Bering Glacier: Interdisciplinary Studies of Earth's Largest Temperate Surging Glacier

edited by

Robert A. Shuchman
Michigan Tech Research Institute
3600 Green Court, Suite 100
Ann Arbor, Michigan 48105, USA

Edward G. Josberger
U.S. Geological Survey
Washington Water Science Center
934 Broadway, Suite 300
Tacoma, Washington 98402, USA

Special Paper 462

3300 Penrose Place, P.O. Box 9140 • Boulder, Colorado 80301-9140 USA

2010

Published by The Geological Society of America, Inc.
3300 Penrose Place, P.O. Box 9140, Boulder, Colorado 80301-9140, USA
www.geosociety.org

Printed in U.S.A.

GSA Books Science Editors: Marion E. Bickford and Donald I. Siegel

Library of Congress Cataloging-in-Publication Data

Bering Glacier : interdisciplinary studies of Earth's largest temperate surging glacier / edited by Robert A. Shuchman, Edward G. Josberger.
p. cm. -- (Special paper ; 462)
Includes bibliographical references.
ISBN 978-0-8137-2462-1 (pbk.)
1. Bering Glacier (Alaska) 2. Surging glaciers--Alaska. 3. Bering Glacier Region (Alaska)--Climate. 4. Geology--Alaska--Bering Glacier Region. I. Shuchman, Robert Allan. II. Josberger, Edward George

GB2425.A4B47 2010
551.31'209798--dc22

2009046633

Cover: Large, front-cover photo of glacier by Robert A. Shuchman. Smaller front- and back-cover photos contributed by Bering Camp participants.

10 9 8 7 6 5 4 3 2 1

We dedicate this book to Austin Post in acknowledgement of his landmark studies of Bering Glacier.

Contents

Dedication iii

Reflections from Austin Post vii

Preface ix

A Wildlife Biologist's Perspective xi

Acknowledgments xiii

1. ***The raw beauty of Bering Glacier*** 1
   Kristine J. Crossen and Christopher Noyles

2. ***Introduction to the Bering Glacier System, Alaska/Canada: Early observations and scientific investigations, and key geographic features*** 13
   Bruce F. Molnia and Austin Post

3. ***Remote sensing of the Bering Glacier Region*** 43
   Robert A. Shuchman, Edward G. Josberger, Liza K. Jenkins, John F. Payne, Charles R. Hatt, and Lucas Spaete

4. ***Hydrography and circulation of ice-marginal lakes at Bering Glacier, Alaska, USA*** 67
   Edward G. Josberger, Robert A. Shuchman, Guy A. Meadows, Sean Savage, and John Payne

5. ***Bering Glacier ablation measurements*** 83
   Robert A. Shuchman, Edward G. Josberger, Charles R. Hatt, Christopher Roussi, P. Jay Fleisher, and Scott Guyer

6. ***Hydrologic processes of Bering Glacier and Vitus Lake, Alaska*** 105
   Edward G. Josberger, Robert A. Shuchman, Guy A. Meadows, Liza K. Jenkins, and Lorelle A. Meadows

7. ***Botanical inventory of the Bering Glacier Region, Alaska*** 129
   Marilyn Barker

8. ***Biogeography and ecological succession in freshwater fish assemblages of the Bering Glacier Region, Alaska*** 167
   Heidi L. Weigner and Frank A. von Hippel

*9. Harbor seal (***Phoca vitulina richardii***) use of the Bering Glacier habitat: Implications for management* 181
Danielle M. Savarese and Jennifer M. Burns

10. The 1993–1995 surge and foreland modification, Bering Glacier, Alaska 193
P. Jay Fleisher, Palmer K. Bailey, Eric M. Natel, Ernie H. Muller, Don H. Cadwell, and Andrew Russell

11. Structural geology and glacier dynamics, Bering and Steller Glaciers, Alaska 217
Ronald L. Bruhn, Richard R. Forster, Andrew L.J. Ford, Terry L. Pavlis, and Michael Vorkink

12. Holocene history revealed by post-surge retreat: Bering Glacier forelands, Alaska 235
Kristine J. Crossen and Thomas V. Lowell

13. Faunal analysis of late Pleistocene–early Holocene invertebrates provides evidence for paleoenvironments of a Gulf of Alaska shoreline inland of the present Bering Glacier margin 251
Anne D. Pasch, Nora R. Foster, and Gail V. Irvine

14. Holocene sea-level changes and earthquakes around Bering Glacier 275
Ian Shennan and Sarah Hamilton

15. Surges of the Bering Glacier 291
Bruce F. Molnia and Austin Post

16. Subarctic hunters to cold warriors: The human history of the Bering Glacier Region 317
John W. Jangala

New Investigator Activities at Bering Glacier 325

17. Acoustic and seismic observations of calving events at Bering Glacier, Alaska 327
Joshua P. Richardson, Katelyn A. FitzGerald, Gregory P. Waite, and Wayne D. Pennington

18. Ice-generated seismic events observed at the Bering Glacier 337
Katelyn A. FitzGerald, Joshua P. Richardson, and Wayne D. Pennington

19. Mapping the fresh-water–salt-water interface in the terminal moraine of the Bering Glacier . 341
Austin B. Andrus, Kevin A. Endsley, Silvia Espino, and John S. Gierke

20. Satellite-derived turbidity monitoring in the ice marginal lakes at Bering Glacier 351
Liza K. Jenkins

21. Investigating biological productivity in Vitus Lake, Bering Glacier, Alaska 361
Nancy A. Auer

22. Bryophytes and bryophyte ecology of the Bering Glacier Region 365
Nancy G. Slack and Diana G. Horton

23. Birds of the Bering Glacier Region: A preliminary survey of resident and migratory birds 373
Carol Griswold

24. Bering Glacier, climate change, and the Southern University research experience 381
Michael Stubblefield, Revathi Hines, and Lionel D. Lyles

Reflections from Austin Post

Bering Glacier is the largest and longest glacier in continental North America, with an area of approximately 5000 km^2 and a length of 190 km. The Bering Glacier System, which includes the Steller Glacier and the vast ice- and snow-filled Bagley Ice Valley and its tributaries, covers more than 6% of the glacier covered area of Alaska and may contain 15–20% of Alaska's total glacier ice; hence it is a major component of the marine and terrestrial ecosystems of the Gulf of Alaska. It is also the largest surging glacier in North America.

Don J. Miller of the U.S. Geological Survey (USGS) first investigated the Bering Glacier region geology in the 1940's, generating benchmark mapping information. I first observed Bering Glacier in 1960 in a pioneering aerial photographic study, annually observing and recording the conditions of hundreds of Northwest North American glaciers. This research was funded by two National Science Foundation grants to the University of Washington, Seattle. Between 1964 and 1983, I continued this research under the direction of Mark F. Meier, USGS, Project Office Glaciology, Tacoma, Washington. The Bering Glacier was observed and conditions were recorded most of those years.

My most important contributions to understanding the complexities of the Bering Glacier was the observation and reporting of the previously unrecognized surging mechanism of many Alaska and Northwest Canada glaciers, including the Bering. These were reported in "Distribution of Surging Glaciers in Western North America " (Austin Post, Journal of Glaciology, vol. 8, no. 53, 1969). Bering was surging throughout most of its length in 1960, and with more than 100 km of its surface broken by innumerable seracs and crevasses, it presented a most incredible sight.

The proposed origin of medial moraine folds on the terminal lobe of the glacier were published in my report "Periodic Surge Origin of Folded Medial Moraines on Bering Piedmont Glacier, Alaska" (Journal of Glaciology, vol. 11, no. 62, 1972).

The annual aerial photography flights also presented a classic opportunity to observe and record changes in following years in which rapid retreat from advanced surge positions took place and were replaced by terminal lakes. New surges, which took place in 1966, 1993, and 1994, failed to reach former extended positions.

Beginning in 1990 and continuing through 1994, under the direction of Bruce Molnia, field parties were established under the USGS "Volunteers for Science" program of which I was a member. This was to facilitate field investigations of the glacier, including bathymetry and sub-bottom profiling of the terminal lakes, carbon-14 dating of living and overridden forests, and periodic high and low aerial photography, establishing terrestrial photo points and mapping changes.

After the 1994 surge, the Bureau of Land Management (BLM), Alaska, which manages the Bering Glacier Region, took over the "Moose International" field camp to continue the important studies initiated by the USGS. Since then, the BLM has expanded the scope of the studies to include not only glaciological studies but also hydrographic surveys of the ice marginal lakes, botanical surveys, paleo-sea level determinations, marine mammal population estimates, and fish surveys. In partnership with the USGS, the BLM has also developed new remote sensing tools to monitor Bering Glacier by working with academia and other institutions.

The studies at Bering Glacier from the BLM field camp have evolved into an international, comprehensive, multidisciplinary program that provides new understanding of this dynamic area. This monograph captures in a single volume these new scientific findings from a wide range of disciplines and preserves data at a single location for future researchers. It is fitting that this monograph is being published as part of the fourth International Polar Year, as it makes an important contribution.

Austin Post,
USGS, Retired

Preface

This monograph presents the results of a comprehensive and diverse series of field studies and science investigations at the Bering Glacier. The results reported in this monograph are from a wide range of disciplines that include: glaciology, geology, paleogeology, hydrology, limnology, oceanography, tectonics, geomorphology, geophysics, meteorology, remote sensing, climate change, anthropology, and ecological studies pertaining to vegetation, fish, and marine mammals. The compilation of these individual studies into a single publication allows for a more complete understanding of how the approximately 5,000 km^2 Bering Glacier System plays a major role in the greater southeastern coastal region of Alaska, and through its wastage, its impact on the circulation of the Northeast Pacific Ocean and on global sea level.

It is the surging dynamics of the Bering Glacier system that has generated the interest that has driven the science investigations reported in this monograph. The Bering Glacier, which has surged at least six times in the last 150 years, is perhaps one of the most dynamic places on earth. The cycle of glacier surges that occurs every twenty to twenty-five years and began centuries ago, accelerates time scales, allowing for physical, hydrographic, climate change, and ecological observations to be made within a lifetime. As the glacier rapidly advances and retreats, sometimes more than 10 km in a short period of time, the physical and biological environment is in constant change. Lakes, streams and rivers form, then dry up or become abandoned as the ice front advances or retreats. Succession forests develop and then are rapidly covered by advancing ice. This allows investigators, such as the authors of this monograph, to capture the geomorphology, paleogeology, hydrography, or ecology of a given region within the Bering Glacier system as a snapshot in time.

The Bering Glacier System is located within one of the most active tectonic regions of the world. Nearby, the Pacific and North American plates collide, forming a convergence zone. The great Alaska earthquake of 1964, with its epicenter located just to the west of Bering Glacier, resulted in a lowering of the area by approximately 1 m. Post-Pleistocene and post-Little Ice Age coastal rebound enhances the crust's uplift. Since the end of the Little Ice Age, isostatic uplift has been approximately 1 m/century. Hence, the Bering area is dynamic and extremely active. Chapters in this monograph report on the tectonic activity today and investigations also use the sediment record to understand historical uplift and subsidence in the area.

As a result of the dynamics of the region, and as reported within this volume, one can find within the Bering Glacier region multi-generation ancient forests that lived 4,000 to 1,200 years ago, were covered by ice advance, and were recently exposed. These forest remnants give insight into the past history of the glacier. Live hemlock and spruce forests with trees as much as 700 to 800 years old are located just beyond the Little Ice Age moraine on the east side of the glacier. On the stagnant medial moraine between the Bering and Steller Lobes of the glacier one can find 200- to 300-year-old trees. The chapter on vegetation investigates successional vegetative species that inhabit an area upon ice retreat and help us better understand the early terrestrial ecology of a region after ice retreat. The proximity to seed sources allows for rapid generation of vegetation on newly exposed glacial till. The trees growing on the glacial till in the moraine show areas of thermokarst development, indicative of the changing climate. These thermokarst areas are unique and help make the Bering Glacier region one of the most unique places on earth. As reported by various authors in the monograph, radiocarbon dating of forest material, shells, and fresh water diatoms coupled with geomorphology and glaciology investigations provide a good picture of the Bering Glacier's activity during the

Holocene. The Bering Glacier region has one of the few well documented Holocene histories from both a climate and a landform perspective.

The rapid advance and retreat of the glacier and the resulting formation of lakes and streams has created a unique ecosystem in the region. As reported in the monograph, the fish populations have established themselves quite quickly in newly formed water bodies that are isolated from species competition and in the case of Sticklebacks, have undergone significant and rapid genetic evolution. Fish species such as dolly varden, which are found in Berg Lake, an ice marginal lake at the terminus of the Stellar Lobe, raise interesting questions as to how these fish came to inhabit the lake. Were they transported in egg form by birds as part of the natural selection process, or was it an anthropogenic activity such as transport on the floats of light aircraft?

Ice advance is also a controlling factor for the resident harbor seal population at Bering Glacier. The seals use icebergs calved by the glacier to outhaul free from predators. During glacier surges, seal populations drastically decline in Vitus Lake, the marginal lake at the glacier terminus. Seal populations within the Bering region would be the greatest prior to a surge, as is the case today, and would be at a minimum during and immediately after a surge event.

The hydrography and resulting ecology of Vitus Lake is also highly variable depending on the time since the last surge. The size and depth of the lake increases, as does the salt content, as a function of the time since last ice advance. Vitus Lake goes from predominantly fresh water after the end of a surge and the beginning retreat to more marine conditions as freshwater discharge decreases. A chapter on the hydrodynamics of the tidally influenced Vitus Lake documents changes in the salinity of the lake and formulates a hydrodynamic model for the interchange of sea water and the glacier-fed lake. The model is then used to simulate conditions under future scenarios, including a warmer climate. Vitus Lake, as discussed in this monograph, is presently classified as oligotrophic, with lower food web aquatic activity constrained to isolated shallow water near shore. The lake also has a large amount of sediment (rock flour) that is deposited into the lake each year as a result of glacier erosion. The remote sensing chapter demonstrates techniques that track the glacier melt water discharge into Vitus via changes in turbidity.

The Bering Glacier system is undergoing rapid melting, approximately 10 m/year at the terminus and 3.5 m/year near the snow equilibrium line. The glacier ablation studies, as reported in detail in this monograph, correlate melt primarily to air temperature and secondarily to other meteorological factors including wind speed. The high melt rates coupled with the large areal extent of Bering Glacier results in upwards of 30 km^3 or 30 gigatons of freshwater discharged into the Gulf of Alaska each year. These observations from the Bering Glacier System can be extrapolated to other regions within Alaska and throughout the world to better estimate sea level rise as a result of changing climate.

The dynamics of this region have also played a role in the activity of the early Native Americans. A century ago, Bering Glacier was approximately 200 m thicker than it is today. A thicker glacier would allow for easier traverses into the interior of Alaska. One chapter in the monograph addresses the use of the Bering Glacier by early man. Habitation sites dating from eight- to ten-thousand years ago likely exist within the Bering region. Based on work reported within this collection of investigations, it is hypothesized that the sites are situated 60 to 100 meters above sea level because the entire area is being uplifted continuously as previously discussed.

This monograph, which reports on the interdisciplinary investigations that have been conducted at Bering Glacier, clearly demonstrates how each individual study is useful as a stand-alone chapter and provides insight into understanding an aspect of the Bering Glacier as an ecosystem. And, the collective set of investigations provides an even greater understanding of this unique place on the planet.

Senior Editors:
Dr. Robert A. Shuchman
Dr. Edward G. Josberger

Contributing Editors:
Dr. Bruce F. Molnia
Dr. John F. Payne
Ms. Liza K. Jenkins

A Wildlife Biologist's Perspective

The sun reflected from the ice face, glittering and sparkling flashes that made my eyes water. In contrast to this burning whiteness, the shadows were as intensely blue as the cloudless sky. I could hear a steady trickle of water, plinking into the pool before me and roaring in the distance as the pools formed streams and rivers, rushing to the sea. In my chest, I felt a bass resonance with the growling, shifting mass. Despite the stark sunlight and ambient temperature, a piercing chill brushed my face with each breeze. I was nearly overcome by the sensory overload and marveled at this thing before me: the Bering Glacier. The glacier's awe-inspiring presence is only magnified by its story.

The calendar date was sometime in early July 1993. I had been asked by the U.S. Geological Survey (USGS) to join a small group that was studying the "surge" of the Bering Glacier. The small ragtag research camp sat on the edge of a brown colored lake that really didn't look like a lake since it was filled with moving ice flows. The brown water, I was told, was coming from under the glacier (after all, what do I know about glaciers—I'm a wildlife biologist), and the ice flows acted as though they were in a demolition derby, bent on breaking free in the wide open of the Gulf of Alaska. Standing on the shore of Vitus Lake, near what is known as the narrow channel, I was looking up at the towering blue face of a glacier breaking free of the bedrock across which it would normally be grinding at a snail's pace. The very small space between the bedrock and the glacier ice was now filled with water, literally lifting the massive structure up and allowing it to "surge" forward as though it were on a well-oiled surface.

This was not just any glacier: the Bering Glacier is the largest and longest glacier in continental North America, with an area of approximately 5175 km^2, and a length of 190 km. It is also the largest surging glacier on Earth and may contain 20% of Alaska's ice. The sheer size of this glacier and its dynamic nature provide opportunities for studies and research rarely found on the planet. These opportunities have led to a scientific program that incorporates elements of both physical and biological sciences, and provides a forum for integrating knowledge across the varied scientific disciplines. The reader of this monograph must know that the knowledge gained by the authors and their teams have not been gathered in the solitary confines of their individual expertise. Rather, they have worked together, sometimes grudgingly, to form partnerships and friendships, forging foundations for a holistic approach to understanding the environments of the Bering Glacier. However, this is just the beginning. This introduction will lead the reader through a brief history of studying the Bering Glacier, introduce programs, and end with the value of integrating science in order to better understand the forces imposed upon our planet. I will conclude with more reflection and a glimpse into the future.

The Bering Glacier is named for the Danish explorer, Vitus Bering, who explored the Alaska Gulf Coast in July 1741. He and naturalist Georg Wilhelm Steller described high mountains and snowy peaks from an area south of Controller Bay. Israel Russell, from the University of Michigan, in his book, *Glaciers of North America* (1897) stated that the Bering Glacier has been seen by vessels passing along the coast, but no one had explored the glacier. Around the time of Vitus Bering's exploration, there were three cultural groups in the Bering Glacier region, but no descriptions have been found that are specific to the Bering Glacier. At the dawn of the twentieth century, descriptions of the Bering Glacier began to surface. Early topographers developed the first crude maps of the region, with the first quality topographic maps appearing after World War II. The dynamic nature of this glacier makes mapping efforts dated almost upon release.

Scientific interest in the Bering Glacier tended to focus on the physical aspects of the glacier, its movements and periodic surges until the mid-1990s. Following the surge event of 1993–95, the USGS led a considerable effort to determine the physical surge mechanisms and glacier recovery post surge. The Bureau of Land Management (BLM), until the surge event, had little interest in the area even though the majority of the glacier and much of the adjacent land is under BLM administration. When USGS surge event studies began to sunset, the BLM began surveying the region for biological diversity and collecting information prior to initiating land use planning efforts. By 1998, a partnership between BLM resource specialists and USGS glaciologists was beginning to document rapid changes in both the physical and biological parameters of the glacier. The evolution of this partnership led to the formation of an integrated science program that includes the wide variety of science disciplines outlined in this monograph. Today, the program has developed into the opportunity for seemingly unrelated physical, biological, and social science disciplines to discover the benefits of viewing science through a holistic approach rather than a single discipline. This monograph provides a "status of our knowledge" of the Bering Glacier, not as separate journal publications, but as a whole. It incorporates anthropology, paleoecology, hydrology, geology, remote sensing, terrestrial and fisheries biology, botany, oceanography, marine biology, geography, and, of course, glaciology. The BLM has also embraced the idea of science integration in the East Alaska Resource Management Plan by designating the Bering Glacier as a Research Natural Area so science related activities can continue.

Entry into the twenty-first century has brought new challenges to the steps of science. It has been known for decades that Earth has entered into a warming phase. The Arctic and sub-Arctic are experiencing firsthand the accelerating processes of climate change. The Arctic ice cap has thinned and its summer retreat is at a previously unknown extent. Both high latitude and high altitude glaciers are melting at accelerating rates, contributing to sea level rise and ocean chemical changes that may not be conducive to current sea life. All of this brings changes to the environmental conditions most of us have come to view as static, or gently fluctuating through normal cycles. Addressing these types of changes through the traditional approach of a single scientific discipline will not provide the answers we all will need to adapt to change. In fact, a traditional approach may limit our ability to predict changes outside of a specific discipline. Bringing science together, like the Bering Glacier Integrated Science Program, serves as a strong base for modeling future scenarios and predicting with far greater confidence environmental conditions of the future, thus increasing our ability to adapt to changing conditions.

I have stood before this incredible wonder of nature in awe of its raw beauty and spectacular setting, with the Gulf of Alaska on one side and the Bagley Ice Field on the other. Standing atop the Grindell Hills on a clear day with the Bering Glacier surrounding me, I wished every day could be this perfect in this imperfect world. I have experienced the power of ice and water as the termination of the surge began in July 1994, watching house-sized pieces of ice jetting from the front of the glacier into newly forming lakes. I have watched and listened as raging, immeasurable amounts of water carried the ice toward the Gulf of Alaska. The work we began on the Bering Glacier enabled me to catch a glimpse of the hidden treasures I sensed the first day I stood before it. This is an exciting time in our planet's evolution, and I consider myself fortunate to have participated in this study. I hope the sense of excitement this project generated has been captured in this monograph and that those who read it are inspired to continue our efforts.

John F. Payne
Bureau of Land Management

Acknowledgments

The editors want to express their gratitude to the many people who made this monograph possible. First and foremost are the monograph's authors who spent a considerable amount of time preparing and revising their material. Next are the monograph's reviewers and editorial board whose comments greatly contributed to improving the overall quality of the material.

The development of the monograph was partially supported and certainly encouraged by the Bureau of Land Management (BLM), which currently has jurisdiction over the Bering Glacier Region and operates the field camp each summer. Michigan Technological University (Michigan Tech) and the U.S. Geological Survey are thanked for their financial contributions to the preparation and printing of this monograph.

Editorial Board:
Robert A. Shuchman, Michigan Tech Research Institute
Edward G. Josberger, U.S. Geological Survey
John F. Payne, North Slope Science Initiative and the Bureau of Land Management
Bruce F. Molnia, U.S. Geological Survey
Liza K. Jenkins, Michigan Tech Research Institute
Kristine J. Crossen, University of Alaska

The scientific discoveries reported in this monograph would not have occurred without the management, oversight, and logistical support provided by BLM. Without the generous helicopter time and support of boat operations on Vitus and Berg Lakes, the field activities could not have been performed. The BLM Bering Glacier Field Camp allowed investigators to concentrate on their research in a safe setting. Special thanks go to Scott Guyer, John Payne, Chris Noyles, and Nathan Rathbun of BLM for running Bering Camp each summer. We would also like to acknowledge Vanessa Rathbun from the BLM for her help with the book cover design.

Additional thanks are given to Gail Raney, Steve Raney, and John Tucker of Fishing and Flying, Cordova, Alaska, for their role in transport of scientists and gear to various parts of the glacier. Finally, we would like to thank the family of Larry Hancock for the flying support he provided to the Bering Camp operation. Unfortunately, Larry died in an aviation accident in 2006.

The Geological Society of America
Special Paper 462
2010

The raw beauty of Bering Glacier

Kristine J. Crossen
Department of Geological Sciences, University of Alaska, 3211 Providence Drive, Anchorage, Alaska 99508, USA

Christopher Noyles
U.S. Bureau of Land Management, 222 W. 7th Avenue, Anchorage, Alaska 99513, USA

The Bering Glacier system is the largest temperate surging glacier in the world and contains ~20% of the glacial ice in Alaska. The Bering and Steller Glaciers flow out of the Bagley Ice Field (Bagley Ice Valley) and terminate in a 60-km-wide Piedmont Lobe that fans out below the Saint Elias Mountains. The Bering and Steller Lobes are separated by a large debris-covered medial moraine that has been deformed into zigzag patterns during surge events. The Steller Lobe terminates in Berg Lake, which empties into the Bering River, and the Bering Lobe terminates in Vitus Lake, which empties into the Seal River.

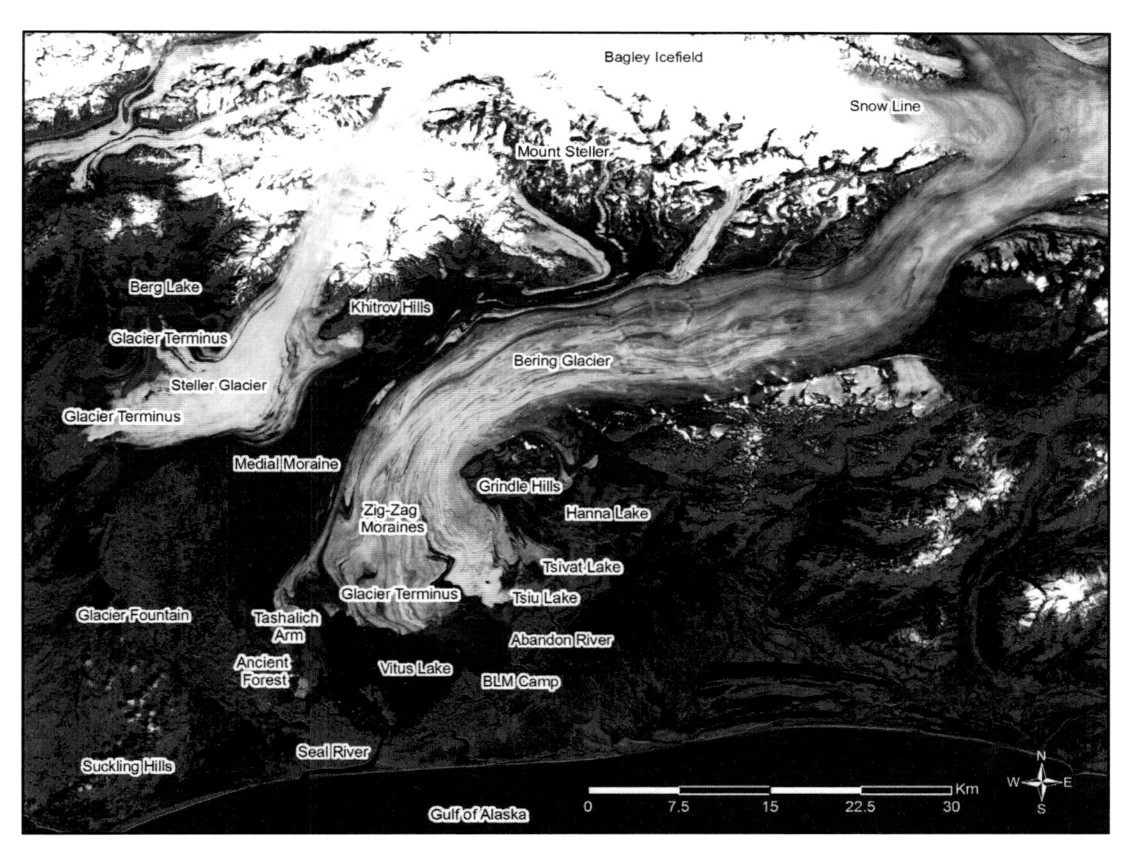

Figure 1. Overview map

Crossen, K.J., and Noyles, C., 2010, The raw beauty of Bering Glacier, *in* Shuchman, R.A., and Josberger, E.G., eds., Bering Glacier: Interdisciplinary Studies of Earth's Largest Temperate Surging Glacier: Geological Society of America Special Paper 462, p. 1–12, doi: 10.1130/2010.2462(01). For permission to copy, contact editing@geosociety.org.

Figure 2. Bagley Ice Field

The Bagley Ice Field forms a high plateau that feeds the Bering and Steller Glaciers. Thick snowpack accumulates yearly until it compacts and recrystallizes into glacial ice. The equilibrium line is located at 1200 m elevation, above which the glacier is covered with fresh snow year round. Here the ice rounds a sharp corner just below the equilibrium line as it leaves the Bagley Ice Field and descends to the upper reaches of Bering Glacier.

Figure 3. Ablatometer

Glaciologists calculate the health of a glacier by measuring both the amount of mass added each year by snowfall (accumulation) and the amount lost to melting (ablation). Greater accumulation means that a glacier becomes both thicker and longer, and its terminus will advance. Greater ablation means that melting will dominate, and the glacier becomes both thinner and shorter, causing the terminus to retreat. Scientists on Bering Glacier use an ablatometer, which bounces radio waves off the ice to measure summer ablation. (See Shuchman et al., this volume.)

Figure 4. Crevasses

Crevasses are deep cracks that descend into the ice. They form in response to the ice sliding downslope over an uneven bed. The crevasses in the foreground cut across a dirty medial moraine and are filled with fresh snow. The circular crevasses in the center suggest a small depression beneath the ice.

Figure 5. Moulin

Summer melting causes large supraglacial streams to form on the ice surface. When the running water intersects a crevasse, it cascades into the ice, carving a circular opening called a *moulin*. Studies have shown that moulins may reach the base of the glacier or may run through tunnels inside the ice. These openings in the glacier are dangerous because of slippery slopes and rushing water. (See Shuchman et al., this volume.)

Figure 3.

Figure 2.

Figure 4.

Figure 5.

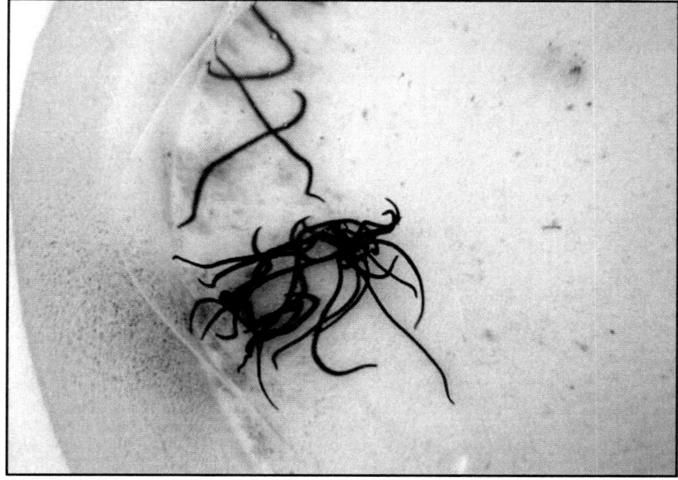

Figure 6.

Figure 6. Ice worms

Small meltwater pools form in the summer on the lower reaches of the glacier. Ice worms 2–3 cm (1 in.) long and no thicker than sewing thread can be found in the shallow water. They live between grains of ice and emerge to feed on algae or pollen found on the glacier surface. Their systems contain a type of antifreeze that allows them to live in this inhospitable environment.

Figure 7. Berg Lake

The northwestern part of the Piedmont Lobe terminates in Berg Lake. There, the vertical face of Steller Glacier calves to form icebergs. Berg Lake exhibits raised shorelines as much as 30 m (100 ft) above the current lake level, which were formed when thicker ice dammed higher water levels. (See Molnia and Post, this volume.)

Figure 8. Steller Glacier terminus

When the face of Steller Glacier calves into Berg Lake, the icebergs displace the lake water and generate tsunami-like waves. This lifts the bergs and strands them along the rocky lakeshore, where their translucent blue color is apparent.

Figure 9. Glacier fountain

South of Berg Lake, the terminus of the glacier becomes hidden by vegetation. But in the midst of a small basin, a glacial fountain rises to the surface and builds a terrace of ice crystals. These fountains originate when supercooled water (–2 °C) rises from an overdeepening beneath the ice, and crystals form as pressure is reduced.

Figure 7.

Figure 8.

Figure 9.

Figure 10.

Figure 10. Debris-covered medial moraine

A substantial medial moraine separates the lobes of Steller Glacier (in the foreground) from the Bering Glacier (in the background). The morainal debris is derived by rockfall from mountain slopes above the glacier and is transported along its edge as a lateral moraine. As the two major lobes join, the lateral moraines coalesce to form a medial moraine in the center of the larger ice body.

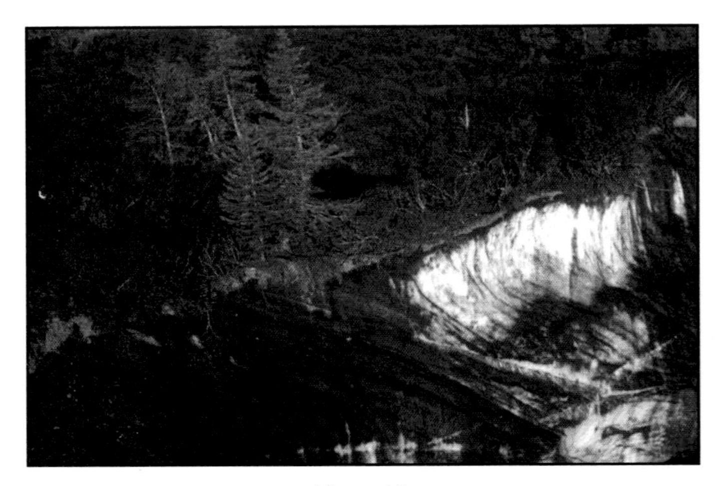

Figure 11.

Figure 11. Forest-covered medial moraine

The medial moraine is cored by ice and collapses into large kettles as the underlying ice melts. This debris cover is thick enough to support living forests, including spruce trees. Other supraglacial forests exist in Alaska at the Malaspina and Matanuska Glaciers.

Figure 12.

Figure 12. Zigzag moraine

The medial moraine along the edge of the Bering Lobe is deformed into a zigzag pattern by glacial surges. Surges occur when pressurized water trapped below the ice causes the glacier surface to rise and advance rapidly. Bering Glacier is well known for its surge events, which have occurred irregularly every 25–30 yr since 1900, most recently between 1993 and 1995. (See Fleisher et al., this volume.)

Figure 13. Bering Glacier and Vitus Lake

Bering Glacier calves along a 25-km-long terminus into Vitus Lake, where water depths range from 40 to 250 m along

Figure 13.

Figure 14.

Figure 15.

the ice front. To the north, Mt. Steller rises above the glacier. The lower dark peaks form the Grindle Hills, which are nunataks surrounded by ice. Since the end of the 1995 surge event, the glacier terminus has retreated as much as 5 km. As much as 100 km^3 of ice has been lost in the past 50 yr. (See Shuchman et al., this volume.)

Figure 14. Blue Ice

Tashalich Arm is the western embayment of Vitus Lake. Here the ice front calves along a 25-m-high cliff into the 180-m-deep lake. The calving exposes a fresh section of the ice, which is deep blue in color. This color fades rapidly after the calving event. The blue color is due to the reflection of blue wavelength light as the other wavelengths are absorbed into the ice crystals.

Figure 15. Iceberg waterlines

Most of the volume of an iceberg is submerged. If the ice melts or breaks off below the lake surface, the remaining berg floats higher in the water. This creates a series of waterlines along the iceberg margin.

Figure 16. Unvegetated surge margin

The 1993–1995 surge event caused Bering Glacier to advance as much as 1–7 m/d, overriding the lake bed and the Taggland Peninsula. Following the surge, the ice has retreated rapidly from the overrun areas, and the surge margin is clearly delineated by the dramatic change in vegetation.

Figure 17. Ice cave

Ice caves are commonly found along the glacier margin, often formed by streams that flowed under the ice. The walls of the caves are sculpted by the wind and exhibit indented *sun cups* caused by melting and sublimation.

Figure 18. Ancient Forest site

The Ancient Forest site melted out of the ice in 1998. Trees, peat, cones, and needles indicate that a long-lived maritime forest occupied this location from ca. 200 B.C. to A.D. 600 prior to ice advance over the area. (See Crossen and Lowell, this volume.)

Figure 19. Seal River

The 9-km-long Seal River drains Vitus Lake and discharges 8–12 km^3 of fresh water into the Pacific Ocean each year. The lake is dammed behind the moraine ridge built by Bering Glacier when it reached its maximum Little Ice Age position about A.D. 1890. The moraine forms the south shore of the lake in this area. (See Shuchman et al., this volume.)

Figure 16.

Figure 19.

Figure 17.

Figure 18.

Figures 20 and 21. Seals on bergs

The Seal River is named for the numerous seals that find shelter in Vitus Lake. They swim up the Seal River into the fresh-water lake and outhaul on the icebergs, where they are protected from orcas and other predators. They feed on salmon and other fish in the lake. (See Savarese and Burns, this volume.)

Figure 22. Series of eastern lakes

The northeastern side of Bering Glacier is characterized by a series of small lake basins. These fresh-water lakes have alternately been exposed and overrun by the ice numerous times in the past. A spectacular outburst flood from July to September 1993 released 75 m^3 of sediment, enough to fill the entire basin of Tsivat Lake. The lakes in order, from near to far, include an open-water Hanna Lake, the sediment-filled basin of Tsivat Lake, and Vitus Lake in the distance. (See Fleisher et al., this volume.)

Figure 23. Tsui Lake

Tsui Lake lies between Tsivat and Vitus lakes and was overrun by ice advance during the 1993 surge event. A calving margin occupied the lake until 2006, when the ice dam separating Tsui Lake from Vitus Lake was breached and the lake drained in about two weeks. The dry lake bed was exposed and the water level dropped 17 m so that the river in the background became abandoned. (See Fleisher et al., this volume.)

Figure 24. BLM Bering Glacier Camp

The U.S. Bureau of Land Management Bering Glacier Camp on the east shore of Vitus Lake is home to 30 or more scientists each summer, who study this remarkable natural laboratory. Current studies include the tectonics of this active area; the

Figure 20.

Figure 21.

Figure 22.

Figure 23.

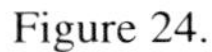

Figure 24.

Figure 25.

velocity and melting of the ice; the water characteristics of the lakes; the stratigraphy and fossils of the glacial deposits; and the ecology of the seals, fish, and plants in this newly deglaciated terrain.

Figure 25. Harbor seal

Harbor seals enter the fresh water of Vitus Lake during the summer to escape predators. This seal is resting among the fireweed after scientists had captured, weighed, and measured it. (See Savarese and Burns, this volume.)

Figure 26. Mountain goat

Mountain goats are found on the mountaintops and nunataks (peaks surrounded by ice) near Bering Glacier. This lone male goat followed the scientists all day as they hiked around the mountain.

Figure 27. Bear tracks

Brown bears are commonly sighted from the air and sometimes seen on the ground near Bering Glacier. Most often, scientists see evidence of the bruins in their tracks along shorelines.

Figure 26.

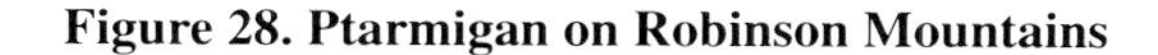

Figure 27.

Figure 28. Ptarmigan on Robinson Mountains

Rock ptarmigan graze among the red stalks of the Alaska spiraea in the high mountains during the summer. This hardy mother and her fuzzy baby will transform for the winter to white plumage and feathered feet.

Figure 29. Khitrov Hills

Plants inhabit a wide variety of environments from sea level to mountaintops surrounding Bering Glacier. From this view atop the tundra-carpeted Khitrov Hills, you can see down to Bering Glacier on one side and Steller Glacier on the other.

Figure 30. Weeping Peat Island

Following the 1993–1995 surge event, Bering Glacier has steadily retreated to uncover fresh landscapes each summer. Early colonizing plants like dwarf fireweed move into newly deglaciated terrain, like this one on Weeping Peat Island along Tsui Lake.

Figure 31. Chukchi primrose

The Chukchi primrose lives in wet stream beds surrounded by moss. This tiny plant, only 10 cm (4 in.) tall, produces spectacular fuchsia blooms on a single stalk. (See Barker, this volume.)

Figure 32. Pacific buttercup

The rare Pacific buttercup prefers moist sites and woodlands. This early bloomer has red-tinged petals and leaves that help produce heat for the plant. (See Barker, this volume.)

Figure 28.

Figure 29.

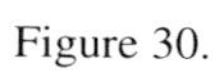

Figure 30.

Figure 31.

Figure 32.

Figure 33.

Figure 33. Alaska poppy

The yellow Alaska poppy prospers amid the loose rock on this scree slope in the Robinson Mountains. (See Barker, this volume.)

Figure 34. Arctic willow

This tiny Arctic willow is only 5 cm (2 in.) high. Its fuzzy seeds cover the plant and wait to be blown to new sites in the Robinson Mountains. (See Barker, this volume.)

Figure 35. Purple saxifrage

This purple saxifrage, growing in the cracks of sandstone, illustrates why the name of this tiny plant means "rock breaker." It blooms so early that its iridescent pink flowers often peek through the snow. (See Barker, this volume.)

Figure 34.

Photographic Contributions by:
Marilyn Barker, University of Alaska
Ann Claerbout, U.S. Bureau of Land Management
Kristine J. Crossen, University of Alaska
Jay Fleisher, State University of New York
Scott Guyer, U.S. Bureau of Land Management
Liza Jenkins, Michigan Tech Research Institute
Christopher Noyles, U.S. Bureau of Land Management
Danielle Savarese, LGL Alaska Research Associates
Robert Shuchman, Michigan Tech Research Institute

REFERENCES CITED

Barker, M., 2010, this volume, Botanical Inventory of the Bering Glacier Region, Alaska, *in* Shuchman, R.A., and Josberger, E.G., eds., Bering Glacier: Interdisciplinary Studies of Earth's Largest Temperate

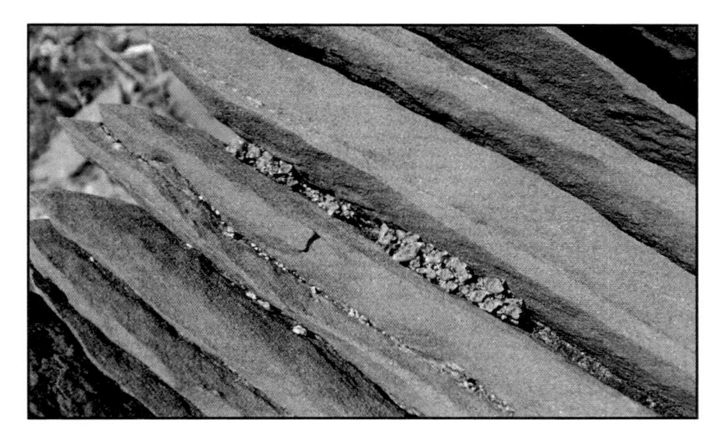

Figure 35.

The Geological Society of America
Special Paper 462
2010

Introduction to the Bering Glacier System, Alaska/Canada: Early observations and scientific investigations, and key geographic features

Bruce F. Molnia
U.S. Geological Survey, 926A National Center, 12201 Sunrise Valley Drive, Reston, Virginia 20192, USA

Austin Post
U.S. Geological Survey (Retired), 2014 Bradley Street, Dupont, Washington 98327-7712, USA

ABSTRACT

The purpose of this chapter is to introduce the reader to the Bering Glacier and the Bering Glacier System. This will be done by (1) providing a summary of the early observations and geographic descriptions of Bering Glacier, (2) identifying scientific studies that have provided insights to the unique character of the Bering Glacier System and its unique surroundings, and (3) presenting descriptions of key geographic features that are part of the system and its surroundings. The Bering Glacier System is the largest glacier in continental North America and the largest temperate surging glacier on Earth.

INTRODUCTION

Bering Glacier System (Fig. 1) is the name that the U.S. Board on Geographic Names (USBGN) approved in 1997 for what is commonly referred to as the *Bering Glacier*. The USBGN description states that the system "comprises a complex of large glaciers, ice valleys, ice streams and smaller tributaries" and is "named for the Bering Glacier, the most prominent feature of this large glacial system." As described, the system "[e]xtends from the eastern edge of the Chugach National Forest, S to the Gulf of Alaska and E into Canada…." (USBGN Decision 1729785, April 08, 1997). Although in the glaciological sense, no ice streams are present, and the extension into Canada is actually the place of origin of the system, the USBGN description of a connected network of large glaciers, ice valleys, and small tributaries effectively captures the essence of the Bering Glacier System. Officially, the name *Bering Glacier* is confined to two important components of the system, the Bering Piedmont Glacier and the Central Valley Reach. The Bering Piedmont Glacier is composed of the Bering Lobe, the Steller Lobe, and the Central Medial Moraine Band. Although technically incorrect, in common usage, the terms "Bering Glacier" and "Bering Glacier System" are used interchangeably.

Among the reasons that this volume focuses on the Bering Glacier System are:

- It is the world's largest temperate glacier, with an area of ~5,175 km^2.
- It is the longest valley glacier in the Northern Hemisphere, with a length of >190 km.
- It is the world's largest nonpolar surging glacier and likely the world's largest surging glacier.
- It dwarfs all but the largest valley glaciers in Antarctica and Greenland in area, length, and ice volume.
- Alone, its area represents nearly 7% of the glacier-covered area of Alaska.
- Alone, it contains ~15% of the glacier ice in Alaska.
- Integrated studies have documented many aspects of its behavior on time scales ranging from millennial to decadal to annual.

Molnia, B.F., and Post, A., 2010, Introduction to the Bering Glacier System, Alaska/Canada: Early observations and scientific investigations, and key geographic features, *in* Shuchman, R.A., and Josberger, E.G., eds., Bering Glacier: Interdisciplinary Studies of Earth's Largest Temperate Surging Glacier: Geological Society of America Special Paper 462, p. 13–42, doi: 10.1130/2010.2462(02). For permission to copy, contact editing@geosociety.org.

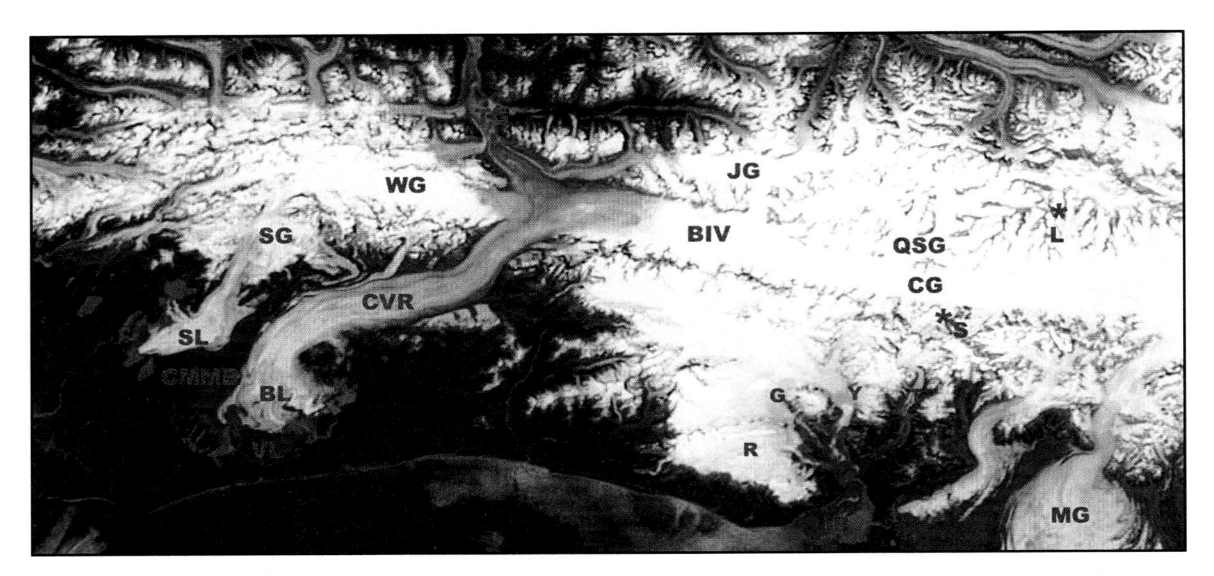

Figure 1. Moderate Resolution Imaging Spectroradiometer (MODIS) image, showing the principal components of the Bering Glacier System and selected adjacent features. BIV—Bagley Ice Valley; BL—Bering Lobe; CG—Columbus Glacier; CMMB—Central Medial Moraine Band; K—Kaliakh Lobe; CVR—Central Valley Reach; IB—Icy Bay; G—Guyot Glacier; Y—Yahtse Glacier; JG—Jefferies Glacier; QSG—Quintino Sella Glacier; SG—Steller Glacier; SL—Steller Lobe; TG—Tana Glacier; VL—Vitus Lake; WG—Waxell Glacier; R—Robinson Mountains; *L—Mount Logan; *S—Mount Saint Elias; MG—Malaspina Glacier. Collectively, the Bering Lobe, the Steller Lobe, and the Central Medial Moraine Band constitute the Bering Piedmont Glacier. The image is oriented with north up, and covers an area that is ~225 km from east to west and ~105 km from south to north. MODIS, a key instrument aboard the Terra (EOS AM) and Aqua (EOS PM) satellites, provides excellent synoptic images of geographic features as large as the Bering Glacier. This image has 250 m pixels.

To assist the nonscientist to understand the unique character of Bering Glacier, the terms *temperate glacier* and *surging glacier* are defined below:

Temperate glacier—Bering Glacier is a temperate glacier. The term *temperate glacier* refers to the glacier's thermal regime. A temperate glacier is one in which the ice temperature is warm enough so that liquid water coexists with glacier ice (frozen water) for all or part of the year. Therefore the temperature of the ice is a fraction of a degree below the melting point of water. Consequently, a small amount of climate change can have a large impact on the stability of the glacier. Temperate glaciers often have major conduit systems with liquid water on (supraglacial), in (englacial), and under (subglacial) them.

Surging glacier—Bering Glacier is a surging glacier. Most glaciers are nonsurging glaciers. Most have relatively constant average annual flow rates. Variations that occur are generally predictable on a seasonal basis. Velocities are low, typically averaging under a meter/day. Surging glaciers exhibit substantial variations and major flow irregularities. Some glaciers have dramatic annual and subannual velocity changes, often characterized by brief velocity increases of at least an order of magnitude. These glaciers are called surging or surge-type glaciers. In a few instances, surging glaciers have had documented velocities of 30–300 m/day. Bering Glacier is the largest surging glacier in Alaska. A separate chapter describes its twentieth century surge history. Some surges involve large volumes of ice displacement and often are characterized by rapid advance of the glacier terminus.

EARLY GEOGRAPHIC INVESTIGATIONS OF THE BERING GLACIER SYSTEM

Oral traditional knowledge suggests that Bering Glacier was a traditional route used by Alaskan Natives to traverse the Chugach Mountains from the coast to interior Alaska. However, little actual detail is known. The first documented sighting of Bering Glacier did not occur until 1839. That summer, Englishman Sir Edward Belcher (1843) observed the fractured terminus of a previously unknown large glacier as he sailed along the coast of the Gulf of Alaska, west of Yakutat Bay. His written description and sketches of the glacier's surface suggest that it was Bering Glacier that he described, and that it was surging at the time of his observation.

In 1880, the U.S. Coast and Geodetic Survey (USC&GS) adopted the name *Bering Glacier* for the same large ice feature observed by Belcher, more than 40 a earlier. The USC&GS chose the name to commemorate Captain Commander Ivan Ivanovich (Vitus) Bering, the Danish explorer employed by the Russian Government, who "discovered" Alaska in 1741.

Lieutenant H.W. Seaton-Karr was the first early observer to observe the eastern end of the Bering Glacier System and appreciate Bering Glacier's true size and complexity. In July 1886, Seaton-Karr was a member of a *New York Times*–sponsored climbing expedition. The expedition made an unsuccessful attempt to reach the summit of Mount Saint Elias. From a viewpoint on the flank of the mountain, Seaton-Karr (1887) observed (p. 109–110):

In this direction (west) the "foothills" of Elias stood like islands in the enormous expanse of glacier stretching prairie-like as far as the eye could penetrate through the crystalline air toward the country of the Atna or Copper River.

Several weeks later, as he sailed along the Gulf of Alaska coast, >100 km west of Mount Saint Elias, he wrote (p. 139–140):

I had understood that with Icy Cape, [Author's note: Icy Cape is the southwestern margin of Icy Bay.] the last ice along the coast was left behind. But looming twenty miles or so to the westward appears another vast ice-plain . . . which sweeps down and opens fan-like on the ocean, where the coast range of "foot-hills" comes to an end. It is evidently the opening or outlet of the vast glacier-desert or ice-lake which we saw from the slopes of Mount St. Elias, lying to the northwest of the mountain. Its birthplace is an icy range that forms an enlarged continuation of the great western ridge of Elias. It is not marked or mentioned by the early navigators, all of whom mistook the true nature of these stupendous glaciers.

This was the first suggestion that the Bering Glacier of the USC&GS and the eastern expanse of glacier that Seaton-Karr had previously described were part of the same system.

Through the end of the nineteenth century, Bering Glacier was unexplored, with only the eastern piedmont lobe having been seen from offshore. As depicted on late-nineteenth century USC&GS charts and U.S. Geological Survey (USGS) topographic maps (Fig. 2), it is clear that late nineteenth and early twentieth century cartographers had practically no knowledge of the size, shape, or geographic location of the glacier's features and sources. More than half a century passed following Seaton-Karr's observations before the relationship of the components of the Bering Glacier System was clearly mapped and understood.

In 1932 the USBGN further expanded the use of the name *Bering Glacier* not only to the terminus, but to the 78-km-long segment of the glacier that extended from the terminus to its "head" east of the eastern end of Waxell Ridge (Orth, 1971). Today, this is the head of the Central Valley Reach valley glacier segment of the Bering Glacier.

The first geological visit to the Bering Glacier region was made by G.C. Martin (1905), a USGS geologist who was mapping the Bering River Coal Fields, a large undeveloped coal reserve on the west side of the glacier, adjacent to the Steller Lobe terminus region. Martin (p. 17) described the Bering as

a large glacier of the piedmont type. . . . It is a large field of stagnant ice which has over flooded the eastern extension of the zone of coastal foothills, a great many valley glaciers coming from the Chugach Range enter it as tributaries. It is fringed along its southwestern border by a wide moraine, while the ice itself is thickly covered by rock debris for a distance of several miles from its front, and, as is in the case of the Malaspina Glacier, this covering is so thick that it is often impossible to determine the margin of the glacier.

Martin and his field party made a topographic map of the region and took the first known photographs of the Bering Glacier. The topographic map and several of the photographs (Fig. 3) document that at the beginning of the twentieth century, retreat of the Steller Lobe was under way. Pre-1905 terminus retreat had produced five small glacier-dammed lakes, Berg Lake 1 through Berg Lake 5, each filling a basin between part of the Steller margin and the adjacent mountains.

In 1908, Martin published a second description of Bering Glacier (Martin, 1908), this time accompanied by a 1:200,000-scale topographic reconnaissance map and a detailed 1:62,500-scale topographic map, both dated 1907. A geologic map (Fig. 4) of a part of the area adjacent to the Steller Lobe was also produced (Martin, et al., 1907). Martin (1908, p. 16) wrote:

Bering Glacier . . . is a huge, even-surfaced, stagnant mass of glacial ice, which is fed by many valley glaciers coming from the high mountains north of it. It is a piedmont glacier of the same general character and of about the same size as Malaspina Glacier. Portions of its surface form a good highway to some of the coal camps in the east end of the Bering River coal field but most of it is so covered by irregular masses of rock and gravel and so much crevassed that travel over its surface is difficult and dangerous.

The geographic relationship between the major components of the Bering Glacier System was first observed and photographed from the air in 1938 by Bradford Washburn. Washburn (Sfraga, 2004) explored the mountains adjacent to the coastal region of the Gulf of Alaska on behalf of the National Geographic Society (NGS) and took nearly 50 aerial photographs (Fig. 5) that clearly provided the first extensive recording of the Bering Glacier System's extent and physical characteristics.

SCIENTIFIC INVESTIGATIONS OF THE BERING GLACIER SYSTEM

Field Research

Following the Second World War, USGS geologist Don J. Miller (1958) mapped the geology of the Bering Glacier area and provided information about many of the associated glaciers. These studies were continued, following Miller's death, by USGS geologist George Plafker.

The first purely scientific aerial investigation of Bering Glacier was a multiyear, systematic study by coauthor Austin Post that began in 1960. This National Science Foundation–funded project was a first-of-its-kind, consisting of annual collection and analysis of detailed aerial photographs of several hundred western North American glaciers, including essentially all of the Bering Glacier System. This study, conducted between 1960 and 1963, was administered by the University of Washington, Seattle, and directed by Philip Church. Beginning in 1964 and continuing through 1982, funding for Post's study was provided by the USGS Project Office of Glaciology, under the direction of Mark F. Meier.

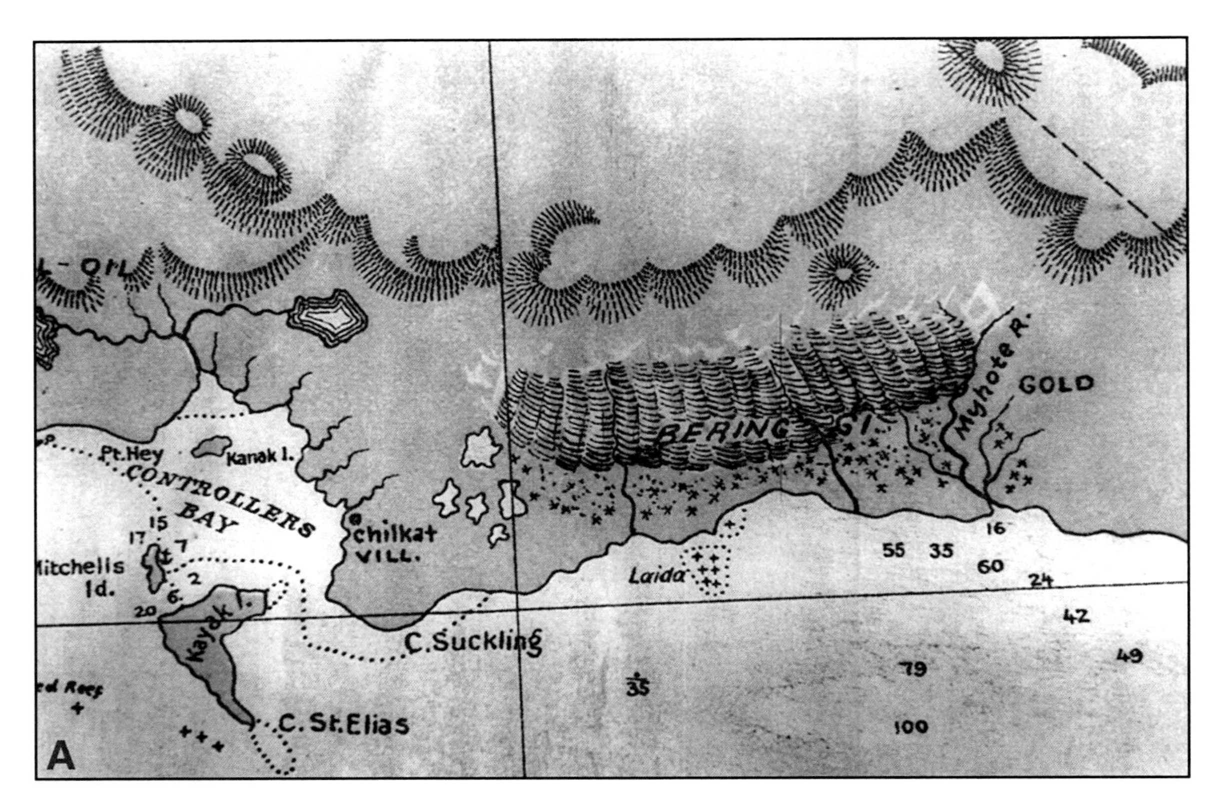

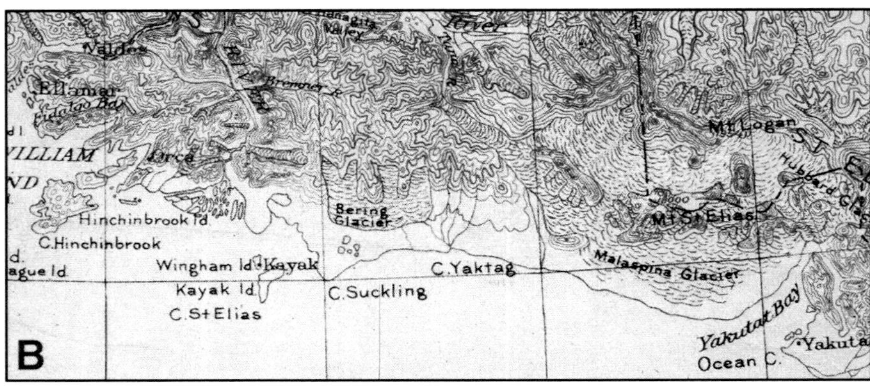

Figure 2. (A) Part of an 1880 U.S. Coast and Geodetic Survey (USC&GS) nautical chart of the eastern Gulf of Alaska, showing the area adjacent to Bering Glacier. The image is oriented with north up, and covers an area that is ~80 km from east to west and ~50 km from south to north. (B) Part of a 1900 U.S. Geological Survey (USGS) 1:500,000-scale topographic map of coastal south-central Alaska, showing the area from Yakutat to Prince William Sound. As depicted on both the nautical chart and the topographic map, there is no connection between Bering Glacier's Piedmont Lobe and its eastern accumulation area. Both the nautical chart and the map clearly illustrate the scarcity of information that was available about the geography of Bering Glacier in the latter part of the nineteenth century and at the beginning of the twentieth century. The image is oriented with north up, and covers an area that is ~340 km from east to west and ~115 km from south to north.

Between the late 1980s and 2002, the USGS, under the direction of the first author, conducted annual scientific investigations of the glaciology and glacial geology of both the Bering and Steller Lobes and the adjacent marginal lakes and foreland. The USGS studies were responsible for providing a comprehensive understanding of many aspects of the dynamics, morphology, and history of the Bering Glacier System. Activities included library and archive research; determination of past surge history; analysis of historical photography from identifiable photo stations (earliest photographs date from ~1905); reconnaissance field mapping by the authors (1988–2002); time lapse photography from in situ camera systems during the 1993–1995 surge; measurement of discharge and flow from telemetering stage recorders (1991–1995);

Figure 3. (A) South-looking 1905 photograph of the western part of the terminus of Steller Glacier and First Berg Lake with Bering Glacier in background. (Martin photo no. 246). (B) East-looking 1905 photograph of the western part of the terminus of Steller Glacier and First Berg Lake, taken from an elevation of ~500 m on Carbon Mountain, the ridge that bounds the northwest side of Berg Lake (Martin photo no. 245). Both photographs are by G.C. Martin, courtesy of the U.S. Geological Survey Photo Library, Denver, Colorado.

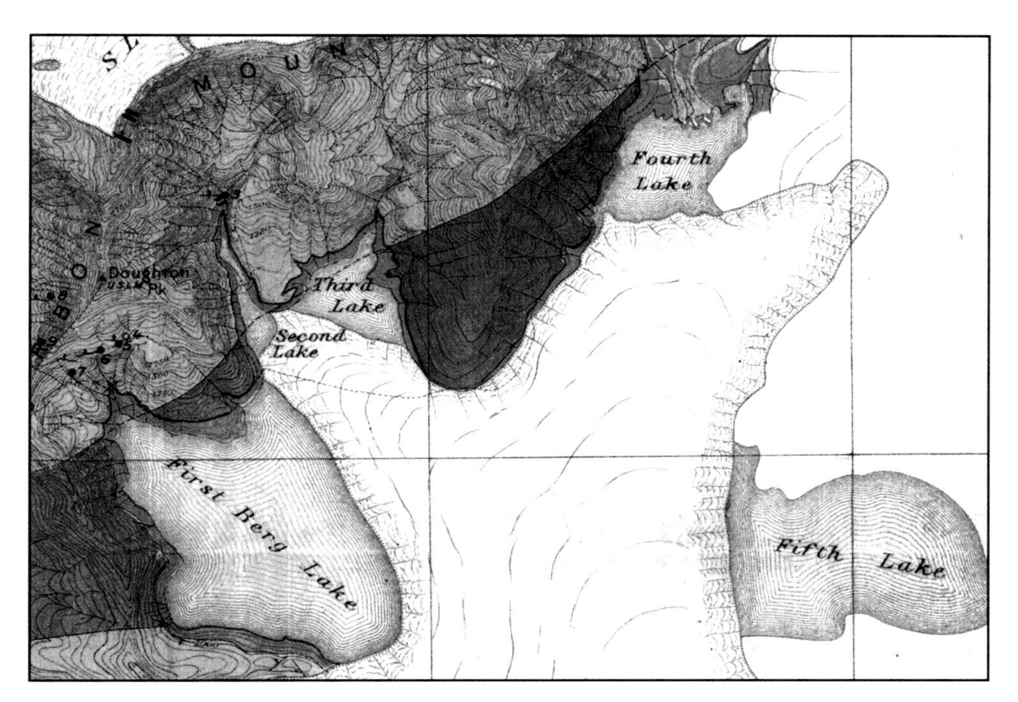

Figure 4. Portion of a 1907 1:62,500-scale USGS *Geologic Map and Sections of the Controller Bay Region, Alaska* by G.C. Martin, assisted by A.G. Maddren, Sidney Paige, and C.E. Weaver. The map shows that by the beginning of the twentieth century, five separate Berg Lakes had developed around the retreating terminus of the Steller Glacier. The map is oriented with north up, and covers an area that is ~6 km from east to west and ~5 km from south to north.

ice-penetrating radar surveys to determine bed depth and ice thickness (1990–1993); high resolution marine seismic reflection surveys of Vitus Lake, the glacier's southeastern ice marginal lake (1991 and 1993); seismic refraction surveys of the glacier's outwash plain (1991–1993); dynamics of 1994 floods at Berg Lake and Weeping Peat Island; dendrochronological and tree coring studies (1976–2002); monitoring of glacier terminus position stakes and shoreline erosion stakes at selected sites during the 1993–1995 surge; precision location of geographic features using differential global positioning system (GPS) (1992–2001); and sampling for water chemistry and suspended sediment load (1976–1980, 1993, 1995). Results of many of these investigations are published in Molnia and Post (1995) and Molnia (2008).

Beginning ~1990, annual field research was undertaken in the Tsivat and Tsiu Lakes area of the eastern piedmont lobe by field parties led by P.J. Flesher, State University of New York College at Oneonta, and Ernest Muller, Syracuse University, Syracuse, New York. Detailed observations of these lakes and their recent changes are presented in a later chapter of this volume.

Figure 5. (A) West-looking 1938 summer photograph by Bradford Washburn, showing the eastern terminus of Bering Glacier. Vitus Lake is present but is much smaller than in the latter part of the twentieth century. The sediment-laden Seal River exits the lake at the middle of the left edge of the image. The river in the foreground is Midtimber River. Subsequent lowering of Vitus Lake's level resulted in the cessation of flow into the outlet of Midtimber River. The end of the distant ridge, in the upper center of the photograph, is ~75 km west of the southeast margin of Vitus Lake. Unnumbered Washburn photo. (B) North-northeast-looking 1938 photograph by Bradford Washburn, showing the western terminus of Steller Glacier and Berg Lake. Berg Lake is much smaller than in the latter part of the twentieth century. By 1938, First, Second, and Third Berg Lakes had expanded and had become one, Fourth Lake had drained, and Fifth Lake has expanded in area but is separate from all the other Berg Lakes. Fifth Lake remains dammed by the eastern side of the Steller Glacier terminus. Washburn photo number 2274. Washburn photographs are used with the permission of the National Snow and Ice Data Center, Boulder, Colorado.

Following the 1993–1995 surge, the U.S. Bureau of Land Management (BLM) took an active role in developing and managing interdisciplinary studies at the Bering Glacier. Examples included resource-management, fisheries, botanical, hydrologic, and marine mammal investigations. Many of these specific disciplinary investigations are described in later chapters of this book.

Remote Sensing Data

The Bering Glacier area is covered by an extensive photographic and remotely sensed database that spans eight decades. Data collections include >75 vertical and oblique aerial photographic data sets, including the Washburn 1938 photographs and many of the Post photographs made between 1960 and 1982. In all, this photographic data set contains >10,000 aerial photographs (1938–2008); >7500 ground-based photographs of Bering Glacier System features by early observers and the authors (1905–2008); >50 Landsat Multispectral Scanner (MSS) and Thematic Mapper (TM) and Enhanced Thematic Mapper (ETM) images (1972–2007); ~150 digital satellite, Space Shuttle, and airborne radar images (1978–2007); and ~45 h of airborne video (1989–2007). These data provide a comprehensive database and baseline to enhance and extend the results of the numerous field studies. Field observations and ice-surface measurements have provided ground truth and complementary information that enhance our ability to interpret the extensive remotely sensed database.

In 2008 the U.S. Government released two declassified high-resolution electro-optical digital images of the lower Bering Lobe and adjacent Vitus Lake, one from September 1996 and the other from May 2005. The images, collected by National Technical Systems, were released at a resolution of 1 m (Fig. 7).

Topographic Mapping

In addition to Martin's early twentieth century map of the Bering terminus area, a complete topographic and image map base exists for monitoring and recording changes of the Bering Glacier System. Topographic mapping by USGS field parties began during the first half-decade of the twentieth century. Subsequently, the USGS (1959, 1984) prepared 1:250,000-scale and 1:63,360-scale topographic maps of the entire glacier using 1957 and 1972 aerial photography. Using June 19, 1991, Landsat TM data, the BLM produced a satellite image base map at a scale of 1:100,000 and a set of individual 1:63,360-scale quadrangle maps that provide a detailed view of the Lower Bering Glacier prior to the 1993–1995 surge.

GEOGRAPHIC FEATURES OF THE BERING GLACIER SYSTEM AND SURROUNDING AREAS

This section presents a brief introduction to a number of geographic features that constitute key components of the Bering

Glacier System and its surroundings. Whereas some features are mentioned only in this section, others are mentioned in nearly every chapter of this volume. Collectively, the following summary will permit readers to understand the geographic fabric that is the Bering Glacier System. The geographic features presented are not meant to be all-inclusive. Rather, those selected provide the reader with basic nomenclature to understand major ice, bedrock, and hydrologic components of the Bering Glacier System.

The entire Bering Glacier System lies within 75 km of the Gulf of Alaska coast (Fig. 1) in the eastern Chugach and western Saint Elias Mountains. Its accumulation area, including the Bagley Ice Valley and its associated tributaries, covers an area of ~3,210 km^2. Its ablation area, which includes the valley glacier segments of the Bering and Steller Glaciers and the complex terminal piedmont lobe, covers an area of ~1,965 km^2.

Geographic features will be presented as follows: (1) valley and ice features of the northern Bering Glacier System and its surrounding area, (2) ice features of the southern Bering Glacier System and its surrounding area, (3) bedrock features of the area within and surrounding the Bering Glacier System, (4) lakes and streams associated with the Bering Glacier System, (5) Bering Glacier Foreland, and (6) Bering Trough.

Valley and Ice Features of the Northern Bering Glacier System and Its Surrounding Area

Bagley Ice Valley

The Bagley Ice Valley, with a length of ~130 km, is a 7.5–12.5 km-wide, ice-filled, nearly straight valley, bordered by the Chugach and Saint Elias Mountains. In actuality, most of the valley is a fault trench, part of the Chugach–Saint Elias Fault System. The Bagley Ice Valley was named for James W. Bagley, a USGS topographic engineer who developed the Bagley T-3 camera and mapped in Alaska prior to the First World War. The eastern part of the Ice Valley begins at the confluence of the Columbus and Quintino Sella Glaciers, which is ~12 km northwest of Mount Huxley, at an elevation of ~1,800 m. From east to west, this valley is filled with an unnamed valley glacier, Waxell Glacier, and the upper Steller Glacier. Combined, these three glaciers make up the bulk of the accumulation area of the Bering Glacier System. Prior to 1997, the Bagley Icefield was used to describe the area that extended from the head of the Steller Glacier, east for nearly 130 km, to the head of the Columbus Glacier (Orth, 1967). [*Editor's note:* Although the name was officially changed by the BGN and the eastern limit defined by each term is different, some of the authors of this volume continue to use the terms Bagley Icefield or Bagley Ice Field. Therefore, in this volume, the names Bagley Ice Valley and the Bagley Icefield (Ice Field) will be used interchangeably.]

Unnamed Valley Glacier

The ~65-km-long west flowing unnamed valley glacier fills the eastern half of the Bagley Ice Valley. It originates west of the U.S.-Canada border at an elevation of ~1800 m. It terminates at the head of Bering Glacier's valley glacier segment, the Central Valley Reach, at an elevation of ~1000 m. The western part of this unnamed valley glacier becomes involved in surges of the Bering Lobe. During the 1993–1995 surge it was a major source of ice that was transported to the terminus region.

Waxell Glacier

East-flowing Waxell Glacier fills the central portion of the Bagley Ice Valley. With a length of ~35 km, it originates north of eastern Waxell Ridge and flows east to an elevation of ~1000 m, where it joins Bering Glacier at the Bering-Tana Glacier Divide, at the head of the Central Valley Reach. Together, Waxell Glacier and the unnamed valley glacier are the primary source of Bering Glacier's Central Valley Reach and Bering Lobe.

Steller Glacier

The upper part of the west-flowing Steller Glacier, extending to its divide with the Martin River Glacier, comprises the western portion of the Bagley Ice Valley. It originates in the eastern Chugach Mountains, north of western Waxell Ridge. The width of the glacier varies from 5.0 to 9.5 km. It has no distributaries. Steller Glacier undergoes frequent short-lived pulses of increased flow. With a frequency of ~10 a, the last event occurred in 2000 and 2001. According to Orth (1967), the glacier's name was derived in 1950 from Mount Steller. Initially, the name was applied only to the lower ~20 km of the glacier, with the rest of Steller Glacier being considered part of what was previously called the *Bagley Icefield.*

Quintino Sella Glacier

Quintino Sella Glacier originates in Canada and flows south for ~20 km until it merges with Columbus Glacier, west of Table Mountain, ~10 km west of the U.S.–Canada border. This confluence marks the eastern limit of the Bagley Ice Valley. It ranges in elevation from >5000 m to <2000 m, and it lies entirely within Bering Glacier's accumulation area. Its greatest width is ~8 km. Quintino Sella Glacier was named in 1897 for the pioneer of Italian alpinism by the Duke of the Abruzzi during his successful first assent of Mount Saint Elias (de Fillippi, 1899).

Columbus Glacier

Columbus Glacier, entirely within Bering Glacier's accumulation area, originates in Canada and flows to the west, crossing the U.S.-Canada border, and joining with Quintino Sella Glacier. It ranges in elevation from >5000 m to <2000 m. Its greatest width is ~10 km. Columbus Glacier was named on July 31, 1897, for Christopher Columbus by the Duke of the Abruzzi (de Fillippi, 1899).

Jefferies Glacier

Part of the eastern Jefferies Glacier is a tributary to upper Bagley Ice Valley. Jefferies Glacier is a westward flowing valley glacier, lying in the basin directly north of the Bagley Ice Valley. It originates west of the northern part of Quintino Sella Glacier at an elevation of ~2600 m. Jefferies Glacier has a length of ~80 km and a maximum width of ~8 km. Approximately 45 km west of the U.S.-Canada border, part of this glacier overtops its divide and flows to the south, contributing ice to the upper Bagley Ice Valley. The rest of Jefferies Glacier flows to the west and is a major source of Tana Glacier, external to the Bering Glacier System.

Tana Glacier

Tana Glacier is a north-flowing glacier north and west of Jefferies Glacier. The location of its medial moraines suggests that it receives about half of its flow from Waxell Glacier. Tana Glacier flows ~40 km northwest from Natural Arch to a stagnant, slowly receding moraine covered, thermokarst lake–filled terminus from which the Tana River discharges. The periodic surges of Bering Glacier do not affect Tana Glacier.

Bagley Ice Valley Equilibrium Line

Bering Glacier's equilibrium line is in the eastern Bagley Ice Valley northeast of Mount Miller. Since the 1950s the equilibrium line altitude (ELA) has risen from <700 m to ~1200 m. During the late 1980s the ELA ranged from ~900 to 1050 m. Meier and Post (1962) computed the accumulation area ratio (AAR) for Bering Glacier to be 0.63. Viens (1995) calculated an AAR of 0.62 for the entire Bering Glacier System, an AAR of 0.66 for the Stellar Glacier, and 0.614 for the remainder of the glacier system. The entire piedmont lobe lies below the ELA.

Ice Features of the Southern Bering Glacier System and Its Surrounding Area

Central Valley Reach

Located south of Waxell Ridge and north of the Grindle and Sint Hills, the Central Valley Reach is the ~45-km-long by ~13-km-wide valley glacier segment of the Bering Glacier. With an area of ~450 km^2, it represents a major portion of the glacier system's ablation area. The Central Valley Reach originates, and the Bagley Ice Valley terminates, at the "Gateway," the bedrock throat between Natural Arch on the south and Juniper Island on the north. Ice-penetrating radar (IPR) studies (Trabant et al., 1991; Trabant and Molnia, 1992) reveal that in places the bed of the Central Valley Reach is below sea level and that maximum glacier ice thicknesses exceed 800 m. Surface morphology and irregularities suggest that deep bedrock channels are present and that bedrock highs and ridges exist at depth. Continued retreat of Bering Glacier may uncover this huge basin, exposing its complex topography.

Kaliakh Lobe

Kaliakh Lobe (Fig. 1) is a ~12-km-long by ~3-km-wide southeast-flowing distributary that exits the east side of the Central Valley Reach. It terminates southwest of the Sint Hills in an ice-marginal lake that is the source of the Kosakuts River. After flowing south for ~5 km, the Kosakuts River flows into the Kaliakh River, which drains into the Gulf of Alaska, southwest of Kulthieth Mountain. The Kaliakh River was formerly an outlet stream, draining the northeastern margin of the Bering Lobe. Aside from the Seal River, which is the primary drainage way for the entire Bering Lobe, the Kosakuts and Kaliakh Rivers are the only streams that currently transport eastern Bering Glacier System discharge into the Gulf of Alaska.

Bering Piedmont Glacier

Bering Piedmont Glacier (Fig. 6) consists of three components: the Bering Lobe, the Steller Lobe, and the intervening Central Medial Moraine Band. This semicircular piedmont glacier has an area of ~900 km^2. The distance around the perimeter, from Berg Lake on the northwest to the Grindle Hills on the east, is ~80 km. The entire Piedmont Lobe has a maximum elevation of ~550 m and lies completely within the area of ablation. It is about one-third smaller than Malaspina Glacier's piedmont lobe and constitutes ~17.60% of the area of the entire Bering Glacier System. The entire Piedmont Lobe lies well below the equilibrium line and is subject to severe ablation. In addition, numerous terminal lakes are forming and rapidly increasing in size as the glacier recedes. Lake formation has accelerated both the rate of ice melting and terminus retreat. The eastern part of the glacier is subject to large-scale periodic surges.

Bering Lobe. Bering Lobe (Fig. 7) constitutes the eastern part of the Piedmont Lobe and makes up ~6.76% of the area of the entire Bering Glacier System. Bering Lobe has an area of ~350 km^2 and drains the main glacier by way of the Central Valley Reach. IPR studies conducted in 1991 by the USGS (Trabant et al., 1991; Trabant and Molnia, 1992) to determine its ice thickness, morphology, and depth to the glacier's bed, revealed that in places the glacier's bed extended down to 350 m below sea level. The maximum glacier ice thickness exceeds 800 m. The IPR studies suggested that much of the eastern piedmont lobe occupies a deep, subsea-level basin or series of channels. Depths to bedrock under the Bering Lobe were compatible with or even deeper than the bedrock depths in Vitus Lake. There, Molnia et al. (1996) used high-resolution seismic reflection profiling to document that Vitus Lake occupies a subsea-level basin (see below). The IPR and seismic surveys probably investigated different areas of the same basin. The Bering Lobe surges with a frequency of 20–30 a. The last major surge ended in 1995. A minor surge occurred in 2009.

Steller Lobe. The Steller Lobe (Fig. 8) consists of relatively debris-free active ice and makes up ~3.48% of the area of the entire Bering Glacier System. Its terminus bifurcates, with the northern portion flowing in a northerly direction and terminating in Berg Lake. The southern portion flows to the west and forms the eastern shoreline of several ice-marginal lakes north of the Suckling Hills, including Starodubtsov Lake and Lakes Roselius and Ivanov. At its widest the width of the Steller Lobe approaches 12 km. Tributaries of the 180 km^2 Steller Lobe and Steller Glacier cover an area of 644 km^2. This constitutes ~12.44% of the area of the entire Bering Glacier System. Aside from ~20 IPR measurements made near the eastern edge of the Steller Lobe in 1991, northeast of Nichawak Mountain, the thickness of most of the Steller Lobe has not been determined. The 1991 measurements

Figure 6. Two remotely sensed views of the Bering Piedmont Glacier. The piedmont lobe consists of three components: the Bering Lobe, the Steller Lobe, and the intervening Central Medial Moraine Band (see Fig. 1). This semicircular piedmont glacier has an area of ~900 km^2. (A) This August 1989 airborne x-band synthetic aperture radar (SAR) image shows the terminus of Bering Glacier 4 a prior to the onset of the 1993 surge and >20 a following the previous surge. By 1989, parts of the terminus had retreated >6 km from the maximum position attained at the end of the last surge in 1967. The image is oriented with northeast up, and covers an area that is ~55 km from east to west and ~40 km from south to north. (B) This 10 September 2001 false color Landsat Thematic Mapper (TM+), bands 6, 4, 2 false color image shows the retreating terminus of the glacier ~6 a following the end of the 1993–1995 surge. By 2001, retreat of the terminus had exposed about half of the area of Vitus Lake, covered by surge displaced ice. In this image, glacier ice and snow are blue, liquid water is black, nonvegetated sediment is red, and vegetation is green and yellow green. The image is oriented with north up, and covers an area that is ~80 km from east to west and ~60 km from south to north.

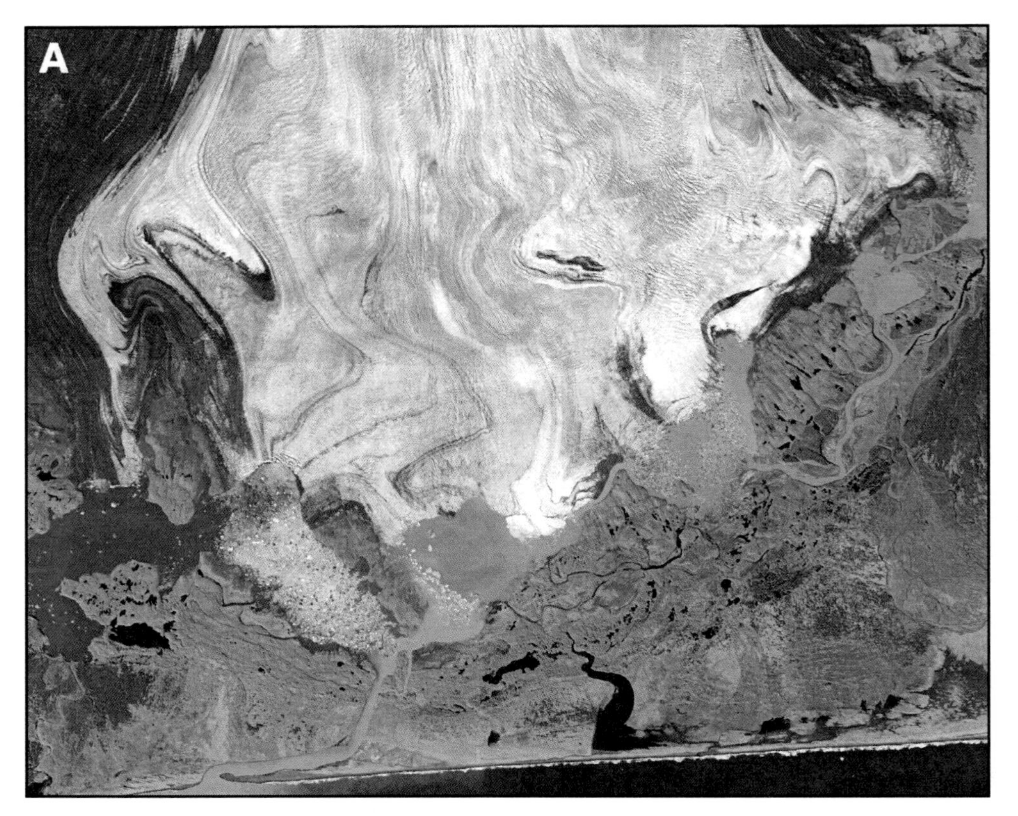

Figure 7. In 2008 the U.S. Government released two declassified high-resolution electro-optical digital images of the lower Bering Lobe and adjacent Vitus Lake. The images, collected by National Technical Systems in September 1996 and May 2005, were released at 1 m resolution. The images are both oriented with north up, and cover an area that is ~36 km from east to west and ~24 km from south to north. (A) This September 1996 image was collected about a year after the end of Bering Glacier's last major surge of the twentieth century. The surface of the glacier still shows evidence of the intensive crevassing that was related to its surge ice displacements and rapid movement. West of the island of Beringia Novaya, a semicircular calving embayment has formed. There, the westernmost calving area seen on the image, intensive calving has already resulted in ~800 m of terminus retreat. To the east, three other sites are actively calving icebergs. (B) This May 2005 image was collected about a decade after the end of Bering Glacier's last major surge. The glacier terminus has retreated as much as 7 km from its 1996 location. The only place where the terminus is within a kilometer of its 1996 position is at the easternmost part of the terminus. Only a few areas are actively calving icebergs. Two semicircular embayments are visible, one on either side of the Taggland Peninsula. The one on the west side is at the head of Tashalich Arm. The one on the east side is at the northwestern corner of Vitus Lake.

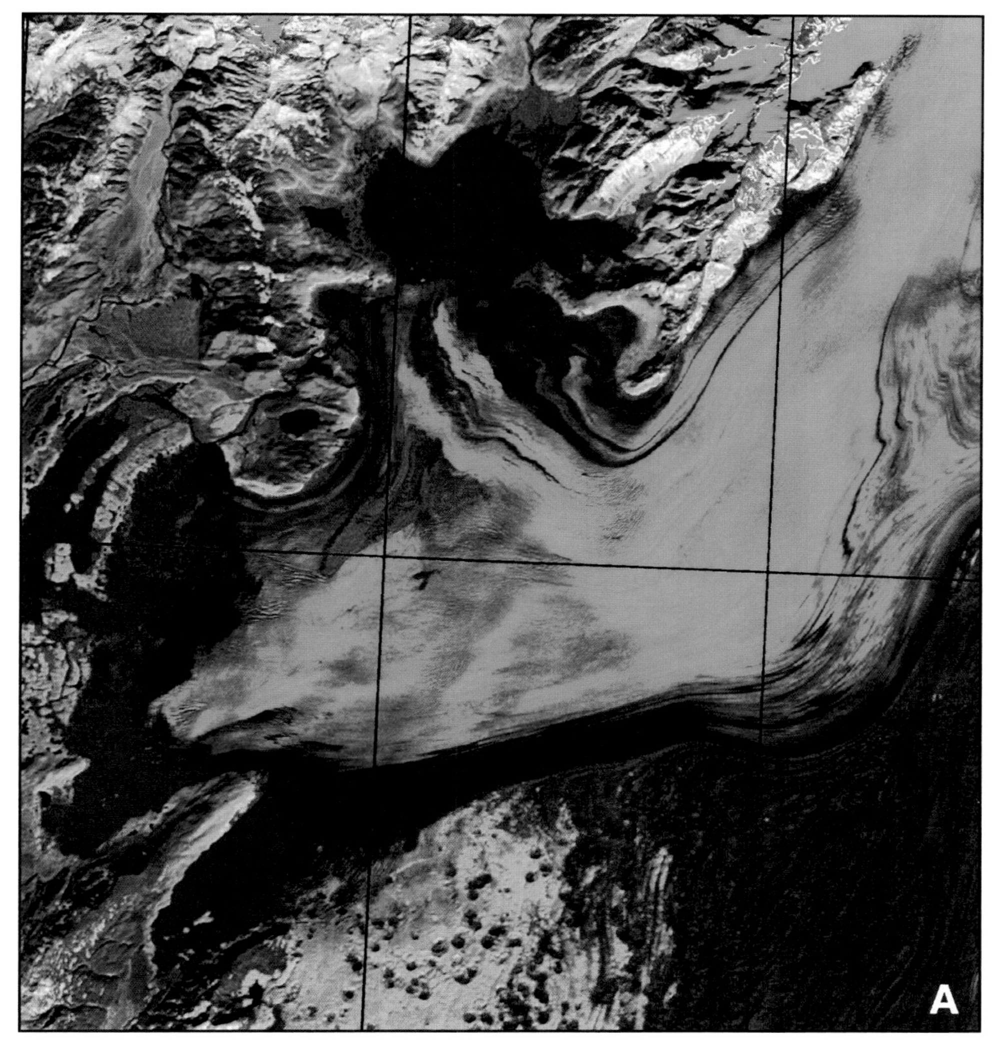

Figure 8. Two views of the Steller Lobe of the Bering Piedmont Glacier. The Steller Lobe has an area of 180 km^2. (A) This subscene of the September 2001 Landsat image shown in Figure 6B focuses on the details of Steller Lobe's bifurcating terminus. The northern lobe serves as the ice dam for Berg Lake, whereas the southern lobe dams Starodubtsov Lake, Lake Ivanov, and Lake Roselius. The bright reddish triangular area between the lobes is a mass of sediment that originated as a landslide or ice-rock avalanche many kilometers up-glacier. As it has been transported down-glacier, its original shape has been significantly distorted. Presently, the northern part of the slide mass is flowing toward Berg Lake, whereas the southern part is moving toward Lake Ivanov. The southernmost margin of the Steller Lobe is covered by an accumulation of ungulatory medial moraine sediment deposits. Nearly all of this sediment has originated in the Khitrov Hills. The image is oriented with north up, and covers an area that is ~22 km from east to west and ~27 km from south to north. (B) Northeast-looking, 11 August 2007 oblique aerial photograph, showing the Steller Lobe terminus in Lake Ivanov (lower right) and Starodubtsov Lake (lower left). A channel is developing that will soon connect the two lakes. It lengthens by the breaching of ice divides between thermokarst pits (ice sinkholes) and the capture of these thermokarst pits. The Steller Lobe terminus is beginning to float and produce large tabular icebergs through disarticulation. A channel is visible in the middle of the right edge of the photograph. As it lengthens into the Steller Lobe, this channel will connect Lake Ivanov to Lake Roselius. The image covers an area that is ~35 km from east to west and ~65 km from south to north. USGS photograph by Bruce F. Molnia (photo no. BFM-2007–0622a).

determined ice thicknesses between 136 and 349 m and depths to bedrock of between 3 and 149 m below sea level.

Central Medial Moraine Band. Post (1972) described the Central Medial Moraine Band (Fig. 9) as a very large debris band composed of repeatedly folded moraines that extend across the center of Bering Lobe. The Central Medial Moraine Band makes up ~7.36% of the area of the entire Bering Glacier System. Based on IPR thickness measurements and bed depths, the southern portion of the band appears to be composed of thin, debris-covered ice that overlies generally shallow bedrock between the thicker Bering and Steller Lobes. At one locality within the Central Medial Moraine Band, IPR measurements showed that bedrock was ~20 m above sea level and was covered by <100 m of ice. Maximum ice thicknesses exceed 500 m. The folded moraines

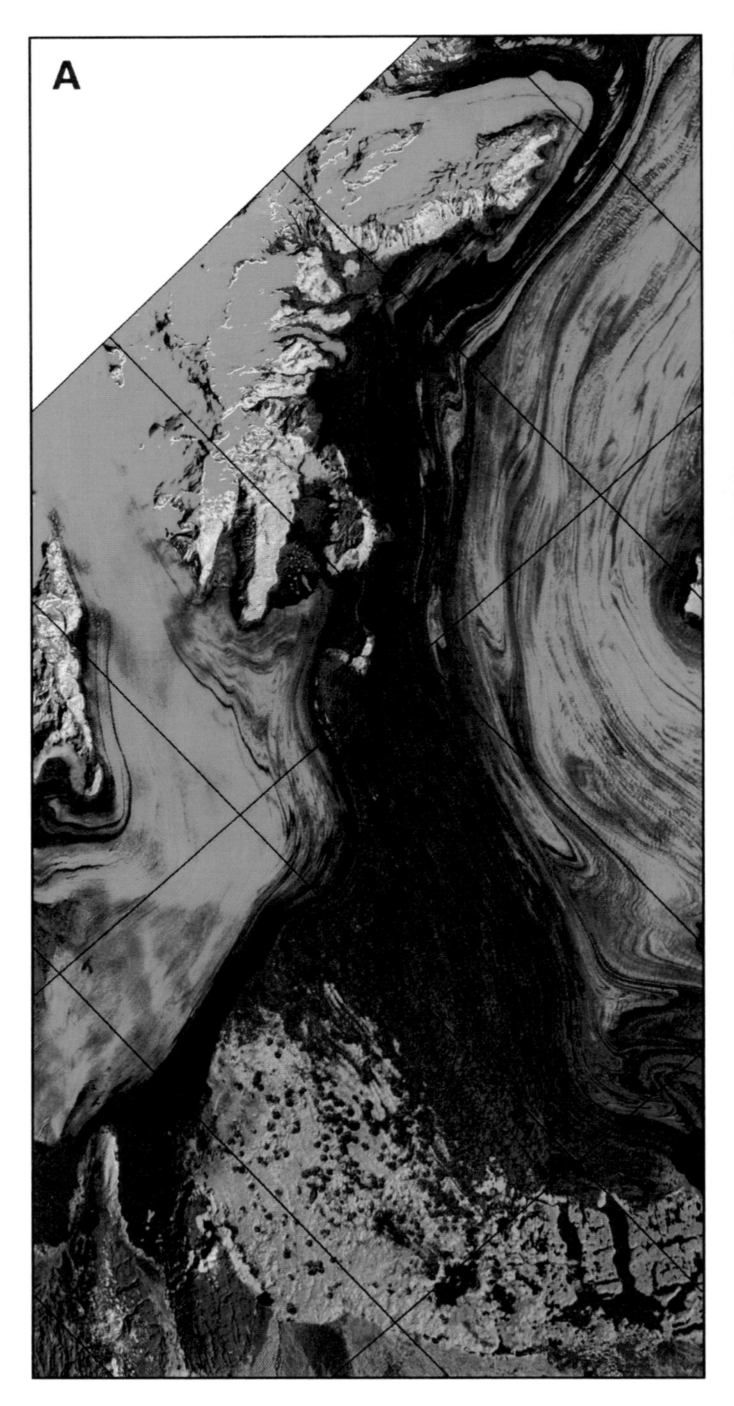

Figure 9. Two views of the Central Medial Moraine Band of the Bering Piedmont Glacier. This moraine band has an area of ~400 km^2. (A) This subscene of the September 2001 Landsat image shown in Figure 6B depicts the entire band. It can be seen originating from the southwest margin of Waxell Ridge and from the southeast margin of the Khitrov Hills. The complex character of its folded loop moraines can be seen along its entire length. The southern one-third of the Central Medial Moraine Band is characterized by hundreds of thermokarst pits. At many places they can be seen to have coalesced, forming irregularly shaped depressions and lakes. The chartreuse green area at the southern margin of the moraine band is covered with vegetation. The image is oriented with northeast up and covers an area ~55 km from northeast to southwest and ~25 km from southeast to northwest. (B) June 1938 northwest-looking oblique aerial photograph by Bradford Washburn (Washburn photo no. 1094), showing the head of the band. The intricately folded character of individual nested folded loop moraines can be seen. The folds are produced during Bering Glacier's surges. Individual moraines may be as high as 20 m. At its maximum, the Central Medial Moraine Band is ~20 km wide. This Washburn photograph is used with the permission of the National Snow and Ice Data Center, Boulder, Colorado.

of the Central Medial Moraine Band are derived from tributary glaciers flowing into the Central Valley Reach from the south side of Waxell Ridge and the southeastern Khitrov Hills. There, drag folds are formed in these moraines owing to differential flow that takes place during periodic surges. A new fold is formed with each surge. Nearly two dozen such folds are identifiable in the Central Medial Moraine Band. With a surge frequency of 20–30 a, as many as 600 a of surge history may have been recorded in the Central Medial Moraine Band. Including the stagnant, highly compressed folds of the terminus portion of the Central Medial Moraine Band, the record may span >2000 a. This stagnant marginal ice between Mount Campbell and the Suckling Hills likely contains the oldest ice in the Bering Glacier System. Much of the southern part of the Central Medial Moraine Band is covered by vegetation (Fig. 10).

Bedrock Features of the Area within and Surrounding the Bering Glacier System

Mountains and Peaks

The highest peaks in the Saint Elias Mountains segment of the Bering Glacier System drainage are Mount Logan (5959 m)

Figure 10. Two oblique aerial photographs of part of the southern Central Medial Moraine Band. Much of this area is covered by vegetation. Much of the vegetation sits directly on glacier ice and is rooted in glacial till. (A) Summer 1991 oblique, low-altitude aerial photograph of part of the southern area of the Central Medial Moraine Band. Here, vegetation, dominated by alder, covers much of the glacier's surface. Many nearly vertical exposures of bare glacier ice can be seen. Differential melting has resulted in the formation of several large circular thermokarst pits. Here, the largest is nearly 100 m long and 20 m deep. Lakes and drainageways are visible in many of these pits. Typically, glacier till thicknesses are <1 m. USGS photograph by Bruce F. Molnia (photo no. BFM-1991–99). (B) Summer 1991 oblique, very low altitude aerial photograph of the interior of a single thermokarst pit in the southern Central Medial Moraine Band. Here, dozens of alder bushes are sloughed off the surface by differential melting and slump down the walls of the thermokarst pit. In places the exposed ice is >25 m high. A meandering channel can be seen on the floor of the pit. USGS photograph by Bruce F. Molnia (photo no. BFM-1991–100).

and Mount Saint Elias (5489 m). These are the second and fourth highest peaks in North America. Mount Miller (3266 m) is the highest peak of Barkley Ridge, a western outlier of the Saint Elias Mountains. The highest peaks in the Chugach Mountains segment of the Bering Glacier System are Mount Tom White (3418 m) and Mount Steller (3237 m). Mount Steller is the highest point on Waxell Ridge.

Robinson Mountains. The Robinson Mountains are a 70-km-long, east-west–trending mountain range that extends from west of Icy Bay to the eastern margin of the Bering Lobe. The eastern third of the mountains is covered by the Guyot and Yahtse Glaciers. Maximum elevations are <2000 m. A line of cirques on the northwestern third of the mountains, northeast of McIntosh Peak, contribute glacier ice directly to the Central Valley Reach. Donald Ridge, a 350-m-high, northeast-southwest–oriented ridge and the Grindle Hills make up the western extent of the Robinson Mountains, west of the Kaliakh River. The Grindle Hills (Fig. 11), on the northeast corner of the Bering Lobe, are a U-shaped, 5-km-long, northwest-facing nunatak, composed predominantly of Yakataga Formation sedimentary rocks. Maximum elevations approach 800 m.

Figure 11. Two 11 August 2007 west-southwest oblique aerial photographs, showing the Grindle Hills and vicinity. Composed primarily of the Yakataga Formation, the morphology of the Grindle Hills suggests that it was a large Pleistocene complex cirque. It contains several ancestral tarn lakes. Today, it is a complex nunatak that projects above the Bering Lobe. A small striated tongue of the Bering Lobe enters its breached western side. Maximum elevations are ~750 m. (A) Distant view showing the Grindle Hills and the adjacent Bering Lobe. (B) Close-up of the Grindle Hills. USGS photographs by Bruce F. Molnia (photo nos. BFM-2007–2775 and 2007–2779).

Hills and Ridges

In addition to the major peaks, several lesser ridges are associated with the Bering Glacier System. From west to east, these are the Plenisner Hills (west of the Steller Glacier), Khitrov Hills (between the Steller Glacier and the Central Valley Reach), Grindle Hills (north of the Bering Lobe), and Sint Hills (south of the Central Valley Reach). With the exception of the Grindle Hills and Barkley Ridge, all of these features are named after members of Vitus Bering's 1741 crew.

Waxell Ridge. Waxell Ridge (Fig. 12), with a length of 40 km, separates the Central Valley Reach from the Bagley Ice Valley. Its high point is the summit of Mount Steller. Many cirque and small valley glaciers descend from the south side of the ridge and the flanks of Mount Steller and flow south. About half a dozen of these small tributaries to the Central Valley Reach are named for members of Bering's crew: Hesselberg Glacier, named for the oldest crew member; Yushin Glacier, named for the navigator who kept detailed journals of the voyage; Betge Glacier, named for the ship's surgeon; and Ovtsin Glacier, named for Bering's lieutenant. During the last half of the twentieth and the early twenty-first century, every one of these glaciers has thinned and retreated. On the south flank of Mount Steller, glaciers originate in four cirque basins. Prior to 1950, a single large glacier exited each cirque basin and descended to the base of the mountain, connecting to one of two debris-covered valley glaciers that flowed into Bering Glacier. Today, many of these former tributaries no longer reach the Bering. In 2005 and 2006, three very large ice and rock avalanches occurred from the crest of Waxell Ridge (Fig. 13).

Barkley Ridge. Barkley Ridge is the USBGN name assigned to the western ~12 km of a series of discontinuous ridges and nunataks that constitute the southern wall of the Bagley Ice Valley. Mount Miller is the only named peak. North of Mount Miller, Sorokin Glacier, a 2-km-wide by 11-km-long valley glacier, flows westwardly and joins the Central Valley Reach southwest of Natural Arch. Sorokin was Bering's sail maker. The western part of Barkley Ridge is separated from the Robinson Mountains by Kuleska Glacier, a 17-km-long, debris-covered valley glacier that heads near Mount Miller. Kuleska was Bering's blacksmith. The remainder of the ridge complex is subparallel to the Juniper Island ridges and nunataks, and also extends eastward for nearly 60 km. Maximum elevations increase to the east. Near the Canadian border, maximum ridge crest elevations are ~3000 m. More than 50 small cirque glaciers flow from the ridge northward toward the Bering Glacier. Some no longer make contact.

Juniper Island. Juniper Island (Fig. 14) is the USBGN name for the western end of a series of discontinuous ridges and nunataks that make up the northern wall of the Bagley Ice Valley. The ridge complex extends eastward for nearly 60 km. Elevations increase from ~1500 m at Juniper Island to ~2500 m at the eastern end of the ridge complex. Only a few small cirque glaciers flow from the ridge toward Bering Glacier. Nearly a dozen cirque and valley glaciers flow from the north side of the ridge to the Jefferies Glacier.

Khitrov Hills. The Khitrov Hills are a ~15-km-long arcuate ridge that separates Steller Glacier from the Medial Moraine Band. The hills have maximum elevations of ~1000 m and are south-southwest of Waxell Ridge. A number of ice-marginal lakes separate adjacent glacier margins from the bedrock slopes of the hills. The two largest of these lakes, each >1 km in length, lie at the south end of the ridge.

Ridges Adjacent to the Steller Lobe. Three small mountains are adjacent to the southwestern margin of the Steller Lobe. The three—Mount Campbell, Nichawak Mountain, and Gandil Mountain—were ice-marginal ridges during Steller Glacier's Little Ice Age maximum extent. Mount Campbell has a length of ~3 km and a maximum elevation of ~550 m. Nichawak Mountain has a length of ~5 km and a maximum elevation of ~500 m. These two ridges are separated by the Campbell River. Gandil Mountain has a length of ~1.5 km and a maximum elevation of ~280 m. It is separated from Nichawak Mountain by the Nichawak River.

Lakes Associated with the Bering Glacier System

Terminal (Ice-Marginal) Lakes

Soon after the first decade of the twentieth century, loss of ice by melting along the piedmont lobe margin resulted in the development of a series of lakes. Field (1975) reported that between 1905 and 1957, as the glacier retreated back from its Little Ice Age maximum moraine, the percentage of the perimeter of Bering Glacier ending in lakes increased from ~10% to ~54%. These lakes and several other Bering Glacier features have been used as targets to compare the use of spaceborne ERS-1 and SIR-C synthetic aperture radar (SAR) with airborne SAR in identifying different types of glacial features (Molnia and Molnia, 1995).

By the early 1920s, with continued melting and retreat, numerous small lakes expanded and united, forming Vitus Lake (Molnia and Post, 1995) on the southern margin of the Bering Lobe and the Berg Lakes on the northwestern margin of the Steller Lobe. Retreat of the piedmont lobe's margin has resulted in their continued expansion. Soon after the first decade of the twentieth century, loss of ice by melting along the piedmont lobe margin resulted in the development of other lakes along much of the glacier's perimeter. At various times, lake expansion has been temporarily interrupted by a series of twentieth century surges.

Lakes Associated with the Bering Lobe

Vitus Lake

Vitus Lake (Fig. 15), the largest terminal lake, presently has a length of 23 km, a maximum width of 8.5 km, and an area of ~160 km^2. Following the 1993–1995 surge, the area of Vitus Lake was reduced to ~50 km^2. Subsequent retreat of the Bering Lobe margin has resulted in Vitus Lake's expansion close to its pre-surge dimensions.

Pre–1993–1995 surge bathymetric surveys of the lake by the authors and Carlson in 1991 and 1993 revealed complex bottom

Figure 12. Two 11 August 2007 oblique aerial photographs, showing most of Waxell Ridge, including Mount Steller. (A) View to the northeast, showing the easternmost part of Waxell Ridge. Rather than having a linear crest, Waxell Ridge is composed of a complex series of connected horns and cirques with associated valleys. At least five glaciers flow from this part of the ridge. Bagley Ice Valley is to the upper left. (B) View to the north, showing the central part of Waxell Ridge, including Mount Steller. The large sediment cover on the surface of the glaciers that descend from Mount Steller is indicative of the intensive amount of ice and rock avalanche and landslide activity that occurs here. As was the case with the eastern part of Waxell Ridge, central Waxell Ridge is a complex series of connected horns, cirques, and valleys. Although not precisely aligned, the left side of photograph 12A and the right side of photograph 12B join to form a panorama. USGS photographs by Bruce F. Molnia (photo nos. BFM-2007–2616 and BFM-2007–2606).

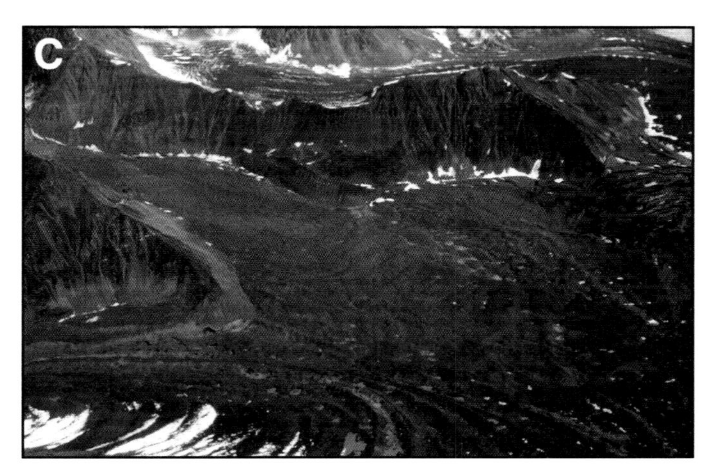

Figure 13. Three 11 August 2007 oblique aerial photographs, showing the Mount Steller landslide–rock avalanche. This large Waxell Ridge mass wasting event occurred 14 September 2005. (A) North-looking view of Mount Steller, showing the large breached cirque that composes much of its south side. The smooth, lighter-colored, steep upper part of the basin was the area of origin of the 2005 slide. The haze surrounding the summit area consists of fine-grained dust that is being blown from the slide source area. Prior to the September 2005 event, a cirque glacier filled the floor of the eastern part of the cirque. The slide mass removed the eastern part of the glacier from the cirque floor and deposited it below as a reconstituted glacier. (B) North view of the central part of Mount Steller, showing the path of the 2005 slide. The glacier ice that was ripped from the cirque floor was deposited as a debris-covered, reconstituted glacier. (C) Northeast look at the toe of the slide. The deposit that constitutes the toe is composed of a mixture of rock, glacier ice, and snow. It was deposited on top of a series of hummocky moraines. As the snow and ice component of the toe has melted, the toe has thinned >10 m, and the preexisting hummocky topography is beginning to show through the debris. USGS photographs by Bruce F. Molnia (photo nos. BFM-2007–0534, BFM-2007–784, and BFM-2007–2613).

Figure 14. East-looking 27 July 2006 oblique aerial photograph, showing the western end of Juniper Island (center). Juniper Island separates the Bagley Ice Valley (right) from Jefferies Glacier (left). The ridge is composed of sedimentary and metasedimentary rocks. Both sides of Juniper Island are mantled with thick covers of lateral moraine. These sediments were deposited as the glaciers on both sides thinned. The mountain at the right edge of the photograph is Mount Saint Elias. USGS photograph by Bruce F. Molnia (photo no. BFM-2006–1536).

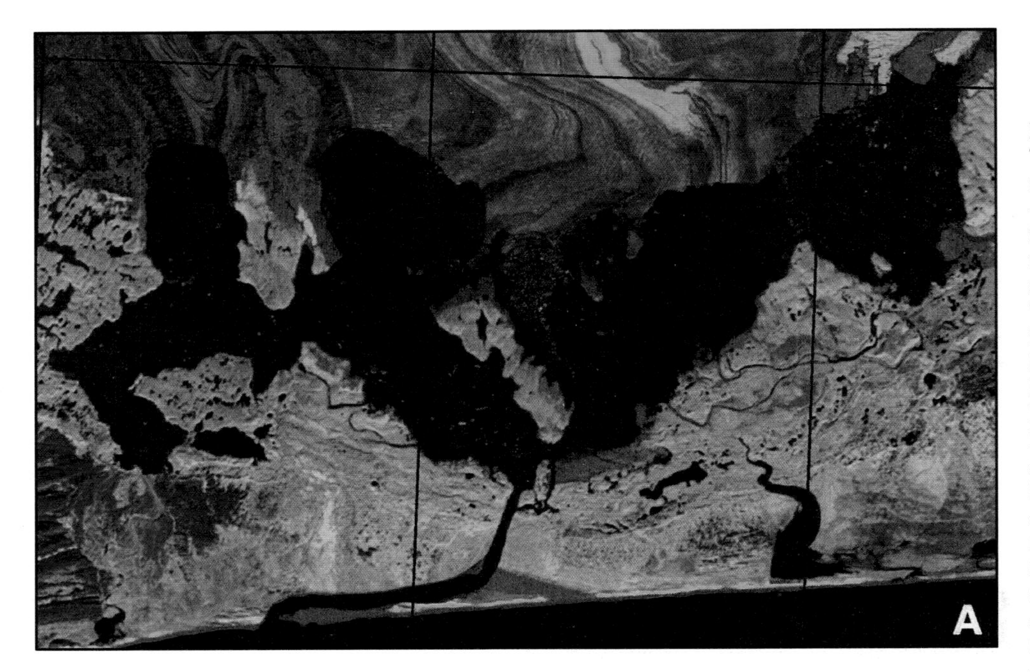

Figure 15. Two views of Vitus Lake. This lake has a length of 23 km, a maximum width of 8.5 km, and an area of ~160 km^2. (A) Subscene of the 10 September 2001 false color Landsat Thematic Mapper (TM+), band 6, 4, 2 false color image, showing all of Vitus Lake. The drainageway at the bottom of the image that connects Vitus Lake to the Pacific Ocean is the Seal River. The large island in the center of the lake is Beringia Novaya. To its right is Pointed Island, with a length of ~1.0 km. Northeast of Pointed Island is the larger of two islands, called the Wallypogs. The second Wallypog can be seen just emerging from under the retreating margin of Bering Glacier between Pointed Island and the larger Wallypog. The image is oriented with north up and covers an area that is ~25 km from east to west and ~20 km from south to north. (B) Southeast-looking oblique aerial photograph taken 11 August 2007, showing nearly all of Vitus Lake. Compare the distance between the ice margins in 2007 with that in 2001. In places, the terminus has retreated as much as 5 km during the nearly 6 a between the time when Figure 15A and Figure 15B were taken. At the right middle edge of the photograph is the Taggland Peninsula. Note the boundary between the vegetation-covered right side of the landmass and the bare left side. The color boundary marks the limit of the area covered by the terminus of Bering Glacier at the end of the 1996 surge. From left (east) to right (west), the islands visible in Vitus Lake are Whaleback Island, the Wollypogs, Pointed Island, and Beringia Novaya. USGS photograph by Bruce F. Molnia (photo no. BFM-2007–2797).

morphology, with at least four deep basins with water depths of 165 m, 135 m, 135 m, and 85 m. Two seismic reflection surveys (Carlson et al., 1993; Molnia et al., 1996) revealed maximum glacial-lacustrine sediment-fill thicknesses of >110 m, with depths to bedrock basement reaching a maximum of 275 m below sea level (Molnia et al., 1996). Taxpayers Bay, the southeasternmost triangular shaped, 4.2 km by 4.5 km embayment in the shoreline of Vitus Lake, is virtually flat-floored with a depth of ~40 m. It also has bedrock depths >200 m.

Six Islands project above the surface of Vitus Lake. From east to west they are Tsistus (Arrowhead) Island, Whaleback Island, The two Wollypogs, Pointed Island, and Beringia Novaya (Long Island). All of the islands are composed of stratified glacial-lacustrine sediment and glacial-marine sediment and are covered by glacial till. They are typically flat surfaced with maximum elevations of <20 m. Tsistus (Arrowhead) Island (Fig. 16) is ~1.7 km long and a maximum of 1.2 km wide. It is heart shaped, with its tapered end facing southeast. Whale-

Figure 16. Two oblique aerial photographic views of Tsistus (Arrowhead) Island. (A) View taken 25 June 1993 of the ~1.7-km-long island as it appeared after emerging from under the retreating Bering Glacier following the 1970s surge. The island was covered by the glacier for >15 a, emergent for <1 a, and then quickly covered by the 1993–1995 surge. (B) August 1994 view of the island after all but the southern end of the island was covered with surge-advanced ice. USGS photographs by Bruce F. Molnia (Photo nos. BFM-1993–2156 and BFM-1994–1497).

back Island, The Wollypogs, and Pointed Island all sit on top of a 3.5-km-wide bathymetric high in the middle of Vitus Lake. Maximum water depths here are <20 m. Whaleback Island is ~0.5 km long and a maximum of 0.2 km wide. It is oriented north-south and bulges toward the east. The larger of the two Wollypogs is ~0.3 km long and a maximum of 0.12 km wide. Pointed Island is ~1.2 km long and a maximum of 0.34 km wide. As its name implies, the northern half of the island tapers and ends in a narrow point. Beringia Novaya (Fig. 17) is ~4.5 km long. Its width ranges from a maximum of 1.7 km to a minimum of 0.42 km.

Northwest of Beringia Novaya is a 6.5-km-long, north-south–oriented peninsula of land called Taggland. It is up to 1.9 km wide. At its closest, it is 2.2 km west of Beringia Novaya. IPR studies conducted 2 km north of the north end of Taggland suggest that the depth to the glacier's bed there is below sea level. Hence, continued retreat of Bering Glacier may reveal that Taggland is an island.

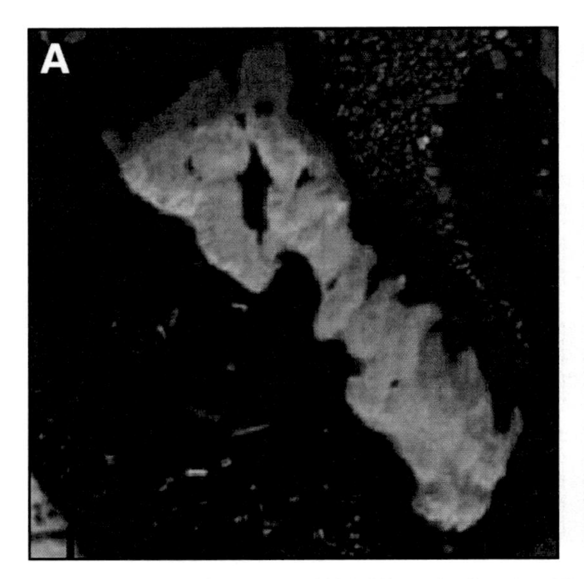

Figure 17. (A) 10 September 2001 false color Landsat Thematic Mapper (TM+), bands 6, 4, 2 false color image showing ~4.5 km long Beringia Novaya. The red areas on the top (north) and right (east) sides of the island are vegetation free and were covered by glacier ice at the end of the last surge in 1996. North is to the top. (B) October 1974 south-looking oblique aerial photography showing most of the island and adjacent Vitus Lake soon after Beringia Novaya emerged from under the retreating terminus of Bering Glacier following the 1960s surge. Note that the only vegetation present on the island is at the south end, clearly defining the maximum extent of the last surge. Also note the numerous grooves that were cut into the surface of the island by the recent advance. Vitus Lake's level was several meters higher in 1974 than following the 1993–1995 surge. USGS photograph by Bruce F. Molnia (BFM-1974-1263).

South of the Taggland Peninsula, Vitus Lake narrows to a width of 2.1 km, its narrowest anywhere. Between the south end of Taggland and the mainland to the south, water depths are <50 m. They are significantly deeper both to the east and west.

Tashalich Arm (Fig. 18), on the west side of Vitus Lake, is 7.0 km long and as wide as 3.5 km. It is steep walled and very deep, with maximum water depths >180 m. The west shore of Tashalich Arm is the northwesternmost extension of Vitus Lake. The surface topography of the adjacent Bering Lobe suggests that this deep channel continues north under the glacier as far as the south end of the Khitrov Hills. Judging from adjacent surface topography, several more channels and islands will probably be uncovered as the ice recedes. South of Tashalich Arm is Laurie Bay, a 6.2-km-long L-shaped basin with a maximum water depth of >80 m. Its maximum width is 2.1 km.

Figure 18. South-southeast-looking 11 August 2007 oblique aerial photograph of Tashalich Arm. The arm is a maximum of 2.4 km wide and nearly 6.5 km long. Tashalich Arm is the deepest part of Vitus Lake, with maximum depths exceeding 185 m. The gray to green transition marks the maximum extent that ice advanced on either side of the Arm during the 1993–1995 surge. The part of the glacier that filled the Arm during the surge extended nearly 1 km beyond the maximum land extent of the ice. Note the arcuate embayment at the head of the arm and the abundant calving that was under way. USGS photograph by Bruce F. Molnia (photo no. BFM-2000–2796).

Eastern Marginal Lakes

Prior to the 1993–1995 surge, two ice marginal lakes, Tsiu and Tsivat Lakes, were adjacent to the southeastern part of the Bering Lobe. They were separated by Weeping Peat Island and connected by Tsivat Outlet. In July 1994 a large outburst flood that originated on the east side of Weeping Peat Island deposited enough sediment to first fill Tsivat Lake and then later, most of Tsiu Lake. Prior to the surge, Weeping Peat Island was 2.9 km long by 2.5 km wide with a trapezoidal shape. Early in the surge, the terminus of the Bering Lobe advanced over Peat Falls Island, covered the northern part of Weeping Peat Island, and rapidly flowed into the northern parts of both lakes.

Tsiu Lake. Pre-surge Tsiu Lake, the westernmost of the two lakes, was ~5.0 km long and ~1.2 km wide. It was oriented northwest to southeast, with maximum water depths >40 m. Peat Falls Island defined its northern margin. It drained into Vitus Lake to the west through a 200-m-wide outlet channel east of Arrowhead Island. The terminus of Bering Glacier constituted the north wall of the channel. By late 1993, this channel was blocked by the advancing terminus of the glacier. As a result, the lake was forced to develop a new drainageway to the east. The original western outlet remained ice-dammed until August 2006, when it reopened subglacially. Much of the lake was filled with sediment during the 1994–1995 outburst flood.

Tsivat Lake. Pre-surge Tsivat Lake, the easternmost of the two lakes, was ~3.5 km long and ~2.7 km wide. Its maximum water depths were >50 m. The terminus of Bering Glacier was the west margin of the lake. It drained into Tsiu Lake to the west, through a 200-m-wide outlet channel, Tsiu Outlet, on the southeast side of Weeping Peat Island. The 1993 closure of Tsiu Lake's western outlet resulted in Tsivat Lake's developing another outlet, draining through its southeastern margin into the Abandoned River. Like Tsiu Lake, much of Tsivat Lake was filled with sediment during the 1994–1995 outburst flood.

Hanna Lake. Hanna Lake lies in a 2.3-km-long by 1.7-km-wide bedrock basin >3 km east of Bering Glacier and >4 km north of Tsivat Lake. It drains to the south through a small unnamed lake before its waters enter Tsivat Lake.

Lakes Associated with the Medial Moraine Band

Several hundred lakes exist in the essentially stagnant Central Medial Moraine Band. Two, the Parallel Linear Lakes (Fig. 19), cut across the outer Central Medial Moraine Band and are the product of more than half a century of differential melting. The remaining lakes, mostly thermokarst lakes and pits, are the product of in situ differential melting.

Parallel Linear Lakes

About 2 km west of Tashalich Arm is the northeast end of the eastern of two northeast-southwest–trending Parallel Linear Lakes. This lake, Lynne Canal, is ~5.5 km long and ~0.6 km wide. Its maximum water depths are between 40 and 50 m. About 1.2 km to its west is the second lake, Loch Rina, which is ~4 km long and also ~0.6 km wide. Its water depths were never determined. The two lakes lie along the eastern margin of the Medial

Figure 19. Two views of the Parallel Linear Lakes. The lakes, each ~5.2 km long, are the product of thermokarst erosion. (A) South-looking 11 August 2007 oblique aerial photograph, showing the two large connected lakes that drain to the southeast (upper left) into Vitus Lake. Since the early part of the twentieth century the lakes have slowly developed by the melting of subsurface ice and the joining of thermokarst pits. North is at the bottom of the image. USGS photograph by Bruce F. Molnia (photo no. BFM-2007–2814a). (B) A 10 September 2001 false color Landsat Thematic Mapper (TM+), bands 6, 4, 2 false color image, showing essentially the same area as seen in Figure 19A. During the 6 a, 1 mo., and 3 d between images, only small changes have occurred. The most obvious are at the northern end of the two lakes, where they contact the peninsula that separates them. There, both lakes have lengthened a few tens of meters by expanding into the base of the peninsula.

Moraine Band. They began to develop in the 1940s, initially by the coalescing of thermokarst pits. They have continued to lengthen and widen through the early twenty-first century.

Lakes Associated with Steller Lobe

As is case with the Bering Lobe, much of the western part of the terminus of the Steller Lobe is fronted by marginal lakes. Aside from Berg Lake, which was already present at the beginning of the twentieth century (although then much smaller), the other western marginal lakes formed much later, mostly through Steller Glacier terminus retreat during the second half of the twentieth century. By the beginning of the twenty-first century, three large lakes, Lake Roselius, Lake Ivanov, and Starodubtsov Lake, and a number of small unnamed lakes, fronted the western margin of the Steller Glacier.

Berg Lake

On the northwestern side of the piedmont lobe, Berg Lake (Fig. 20), with an area of ~30 km^2, is the largest of several ice-marginal lakes formed by the retreat of the Steller Lobe. During the Little Ice Age the basin presently occupied by Berg Lake was filled by the Steller Glacier, with its terminus pushing up the slopes of Carbon Mountain, the ~1000-m-high ridge that makes up much of the northern margin of the Berg Lake basin.

When Martin (1908) visited Berg Lake in 1905, its level was ~50 m higher than present. He mapped five separate small lakes, many with evidence of fluctuating water levels. He reported maximum water depths of ~250 m and noted that the water level had recently been as much as 100 m higher. The existence of these five lakes suggests that by 1905 the Steller Lobe was already in retreat from its Little Ice Age maximum position. By the early 1940s, continued retreat of the Steller Lobe resulted in the five lakes coalescing into one. Retreat of the Steller Lobe has been interrupted by several small pulses or mini-surges, including the one that occurred in 2000 and 2001. Post and Mayo (1971) note that in 1970 the lake was 207 m deep, ~40 m lower than in 1905. Prior to the twentieth century, the level was slightly higher than 300 m. Post and Mayo (1971) note that the total lake area had increased from ~12 km^2 in 1905 to ~28 km^2 in 1970. These changes are the result of the ongoing thinning and retreat of the Steller Lobe. By 2005 the area had increased to nearly 35 km^2.

Berg Lake has a history of at least three late twentieth century outburst floods. In 1984, 1986, and again in 1994, Berg Lake partially drained, causing major flooding in the Bering River drainage and severely impacting wildlife habitats. The May 1994 outburst flood event resulted from the catastrophic draining of Berg Lake through a ~500-m-long subglacial channel through the northwestern margin of the Steller Glacier. It resulted in >50 m of lowering of the lake level in a 72 h period. An estimated 5.5×10^9 m^3 of water escaped and drained through the Bering River, completely inundating the floor of the Bering River valley from valley wall to valley wall. As May is the time of moose calving and significant avian nesting activity, especially for Dusky Canada Geese and Trumpeter Swans, the flood had a major impact on the local ecosystem. Outburst floods, such as these, were predicted by Post and Mayo (1971), who determined that continued thinning of the glacier would result in a cycle of outburst flooding.

Starodubtsov Lake

Starodubtsov Lake (Figs. 8B, 21), with a length of 7.5 km and a width of 2.2 km, lies north of Lake Ivanov and southwest

Figure 20. A 3 October 2000 ground-based, southeast-looking photographic view of the terminus of Steller Glacier in Berg Lake. In 1994 a major draining of the lake occurred as lake water cut an englacial channel through the terminus. The northern part of the terminus, the part closest to the channel, was temporarily separated from the rest of the glacier. The ice mass seen here was pushed forward by a 2000–2001 advance of the Steller Glacier. Although it made contact with the shoreline of Berg Lake seen in the foreground, it failed to dam Berg Lake and cause a rise in lake level. Unnumbered BLM photograph by John Payne.

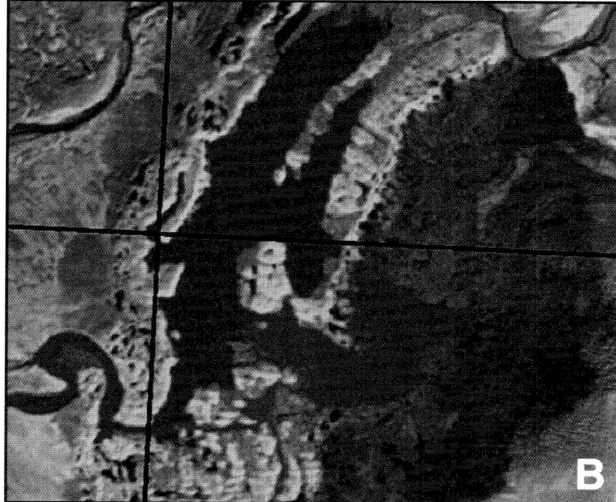

Figure 21. Two views of Bering Glacier's west ice-marginal lakes. (A) An 11 August 2007 northeast-looking oblique aerial photographic view of Lake Roselius (right), Lake Ivanov (left center), Starodubtsov Lake (upper left edge), and Berg Lake (top, just left of center). All are ice marginal lakes that expanded through retreat of the margins of the Steller Lobe. Lake Roselius lies between Mount Campbell, to the east, and Nichawak Mountain to the west. Lowering of its ice dam has resulted in a significant decrease in lake area. A meandering channel is developing at the north end of Nichawak Mountain that will connect Lake Roselius to Lake Ivanov. USGS photograph by Bruce F. Molnia (photo no. BFM-2007–2827). (B) A 10 September 2001 false color Landsat Thematic Mapper (TM+), bands 6, 4, 2 false color image showing all of Starodubtsov Lake.

of Berg Lake. It is bounded on the east and west by moraine and outwash sediment and on the east by the glacier's terminus. It is arcuate in shape, with depths <10 m. It drains to the west through a narrow passage that becomes the head of the Gandil River.

Lake Ivanov

V-shaped Lake Ivanov (Figs. 8B, 21), with a length of 4.1 km and a width of 2.6 km, is bounded on the south by Nichawak Mountain. Its western and northern perimeters are composed of morainal material, whereas its eastern margin is glacier ice. Its water is ~12 m deep. Lake Ivanov drains to the north through a narrow passage and connects with Starodubtsov Lake.

Lake Roselius

Lying between Mount Campbell and Nichawak Mountain, Lake Roselius (Fig. 21) is 3.9 km long by 2.4 km wide and elliptical in shape. Its maximum depths are ~40 m. The western margin of the lake is composed of morainal material, and the eastern margin is glacier ice. Lake Roselius drains to the southwest through the Campbell River.

Hohlinsock Lake

At the northwestern margin of the Central Valley Reach, this small glacial lake was named by the second author for its fancied resemblance to a large hole in the heel of a striped stocking.

Lake Sofron

At the south end of the Khitrov Hills, Lake Sofron is subject to infrequent abrupt dumping, which greatly reduces its area. As the Stellar Glacier thins, the level of the lake lowers. Presumably, with time, this lake will greatly expand as the Steller terminus retreats.

Small Unnamed Lakes

At least three small unnamed lakes exist adjacent to the western margin of the Steller Lobe. The southernmost of these, a small lake on the northeast side of Mount Campbell, is ~0.6 km in maximum dimension. Aside from the ice margin on the east, it lies in a bedrock basin. All of the other small lakes are moraine dammed or ice dammed.

Current and Former Outlet Streams Associated with the Bering Glacier System

At the beginning of the twentieth century, Bering Glacier's Piedmont Lobe had nearly a dozen major drainage outlets, arranged in a radial pattern along its margin. These included, from east to west, the Kaliakh River, the many branches of the Tsiu and Tsivat Rivers, Midtimber River, Seal River, Tashalich River, Campbell River, Kiklukh River, Oaklee River, Edwardes River, Nichawak River, Gandil River, Bering River, and the unnamed river that drains Berg Lake. A significant retreat of the Piedmont Lobe's terminus began sometime between 1905 and 1910. Since then, Bering Glacier has lost nearly 100 km^2 of ice from the terminus area of the Piedmont Lobe.

In the century since Bering Glacier's terminus began to retreat from its Little Ice Age maximum position, the retreat has resulted in the loss of contact between the glacier and many of its former outlet streams. This coincided with the development of several marginal lakes that drain large regions of the glacier's perimeter. Today, all of the Piedmont Lobe's discharge flows through four rivers: Seal River, Campbell River, Gandil River, and the unnamed river that drains Berg Lake.

Current Outlet Streams

Seal River

Seal River (Fig. 22) is a 6-km-long, low gradient, tidally dominated stream that drains Vitus Lake and flows diagonally across the Bering Foreland from east to west. On 28 June 1991 the discharge of the Seal River was measured where it exits Vitus Lake and was found to be 1.4×10^3 m^3/s. At the beginning of the twenty-first century, >90% of the drainage of the Bering Lobe and the Central Medial Moraine Band flowed through Vitus Lake, entering the Gulf of Alaska through the Seal River. Lake expansion and stream capture resulted in the Seal River becoming the dominant drainage outlet for the entire Bering Glacier System prior to 1950 (Molnia and Post, 1995).

Campbell River

Campbell River is an ~20-km-long, low gradient stream that drains Lake Roselius and flows into Controller Bay, west of Kanak Island. Most of its course is through a marshy wetland. Its lower reaches are tidally influenced.

Gandil River

Gandil River is an ~18-km-long, low gradient stream that drains Starodubtsov Lake and flows into Bering River west of Gandil Mountain. At various times its course west of Starodubtsov Lake has widened to form a lake of varying size. In 1992 it was as much as 6 km long by as much as 1.6 km wide. In 2001 it was ~3 km long by ~1 km wide.

Unnamed River that Drains Berg Lake

An unnamed river drains Berg Lake. Informally it is known as *Berg Lake outlet stream.* It exits the southwest corner of Berg Lake and then flows through a steep bedrock gorge before entering the northeast corner of Starodubtsov Lake. Its total length is ~5 km. From Berg Lake to Starodubtsov Lake its elevation decreases by ~60 m.

Former Outlet Streams

More than a dozen streams drained the perimeter of Bering Glacier at the beginning of the twentieth century (Fig. 23). These streams were the many branches of the Tsiu and Tsivat Rivers, Midtimber River, Seal River, Tashalich River, Kiklukh River, Oaklee River, Edwardes River, Nichawak River, and Bering River. All are detached from Bering Glacier and no longer receive any of its meltwaters. However, during outburst flood conditions, such as during the 1994 flood of Bering River, these streams were inundated with floodwater, and they served as the conduits for these floodwaters to the Gulf of Alaska.

Kaliakh River

Formerly the northeasternmost outlet stream of the Bering Lobe, the Kaliakh River formerly drained the area adjacent to the southeastern Grindle Hills and Donald Ridge. With a length of ~25 km, the river now receives most of its discharge from the Kosakuts River, formerly a minor tributary. In 1996 the lead author investigated the former points of origin of the Kaliakh, Tsivat, and Tsiu Rivers. In each case the heads of the former channels were adjacent to the Little Ice Age maximum terminus position of the glacier on its former outwash plain. All were several kilometers from the closest glacier ice and 15–25 m above the level of the Bering Lobe's marginal lakes.

East and West Tsivat River

Today, nearly a dozen small streams coalesce to form each of the two major channels of the ~25-km-long Tsivat River. A century ago these streams were part of a major braided stream network that drained the central part of the eastern Bering Lobe. During the second half of the twentieth century the mouth of the Tsivat River was sealed by a major washover, and its flow was diverted to the southwest where it joined with that of the Tsiu River.

Tsiu River

The Tsiu River consists of a network of small channels that currently originate in the Bering Foreland, west of the Tsivat River. A century ago these streams were part of a major braided stream network that drained the central part of the eastern Bering Lobe.

Midtimber River (Now Midtimber Lake)

Midtimber Lake is the remnant of what formerly was the Midtimber River. During the first half of the twentieth century the Midtimber River originated at the margin of Bering Glacier, one of more than a dozen rivers that radially drained away from the glacier. By the beginning of the second half of the twentieth century, terminus retreat and a lowering of the level of Vitus Lake beheaded this river, isolating its channel from a water source. Washovers and storm surges filled the former mouth of the river with sediment, leaving the south end of the lake >50 m from the high tide line.

Tashalich River

The Tashalich River is a meandering stream with a length of ~10 km. Historically, it drained the central and southwestern parts of the glacier margin, extending from what is now Laurie Bay to the Gulf of Alaska. When observed in 2007, the northernmost extent of the stream was ~600 m from the south shore of Laurie Bay, whereas the southern extent of its channel failed to reach the beach adjacent to the Gulf of Alaska.

Kiklukh River

The Kiklukh River is a meandering stream with a length of >20 km. Historically, it drained the southwestern part of the

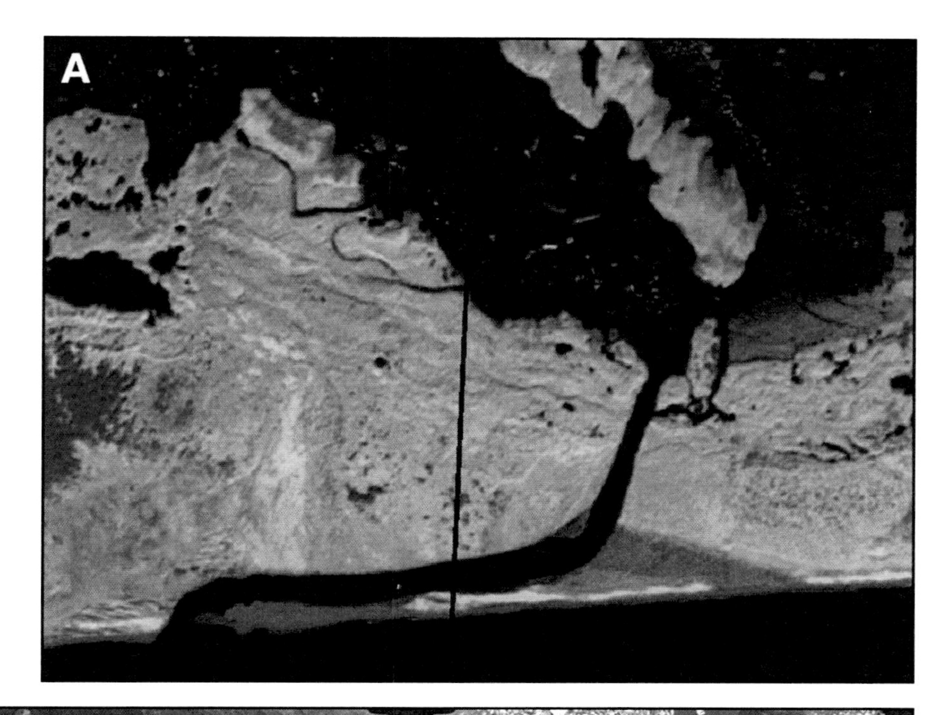

Figure 22. Two views of the Seal River, the 6-km-long, low-gradient, tidally dominated stream that drains Vitus Lake. Seal River is the primary distributary for the entire Bearing Glacier System. (A) A 10 September 2001 false color Landsat Thematic Mapper (TM+), bands 6, 4, 2 false color image, showing all of the river and adjacent southern Vitus Lake. (B) A 7 September 1994 vertical aerial photograph showing the entire length of the Seal River and the large sediment plume entering the Gulf of Alaska from the river's mouth. Suspended sediment loads carried during the 1993–1995 surge frequently exceeded 1 g/L. For a number of years prior to the surge, the suspended sediment load was only a few milligrams per liter. (BLM Photograph no. 94V4–85).

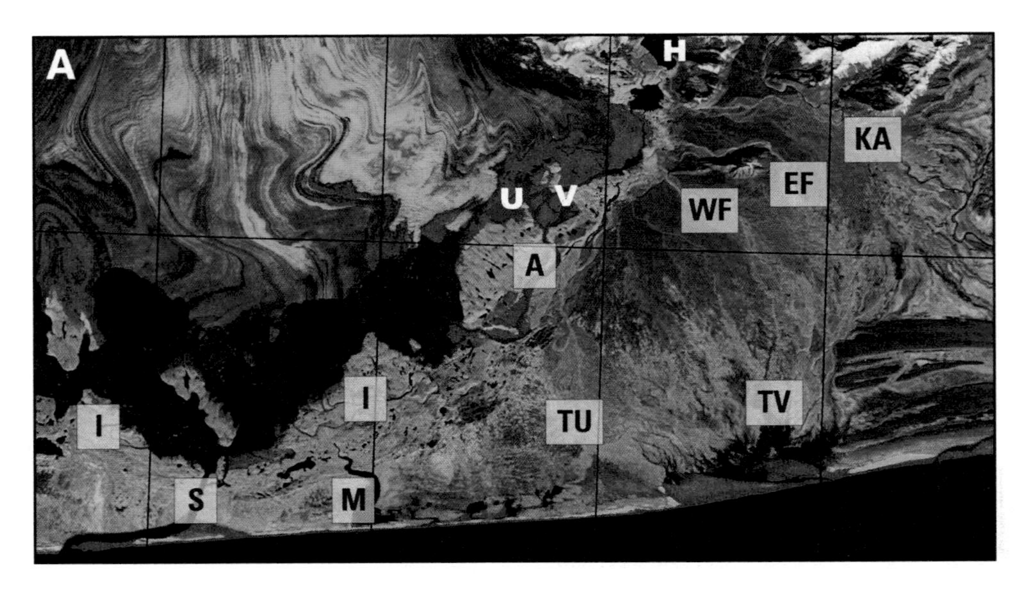

Figure 23. Two annotated 10 September 2001 false color Landsat Thematic Mapper (TM+), bands 6, 4, 2 false color image subscenes, showing more than a dozen current and former outlet streams that drain or have drained the perimeter of Being Glacier. (A: top image) Image showing the area from just west of the mouth of the Seal River to east of the mouth of the Tsivat River. From east to west the current and former drainages are Kaliakh River (KA), East Fork of the Tsivat River (EF), West Fork of the Tsivat River (WF), Mouth of the Tsivat River (TV), Tsiu River (TU), Abandoned River (A), Midtimber River, now Midtimber Lake (M), and Seal River (S). Two ice-marginal stream channels that developed during the 1972 surge are labeled with an I. The image is oriented with north up, and covers an area that is ~40 km from east to west and ~20 km from south to north. (B: bottom image) Image showing the area west of the mouth of the Seal River. Beginning in the east and going around the glacier in a clockwise direction, the current and former drainages are Tashalich River (TA), Kiklukh River (KI), Oaklee River (O), Edwardes River (E), Campbell River (CA), Nichawak River (N), Gandil River (G), Bering River (B), and the unnamed river that drains Berg Lake (U). Stillwater Creek (S) and Canyon Creek (CC) are tributaries to the Bering River.

glacier margin, extending from the Little Ice Age moraine to the Gulf of Alaska. Today it extends from just southeast of the Parallel Linear Lakes to the marshes associated with the western beach ridges, directly east of the Suckling Hills. Its mouth does not reach the Gulf of Alaska.

Oaklee River

The Oaklee River is a meandering stream with a length of >20 km. It is west of the Suckling Hills and consists of a network of anastomosing channels. Historically it drained much of the eastern and central part of the Central Medial Moraine Band. Today it extends from southwest of the Parallel Linear Lakes to the marsh and wetland area associated with the western beach ridges, directly east of the Suckling Hills, north of Controller Bay. Its mouth does not reach Controller Bay.

Edwardes River

The Edwardes River is an anastomosing stream with a length of >12 km. It is northwest of the Suckling Hills. Historically it drained the westernmost Central Medial Moraine Band. Today it extends from south of Mount Campbell to the marsh and wetland area north of Controller Bay. Its mouth does not reach the Gulf of Alaska.

Nichawak River

The Nichawak River drains from south of Gandil Mountain to the Bering River. Its length is ~10 km. Historically it drained the southwesternmost part of the Steller Lobe.

Bering River

Today, the Bering River receives most of its discharge from outside the Bering Glacier System. Much of its water originates as meltwater from Kushtaka Glacier, flowing through Kushtaka Lake and Stillwater Creek before entering Bering River. Formerly, Bering River was the primary distributary for the northern part of the Steller Lobe. It also served as the pathway for Berg Lake's discharge into Controller Bay. During the twentieth century the retreat of the terminus of the central Steller Lobe and the growth of Starodubtsov Lake funneled most of this meltwater south of Bering River and into the Gandil River drainage. During the 1994 Berg Lake flood the volume of water flowing into Starodubtsov Lake water exceeded its capacity, and the overflow spread westward before flowing into Bering River. When photographed at near peak flood stage, the Bering River valley was covered with floodwaters from the foot of the Don Miller Hills on the west to Nichawak Mountain and Mount Campbell on the east.

Abandoned River

The Abandoned River is a former ice-marginal stream that developed to accommodate the Bering Lobe's discharge both during and following an early to mid-twentieth century surge. It was reactivated in subsequent surges, including, most recently, the 1993–1995 surge. In 1993, advancing ice blocked the normal drainage pathway, east of Arrowhead Island, of waters from the eastern marginal lakes. As was the case in previous surges, the discharge from the eastern Bering Lobe was forced to flow through the Abandoned River into Taxpayers Bay and Vitus Lake. It took until 2006 for the Arrowhead Island passageway to reopen and for the Abandoned River to once again be abandoned. The landmass between, which the Abandoned River separated from the adjacent foreland, is known as *Bentwood Island.*

Bering Foreland

Vitus Lake is separated from the Gulf of Alaska by the Bering Foreland, a 3-km-wide beach–outwash plain–moraine complex. The sand beach bounding the southern edge of the foreland is as wide as 475 m and is composed of reworked outwash-plain sand and gravel, and sand contributed by longshore transport from the east. No bedrock is exposed at the surface in any part of the Bering Foreland between Cape Yakataga and the Suckling Hills. In fact, with the exception of one small exposure of the Yakataga Formation, exposed along the east side of Tashalich Arm in 1992, there are no known bedrock outcrops in the Vitus Lake area. No shallow bedrock was identified in seismic refraction profiling of the beach in 1992.

The thickness of the foreland sediment wedge is several hundred meters. At two locations, southwest of the Parallel Linear Lakes and directly east of the mouth of the Tsiu River, are series of subparallel former beach ridges separated by marsh-filled swales. The western ridge complex is a maximum of ~8 km long and ~3.5 km wide with at least eight ridge segments. The eastern ridge complex is also a maximum of ~8 km long and ~3.5 km wide with as many as 10 ridge segments. Radiocarbon dating of organic material contained in these segments by the first author determined that all ridge segments were younger than 600 a old.

Bering Trough

Offshore of Bering Glacier in the Gulf of Alaska is the Bering Trough (Fig. 24), a sediment-floored, 45-km-long submarine valley (Carlson et al., 1982). It is the result of Pleistocene, if not even earlier, erosion. The Bering Trough has maximum water depths of 321 m, maximum bedrock depths of ~500 m, and a maximum width of 25 km. Vitus Lake acoustic basement depths and IPR measurements of glacier-bed depths suggest that the sub-sea-level basins and channels that underlie the Bering Glacier and Vitus Lake are continuations of the Bering Trough.

DISCUSSION—SPECULATION ON THE FUTURE OF THE BERING GLACIER SYSTEM

Not withstanding the complexity introduced by its dynamic surge history, Bering Glacier has been retreating and thinning for more than a century. As Alaskan temperatures rise, the ongoing melting, thinning, and retreat of the Bering and Steller Lobes will continue, if not accelerate. Bering Glacier's major terminal margin lakes—Vitus, Berg, and Starodubtsov Lakes

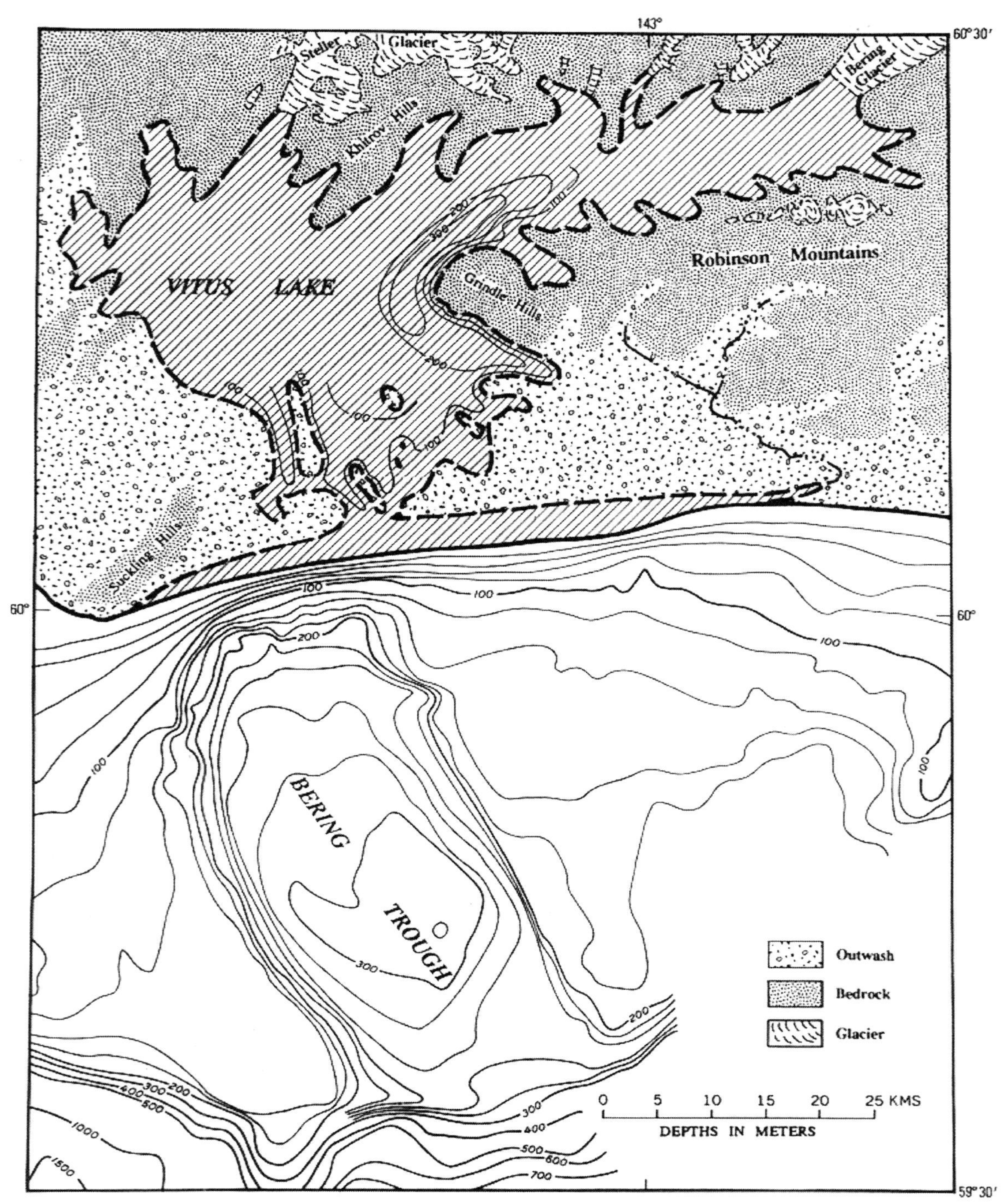

Figure 24. Cartoon showing the relationship between Bering Trough and the basin system that underlies the Piedmont Lobe of Bering Glacier. Gulf of Alaska bathymetry shows the location, morphology, and depths associated with Bering Trough. Vitus Lake is drawn to show its pre-1993 surge bathymetry (contours adjacent to its southern margin) and an approximate maximum size that it could attain following a catastrophic retreat of the Bering Glacier. Depths adjacent to the Grindle Hills are depths below present sea level to bedrock measured by radio-echosounding traverses. The area depicted by the diagonal pattern is the area where the bed of Bering Glacier is below sea level.

and Lakes Roselius and Ivanov—will continue to expand in length, width, and area. As we are already seeing, the Piedmont Lobe termini are rapidly thinning. The resulting flotation of the termini is accelerating, thus increasing the rate of retreat. Continued retreat will result in the joining of the large marginal lakes surrounding the terminus and ultimately in the formation of a single, large connected open body of water surrounding the entire southern perimeter of Bering glacier. Although IPR ice thickness soundings are limited, they suggest that all of the Bering Lobe and an unknown but extensive part of the Central Valley Reach extend a significant distance below sea level. Ongoing retreat could result in this area becoming ice free, with ice being replaced by deep water. Ultimately, this could result in a 60-km-long by 40-km-wide Vitus Lake. A similar retreat of parts of the Steller Lobe will result in an even larger Vitus Lake, producing one of Alaska's largest lakes.

As Vitus Lake grows, its increased size will capture more of the glacier's sediment. This will reduce the volume of sediment available to recharge and maintain the beach and foreland that separate Bering Glacier from the Gulf of Alaska. Consequently, the foreland and beach complex adjacent to the Gulf of Alaska shoreline will be more vulnerable to erosion and breaching.

Ultimately, Vitus Lake could become an extensive saltwater bay, embayment, or inlet extending nearly 70 km into the Chugach Mountains (Fig. 25). Based on IPR data, Vitus Lake could approach the size of Yakutat Bay or Glacier Bay.

SUMMARY

This chapter serves to introduce the reader to the Bering Glacier and the Bering Glacier System. It provides a summary of the early observations and geographic descriptions of Bering Glacier, identifies scientific studies that provide insights into the unique character of the Bering Glacier System and its unique surroundings, and presents descriptions of key geographic features that are part of the system and its surroundings. The Bering Glacier System is the largest (~5175 km^2) and longest (>190 km) glacier in continental North America and the largest temperate surging glacier on Earth. It is the world's largest nonpolar surging glacier and likely the world's largest surging glacier. Bering

Figure 25. Painting by artist Michael Carroll, showing his conceptualized view of what a new Chugach Mountains landscape might look like after a catastrophic retreat of the Bering Lobe–Bering Glacier from a maximum position attained following a major surge advance. Although not topographically or glacially precise, Carroll's painting presents a radically different coastline than exists today. The painting presents three tiers of water and ice. Blue-black water represents the ocean and Vitus Lake. Medium blue ice is a "current" glacier and terminus position. In terms of accuracy, it is close to being correct in western Vitus Lake, but too far advanced in eastern Vitus Lake. White ice is a significantly retreated future position of Bering Glacier. Carroll's painting shows open ocean–bay right up to two tidewater glacier tongues. The Taggland Peninsula has become an island. A number of newly emergent islands are shown. The light blue haze along the foothills represents the current height of Bering Glacier. (Painting in the collection of the first author.)

Glacier dwarfs all but the largest valley glaciers in Antarctica and Greenland in area, length, and ice volume. Alone, its area represents nearly 7% of the glacier-covered area of Alaska, and it contains ~15% of the glacier ice in Alaska. Integrated studies have documented many aspects of its behavior on time scales ranging from millennial to decadal to annual.

REFERENCES CITED

Belcher, E.B., 1843, Narrative of a voyage around the world performed in Her Majesty's ship *Sulphur* during the years 1836–1842: London, Henry Colburn Publisher, v. 1, p. 75–79.

Carlson, P.R., Bruns, T.R., Molnia, B.F., and Schwab, W.C., 1982, Submarine valleys in the northeastern Gulf of Alaska: Characteristics and probable origin: Marine Geology, v. 47, no. 3/4, p. 217–242.

Carlson, P.R., Tagg, R.A., and Molnia, B.F., 1993, Acoustic profiles of sediment in a melt-water lake adjacent to the Berin Glacier, Alaska, RV Karluk cruise K2-91-YB, July 1–7, 1991: U.S. Geological Survey Open-File Report 93-266, 26 p.

de Fillippi, Fillippo, 1899, The ascent of Mount St. Elias [Alaska] by H.R.H. Prince Luigi Amedeo Di Savoia Duke of the Abruzzi: New York, Frederick A. Stokes Co., 242 p., 4 mosaic plates, 2 maps.

Field, W.O., 1975, Glaciers of the Chugach Mountains, *in* Field, W.O., ed., Mountain Glaciers of the Northern Hemisphere: Hanover, New Hampshire, U.S. Army Corps of Engineers, Cold Regions Research and Engineering Laboratory, p. 299–492.

Martin, G.C., 1905, The Petroleum fields of the Pacific Coast of Alaska with an account of the Bering River's coal deposits: U.S. Geological Survey Bulletin 250, 64 p.

Martin, G.C., 1908, Geology and mineral resources of the Controller Bay Region, Alaska: U.S. Geological Survey Bulletin 355, 141 p.

Martin, G.C., Maddren, A.G., Paige, S., and Weaver, C.E., 1907, Geological map and sections of the Controller Bay Region, Alaska: U.S. Geological Survey, scale 1:100,000, 1 sheet (released as sheet V *in* Martin, G.C., 1908, Geology and Mineral Resources of the Controller Bay Region, Alaska: U.S. Geological Survey Bulletin 355, 141 p.)

Meier, M.F., and Post, A., 1962, Recent variations in mass net budgets of glaciers in western North America: International Association of Scientific Hydrology, Obergurgl Symposium, Publication 58, p. 63–77.

Miller, D.J., 1958, Anomalous glacial history of the northeastern Gulf of Alaska region: Geological Society of America Bulletin, v. 69, p. 1613–1614.

Molnia, B.F., 2008, Glaciers of Alaska, with sections on Columbia and Hubbard tidewater glaciers by Krimmel, R.M.; The 1986 and 2002 temporary closures of Russell Fiord by the Hubbard Glacier by Molnia, B.F., Trabant, D.C., March, R.S., and Krimmel, R.M.; and Geospatial inventory and analysis of glaciers: A case study for the eastern Alaska Range by Manley, W.F., *in* Williams, R.S., Jr., and Ferrigno, J.G., eds., Satellite Image Atlas of Glaciers of the World: U.S. Geological Survey Professional Paper 1386-K, 705 p.

Molnia, B.F., and Molnia, M.I., 1995, Comparison of spaceborne ERS-1 and SIR-C synthetic aperture radar (SAR) with airborne SAR of the Bering Glacier, Alaska: Geological Society of America Abstracts with Programs, v. 28, p. 457.

Molnia, B.F., and Post, A., 1995, Holocene history of Bering Glacier, Alaska: A prelude to the 1993–1994 surge: Physical Geography, v. 16, p. 87–117.

Molnia, B.F., Post, A., and Carlson, P.R., 1996, 20th-century glacial-marine sedimentation in Vitus Lake, Bering Glacier, Alaska, U.S.A.: Annals of Glaciology, v. 22, p. 205–210.

Orth, D.J., 1967, Dictionary of Alaska place names: U.S. Geological Survey Professional Paper 567, 1084 p., 12 maps.

Orth, D.J., 1971, Dictionary of Alaska place names: U.S. Geological Survey Professional Paper 567, 1084 p., 12 maps, scale 1:2,500,000.

Post, A., 1972, Periodic surge origin of folded medial moraines on Bering Piedmont Glacier, Alaska: Journal of Glaciology, v. 11, p. 219–226.

Post, A., and Mayo, L.R., 1971, Glacier dammed lakes and outburst floods in Alaska: U.S. Geological Survey Hydrological Investigations Atlas 455, scale 1:1,000,000, 3 sheets, 10 p.

Seaton-Karr, H.W., 1887, Shores and Alps of Alaska: London, Low, Marston, Searle, and Rivington, 248 p.

Sfraga, M., 2004, Bradford Washington—A life of exploration: Corvallis, Oregon State University Press, 280 p.

Trabant, D.C., and Molnia, B.F., 1992, Ice thickness measurements of Bering Glacier, Alaska, and their relation to satellite and airborne SAR image patterns: Eos (Transactions, American Geophysical Union), v. 73, p. 181, supplement (October 27).

Trabant, D.C., Molnia, B.F., and Post, A., 1991, Bering Glacier, Alaska—Bed configuration and potential for calving retreat: Eos (Transactions, American Geophysical Union), v. 72, (44-supplement), p. 159.

U.S. Geological Survey, 1959, Minor revisions, 1983, Bering Glacier, Alaska, 1:250,000-scale topographic map, 1 sheet.

U.S. Geological Survey, 1984, Bering Glacier, Alaska: 1:63,360-scale topographic map, 1 sheet.

Viens, R.J., 1995, Dynamics and mass balance of temperate tidewater calving glaciers of Southern Alaska [M.S. thesis]: University of Washington, Seattle, 149 p.

Manuscript Accepted by the Society 02 June 2009

The Geological Society of America
Special Paper 462
2010

Remote sensing of the Bering Glacier Region

Robert A. Shuchman*
Michigan Tech Research Institute, 3600 Green Court, Suite 100, Ann Arbor, Michigan 48105, USA

Edward G. Josberger
United States Geological Survey, Washington Water Science Center, 934 Broadway, Suite 300, Tacoma, Washington 98402, USA

Liza K. Jenkins
Michigan Tech Research Institute, 3600 Green Court, Suite 100, Ann Arbor, Michigan 48105, USA

John F. Payne
North Slope Science Initiative, c/o Alaska State Office (910), Bureau of Land Management, 222 West 7th Avenue #13, Anchorage, Alaska 99513, USA

Charles R. Hatt
Lucas Spaete
Michigan Tech Research Institute, 3600 Green Court, Suite 100, Ann Arbor, Michigan 48105, USA

ABSTRACT

Satellite remote sensing is an invaluable tool for monitoring and characterizing the Bering Glacier System. Applications of glacier remote sensing include, but are not limited to, mapping extent and features, ice velocities through sequential observations, glacier terminus locations, snow line location, glacier albedo, changes in glacier volume, iceberg surveys and calving rates, hydrographic and water quality parameters in ice marginal lakes, and land-cover classification maps. Historical remote sensing images provide a much needed geospatial time record of the dynamic changes that Bering Glacier has undergone, including changes from its surge behavior and response to climate change. Remote sensing images dating back to the early 1990s have been used to map the glacier terminus retreat of ~5 to 7 km, which has resulted in Vitus Lake increasing in volume 9.4 km^3 (~260%) from 1995 to 2006. Using elevation data obtained from remote sensing and GPS surface points, we have determined that the glacier elevation has decreased by ~150 m at the terminus and 30 m at the equilibrium line (~1300 m) since 1972. Satellite observations have recorded the upward migration in altitude of the equilibrium line to its present (2006) position (slightly >1200 m). The decrease in glacier volume, obtained using remote sensing–derived elevation data, from 1957 to 2004 is estimated at ~104 km^3. Remote sensing data also have mapped the sediment-rich (rock flour) water flowing into Vitus Lake, providing insight into the hydrologic circulation of the Bering Glacier System, showing major glacier discharge from the Abandoned River, Arrowhead Point, and Lamire Bay in the area of Vitus Lake west of Taggland.

*shuchman@mtu.edu

Shuchman, R.A., Josberger, E.G., Jenkins, L.K., Payne, J.F., Hatt, C.R., and Spaete, L., 2010, Remote sensing of the Bering Glacier region, *in* Shuchman, R.A., and Josberger, E.G., eds., Bering Glacier: Interdisciplinary Studies of Earth's Largest Temperate Surging Glacier: Geological Society of America Special Paper 462, p. 43–65, doi: 10.1130/2010.2462(03). For permission to copy, contact editing@geosociety.org.

INTRODUCTION AND BACKGROUND

The Bering Glacier is the largest and longest glacier in continental North America, with an area of ~5,175 km^2 and a length of 190 km. It is also the largest surging glacier in the world, having surged at least five times during the twentieth century. The last great surge occurred in 1993–1995. Bering Glacier alone covers >6% of the glacier-covered area of Alaska and may contain 15–20% of Alaska's total glacier ice (Molnia and Post, 1995). The entire glacier lies within 100 km of the Gulf of Alaska. The rapid ongoing retreat of the glacier and expansion of Vitus Lake at the glacier terminus have provided opportunities for establishment of new habitat and the introduction of new flora and fauna. The post-surge retreat of Bering Glacier has created a dynamic landscape of reticulated and fluted surfaces with subtidal invertebrate fossils, lake sediments, and previously overrun forests.

Given the current climatic change scenario of warmer summers, Bering Glacier is retreating rapidly, resulting in a dramatically different and changing landscape. Remote sensing observations have and will continue to play an active and critical role in monitoring the baseline characteristics and dynamics of the vast expanse of the Bering Glacier System. The use of remote sensing provides a synoptic, historic record of the glacier's changing landscape.

In this chapter we first give a general overview of remote sensing capabilities for observing the cryosphere and then describe specific applications of remote sensing to the Bering Glacier System. We will show through numerous examples that remote sensing data, either solely or when combined with in situ measurements in a geographic information system (GIS), provide information on terminus location, glacier movement, iceberg and calving rate surveys, Vitus Lake frontal boundaries, land cover classification, snow line delineation, Vitus Lake and Berg Lake volumes, water quality, glacier surface features (crevasses, moraines, etc.), harbor seal monitoring (resolution dependent), and height information in the form of digital elevation models (DEMs) to study changes in glacier mass-volume over time. The historical record of electro-optical (EO) and RADAR imagery, more than four decades in length, provides a historic record of the glacier's changing landscape in response to its surge characteristics and climate driven changes.

WHY REMOTE SENSING?

Bering Glacier occupies a vast area, and only remote sensing systems have the potential to capture the synoptic view of the entire landscape required to understand this complex system. Given the geographic extent of the glacier, in situ observations alone cannot adequately characterize glacier dynamics. Remote sensing provides additional information that can supplement more intensive sampling efforts and help extrapolate findings.

Remote sensing is a proven tool for determining and quantifying landscape change. Land-cover-change maps of the Bering region can be generated at reduced costs in comparison with traditional field-data-collection methods. This is particularly true at Bering Glacier because most field verification is done via helicopter deployment or by boat.

Satellite remote sensing data are collected year round and can provide valuable information about the system when the glacier is inaccessible to ground sampling owing to extreme weather conditions. During the winter months it is not uncommon to record wind speeds >35 m/s (~70 knots) at the Bering Glacier field camp weather station as strong low pressure systems move in from the Gulf of Alaska.

The Bering Glacier is 124 km from Cordova and 208 km from Yakutat, the nearest points of civilization. The glacier's remote location limits on-the-ground access year round, and field sampling events require extensive logistical considerations. Thus, remote sensing can offer cost-effective insight into areas that have not been or cannot be accessed.

Remote sensing can be defined as observing the properties of an object through a nondirect (remote) measurement. A variety of remote sensing systems can be used to provide information on the Bering Glacier region. Remote sensors can be active (providing their own source of illumination) or passive (detecting radiation that is either reflected or emitted from an object). Remote sensors can operate in all parts of the electromagnetic spectrum but typically in the visible (0.4–0.7 μm), near-infrared (0.7–2.5 μm), mid-infrared (3–5 μm), infrared (8–14 μm), and microwave (1 mm to 1 m/300MHz to 300 GHz) range. EO sensors are those that operate in the 0.4–14 μm range, but they cannot image through extensive cloud cover, whereas microwave sensors operate at longer wavelengths (lower frequencies) and have the ability to sense objects through clouds, rain, and snow.

Remote sensors have a range of spectral, spatial, and temporal resolution. Spectral resolution refers to the number of different wavelength bands the sensor employs to image the earth. Spatial resolution is the spot size the sensor images on the ground surface. The higher or finer the spatial resolution, the more detail is discernible. Temporal resolution refers to how often a sensor passes over a given point on the ground. This is typically referred to as the repeat cycle.

An example of an active EO satellite sensor is the laser profiler on ICESat, whereas passive EO satellite sensors include Landsat, ASTER, MODIS, MERIS, and the high-spatial-resolution commercial systems such as QuickBird, Worldview, and Ikonos. The source of this EO passive sensing is either reflected sunlight or thermal (temperature) emissions. Spatial resolution for an EO sensor is a function of the size of the optics and the distance from the sensor to the object being observed. The ground resolution of EO satellite sensors varies from approximately 0.5 m to 1 km, depending on the platform. The reflectance in the EO portion of the electromagnetic spectrum for various snow and glacier ice features is shown in Figure 1 (modified from König et al., 2001). Also shown in Figure 1 are the satellite bands used by Landsat, MODIS, and ASTER.

The albedo, or amount of reflected solar energy, can be determined using EO satellite sensors. Albedo is an important

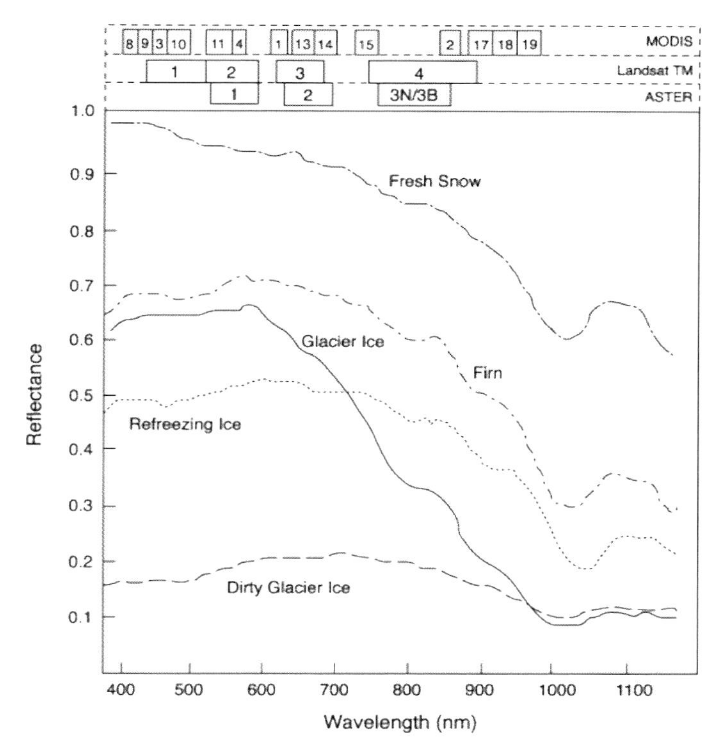

Figure 1. Spectral reflectance values for various snow and glacier ice types and the corresponding satellite bands used by MODIS, Landsat TM, and ASTER. Figure modified from König et al. (2001).

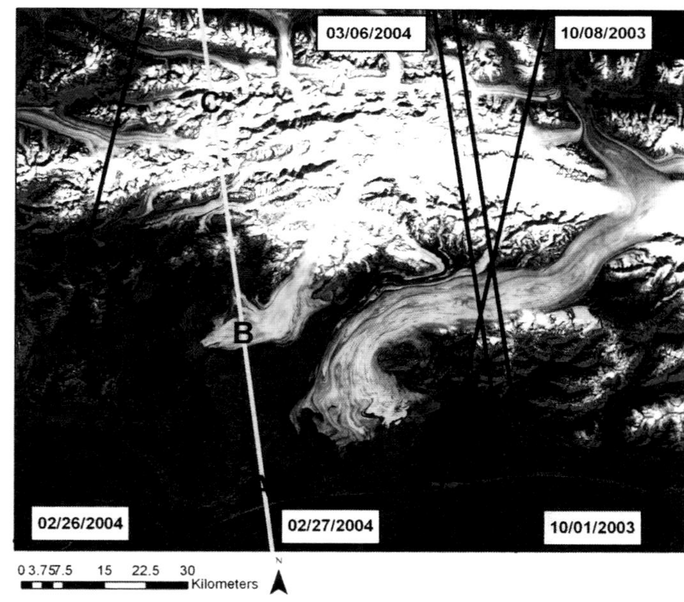

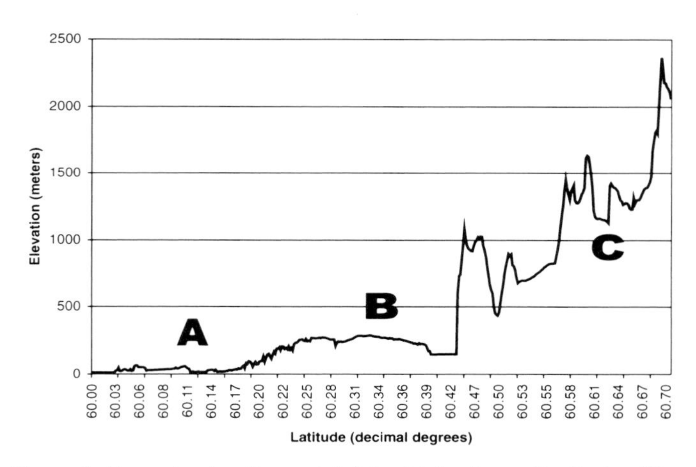

Figure 2. Example elevation point data obtained over the Bering Glacier by the NASA ICESat system (top) with a corresponding elevation profile for the 27 February 2004 data acquisition (bottom). Points A, B, and C on the satellite image correspond to points A, B, and C on the graph.

parameter used in modeling the accumulation and ablation of snow and ice. Thus a model for the ablation of Bering Glacier would utilize a spatially varying albedo that would be determined from a satellite sensor, typically Landsat or MODIS. Snow covered ice would have a high albedo, bare ice would have a lower albedo, and debris covered ice would have an even lower albedo.

DEMs, which can be obtained through remote sensing techniques or ground surveys, are vital data sets needed to understand the change in glacier mass, determine glacier flow, and distribute meteorological variables on the glacier for glacier melt modeling. EO sensors that have forward and backward looking instruments, or have 60% image overlap, produce stereo images that are used to derive elevation data. Space borne lasers (such as in the Geoscience Laser Altimeter System [GLAS] on ICESat) also produce height information by transmitting a pulse of electromagnetic energy and accurately recording the time it takes to strike an object and be reflected back to the sensor. The height information produced by a laser is a profile (single elevation value often directly under the spacecraft) and has a height resolution of ~10 cm (Zwally et al., 2002). Example GLAS passes collected over the Bering Glacier region are shown in Figure 2.

Microwave satellite sensors also can be active or passive, and each provides a unique view of snow and ice, each independent of solar illumination and cloud cover. The Special Scanning Microwave Radiometer, SSMI, is a passive microwave sensor on the Defense Mapping Satellite System (DMSP), a version of which has been observing the earth continuously since 1979. It measures the electromagnetic energy emitted by an object, usually at different frequencies, to observe what is referred to as microwave brightness temperatures. Snow and ice have different brightness temperatures than bare earth vegetation and open water. Microwave temperatures also vary as a function of the snow depth and the liquid water content; hence it can provide critical information for snowmelt runoff modeling. The spatial resolution of the SSMI for the frequencies used in cryospheric investigations ranges from 12.5 to 50 km and therefore provides information only for large-scale features, such as the Bagley Ice Field (the source of ice that feeds the Bering Glacier).

Synthetic Aperture Radar (SAR) is an active microwave satellite sensor that operates in the 1–10 GHz frequency range and has a ground resolution of 30 m or less. Its all-weather capability and high resolution make it an ideal sensor for the repeated monitoring of glacier features such as terminus position, snow line position, and iceberg distribution. SAR observations also can be used to derive DEMs in one of two ways. SAR imagery of the same spot can be collected at two viewing angles, thus creating stereo pairs similar to EO sensors. An alternative and more accurate approach utilizes a method referred to as interferometric SAR or InSAR. InSAR images an area with two very slightly varying imaging geometries and compares the phase difference between the two returning radar pulses to obtain the required height information (Lu et al., 2007). InSAR can provide very fine height resolution data on the order of 1 to 3 m. InSAR techniques (Fatland and Lingle, 2002) also can be used to obtain glacier movement (velocity and accelerations). An area of the glacier is imaged during 1–3 d intervals, and again the phases of the radar returns are compared, but in this case to determine movement of the surface between frames.

It is important to note that not all elevation data sets derived from remote sensing systems can be classified as DEMs. If the surface that is being measured by the remote sensing systems is not bare earth (i.e., forested), then the resulting elevation data set must be considered a Digital Surface Model (DSM). The Bering Glacier can be considered bare earth, so the resulting elevation data sets of the area are considered DEMs.

Satellite remote sensors often trade off spatial resolution for ground imaging coverage or what is termed *swath width*. For example, QuickBird, an EO sensor, produces 60 cm spatial resolution imagery, but its swath is a mere 16 km. Alternatively MODIS images a swath on the Earth of ~2300 km at a resolution of 250 m. Thus the tradeoff of fine detail of the Earth's surface is confined to a small area, or a synoptic view at a coarser scale. Satellite orbits also play a role in imaging glaciers such as the Bering. Polar orbits with altitudes above the Earth of ~800 km having inclination angles (angles from the equator) of 70° or greater provide for revisit times to image the Bering Glacier region such that temporal changes can be observed.

Table 1 summarizes the satellite sensors used to study Bering Glacier. Presented for each sensor is the type (EO or microwave), the temporal sampling (historical coverage in time), spatial and spectral resolution, swath width and revisit time, glacier information obtainable from the given sensor, and a "remarks" column that presents pros and cons. As the table indicates, passive EO satellite data when cloud free provide the greatest amount of information about the glacier. The level of detail obtainable from the remote sensing data is a function of the spatial resolution and the swath width. Examination of the table also reveals that the all-weather day or night operation of SAR can image the glacier system at repeatable time intervals useful for determining the snow equilibrium line and iceberg and calving rates. SAR data exist from the early 1990s and thus are useful for Bering Glacier pre-surge bench marking.

In addition to satellites, aerial photography or video has been shown to be a useful tool for characterizing glacial properties. For example, historical photography dating back to the early 1930s (Molnia, 2001) can be used to map historic terminus positions. Aerial photography typically has a spatial resolution on the order of 1 to 5 ft, which is very useful for identifying surface glacier features such as moraines, crevasses, moulins, melt ponds, and marine animals such as harbor seals that have out-hauled on icebergs in Vitus Lake. The higher resolution provided by aerial photography is also useful in generating glacier surface movements, because the same feature in time lapse imagery can be more easily identified.

REVIEW OF GLACIER REMOTE SENSING STUDIES

A number of investigators (König et al., 2001; Zwally et al., 2002; Winther et al., 2005; Chen et al., 2006) have demonstrated the utility of using remote sensing satellites for glacier feature characterization. For example, Winther et al. (2005) in their chapter entitled "Remote sensing of glaciers and ice sheets" summarize the use of MODIS, ENVISAT, RADARSAT, MERIS, and ICESat to map snow equilibrium line, surface albedo, characteristic glacier surface zone, glacier velocity, glacier mapping, change detection, surface features, and snow pack characteristics. Winther's group gives special attention to exploiting visible EO data when cloud-free conditions exist and then relying on the all-weather microwave radar data when the glacier is under cloud cover, or during the winter when solar illumination is at a minimum.

Chen et al. (2006) use satellite gravity measurements from the Gravity Recovery and Climate Experiment (GRACE) as an indication of mass change to study potential long-term mountain glacier melting in southern Alaska and elsewhere. The first 3.5 a of GRACE monthly gravity data, spanning April 2002 to November 2005, shows a prominent glacial melting trend in the coastal Alaskan region around the Gulf of Alaska, which correlates well with limited U.S. Geological Survey (USGS) mass balance measurements from Bidlake et al. (2007).

Zwally et al. (2002) utilized the laser altimeter on the NASA's ICESat satellite to generate high resolution (~5 cm accuracy) profiles of various glaciers around the world, including partial coverage of the Bering Glacier System. The profiles collected over various years allow for the estimation of volumetric-change maps to be generated. Unfortunately, the laser profiling system on board the spacecraft is nearing the end of its operational effectiveness.

BERING GLACIER REMOTE SENSING STUDIES

Specific Bering Glacier remote sensing studies that utilize the satellite sensors discussed above are presented in the following subsections and include synoptic landscape mapping, land cover classification, glacier terminus monitoring and mapping, Vitus Lake area and volume change analysis, snow line mapping,

TABLE 1. SATELLITE SENSORS USED TO STUDY BERING GLACIER

Satellite sensor	Sensor type	Temporal range of data	Revisit time and swath width	Spectral resolution	Spatial resolution	Glacier applications	Remarks
MODIS Aqua Terra http://modis.gsfc.nasa.gov/	Passive EO	2000–present 2002–present	1–2 day revisit 2300 km	36 bands 0.4–14.5 μm	250 m 500 m 000 m	Snowline, terminus positions, water quality, albedo derived product, vegetation map, land cover, snow accumulation	Pros: 40 data products that combine different spectral, temporal, and spatial resolutions. Perhaps the most relevant to the Bering Glacier area is the 8 day reflectance product that has a spatial resolution of 250 m and combines 8 days of reflectance data into one image, making it ideal for the usual cloudiness of the area. Free and collected every day. Cons: Spatial resolution limits detailed glacier feature mapping.
Landsat ETM+ http://landsat.gsfc.nasa.gov/	Passive EO	1999–present	16-day revisit 183 km	8 bands 0.4–12.5 μm	15 m pan 30 m 60 m thermal	Snowline, terminus positions, land cover mapping, vegetation map, glacier features	Pros: Historical temporal extent, multispectral allows for observing contrasts, resolution. Cons: The long repeat time coupled with the usual cloudiness of the area makes it hard to find useful data, especially when you want it. The data are expensive. Problems with Scan Line Correction (SLC). Missing data.
Landsat MSS/TM http://landsat.gsfc.nasa.gov/	Passive EO	1973–present	16-day revisit 183 km	7 bands 0.4–12.5 μm	30 m 120 m thermal	Snowline, terminus positions, land cover mapping, Vitus Lake frontal boundaries, albedo	Pros: Historical temporal extent, multispectral allows for observing contrasts, resolution. Cons: The long repeat time coupled with the usual cloudiness of the area makes it hard to find useful data, especially when you want it. The data are relatively expensive.
ASTER http://asterweb.jpl.nasa.gov/	Passive EO	2000–present	4–16 day revisit 60 km	15 bands 0.5–12 μm	15 m VNIR 30 m SWIR 90 m TIR	Snowline, terminus positions, DEM generation, land cover, vegetation map, glacier surface features, glacier velocity, water quality	Pros: Multispectral allows for observing contrasts, resolution, DEM creation, inexpensive to free. Cons: Temporal extent. The long repeat time coupled with the usual cloudiness of the area makes it hard to find useful data on demand.
Commercial Digital Globe, Quickbird, and Worldview-1/2 / Geoeye, Ikonos http://www.digitalglobe.com/index.php/6 http://www.geoeye.com/CorpSite/products/imagery-sources/Default.aspx#ikonos	Passive EO	2001–present	3–7 day revisit Min 16.5 km	5 bands 0.45–0.9 μm	0.6 m pan 2.4 m	Glacier velocity, glacier surface features, iceberg and calving monitoring, seal monitoring, water quality	Pros: Spatial resolution, revisit interval. Cons: Expensive, same problems as other optical sensors
ICEsat/GLAS http://icesat.gsfc.nasa.gov/	Active EO	2003–present	91-day revisit 70 m	0.8 nm	70 m horizontal 3 cm vertical	Topography, ablation, sea ice thickness	Pros: Sensor designed specifically for ice sheet monitoring, precise. Cons: Limited coverage and cloud cover.

(*Continued*)

TABLE 1. SATELLITE SENSORS USED TO STUDY BERING GLACIER (*Continued*)

Satellite sensor	Sensor type	Temporal range of data	Revisit time and swath width	Spectral resolution	Spatial resolution	Glacier applications	Remarks
ERS 1/2 http://earth.esa.int/ers/0	SAR	1991–present	35-day revisit 100 km	5.3 GHz C-band	25 m	Snowline, terminus positions, iceberg monitoring, glacier velocity, DEMs	Pros: Historical temporal extent, acquisition regardless of atmospheric and sun illumination conditions, wavelength allows penetration through snow cover allowing for glacier measurements year round, inexpensive. Cons: Moderate resolution, backscatter varies as function of liquid water content of glacier surfaces.
RADARSAT 1/2 http://gs.mdacorporation.com/	SAR	1995–present	Approx. 6-day revisit 50–500 km	5.6 GHz C-band	8–100 m	Snowline, terminus positions, iceberg monitoring, glacier velocity, DEMs	Pros: Historical temporal extent, acquisition regardless of atmospheric and sun illumination conditions, wavelength allows penetration through snow cover allowing for glacier measurements year round, inexpensive. Cons: Moderate resolution, backscatter varies as function of liquid water content of glacier surfaces.
Envisat http://earth.esa.int/ers/	SAR	2002–present	35-day revisit 100–400 km	5.6 GHz C-band	10–25 m	Snowline, terminus positions, iceberg monitoring, glacier velocity, DEMs	Pros: Historical temporal extent, acquisition regardless of atmospheric and sun illumination conditions, wavelength allows penetration through snow cover allowing for glacier measurements year round, inexpensive. Cons: Moderate resolution, backscatter varies as function of liquid water content of glacier surfaces.
SSM/I http://www.ssmi.com/	Passive microwave radiometric	1987–present	Daily revisit 400 km	19.35, 22.2, 37.0, 85.5 GHz	12.5 km 25 km	Snowfield assessment, sea ice concentrationsnowmelt	Pros: Synoptic daily coverage allows for determination of freeze up and thaw. Cons: Coarse resolution.
ALOS PALSAR http://www.palsar.ersdac.or/jp/e	Polarimetric SAR	2004–present	46-day revisit, 25–75 km depending on polarization	1.3 GHz, L-band	10 and 100 m	Snowline, terminus positions, iceberg moniotiring, glacier velocity, DEMS	Pros: Fully polarmetic, longer wavelength, better penetration Cons: Limited swath width

time series environmental monitoring, elevation mapping, yearly melt estimates using in situ ablation and remote sensing data, iceberg and calving rate estimates, water quality measurements, glacier surface feature mapping, harbor seal population estimates, and albedo estimates.

Synoptic Landscape Mapping

One of the most fundamental uses of remote sensing imagery is to obtain a synoptic view of a large geographic area. Viewing the earth from space provides an overhead perspective that reveals patterns or features in the landscape that may not be otherwise readily apparent. These patterns or features can also be coarsely delineated to provide a starting point for further research activities or to provide information on historic and current landscape processes.

Figure 3 demonstrates the ability of the Landsat satellite to map geologic and glacial features in the Bering Glacier region. These landforms are typical features associated with a glacial environment. Moraines, glacier lobes, glacier termini, and ice marginal lakes are all easily identifiable from the August 2006 Landsat image in Figure 3 (top). The vegetation-free medial moraine separating the Bering and Steller Lobes is dark and is identified and labeled in the figure in orange. Glacier ice is labeled red, glacier termini pink, and ice marginal lakes blue. Other identifiable features include the farthest historical seaward extent of the glacier, outwash fans, and beach ridges.

Patterns discernible from a synoptic view can provide additional information about the environment, including processes occurring over shorter time periods. For example, the deformed medial moraines shown in yellow in Figure 3 (top) are evidence of the glacier's recurring periods of surge and retreat. From these folded moraines, Molnia and Post (1995) estimated that Bering Glacier has surged five times in the previous century. The Landsat data in Figure 3 are shown using two different band combinations. In Figure 3 the top image utilizes a near-IR, red, and green band combination, whereas the lower image utilizes a red, green, and blue combination. The top image is a false-color image because it depicts the landscape in colors that differ from the human perception of the land. For example, in this image vegetation is shown in red instead of green. The bottom image is a natural-color image, and objects in this image appear as they would to the human eye. Thus, changing the band combination changes the portion of the electromagnetic spectrum through which we are viewing the landscape. For this reason the different combinations of the Landsat bands show different versions of the landscape. The false color image (near-IR, red, and green) best delineates the glacial features presented.

Land Cover Classification

A land cover map created from a 19 June 1991 Landsat data set is shown in Figure 4. This map was made by U.S. Bureau of Land Management (BLM) and Ducks Unlimited scientists using a supervised pixel-by-pixel classification procedure by which training sets representing each of the 15 classes are located and characterized using helicopter or ground surveillance. The computer then statistically characterizes (creates a vector) for each of the training sets with respect to the spectral signature from the four satellite channels. Each Landsat pixel, represented by the four spectral signatures (blue, green, red, near-IR), is then best mapped into one of the identified classes. The classification map shown in Figure 4 has been smoothed (low pass filtered) to reduce the "salt-and-pepper" nature of the image. This land cover map shows the diversity of vegetation types surrounding Bering Glacier. The dominant land cover types other than ice, snow, and water include closed and open needleleaf forest, open and closed alder-willow, closed mixed forest, herb-graminod, aquatic emergent, salmonberry-fern, and sparse vegetation. The reader is encouraged to read the chapter by Barker (this volume) for additional insight into the diverse vegetation community of the Bering Glacier region.

A new method for generating land cover maps using multispectral images has emerged that produces highly accurate classifications without the salt-and-pepper look of traditional approaches. This new technique, referred to as object-based algorithms, mimics the thought process of the human mind by taking into consideration the size, shape, and context of a cover type, in addition to its spectral properties. The Bering Glacier region should be remapped with respect to vegetation cover using this new improved procedure. The new map should be compared to the 1991 classification and changes noted. The changes should reflect the dynamics of both the glacier and the surrounding forelands.

Terminus Mapping

Both high resolution EO sensors such as Landsat and the microwave sensor SAR are useful for mapping glacier terminus locations. Figure 5 shows the change in terminus location of the Bering Lobe, derived by using both Landsat and SAR imagery. The pre-surge (1992) and post-surge (1995) positions are mapped on the Landsat image. The terminus has retreated (post-surge) on average by ~0.38 km/a with a maximum of 0.47 km of retreat between 1999 and 2000, and with a minimum of 0.21 km of retreat between 2002 and 2003, since the end of the surge (1995). Significant retreat of 0.43 km/a also occurred between 2004 and 2005; a record high for average air temperature occurred in the summer of 2004. The historical satellite observations also reveal that the terminus of the Bering Lobe in 2007 was near the 1992 pre-surge location of the glacier. In general, Landsat and SAR results produce comparable terminus mapping results.

Vitus Lake Area and Volume Change

The time history of glacier terminus positions obtained from satellite observations clearly shows that the Bering Glacier terminus retreated approximately 5 to 7 km since the end of

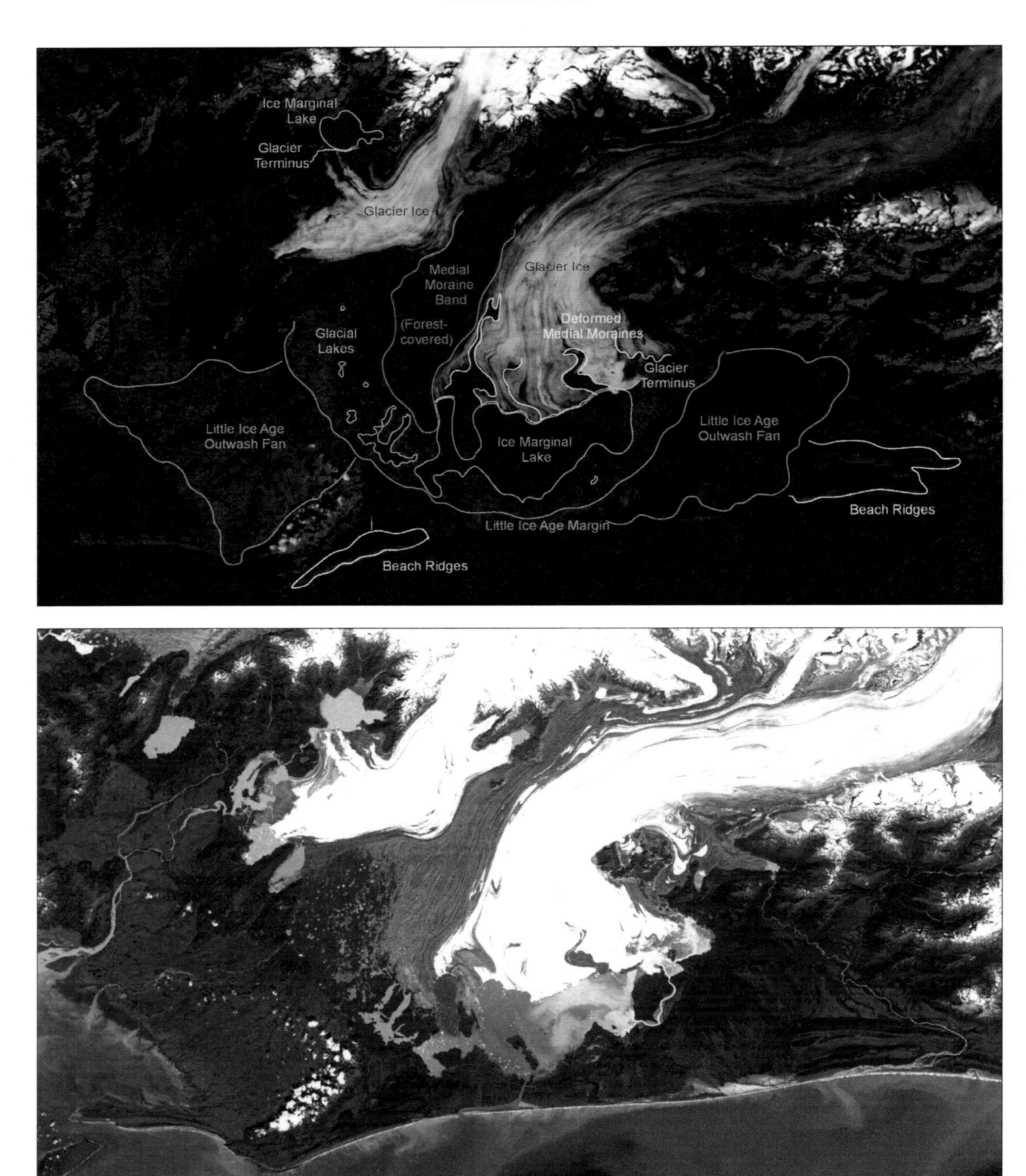

Figure 3. Geologic features from the Bering Glacier region are easily discernible from images obtained from remote sensing. A Landsat image from 7 August 2006 has been used to synoptically identify several glacial landforms in the Bering Glacier region. In this figure (top), the farthest forward extent of Bering Glacier is shown in green and labeled "Little Ice Age margin." The outwash fans and the beach ridges associated with this margin are shown in light blue and white, respectively. The medial moraine separating the Bering and Steller Lobes is shown in orange, the deformed medial moraines in yellow, the ice marginal lakes in blue, the glacier terminus in pink, and the glacier ice of the Bering and Steller Lobes is labeled in red. The Landsat data in this figure are shown using two different band combinations. The top image utilizes a near-IR, red, and green (false color) band combination, whereas the lower image utilizes a red, green, and blue (natural color) combination. The false color image best delineates the glacial features presented.

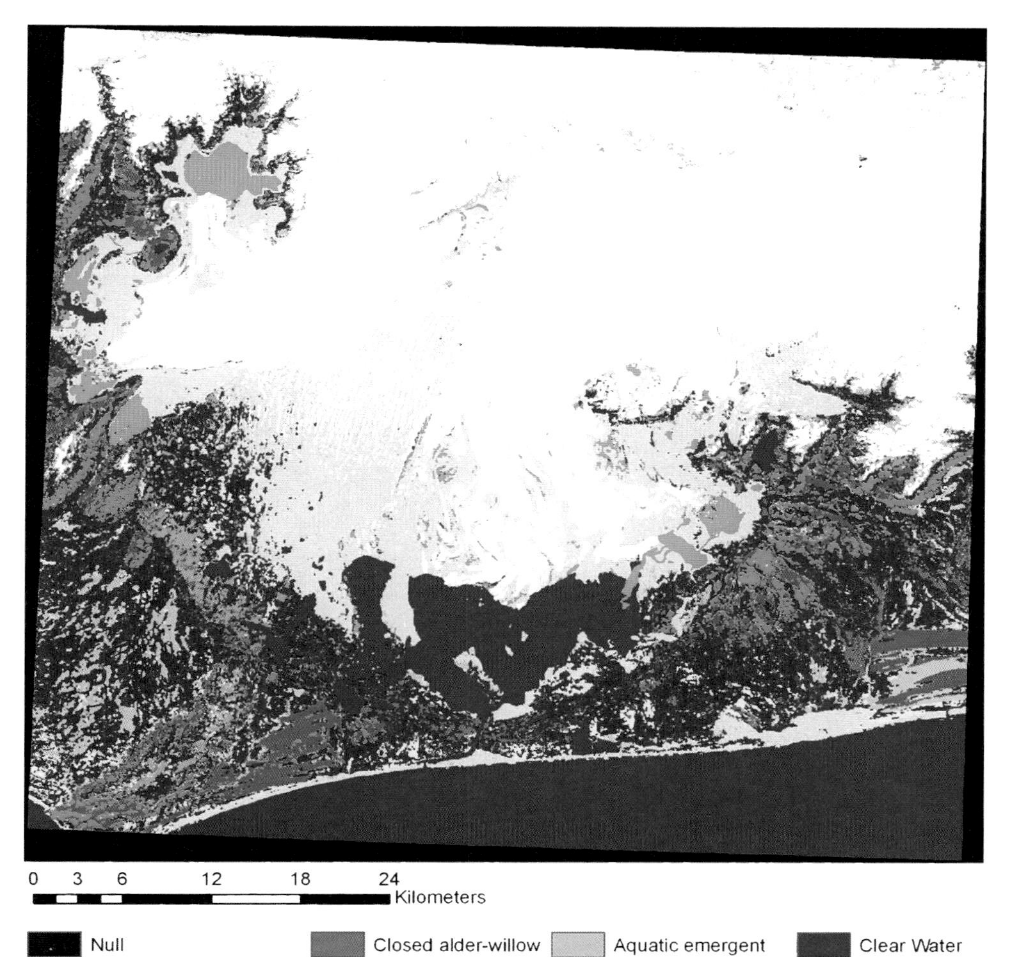

Figure 4. This land cover map has been modified (smoothed with a low pass filter and classes combined) from the 1991 U.S. Bureau of Land Management (BLM) and Ducks Unlimited classification. The land cover maps show the diversity of the vegetation types surrounding the Bering Glacier. The dominant land cover types other than ice, snow, and water include closed and open needleleaf forest, open and closed alder-willow, closed mixed forest, herb-graminoid, aquatic emergent, salmonberry-fern, and sparse vegetation.

the 1995 surge to 2007, which has resulted in a large expansion of Vitus Lake.

We mapped the bathymetry from a small boat using a portable fathometer (depth sounder) that simultaneously logged global positioning system (GPS) position depth and temperature once every second, which, at the typical speed of the survey boat (3.6 m/s), resulted in one data point approximately every 3.6 m. It took three field seasons, 2002–2004, to completely map Vitus Lake, and a modified kriging interpolation technique was used to convert the track data to a uniform grid with a 15 m posting (Josberger et al., 2006).

To determine the increase in lake volume, we combined the bathymetry obtained by the University of Michigan with the satellite-derived expanding lake area in a geographic information system (GIS). Figure 6 shows the bathymetry, and Table 2 gives the area, volume, and percentage change per year for the lake. From 1995 to 2006, Vitus Lake has increased in volume to ~9.4 km^3 (a 261.3% change) with a total area of 130.8 km^2 (a 124.1% change).

The ever increasing volume of Vitus Lake has the potential for altering the ecosystem. As reported by Josberger et al. (this volume) and Josberger et al. (2006), the water quality parameters (temperature, density, salinity, dissolved oxygen, pH, and turbidity) are highly complex and have changed over the 5 a observation period. Since the surge ended, the harbor seal population in Vitus Lake has steadily increased to ~1000 individuals (Saverese and Burns, this volume), which is an indicator of a changing aquatic ecosystem.

Equilibrium Line Mapping

The equilibrium line is the location on the glacier below which all the snow that accumulated during the previous winter completely melts, leaving firn or bare glacier ice visible. Hence

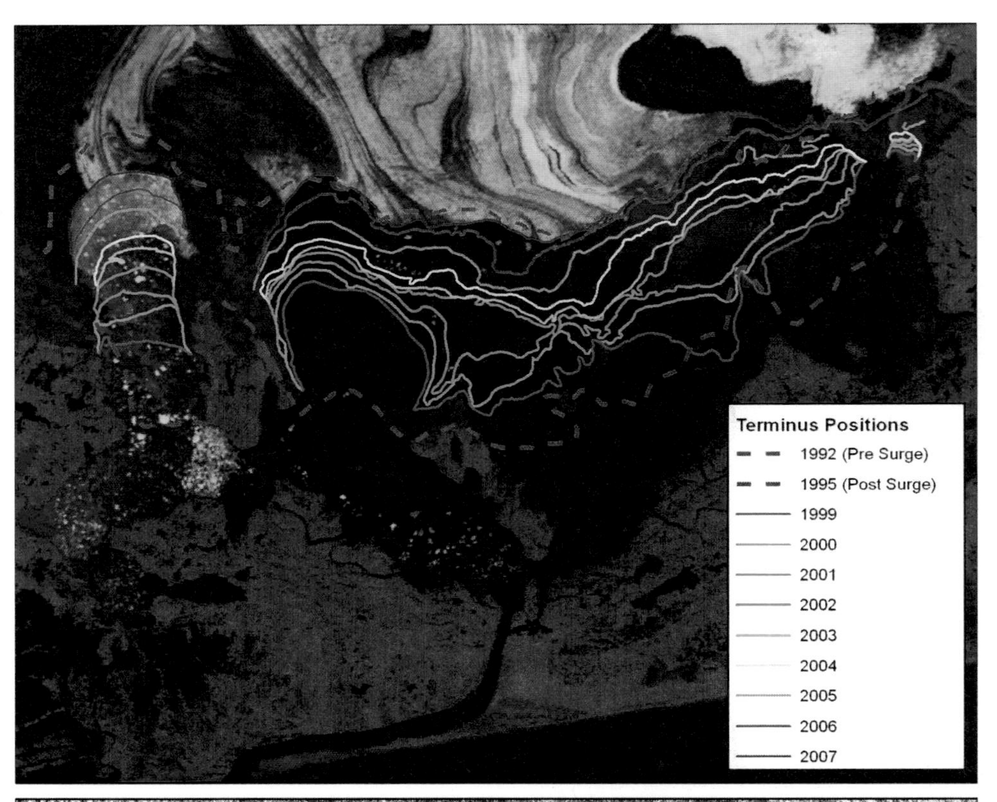

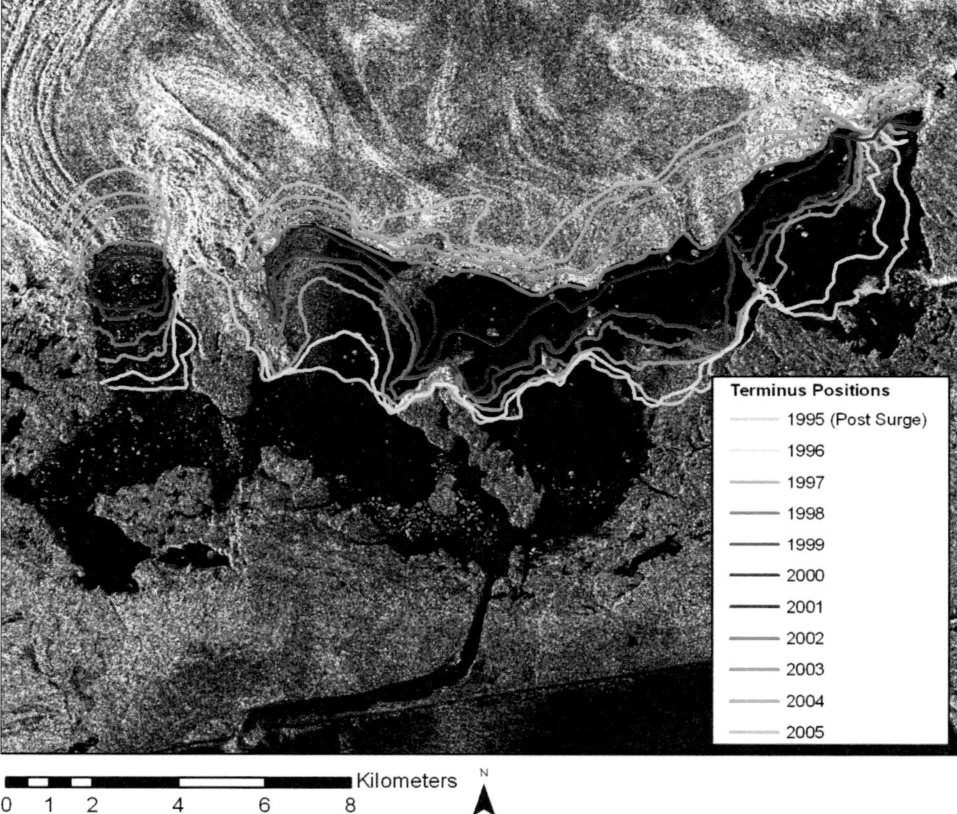

Figure 5. Bering Glacier terminus positions, derived using Landsat imagery (top) and ERS-2 SAR imagery (bottom). These maps show that Landsat and SAR produce comparable terminus mapping results. Note that the glacier has receded ~4.8 km since the 1995 surge and is at the position of the pre-surge terminus location.

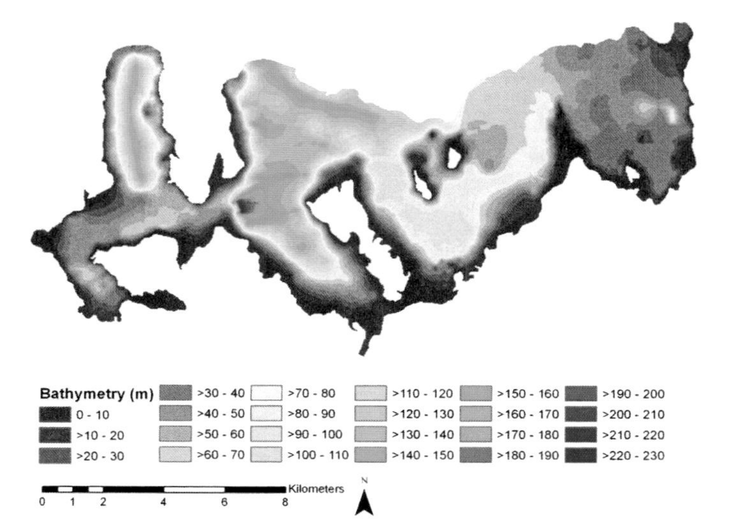

Figure 6. Vitus Lake bathymetry was obtained in 2002–2004 by the University of Michigan field deployable GPS-enabled fathometer system. The maximum lake depth is >230 m and is represented by the dark blue color.

TABLE 2. VITUS LAKE VOLUME AND AREA CHANGES

Year	Volume			Area		
	km³	Percentage change		km²	Percentage Change	
		Annual	Cumulative		Annual	Cumulative
1995	2.6	—	—	58.4	—	—
1999	4.4	67.0	67.0	80.6	38.0	38.0
2000	4.9	12.2	87.3	89.0	10.4	52.4
2001	5.4	10.4	106.8	96.2	8.1	64.7
2002	6.2	14.9	137.5	105.8	9.9	81.1
2003	6.4	4.0	147.0	108.8	2.9	86.3
2004	6.9	6.5	163.1	114.0	4.8	95.3
2005	8.2	18.8	212.6	122.5	7.4	109.8
2006	9.4	15.6	261.3	130.8	6.8	124.1

Note: Changes were derived using Vitus Lake bathymetry and satellite-derived terminus locations. Vitus Lake is presently ~9.4 km³ in volume and ~130 km² in area.

the equilibrium line altitude (ELA) depends on both the winter snowfall and the summer temperature. Alaska mean summer temperature has increased ~0.55 °C from 1977 to 2004 (Molnia 2007), which has attributed in part to the ELA rising to the 2006 altitude of 1280 m above sea level. Historical remote sensing images can be used to quantify the change in altitude of the ELA, as shown in Figure 7, which shows the ELA for 1995, 1999, 2000, 2002, and 2006 derived from EO satellites. For the 10 a period 1995–2006, the ELA has increased by ~200 m with a horizontal translation of 10 km. Also shown in Figure 7 are elevation contours, shown in gray, derived from a stereo pair of 2004 ASTER images.

The migration of the snowline on the glacier over the course of the melt season is a key parameter with which to calibrate snow-glacier ablation models, and the MODIS satellite system is particularly useful for this mapping endeavor. Recall that MODIS is a moderate spatial resolution (250 m) multispectral EO satellite, with data provided by NASA free of charge. It has a 2500 km swath (1500 of which has usable viewing geometry) with a late morning and early afternoon (Terra and Aqua) revisit interval. An 8 d composite product using best available data within the eight-day time frame makes the system useful for Bering mapping, given the cloudiness of this area.

Figure 8 shows six late August MODIS scenes, from 2001 to 2006. The snowline, and likely the ELA because it is late in the ablation season, is shown by the letters A through F. An annual comparison of the Bering Glacier region for the August time frame is shown in Figure 8. Note from the points labeled A through F that the snowline has clearly moved to higher altitudes during this 5 a observation period.

The MODIS data can also be used to observe intra-annual variability. Figure 9 shows three 8 d composites spanning a time period from 14 September 2004 to 6 October 2004. These images show the ELA well up into the Bagley Ice field (A) in mid-September of 2004. The MODIS composite of 22–29 September (B) shows the cloud cover associated with an early winter storm, and (C) shows the new snowline after the extensive snowfall event. Information about the first snowfall can be useful in selecting which satellite image to use for determining the snowline. It is difficult to determine the snowline from a single EO or radar image. In the EO bands the reflectivity of snow versus glacier surface is hard to interpret in the fall owing to the extensive meltwater present (Fig. 1). The microwave radar signature is also attenuated because of the large amount of liquid water present. By combining, for example, MODIS data with SAR data, a more discernible snowline can be obtained.

Elevation Mapping and Glacier Volume Estimation

Remote sensing satellite data can provide detailed height information. Forward and backward looking pairs of ASTER images can create a DEM from a stereographic analysis that is part of the Japanese software package SilcAst. The resulting DEM has 15 m postings in the horizontal and vertical directions. Figure 7 presents contours derived from ASTER observations superimposed on the upper Bering Glacier and a part of the Bagley Ice Field for 2004.

In order to compute the volumetric change of the Bering Lobe, we georegistered and compared the remote sensing-derived DEM to historical elevation data sets. The top panel in Figure 10 gives elevation profiles starting in 1900 through 2004. The profiles are from the center line of the glacier, shown in yellow in the lower panel of Figure 10. The 2004 elevation data were obtained from the aforementioned ASTER DEM analysis, whereas the other elevation data were provided by the USGS (Molnia and Post, 1995). Note that extensive wastage has occurred at the lower elevations, whereas at elevations of 800–1000 m the elevation values are similar, within the accuracy of the observations. To further validate the 2004 remote sensing–derived elevations,

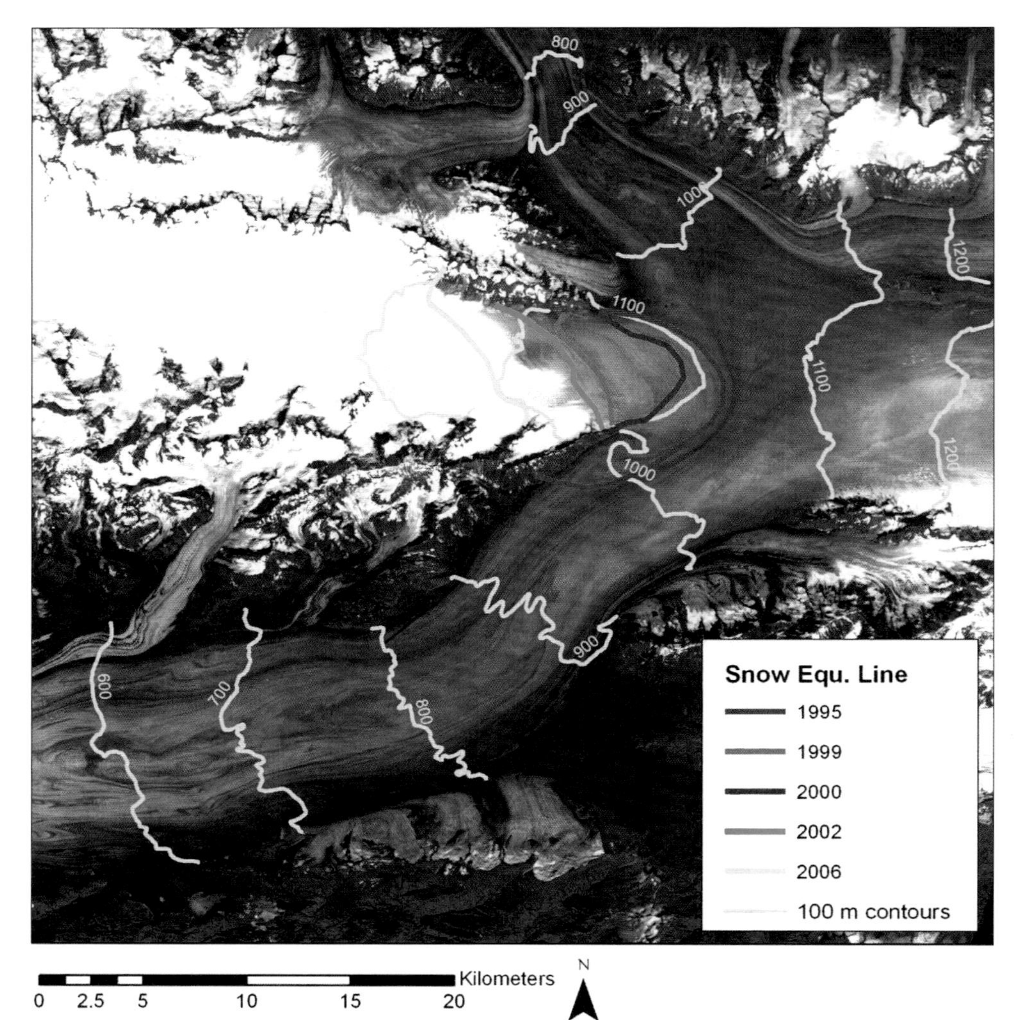

Figure 7. The snow equilibrium line, mapped using remote sensing satellite data, shows that the snow equilibrium line has migrated up-glacier. The purple line shows the location of the snow equilibrium line in 1995, orange 1999, red 2000, green 2002, and yellow 2006. Also included are elevation contours, shown in gray, obtained from a stereo pair of 2004 ASTER scenes. A 2003 ASTER satellite image has been used as a base map showing a snow equilibrium line. The 2006 equilibrium line has migrated up glacier ~200 m in elevation since 1995.

a limited set of GPS elevations was collected during the summer 2006 field season. These GPS elevation values, when compared to the ASTER elevations, showed comparative heights within 15 m, the height resolution of the ASTER derived DEM. It is interesting to note that the glacier has lost ~150 m of elevation near the terminus and ~30 m at 300 m below the location of the 2004 equilibrium line for the period 1957–2004.

The entire volume of the glacier was estimated by combining the radar sounding–determined bedrock depth data from Molnia and Post (1995) with the glacier surface altitude data determined by the 1972 aerial photography–derived USGS DEM map. By subtracting the bedrock elevations from the glacier elevations, the ice depth at each pixel was found. Integrating these values over every pixel revealed that the total glacial volume in 1972 was ~662 km^3.

The change in glacier volume was calculated using a comparison of the ASTER-derived DEM values with historical data reported by Molnia and Post (1995). By subtracting the 1957 historical elevation data from the 2004 ASTER-derived elevations, the total vertical melt along a transect through the center of the glacier (Fig. 10) was calculated for this period. By dividing the glacier into discrete sections, calculating the area for each section using the DEM data, and multiplying the area of each section by the previously calculated vertical melt, it was found that the total volume change of the Bering Lobe from 1957 to 2004 was ~104 km^3. If the last surge (1993–1995) is ignored in this analysis, and assuming an average melt rate each year from 1957 to 2004, the two previous calculations indicate that during this time period the glacier melted, on average, 2 km^3 per year and that the glacier lost 13% of its total mass.

Another method for elevation data extraction is the use of stereo-pair aerial photographs. In August 2006, black and white aerial photography was flown along the glacier terminus. These data were used to obtain point elevation measurements using parallax techniques. Past experience indicates that parallax measurements can provide spot heights to within plus or minus 1/2000 of the flying height. The 2006 images were taken from an altitude of ~4000 m, which suggests an approximate error

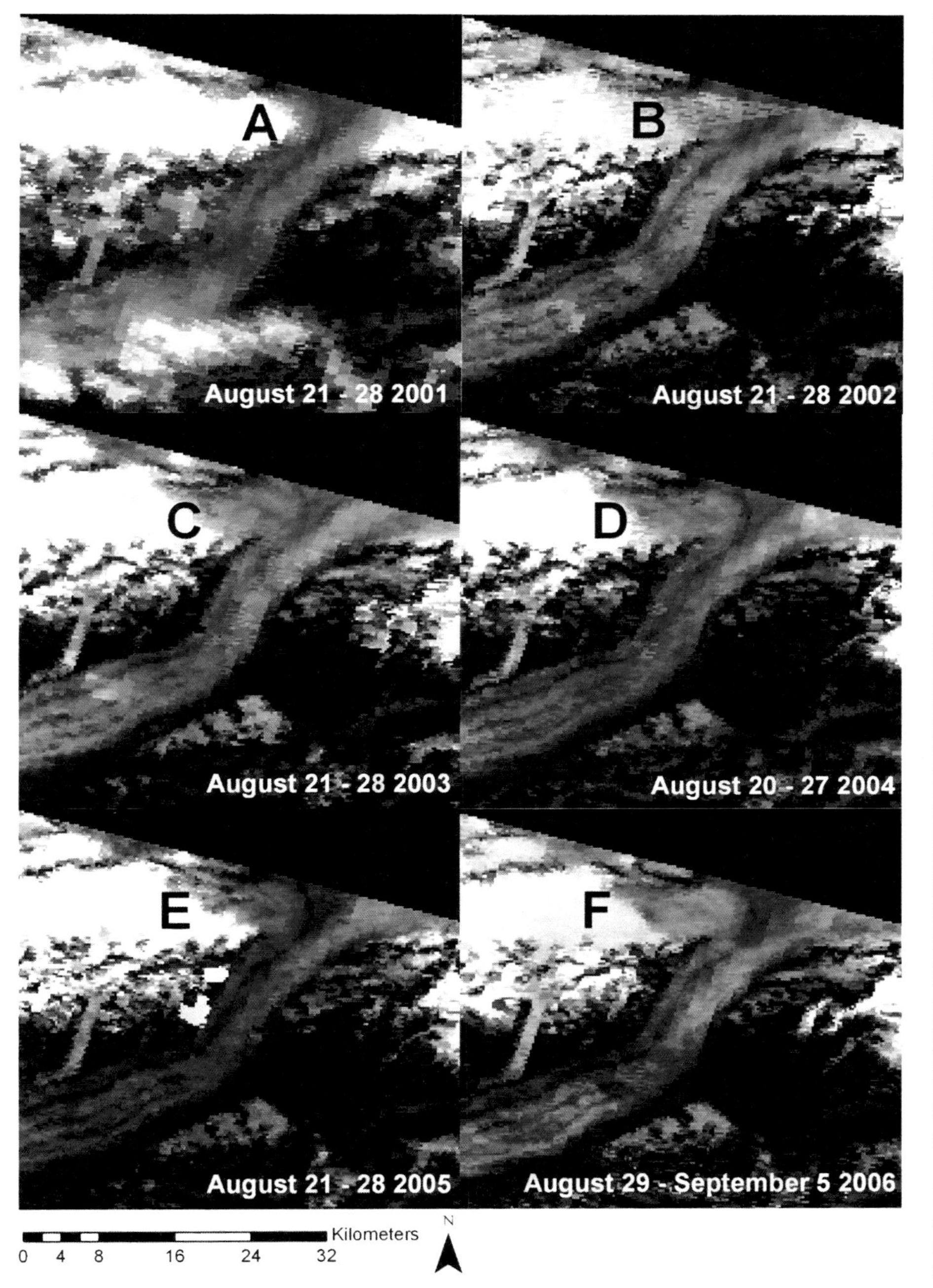

Figure 8. Six MODIS 8 d composites show the inter-annual changes occurring at higher elevations on the Bering Glacier. The imagery shows that during this 5 a observation period the snowline has moved to higher altitudes (~100 m).

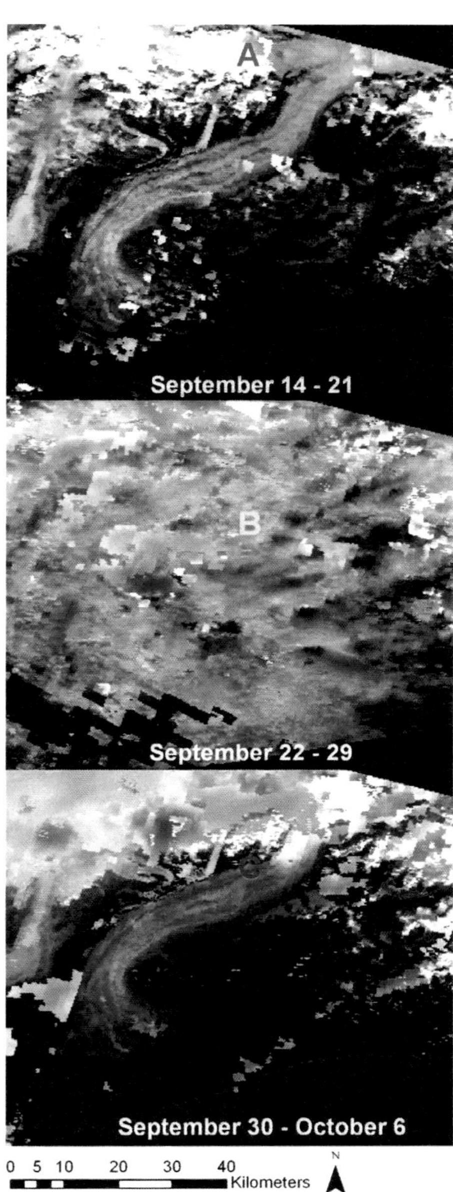

Figure 9. Three MODIS 8 d composites, spanning a time period from 14 September 2004 to 6 October 2004, show changing weather conditions on the glacier, and specifically the documentation of the first snowfall of the season. (A) End-of-season snow equilibrium line; (B) first snowfall; (C) new snowline.

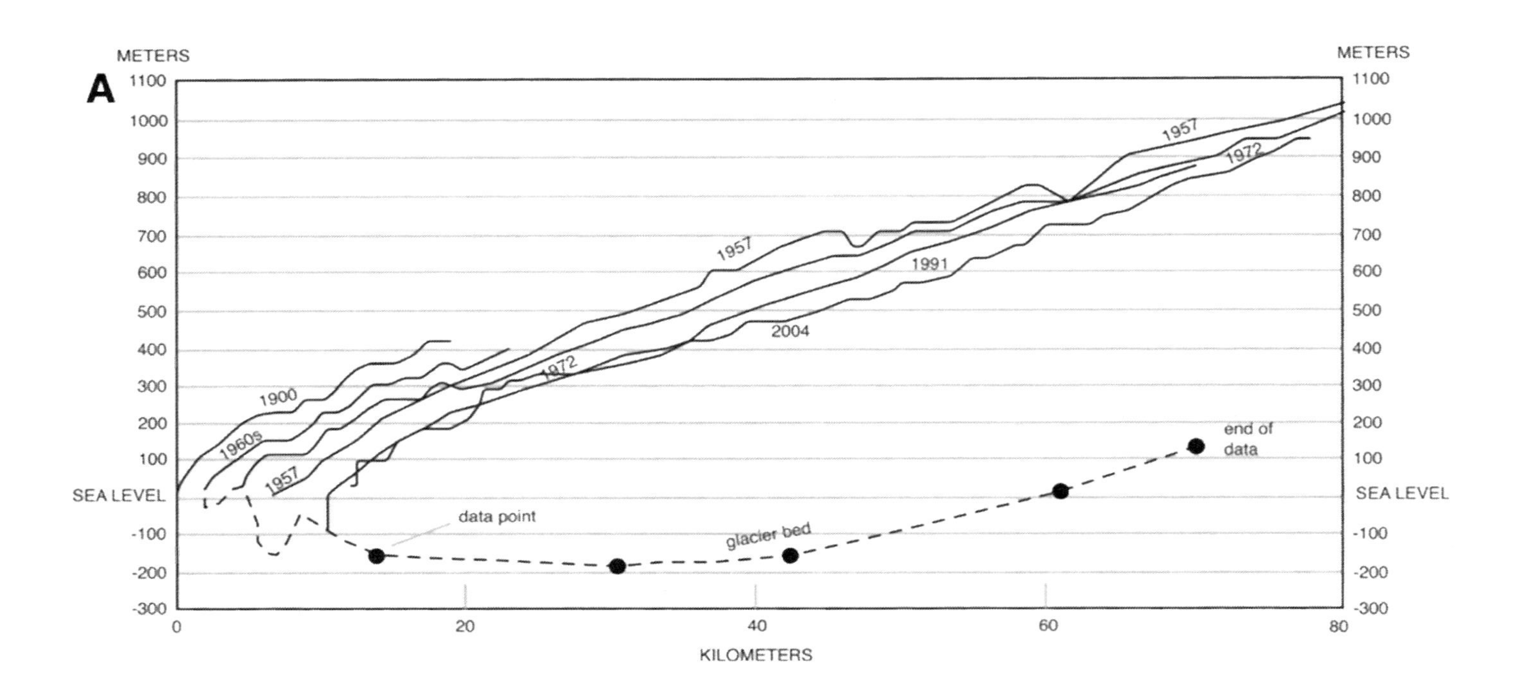

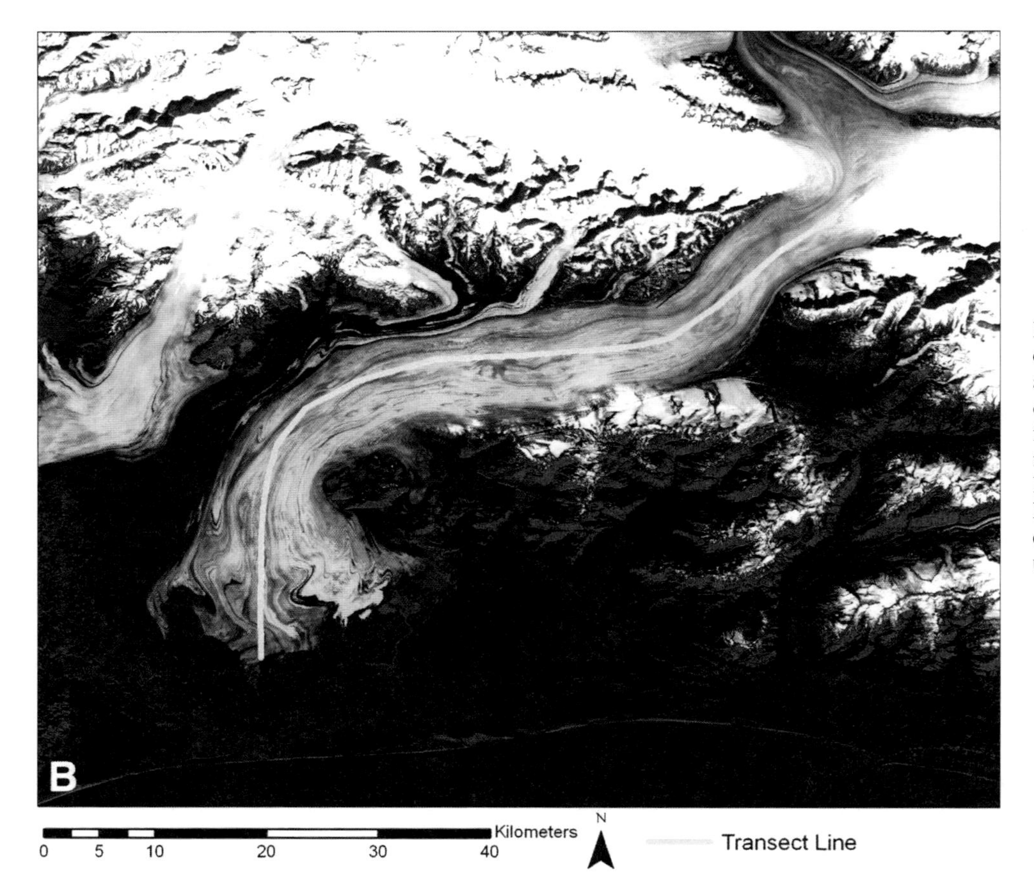

Figure 10. (Top) Changes in observed elevation of the Bering Glacier surface applicable to ablation or melting that has occurred since 1900. Figure modified from Molnia and Post (1995), with permission from Post. (Bottom) Transect line used to determine elevation profile. Note that the glacier has decreased in elevation by ~200 m from 1957 at 40 km up-glacier (500 m elevation).

of ±2 m. A conservative estimate of the error in these measurements is ±3 m.

Total Summer Melt Estimate Using In Situ Ablation and Remote Sensing

The synergy of remote sensing observations and in situ measurements is a powerful tool for estimating annual glacier ablation. We used five seasonal melt rate measurements at ascending positions on the glacier, starting at the terminus and ending near the equilibrium line (Fig. 10). The melt rates obtained (Table 3) for the 2005 season show ablation rates of 5.08 cm/d at the terminus and 3.88 cm/d slightly below the equilibrium line. These data points determined the ablation rate as a function of altitude for the entire glacier.

A 2005 Landsat scene was then used to create a minimum and maximum estimate for the ice/snow area of the Glacier (Bering Lobe) that melts into Lake Vitus (minimum determined as 1550 km^2, and maximum as 2200 km^2). Using digital elevation data obtained from the ASTER image pair, melt rates were

TABLE 3. MELT RATES OBTAINED FOR THE 2005 SEASON

Site	Starting latitude (°N)	Starting longitude (°W)	Starting date	Starting elevation (ft)	Ending latitude (°N)	Ending longitude (°W)	Ending date	Total melt (cm)	Total days	Melt per day (cm)	Horizontal migration (m)
S01	60.384	143.809	6/16/2005	596	60.385	143.809	9/20/2005	514	96	5.35	37.42
S02	60.397	143.666	6/16/2005	1483	60.397	143.667	9/20/2005	500	96	5.21	112.34
B01	60.163	143.410	5/11/2005	199	60.163	143.410	9/19/2005	665	131	5.08	4.61
B02	60.278	143.376	5/11/2005	850	60.278	143.376	9/19/2005	534	131	4.08	10.39
B03	60.358	143.268	6/16/2005	1532	60.358	143.268	9/20/2005	528	96	5.50	4.15
B05	60.490	142.613	6/16/2005	3203	60.488	142.617	9/20/2005	488	96	5.08	268.69
B06	60.575	142.770	6/16/2005	4107	60.575	142.769	9/20/2005	372	96	3.88	41.47

Note: These melt rates show ablation rates of 5.08 cm/day at the terminus and 3.88 cm/day near the snow equilibrium line (top). Site locations are shown in the Landsat image for reference (below).

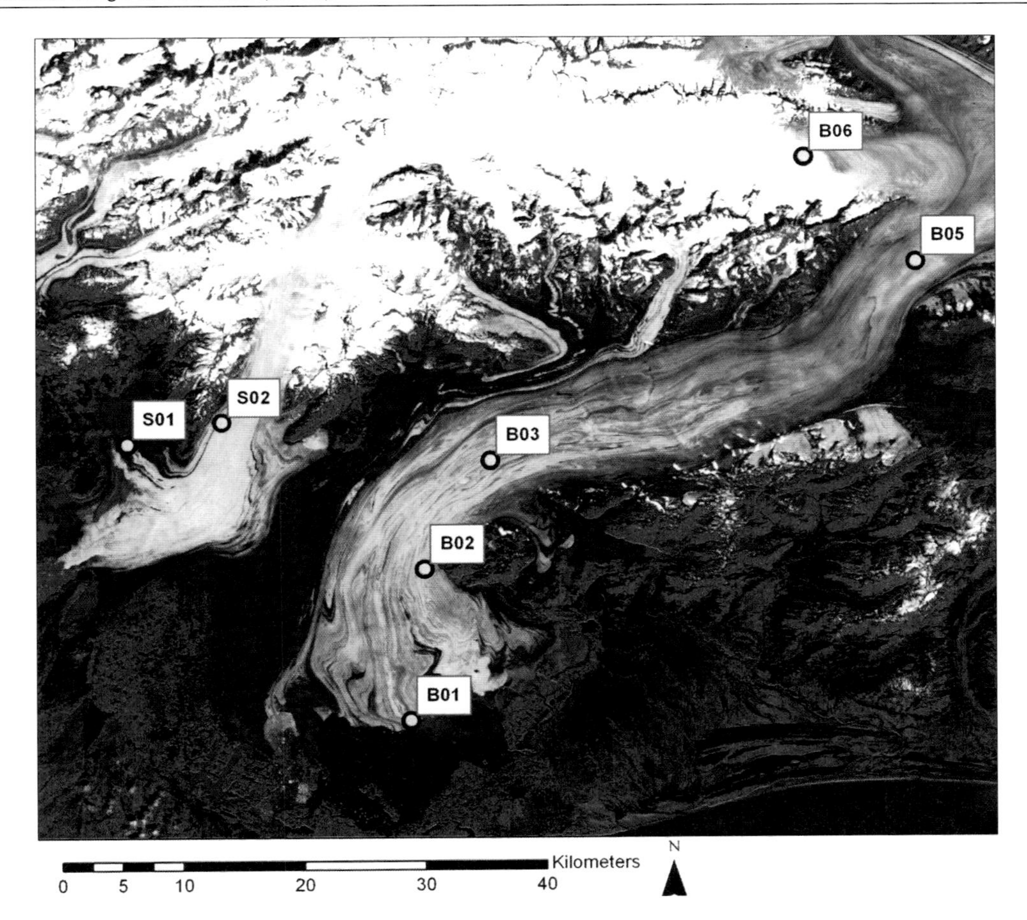

calculated for each pixel within both the minimum and maximum area estimate. Using these data, it was found that between 6.5 km³ and 9.1 km³ of ice from the glacier ablated during the 120 days from 1 May to 1 September 2005. The water equivalent of this ice is ~5.9–8.3 km³. Since the volume increase of Vitus Lake obtained from the USGS water level gauge was shown to be a maximum 0.12 km³ in 2005 (from bathymetric mapping and water height measurements), ~5.8–8.2 km³ of fresh water flowed into the Gulf of Alaska during that year—a tremendous amount of water. By comparison, the Copper River, the largest watershed in south-central Alaska and the third largest watershed in Alaska, discharges ~60 km³ of water into the ocean each year (USGS, 2007).

Iceberg and Calving Rate Estimates

During the twentieth century, Bering Glacier has surged five times; each surge has been followed by a period of rapid terminus retreat (Fig. 5). As reported by Molnia (2006, personal commun.) each post-surge-retreat phase began with active calving when the terminus was grounded and the ice cliff was ~100 m high. As the Bering Lobe rapidly thins by as much as 6–10 m/a (Shuchman et al., this volume) the calving slows as the terminus approaches flotation thickness, and disarticulation becomes the dominant process responsible for terminus retreat. Disarticulation is the passive separation of large tabular pieces of ice from the Bering terminus.

Thus, iceberg density and calving rates can provide insight into glacier dynamics, and the number and size of icebergs in Vitus Lake may be an important constituent of the ecosystem required by harbor seals, whose numbers now exceed 1000 (Savarese and Burns, this volume) individuals. Remote sensing systems with spatial resolutions of 25 m or finer can resolve the larger icebergs in Vitus Lake. SAR sensors are particularly useful owing to their all-weather imaging capability. Figure 11 is a sequence of four ERS-2 SAR satellite images collected in February, May, July, and August of 2004, showing icebergs in Vitus Lake. The table in Figure 12 summarizes the count of icebergs in the three parts of Vitus Lake for each of the four SAR images as well as one Landsat image. Comparisons of the 2000, 2002, and 2004 ERS-2 SAR counts show a decrease in iceberg totals, supporting Molnia's disarticulation hypothesis. Qualitative examination of the Landsat scene also supports the large tabular iceberg argument, particularly in the central part of Vitus Lake where the terminus height is now low. In the table, Landsat iceberg counts for zones 2 and 3 are similar to the May SAR derived counts. However, zone 1 counts vary between the Landsat and SAR images owing to dissimilar ice characteristics within the zone. When ice characteristics are similar the SAR and Landsat data produce similar estimates. Comparison of the four estimates

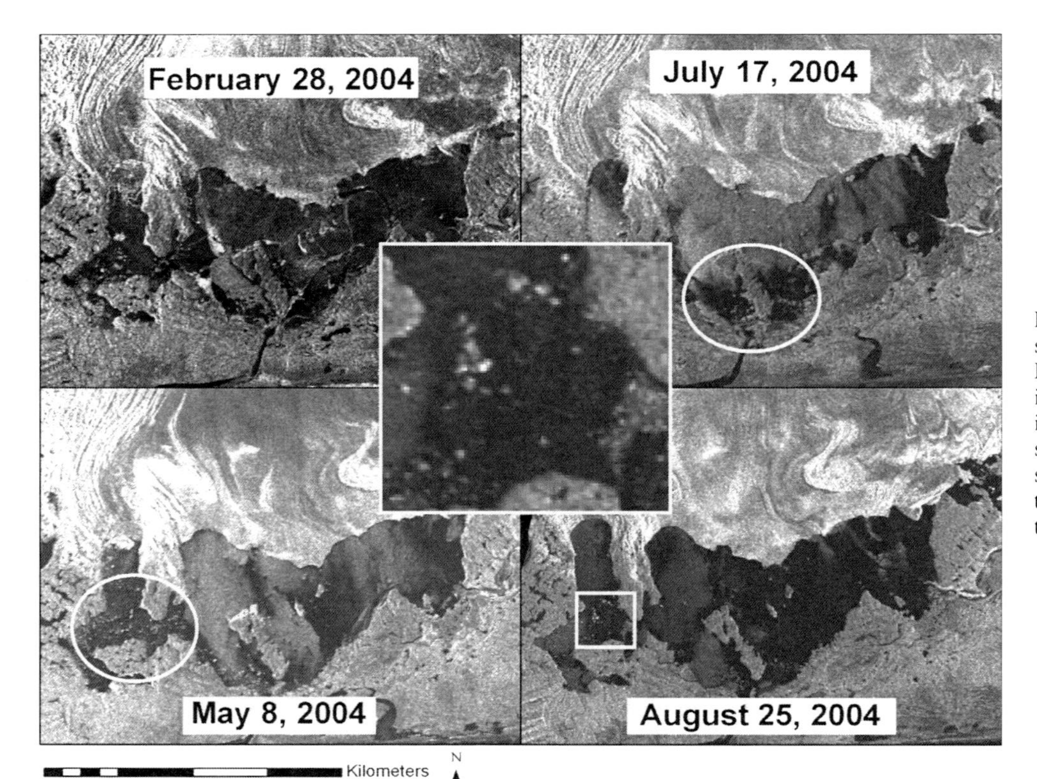

Figure 11. Sequence of four ERS-2 SAR satellite scenes collected 28 February, 8 May, 17 July, and 25 August 2004, showing icebergs visible in Vitus Lake. Circles indicate visible clusters of icebergs. The square in the 25 August image corresponds to the zoom box in the middle of the figure; icebergs are clearly seen. Notice ice cover in the 28 February image.

Sensor	Dates	Zone 1	Zone 2	Zone 3
SAR	Aug 00	109	70	205
SAR	Aug 02	140	95	109
SAR	Feb 04	Frozen	Frozen	Frozen
SAR	May 04	194	48	38
SAR	Jul 04	30	19	43
SAR	Aug 04	51	37	53
Landsat	Apr 04	90	52	36

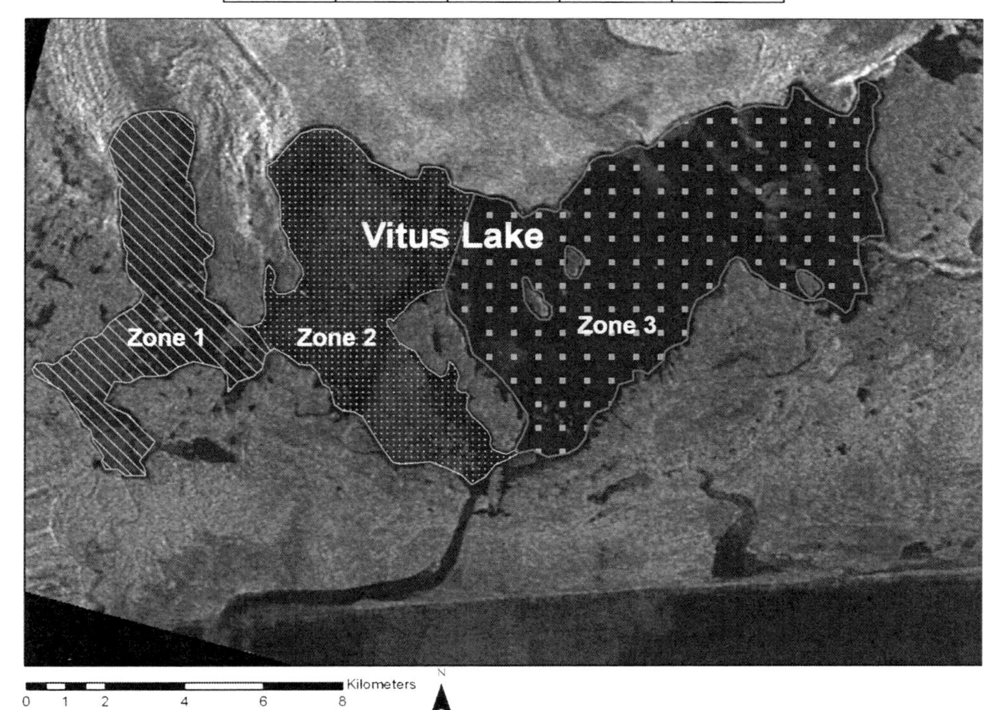

Figure 12. Iceberg counts for 2000, 2002, and four 2004 ERS-2 SAR satellite scenes and a 2004 Landsat scene (top). The zones used for iceberg delineation are shown in the reference image (bottom). For 2004, notice the large number of icebergs in May and April as a result of lake-ice breakup. The August estimates (2000, 2002, and 2004) show variability as a result of average summer temperature and height of the glacier terminus.

collected during the different seasons suggests that iceberg analysis should be conducted during the summer and fall seasons to minimize the effects of lake-ice breakup.

Albedo Estimations

Albedo is the ratio of reflected to incident light on a surface. It can be measured at individual wavelengths or integrated over a range of wavelengths. Albedo is an important parameter for understanding heat transfer in environmental systems, because it determines how much heat a system absorbs from the sun. Using a Landsat image from 7 August 2006, we were able to create an estimate of albedo for the entire Bering Glacier (Fig. 13) using an algorithm described by Greuell and Oerlemans (2004). Most of the glacial ice has albedo values between 0.25 and 0.65, which are typical values (Fig. 1). Snow has values near 1.

Using these estimates, along with coincident melt and meteorological observations, it is possible to create a heat transfer model for the Bering Glacier. Wind speed and temperature data contain information necessary to quantify convective heat transfer. These data can be combined with the albedo imagery, which contains radiative-heat-transfer information, to create a comprehensive glacial melt model. This would enable an accurate estimation of summer melt on a pixel-by-pixel basis for the entire glacier.

Water Quality Estimates

High resolution multispectral EO satellite data from Landsat (30 m resolution) or QuickBird (2.4 m) can provide useful information on the water quality of the lakes within the Bering Glacier System. MODIS data lack sufficient resolution (1 km) to resolve the smaller Bering Glacier lakes and the frontal features within Vitus Lake.

Water discharge from the glacier is typically sediment laden, with concentrations as high as 2800 mg/L (Josberger et al., 2006). An in situ water-quality instrument (ALWAS—Automated Lagrangian Water Quality Assessment System) has recorded turbidity ranges from zero NTU (nephelometric turbidity unit) to ~1,000 NTU in Vitus Lake (Liversedge, 2007). Typical values of

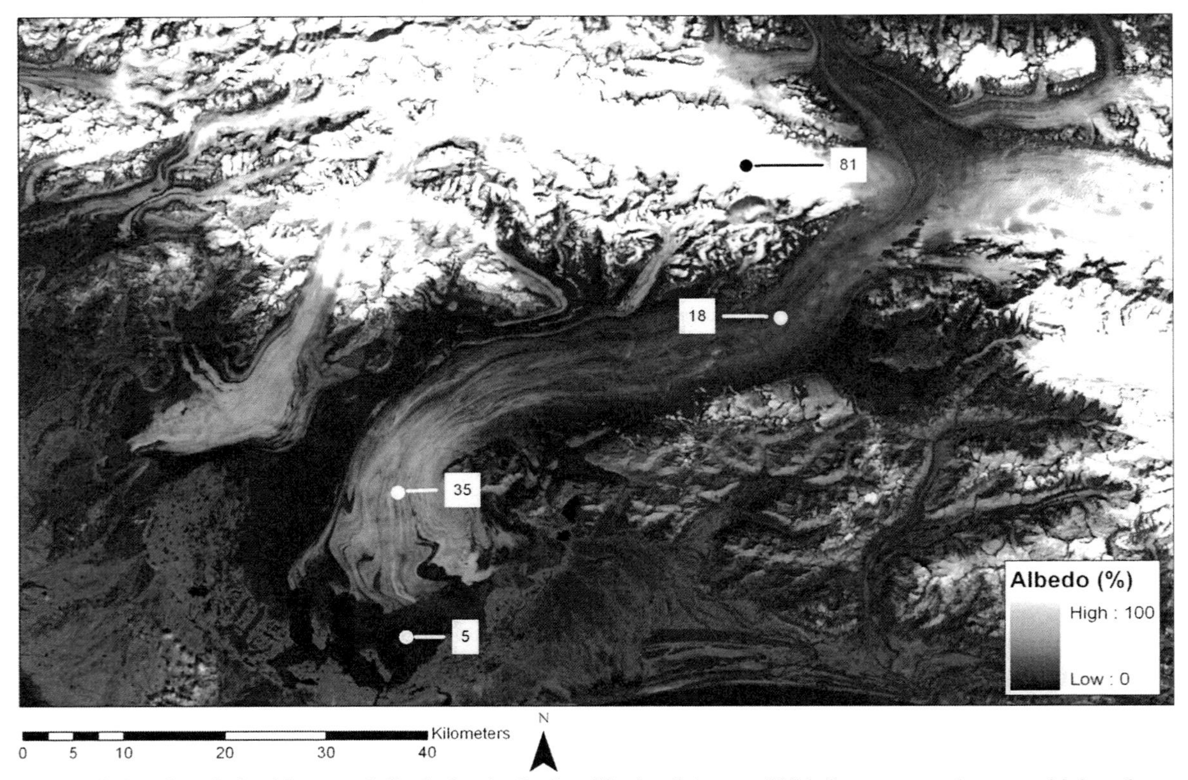

Figure 13. Landsat-derived image of albedo for the Bering Glacier, 7 August 2006. Snow-covered areas at higher elevations have high albedo values near 100%. Clean glacial ice has albedo values between 25% and 65%. Albedo measurements can be used, in conjunction with temperature and wind speed data, to create glacial melt models. Note that Vitus Lake is ~5% in areas where glacial rock flour is not present.

the sediment-laden water are 250 NTU or higher. Because surface water with large turbidity values has higher reflectance values, high turbidity areas are easily identified on the satellite images owing to the increased reflectivity. Figure 14 shows three Landsat images of Vitus Lake collected in September 2005, August 2006, and August 2007. These Landsat images show different amounts of sediment loading across the lake.

A model-based turbidity algorithm was developed by Liversedge (2007) for use in glacially influenced lakes and specifically in ice-marginal lakes at the Bering Glacier. This algorithm uses the red and near-infrared portions of the electromagnetic spectrum in a multiple liner regression equation to predict turbidity concentrations with an r^2 of 0.956 and an adjusted r^2 of 0.949. Inter- and intra-annual turbidity maps generated using this algorithm can be used to examine glacial processes and predict glacial trends such as surge and outburst flood events.

A QuickBird image of the Abandoned River collected in August 2005 is shown in Figure 15 and illustrates that the increased resolution of 1 m facilitates the mapping of the various turbidity concentrations within Vitus Lake. Examination of the high resolution QuickBird image also can be used to show the complex surface circulation of Vitus Lake water near the vicinity of the mouth of the Abandoned River. At the mouth of the river, sediment-rich water (concentrations of 2650–2800 mg/L) has a maximum velocity of ~3–4 m/s, with a total discharge of ~600 m^3/s (Josberger et al., 2006). This Abandoned River delta forces water from the River to flow north, south, along the shoreline, and also directly into Vitus Lake. The cold, sediment-laden water has a significantly higher density than the surrounding Vitus Lake water and rapidly sinks, leaving a much reduced surface signature. The interaction of these distinctly different water masses is expressed as an eddy observed in the upper left part of Figure 15.

Surface Glacier Feature Mapping

High resolution EO satellite data and aerial photography are useful tools to characterize surface features on the glacier. Additionally, high-resolution, time-lapsed images can be used to obtain glacier velocity and deformation information. A QuickBird image of the Steller Lobe terminus (Fig. 16) demonstrates the utility of high resolution satellite data to resolve surface features. The high resolution image with a ground spatial distance (resolution) of ~1 m was collected on 24 June 2004. For comparison, a Landsat image collected on 10 September 2001 is also included in the figure. The QuickBird image shows part of the Steller Lobe. Crevasses filled with residual snow (A), rock debris

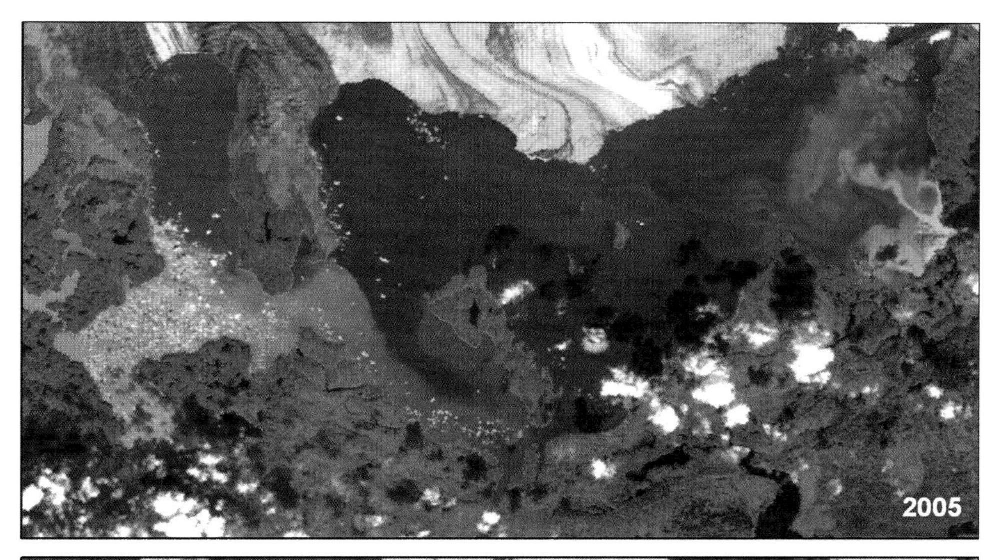

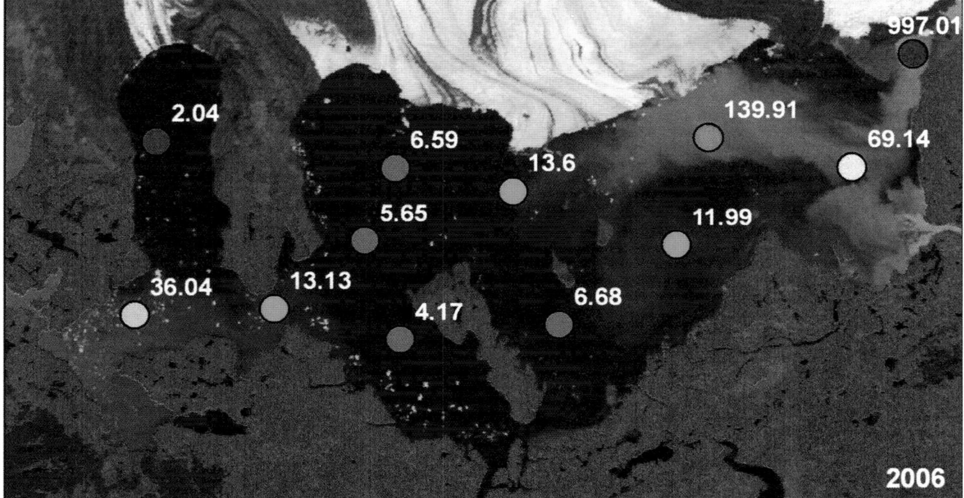

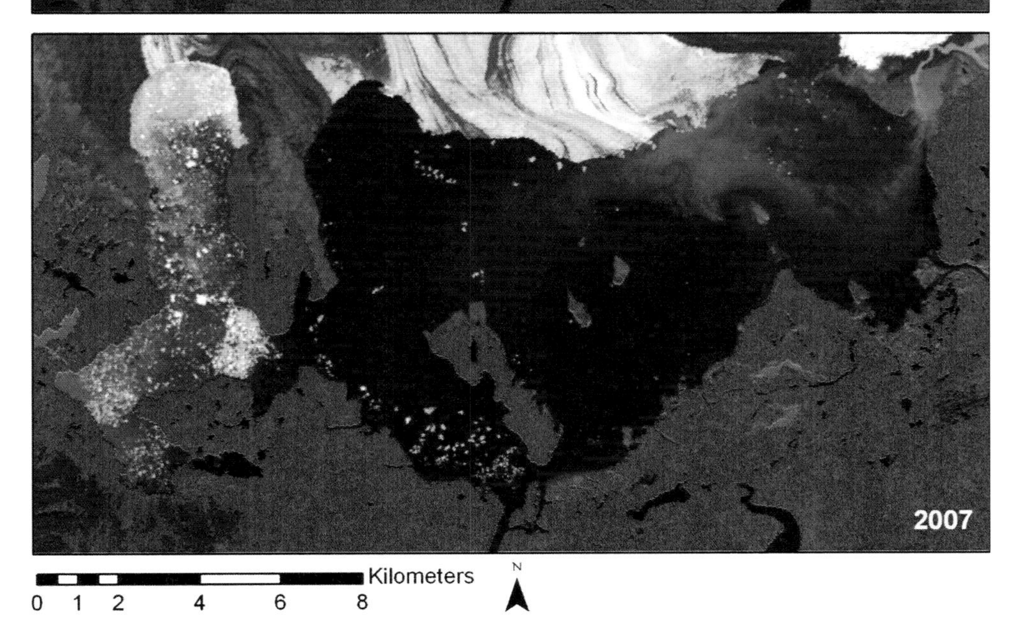

Figure 14. Three Landsat images (September 2005 Landsat 5 (top), August 2006 Landsat 7 (middle), and August 2007 Landsat 7 (bottom) show different spatial patterns of sediment loading in Vitus Lake. In situ surface turbidity values collected August 2006 using the ALWAS water quality instrumentation is shown (middle), reflecting good correlation between high reflectance satellite data and high turbidity values.

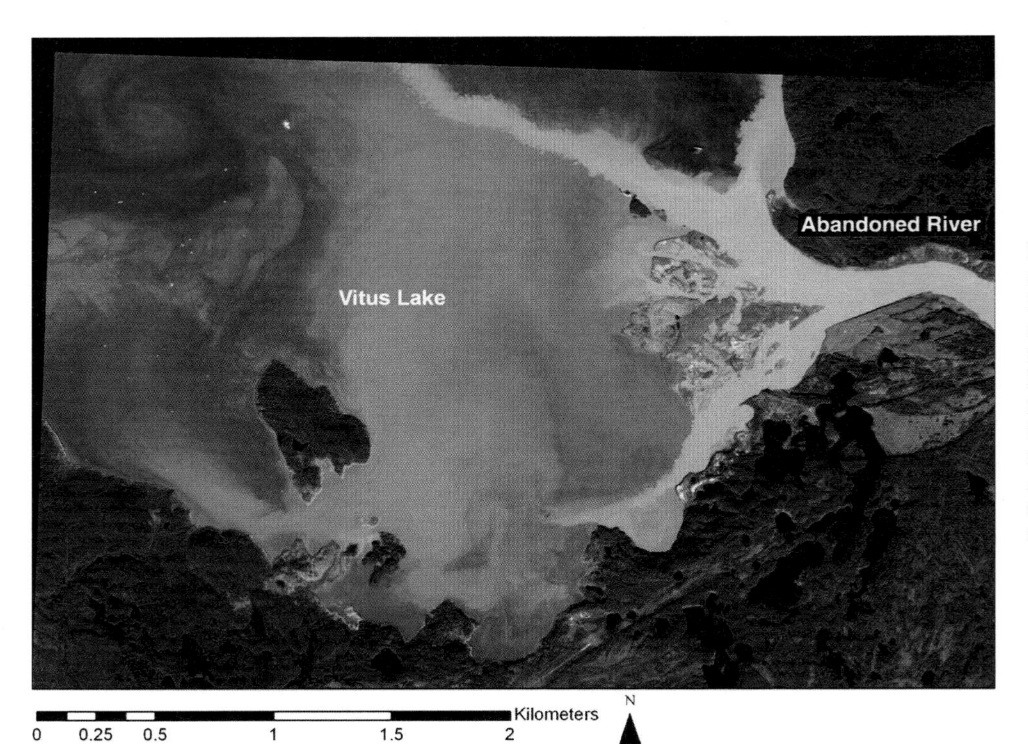

Figure 15. Increased spatial resolution of QuickBird satellite data (image collected 9 August 2005) facilitates the mapping and delineation of different turbidity concentrations in Vitus Lake. Note the eddy in the upper left corner of the image. This eddy is indicative of the interaction of distinctly different water masses converging at the mouth of the Abandoned River. The brighter water signatures are indicative of increased sediment concentration.

in lateral moraines (B), and melt ponds (C) are visible. The orientations of the crevasses, which are discernible on the image, provide insight into the flow of the ice. Most of these features are not visible in the Landsat image.

Seal Population Estimates

The harbor seal *(Phoca vitulina richardsii)* population in Vitus Lake continues to grow since the last surge. Saverese and Burns (this volume) estimate the current population to be ~1000 individuals. We attempted to identify and count the Bering Glacier resident harbor seal population with QuickBird commercial satellite imagery. The QuickBird system has a 60-cm-resolution panchromatic sensor (black and white) and a 2.44 m resolution for the visible and near infrared sensors. Figure 17 shows an image created by fusing (pixel sharpening) the 60 cm black and white data with the 2.44 m color data to create an overall 60-cm-resolution natural color image. The QuickBird data shown in Figure 16 were acquired 9 August 2005 with near coincident photography collected from a helicopter.

As indicated on the helicopter photograph, the seals outhaul on the icebergs free from potential predators. Our analysis of the QuickBird data showed that the individual 1- to 2-m-long seals are not discernible on either the fused natural color image (Fig. 16) or the panchromatic image. However, our analysis did show that several of the icebergs have lower reflectivity, possibly indicating a congregation of individual seals. A mixed pixel technique in which the combined reflectivity is calculated for the ice plus seals may be useful for seal population estimates. The photograph taken from the helicopter, in which the seals are easily discernible, further demonstrates the utility of high resolution aerial photography for some glacier monitoring applications.

Long-term Glacier Monitoring

In addition to the systems previously described, a recent (2009) publically available dataset from U.S. National Imagery Systems is now accessible for long-term glacier monitoring. These datasets can be accessed via http://gfl.usgs.gov with additional datasets located at http://eros.usgs.gov/#/Find_Data/Products_and_Data_Available/Declassified_Satellite_Imagery_-_1 and http://eros.usgs.gov/#/Find_Data/Products_and_Data_Available/Declassified_Satellite_Imagery_-_2. As more data becomes available from U.S. National Imagery Systems, additional temporal monitoring of glacier systems will be possible.

Figure 18 shows a September 1996 image (recent post-surge) and a May 2005 image from U.S. National Imagery Systems that shows the extensive retreat of the Bering Glacier terminus and resulting increase in Vitus Lake area for the nine-year interval. Also note the large number of icebergs that are present in the September 1996 image, and the large amount of sediment in Seal River for 1996, but not in 2005. These images, which demonstrate the utility of decadal monitoring via remote sensing, are part of the USGS Global Fiducials Library (GFL), which is an archive of images from U.S. National Imagery Systems that represents a long-term periodic record for selected scientifically

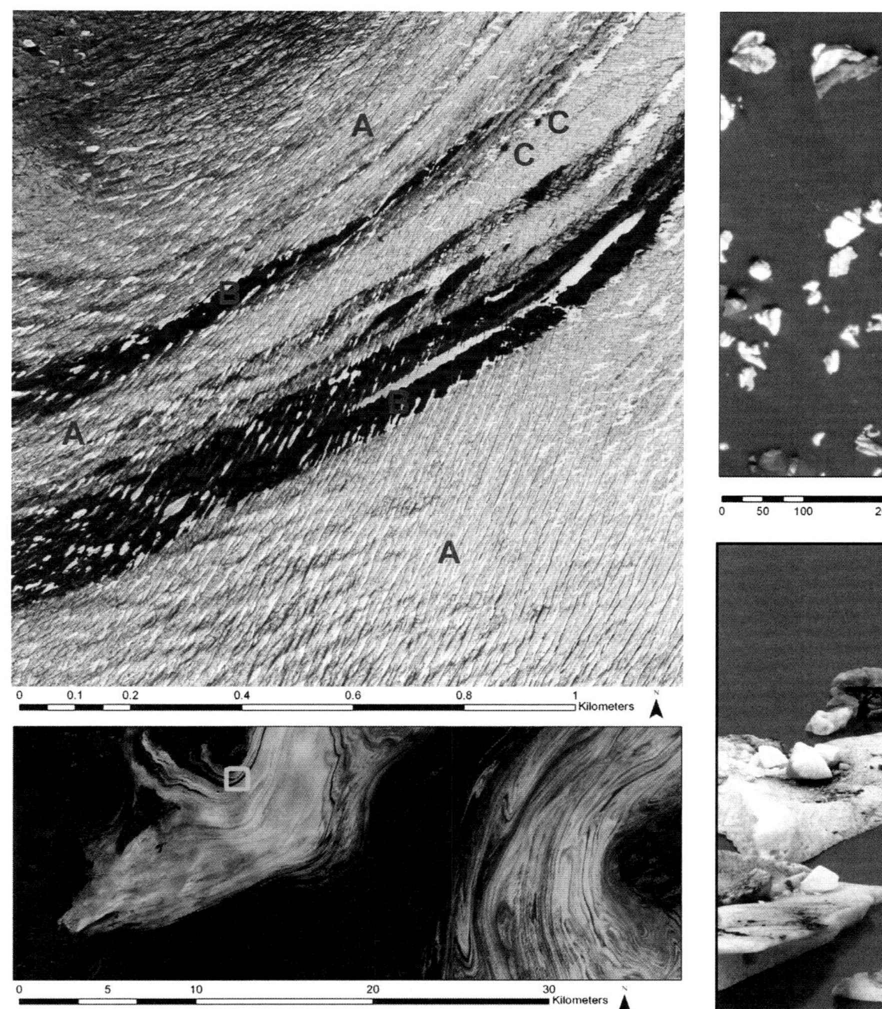

Figure 16. High resolution data from QuickBird (60 cm GSD) of the Bering Glacier Steller Lobe, collected 26 June 2006 (top), showing crevasses with residual snow (A), rock debris representing the lateral moraines (B), and melt ponds (C). A 10 September 2001 Landsat scene (bottom) is shown for context.

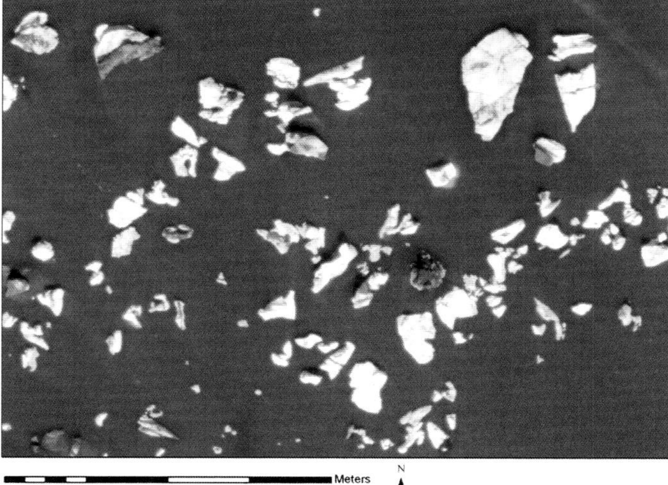

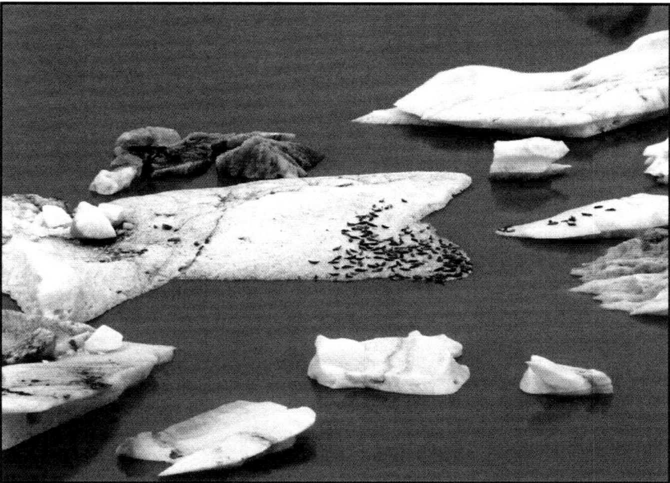

Figure 17. Remote sensing can be used to identify and count the resident Bering Glacier harbor seal *(Phoca vitulina richardsii)* population. It was determined that the seals cannot be resolved using QuickBird satellite data (top, image collected 9 August 2005), but can be resolved using helicopter (bottom) and aerial photography. A mixed pixel technique, in which the combined reflectivity is calculated for the ice plus seals, could prove to be useful for delineation of the seal population using QuickBird.

important sites such as Bering Glacier (see http://gfl.usgs.gov for additional details).

CONCLUDING REMARKS

The Bering Glacier occupies a vast area (5200 km^2 with a length >190 km), and remote sensing systems have demonstrated that they can capture a synoptic and historical view of the landscape. Remote sensing EO and RADAR technologies, when combined with in situ field measurements and GIS, have proven to be powerful tools for environmental monitoring at the Bering Glacier. Applications include glacier movement, terminus locations, iceberg and calving surveys, Vitus Lake frontal boundaries, land cover classification, snow line delineation, Vitus Lake volume analysis, generation of DEMs for glacier volume change, and albedo measurements.

Each remote sensing system used in the Bering Glacier analysis has strengths and weaknesses, and no one sensor can be used for every aspect of monitoring owing to differences in spatial resolution, spectral resolution, temporal resolution, and atmospheric conditions. Combining data derived from different remote sensors with field-based measurements can be a powerful tool for environmental monitoring at the Bering Glacier. For example, SAR/MODIS data fusion can be used to generate improved information on the snowline and terminus.

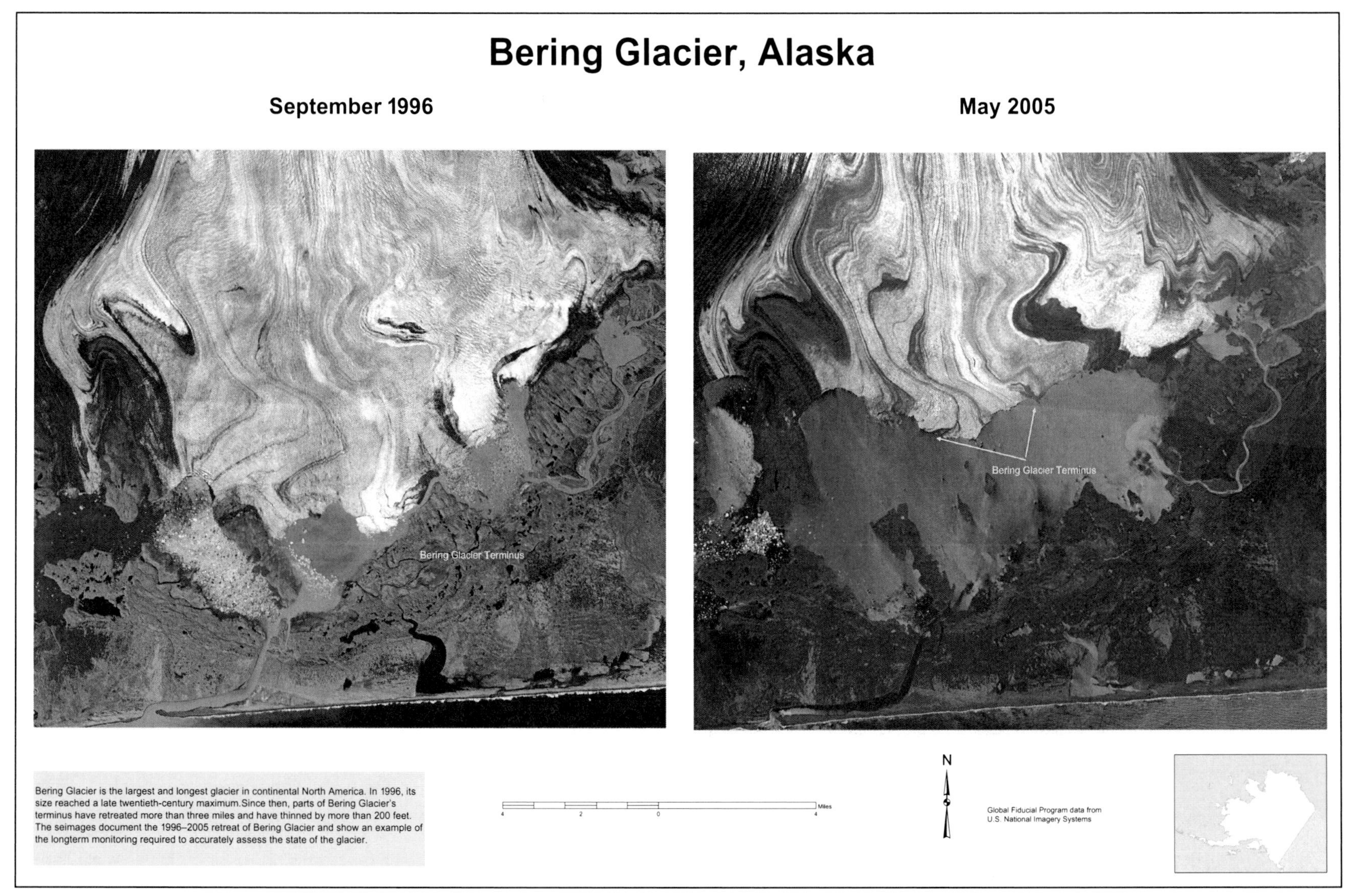

Figure 18. These images from U.S. National Imaging Systems document the 1996–2005 retreat of Bering Glacier and show an example of the long term monitoring required to accurately assess the state of the glacier.

Additional Bering Glacier remote sensing studies are recommended. These include completion of the snow equilibrium line analysis, generation of additional DEMs of the Bering Glacier System using SAR, ASTER, and ICESAT profile data, generation of and updates to land cover maps for the Bering Glacier region using an object based classification system such as Definiens, quantification of iceberg density in Lake Vitus on a yearly basis using a combination of EO and SAR data sets, exploration of the use of MODIS for the water quality assessment of Vitus Lake, and further examination of the use of high resolution commercial satellite data, such as QuickBird, to estimate seal populations.

The methodologies developed and demonstrated in Bering Glacier remote sensing data sets are applicable to other glacier systems throughout the world. The results presented here are consistent with other remote sensing glacier examinations.

ACKNOWLEDGMENTS

This activity was partially funded under U.S. BLM grant no. LAA-01-0018. The authors would like to acknowledge Chris Noyles and Scott Guyer of BLM for their help in collecting the in situ observations. Rick Powell of Michigan Tech Research Institute (MTRI) is thanked for providing the ASTER derived DEM information as well as acquiring the high resolution QuickBird data. Kristine Crossen of the University of Alaska, Anchorage is thanked for her assistance is delineating geologic landforms in the Bering region.

REFERENCES CITED

Bidlake, W.R., Josberger, E.G., and Savoca, M.E., 2007, Water, Ice, and Meteorological Measurements at South Cascade Glacier, Washington, Balance Years 2004 and 2005: U.S. Geological Survey Scientific Investigations Report 2007–5055, 69 p.

Chen, J.L., Tapley, B.D., and Wilson, C.R., 2006, Alaskan mountain glacial melting observed by satellite gravimetry: Science, v. 248, p. 368–378.

Fatland, D.R., and Lingle, C.S., 2002, InSAR observations of the 1993–95 Bering Glacier (Alaska, U.S.A.) surge and a surge hypothesis: Journal of Glaciology, v. 48, p. 439–451, doi: 10.3189/172756502781831296.

Greuell, W., and Oerlemans, J., 2004, Narrowband-to-broadband albedo conversion for glacier ice and snow: Equations based on modeling and ranges of validity of the equations: Remote Sensing of Environment, v. 89, p. 95–105, doi: 10.1016/j.rse.2003.10.010.

Josberger, E.G., Shuchman, R.A., Meadows, G.A., Savage, S., and Payne, J., 2006, Hydrography and circulation of ice-marginal lakes at Bering Glacier, Alaska, U.S.A.: Arctic, Antarctic, and Alpine Research, v. 38, p. 547–560, doi: 10.1657/1523-0430(2006)38[547:HACOIL]2.0.CO;2.

König, M., Winther, J.-G., and Isaksson, E., 2001, Measuring snow and glacier ice properties from satellite: Reviews of Geophysics, v. 39, p. 1–28, doi: 10.1029/1999RG000076.

Liversedge, L., 2007, Turbidity mapping and prediction in ice marginal lakes at the Bering Glacier System, Alaska [M.S. thesis]: Ann Arbor, University of Michigan, School of Natural Resources and Environment, 50 p.

Lu, Z., Kwoun, O., and Rykhus, R., 2007, Interferometric Synthetic Aperture Radar (InSAR): Its past, present, and future: Photogrammetric Engineering and Remote Sensing, v. 73, p. 217–221.

Molnia, B.F., 2001, Glaciers of Alaska: Alaska Geographic, v. 28, no. 2, 112 p.

Molnia, B.F., 2007, Late nineteenth to early twenty-first century behavior of Alaskan glaciers as indicators of changing regional climate: Global and Planetary Change, v. 56, p. 23–56, doi: 10.1016/j.gloplacha.2006.07.011.

Molnia, B.F., and Post, A., 1995, Holocene history of Bering Glacier, Alaska: A prelude to the 1993–1994 surge: Physical Geography, v. 16, p. 87–117.

U.S. Geological Survey, 2007, National Water Information System, Real-Time Water Data for Alaska: http://waterdata.usgs.gov/ak/nwis/uv.

Winther, J.-G., Bindschadler, R., König, M., and Scherer, D., 2005, Remote sensing of glaciers and ice sheets: Remote Sensing in Northern Hydrology: Measuring_Environmental Change, Geophysical Monograph Series, C.R. Duguay and A. Pietroniro, Washington, D.C., American Geophysical Union, v. 163, p. 39–62.

Zwally, H.J., Schutz, B., Abdalati, W., Abshire, J., Bentley, D., Brenner, A., Bufton, J., Dezio, J., Hancock, D., Harding, D., Herring, T., Minster, B., Quinn, K., Palm, S., Spinhirne, J., and Thomas, R., 2002, ICESat's laser measurements of polar ice, atmosphere, ocean, and land: Journal of Geodynamics, v. 34, p. 405–445, doi: 10.1016/S0264-3707(02)00042-X.

Manuscript Accepted by the Society 02 June 2009

Printed in the USA

The Geological Society of America
Special Paper 462
2010

Hydrography and circulation of ice-marginal lakes at Bering Glacier, Alaska, USA

Edward G. Josberger*
U.S. Geological Survey, Washington Water Science Center, 1201 Pacific Avenue, Suite 600, Tacoma, Washington 98402, USA

Robert A. Shuchman
Michigan Tech Research Institute, MTRI, 3600 Green Court, Suite 100, Ann Arbor, Michigan 48105, USA

Guy A. Meadows
Department of Naval Architecture and Marine Engineering, University of Michigan, Ann Arbor, Michigan 48109, USA

Sean Savage
Altarum Institute, P.O. Box 134001, Ann Arbor, Michigan 48113-4001, USA

John Payne
U.S. Bureau of Land Management, 222 W 7th Ave, Anchorage, Alaska 99513, USA

ABSTRACT

An extensive suite of physical oceanographic, remotely sensed, and water quality measurements, collected from 2001 through 2004 in two ice-marginal lakes at Bering Glacier, Alaska—Berg Lake and Vitus Lake—shows that each lake has a unique circulation controlled by specific physical forcing within the glacial system. Conductivity profiles from Berg Lake, perched 135 m above sea level (a.s.l.), show no salt in the lake, but the temperature profiles indicate an apparently unstable situation: the 4 °C density maximum lies at 10 m depth, not at the bottom of the lake (90 m depth). Subglacial discharge from the Steller Glacier into the bottom of the lake must inject a suspended sediment load sufficient to marginally stabilize the water column throughout the lake. In Vitus Lake, terminus positions derived from satellite imagery show that the glacier terminus rapidly retreated from 1995 to the present, resulting in a substantial expansion of the volume of Vitus Lake. Conductivity and temperature profiles from the tidally influenced Vitus Lake show a complex four-layer system with diluted (~50%) seawater in the bottom of the lake. This lake has a complex vertical structure that is the result of convection generated by ice melting in salt water, stratification within the lake, and fresh water entering the lake from beneath the glacier and surface runoff. Four consecutive years, 2001 through 2004, of these observations in Vitus Lake show little change in the deep temperature and salinity conditions,

This paper was originally published as Josberger, E.G., Shuchman, R.A., Meadows, G.A., Savage, S., and Payne, J., 2006, Hydrography and circulation of ice-marginal lakes at Bering Glacier, Alaska, USA: Arctic, Antarctic and Alpine Research, v. 38, no. 4, p. 547–560.
*ejosberg@usgs.gov

Josberger, E.G., Shuchman, R.A., Meadows, G.A., Savage, S., and Payne, J., 2010, Hydrography and circulation of ice-marginal lakes at Bering Glacier, Alaska, USA, *in* Shuchman, R.A., and Josberger, E.G., eds., Bering Glacier: Interdisciplinary Studies of Earth's Largest Temperate Surging Glacier: Geological Society of America Special Paper 462, p. 67–81, doi: 10.1130/2010.2462(04). For permission to copy, contact editing@geosociety.org.

indicating limited deep water renewal. The combination of the lake level measurements with discharge measurements, through a tidal cycle, by an Acoustic Doppler Current Profiler (ADCP) deployed in the Seal River, which drains the entire Bering system, showed a strong tidal influence but no seawater entry into Vitus Lake. The ADCP measurements, combined with lake level measurements, established a relationship between lake level and discharge, which when integrated over a tidal cycle gave a tidally averaged discharge ranging from 1310 to 1510 m^3 s^{-1}.

INTRODUCTION

As a result of global warming, glaciers are receding and thinning worldwide (Dyurgerov and Meier, 1997; Arendt et al., 2002; Meier and Dyurgerov, 2002). Consequently, existing ice-marginal lakes are expanding and new ones are forming; examples include Mendenhall Lake, Alaska (Motyka et al., 2002) and Tasman Lake, New Zealand (Purdie and Fitzharris, 1999). As these lakes evolve, the associated lacustrine ecosystems will be controlled by the physical conditions that occur in each lake. The ice-marginal lakes of Bering Glacier provide a unique natural laboratory for the study of ice-marginal lakes, because the region contains numerous lakes of varying sizes and characteristics. In addition, Bering Glacier is a surging glacier (Post, 1969). The last surge ended in 1995, which resulted in an extended glacier position that was more out of equilibrium with the local climate than would have occurred if the glacier had not surged. Also, in recent years, south-central Alaska has undergone exceptionally warm summers (Alaska Climate Research Center). All of these factors have combined to produce extremely rapid thinning and retreat of the glacier. Hence, the ice-marginal lakes at Bering Glacier are expanding at a rate faster than would normally be expected, and as such, they act as high-speed analogs for other ice-marginal lakes.

Bering Glacier is the largest and longest glacier in continental North America, with an area of ~5175 km^2 and a length of 190 km. The entire glacier lies within 100 km of the Gulf of Alaska. Figure 1 shows the extent of Bering Glacier and the location of some of the landmarks mentioned in this paper. Bering Glacier alone is >6% of the glacier-covered area of Alaska and may contain 15%–20% of Alaska's total glacier ice (Molnia, 2000). Because of its size and large fresh water discharge, it is a major component of the marine and terrestrial ecosystems of the Gulf of Alaska, and its discharge is an important driving mechanism of the circulation in the Gulf of Alaska (Royer, 1979).

Bering Glacier exhibits surging behavior: a short period (1–2 a) of rapid advance followed by a longer period (decades) of gradual retreat. It is the largest surging glacier in America, having surged at least five times in the twentieth century, most recently from 1993 through 1995. Molnia and Post (1995) give a thorough review of the Holocene history of Bering Glacier up to the 1993–1995 surge. The latest surge reversed the retreat of Bering Glacier, which had been taking place since the previous surge in 1968. However, Muller and Fleisher (1995) correctly anticipated that this advance would be only temporary and that the glacier would once again retreat after the surge had subsided. Our observations show that Bering Glacier is thinning and the terminus is retreating, continually increasing the size of Vitus Lake. Our measurements describe the retreat of the Bering Glacier terminus and the current hydrologic conditions that result from the complex interaction of tidewater glacier dynamics with both fresh water and seawater in a complicated bathymetric setting.

The entire Bering Glacier terminus region is also of great interest to the U.S. Bureau of Land Management (BLM), which is mandated with managing this area under the pressure of competing interests. Anthropogenic activity in this region continues to increase, with observable impacts on pristine and unique natural habitats. The last 100 yr have brought significant changes to the number of people and their methods of access to the Bering Glacier area. In the early 1900s most of the people visiting the glacier were subsistence hunters, fishermen, trappers, and gold prospectors. Oil exploration in the 1960s added the development of temporary roads. The passage of the Alaska Native Claims Settlement Act in 1971 began the process of conveying land under BLM administration to Alaska Natives. As part of this process, mineral rights in land near Berg Lake were conveyed to a local Native corporation, which in turn sold the rights to an Asian corporation interested in developing the coal and oil potential. The 1990s brought sport fishing and big-game hunting cabins and lodges into the area. Two public-use cabins on the shores of Vitus Lake, built in 2002, have resulted in increased ecotourism in the area. Understanding the physical changes occurring in the ice-marginal lakes of the Bering Glacier system is necessary for the BLM to formulate a management plan for this biologically diverse and environmentally significant wilderness area.

Berg Lake and Vitus Lake are of particular interest, because their sizes and circulations are dominated by glacier advance or retreat, ice melting, subglacial and surface runoff, iceberg calving, terminus disarticulation, and a large sediment flux. Each lake has a significant portion of its margin at a calving glacier ice wall that extends to the deepest parts of the lakes. Furthermore, each lake receives surface runoff and subglacial discharge that probably carry a large sediment load. The lakes are different in that Vitus Lake is at sea level and is brackish, whereas Berg Lake is 135 m above sea level (a.s.l.) and is entirely fresh. An ice-penetrating-radar study, carried out by D. Trabant, U.S. Geological Survey (USGS), and reported by Molnia (1993), showed that the bedrock below the medial moraine area between the Steller and the Bering Lobes rises up hydrologically to separate the two lakes.

This paper presents the results of a bathymetric and hydrographic study of Berg and Vitus lakes that was carried out to determine the primary mechanisms that control the circulation, exchanges, and aquatic conditions in each lake. The data provide baseline information for future studies for understanding changes in the physical setting as well as changes in fish and marine mammal distributions and populations. Indeed, Savarese (2004) showed that the number of harbor seals in Vitus Lake can reach a population of 1000. The hydrographic measurements were made in August 2001, 2002, 2003, and 2004. Analysis of these data has revealed surprising convective mechanisms driven by the effects of suspended sediment, and the interaction of melting ice with both fresh water and salt water.

BATHYMETRIC DATA COLLECTION AND ANALYSIS

Data Collection System

Bathymetric data for Berg Lake and Vitus Lake were collected using a portable bathymetric survey system. This system includes an integrated survey-grade differential global positioning system (DGPS), fathometer, and surface temperature sensor. The fathometer transducer transmitted a 130 dB conical beam at 200 kHz with a width of ± 20°, and a laptop computer integrated and recorded the data. The complete system is roughly the size of an average suitcase and is easily transported to remote locations; we deployed the system in a 5 m inflatable survey boat.

Bathymetric Data Collection

The bathymetric survey system was configured to take spatially referenced depth and temperature measurements at a time interval of once per second, which, at the typical speed of the survey boat (3.6 m s^{-1}), gave one data point approximately every 3.6 m. The comparatively small size, shallow water, and simple shape of Berg Lake simplified the bathymetric data collection, and the entire survey was completed in 2 d during 2001, resulting in >12,000 data points. In contrast, because of the size, complex shape, and icebergs blocking parts of Vitus Lake, the complete survey required three field seasons—2002, 2003, and 2004. The resulting bathymetric data set contains ~100,000 data points.

There are several potential sources of error in the bathymetric data. In the deep basins the fathometer would "lose" the bottom. This was particularly true of Vitus Lake, where depths can exceed 150 m and where the bottom in deep basins consists of a layer of loosely consolidated fine-grained glacial sediment that attenuates rather than reflects the acoustic signal. We obtained samples of the bottom sediment using oceanographic Niskin

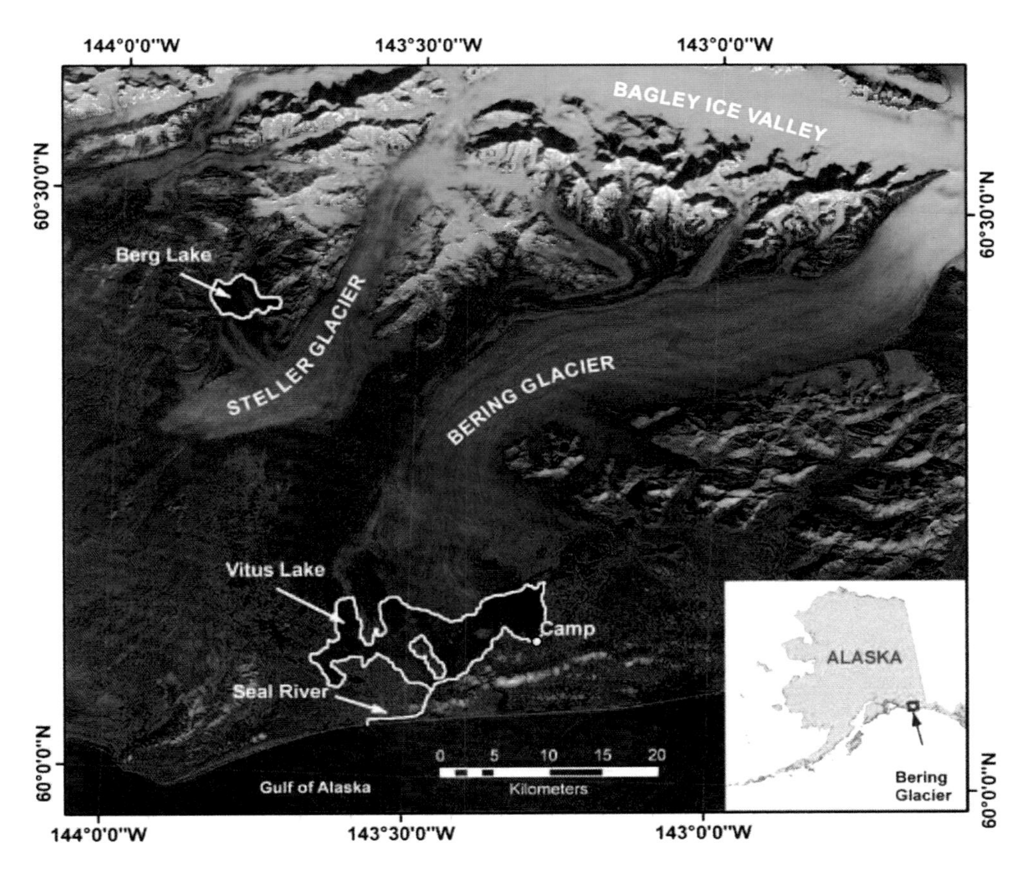

Figure 1. Landsat 7 pseudo-color image from 29 September 2002, showing the Bering Glacier System and the location of several physiographic features cited in this paper.

bottles and also found bottom sediment adhering to the conductivity-temperature-depth (CTD) instrument after a deep cast. Where submerged portions of icebergs obscured the view of the lake bottom, a "double bottom" was detected, including a weak signal indicating the response from the ice and a stronger signal indicating response from the lake bottom.

Following data collection, we identified and corrected erroneous data points in two ways. If an error was due to depths exceeding the maximum detectable depth by the fathometer, the local maximum known depth was assigned to that location. Likewise, where a double bottom was identified, the deeper measurement was assigned. In contrast, if the depth value is invalid because the transition between lake bottom and open water was too gradual, a manual linear interpolation was applied between the nearest points of known depth (commonly over short distances). If the source of error could not be identified or easily fixed, the data point was removed from the data set.

Bathymetric Interpolation

The objective of collecting bathymetric data is to generate a uniformly gridded surface that accurately represents the bottom of both Berg and Vitus Lakes. Combining the bathymetric data with other spatial data, CTD observations, for example, in a geographic information system (GIS) results in a powerful tool for organizing and analyzing the large amount of spatial and temporal data that has been acquired from Bering Glacier lakes. The derivation of a uniformly gridded surface from the survey transects requires an interpolation technique that can adapt to highly convoluted shorelines and spatially varying depth samples along irregularly spaced transects. We used ordinary kriging, which offers the advantage of customizing the interpolation weights to the data set as well as providing an estimate of the error in the predicted depth. Landsat 7 imagery provided the lake boundary. The depth was set to zero along the digitized shoreline, whereas along the glacier terminus, depths were set to the nearest measured depths.

A separate semivariogram model was developed (optimized, using the Earth System Research Institute's [ESRI's] Geostatistical Analyst for ArcGIS 9.0) for each lake, based on the 12,000 sample points in Berg Lake and the 100,000 sample points in Vitus Lake. The smaller data set and simple shape of Berg Lake resulted in a straightforward kriging based on a spherical semivariogram model with a major range of 3715 m, a partial sill of 771 m^2, a nugget of 21.9 m^2, a lag interval of 551 m with 12 lags, and a search neighborhood that included at least five points in each quadrant (areas bounded by the cardinal directions), yielding a grid with a 15-m posting. Although the kriging algorithm could accommodate the entire 100,000 point data set from Vitus Lake, the initial interpolation produced a surface in which the survey transects were readily apparent as discrete breaks in the surface. We found that using every 10th sample point produced a natural appearing surface with a 30-m posting. The interpolation was based on an exponential semivariogram model with a major range of 7597 m, a partial sill of 1878.3 m^2, a nugget of 188.54 m^2, a lag interval of 1019.4 m with 12 lags, and a search neighborhood that included at least five points in each quadrant.

Hydrographic and Water Quality Measurements

We used a Seabird oceanographic CTD to measure the depth profiles of salinity and temperature at various locations in each lake. In 2003 we measured the vertical distribution of pH, temperature, dissolved oxygen (DO), turbidity, total dissolved solids (TDS), and conductivity with a Horiba U-20 XD Series water quality monitoring probe. A vertical water quality profile was generated by sampling every 1–2 m to a depth of 10 m and then sampling every 5 m to the bottom of the lake. These measurements, which are temperature corrected, were limited to a depth of 100 m owing to the length of the instrument cable. Hand-held GPS receivers gave the location of each CTD and water-quality-parameter profile, and these spatially located data were incorporated into the Bering Glacier GIS. In 2003 and 2004 we used a USGS Acoustic Doppler Current Profiler (ADCP) to measure the discharge of the Seal River, which drains the entire Bering Glacier System into the Gulf of Alaska, and the Abandoned River, which emerges from the glacier terminus near the east end of Vitus Lake. Two Ott water level gauges measured the water level in Seal River and at the east end of Vitus Lake near the BLM Bering Camp. For measurements in Berg Lake, a helicopter transported all of the equipment, including the boat, to the lake. The observations in Vitus Lake were carried out from inflatable boats from the BLM field camp. Standard meteorological data—air temperature, wind velocity, barometric pressure, and relative humidity—were collected throughout each summer season at a weather station at the Bering Camp airstrip.

BERG LAKE

Berg Lake is perched at ~135 m a.s.l. behind an ice margin that is part of the Steller Glacier terminus (Figs. 1, 2). The ice front is oriented nearly east-west and is ~3 km long. The ice cliff is ~60 m above lake level, and the western portion is the most dynamic with active calving and ice velocities toward the lake of ~1 m d^{-1}. Numerous small streams in the valleys to the north drain into Berg Lake. The only outlet stream flows to the west between the ice margin and a steep bedrock shore. Hence, a small glacier advance can dam the stream, causing a rapid rise in lake level.

The lake level and size have varied considerably over the past century. Early maps of the region (Molnia and Post, 1995) show the Steller Glacier terminus much farther to the north, filling the present-day Berg Lake, creating several ice-dammed lakes backed up in valleys that now contain streams. Strandlines on the mountains surrounding the lake indicate that the lake level has been as high as 140 m above its present level. A comparison of USGS vertical photographs acquired during the 1993–1995 surge period of Bering Glacier, and in 2002, shows

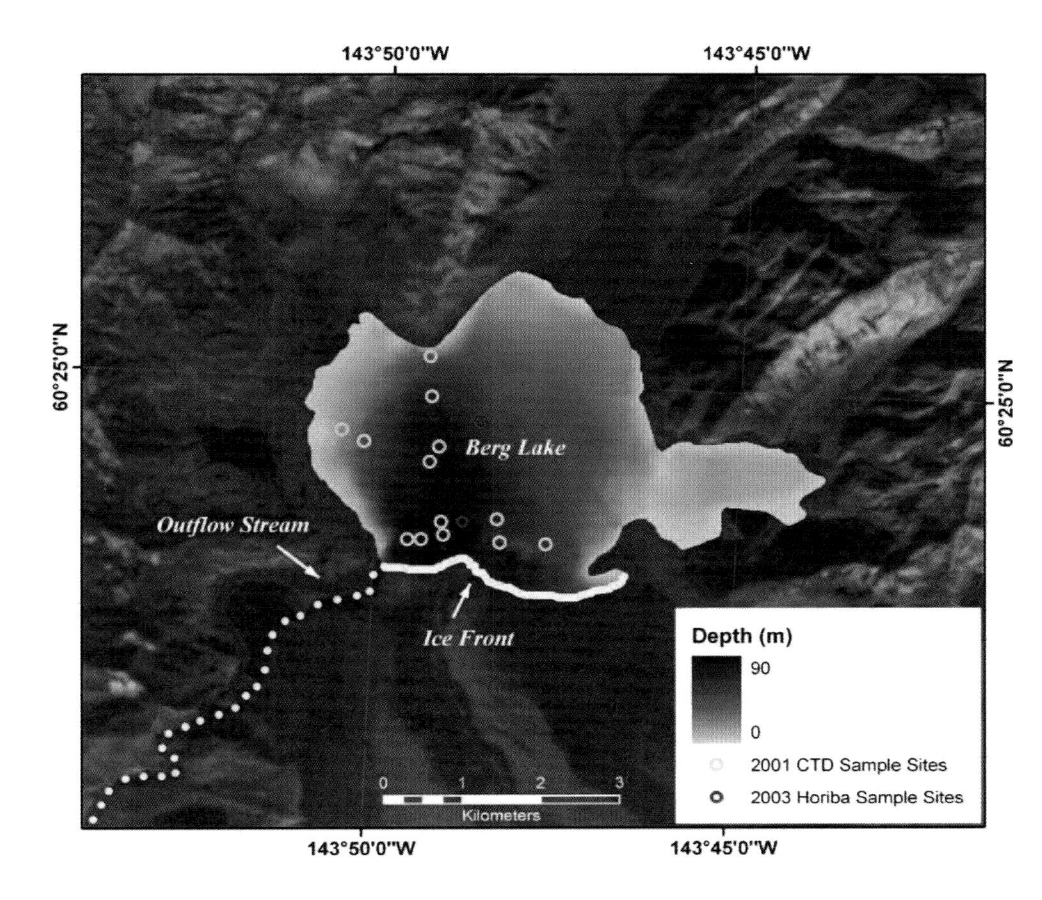

Figure 2. Bathymetry of Berg Lake and locations of the 2001 CTD and 2003 Horiba sampling sites overlaid on a Landsat 7 pseudo-color image from 29 September 2002.

that the Steller Glacier terminus was 500 m farther north than its current position. In this advanced position the terminus dammed the outlet stream at the southwest corner of the lake between the terminus and a bedrock promontory. From mid-March 1993 to mid-May 1994 the lake level rose from 140 m, near its current level, to 198 m. The failure of the ice dam in late May 1994 produced an outburst flood that inundated ~300–400 km^2 to the west. Photographs from 7 September 1994 show the lake level near 2001–2003 levels.

Berg Lake Bathymetry

Because of the small size of Berg Lake and its widely scattered icebergs, we were able to carry out the bathymetric survey in 2 d, 7–8 August 2001. Figure 2 shows the color-coded bathymetry of the lake merged with part of a Landsat 7 image. The deepest part of the lake is ~90 m, in the southwest part of the lake, directly adjacent to the most actively calving and fastest flowing part of the glacier terminus: a comparison of this bathymetry with that obtained by Austin Post (2002, personal commun.) of the USGS in 1996 shows that no significant changes have occurred in either lake extent or in depth since that time. With the bathymetry processed into a grid, as described previously, we computed the 2001 volume of Berg Lake to be 0.48 km^3.

Berg Lake Hydrography

Surface Temperature

Situated in an amphitheater of mountains, open to the south, Berg Lake receives considerable solar radiation through the summer that warms its surface layers. The surface temperatures, which were measured by the temperature sensor on the acoustic depth sounder, reached nearly 14 °C in the shallow bays away from the glacier terminus, particularly on the east side. These regions may be important fish habitats.

Vertical Temperature Structure

To determine the vertical temperature structure in Berg Lake we made nine CTD casts, the locations of which are denoted by the yellow circles in Figure 2. Figure 3 gives the profiles. As expected, because the lake is well above sea level, we detected no appreciable salt in any of the casts. The general temperature structure found in Berg Lake consists of three layers. The upper layer extends from the surface to ~8 m, where the temperature equals the temperature of maximum density for fresh water, 4 °C. In this layer the temperature decreases uniformly and rapidly with increasing depth, and the resulting strong stratification and turbid water act to confine the solar heating to this layer. The middle layer spans the depth region from 8 to 50 m, where the

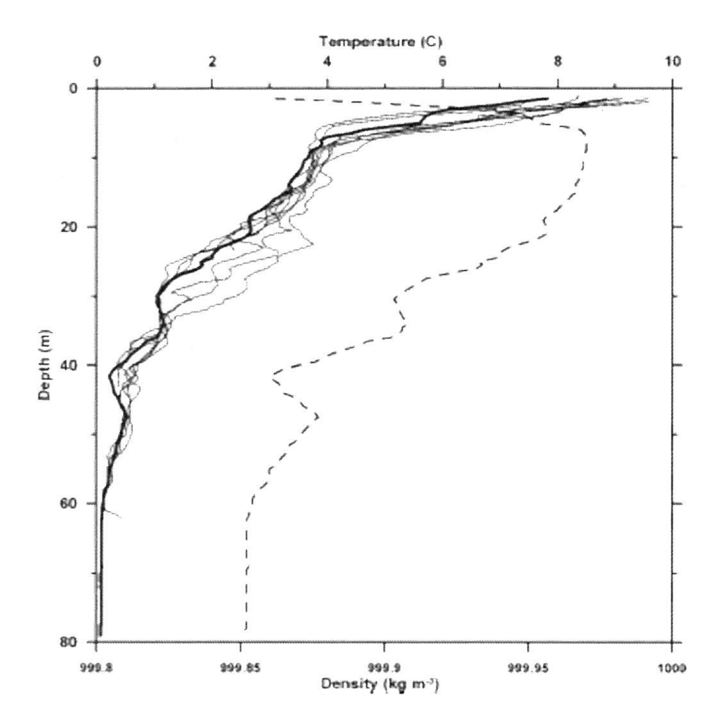

Figure 3. Temperature and density profiles from Berg Lake. The darker line is the profile from the deepest cast, the dashed line is the density profile from this cast using the equation of state for fresh water, and the fine lines are the temperature profiles from all of the casts.

temperature decreases from 4 °C to nearly 0 °C, with high variability between casts. This layer contains numerous steps and inversions, with a vertical scale of 1–10 m, that are characteristic of convective overturning, and the temperature can fluctuate by as much as 1 °C about its mean value, particularly between 20 and 30 m in depth. The third layer extends from 50 m to the bottom, at 80 m, where the temperature remains nearly constant at slightly above 0 °C.

At first inspection, these temperature profiles indicate an unstable situation with the densest water, that near 4 °C, overlying less dense water colder than 4 °C, as the dashed line in Figure 3 shows. The stabilizing factor is the discharge of sediment-rich water from beneath the glacier to the bottom of the lake. The water quality measurements from Berg Lake, discussed in the next section, show that turbidity increases with increasing depth. Hence the subglacial discharge that enters Berg Lake from beneath the Steller Lobe must contain a suspended sediment load that is sufficient to stabilize the water column. As the sediment settles out, convective instabilities form and overturning occurs, producing the temporal and spatial variability observed in the temperature profiles. Also, the bottom of Berg Lake contains fine-grained soft, sticky mud, which we found clinging to the CTD upon retrieving the instrument after each deep cast.

An additional component of the circulation in Berg Lake is the vertical convection generated by melting of the ice wall that forms its southern boundary and icebergs. The flow may be upward or downward, depending on the water temperature within a few meters of the ice (Josberger and Martin, 1981), a result of the 4 °C density maximum. For water temperatures between 0 and 4 °C the meltwater is lighter than that at 4 °C, resulting in only an upward flow. Water temperatures >8 °C produce only negatively buoyant meltwater and downward flow. For temperatures between 4 and 8 °C, the buoyancy distribution across the boundary layer is both positive and negative, resulting in bidirectional flow. Given the vertical temperature distributions observed in Berg Lake (Fig. 3), the warm upper layer will cause rapid melting, and the convection will be characterized by upward and downward flows. Below 10 m, the melt rate will decrease with decreasing water temperature, and only downward convection will take place.

Berg Lake Water Quality Measurements

In 2003 we measured water quality at two sites in Berg Lake, shown in Figure 2, and the profiles are shown in Figure 4. The observations show a relatively high pH value of 9, which is constant with depth. The conductivity, which is influenced by sediment in the water, is close to zero at the surface and reaches a maximum at 10 m. The turbidity values (NTU), which also reflect sediment in the water column, are quite high (250 NTU or greater) and increase with depth, a result of the previously described input of sediment laden water entering the lake from beneath the base of the glacier. The dissolved oxygen profiles show near saturation values in the surface layers that decrease rapidly to ~6 mg L^{-1} at 10 m and then slowly decrease with increasing depth. The temperature profiles, to the accuracy of the instrument, indicate no supercooled water at depth, which is consistent with the CTD temperatures from the previous year. The TDS profile for the site away from the ice, in general, follows the conductivity profile. The TDS profile near the glacier terminus also follows the conductivity profile, except for a minimum observed at a depth of 40–55 m. The cause of this minimum is not understood, but there are similar perturbations in the temperature and salinity profiles, which may be the result of the previously described circulation.

VITUS LAKE

Vitus Lake (Figs. 5, 6) undergoes rapid changes in size, volume, and water characteristics with the Bering Glacier surge and retreat cycle. We found that the lake, situated nearly at sea level, contains seawater from the Gulf of Alaska, and it is a holding basin that receives most of the water from the Bering Glacier System (Merrand and Hallet, 1996). Although Fleisher et al. (1998) describe some glacier discharge that enters the Tsiviat Lake Basin, east of Vitus Lake, Vitus Lake receives a large number of icebergs from the calving glacier terminus, subglacial discharge, and surface runoff from the glacier as well as runoff from the surrounding land. There is a large contribution of sediment-rich water at the surface from the Abandoned River at the far east end of the lake. The system drains into the Gulf of Alaska through the 8-km-long, 10-m-deep Seal River. As Bering Glacier

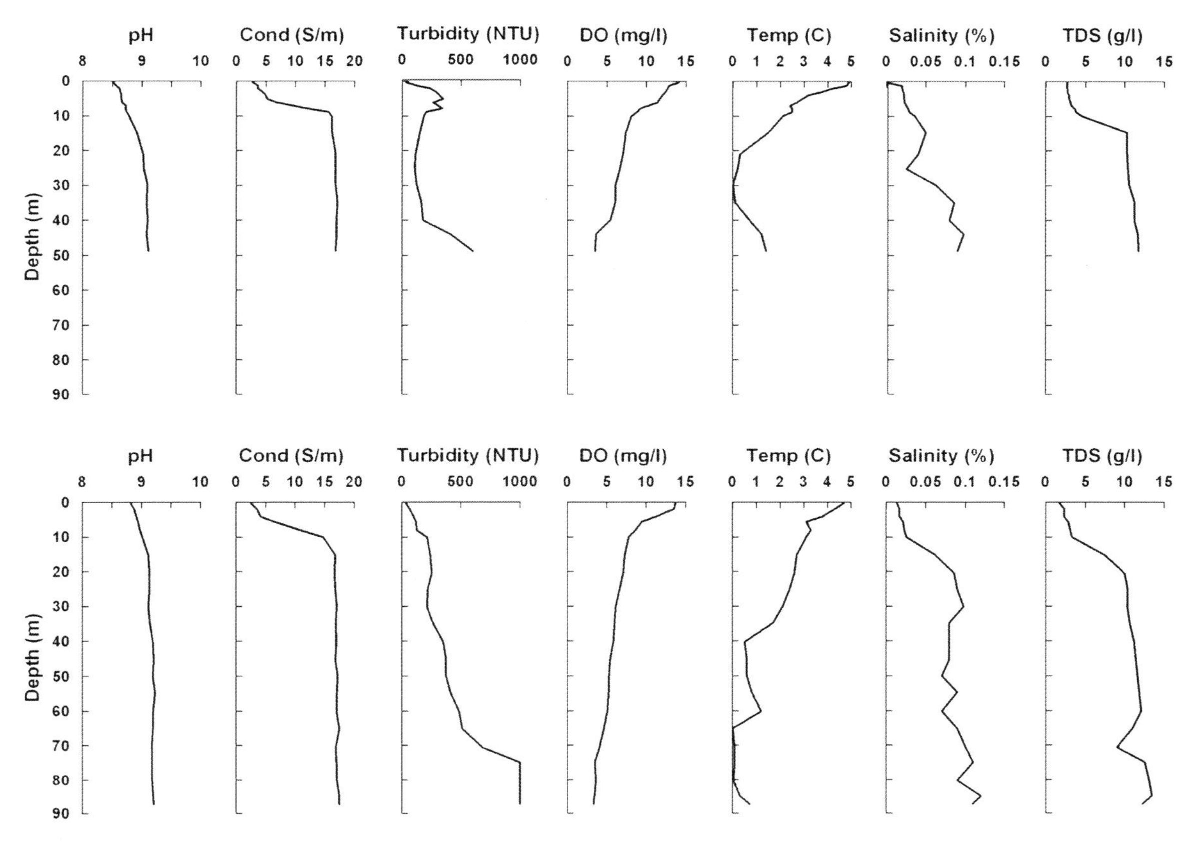

Figure 4. Water quality profiles for two deep casts in Berg Lake from 2003. Measurements in the upper set were taken near the middle of the lake; those in the lower set were taken from closer to the ice edge in southwest Berg Lake. See Figure 2.

rapidly retreats, Vitus Lake is rapidly expanding. Our bathymetric measurements act as a baseline measurement with which to gauge sedimentation rates as Vitus Lake expands.

Vitus Lake Bathymetry

Vitus Lake continues to expand rapidly in both area and volume since the last surge ended in 1995. Owing to the size and difficulty of navigating in the iceberg-clogged regions of the lake, it took three field seasons to fully map the bathymetry of Vitus Lake. Figure 5 shows the bathymetry of Vitus Lake and the retreat of the Bering Glacier terminus from 1995 to 2004, superimposed on a Landsat 7 false-color image acquired 25 April 2003. A comparison of this bathymetry to presurge measurements by Austin Post (2005, personal commun.) shows that the two are in general agreement. The deepest water in the lake is generally found adjacent to the glacier terminus, particularly in the central basin called McMurdo Sound, and farther west in Tashalich Arm, where the depths reach and exceed 150 m. A sill at 50 m restricts the circulation between the deep parts of Tashalich Arm from the rest of the lake, which results in very different deep water properties.

With a continuous surface representation of the bathymetry of Vitus Lake, calculating water volume is straightforward. Assuming that the bathymetry of the lake is relatively static over the 3 a survey period, volume change can be estimated as a function of time as the glacier terminus retreats, derived from Landsat 7 imagery, generally from late summer or early fall of each year. Table 1 gives the lake area and volume estimates from 1995 through 2004.

Between 1995 and 2004, the volume of water in Vitus Lake increased by >260%, and the lake area increased by 195%. The annual rate of increase in water volume has increased each year between 1995 and 2002, assuming the annual rate of change between 1995 and 1999 is constant as the glacier terminus retreated. Because the time interval between the 2002 and 2003 data points is only 5 mo., the rate of change appears to have decreased slightly.

Vitus Lake Hydrography

The surface temperatures for most of Vitus Lake range from near 0 to +2 °C, except in an isolated bay at the far southwest end of the lake where, in the summer of 2001, the temperature reached 7 °C. The surface temperatures are considerably colder than those observed in Berg Lake, probably resulting from large amounts of glacial meltwater entering the lake at the surface and

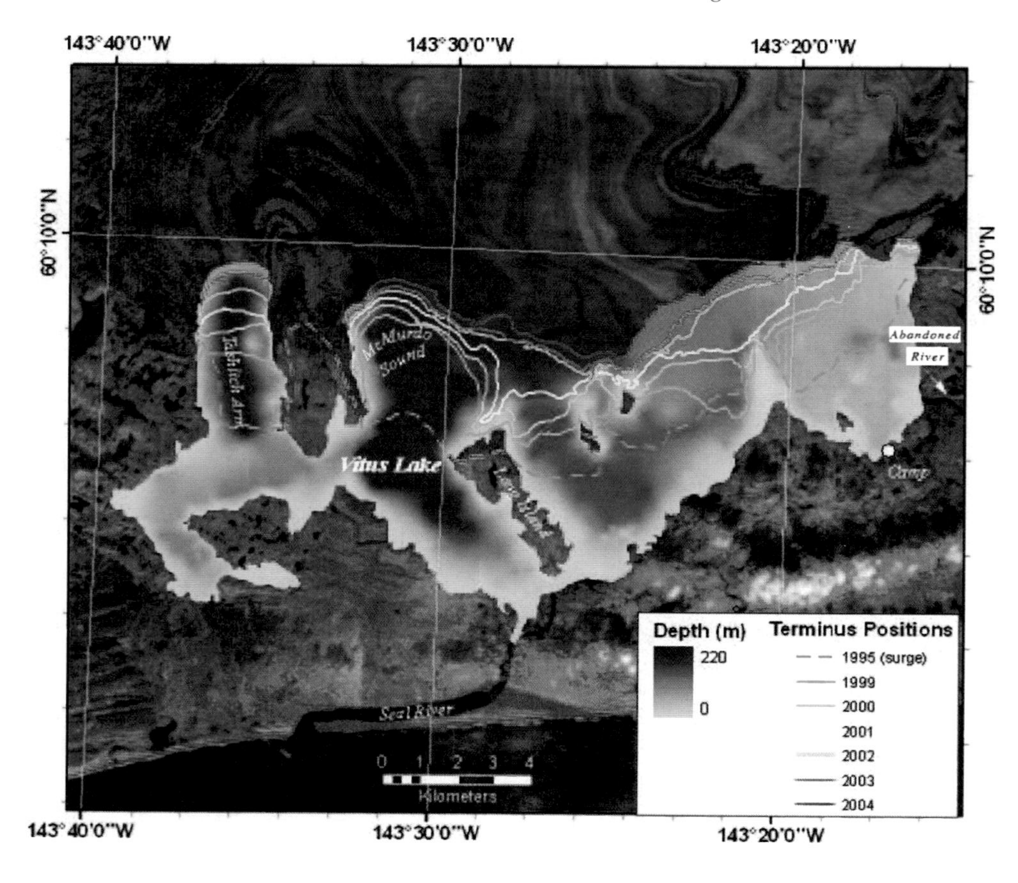

Figure 5. Terminus retreat of Bering Glacier from 1995 to 2004 and the bathymetry of Vitus Lake superimposed on a Landsat 7 pseudo-color image from 29 September 2002.

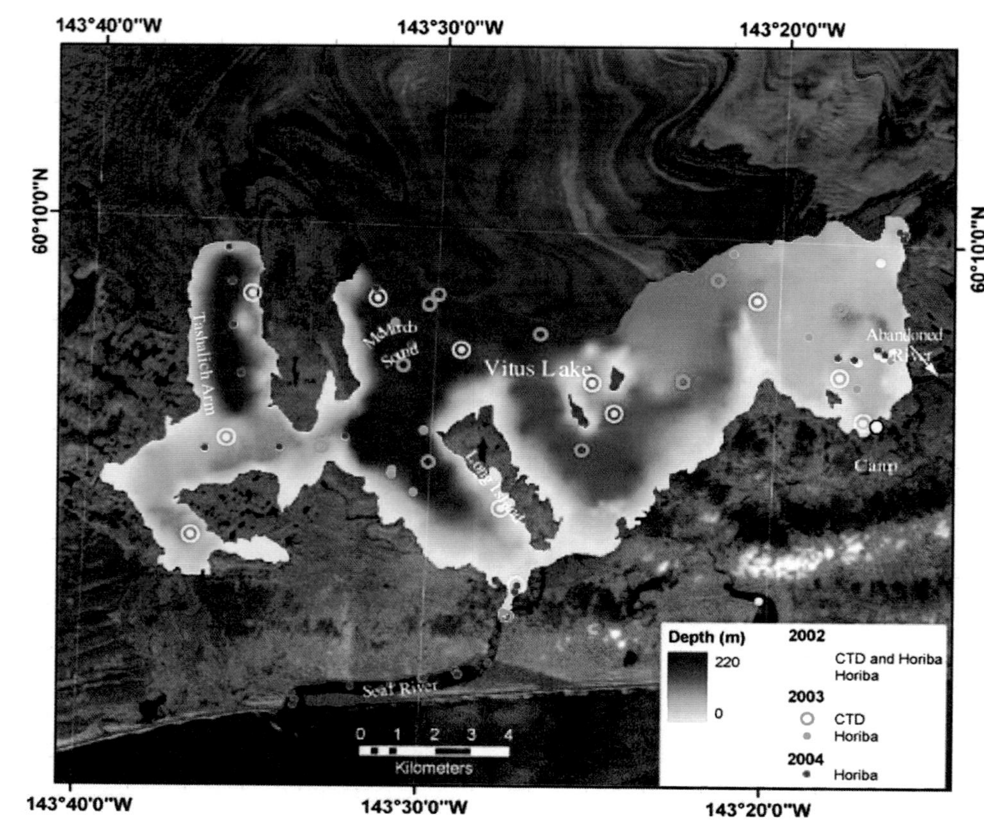

Figure 6. CTD and Horiba sampling locations in Vitus Lake for 2001 through 2004 superimposed on the bathymetry, with a pseudo-color Landsat 7 image from 29 September 2002 as background.

TABLE 1. VOLUME AND AREA OF VITUS LAKE DURING THE STUDY PERIOD

Year	Volume			Area		
	km³	Annual (% change)	Cumulative (% change)	km²	Annual (% change)	Cumulative (% change)
1995	2.6	—	—	58.4	—	—
1999	4.4	67.0	67.0	80.6	38.0	38.0
2000	4.9	12.2	87.3	89.0	10.4	52.4
2001	5.4	10.4	106.8	96.2	8.1	64.7
2002	6.2	14.9	137/5	105.8	9.9	81.1
2003	6.4	4.0	147.0	108.8	2.9	86.3
2004	6.9	6.5	163.1	114.0	4.8	95.3

Note: Elapsed time between 2002 and 2003 was only 5 mo.

at depth. In addition, the Bering Lobe reflects rather than absorbs solar radiation, as is the situation at Berg Lake, which tends to result in a cold katabatic flow down-glacier onto the lake.

The CTD measurements in Vitus Lake, at the locations shown in Figure 6, revealed a complex vertical structure that is the result of the interaction of fresh water, seawater, and glacier ice in a complex bathymetric setting. Figure 7A shows three CTD casts in the deepest part of the main basin of the lake, McMurdo Sound. The cast from 2001 was from the east side of Long Island (Fig. 5) because dense concentrations of icebergs prevented us from entering McMurdo Sound, which we were able to enter in 2002, 2003, and 2004. The hydrographic conditions below ~60 m have remained remarkably uniform over 3 a, with a small change observed in 2004, which will be discussed later.

The basic structure of the water column consists of four distinctive layers, which, beginning at the top, are:

Layer 1: A fresh surface layer from 0 m to ~30 m, with temperatures ranging from 0 to 2.5 °C, and salinities ranging from 2 to 2.5 practical salinity units (psu).

Layer 2: A very cold intermediate layer from 30 m to ~55 m, where the temperature is 0 °C or less, and the salinity rapidly increases from 2 to ~18 psu.

Layer 3: A layer with remarkably uniform temperatures and salinities from ~55 m down to near the bottom. The salinity is near 18 psu, approximately one-half the salinity of the Gulf of Alaska seawater, and the temperature ranges from 1 to 1.4 °C.

Layer 4: A turbid bottom layer, as thick as 10 m, which is composed of a high concentration of suspended fine silt, from 30 to 300 g L^{-1}.

This basic structure is found throughout the lake, and its structure in shallower regions is a truncated version of that observed in the deep areas, except for Tashalich Arm, where the aforementioned sill restricts deep water exchange below 50 m; this will be discussed later. The most spatially and temporally variable layer is the surface layer, which is modified by solar radiation, precipitation, and surface runoff.

In the intermediate layer the salinity distribution produces a highly stratified layer, which inhibits vertical exchange between the surface water and the deep water, thus isolating the surface layer from the deep layer. The temperature-salinity characteristics of the intermediate layer result from the combination of dilution of the deep layer with cold fresh water and the melting of glacier ice. When fresh water near 0 °C mixes with deep water, the case for subglacial discharge, the resulting mixture must lie on the dilution line shown on the temperature-salinity (T-S) diagrams (Fig. 7B). Greisman (1979) and Gade (1979) show that when ice melts in salt water, the resulting mixture lies on a different line, called the melt line, the slope of which depends on the water temperature and salinity. For the cold saline deep layer, the melt line is steep (Fig. 7B), because the heat necessary to melt ice in cold water requires cooling a large amount of water.

As Figure 7B shows, the T-S properties for the intermediate layer lie between the dilution line and the melt line. The water at the bottom of the intermediate layer has the characteristics of melting, whereas water in the upper part of the layer takes on more of the characteristics of dilution. The T-S diagrams also indicate a change in the relative proportion between dilution and melting for the different years. In 2001 the intermediate layer was more characteristic of melting than of dilution for both 2002 and 2003. The temperature for 2001 was below 0 °C from 30 to 50 m, indicating a greater impact from melting for that year. This trend may be the result of two or more factors. First, the lake greatly expanded in volume over these years. Second, the number and concentration of icebergs in the lake decreased, which suggests a reduction in the amount of upwelling generated by deep melting.

The convection in Vitus Lake is also generated and controlled by dilution and melting processes. As shown by Walters et al. (1988) and Motyka et al. (2003), temperate tidewater glaciers in this region generate large subglacial discharges that rise in a plume next to the glacier terminus. Also, the discharge typically occurs in the deeper regions of the lake, entraining water from the deep saline layer as it rises. This results in a mixture that is denser than the surface layer and flows horizontally away from the glacier terminus as it encounters the strong pycnocline at the bottom of the surface layer (Huppert and Josberger, 1980; Neshyba and Josberger, 1980; Josberger and Martin, 1981). In our surveys we did not notice specific upwelling plumes along the ice front, which suggests that if plumes exist they do not reach the surface but rather flow out laterally when they encounter the strong capping pycnocline.

Vitus Lake Water Quality Measurements

Figure 6 also shows the location of the water quality profiles made in 2002, 2003, and 2004. Figures 8 and 9 show a representative water quality data set, DO profiles, pH, TDS, and turbidity measurements plotted with co-located profiles of temperature and salinity, one from McMurdo Sound and one from Tashalich Arm. In McMurdo Sound the pH is uniform with depth, at a value just under 6. However, TDS follow the salinity distribution, at a minimum just under 3 g L^{-1} in the surface layers and increasing to a maximum of 19 g L^{-1}. The DO has a surface value

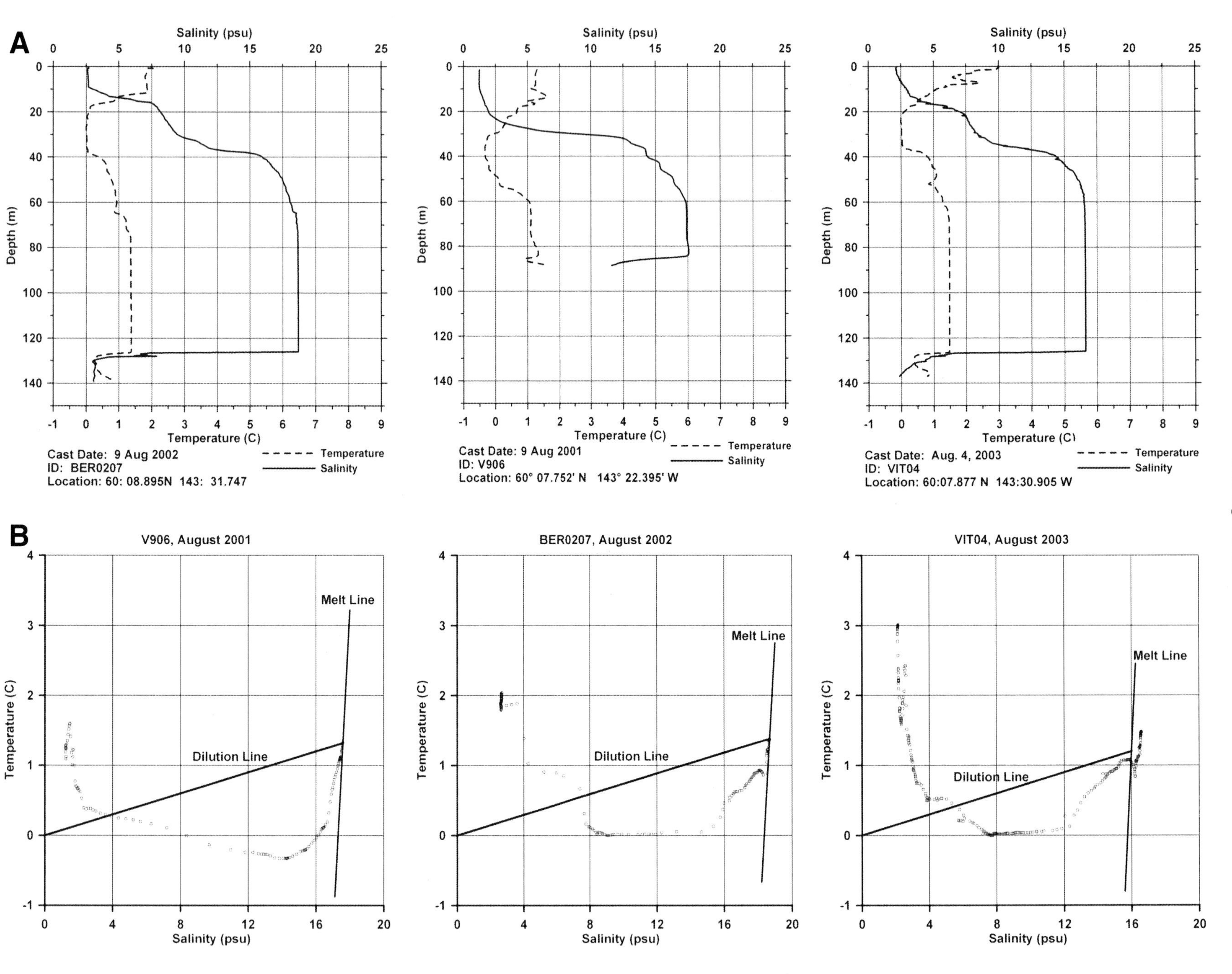

Figure 7. (A) Deep CTD profiles from Vitus Lake for 2001, 2002, and 2003. (B) Temperature and salinity (T-S) diagrams from Vitus Lake for 2001, 2002, and 2003, showing the internal water mass structure and its relation to the dilution line and the melt line.

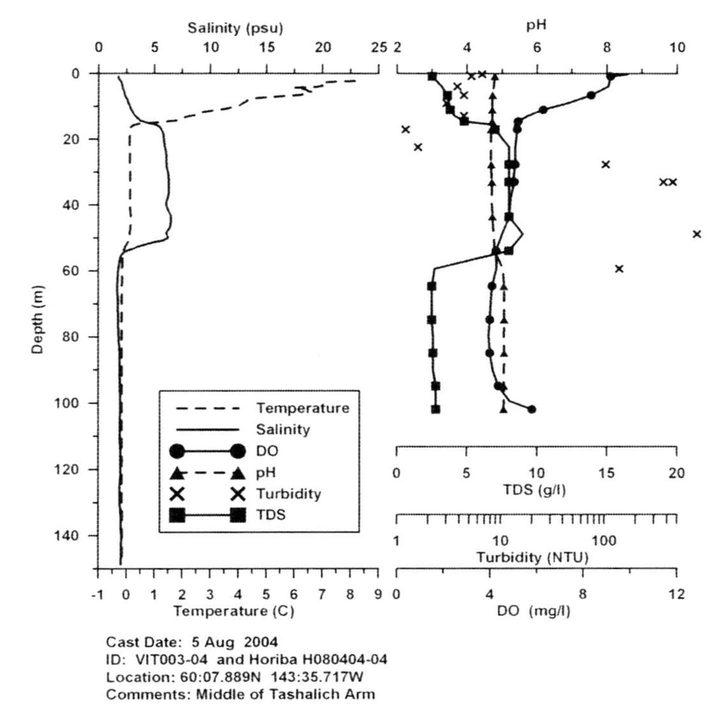

Figure 8. Vertical profile of CTD sampling in Tashalich Arm plotted along with coincident Horiba parameters (pH, DO, turbidity, and TDS).

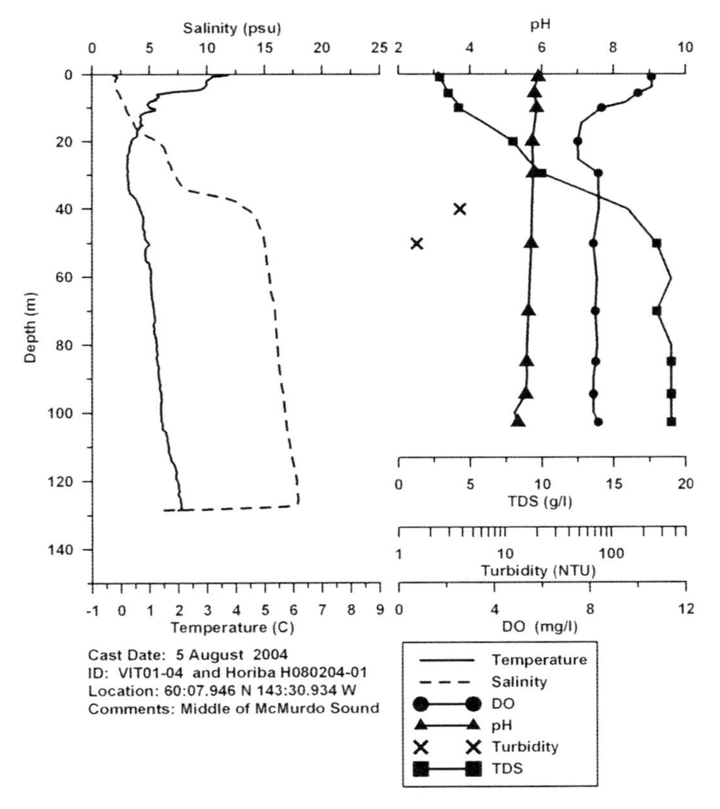

Figure 9. Vertical profile of CTD sampling in McMurdo Sound plotted along with coincident Horiba parameters (pH, DO, turbidity, and TDS).

of ~11 mg L^{-1}, then decreases to 7.5 mg L^{-1} at a depth of 20 m, then slightly increases to 8.2 mg L^{-1} for the rest of the profile. We obtained only two NTU values for McMurdo Sound because of sensor problems, but those values at 40 and 50 m depth indicate clear, relatively sediment-free, highly oxygenated water, which could support a robust set of aquatic organisms, as indicated by the rapidly growing population of harbor seals. This clear water is a result of the removal of the suspended sediments by flocculation when it mixes with the saline water in Vitus Lake.

Tashalich Arm is the deepest basin in Vitus Lake, with water depths exceeding 150 m, the length of line used to lower the CTD instrument. This arm is frequently filled with large icebergs that calve off the western part of the Bering Lobe. As observed in 2004, catastrophic calving events across the entire terminus can rapidly fill the fjord with icebergs of all sizes. Prior to this event, we noted water lines that result from wave-induced heat transfer (Josberger, 1978) on the terminus face bowing upward, indicating that the ice was deforming as a result of flotation. Waves from the calving event of 2004 stranded floating ice blocks, with a characteristic size of 1 m, high and dry on the beach above the water level.

Temperature, salinity, and water quality profiles in Tashalich Arm (Fig. 8) are dramatically different from those observed in McMurdo Sound (Fig. 9), especially below 18 m. For the surface layer, above 18 m, both parts of the lake have similar properties, but from this depth to 55 m, the sill depth, the temperature and salinity are nearly constant at +0.1 °C and 6.1 psu in Tashalich Arm, whereas the salinity in McMurdo Sound rapidly increases to 15 psu and greater. The deep water of Tashalich Arm, >55 m, is isothermal at –0.16 °C and isosaline at 1.9 psu, which is water at its salinity-determined freezing point. This is the result of the large amounts of ice at depth, large icebergs, a deep terminus, and little exchange of water with the main basin. The large flux of ice into Tashalich Arm and the likelihood of a great deal of subglacial discharge will generate only outflow over the sill, thus preventing the more saline water of Vitus Lake from entering the arm.

Figure 8 also demonstrates the impact of the sill, and little exchange with the main part of the lake. The dissolved oxygen has a surface value of ~8.5 mg L^{-1}, which decreases rapidly to 5 mg L^{-1}, and then remains at this value to 100 m. In the main basin, DO remains near 8 mg L^{-1} at all depths. Likewise, TDS values are quite low at depth in Tashalich Arm (~2.5 g L^{-1}) but are near 18 g L^{-1} in McMurdo Sound. The pH values for Tashalich Arm are approximately uniform with depth at a value of 5, whereas in the main basin the pH is at 6. (Recall that the pH for Berg Lake was near 9.) The turbidity values for Tashalich Arm indicate the surface water to be relatively clear of glacial rock flour, but the turbidity increases in the isohaline layer to values of 20 NTU.

Lake Level and ADCP Measurements

To determine the conditions necessary for seawater to enter Vitus Lake through Seal River, the only reasonably possible source, we measured water levels, using an arbitrary datum, in Seal River and Vitus Lake and carried out ADCP discharge measurements in Seal River. Figure 10 shows the measured time series of water levels from both sites, Seal River and Vitus Lake, and the U.S. National Oceanic and Atmospheric Administration (NOAA) tidal predictions for Yakutat, 220 km to the east, for the period 1–12 August 2003. The tide at Yakutat is a combination of semidiurnal (~12 h) and diurnal (~24 h) components with a mean range of 2.38 m and a spring range of 3.07 m.

For Seal River, both the semidiurnal and diurnal components are clearly evident, although the tidal range is attenuated to ~2 ft. (Water levels are in feet because the NOAA tidal predictions are given in feet, and our instruments record in feet.) The peaks and troughs in the record from Seal River are nearly in phase with the NOAA predictions for high and low water at nearby Wingham Island, 50 km to the west. At the east end of the lake the tidal range is further attenuated to ~1 ft, and the semidiurnal portion of the signal is virtually gone. The diurnal signal has more of a sawtooth shape where the lake level rises rapidly and then slowly decreases. Toward the end of the record the average lake level rose in response to the increase in tidal level. Precipitation was not the cause of the rise in lake level; records from the BLM Camp weather station show that the last significant precipitation event was on 30 July 2003, when 3.8 cm of rain fell.

To measure the discharge of both Seal River and the Abandoned River we used a USGS ADCP, deployed from an inflatable boat. At Seal River on 4 August 2003 we carried out 16 transects across the river, which spanned a flood tide (Fig. 11). On 7 August we carried out an additional eight transects during the early part of an ebb tide. The flow in Seal River strongly depends on the tidal level in the Gulf of Alaska. Figure 11 shows that the river discharge slowed to near zero at high tide and reached a maximum of 2039 $m^3 s^{-1}$ at low tide. Observations near high tide (9.6 ft [2.93 m], 1925 Alaska Daylight Time [ADT], 5 August 2003, at Wingham Island) showed that the surface flow in Seal River did reverse and flowed into the lake at the time of high tide. We measured the surface velocity by tracking the drift of our inflatable boat with a GPS receiver. The river center line surface velocities were 1.5 km h^{-1}, and there was no wind during the drift period. Five CTD casts over the time span from 1850 to 2000 ADT, while the river flow had reversed, found uniform vertical temperature and salinity conditions of 2.5 °C and 2.8 psu, which were characteristic of the upper 10 m of Vitus Lake. We found no evidence of Gulf of Alaska seawater entering the lake at the bottom of the river.

Our measurements show that the discharge from Vitus Lake to the Gulf of Alaska through the Seal River occurs as a tidally driven series of pulses (Fig. 12). To compute the average discharge of Seal River averaged over a tidal cycle, we used the time series of water level measurements from Seal River with the ADCP flow measurements to establish a tidally varying relationship between river height and discharge (Fig. 11). This is analogous to a conventional hydrologic rating curve except that it is inverted; when the water level increases, the flow decreases, and conversely. As shown in Figure 12, we used our rating curve to

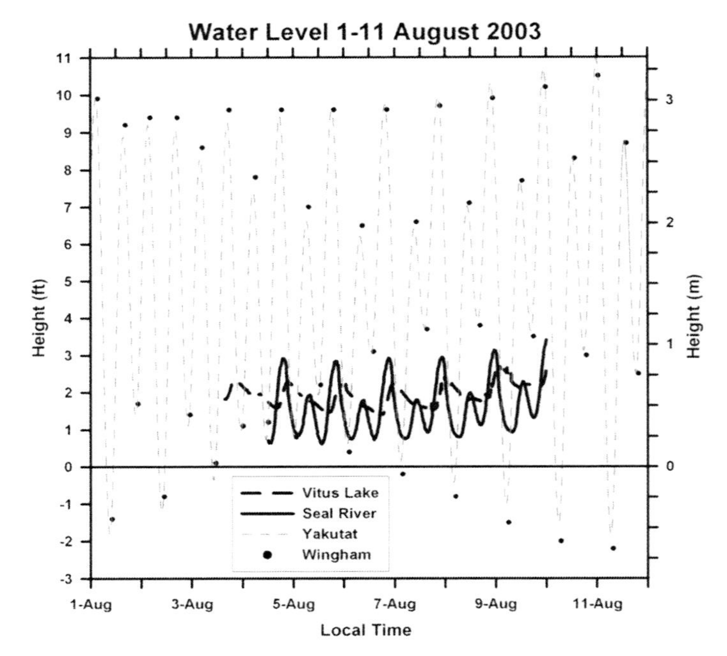

Figure 10. Times series plot of the lake level variations from Seal River and Vitus Lake sites, and the NOAA tidal predictions for Yakutat and Wingham Island. Note that the Seal River and Vitus Lake time series are plotted at an arbitrary height.

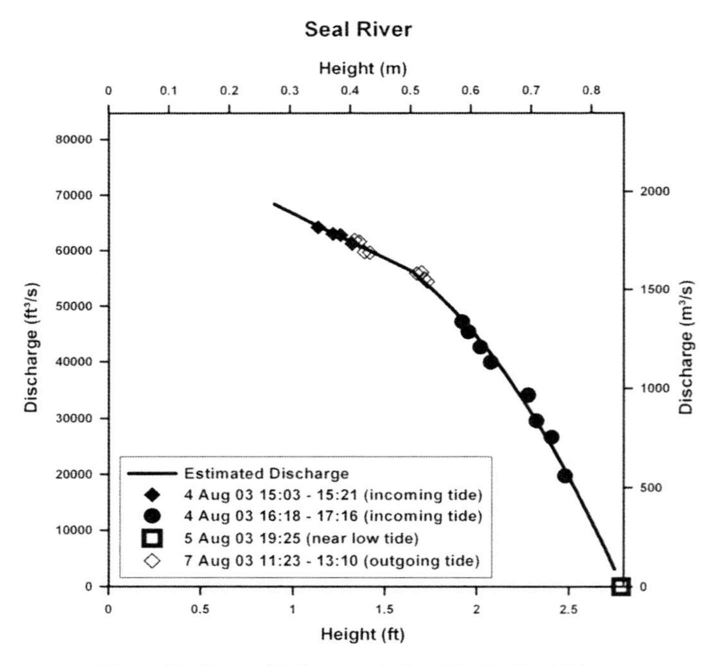

Figure 11. Stage-discharge relationship for Seal River.

compute a time series of discharge, and then numerically integrated the discharge time series over the tidal cycle to obtain the average. The average flow over a tidal cycle ranged from 1310 to 1510 $m^3 s^{-1}$. The average discharge shows a decreasing trend, which is a result of increasing tidal heights (Fig. 10). Using these

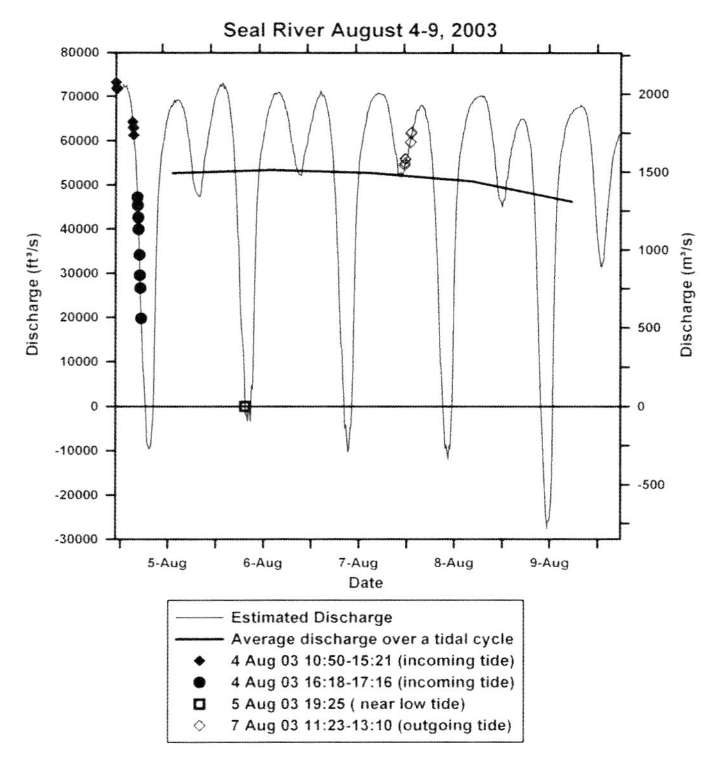

Figure 12. Tidally varying discharge and discharge averaged over the tidal cycle for Seal River computed from the river stage and discharge relationship. The symbols show when the discharge measurements were made through the tidal cycle.

values for discharge and the measured lake volume gives an estimate for the residence time of ~2 mo.

At Abandoned River we made four discharge transects on 5 August 2003 during the period 1415–1450 ADT, and the average discharge was 585 $m^3 s^{-1} \pm 12 m^3 s^{-1}$. We made the measurements at the mouth of the river rather than in the river, because river conditions were too hazardous. The surface velocity was ~10 km h^{-1} with 0.3–0.6 m standing waves. By doing so we missed a small branch of the river that was too shallow to measure; we estimate that the flow in this branch was <5% of the measured flow.

For 2002 and 2003, we took surface water samples in the main stream of Abandoned River to determine the sediment load. The USGS Cascade Volcano Observatory analyzed these samples, and the sediment concentration was found to be 2520 and 2829 mg L^{-1}, respectively. When combined with the ADCP measurements, these results yield a sediment flux of ~127,000 metric tons per day (t d^{-1}) (or with a nominal sediment density of three times that of water, the yield is ~$4.3 \times 10^4 m^3 d^{-1}$) into Vitus Lake during summer discharge conditions. Furthermore, this value represents a minimum flux on this day, as the sediment concentration increases with depth. Fleisher et al. (2003) estimate the average sediment flux from this section of the glacier to be 227 $\times 10^6 m^3$ for the 10 a period 1991–2000, for a daily average of $6.2 \times 10^3 m^3 d^{-1}$. Given that our determination was made during peak summer flows, and that of Fleisher et al. (2003) included the winter months, when the flow is low, these sediment flux values are in reasonable agreement.

As Merrand and Hallet (1996) describe, most of the sediment flux into Vitus Lake remains in the lake. The large grains settle out immediately at the mouth of the river, whereas the very fine grained material spreads throughout the lake where salt, even in low concentrations, causes the sediment to flocculate and settle in the deep basins of Vitus Lake. This process, for all sediment sources, produces the 10-m-thick bottom layer in the deeper basins of the lake that we observed in the CTD measurements as layer 4.

When compared with the flow of the Abandoned River into Vitus Lake, the outflow by Seal River is 2.5–3 times greater. Hence, there must be significant inflow from other sources to maintain a nearly constant lake level. Immediately prior to and during these measurements there was no rainfall. Hence, a subglacial discharge of ~1000 $m^3 s^{-1}$ must be occurring along the extensive terminus front, probably from a series of subglacial channels (Fleisher et al., 1998). Walters et al. (1988) determined the seasonal subglacial discharge from nearby Columbia Glacier, also a tidewater glacier. There, the discharge for the summer reached 300 $m^3 s^{-1}$ and approached zero during the winter, when the hydrologic system is frozen.

Seawater Entry into Vitus Lake

The presence of salt in Vitus Lake is not surprising, considering that the lake is nearly at sea level. The intriguing question is, What circumstances allow Gulf of Alaska seawater, with a typical salinity of 32 psu, to enter the lake and form the deep water with a salinity of ~18 psu? Certainly, high tides are necessary to allow salt water to enter the lake. Observations in Seal River on 5 August 2003, at the time of a +2.9 m tide at Wingham Island, show a 2 km h^{-1} flow into the lake. However, CTD observations showed that the entering water had the same salinity and temperature as the upper 10 m of the lake; hence the inflowing water had pooled at the mouth of the river and was returning. Furthermore, it would take seawater more than 3 h to traverse the 9-km-long Seal River and reach the lake. In this amount of time the tide would have ebbed sufficiently, and the flow once again would be out of the lake.

Seawater entry is highly episodic, and most likely during the winter, when the melting of snow and ice ceases, the flux of fresh water entering the lake virtually ceases. However, the 3 yr CTD record shows a remarkably well mixed deep layer that is nearly constant over 2 yr. The complete absence of vertical structure indicates that the current state is probably the result of a single mixing event rather than numerous events that would have certainly generated vertical variations in both the salinity and the temperature.

The most likely singular event that may have produced the deep water is the impact of the remnants of typhoon Oscar on the Bering region of the Alaska coast in the fall of 1995, just after the surge had ended. At this time Vitus Lake was at a minimum size, at one-half its 2003 volume. Anecdotal reports from aircraft pilots familiar with the area reported that the land separating Vitus Lake from the Gulf of Alaska was completely inundated by the associated storm surge. It is possible that other mechanisms may introduce seawater into Vitus Lake, but we believe that the most plausible explanation is reverse flow through the Seal River during extreme events.

CONCLUDING REMARKS

This study has used hydrographic observations to define the primary mechanisms that drive the circulation and exchanges in the ice-marginal lakes at Bering Glacier, Berg Lake and Vitus Lake. Striking differences were found between each lake. In Berg Lake, we have found that the circulation is driven by the contribution of fine-grained sediments from subglacial water entering the lake. For Vitus Lake, salt from Gulf of Alaska seawater, a complicated bathymetry, sediment input, and ocean tides act to produce a highly complex circulation. The presence of salt in Vitus Lake acts to confine the input of fine-grained sediment in the lake by promoting flocculation and settling, resulting in a clear, blue lake. In contrast, Berg Lake, which has no salt, is quite turbid, and significant amounts of the fine-grained sediment input leave the lake via the outlet stream. Berg Lake has remained nearly constant in size, whereas Vitus Lake has rapidly increased in both size and volume and is likely to continue to do so. The surface temperatures in Vitus Lake rarely exceeded 3 °C, whereas in Berg Lake surface temperatures reached 14 °C in the shallow bays away from the glacier terminus, particularly on the east side; these regions may be important fish habitats. With the diverse conditions observed in these two lakes, they represent analogs to other ice-marginal lakes during a period of deglaciation.

Wang et. al. (2004) stated: "The freshwater discharge into the Gulf of Alaska (GOA) has an important effect on coastal circulation." These authors also estimate that the mean annual total freshwater discharge into the Gulf of Alaska is between 19,000 and 31,000 $m^3 s^{-1}$. Given this recent estimate for the entirety of the Gulf of Alaska, our measurement of the relative contribution of the Bering Glacier System, through the Seal River discharge, is significant. The discharge of the Bering Glacier System, at peak summer melt, is on the order of 5%–7% of the total average annual flow into the entire Gulf of Alaska (Schumacher and Reed, 1980; Royer, 1979, 1981). Hence, the Bering Glacier discharge is a significant driving mechanism of the local circulation in the Gulf of Alaska.

The calculated sediment flux into Vitus Lake through the glacial discharge of the Abandoned River was on the order of 127,000 t d^{-1}. Given this large flux of glacial sediment, significant bathymetric changes to the lake and its associated ecosystem should be anticipated. Fortunately, a previous hydrographic survey of Vitus Lake was completed prior to the surge of 1993 (Austin Post, 2005, personal commun.). Owing to the recent and rapid retreat of the post-surge ice edge, it now appears that the size and extent of Vitus Lake now is similar to its condition at the time of Post's survey. Similarly, the recorded bathymetry is also comparable. It is anticipated that the large sediment flux will begin to modify this bathymetry. The bulk of the sediment load from the Abandoned River is deposited in the easternmost basin of Vitus Lake and is blocked to the west by the presence of a north-south–trending subsurface ridge.

Given current climatic change scenarios, as exemplified by recent record-setting hot summers in south-central Alaska, we expect that the Bering Glacier System may undergo dramatic changes in the next few decades. The strong negative glacier mass balance in the region (Arendt et al., 2002) and the resulting glacier thinning may extend the time between surges by Bering Glacier or possibly eliminate the surge behavior entirely. In either case, Bering Glacier will continue to retreat to the point at which the bedrock topography is at sea level. Ice-penetrating-radar measurements reported by Molnia and Post (1995) show that bedrock topography reaches sea level ~40 km up-valley from its 2004 position. Our measurements show that the average rate of terminus retreat is ~1 km a^{-1}. Hence, we expect that in the coming decades Vitus Lake will greatly expand in size and volume. Increasing lake volume will alter the circulation in the lake; the impact of seawater intrusions will be reduced. Furthermore, there will be relatively less ice in the lake, which will reduce the amount of melt-driven vertical convection. These changes will have unknown ramifications on the extent and diversity of the ecosystems in and around Vitus Lake.

The hydrography of Berg and Vitus Lakes and their associated ecosystems are dominated by the dynamic processes of Bering Glacier. The hydrological data collected and reported in this paper serve as a baseline for understanding future changes in this rapidly evolving system. The established GIS framework will continue to serve as a tool for providing insight into environmental habitat changes from glacier dynamics as the hydrological system is monitored on an annual basis.

ACKNOWLEDGMENTS

Partial support for this program was provided to the U.S. Geological Survey and the Altarum Institute by the U.S. Bureau of Land Management. Support to the Altarum Institute was provided under grant no. LAA-01-0018. Many thanks are due Lorelle Meadows, Katie Harding, and Lauren Russell of the University of Michigan Naval Architecture and Marine Engineering Department for assistance in data collection, preparation, and analysis. The authors also would like to thank Scott Guyer and Chris Noyles of the Bureau of Land Management for their assistance in collecting the field observations.

REFERENCES CITED

Alaska Climate Research Center, University of Alaska, Geophysical Institute, http://climate.gi.alaska.edu/AKCityClimo/AK_Climate_Sum.html.

Arendt, A.A., Echelmeyer, K.A., Harrison, W.D., Lingle, C.S., and Valentine, V.B., 2002, Rapid wastage of Alaska glaciers and their contribution to rising sea level: Science, v. 297, p. 382–386, doi: 10.1126/science.1072497.

Dyurgerov, M.B., and Meier, M.F., 1997, Year to year fluctuations of global mass balance of small glaciers and their contribution to sea-level changes: Arctic and Alpine Research, v. 29, p. 392–402, doi: 10.2307/1551987.

Fleisher, P.J., Cadwell, D.H., and Muller, E.H., 1998, Tsivat Basin conduit system persists through two surges, Bering piedmont glacier, Alaska: Geological Society of America Bulletin, v. 110, p. 877–887, doi: 10.1130/0016-7606(1998)110<0877:TBCSPT>2.3.CO;2.

Fleisher, P.J., Bailey, P.K., and Cadwell, D.H., 2003, A decade of sedimentation in ice-contact, proglacial lakes, Bering Glacier, Alaska: Sedimentary Geology, v. 160, p. 309–324, doi: 10.1016/S0037-0738(03)00089-7.

Gade, H.G., 1979, Melting of ice in sea water: A primitive model with application to the Antarctic ice shelf and icebergs: Journal of Physical Oceanography, v. 9, p. 189–198, doi: 10.1175/1520-0485(1979)009<0189:MOIISW>2.0.CO;2.

Greisman, P., 1979, On upwelling driven by the melt of ice shelves and tidewater glaciers: Deep-Sea Research, v. 26A, p. 1050–1065.

Huppert, H.E., and Josberger, E.G., 1980, The melting of ice in cold stratified water: Journal of Physical Oceanography, v. 10, p. 953–960, doi: 10.1175/1520-0485(1980)010<0953:TMOIIC>2.0.CO;2.

Josberger, E.G., 1978, A laboratory and field study of iceberg deterioration, *in* Proceedings of the First International Conference on Iceberg Utilization, Ames, Iowa State University: Elmsford, New York, Pergamon Press, p. 245–264.

Josberger, E.G., and Martin, S., 1981, A laboratory and theoretical study of the boundary layer adjacent to a vertical melting ice wall in salt water: Journal of Fluid Mechanics, v. 111, p. 439–473, doi: 10.1017/S0022112081002450.

Josberger, E.G., and Neshyba, S., 1980, Iceberg melt-driven convection inferred from field measurements of temperature: Annals of Glaciology, v. 1, p. 113–118.

Meier, M.F., and Dyurgerov, M.B., 2002, How Alaska affects the world: Science, v. 297, p. 350–351, doi: 10.1126/science.1073591.

Merrand, Y., and Hallet, B., 1996, Water and sediment discharge from a large surging glacier: Bering Glacier, Alaska, USA, summer 1994: Annals of Glaciology, v. 22, p. 233–240.

Molnia, B.F., 1993, Major surge of the Bering Glacier: Eos (Transactions, American Geophysical Union), v. 74, p. 322.

Molnia, B.F., 2000, Glaciers of Alaska: Alaska Geographical Society, v. 28, no. 2, 112 p.

Molnia, B.F., and Post, A., 1995, Holocene history of Bering Glacier, Alaska: A prelude to the 1993–1995 surge: Physical Geography, v. 16, p. 87–117.

Motyka, R.J., O'Neel, S., Conner, C.L., and Echelmeyer, K.A., 2002, Twentieth century thinning of Mendenhall Glacier, Alaska, and its relationship to climate, lake calving and glacier run-off: Global and Planetary Change, v. 35, p. 93–112, doi: 10.1016/S0921-8181(02)00138-8.

Motyka, R.J., Hunter, L., Echelmeyer, K.A., and Conner, C., 2003, Submarine melting at the terminus of a temperate tidewater glacier, LeConte Glacier, Alaska, U.S.A.: Annals of Glaciology, v. 36, p. 57–65, doi: 10.3189/172756403781816374.

Muller, E.H., and Fleisher, J.P., 1995, Surging history and potential for renewed retreat: Bering Glacier Alaska, U.S.A.: Arctic and Alpine Research, v. 27, p. 81–88, doi: 10.2307/1552070.

Neshyba, S., and Josberger, E.G., 1980, An estimation of Antarctic iceberg melt rate: Journal of Physical Oceanography, v. 10, p. 1681–1685, doi: 10.1175/1520-0485(1980)010<1681:OTEOAI>2.0.CO;2.

Post, A.S., 1969, Distribution of surging glaciers in western North America: Journal of Glaciology, v. 8, p. 229–240.

Purdie, J., and Fitzharris, B., 1999, Processes and rates of ice loss at the terminus of Tasman Glacier, New Zealand: Global and Planetary Change, v. 22, p. 79–91, doi: 10.1016/S0921-8181(99)00027-2.

Royer, T.C., 1979, On the effect of precipitation and runoff on coastal circulation in the Gulf of Alaska: Journal of Physical Oceanography, v. 9, p. 555–563, doi: 10.1175/1520-0485(1979)009<0555:OTEOPA>2.0.CO;2.

Royer, T.C., 1981, Baroclinic transport in the Gulf of Alaska. II. Fresh water driven coastal current: Journal of Marine Research, v. 39, p. 251–266.

Savarese, D.M., 2004, Seasonal trends in harbor seal *(Phoca vitulina richardsii)* abundance at the Bering Glacier in south central Alaska [M.S. thesis]: Anchorage, University of Alaska, 70 p.

Schumacher, J.D., and Reed, R.K., 1980, Coastal flow in the northwest Gulf of Alaska: The Kenai Current: Journal of Geophysical Research, v. 85, p. 6680–6688, doi: 10.1029/JC085iC11p06680.

Walters, R.A., Josberger, E.G., and Driedger, C.L., 1988, Columbia Bay, Alaska: An "upside down" estuary: Estuarine, Coastal and Shelf Science, v. 26, p. 607–617, doi: 10.1016/0272-7714(88)90037-6.

Wang, J., Meibing, J., Musgrave, D.L., and Moto, I., 2004, A hydrological digital elevation model for fresh water discharge into the Gulf of Alaska: Journal of Geophysical Research, v. 109, C07009, doi: 10.1029/2002JC001430.

Manuscript Accepted by the Society 02 June 2009

The Geological Society of America
Special Paper 462
2010

Bering Glacier ablation measurements

Robert A. Shuchman*
Michigan Tech Research Institute, 3600 Green Court, Suite 100, Ann Arbor, Michigan 48105, USA

Edward G. Josberger
U.S. Geological Survey, Washington Water Science Center, 934 Broadway, Suite 300, Tacoma, Washington 98402, USA

Charles R. Hatt
Christopher Roussi
Michigan Tech Research Institute, 3600 Green Court, Suite 100, Ann Arbor, Michigan 48105, USA

P. Jay Fleisher
State University of New York at Oneonta, Oneonta, New York 13820, USA

Scott Guyer
Bureau of Land Management, 222 West 7th Avenue, #13, Anchorage, Alaska 99513, USA

ABSTRACT

Bering Glacier is rapidly retreating and thinning since it surged in 1993–1995. From 2002 to 2007 we have mapped the terminus position and measured the surface ablation from the terminus region up-glacier to the snowline in the Bagley Ice Field. Since the last surge the terminus has retreated, primarily by calving, ~0.4–0.5 km/a, and the terminus position is at the 1992 pre-surge position. The glacier surface in the terminus region is presently downwasting by melting at ~8–10 m/a and 3.5–6.0 m/a at the approximate altitude of the equilibrium line, 1200 m. The average daily melt for Bering Glacier is ~4–5 cm/d at mid-glacier, and this melt rate appears to be steady, regardless of insulation and/or precipitation. The melt from the Bering Lobe of the glacier system generates between 8 and 15 km^3 of fresh water yearly, which flows directly into the Gulf of Alaska via the Seal River, potentially affecting its circulation and ecosystem. Elevation measurements from 1957 compared with our measurements made in 2004, combined with bed topography from ice penetrating radar, show that the Bering Lobe has lost ~13% of its total mass.

INTRODUCTION

As a result of global warming, the glaciers of coastal south-central Alaska are undergoing extensive melting, also known as ablation. They are rapidly receding and thinning owing to warmer Alaskan summers, with 2004 having been one of the warmest on record (Alaska Climate Research Center, 2007). Few comprehensive melt measurements of these Alaskan glaciers exist, and the nearest glacier to Bering with a long term ablation record (48 a record) is Wolverine Glacier, which is part of the U.S. Geological Survey (USGS) Benchmark Glacier Program (E.G. Josberger, 2007). This unique seasonal-mass-balance USGS record shows that in 1988 the glaciers in this region began a period of rapid melting that has continued through the present. This melting is

*shuchman@mtu.edu.

Shuchman, R.A., Josberger, E.G., Hatt, C.R., Roussi, C., Fleisher, P.J., and Guyer, S., 2010, Bering Glacier ablation measurements, *in* Shuchman, R.A., and Josberger, E.G., eds., Bering Glacier: Interdisciplinary Studies of Earth's Largest Temperate Surging Glacier: Geological Society of America Special Paper 462, p. 83–104, doi: 10.1130/2010.2462(05). For permission to copy, contact editing@geosociety.org.

the result of generally warmer summers rather than decreased snowfall. The extensive melt from glaciers on the south-central coast of Alaska is contributing to a rise in sea level of 0.27 mm/a, which is on the order of the contribution from Greenland glaciers (Dyurgerov and Meier, 1997; Arendt et al., 2002; Meier and Dyurgerov, 2002).

In addition to contributing to sea level rise, glacier runoff plays an important role in the circulation of the Gulf of Alaska. Royer and Grosch (2006) report a thickening and warming of upper layer, low salinity water in the Gulf of Alaska that they attribute to an increase in glacier melt, which will have significant impacts on the ecosystem and quite possibly on global ocean circulation.

From a water resource perspective, fluctuations in glacier melt impact the quality and quantity of the water in a glaciated basin, affecting runoff timing, temperature, and turbidity—critical factors that can influence aquatic ecosystems. If the Bering Glacier continues to retreat, it has the potential for developing a robust salmon fishery, similar to the one in Copper River 100 km to the west. Measurements of the discharge of the Copper River Basin by the USGS from 1988 to 1995 show that the minimum discharge occurs in the months of December through April, an increasing discharge in May and a maximum in July, and then decreases through November (Fig. 1). These discharge data, combined with the climatic history of the region, helps us put into perspective the Bering melt and its discharge into the Gulf of Alaska.

Situated on the Gulf of Alaska, Bering Glacier is the longest and largest glacier in continental North America, with an area of ~5,175 km^2 and a length of 190 km (Molnia and Post, 1995). We have measured summer wastage from 8 to 10 m in the terminus region to 3.5–6 m at an altitude of ~1200 m, which is near the equilibrium line altitude (ELA). At sea level, this rapid melt has resulted in the expansion of the large proglacial lake (Vitus Lake) by calving, as reported by Josberger et al. (2006 and this volume) and Shuchman et al. (this volume). The rapid melt also results in the discharge of large amounts of fresh water into the Gulf of Alaska through the Seal River.

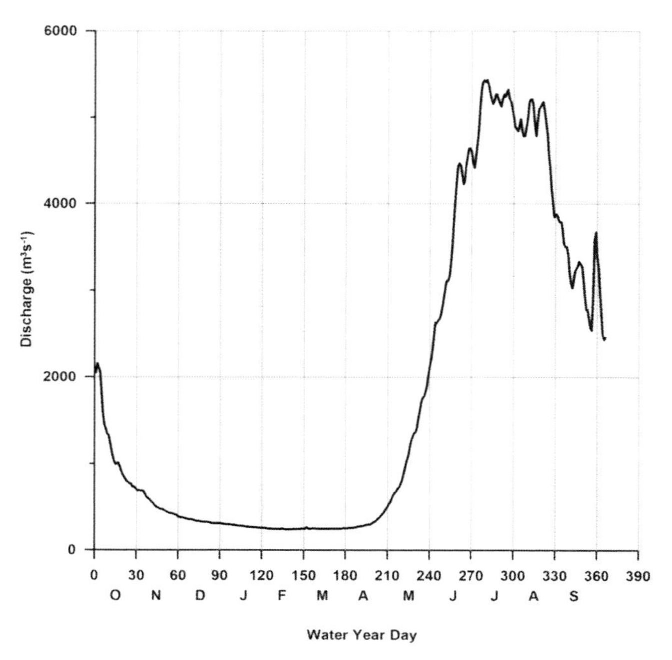

Figure 1. Copper River mean daily discharge from 1988 to 1995, showing peak flow in July and minimum discharge in the months of December through April.

This chapter summarizes the in situ ablation measurements made since 2000 of Bering Glacier, which are combined with local meteorological measurements to derive a melt model for the glaciers in the region. A satellite-derived digital elevation model (DEM) from 2004 is compared with historical elevation profiles derived from USGS topographic maps from the early 1900s, 1957, 1972, and 1995 to estimate the net glacier elevation change since the early 1900s and mass loss since 1957. The melt measurements are also used to estimate yearly fresh water discharge into the Gulf of Alaska. The hourly in situ glacier ablation measurements that were made during the summer melt period, in conjunction with coincident global positioning system (GPS) locations and meteorological observations, are used to better understand the relative importance of temperature, wind, solar radiance, relative humidity, and rainfall on the melting process.

HISTORICAL OBSERVATIONS OF BERING GLACIER

Bering Glacier has undergone at least six major episodes of surging (Molnia and Post, 1995) during the 1900s. Surging is defined by Molnia and Post (1995) as "rapid displacements and movements of large quantities of ice within a glacier, frequently accompanied by a significant advance of the glacier's terminus" (p. 89). Surging generally alternates with long periods of normal flow, stagnation, and recession. Bering Glacier is the largest temperate alpine surging glacier in the world. Since the end of the last surge in 1995, the Bering Lobe portion of the terminus has been retreating 0.4–0.5 km per year, and the present terminus is near the 1992 pre-surge position (Fig. 3, Shuchman et al., this volume).

The USGS has extensively studied the Bering Glacier for the last century. USGS field parties topographically mapped part of the glacier, starting in the early 1900s (Molnia and Post, 1995). After the Second World War, Austin Post, USGS, was instrumental in obtaining aerial photographs of Bering Glacier and other glaciers identified with surging behavior. In the 1990s, USGS scientists documented the 1993–1995 surge and employed ice-penetrating radar (IPR) to map the glacier ice thickness and provide detailed elevation data (Fleisher et al., 1995; Molnia and Post, 1995). These elevation data are invaluable for determining glacier mass changes as well as annual ice loss.

At the east end of the glacier terminus, near Weeping Peat, Peat Falls, and Arrowhead Islands, Jay Fleisher of the State University of New York (SUNY) at Oneonta also has mapped (Fleischer et al., 2006 and this volume) the terminus retreat from 1995 to the present and measured daily melt rates at three sites ~1 km up-glacier. Fleisher's EarthWatch Team found during their 10 day observation period that ablation generally ranged from 4 to 6 cm per day on overcast, cooler days and 7–9 cm per day when the weather was dry and sunny. The Fleisher group further observed that from 1998 annual downwasting varied between 10 and 16 m/a for debris-free ice and 3–9 m/a beneath a well established debris band. These observations indicate that knowledge of the debris distribution on the glacier is important when modeling the melt rate.

IN-SITU ABLATION MEASUREMENTS

A collaborative team, consisting of Michigan Tech Research Institute (MTRI), USGS, and BLM scientists, has measured glacial ablation and movement rates since 2000 on the Bering Glacier. The measurements, conducted from 2000 to 2003, consisted of limited one to two week ablation measurements at approximately 10 sites distributed in the terminus region, at Berg Lake, and no elevation higher than the west end of the Grindle Hills, during the mid-July to mid-August time periods. These measurements were obtained using the standard glaciological procedure of repeatedly measuring the lowering of the glacier surface on a marked plastic stake placed into a hole drilled in the ice. At each site the GPS location, altitude, date, time, and distance to the ice surface were recorded every several days. These observations show that during the summer peak melt period, the glacier ablated at a rate of ~5 cm per day.

From 2004 to the present a more robust set of glacier ablation measurements have been recorded by this collaborative team. In each of the years from 2004 through 2007, between five and eight sites from the terminus to slightly above the snowline have been instrumented in May or June with 1 in (2.54 cm) pipe placed in steam-drilled holes 10 m deep (Fig. 2). These sites were periodically visited to record total melt and position over the melt season. The pipe was typically installed in early June, and the final observations were made in the middle of September, providing approximately four months of observations. Table 1 summarizes the four years of measurements. For each site, a beginning and ending location (latitude, longitude), elevation on the glacier, start and end dates, total melt, total horizontal displacement, and a computed daily melt rate are displayed. Figure 3 is a Landsat image acquired 7 August 2006 that shows the locations of the ablation measurement sites. The annual average summer melt (assuming 150 days of melt) is ~7.5 m near the terminus and 5.7 m near the average equilibrium altitude of 1200 m.

Figure 2. Steam-drilling holes in the glacier ice in June 2004 at Bering Glacier, Alaska. (An installed GASS instrument can be seen in the background.)

Also presented in Table 1 are the horizontal daily displacements of the glacier at each ablation site (cumulative values of <5 m were recorded as "no significant movement"). The displacements are generally much <3 m/d, which is typical of a surging glacier between surge events. A notable exception to the glacier horizontal velocity measurements is the area of the Bering Lobe at sites B04 and B05 (2006). These sites, high on the glacier and in the Bagley Ice Field, at altitudes of 2400 and 3250 ft (732 and 991 m), respectively, show 200–300 m of displacement for the summer of 2006. The displacement at site B05 in the summer of 2005 was only ~10 m. Also, note in Table 1 that site S01 (Steller Lobe terminus), showed significant movement, indicating active ice.

Albedo, the percentage of solar radiation reflected from the surface of the glacier, also influences the average daily melt rates at each site. Using a technique presented in the remote sensing chapter by Shuchman et al. (this volume) the August 2006 Landsat image was used to calculate approximate albedo for each of the 2006 measurement sites (B01–B06, Jay site), whereas an August 2004 Landsat image was used for the 2004 albedo calculations (Table 2). The albedo values ranged from a low of 8% at the Jay site, 41 m elevation, to 74% on the snow-covered B06 site at 1250 m elevation. For 2004 the albedo values had a low of 31% and a high of 80% . In general, lower albedo resulted in greater melt rates. Significant rock debris (i.e., lower albedo) can insulate the ice, but our limited data do not show this effect. The average albedo for the glacier decreased from 2004 to 2006 as a result of the exposure of underlying rock debris.

WEATHER DATA

The long-term climate record for the Bering Glacier region was obtained from the University of Alaska Climate Center (2007). Historical temperature, precipitation, and snowfall depth are available for Cordova and Yakutat, Alaska, locations that bound both the east and west margins of Bering Glacier. The Yakutat data, which are in general agreement with the Cordova observations, are represented in Figures 4 and 5. Figure 4 shows the mean annual temperature for Yakutat in degrees Celsius and Fahrenheit from 1917 to the present, whereas Figure 5 shows monthly values of average, minimum, and maximum temperatures, as well as precipitation, snowfall, and snow depth for the years 2004–2007. Figure 4 shows that the warmest year since 2000 was 2005, with an average annual temperature of 42.1 °F (5.6 °C). In 2006 the average annual temperature decreased to

TABLE 1. SUMMARY OF BERING GLACIER ABLATION-MELT RATES FOR THE YEARS 2004–2007

SITE	Year	Starting latitude	Starting longitude	Starting date	Starting elevation (m)	Ending latitude	Ending longitude	Ending date	Total melt obs (cm)	Total days obs	Melt per day (cm)	Total 150 day melt (m)	Horizontal migration (m/day)
AB01	2004	60.3863	−143.8146	6/3/2004	269	60.3867	−143.8148	9/24/2004	643	113	5.7	8.55	0.44
AB02	2004	60.2802	−143.3755	6/3/2004	280	60.2801	−143.3755	9/24/2004	564	113	5.0	7.50	No significant movement
AB03	2004	60.3771	−143.8041	6/4/2004	222	60.3775	−143.8045	9/24/2004	577	112	5.2	7.80	0.42
AB04	2004	60.1675	−143.4188	6/4/2004	93	60.1675	−143.4188	9/18/2004	531	106	5.0	7.50	No significant movement
AB05	2004	60.2067	−143.5889	6/4/2004	148	60.2066	−143.5889	9/22/2004	777	110	7.1	10.65	No significant movement
S01 Berg	2005	60.3843	−143.8092	6/16/2005	182	60.3846	−143.8092	9/20/2005	514	96	5.35	8.02	2.8
S02 Berg	2005	60.3974	−143.6660	6/16/2005	452	60.3967	−143.6674	9/20/2005	500	96	5.21	7.81	0.43
B01	2005	60.1630	−143.4098	5/11/2005	61	60.1630	−143.4098	9/19/2005	665	131	5.08	7.62	0.28
B02	2005	60.2782	−143.3763	5/11/2005	259	60.2781	−143.3763	9/19/2005	534	131	4.08	6.12	0.86
B03	2005	60.3578	−143.2683	6/16/2005	467	60.3578	−143.2682	9/20/2005	528	96	5.5	8.25	No significant movement
B05	2005	60.4896	−142.6128	6/16/2005	976	60.4883	−142.6168	9/20/2005	488	96	5.08	7.62	0.1
B06	2005	60.5753	−142.7697	6/16/2005	1252	60.5753	−142.7689	9/20/2005	372	96	3.88	5.82	No significant movement
Jay	2006	60.2004	−143.2528	6/14/2006	41	60.2004	−143.2528	7/21/2006	390	37	10.54	15.81	No significant movement
B01	2006	60.1651	−143.4191	6/12/2006	61	60.1651	−143.4191	9/21/2006	632	101	6.26	7.39	No significant movement
B02	2006	60.2779	−143.3779	6/13/2006	255	60.2779	−143.3779	9/21/2006	378	100	3.78	5.67	No significant movement
B03	2006	60.3540	−143.2718	6/13/2006	462	60.3540	−143.2718	9/21/2006	365	100	3.65	5.47	No significant movement
B04	2006	60.3827	−142.9202	6/14/2006	739	60.3821	−142.9237	9/21/2006	383	99	3.87	5.80	2.02
B05	2006	60.4879	−142.6169	6/13/2006	995	60.4863	−142.6219	9/21/2006	420	100	4.20	6.30	3.2
B06	2006	60.5755	−142.7719	6/13/2006	1257	60.5755	−142.7713	9/21/2006	240	100	2.40	3.60	0.64
T01	2007	60.2053	−143.5788	6/6/2007	120	60.2053	−143.5789	9/11/2007	505	97	5.21	7.82	No significant movement
B01	2007	60.1735	−143.4319	6/9/2007	66	60.1735	−143.4319	9/11/2007	442	94	4.71	7.07	No significant movement
B03	2007	60.3475	−143.2787	6/9/2007	453	60.3475	−143.2788	9/11/2007	363	94	3.86	5.79	No significant movement
B04	2007	60.3809	−142.9170	6/9/2007	738	60.3808	−142.9179	9/11/2007	384	94	4.09	6.14	1.51*
B05	2007	60.4849	−142.6143	6/9/2007	972	60.4841	−142.6167	9/11/2007	404	94	4.30	6.45	2.69†
B06	2007	60.5762	−142.7740	6/9/2007	1257	60.5762	−142.7734	9/11/2007	371	94	3.95	5.93	0.63†

Note: See Figure 3 for site locations on Glacier. Datum: WGS 84.
*End date for horizontal migration: 14 June 2007.
†End date for horizontal migration: 6 August 2008.

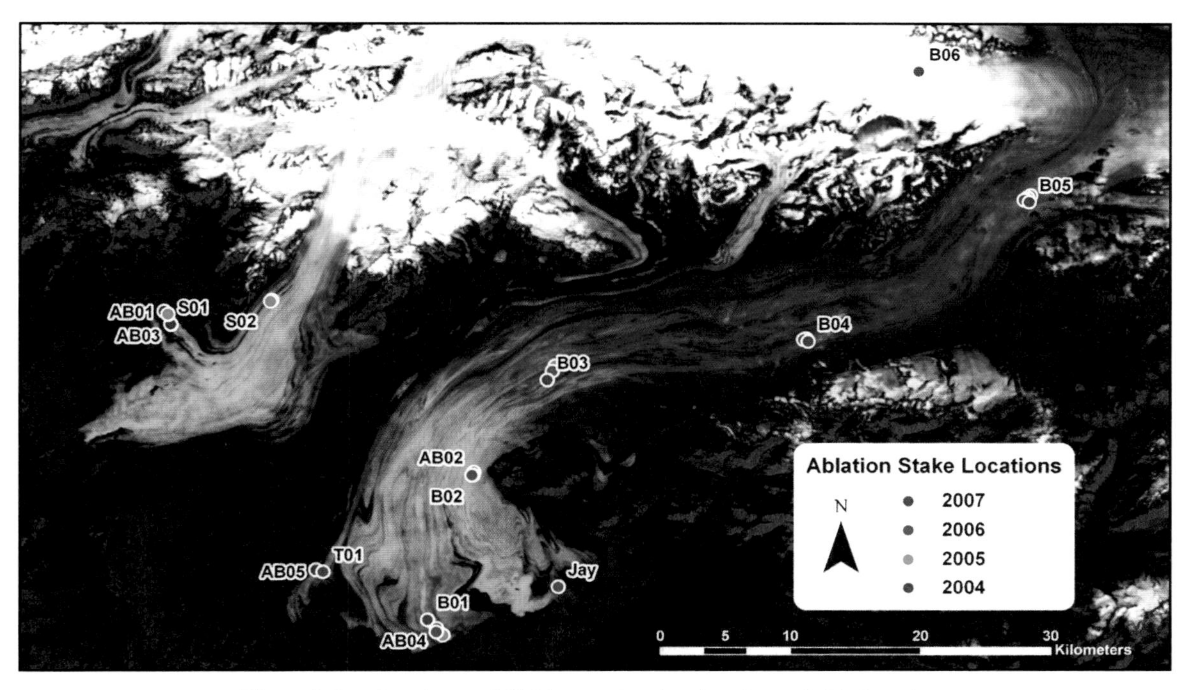

Figure 3. Landsat August 2006 image showing locations of ablation sites.

TABLE 2. REMOTE SENSING–DERIVED ALBEDO VALUES FOR 2004 AND 2006 ABLATION SITES

SITE	Year	Starting latitude	Starting longitude	Starting date	Starting elevation (m)	Ending latitude	Ending longitude	Ending date	Melt per day (cm)	Total days	Albedo (%)
AB05	2004	60.2067	-143.5889	6/4/2004	148	60.2066	-143.5889	9/22/2004	7.06	110	31
AB01	2004	60.3863	-143.8146	6/3/2004	269	60.3867	-143.8148	9/24/2004	5.69	113	42
AB03	2004	60.3771	-143.8041	6/4/2004	222	60.3775	-143.8045	9/24/2004	5.15	112	47
AB04	2004	60.1675	-143.4188	6/4/2004	93	60.1675	-143.4188	9/18/2004	5.01	106	80
AB02	2004	60.2802	-143.3755	6/3/2004	280	60.2801	-143.3755	9/24/2004	4.99	113	71
Jay	2006	60.2004	-143.2528	6/14/2006	41	60.2004	-143.2528	7/21/2006	10.54	37	8
B01	2006	60.1651	-143.4191	6/12/2006	61	60.1651	-143.4191	9/21/2006	6.26	101	33
B05	2006	60.4879	-142.6169	6/13/2006	995	60.4863	-142.6219	9/21/2006	4.20	100	20
B04	2006	60.3827	-142.9202	6/14/2006	739	60.3821	-142.9237	9/21/2006	3.87	99	19
B02	2006	60.2779	-143.3779	6/13/2006	255	60.2779	-143.3779	9/21/2006	3.78	100	42
B03	2006	60.3540	-143.2718	6/13/2006	462	60.3540	-143.2718	9/21/2006	3.65	100	24
B06	2006	60.5755	-142.7719	6/13/2006	1257	60.5755	-142.7713	9/21/2006	2.40	100	74

Note: Datum: WGS 84.

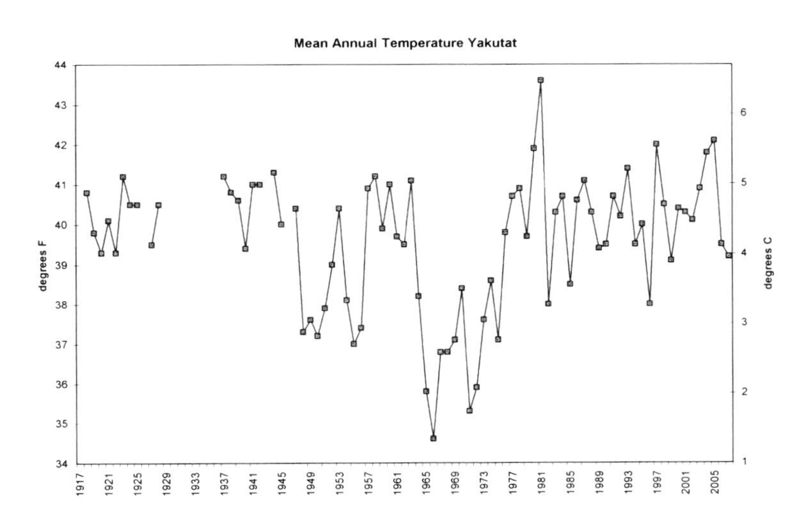

Figure 4. Mean annual temperature for Yakutat, Alaska, from 1918 to 2007. Note that 2004 and 2005 were years of high mean annual temperature.

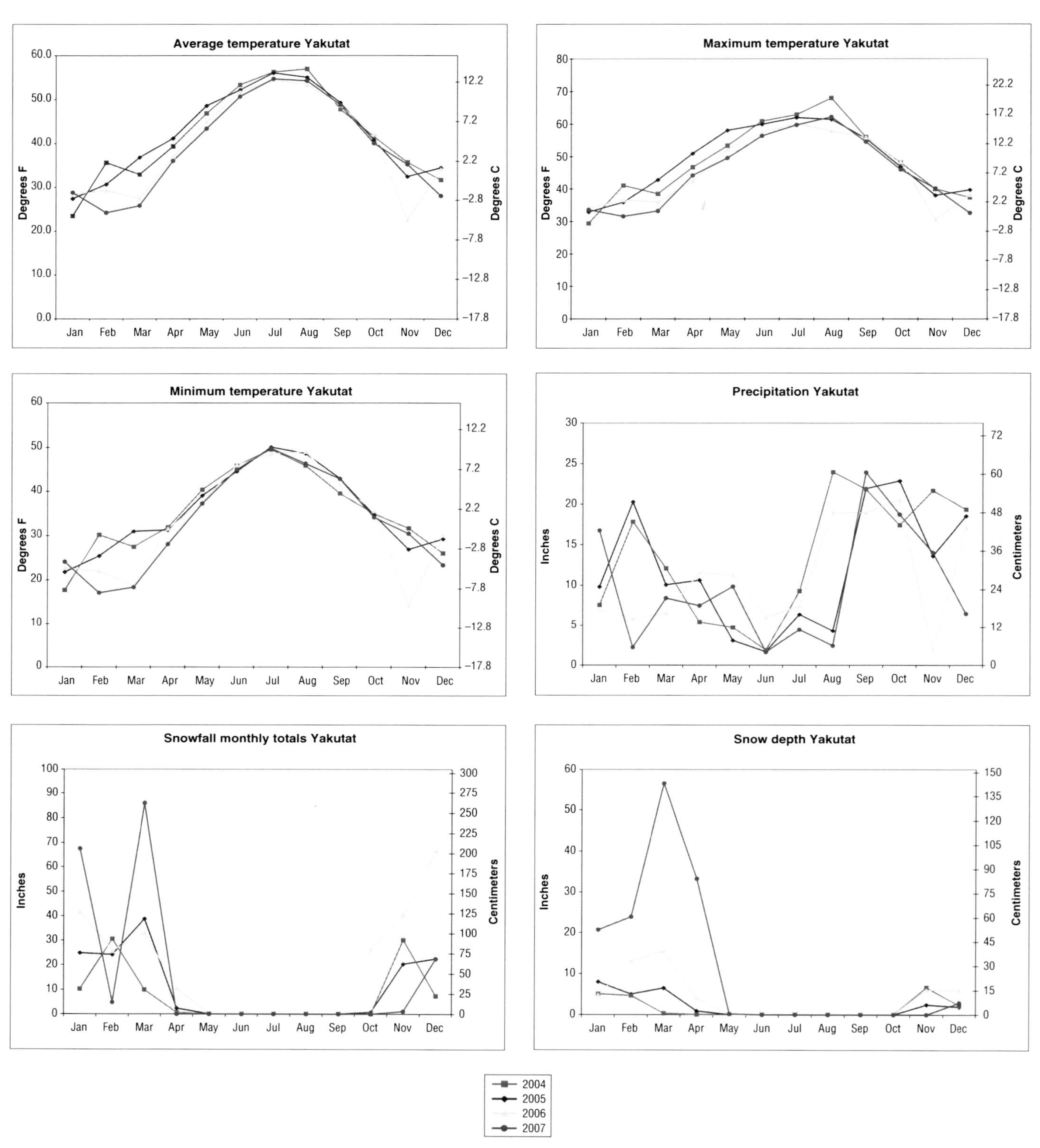

Figure 5. Yakutat, Alaska, monthly values for average, minimum, and maximum temperatures, and precipitation, snowfall, and snow depth for 2004–2007. Note the maximum temperature observed in August 2004.

39.5 °F (4.2 °C), and in 2007 the temperature was slightly colder, 39.2 °F, (4.0 °C), than the previous year. The June, July, and August monthly averages for 2004–2007 show a similar behavior: 2006 and 2007 were cooler than 2004, which was particularly warm. The precipitation data, also shown in Figure 5, are interesting from a snowfall and snow depth perspective. Years 2006 and 2007 had significantly more snow, but the average total precipitation appears to have been similar to the years 2004 and 2005.

A comparison of the April through September average temperatures at Yakutat with the measured ablation rates shows the controlling effect of air temperature. The summers of 2004 and 2005 were exceptionally warm, and the melt rates at corresponding altitudes were the most negative, approaching –10 m in the terminus region. For the summers of 2006 and 2007 the melt rates at all altitudes were 1–2 m less negative, and the average temperature was ~2 °F (1 °C) cooler.

GASS INSTRUMENTATION AND OBSERVATIONS

Bulk ablation and movement data were obtained at five to six sites using a specialized Glacier Ablation Sensing System (GASS) that was designed, fabricated, and deployed by MTRI, USGS, and U.S. Bureau of Land Management (BLM) scientists. GASS recorded on an hourly interval the GPS location, vertical (melt) movement of the glacier, and corresponding meteorological data including temperature, wind speed, humidity, barometric pressure, and upward and downward looking light intensity. The GASS unit (Fig. 6) employed a rechargeable battery and solar panel to provide at least 4.5 months of power, and a micro-processor to manage the power, data collection, and data storage. A Wide Area Augmentation System (WAAS)–enabled GPS recorded position and time for each measurement, and an acoustic ranging sensor measured the absolute distance between the GASS unit and the glacier surface: the difference between successive measurement intervals gives the melt rate. Each unit (up to six) was deployed on previously described ablation stakes ~0.5 m above the ice. The GASS units were visited on approximately a monthly basis to lower the sensor back to the 0.5 m position above the ice, which helps to keep the pole and the sensor perpendicular to the ice surface. The process of installing each GASS unit took ~1.5 h per site, which included steam drilling (Fig. 2) as well as sensor mounting, testing, and instrument activation.

The initial deployment of the GASS units was highly successful. The five instruments operated continuously and nearly without error for almost four months. The solar cell supplied sufficient power even into late summer. The location data provided by the WAAS-enabled GPS units showed an average glacier movement of 49 m over the four month period, which translates into an average velocity of ~0.4 m per day.

The deployment locations of the five sites are shown in Figure 3. Two instruments were deployed in the proximity of the terminus of the Steller Lobe as it flows into Berg Lake (sites AB01 and AB03). Two additional instruments were deployed near the terminus of the Bering Lobe, one at Tashalich Arm (AB05) and the other across Vitus Lake from the BLM summer field camp (AB04). The fifth site was installed near the Grindle Hills (AB02), ~13 km up-glacier from Vitus Lake in the Bering Lobe.

Figure 6. GASS instrument pack with anemometer, mounted on a pole that has been steam-drilled into the glacier.

The instruments were installed in early June 2004, with the initial plan being to retrieve them from the field in early August, near the end of the summer season. From June to August, during peak melt, the positions of the GASS instruments on the poles were adjusted, and any excess pole sections were removed every 3–4 weeks. In early August the data collected to date were downloaded, but given that every instrument was operating almost perfectly, it was decided to leave the instruments on the glacier until the BLM camp (Bering Camp) was broken down for the winter. Typically, the camp is broken down in mid-September, but owing to adverse weather conditions that date was pushed back to early October. Thus, the data record essentially covers the first week in June to the end of September.

As discussed, the GASS instruments utilized an onboard GPS receiver to record time and location every hour during the deployment. The WAAS-enabled GPS system provided geographic location accurate to within ~3 m. In some cases, the amount of weekly, and even cumulative, change in position was not sufficient to exceed the accuracy limitation of the GPS data, which showed how inactive the glacier had become.

Tables 3–7 summarize the average weekly locations reported for the five sites. Weekly averages are reported here to

TABLE 3. WEEKLY AVERAGED GASS POSITION AT SITE AB01

Site AB01—Berg Lake no. 1 (near Berg Lake terminus)					
Week	Begin date	End date	Latitude	Longitude	Distance moved (m)
Initial location	6/3/2004	–	60.386283	–143.814583	–
1	6/3/2004	6/9/2004	60.386270	–143.814618	–
2	6/10/2004	6/16/2004	60.386298	–143.814635	3.3
3	6/17/2004	6/23/2004	60.386325	–143.814630	3
4	6/24/2004	6/30/2004	60.386368	–143.814668	5.2
5	7/1/2004	7/7/2004	60.386377	–143.814654	1.3
6	7/8/2004	7/14/2004	60.386395	–143.814653	2
7	7/15/2004	7/21/2004	60.386423	–143.814670	3.2
8	7/22/2004	7/28/2004	60.386460	–143.814690	4.3
9	7/29/2004	8/4/2004	60.386502	–143.814729	5.2
10	8/5/2004	8/11/2004	60.386514	–143.814721	1.4
11	8/12/2004	8/18/2004	60.386541	–143.814741	3.1
12	8/19/2004	8/25/2004	60.386556	–143.814724	1.9
13	8/26/2004	9/1/2004	60.386578	–143.814734	2.5
14	9/2/2004	9/8/2004	60.386610	–143.814743	3.5
15	9/9/2004	9/15/2004	60.386613	–143.814750	0.6
16	9/16/2004	9/22/2004	60.386636	–143.814752	2.6
17	9/23/2004	9/29/2004	60.386691	–143.814774	6.3
				Total movement	**49.5**

Note: Datum: WGS 84.

TABLE 4. WEEKLY AVERAGED GASS POSITION AT SITE AB02

Site AB02—Grindle Hills					
Week	Begin date	End date	Latitude	Longitude	Distance moved (m)
Initial location	6/3/2004	–	60.280167	–143.375483	–
1	6/3/2004	6/9/2004	60.280166	–143.375446	–
2	6/10/2004	6/16/2004	60.280116	–143.375537	No significant movement
3	6/17/2004	6/23/2004	60.280142	–143.375502	No significant movement
4	6/24/2004	6/30/2004	60.280104	–143.375566	No significant movement
5	7/1/2004	7/7/2004	60.280129	–143.375528	No significant movement
6	7/8/2004	7/14/2004	60.280130	–143.375534	No significant movement
7	7/15/2004	7/21/2004	60.280121	–143.375563	No significant movement
8	7/22/2004	7/28/2004	60.280143	–143.375577	No significant movement
9	7/29/2004	8/4/2004	60.280126	–143.375568	No significant movement
10	8/5/2004	8/11/2004	60.280127	–143.375542	No significant movement
11	8/12/2004	8/18/2004	60.280118	–143.375573	No significant movement
12	8/19/2004	8/25/2004	60.280120	–143.375534	No significant movement
13	8/26/2004	9/1/2004	60.280122	–143.375549	No significant movement
14	9/2/2004	9/8/2004	60.280111	–143.375517	No significant movement
15	9/9/2004	9/15/2004	60.280112	–143.375537	No significant movement
16	9/16/2004	9/22/2004	60.280105	–143.375550	No significant movement
17	9/23/2004	9/29/2004	60.280095	–143.375504	No significant movement
				Total movement*	

Note: Datum: WGS 84.
*Total changes in position were within the accuracy of the GPS device, and it was not possible to determine distance moved.

TABLE 5. WEEKLY AVERAGED GASS POSITION AT SITE AB03

Site AB03—Berg Lake no. 2 (up-glacier from site AB01)					
Week	Begin date	End date	Latitude	Longitude	Distance moved (m)
Initial location	6/4/2004	–	60.377140	–143.804110	–
1	6/4/2004	6/9/2004	60.377143	–143.804122	–
2	6/10/2004	6/16/2004	60.377163	–143.804124	2.3
3	6/17/2004	6/23/2004	60.377198	–143.804170	4.7
4	6/24/2004	6/30/2004	60.377211	–143.804173	1.4
5	7/1/2004	7/7/2004	60.377247	–143.804216	4.7
6	7/8/2004	7/14/2004	60.377251	–143.804234	1.1
7	7/15/2004	7/21/2004	60.377271	–143.804249	2.4
8	7/22/2004	7/28/2004	60.377286	–143.804274	2.1
9	7/29/2004	8/4/2004	60.377316	–143.804304	3.7
10	8/5/2004	8/11/2004	60.377339	–143.804310	2.6
11	8/12/2004	8/18/2004	60.377377	–143.804328	4.4
12	8/19/2004	8/25/2004	60.377403	–143.804377	4
13	8/26/2004	9/1/2004	60.377404	–143.804368	0.5
14	9/2/2004	9/8/2004	60.377432	–143.804386	3.2
15	9/9/2004	9/15/2004	60.377446	–143.804420	2.4
16	9/16/2004	9/22/2004	60.377446	–143.804414	0.3
17	9/23/2004	9/29/2004	60.377507	–143.804464	7.3
				Total movement	**47.1**

Note: Datum: WGS 84.

TABLE 6. WEEKLY AVERAGED GASS POSITION AT SITE AB04

Site AB04—near Vitus Lake terminus					
Week	Begin date	End date	Latitude	Longitude	Distance moved (m)
Initial location	6/4/2004	–	60.167460	–143.418790	–
1	6/4/2004	6/9/2004	60.167478	–143.418776	–
2	6/10/2004	6/16/2004	60.167499	–143.418842	No significant movement
3	6/17/2004	6/23/2004	60.167503	–143.418804	No significant movement
4	6/24/2004	6/30/2004	60.167481	–143.418816	No significant movement
5	7/1/2004	7/7/2004	60.167482	–143.418792	No significant movement
6	7/8/2004	7/14/2004	60.167476	–143.418795	No significant movement
7	7/15/2004	7/21/2004	60.167474	–143.418836	No significant movement
8	7/22/2004	7/28/2004	60.167472	–143.418800	No significant movement
9	7/29/2004	8/4/2004	60.167486	–143.418795	No significant movement
10	8/5/2004	8/11/2004	60.167494	–143.418792	No significant movement
11	8/12/2004	8/18/2004	60.167499	–143.418804	No significant movement
12	8/19/2004	8/25/2004	60.167482	–143.418804	No significant movement
13	8/26/2004	9/1/2004	60.167494	–143.418832	No significant movement
14	9/2/2004	9/8/2004	60.167474	–143.418811	No significant movement
15	9/9/2004	9/15/2004	60.167477	–143.418776	No significant movement
16	9/16/2004	9/22/2004	60.167479	–143.418771	No significant movement
17	9/23/2004	9/29/2004	60.167488	–143.418778	No significant movement
				Total movement*	

Note: Datum: WGS 84.
*Total changes in position were within the accuracy of the GPS device, and it was not possible to determine distance moved.

TABLE 7. WEEKLY AVERAGED GASS POSITION AT SITE AB05

Site AB05—near Vitus Lake terminus					
Week	Begin date	End date	Latitude	Longitude	Distance moved (m)
Initial location	6/4/2004	–	60.206650	–143.588910	–
1	6/4/2004	6/9/2004	60.206646	–143.588880	–
2	6/10/2004	6/16/2004	60.206711	–143.588915	No significant movement
3	6/17/2004	6/23/2004	60.206635	–143.588881	No significant movement
4	6/24/2004	6/30/2004	60.206649	–143.588893	No significant movement
5	7/1/2004	7/7/2004	60.206660	–143.588898	No significant movement
6	7/8/2004	7/14/2004	60.206654	–143.588911	No significant movement
7	7/15/2004	7/21/2004	60.206637	–143.588914	No significant movement
8	7/22/2004	7/28/2004	60.206640	–143.588895	No significant movement
9	7/29/2004	8/4/2004	60.206634	–143.588894	No significant movement
10	8/5/2004	8/11/2004	60.206636	–143.588904	No significant movement
11	8/12/2004	8/18/2004	60.206628	–143.588895	No significant movement
12	8/19/2004	8/25/2004	60.206631	–143.588865	No significant movement
13	8/26/2004	9/1/2004	60.206623	–143.588918	No significant movement
14	9/2/2004	9/8/2004	60.206621	–143.588910	No significant movement
15	9/9/2004	9/15/2004	60.206619	–143.588893	No significant movement
16	9/16/2004	9/22/2004	60.206615	–143.588892	No significant movement
17	9/23/2004	9/29/2004	60.206611	–143.588876	No significant movement
				Total movement*	

Note: Datum: WGS 84.
*Weekly and total changes in position were within the accuracy of the GPS device, and it was not possible to determine distance moved.

reduce positional uncertainty from a short GPS collection period each hour. In 2005, 2006, and 2007, the GPS instrument was programmed to log a position during a 5 min period to ensure higher positional accuracy for the hourly observations. The initial location of each site was obtained at the time of deployment by averaging the GPS position for an hour, the approximate time it took the field team to set up a GASS site. Thus, the reported location is accurate to ~3 m (the published horizontal accuracy of the WAAS GPS).

Figures 7–11 summarize the weekly glacier movement at each of the test sites. An examination of the figures indicates that the average flow rate of the glacier is ~0.4 m per day for the two sites on the Steller Lobe near the Berg Lake terminus (AB01 and AB03). Site AB01, closest to Berg Lake, showed the most movement at 49 m in 114 days, and site AB03 moved ~47 m over about the same period. The sites established on the Bering Lobe (AB02, AB04, and AB05) exhibited much less movement, and, in fact, the total overall displacement was <3 m, which is within the error of the WAAS-enabled GPS instrument. Although two of the sites showed a general trend (movement in a uniform direction), it is not possible to quantitatively determine the glacier movement at those sites from the data acquired by GASS. Therefore, the three sites did not move an appreciable distance during the observational period, indicating the inactive state of the Bering Lobe.

Figure 12 shows the total ablation for each of the five sites from deployment in early June through 24 September 2004: the air temperature (top), wind speed (middle), and precipitation (bottom) as recorded at Bering Camp. After 24 September the response of the acoustic sonar became unstable owing to severe weather and/or power limitations. Note that the maximum ice melt occurred at site AB05, with a value of ~7.8 m. The GASS at site AB01 showed the next highest melt at just over 6.4 m, whereas sites AB02, AB03, and AB04 had melts of 5.6 m, 5.8 m, and 5.3 m, respectively. The average melt for all sites over this period was 6.2 m. Because of a data dropout, which may have been caused by a sensor malfunction, the range information for site AB01 was linearly interpolated between the dates of 1 August and 10 August 2004 (visible as the straight segment in the graph in Fig. 12), using the slope from the range observations over approximately the first 60 days.

Two interesting observations can be made based on examination of the ablation data. First, independent of location, the melt rates are relatively uniform throughout the summer, with an expected gradual decline as fall begins. Second, three of the sites (AB02, AB03, and AB04) showed approximately the same average melt rate, 5.3–5.6 m, whereas sites AB01 and AB05 showed a noticeably higher melt rate. The two higher melt rates occurred at sites with lower albedo as calculated from Landsat-7 images. Sites AB05 and AB01 (Fig. 2) both had extensive rock debris and much lower albedo values (31% and 42%, respectively). Site AB03 (albedo value of 47%) had a slightly higher melt rate than sites AB04 and AB02 (albedo values of 80% and 71%, respectively).

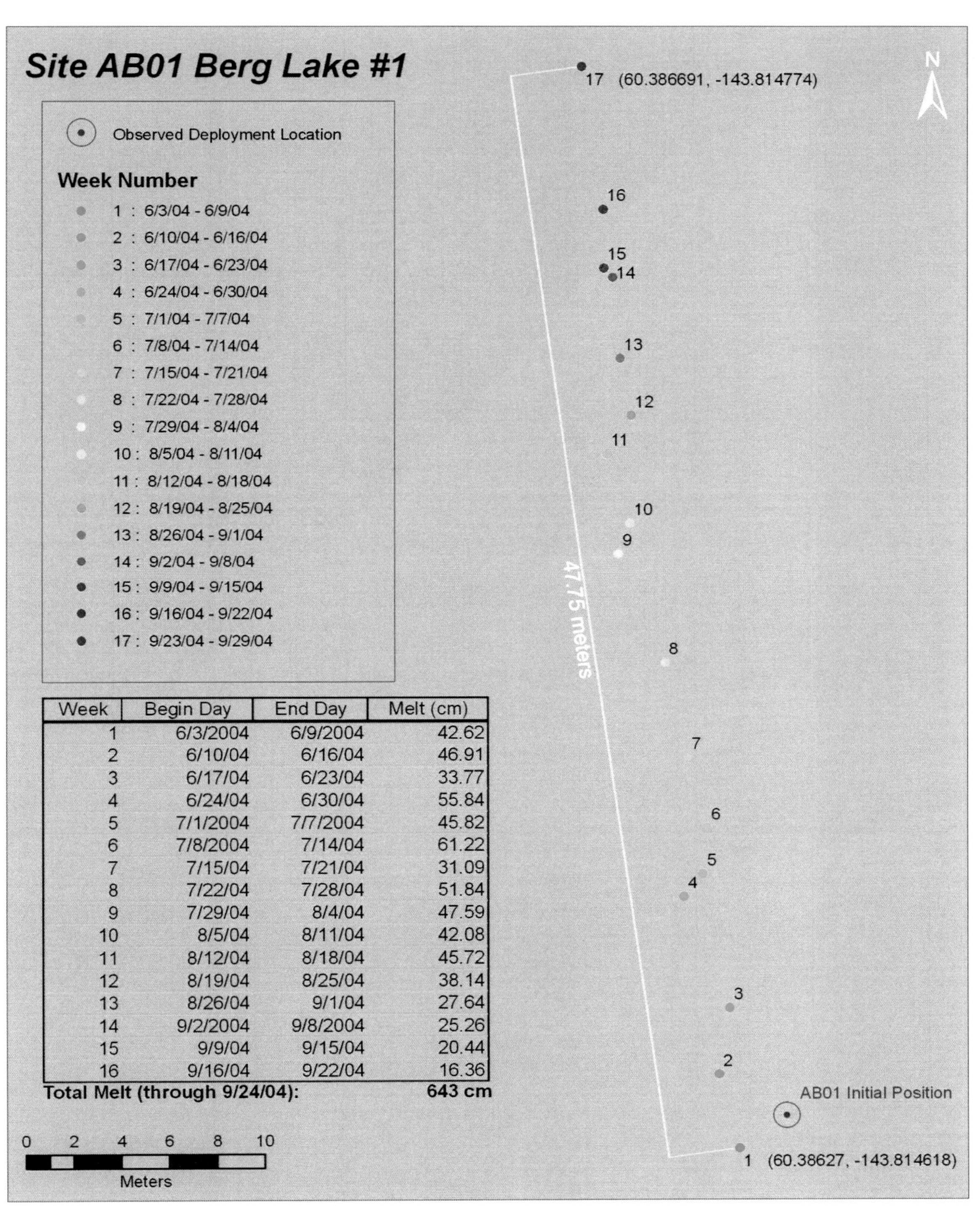

Week	Begin Day	End Day	Melt (cm)
1	6/3/2004	6/9/2004	42.62
2	6/10/04	6/16/04	46.91
3	6/17/04	6/23/04	33.77
4	6/24/04	6/30/04	55.84
5	7/1/2004	7/7/2004	45.82
6	7/8/2004	7/14/04	61.22
7	7/15/04	7/21/04	31.09
8	7/22/04	7/28/04	51.84
9	7/29/04	8/4/04	47.59
10	8/5/04	8/11/04	42.08
11	8/12/04	8/18/04	45.72
12	8/19/04	8/25/04	38.14
13	8/26/04	9/1/04	27.64
14	9/2/2004	9/8/2004	25.26
15	9/9/04	9/15/04	20.44
16	9/16/04	9/22/04	16.36
Total Melt (through 9/24/04):			**643 cm**

Figure 7. Map of weekly average GASS location for site AB01. The background color of this image was generated using the Landsat satellite response and can serve as a relative index of albedo when compared to maps of the other sites.

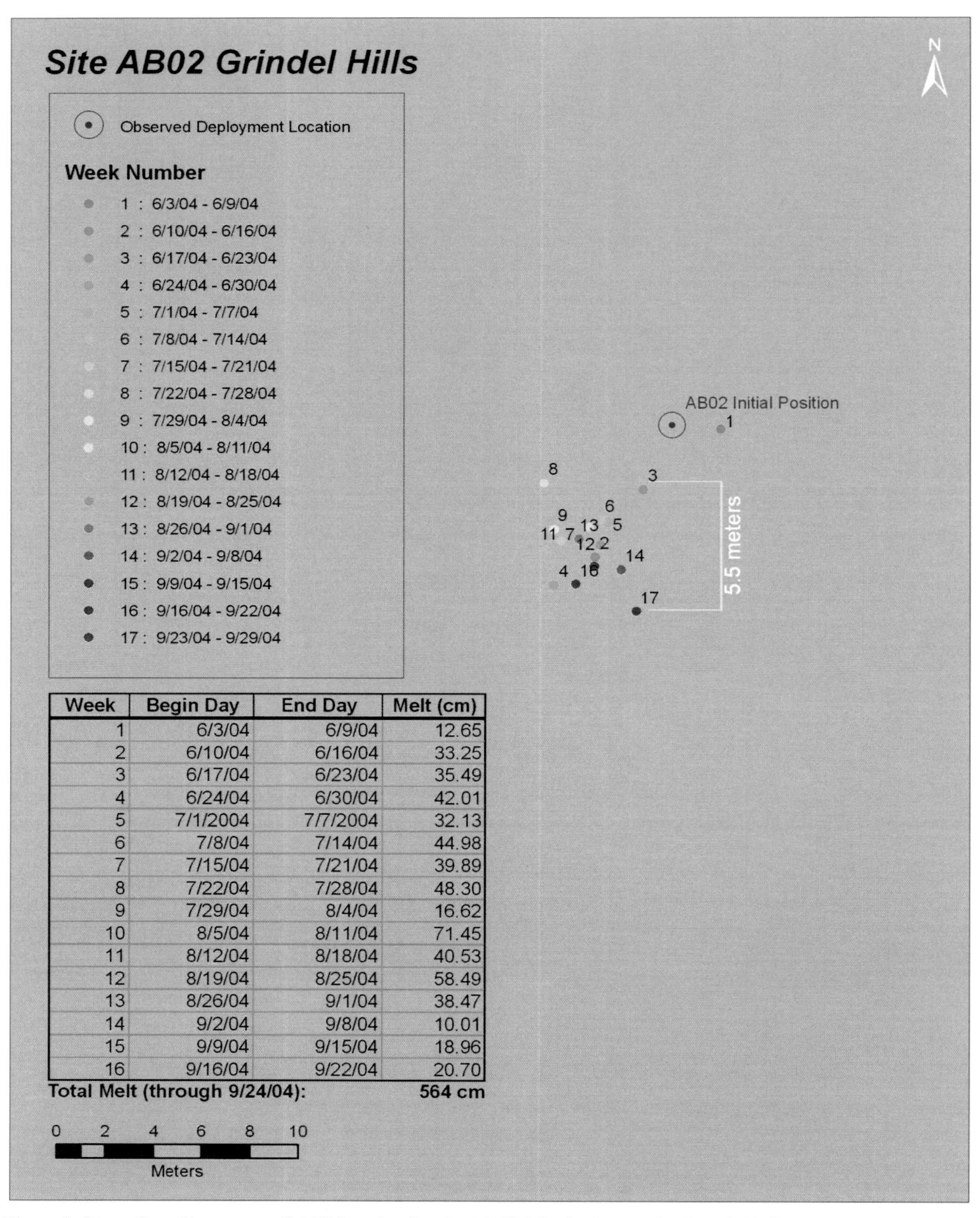

Week	Begin Day	End Day	Melt (cm)
1	6/3/04	6/9/04	12.65
2	6/10/04	6/16/04	33.25
3	6/17/04	6/23/04	35.49
4	6/24/04	6/30/04	42.01
5	7/1/2004	7/7/2004	32.13
6	7/8/04	7/14/04	44.98
7	7/15/04	7/21/04	39.89
8	7/22/04	7/28/04	48.30
9	7/29/04	8/4/04	16.62
10	8/5/04	8/11/04	71.45
11	8/12/04	8/18/04	40.53
12	8/19/04	8/25/04	58.49
13	8/26/04	9/1/04	38.47
14	9/2/04	9/8/04	10.01
15	9/9/04	9/15/04	18.96
16	9/16/04	9/22/04	20.70
Total Melt (through 9/24/04):			564 cm

Figure 8. Map of weekly average GASS location for site AB02. The background color of this image was generated using the Landsat satellite response and can serve as a relative index of albedo when compared to maps of the other sites.

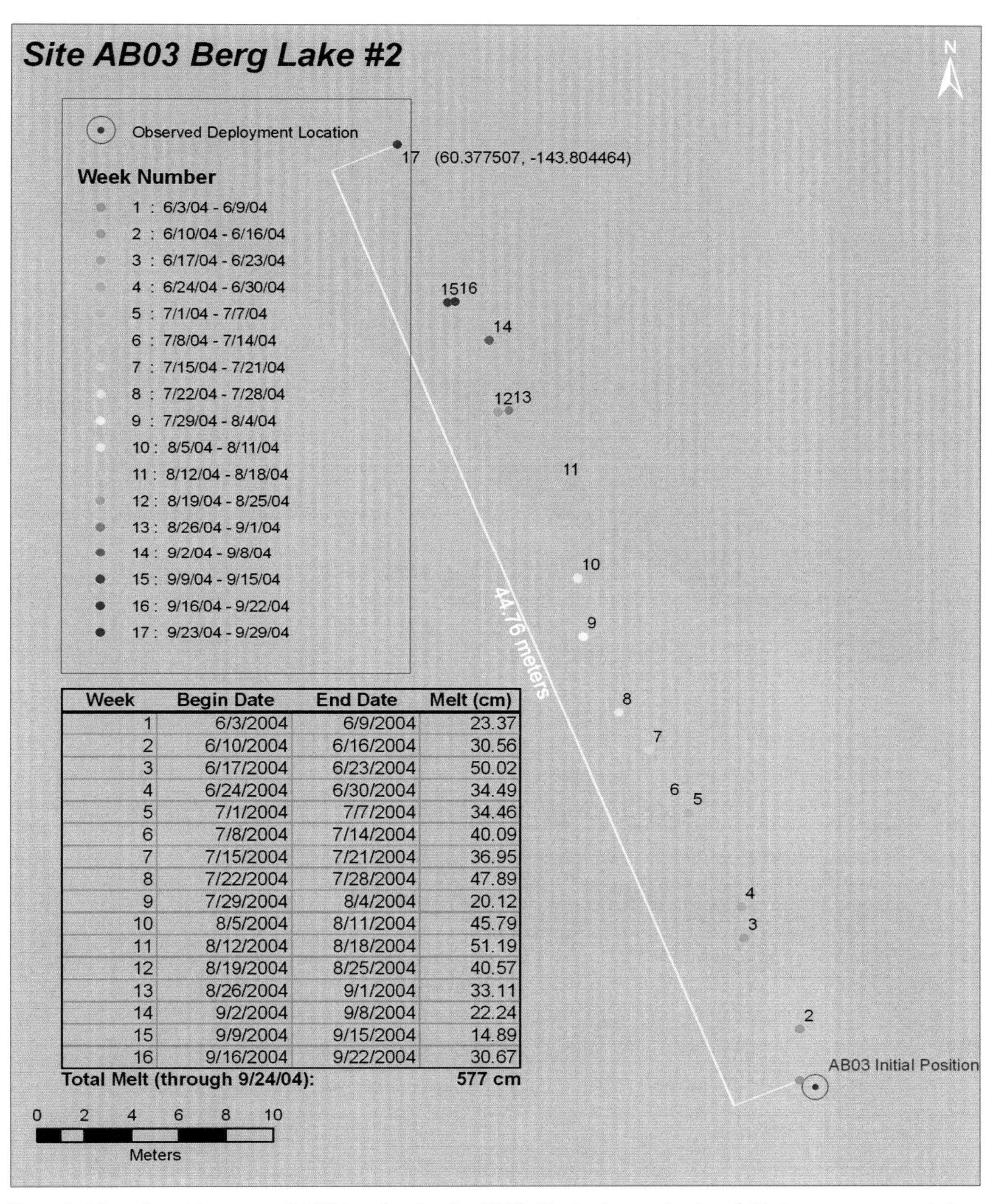

Figure 9. Map of weekly average GASS location for site AB03. The background color of this image was generated using the Landsat satellite response and can serve as a relative index of albedo when compared to maps of the other sites.

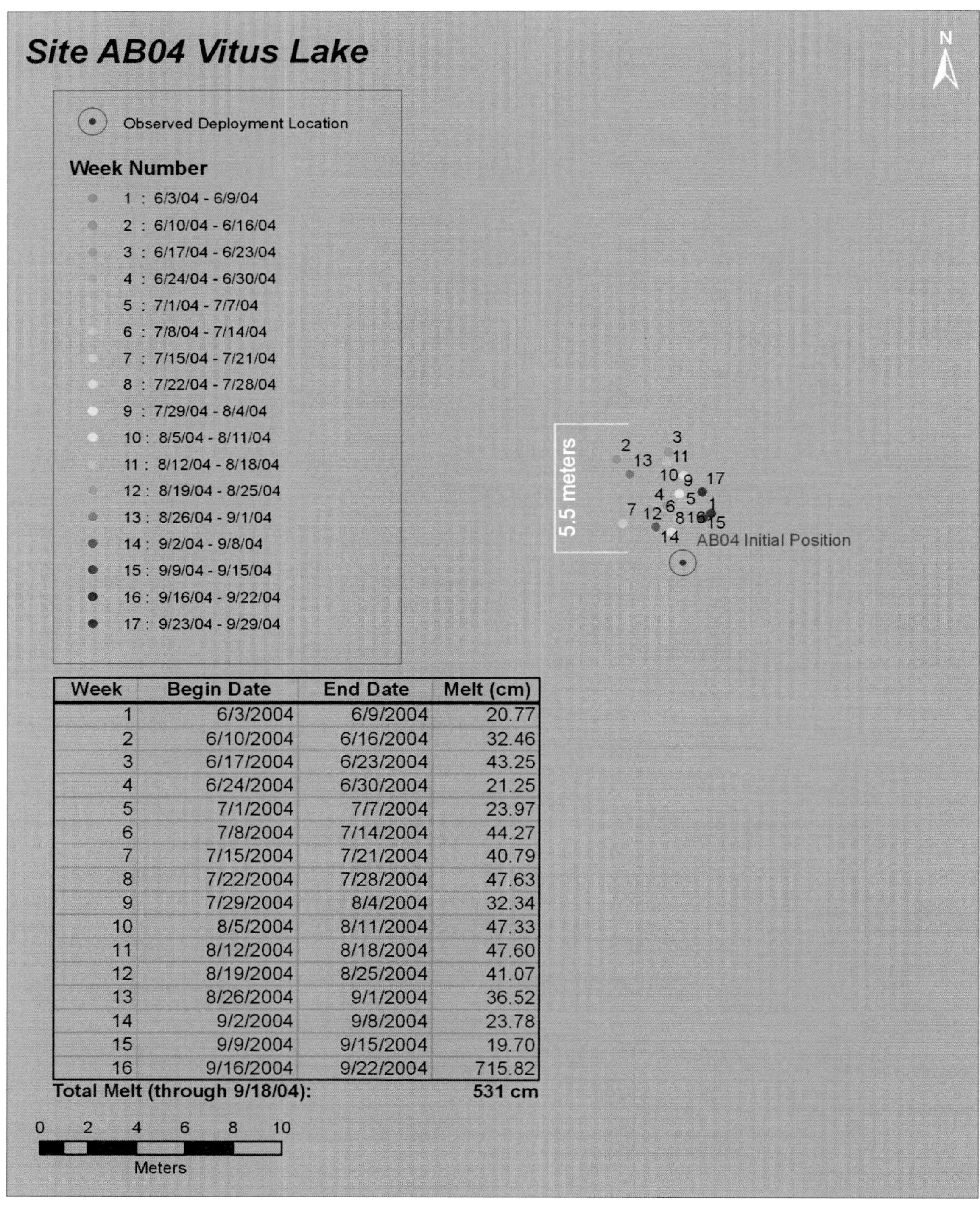

Week	Begin Date	End Date	Melt (cm)
1	6/3/2004	6/9/2004	20.77
2	6/10/2004	6/16/2004	32.46
3	6/17/2004	6/23/2004	43.25
4	6/24/2004	6/30/2004	21.25
5	7/1/2004	7/7/2004	23.97
6	7/8/2004	7/14/2004	44.27
7	7/15/2004	7/21/2004	40.79
8	7/22/2004	7/28/2004	47.63
9	7/29/2004	8/4/2004	32.34
10	8/5/2004	8/11/2004	47.33
11	8/12/2004	8/18/2004	47.60
12	8/19/2004	8/25/2004	41.07
13	8/26/2004	9/1/2004	36.52
14	9/2/2004	9/8/2004	23.78
15	9/9/2004	9/15/2004	19.70
16	9/16/2004	9/22/2004	715.82
Total Melt (through 9/18/04):			**531 cm**

Figure 10. Map of weekly average GASS location for site AB04. The background color of this image was generated using the Landsat satellite response and can serve as a relative index of albedo when compared to maps of the other sites.

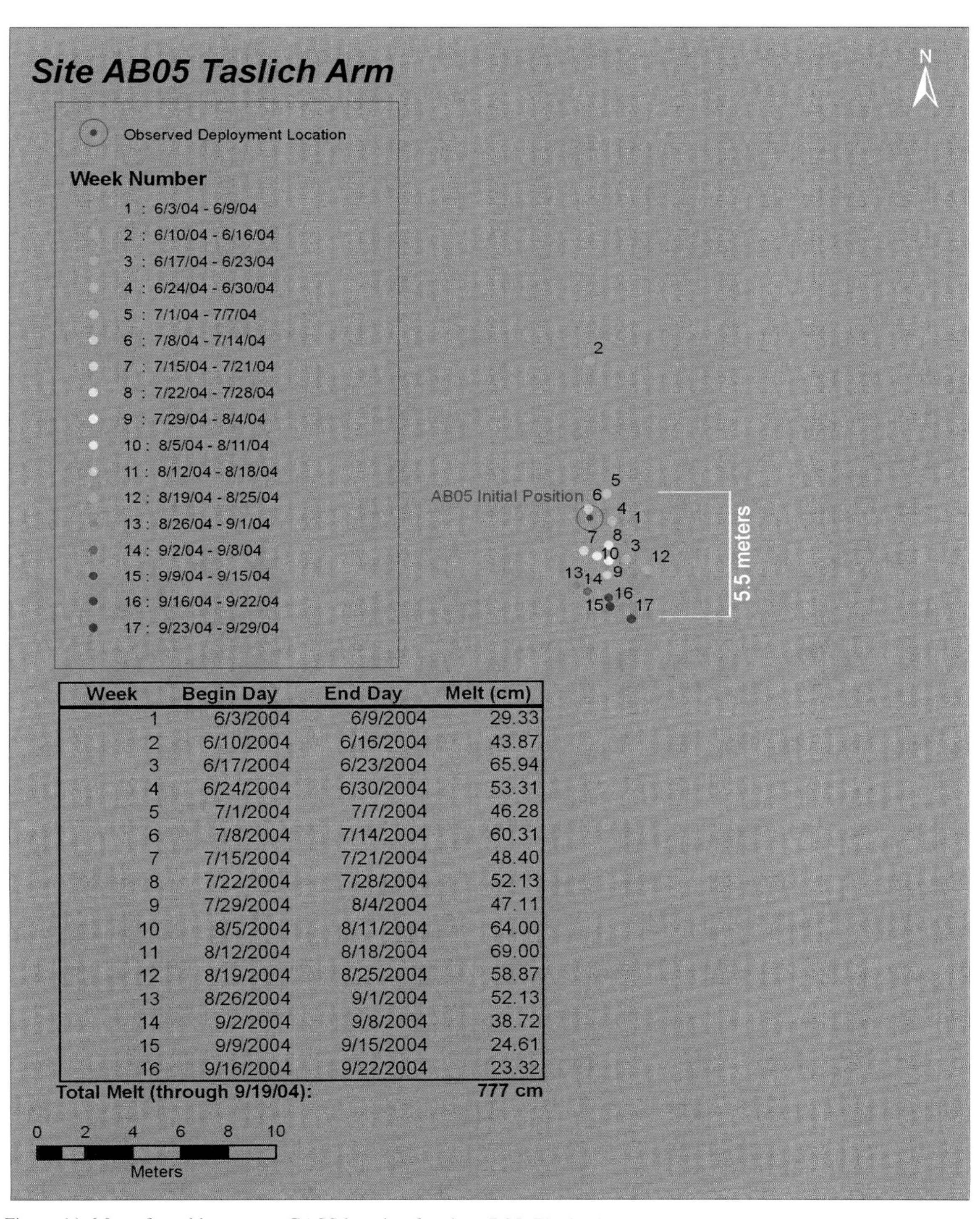

Week	Begin Day	End Day	Melt (cm)
1	6/3/2004	6/9/2004	29.33
2	6/10/2004	6/16/2004	43.87
3	6/17/2004	6/23/2004	65.94
4	6/24/2004	6/30/2004	53.31
5	7/1/2004	7/7/2004	46.28
6	7/8/2004	7/14/2004	60.31
7	7/15/2004	7/21/2004	48.40
8	7/22/2004	7/28/2004	52.13
9	7/29/2004	8/4/2004	47.11
10	8/5/2004	8/11/2004	64.00
11	8/12/2004	8/18/2004	69.00
12	8/19/2004	8/25/2004	58.87
13	8/26/2004	9/1/2004	52.13
14	9/2/2004	9/8/2004	38.72
15	9/9/2004	9/15/2004	24.61
16	9/16/2004	9/22/2004	23.32
Total Melt (through 9/19/04):			**777 cm**

Figure 11. Map of weekly average GASS location for site AB05. The background color of this image was generated using the Landsat satellite response and can serve as a relative index of albedo when compared to maps of the other sites.

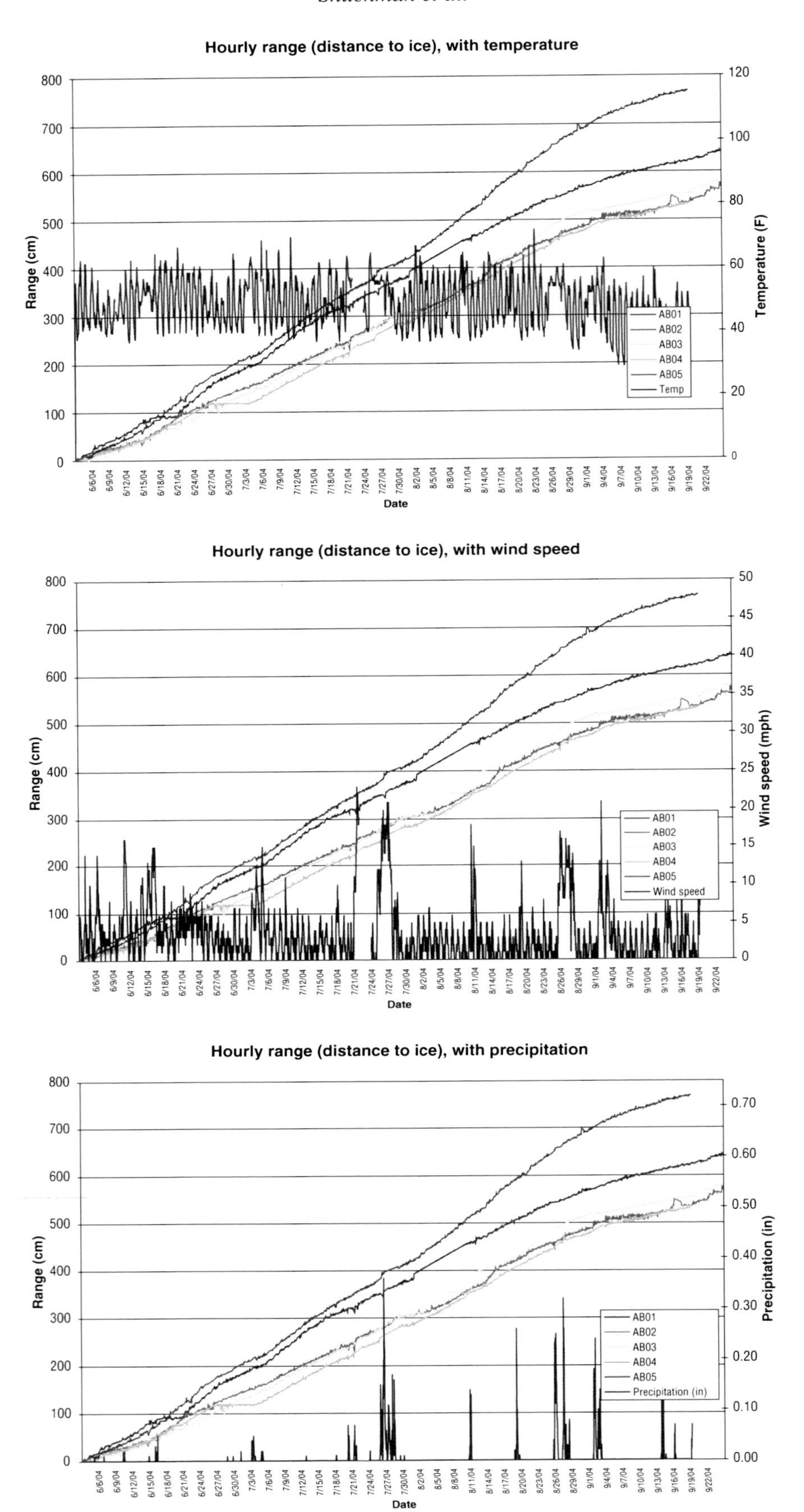

Figure 12. Ablation measurements for 2004 with coincident temperature (top), wind speed (middle), and rain (bottom) meteorological observations.

CONTROLLING FACTORS OF GLACIAL MELT

The observed melt rates exhibited hourly, diurnal, and seasonal fluctuations as a result of changing meteorological conditions; however, the hourly rates are significantly less than the mean daily rates, which resulted in a relatively constant melt rate over the summer. Figure 13 shows hourly and diurnal melt-rate fluctuations along with coincident temperature for a portion of the 2004 site AB01 ablation record. This figure illustrates the close relationship between air temperatures and melt rate on a diurnal time scale. As air temperature increases, there is a corresponding increase in melt rate. The data from the 2005 GASS instruments showed similar results. The GASS sensors deployed in 2006 had battery malfunctions, and therefore resulted in a very limited data collection in May, whereas the 2007 observations support the 2004 and 2005 results.

To determine the most significant control of hourly meteorological factors on melt for Bering Glacier, we carried out a multiple linear regression between the observed 2004 melt rates and the meteorological data at each GASS site. We focused the analysis on air temperature (sensible heat), and the product of air temperature and wind speed (turbulent heat flux), as these parameters are major factors in determining ice ablation. We included solar radiation in a subsequent analysis, described later. The best coefficient of determination (r^2) calculated using this model was ~0.69, which was based on a sensor at site AB04 during June and part of July. These results show that air temperature and wind play a significant role in determining melt rate. However, a more robust set of input parameters that include humidity, upward and downward solar radiation, and precipitation is needed to predict the glacier ablation at a given location.

On a time scale of days, we examined the 2004 GASS record to determine the effect of storms on melt rate. During 2004 there were three storm events when temperature exceeded 11 °C, cloudiness as indicated by a battery voltage of <13 V, wind speed >4m/s, and rain rate >40 mm/hr. The vertical lines in Figure 14 locate the three events in the time series of observations made by the camp weather station. Each parameter was smoothed with a 24 point moving average to eliminate sensor noise. Also included in Figure 14B is the average melt rate of all five GASS sensors, smoothed with a 24 h moving average. The melt rates have a limited response to a storm event, because there are opposing effects on the surface heat budget. Increased cloudiness will reduce the downward shortwave radiation, but warmer temperatures and increased wind speeds will enhance the sensible heat transfer.

The hourly range of ice measurements from the GASS instruments for 2004, 2005, and 2007 revealed that the average daily melt rates are virtually constant through the summer (June through August), regardless of weather conditions. Then, in early fall, when both the solar radiation and the temperature decrease, the melt rate decreases significantly (Fig. 12). An example of the midsummer melt rate uniformity is given in Figure 15, which shows the hourly range of the ice record from site B06 in 2005, for mid-June through mid-July. A linear regression of the range data versus time further demonstrates the independence of melt rate in varying conditions. For the month of records, the correlation coefficient, r^2, equaled 0.9982, virtually a perfect correlation. The regression also showed that the ice melted at ~5.15 cm per day during the observation period. Melt measurements by the EarthWatch program at the southeast corner of the glacier terminus have shown some response to cloudiness. This is likely a result of the lower albedo of the ice and the proximity to non-ice-covered terrain, which could increase the local air temperature.

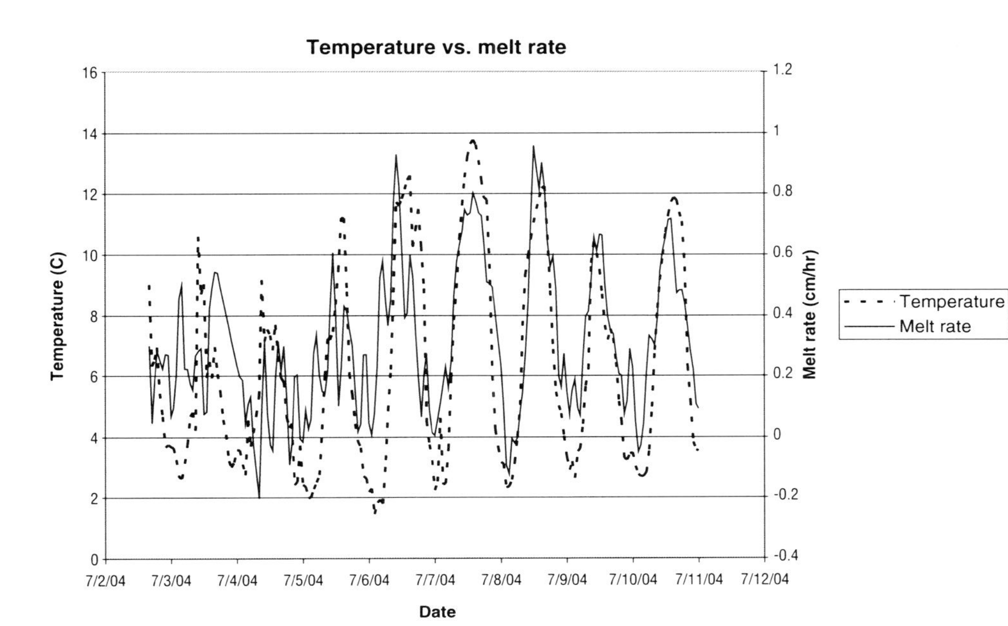

Figure 13. A portion of GASS data, showing representative melt rate and temperature for site AB01. This graph illustrates the strong correlation between melt rate and temperature on a diurnal time scale.

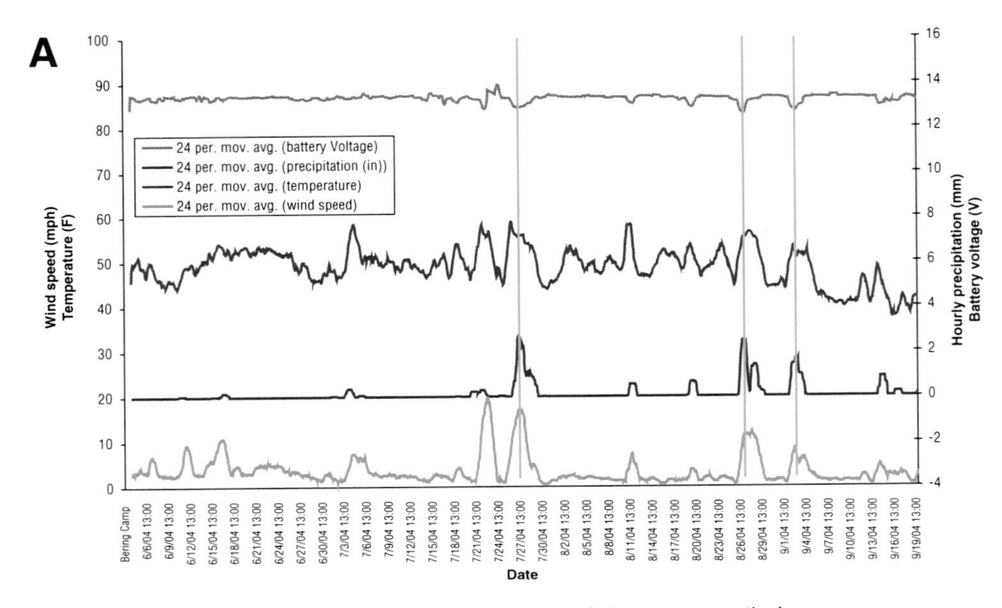

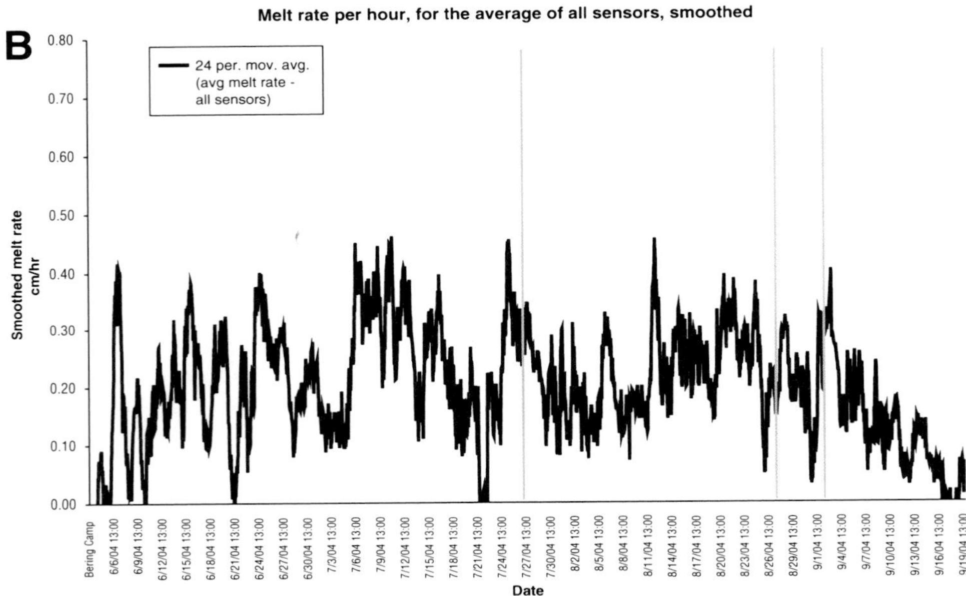

Figure 14. (A) Simultaneous plot of precipitation, battery voltage (proxy for light levels), wind speed, and temperature at Bering Camp. Vertical lines mark light weather events. (B) Melt rate per hour, for the average of all sensors. Note that increased melts do not all correspond to the above identified weather events.

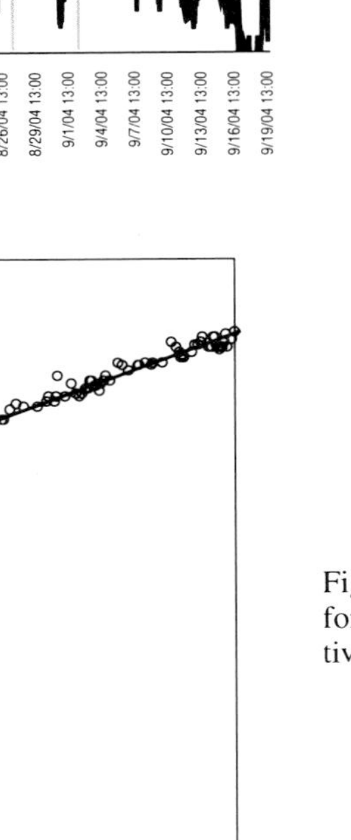

Figure 15. Ablation as a function of time for sensor B06 in 2005, showing a relative uniform melt rate.

In 2005, 2006, and 2007 we extended the ablation measurements to cover not only the terminus region but also the entire glacier up to its source region, the Bagley Ice Field. We measured bulk ablation at six or seven sites at different altitudes on the glacier, ranging from near sea level at the terminus to 1260 m in the Bagley Ice Field. The 2005, 2006, and 2007 bulk ablation measurements were analyzed as a function of altitude and compared with summer ablation measurements from Wolverine Glacier, which is part of the USGS Benchmark Glacier program (Josberger et al., 2007). These measurements extend the altitude range of summer melt from sea level to 500 m; The USGS observations on Wolverine Glacier begin at an altitude of 600 m. The observations (Fig. 16) show that 150 d ablation values fall into three elevation groups. The values are expressed in negative terms, which is the glaciological convention. The three groups are as follows:

1. Less than 200 m, where the ablation is greatest and is also highly variable, ranging from –5 to –7 m. Fleisher et al. (this volume) reports a value of –10 m from a site at the extreme southeastern part of the terminus. The variability is due in part to subtle differences in such factors as ice texture (crystal size and foliation), density of supraglacial debris, aspect (direction of slope exposure), exposure of ice previously saturated by moving englacial meltwater not yet subject to annealing by glacial flow, and proximity of the glacier to different local heat sources. Sites near the flanks of the glacier receive heat absorbed by nearby non-glacier terrain. Temperature measurements in Berg Lake (Josberger et al., 2006) also demonstrate this effect. Sites in the central part of the terminus are not subject to this heat source and melt at a slower rate. The GASS measurements in 2004 also demonstrate this variability.

2. Between 200 and 1000 m; the ablation in this zone is nearly constant with altitude, although there is a uniform offset between 2005 and 2006.

3. Greater than 1000 m; the ablation decreases with altitude, as expected, owing to cooler temperatures and more reflective ice and snow.

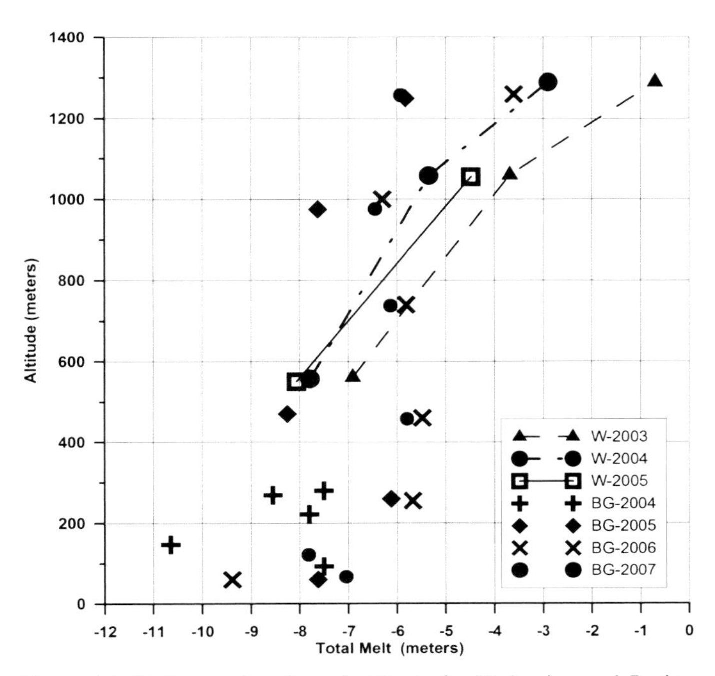

Figure 16. Melt as a function of altitude for Wolverine and Bering Glaciers.

With regard to temperature variability above 200 m, the melt curves, Figure 16, are offset by ~1.2 m, with more melt having occurred in 2005 than in 2006. This is consistent with the weather data (Figs. 4, 5) from Yakutat, which showed that the summer of 2005 was ~2 °F (1 °C) warmer than those of 2006 and 2007. This preliminary finding suggests that we may be able to index the melt to the average temperature and wind speed.

Figure 16 also presents the summer ablation observations from Wolverine Glacier for 2000–2005. Wolverine Glacier is a much smaller glacier situated on the Kenai Peninsula, 400 km west of Bering Glacier. The Wolverine summer balance measurements were recorded at three altitudes, ~550 m, 1000 m, and 1300 m. The summer balance of Wolverine Glacier, as a function of altitude, is nearly linear, and there is a strong suggestion of a change in the slope of the melt rate at or above 1000 m. Although each year is slightly different from the others, the general shape is the same; the balance curves are simply offset from each other. In contrast, the Bering Glacier above 200 m has a melt rate that is uniform up to 1000 m. Above this altitude the change of ablation with altitude is similar to that of the Wolverine Glacier. The melt values from the Bering Glacier span the same range as those from the Wolverine Glacier. Bering melt rates at the highest altitude, 1250 m, are more negative than those measured at Wolverine by ~2–3 m. At ~1000 m, the ablation values are nominally the same, whereas at 600 m the ablation at Wolverine Glacier can be 1–3 m more negative.

YEARLY GLACIER MELT ESTIMATES AND SURFACE ELEVATION CHANGES

The flux of meltwater from glaciers in this region is a critical factor in driving the circulation in the Gulf of Alaska (Royer and Grosch, 2006). To estimate the contribution by Bering Glacier we combined our summer ablation measurements with satellite-derived information on the Bering Lobe, which included glacier extent and a digital elevation model (DEM). We limited the analysis to the Bering Lobe, without regard to the Steller Lobe, because the ablation sites were almost entirely on the Bering Lobe. Furthermore, Josberger et al. (this volume) show that all this meltwater flows into Vitus Lake, where it controls the lake hydrography. Thus, lake level acts as an indicator, responding to the flux of water into and out of the lake. As described in the remote sensing chapter of Shuchman et al. (this volume), the ablation observations were interpolated to generate an equation for melt rate as a function of altitude for each of the four years. August Landsat images were used to estimate minimum and maximum areas for snow- and ice-covered areas of the Bering Lobe (~1550–2200 km^2). Using a geographic information system (GIS), an ASTER satellite–derived digital elevation map was combined with the Landsat satellite–determined glacier area to define the snow- and ice-covered area as a function of altitude. Finally, this information was combined with the ablation rate, as a function of altitude, to yield the total melt of the Bering Lobe. The melt values were then converted to a liquid water equivalent. The estimate assumed 150 days of melt in a given year, which is consistent with the Copper River discharge data as well as the weather history from Yakutat.

Table 8 summarizes the amount of fresh water generated by the Bering Glacier Lobe for the years 2004 through 2007. Minimum and maximum values are presented, based on including or excluding the medial moraine between the Bering Lobe and the Steller Lobe. In 2004 and 2005, the warmer years, the amount of water discharged into the Gulf of Alaska via Seal River from Bering Lobe ranged from 10.6 to 15.0 km^3 (10.6–15.0 Gt). The range for 2006 and 2007, when Yakutat underwent a 1 °C (2 °F) cooler average summer temperature, was significantly less, 8.6–12.7 km^3, and 8.6–12.7 Gt. Meier and Dyurgerov (2002) explain that 1 gigatonne (Gt) is the mass of 1 km^3 of water, which, when added to the ocean, would raise sea level by 1/362 mm.

A similar approach was taken for calculating the 2006 total annual fresh water melt for the entire Bering Glacier watershed. The watershed boundary was obtained from the USGS National Water Quality Assessment Program and was combined with the USGS 60 m DEM. In August 2006, a 16-day MODIS satellite mosaic was used to determine the extent of snow and ice in the entire Bering Glacier watershed, including the Steller Lobe. The total Bering Glacier watershed had an area of ~8990 km^2, whereas the area within the watershed that was determined via the MODIS image to represent snow and ice was ~4737 km^2. These data were then imported into a GIS to calculate total melt over a 150 day period as a function of elevation. The melt-altitude function was obtained from the ablation measurements presented in Figure 16. Three functions were generated: one representing melt from 0 to 200 m, a constant melt for 200–1000 m, and a third function for 1000 m and higher. Using these functions and converting ice to the liquid water equivalent with a ratio of 0.9, the Bering Glacier System produced ~30 km^3 of fresh water for 2006. This does not include contributions from precipitation and pressure melting. Since the ice marginal lakes in the region did not exhibit significant changes in level, it can be assumed that the bulk of this water flowed into the Gulf of Alaska.

We determined the change in Bering Glacier elevation by comparing the elevation data provided by Molnia and Post (1995) with the GPS elevations obtained from the 2006 ablation sites. Figure 17 shows the 2006 data and elevation profiles from past surveys. Examination of Figure 17 reveals that at site B01 (Fig. 3), the surface of the glacier in 2006 was at ~60 m, very near sea level. Note that in 1900 the elevation of the glacier at this location was ~350 m (a loss of 290 m). Table 9 summarizes the change in elevation of the surface of the glacier since 1957 at each of the 2006 ablation sites. Losses of ~200 m were observed at sites B02 and B03, whereas the glacier elevation at sites B04 and B05 have decreased by 60–80 m. For the twentieth century, melt rates in the terminus region were probably at near current levels because the annual average temperatures at Yakutat have not significantly exceeded values over the past 20 a (Fig. 4). The total melt for the past 100 a would have been more than sufficient to completely melt the terminus region. However, the surge mechanism transports large amounts of ice from the accumulation zone to the ablation zone, which maintains the extended position of Bering Glacier.

We estimated the total volumetric change of the Bering Lobe from 1957 to 2004. The total glacier volume was first calculated by combining the IPR data (Molnia and Post, 1995) with the glacier surface altitude data obtained from a 1972 USGS DEM. Ice thickness was determined by subtracting the subglacial elevations from the glacier surface elevations. Integrating ice depth values over the thickness map revealed that the total glacial volume in 1972 was ~660 km^3. Using 2004 and 1957 surface glacier elevation data, the amount of ice ablation was calculated. This was done by dividing the glacier into six stacked rectangles upglacier. We assumed a uniform melt over each rectangle. By multiplying the area of each rectangle by its change in thickness, and summing each of the six volumes, we arrived at a total change in volume of the glacier for the period 1957–2004, which was ~104 km^3. If we assume that the Bering Lobe represents 660 km^3 of ice, 13% of its total mass was lost during this period. It should

TABLE 8. ESTIMATES OF FRESH WATER GENERATED BY THE BERING GLACIER LOBE IN YEARS 2004 THROUGH 2007, ASSUMING 150 DAYS OF MELT

	Minimum glacier water discharge	Maximum glacier water discharge	Minimum precipitation	Maximum precipitation	Minimum total water discharge	Maximum total water discharge
2004	8.47	11.71	1.47	2.28	10.71	15.05
2005	7.46	10.31	2.42	3.76	10.56	14.99
2006	5.94	8.2	2.47	3.82	8.95	12.77
2007	6.35	8.84	1.66	2.58	8.59	12.22

Note: A minimum of 1550 km^2 and a maximum of 2200 km^2 were used as estimates of glacier surface area. Units are km^3. Datum: WGS 84.

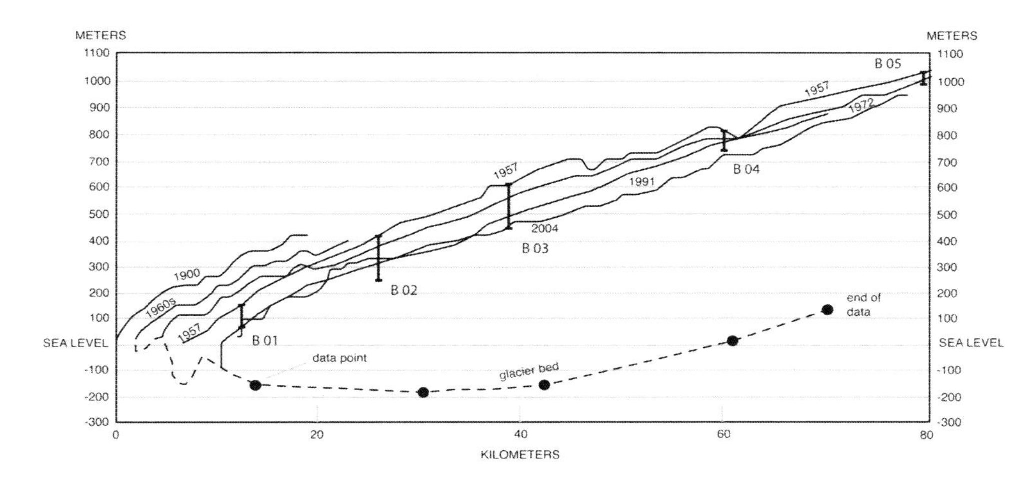

Figure 17. Elevation of Bering Glacier, measured at various times from 1990 to present. Black vertical lines show the differences in elevation between 1957 and 2006 at each ablation stake site. (Adapted from Molnia and Post, 1995).

TABLE 9. APPROXIMATE DIFFERENCES IN ELEVATIONS BETWEEN 1957 AND 2006

Site	Present elevation (m)	Change in elevation since 1957 (m)
B01	61	110
B01	61	300*
B02	255	175
B03	462	170
B04	739	85
B05	995	60

*Change in elevation since 1900.

be noted that this calculation does not take into consideration transfer of ice mass from up-glacier to the Piedmont Lobe or subglacial release of water (jökulhlaup) that interrupted the surges that occurred during this time period.

CONCLUDING REMARKS

Bering Glacier is undergoing rapid melting; in the terminus region of Bering Lobe, at altitudes <200 m, melt rates of 8–10 m per a occur. The melt rates decrease with increasing altitude, becoming 3.5–6.0 m per a at ~1250 m, which is near the snowline. Most estimates of total glacier melt in South Alaska, as reported in the literature (O'Neel and Pfeffer, 2005), are based to varying degrees on the summer balances measured at the Wolverine Glacier (USGS, 2007). Many of the glaciers in this region extend to near sea level, and thus using the Wolverine Glacier measurements in conjunction with these measurements will provide a better estimate of the total fresh water discharge into the Gulf of Alaska.

The melt from the Bering Lobe yields a summer discharge of between 8 and 14 Gt of fresh water. The bulk of this water enters the Gulf of Alaska through the Seal River; however, there may be additional contributions by groundwater flow through the foreland gravel (Andrus et al., this volume). This is likely to be small because Vitus Lake is only 1–2 m above sea level. For the entire Bering Glacier System, which includes the Bering Lobe, Steller Lobe, and perennial snow fields, the summer discharge is ~35 Gt, including precipitation.

The fresh water discharge from this glaciated region is the primary driver of the Alaska Coastal Current (Royer, 1982), which strongly influences the entire marine ecosystem (Royer and Grosch, 2006). Current estimates of the annual fresh water discharge into the Gulf of Alaska are approximately, 725 Gt per year, and this study shows that the Bering System contributes 5%. Increased glacial melt in this region is likely to accelerate the ACC, possibly shifting the rich and diverse marine habitat that supports economically important salmon fisheries and marine mammal populations east and north by as much as 300 km (Royer and Grosch, 2006). Arendt et al. (2002) show that the increased glacier melt has resulted in the annual loss of 100 Gt of ice, which, when compared to our estimates of melt derived discharge from the Bering Lobe alone, ranges from 6 to 13 Gt, or ~10% of the total. We suggest that because of its large contribution to the fresh water budget of the Gulf of Alaska that continued monitoring of Bering Glacier could be used to index future changes.

On a local scale, Vitus Lake, which is the largest pro-glacial lake in the world, is undergoing significant expansion as the glacier retreats. As shown by Josberger et al. (this volume), the injection of glacial meltwater into Vitus Lake controls the hydrography of the lake. The lake is currently oligotrophic, but the present aquatic ecosystem will evolve as the lake ages and the glacier continues to retreat. Indeed, the Vitus Lake harbor seal population has been growing over the past decade and is currently at ~1000 animals (Saverese and Burns, this volume). The increase in seal population is probably related to expanding fish stocks, including salmonids, in Vitus Lake.

ACKNOWLEDGMENTS

Partial support for this program was provided to the U.S. Geological Survey and Michigan Tech Research Institute (MTRI) by the U.S. Bureau of Land Management under grant no. LAA-01-0018. The authors are grateful to Chris Noyles of the BLM for assisting in the installation and removal of the ablation sensors as well as servicing the sensors during each field season. The authors also would like to thank Liza Jenkins and Luke Spaete of MTRI for field assistance in generating the figures and tables used in this chapter, as well as for making editorial suggestions.

REFERENCES CITED

Alaska Climate Research Center, 2007, Climatological Data, Monthly Time Series, from http://climate.gi.alaska.edu/Climate/Location/TimeSeries/.

Arendt, A.A., Echelmeyer, K.A., Harrison, W.D., Lingle, C.S., and Valentine, V.B., 2002, Rapid wastage of Alaskan glaciers and their contribution to rising sea level: Science, v. 297, doi: 10.1126/science.1072497.

Dyurgerov, M.B., and Meier, M.F., 1997, Year-to-year fluctuations of global mass balance of small glaciers and their contribution to sea-level changes: Arctic, Antarctic, and Alpine Research, v. 29, p. 392–402.

Fleisher, J., Muller, E.H., Cadwell, D.H., Rosenfield, C.L., Bailey, P.K., Pelton, J.M., and Puglisi, M.A., 1995, The surging advance of Bering Glacier, Alaska, U.S.A.: A progress report: Journal of Glaciology, v. 41, p. 207–213.

Fleisher, J., Lachniet, M., Muller, E., and Bailey, P., 2006, Subglacial deformation of trees within overridden foreland strata, Bering Glacier, Alaska: Geomorphology, v. 75, p. 201–211.

Josberger, E.G., Shuchman, R.A., Meadows, G.A., Savage, S., and Payne, J., 2006, Hydrography and circulation of ice-marginal lakes at Bering Glacier, Alaska, U.S.A.: Arctic, Antarctic, and Alpine Research, v. 38, p. 547–560, doi: 10.1657/1523-0430(2006)38[547:HACOIL]2.0.CO;2.

Josberger, E., Bidlake, W., March, R., and Kennedy, B., 2007, Glacier mass-balance fluctuations in the Pacific Northwest and Alaska, USA: Annals of Glaciology, v. 46, p. 291–296, doi: 10.3189/172756407782871314.

Meier, M.F., and Dyurgerov, M.B., 2002, How Alaska affects the world: Science, v. 297, p. 350–351, doi: 10.1126/science.1073591.

Molnia, B.F., and Post, A., 1995, Holocene history of Bering Glacier, Alaska: A prelude to the 1993–1995 surge: Physical Geography, v. 16, p. 87–117.

O'Neel, S., and Pfeffer, W., 2005, Geophysical Investigation of Rapid Tidewater Glacier Retreat: IRIS Newsletter, p. 1.

Royer, T.C., 1982, Coastal fresh water discharge in the Northeast Pacific: Journal of Geophysical Research, v. 87, C3, p. 2017–2021, doi: 10.1029/JC087iC03p02017.

Royer, T.C., and Grosch, C.E., 2006, Ocean warming and freshening in northern Gulf of Alaska: Geophysical Research Letters, v. 33, no 16, doi: 10.1029/2006GL026767.

U.S. Geological Survey, 2007, Benchmark Glaciers: from http://ak.water.usgs.gov/glaciology/.

Manuscript Accepted by the Society 02 June 2009

The Geological Society of America
Special Paper 462
2010

Hydrologic processes of Bering Glacier and Vitus Lake, Alaska

Edward G. Josberger*
U.S. Geological Survey, Washington Water Science Center, 934 Broadway, Suite 300, Tacoma, Washington 98402, USA

Robert A. Shuchman
Michigan Tech Research Institute (MTRI), 3600 Green Court, Suite 100, Ann Arbor, Michigan 48105, USA

Guy A. Meadows
Department of Naval Architecture and Marine Engineering, University of Michigan, Ann Arbor, Michigan 48109, USA, and College of Engineering, University of Michigan, Ann Arbor, Michigan 48109, USA

Liza K. Jenkins
Michigan Tech Research Institute (MTRI), 3600 Green Court, Suite 100, Ann Arbor, Michigan 48105, USA

Lorelle A. Meadows
Undergraduate Education, College of Engineering, University of Michigan, 1261 Lurie Engineering Center, Ann Arbor, Michigan 48109-2102, USA

ABSTRACT

Runoff from the mountains and large glaciers on the rim of the Gulf of Alaska is a critical driver for ocean circulation in the gulf and a major contributor to global sea level rise. Bering Glacier is the foremost glacier of this system, with one of the largest proglacial lake-river systems in the world, Vitus Lake, which is linked to the Gulf of Alaska by the Seal River. Vitus Lake, at sea level and >250 m deep in some locations, receives all of the runoff, rainfall, and glacial melt from the Bering Lobe, which then flows into the Gulf of Alaska in the 8-km-long Seal River. Six years of conductivity-temperature-depth (CTD) surveys in Vitus Lake show a highly stratified system with 50% diluted seawater at the bottom. The annual surveys show changes in the deep water temperature and salinity that are the result of seawater intrusions. To understand the complex interaction between lake level and area, glacier discharge, river morphology and flow, sea level fluctuations, and their associated impacts on the lacustrine ecology of Vitus Lake, we developed a hydrodynamic flow model that was calibrated using field measurements of lake level and the flow in Seal River. The model is used to analyze present conditions in Vitus Lake and shows that even with no runoff entering the lake, the distance from the Gulf of Alaska through Seal River to Vitus Lake is too great for typical tidal inflow to reach the lake. Furthermore, the model is used to understand the response of the glacier-lake system to possible future

*ejosberg@usgs.gov

Josberger, E.G., Shuchman, R.A., Meadows, G.A., Jenkins, L.K., and Meadows, L.A., 2010, Hydrologic processes of Bering Glacier and Vitus Lake, Alaska, *in* Shuchman, R.A., and Josberger, E.G., eds., Bering Glacier: Interdisciplinary Studies of Earth's Largest Temperate Surging Glacier: Geological Society of America Special Paper 462, p. 105–127, doi: 10.1130/2010.2462(06). For permission to copy, contact editing@geosociety.org.

scenarios of glacier retreat or advance and changes in runoff. Finally, properly calibrated, such a model would be able to gauge the discharge from the Bering Glacier System by measuring only the lake level.

INTRODUCTION AND HYDROLOGIC SETTING OF BERING GLACIER

The glaciers and ice fields of the Wrangell–Saint Elias mountain ranges along the southern coast of Alaska contain the greatest amount of glacier ice outside of Antarctica and Greenland. This is a result of vigorous winter storms that come ashore from the Gulf of Alaska, depositing thick winter snowpacks that nourish these ice masses. Few measurements of the winter accumulation at high altitude are available. However, Josberger et al. (2007) show that for Wolverine Glacier, a much smaller glacier on the Kenai Peninsula, the winter accumulation can range from 2 to 6 m of water equivalent (mweq) averaged over the area of the glacier. Occasionally during the winter the region receives rain, but rain events are limited to lower elevations and are generally absorbed by the snowpack.

The hydrograph from Copper River, a large glaciated drainage directly east of the Bering System, illustrates the seasonal runoff that is typical of the glaciated basins in this region (Fig. 1, top panel). During the winter months the discharge approaches zero, as the hydrologic system is frozen. Discharge begins to increase in May and reaches a broad peak in late June that extends through August. This broad peak is the result of runoff from melting glaciers that occurs all season long. This is in contrast to drainages that are dominated by the melting of the seasonal snowpack, which results in a sharp runoff spike that diminishes rapidly as the seasonal snowpack disappears.

Numerous studies (Arendt et al., 2002; Meier and Dyurgerov, 2002; Josberger et al., 2007), using a variety of techniques including conventional stake measurements, aircraft lidar, and satellite observations, show that the glaciers in this region are rapidly melting and that the melt rate is increasing. Josberger et al. (2007) show that the recent glacier wastage is occurring uniformly across the entire Northeast Pacific from Alaska to Washington. The nearly 50-year-long mass balance record from the U.S. Geological Survey (USGS) Benchmark Glacier Program shows that since 1989 the three benchmark glaciers, one in each major climatic regime, including Wolverine Glacier on the Kenai Peninsula, have lost mass at a rate of 0.8 mweq per year (USGS Benchmark Glacier Program). For all three Benchmark Glaciers, the summer melt is becoming more negative and becoming more dominant in determining their net mass balance (USGS Benchmark Glacier Program). The contribution to sea level rise from the glaciers in this region is ~0.25 mm a^{-1}, which is equal to the current estimated contribution from Greenland (Krabill et al., 2000).

Bering Glacier is the largest glacier in this region, and it is subject to a-periodic surges that rapidly advance the terminus over the course of a year or two, greatly altering the landscape. The terminus is now retreating from an extended glacial surge that ended in 1995 (Molnia and Post, this volume). An understanding of the myriad processes occurring at Bering—rapid thinning, large subglacial flow of water, and the subsequent collapse of large parts of the terminus—may provide insight into the stability and retreat of the West Antarctic Ice Sheet and the outlet glaciers of Greenland (Fricker et al., 2007). In addition, the runoff from this region into the Gulf of Alaska is an important factor in driving the circulation in the Gulf of Alaska, which supports a rich and economically important marine ecosystem. The recent accelerated glacier melting is now impacting the hydrographic structure in the Gulf of Alaska (Royer and Grosch, 2006). Water column temperatures have increased, and the vertical density stratification has increased.

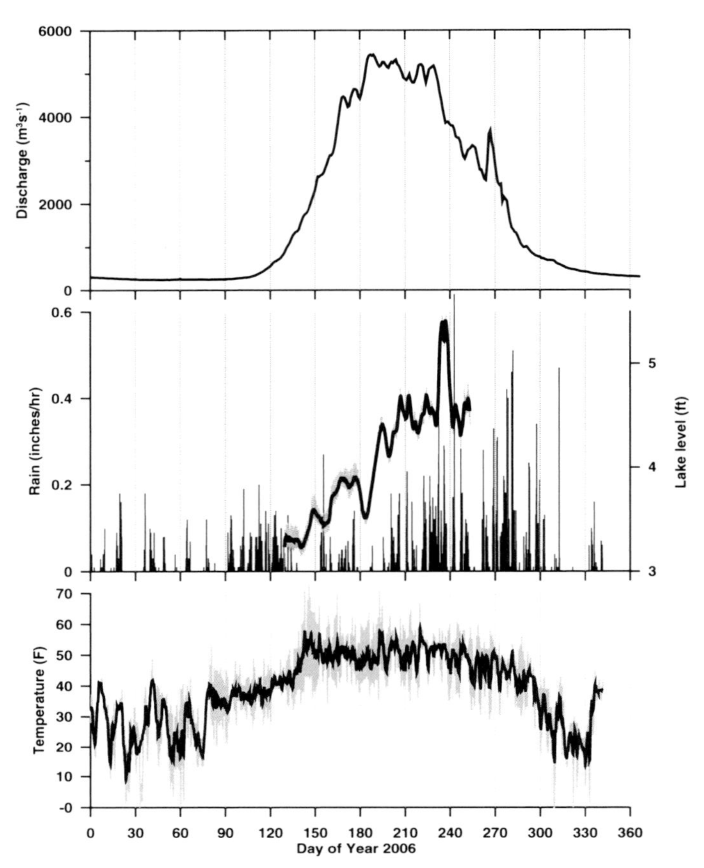

Figure 1. Top panel shows mean daily discharge of Copper River averaged over the period of record, which is 1988 to 1995. The middle panel displays the level of Vitus Lake and the hourly precipitation measured at the BLM field camp weather station for the summer of 2006. The bottom panel shows temperature measured at the BLM field camp weather station. The light gray line is the hourly temperature, and the black line is the 25 h running average temperature.

The Bering Lobe of Bering Glacier (Fig. 2) terminates at sea level in Vitus Lake, which is one of the largest proglacial lakes on Earth (Merrand and Hallet, 1996). This irregularly shaped lake, dotted with islands, measures at this time ~20 km long, following the terminus of Bering Glacier, up to 7 km wide, with water depths that exceed 250 m, well below sea level. Josberger et al. (2006 and this volume) describe its bathymetry and the unusual hydrography of the lake that results from the intrusion of seawater from the Gulf of Alaska into Vitus Lake, where it interacts with the glacial ice and runoff. During the summer, glacier melt, snow melt, and rainfall produce prodigious amounts of water on the glacier surface. Ponds of water are common, as are streams and rivers that carve valleys in the ice. The meltwater drains into crevasses and moulins to become part of the subglacial flow that eventually enters Vitus Lake and then out Seal River into the Gulf of Alaska.

Subglacial discharge may play a significant role in controlling the mass balance of tidewater glaciers, including the Greenland Ice Sheet. For LeConte Glacier, near Juneau, Motyka et al. (2003) found that submarine flow is strongly related to the ice discharge or calving into the fjord. The Greenland Ice Sheet is losing mass at an increasing rate; recent study reveals that the mass loss tripled in the last 11 a (Rignot et al., 2008). This mass loss is a combination of surface melting and accelerated flow into the ocean, which is thought to be a result of surface meltwater reaching the bed. Hence, the results of our Bering Glacier studies will help glaciologists better understand the interactions between glacial submarine runoff, seawater, and fresh water.

An additional component of the hydrology is the calving of icebergs from the terminus into Vitus Lake. Immediately after the surge the terminus consisted of an ice front >50 m high, which was subject to calving. Because of the high melt rates at the sea-level terminus region, 8 m over the course of the ablation period (Shuchman et al., this volume), the ice front has decreased in elevation and now consists of a mixture of ice beaches and lower ice cliffs, <20 m high. In places the terminus is starting to float and rapidly retreat through a process called *disarticulation* (Molnia and Post, this volume). This process starts with large arcuate cracks at the terminus that follow crevasses, and then the entire cracked region separates, producing large tabular icebergs. Lake level fluctuations may also enhance the disarticulation process by flexing the floating portion of the ice sheet. In Tashalich Arm, which is deeper than 150 m, the vertical displacement of horizontal waterline melt features (as described by Josberger and Martin, 1981) indicates isostatic uplift of the ice front. Massive calving events take place in Tashalich Arm. The entire width of the terminus suddenly deteriorates, filling the arm with icebergs and brash ice. The relationship, if any, between the uplift and calving events is not known, but it may be related to migration of the grounding line as the ice thins.

All of the water leaving the Bering drainage, be it snow melt, rainfall, surface glacier melting, submarine glacier melting, or iceberg melting, first enters Vitus Lake and then flows through the 8-km-long Seal River into the Gulf of Alaska. Hence, measurements of the level of Vitus Lake and the flow in Seal River and river flow are essential to understanding the hydrologic cycle of the Bering Glacier System as well as the exchange of water between the Gulf of Alaska and Vitus Lake. This chapter first updates the time series of hydrographic and water quality measurements given in Josberger et al. (2006 and this volume). We derive a model for the flow in Seal River and the lake level that includes the tides in the Gulf of Alaska and the flux of water from the Bering Glacier System. The model is calibrated with the field measurements and is used to understand the factors that cause seawater to enter the lake and the potential changes in the glacier lake river system under a variety of future scenarios. Such information is needed by the U.S. Bureau of Land Management (BLM) in formulating land use policy. Finally, we show how the model could be used to "gauge" the discharge of the entire Bering Glacier System, simply by measuring the lake level through the year.

OBSERVATIONS

The observations used in this chapter are from two sources, in situ measurements on the Bering Glacier, in Vitus Lake, and in Seal River, and a suite of remote sensing observations. All of these observations are described in detail in their respective chapters. This section describes how these observations are intertwined in the glacio-hydrology of the Bering Glacier System.

Glacier Ablation

As described by Shuchman et al. (this volume), the total summer ablation in the terminus region ranges from 6 to 10 m up to altitudes ~1000 m and decreases to 3.5 m in the Bagley Ice Field. Throughout the summer the ablation remains remarkably uniform from day-to-day, independent of weather conditions. For 2004, using the in situ ablation measurements, the glacier- and snow-covered area from satellite observations, and a digital elevation model (DEM) in a geographic information system (GIS), Shuchman et al. (this volume) estimate the average summer meltwater flux to range from 897 to 1273 $m^3 s^{-1}$. Using rainfall measurements from the BLM field camp we estimate that contribution to be 180 $m^3 s^{-1}$, for a total of 1100–1450 $m^3 s^{-1}$. Our Acoustic Doppler Current Profiler (ADCP) discharge measurements, adjusted for changes in lake level, give average flows of 1550 $m^3 s^{-1}$ for 2003, and 2250 $m^3 s^{-1}$ for 2004. Measured in early August, these are near peak flows, which are greater than the mean flow. Hence, our technique for estimating discharge provides reasonable values, and application to other glaciers in this region may provide more accurate runoff estimates.

Vitus Lake

Since originally reported by Josberger et al. (2006), Vitus Lake continues to expand as the terminus retreats from its extended surge position of 1995. Shuchman et al. (this volume) use a combination of Landsat and satellite synthetic aperture radar (SAR) images to map the temporal and spatial evolution of the terminus (Fig. 2). To summarize the retreat, in 1995 Vitus Lake had an area of 58.4 km^2, which grew to 130.8 km^2 in 2006. The annual average retreat rate of the terminus has been 0.4 km a^{-1}.

Similarly, the volume of Vitus Lake has increased nearly fourfold from 2.6 km^3 to 9.4 km^3, from 1995 to 2006. We computed lake volumes using the results of a bathymetric survey from 2003 to 2004, as described by Josberger et al. (2006), and computed lake extent from the satellite observations in a GIS. For years after 2004, when there were no water depth measurements from previously glacier-covered areas, we extrapolated the measured water depths to the terminus position; subsequent measurements in 2007 confirmed this assumption. We recommend periodic surveys of the lake, which not only will chart recently uncovered lake bottom but also will provide estimates of sedimentation rates.

Terminus Retreat

The position of the glacier terminus depends on the calving rate, the difference between the rate of glacier advance and the rate of ice fracturing from the terminus. For Bering Glacier, the calving rate varies both temporally and spatially along the ice front. Figure 3 gives the observed calving rates for 1999–2005 along the transects shown in the figure. We used the fresh water calving law from Warren and Kirkbride (2003) to compute the expected calving rates, which increase linearly with depth. For depth information we used the bathymetric survey described in Josberger et al. (2006 and this volume), which was extrapolated to the glacier terminus for areas that were ice covered during the original survey. In Tashalich Arm, which is the deepest part of the lake, we estimate the water depth to be at least 250 m.

From west to east, the water depth along the terminus generally decreases from >250 m to <20 m. Likewise, the expected calving rate also decreases from ~600 m a^{-1} along the western edge to 100 m a^{-1} along the shallower eastern edge. For comparison, Figure 3 also shows the annual changes in terminus position that occurred along each transect shown in Figure 2, as determined from satellite imagery. Because the ice in the Piedmont Lobe is nearly stagnant (Shuchman et al., this volume), the annual change in terminus position can be considered as the calving rate. The predicted calving rates are in approximate agreement for Tashalich Arm in the west and also for the easternmost transect 15; elsewhere there is considerable scatter (Fig. 3). The geometry of Tashalich Arm is more like the geometry of the calving glaciers used to derive the calving law, which flow in narrow, constrained fjords. The remaining transects are located along the Piedmont Lobe of the terminus, where disarticulation commonly takes place and dominates the calving rate. This process is not included in the calving law of Warren and Kirkbride (2003).

The annual terminus changes show increasing scatter for transects 5–14 (Fig. 3), which coincides with the lake becoming shallower. Transects 5–8, in deep water, all show terminus changes generally less than the expected calving rate, whereas transects 9–14 have annual terminus changes greater than the expected calving rate. For the latter case, the water depths are the shallowest, and the ice is the thinnest. These factors, combined with the very high surface ablation rate in the terminus region, make the glacier highly susceptible to disarticulation. Molnia and Post (this volume), describe this process as the disintegration of a large part of the terminus that results from the glacier approaching flotation thickness as a result of surface melting. The fracturing usually occurs along old crevasse lines. It is unclear why annual terminus changes in the region defined by transects 5–8 show less scatter and generally less than expected calving rates. Perhaps the thicker ice in this region has not yet approached flotation thickness, and it is more resistant to fracturing.

Vitus Lake Hydrography

CTD Observations

For the period 2001 through 2007 we measured the vertical temperature and salinity distributions with a Seabird CTD device at 12 sites in Vitus Lake. Figure 4 shows the CTD profiles from McMurdo Sound, the deepest portion of the main body of Vitus Lake. The basic structure consists of four layers, as described by Josberger et al. (2006 and this volume), with the uppermost layers exhibiting variability resulting from precipitation, wind effects, and surface runoff. Three more years of data, however, show significant changes to the deep layers, which are the result of ice melting in Vitus Lake, meltwater from the Bering basin flowing into the lake with most flowing subglacially and entering at the bottom of the lake, and the input of relatively warm salty water from the Gulf of Alaska via the Seal River and groundwater seepage (Andrus et al., this volume).

Table 1 describes the annual conditions observed in the CTD record, and the following enumeration describes the observed deep temperature and salinity conditions and discusses their significance. For 2001, we were unable to sample the deep water in McMurdo Sound because the glacier covered this region.

- For 2001, 2002, and 2003, the deep water was both isothermal and isohaline; 2002 was the most saline year by as much as 2 practical salinity units (psu). The temperature varied by 0.2 °C.
- In 2004, previously observed vertically uniform conditions were replaced by a nearly linear trend. The bottom salinity and temperature were warmer than the year before. The only source for warmer, more saline water is the Gulf of Alaska.
- In 2005, vertically uniform conditions returned to depths below 100 m, with both a decrease in temperature and salinity when compared with 2004. This suggests little or no new Gulf of Alaska water entering the lake. The vertically uniform conditions probably result from mixing, caused by fresh water flowing out from beneath the glacier and rising in a buoyant plume adjacent to the flakier terminus (Josberger and Martin, 1981).
- In 2006, the deepest 30 m became more saline and warmed to the greatest values observed, 18.77 psu and 3.11 °C, indicating a significant influx of water from the Gulf of Alaska.

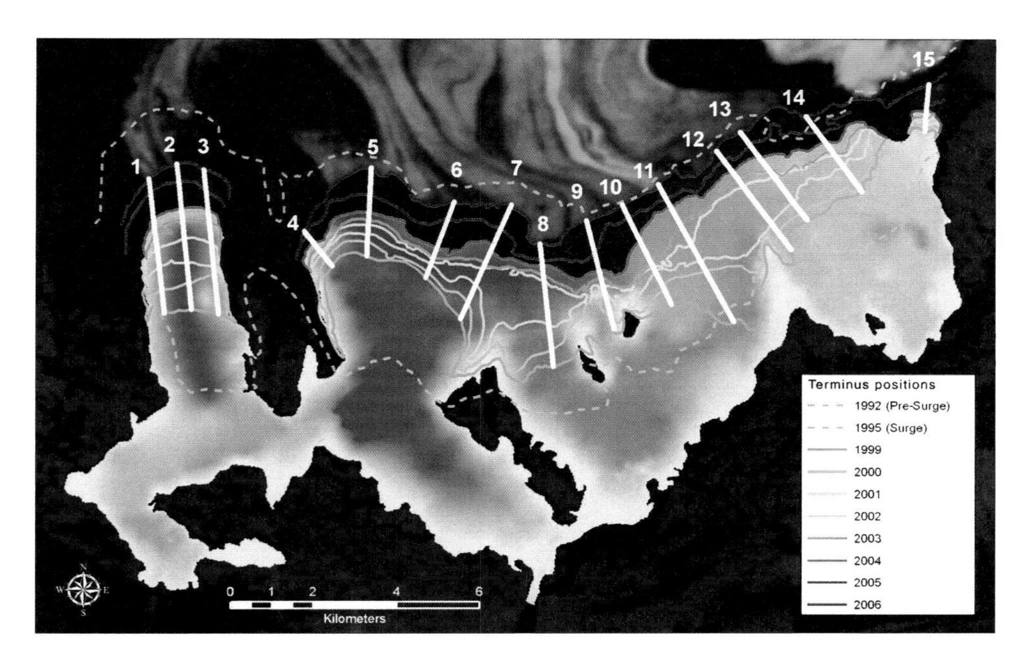

Figure 2. Vitus Lake has increased rapidly in area and volume as the glacier terminus has retreated. This figure shows the terminus positions of the Bering Lobe from 1999 through 2006 and the transect lines along which the changes in terminus position were measured. Pre-surge and surge margins are denoted by the dashed pink and blue lines, respectively.

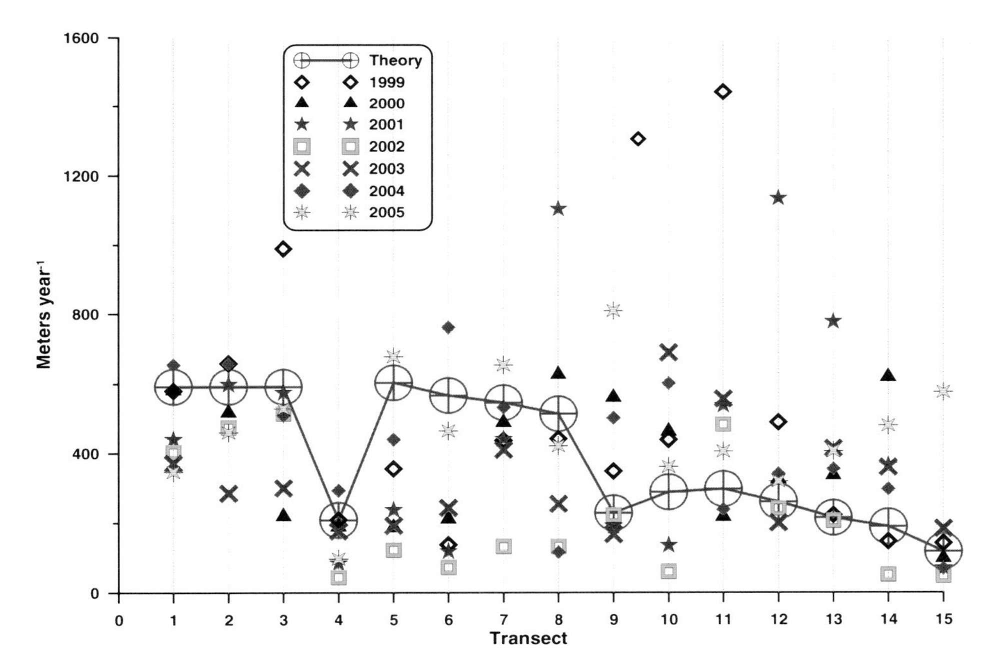

Figure 3. This figure shows annual retreat from 1999 to 2005 of the ice edge at each of the transect lines shown in Figure 2. The blue line indicates the theoretical calving rates from the law of Warren and Kirkbride (2003). Note that the model and observations are in better agreement for the transects in deeper water (transects 1–7).

- Observations for 2007 when compared with 2006 (Fig. 5) show that Vitus Lake was more saline at almost all depths, colder to 80 m, and warmer below 90 m. Again, this was the result of inflow from the Gulf of Alaska. Cooling in the upper layers is the result of upward buoyancy-driven convection, discussed previously.

Table 2 gives the annual change in salt content for the entire lake, the upper layer (1–40 m), and the lower layer (41–125 m). The values were derived by numerically integrating the hypsometric distribution of lake depth at 1 m intervals from our bathymetric data, with the CTD observations for the appropriate year. For 2001, the CTD sampled only the upper 80 m, and we assumed that the salinity was constant from 80 m to 125 m. Because salt is a conservative quantity, with the only source being water in the Gulf of Alaska, increases in total salt content had to be the result of inflow of seawater in the gulf, which has a nominal salinity

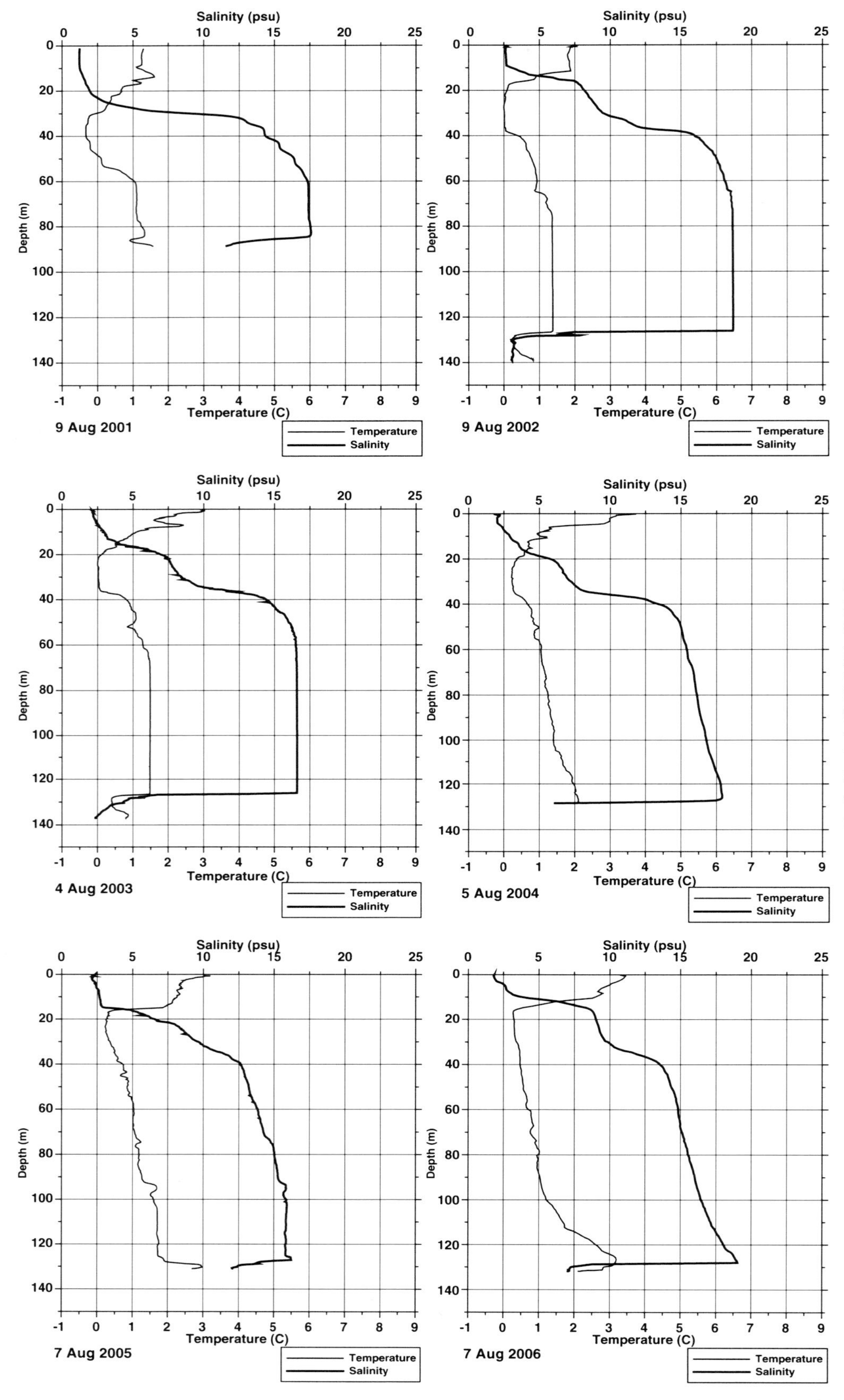

Figure 4. CTD observations from McMurdo Sound, the deepest portion of Vitus Lake, are shown in this figure. One plot is presented from each year from 2001 to 2006. Note that the temperature and salinity annual variations in the near-surface record are the result of wind, precipitation, and surface runoff. The deep record shows significant annual changes resulting from subglacial discharge and potential salt water intrusion from the Gulf of Alaska.

TABLE 1. YEARLY SUMMARY OF DEEP WATER PROPERTIES IN MCMURDO SOUND FROM 2001 TO 2007

Year	Depth range	Salinity (psu)	Temperature (°C)
2001	60–80 m	Isohaline at 17.4	Isothermal at 1.59
2002	70–125 m	Isohaline at 18.67	Isothermal at 1.37
2003	60–125 m	Isohaline at 16.60	Isothermal at 1.48
2004	50–125 m	Trend: 15.02–17.90	Trend: 0.98–2.07
2005	50–100 m	Trend: 13.20–15.90	Trend: 0.84–1.61
	100–125 m	Isohaline at 15.90	Isothermal at 1.61
2006	50–100 m	Trend: 14.50–16.45	Trend: 0.57–1.21
	100–125 m	Trend: 16.45–18.77	Trend: 1.21–3.11
2007	40–125 m	Trend: 15–20	Trend: 0.3

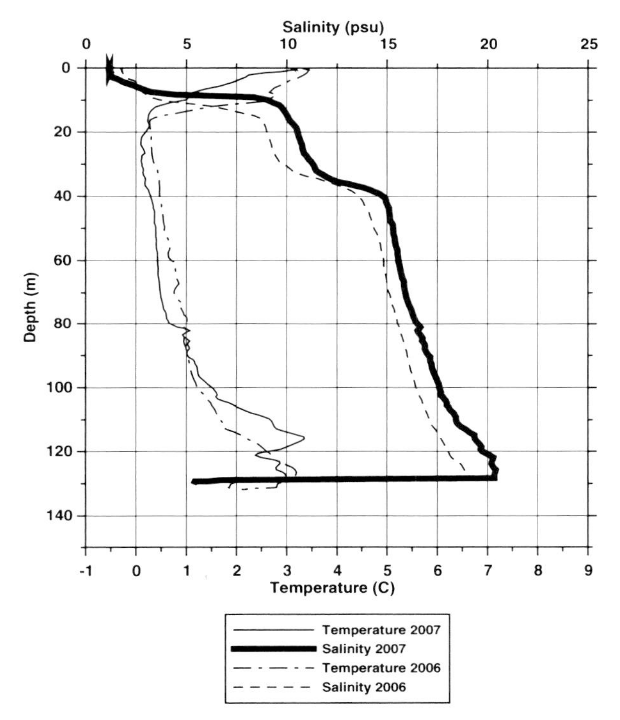

Figure 5. CTD observations from 2007 (heavy lines) compared with the CTD observations from 2006 (thin dashed lines). This figure shows that Vitus Lake was more saline at all depths, colder to 80 m and warmer below 90 m, again the result of inflow from the Gulf of Alaska.

TABLE 2. DIFFERENCE IN AMOUNT OF SALT IN VITUS LAKE FOR ALL DEPTHS, UPPER LAYER, AND DEEP LAYER

	All depths (psu km^3)	Upper 1 to 40 m	Deep 41 to 125 m
2001 to 2002	12.7	3.6	9.1
2002 to 2003	–8.1	–5.7	–2.3
2003 to 2004	–3.0	–1.5	–1.5
2004 to 2005	–1.9	–4.5	2.7
2005 t0 2006	9.3	3.3	6.0
2006 to 2007	8.5	3.1	5.4

of 30 psu. Reduced salinity results from dilution by runoff from the entire basin and, to a lesser extent, ice melting in Vitus Lake. Table 2 shows that the largest influx took place in the winter of 2001–2002, which was followed by a period of decreasing salinity that ended in 2004. Since 2005 the lake has continued to increase its salt content.

Tashalich Arm

CTD observations from Tashalich Arm show strikingly different conditions from those found in the main part of Vitus Lake. This is a result of a sill at ~60 m depth that isolates the deep water in Tashalich Arm. Figure 6 shows the observed water properties for 2004 and 2006. Observations were not made in other years because icebergs and brash ice prevented us from entering the arm. For both 2004 and 2006 the temperature profiles were nearly identical, with relatively warm water in the upper 5 m, ~7 °C that rapidly cooled to just above 0 °C at 15 m. The temperature remains constant to a depth of 60 m, the sill depth, where it decreases to just below 0 °C, and remains at this value to a depth of 145 m, the length of our CTD rope. The salinity structure for both years is similar, but with significant differences. Both years exhibit a surface layer of 2 psu water, below which is a layer of uniformly salty water from ~15 m to 60 m, but this layer was 10 m thicker and 2 psu saltier in 2006. The salinity remained nearly constant at 2 psu below 60 m for both years.

Water Quality Measurements

At the end of the Bering Glacier surge in 1995, Vitus Lake had an area of ~60 km^2 with an estimated volume of 2.6 km^3. By 2006 the area had increased to 130 km^2, and the volume had increased to 9.4 km^3 (Shuchman et al., this volume). This expansion has resulted in a rapidly evolving lake ecosystem that we have documented through a set of water chemistry baseline measurements starting in 2004 and continuing through 2007.

Vitus Lake is very cold and deep, and although it does not contain significant amounts of phytoplankton or zooplankton, it does support significant populations of migrating waterfowl and harbor seals *(Phoca vitulina richardsii)* (Savarese and Burns, this volume). Additionally, fish have been observed throughout the lake as well as a small but significant population of migrating salmon in the Seal River. Vitus Lake does sustain a band of primary productivity in the nearshore area that extends ~5 m from shore (Auer, this volume).

The biology and chemistry of fresh water environments are highly interrelated, and lake water chemistry typically drives the type and abundance of biological organisms present. We have observed the higher-level organisms noted above, but the apparent lack of lower-food-web organisms is perplexing. For this reason we have been monitoring the chemistry of Vitus Lake to help explain food web dynamics, and to document the early colonization of an oligotrophic lake, which is characterized by low primary productivity (the result of low nutrient content) and subsequently low biological activity.

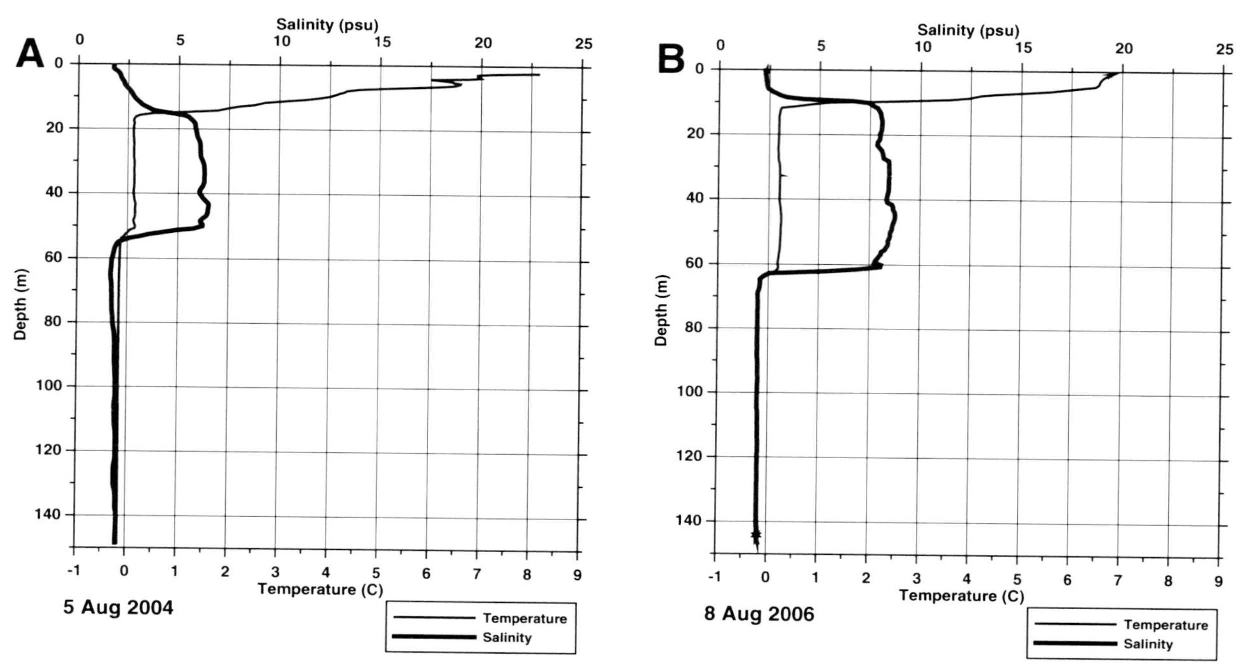

Figure 6. Shown in this figure are CTD casts for Tashalich Arm for 2004 and 2006. Note that the temperature profiles are nearly identical, whereas the salinity profiles for 2006 are thicker and saltier by ~10 m and 2 psu.

We measured an ensemble of water quality parameters during the 2004, 2005, 2006, and 2007 summer field seasons, typically during the end of the July–early August time frame. Profiles data were collected using a Horiba U-22 XD multiparameter probe that measures pH, dissolved oxygen (DO), conductivity, total dissolved solids (TDS), temperature, and turbidity. At the sites shown in Figure 7, we measured water quality properties at the surface, 1 m below the surface, 5 m below the surface, 10 m below the surface, and then every 10 m to the bottom of the lake or until we reached the end of the 100 m cable. We used handheld global positioning system (GPS) receivers to navigate to the established water-quality-sampling sites. Fifteen sites were sampled during the 2004, 2005, 2006, and 2007 summer field seasons.

The Horiba water quality measurements as a function of depth help us to address the variability within Vitus Lake as well as interannual variations. We found that most of the water quality parameters have similar vertical distributions from year to year at each site, with the exception of the high turbidity sites. Figure 8 shows representative vertical profiles for the three years of observations. The profiles from site 7 show that pH is relatively constant with depth, and that DO values are higher at the surface, abruptly decrease at the thermocline, and remain constant at depth; temperature, conductivity, and TDS are consistent with the CTD observations and explanations discussed earlier. The values throughout the profiles indicate water quality that is capable of supporting aquatic life.

The Appendix to this chapter presents the vertical profiles for all sites. These profiles indicate interannual stability in the water quality of Vitus Lake, despite the rapidly changing physical size of the lake. As also seen with the CTD measurements, the temperature profiles show a general pattern of relatively warm surface temperatures that decrease with depth to a minimum at ~20 m. Below this depth the temperatures slowly increase. The warmest surface temperatures, reaching 8 °C, were observed in the western part of Vitus Lake, and they are likely the result of solar heating and limited exchange with the rest of the lake.

To begin to address lower-food-web dynamics within Vitus Lake, we deployed the robot buoy ALWAS (Automated Lagrangian Water Assessment System) in the summers of 2006 and 2007. ALWAS is a relatively inexpensive, free-floating, sail-powered or electrically propelled, water-quality-measuring and watershed-evaluation buoy. It is capable of measuring up to 12 water quality parameters plus the complete GPS position and navigation suite as rapidly as every 40 s. Data are transmitted for real-time viewing and are stored for future retrieval and analysis. The stored data are easily downloaded into geographic database (ESRI shapefile) and spreadsheet formats.

For Vitus Lake, the ALWAS buoy was configured to measure and store on board depth, temperature, conductivity, salinity, total dissolved solids, dissolved oxygen, pH, oxidation-reduction potential, turbidity, chlorophyll-a, nitrates, ammonium, and chlorides. In addition, the ALWAS buoys also provided the full GPS position and navigation suite that includes course and speed over ground.

In 2006 and 2007, we deployed ALWAS concurrently with the CTD and Horiba data collections. Figure 9 presents the pH values obtained by ALWAS during 7–9 August 2006. Note that the pH displays a definite spatial trend across Vitus Lake, with higher observed values in the eastern part of the lake and lower values

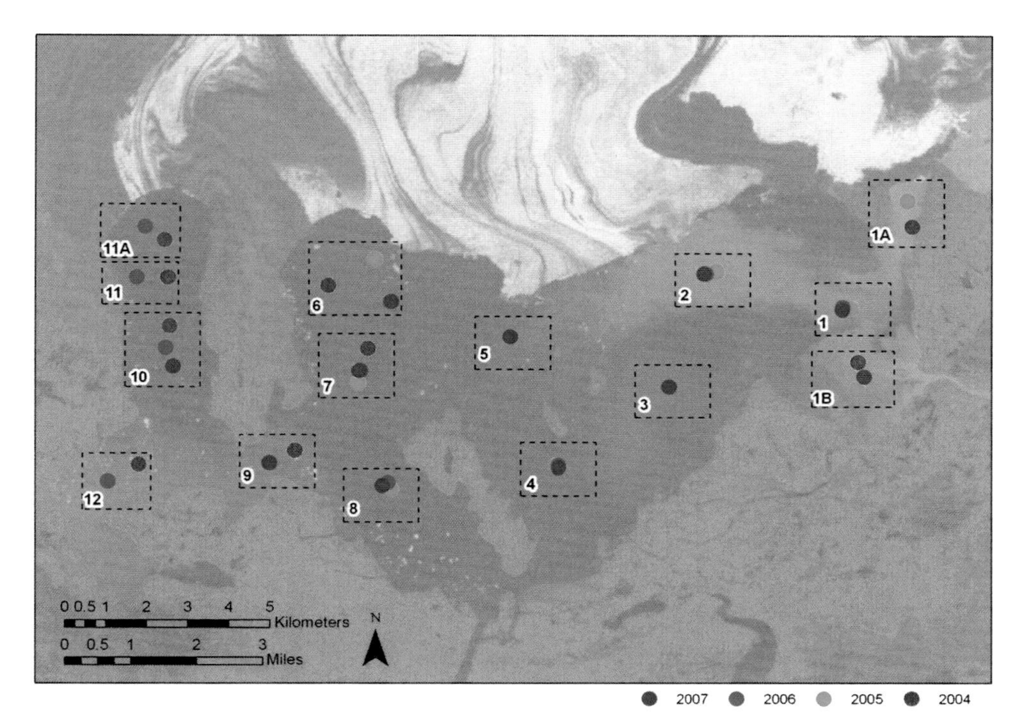

Figure 7. Multiparameter water-quality profile (Horiba) sampling sites for 2004–2007 are shown in this figure. The sites are the same as the CTD sites and are evenly distributed across Vitus Lake.

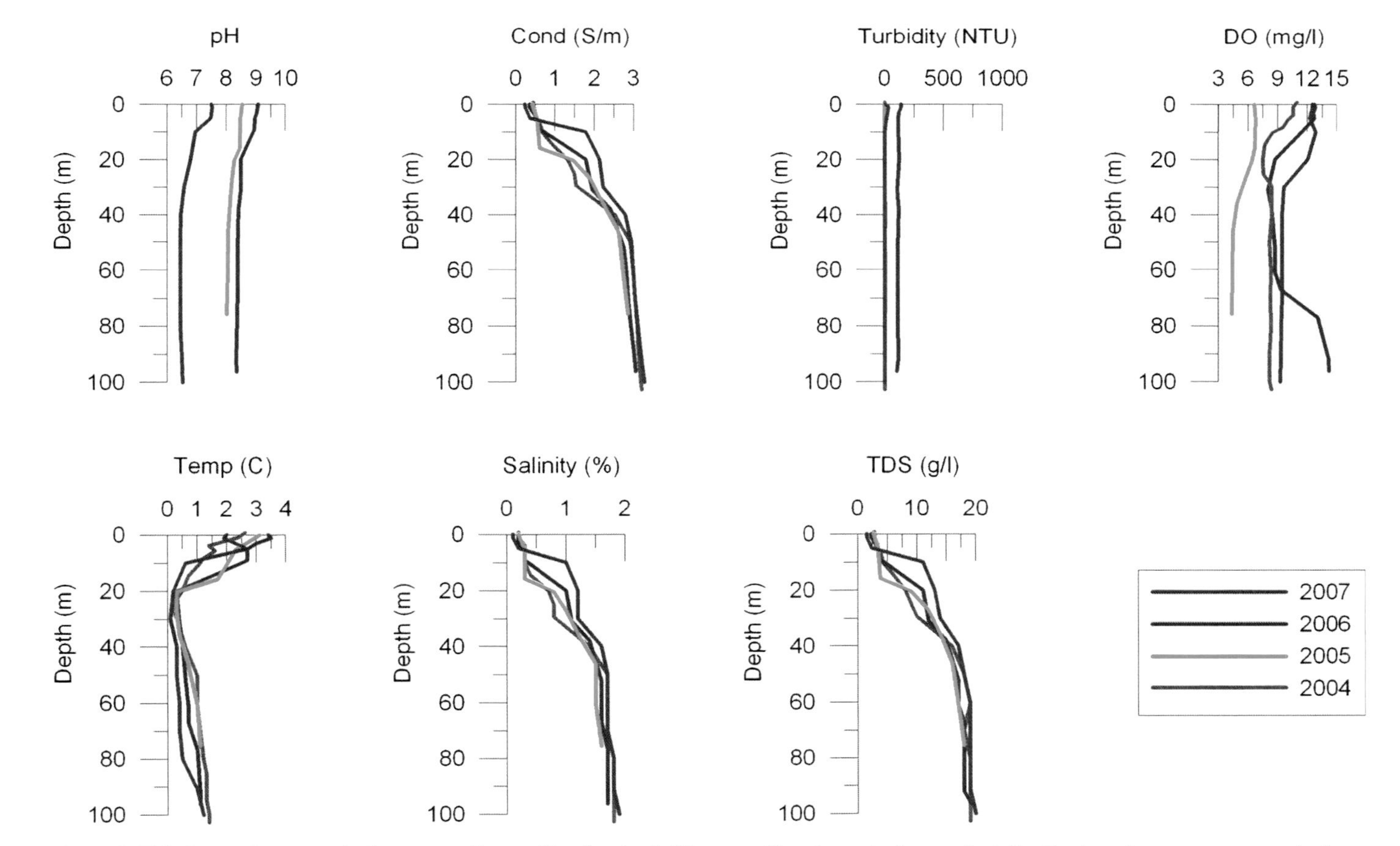

Figure 8. This figure shows vertical water quality profiles for site 7. These profiles show similar vertical distributions from year-to-year, indicating interannual stability. The profiles for the other sampling sites can be found in the Appendix to this chapter.

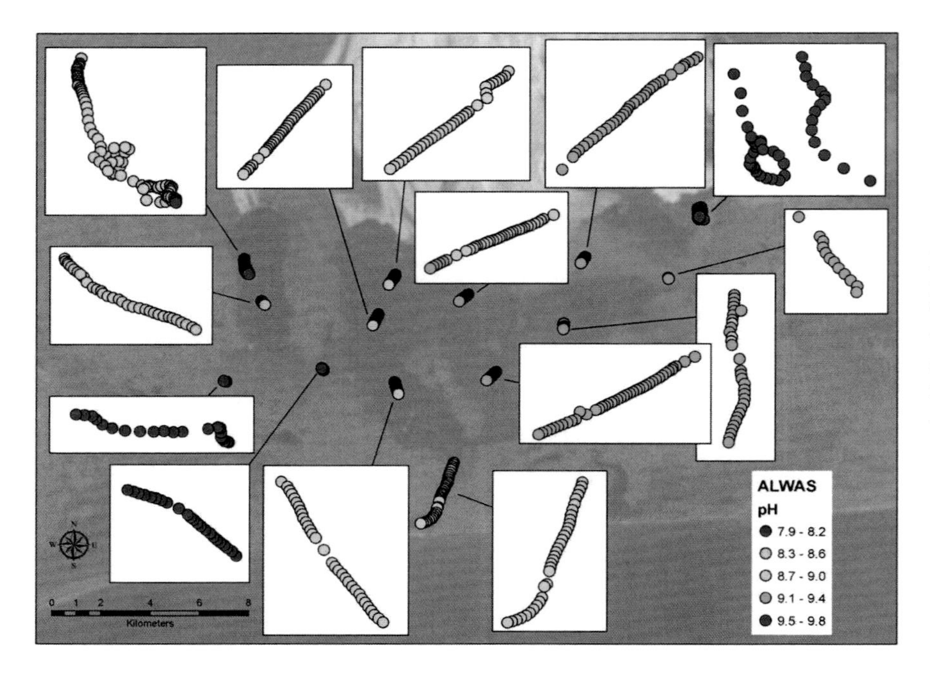

Figure 9. Representative ALWAS map, displaying surface pH values from 2006 for Vitus Lake. This map shows a strong east-west gradient in pH values, most likely resulting from subglacial discharge. The ALWAS statistics for 2006 and 2007 for the measured parameters can be found in the Appendix to this chapter.

in the western part. This east-west gradient was also observed for other water quality parameters besides pH, such as temperature, and to a lesser degree conductivity and turbidity (see Appendix to this chapter for the complete set of ALWAS and Horiba measurements). These gradients are probably driven by glacial water discharge, which is greatest in the eastern part of Vitus Lake. It is important to note that the water chemistry is further confounded by variable wind-driven circulation patterns that affect the surface water quality observations. The observed east-west water quality gradients probably affect biological distribution and abundance within Vitus Lake. Further biological research is needed to determine the extent and strength of these effects.

Lake Level Measurements

The level of Vitus Lake is a key element in the hydrology of the Bering Glacier System as it responds to the tides in the Gulf of Alaska, and the flux of water entering the lake from the entire Bering System. Lake level fluctuations may enhance the disarticulation process by flexing the floating ice sheet. We measured the water level at Seal Rock on the north side of Seal River ~1 km from the northern end of the river and in the bay in front of the BLM field camp at the eastern end of Vitus Lake. We used an OTT water level gauge with an accuracy of 6 mm, and recorded lake levels either once every 15 min or, to extend battery life, once every hour. The measurements took place during 2003, 2004, 2005, 2006, and 2007. For 2003 the measurements spanned 4–10 August, and for the other years measurements began in mid-May or early June and continued through September, when the instruments were recovered.

Figure 1, middle panel, shows the lake level fluctuations as measured in front of the BLM field camp from early May through mid-September 2006. The 25 h running mean average of lake level observations shows a general rise in lake level through the melt season. Also shown in Figure 1 are the meteorological observations from a weather station adjacent to the BLM field camp airstrip. The spikes in rainfall at the end of August were the result of the remnants of a typhoon. During this time the weather station at the BLM field camp measured 4.9 in. of rain from 18 to 21 August, and the Cordova airport measured 7.5 in. of rain over the same period. The lake level record shows that Vitus Lake rises through the summer by ~3 ft (~1 m). The higher frequency fluctuations are the result of rain events, modulated by the complex subglacial hydrology. The highest frequency fluctuations are the result of ocean tides, which are attenuated and phase shifted when compared with the predicted tides at Yakutat. Josberger et al. (2006 and this volume) saw that the tidal signal at Seal Rock is attenuated by 79% and by 93% at the BLM field camp, and the lag at each location is ~100 min and 240 min, respectively. Merrand and Hallet (1996) found similar tidal signals at Seal Rock: peak-to-peak amplitudes of 0.6 m. The observations at the BLM field camp are more indicative of the lake response than those at Seal Rock, because the latter is 1 km from Vitus Lake, and an additional 2.5 km of shallow water on the west side of Long Island is densely packed with grounded icebergs that restrict the flow of water out of the lake. Hence the lake level measurements at the BLM field camp are more representative of the entire lake, and we use them in the calibration of a model for the exchange of water between the Gulf of Alaska and Vitus Lake.

Discharge Measurements

The flow in Seal River is the total amount of liquid water leaving the Bering Lobe, including snowmelt, glacier melt, and precipitation events, but not including evaporation and groundwater. Hence, this flow is a key parameter in constructing a hydrologic balance for the system. The integrated flow was measured with an ADCP mounted on a survey vessel performing multiple transects across the channel. An ADCP determines the water velocity at various depths by measuring the Doppler shift of the returned acoustic signal. An onboard computer manages the data and integrates the velocity distribution across the channel to give the total discharge.

We measured the discharge on 4, 5, and 7 August 2003 and 3 and 4 August 2004 on both a rising and falling tide in the Gulf of Alaska. A total of 24 transects were made in 2003, and 21 transects were made in 2004. The measurements were taken ~1 km south of the lake end of Seal River, across a straight reach. Seal Rock, a large boulder on the west side of Seal River, provided a convenient navigational landmark and a site for water level measurements. The results show that the tide has a strong influence on the short term discharge, whereas the average discharge depends on weather conditions.

The flow in Seal River is strongly influenced by the ocean tides in the Gulf of Alaska. Figure 10 shows two cross channel velocity distributions as measured by the ADCP on 4 August

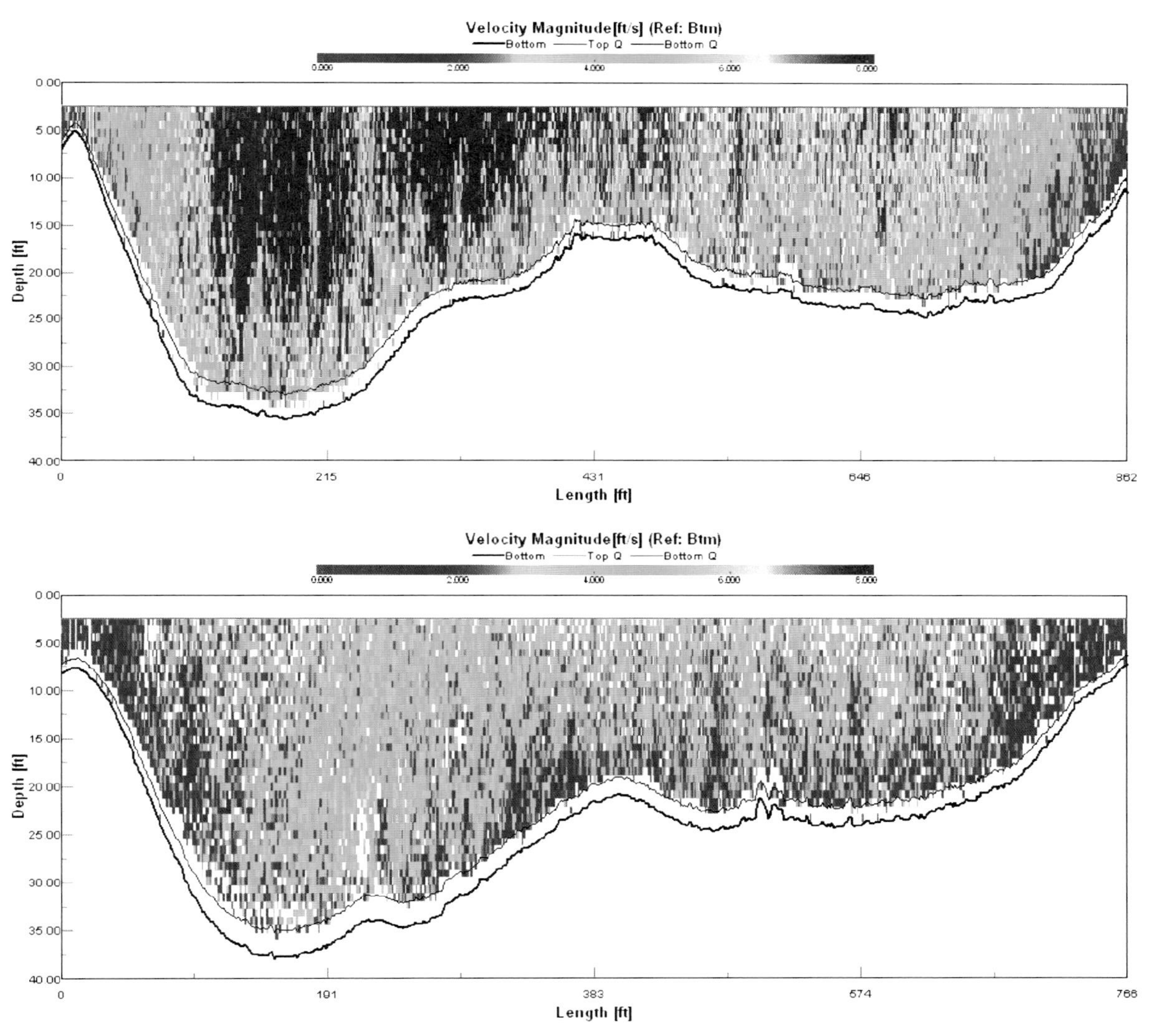

Figure 10. Acoustic Doppler Current Meter (ADCP) velocity cross sections of Seal River, at mid-tide and high-tide in the Gulf of Alaska. The upper plot shows data for 4 August 2004 at transect 001, which was measured at 1133 ADT and which had a flow of 2710 $m^3 s^{-1}$. The lower plot shows data for 4 August 2004 at transect 016, which was measured at 1748 ADT and which had a flow of 1410 $m^3 s^{-1}$. The velocities are color-coded; red indicates speeds of near 2.67 m s^{-1}, whereas blue indicates <0.67 m s^{-1}.

2004 at 1133 Alaska Daylight Time (ADT) and 1748 ADT. The velocities are color coded, in which red indicates speeds of near 8 ft s^{-1} (2.67 m s^{-1}) and blue indicates <2 ft s^{-1} (0.67 m s^{-1}). During this time period the rising tide in the Gulf of Alaska had reduced the flow from 2710 m^3 s^{-1} to 1410 m^3 s^{-1}. At very high tides the flow in the river can halt and even reverse for short periods. Averaged over a tidal cycle, the net discharge from the Seal River was 1550 m^3 s^{-1} in 2003 and 2250 m^3 s^{-1} in 2004, which was one of the warmest summers on record (Alaska Climate Research Center). The average peak discharge from the Copper River in early August is ~5000 m^3 s^{-1}, and the area of the Copper River basin is 62,700 km^2. The Bering Glacier has a drainage area of 9000 km^2, which includes not only the glacier-covered area but also the glacier-free area that drains into Vitus Lake. The discharge measurements, made by Merrand and Hallet (1996) in Seal River during non-surge conditions, were 1550 m^3 s^{-1}.

It is impractical to make continuous measurements of discharge in Seal River, although it is arguably the single most important measurement to be made. Therefore, we seek a surrogate parameter to measure, one that can be related to discharge, which is the level of Vitus Lake relative to sea level. Josberger et al. (2006 and this volume) found a lake level–discharge relationship, based on a few days of measurements in 2003, that was converse to usual stage discharge relationships in streams and rivers: the discharge decreased as the water level rose. The influence of tides limits the application of traditional stream-gauging techniques, which relate the discharge to the level of the stream.

VITUS LAKE–GULF OF ALASKA TIDAL EXCHANGE MODEL

To understand the exchange between the Gulf of Alaska and Vitus Lake, and the relationship between lake level and fresh water influx, we developed an analytic model that allows us to examine the hydrodynamic processes that take place and the factors that govern the interactions. With the model, we can better understand the processes that generated the observed conditions, and to estimate conditions that will occur in the future under different glacier advance or retreat scenarios. The model shows how to use the level of Vitus Lake as a gauge for the fresh water flux entering the lake from the Bering Lobe.

Model Description

Consider a straight channel connecting the open ocean with an enclosed bay that also receives a fresh water influx—in this case, glacier melt and precipitation. Then, the governing hydrodynamic equations are conservation of mass and momentum:

Conservation of Mass

$$A\frac{\partial \eta}{\partial t} = Q_g + Q_s, \qquad (1)$$

where A is area of lake, η is lake level, Q_g is drainage from the Bering Glacier Basin, and Q_s is flow in Seal River. Simply stated, the rate of change of the lake level is the difference between the inflow from the Bering Glacier basin, and the outflow is the net seaward flow in Seal River. In this analysis we assume that the area of Vitus Lake and the width and depth of Seal River, all parameters that we have measured, are constant in time. We used the lake level measurements and the coincident ADCP flow measurements to calculate the amount of water entering the lake from the glacier system. We computed the volumetric rate of change in Vitus Lake from the smoothed 15-min lake-level data measured at the BLM field camp at the time of the ADCP measurements for both 2003 and 2004. The slowly varying Qg is simulated by solving the governing equations for different values of Qg.

Conservation of Momentum

Conservation of momentum relates the flow in Seal River to the tidal elevation, or forcing, at the mouth of the river. We assume a uniform channel, and that friction is proportional to water velocity. Under these assumptions the momentum equation becomes:

$$\frac{\partial u}{\partial t} + g\frac{\partial \eta}{\partial x} = ku, \qquad (2)$$

where u is the velocity, g is the acceleration of gravity, η is the surface elevation, and k is the friction coefficient. We use a linear friction law rather than a quadratic in velocity, as it greatly simplifies the analysis. Hughes and Brighton (1967) show that although friction is expressed as linear in velocity, the friction factor is inherently based on the kinetic energy of the flow and its associated head loss.

Integrating from the ocean to the lake, along the river channel gives:

$$\frac{du}{dt} + \frac{g}{L}(\eta - \eta_0) = ku, \qquad (3)$$

where L is the length of Seal River.

Substitution of the conservation of mass equation (equation 1) into the momentum equation (equation 2) yields:

$$\frac{d^2\eta}{dt^2} + k\frac{d\eta}{dt} + \Omega_0^2\eta_{ocean} + \frac{k}{A}Qg, \qquad (4)$$

where $\Omega_0^2 = \frac{gwd_s}{LA}$, w is the width of Seal River, d_s is the depth of Seal River, A is the area of Vitus Lake, and Qg is the flux of water into the lake from the Bering Lobe.

The result is the equation for a linear oscillator that is damped and forced by the tides. This equation is rich in solutions that correspond to distinct physical mechanisms that occur.

First, there is a "particular" solution that corresponds to a given Qg that is independent of time for the lake level, η_g. This solution represents the balance between the mean flow in the river and the mean pressure gradient that is the result of the lake level elevated above mean sea level:

$$\eta_g = -\frac{KQ_g L}{gwd_s}. \tag{5}$$

Note that this equation is independent of lake area, A. The mean flow only depends on the difference in elevation of the water surface, from one end of the river to the other, averaged over a tidal cycle. Equation 5 can be rearranged to show that Qg could be determined from lake level measurements from a properly calibrated model.

We then look for harmonic solutions that result from the tidal forcing:

$$\eta_{\text{ocean}} = H\cos(\omega t + \phi), \tag{6}$$

where H is the tidal amplitude, ω is the tidal frequency, and ϕ is the phase.

We seek sinusoidal solutions as follows:

$$\eta = A_0 \cos\omega t + B_0 \sin\omega t, \tag{7}$$

where A_0 and B_0 are amplitudes of the sinusoidal response of the lake height fluctuations.

Substitution of equation 7 into equation 4 (and the employment of considerable algebra and trigonometry) gives the lake level as a function of time:

$$\eta = \frac{H\Omega_0^2}{(\Omega_0^2 - \omega^2)^2 + k^2\omega^2}\left[\left(\Omega_0^2 - \omega^2\right)\cos(\omega t) + k\omega\sin(\omega t)\right]. \tag{8}$$

Because this is a linear system, we can use the sum of the tidal components to drive the model. In this case we used the four most dominant solar and lunar tidal components, which are m2, k1, s2, and n2. To compute the flow in Seal River, we substitute the temporal derivative of equation 8 into the continuity equation (equation 1) for each of the four tidal components and sum the results.

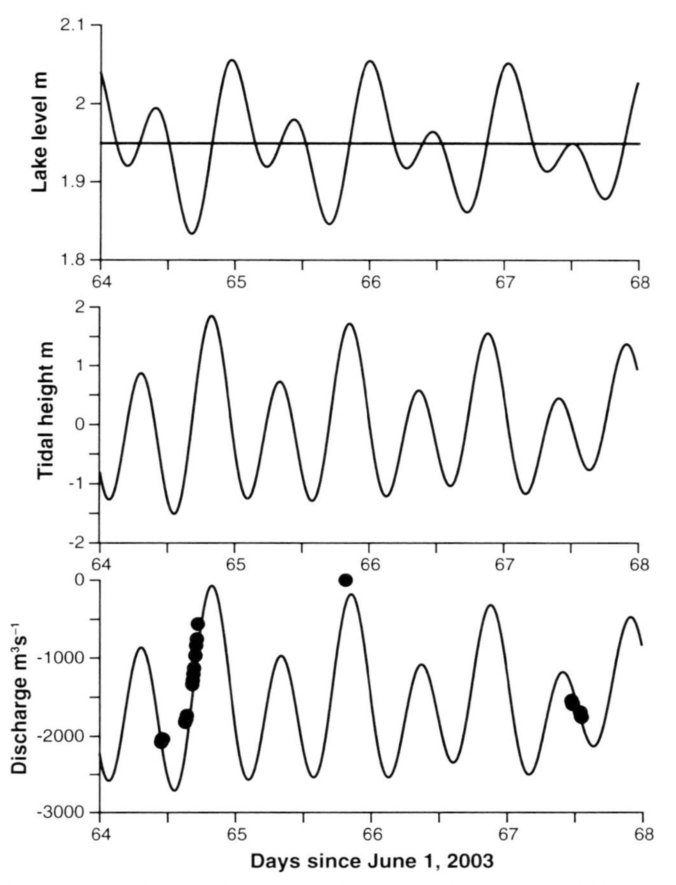

Figure 11. Results from the model, driven by the four primary tidal components, shown in the middle panel, with the friction coefficient, k, set to 0.0029 s^{-1}, and a fresh water flux from the glacier basin of 1550 $m^3 s^{-1}$. The top panel shows the lake response, the straight line is the mean upward displacement of the lake, 1.95 m, and the sinusoidal curve is the temporal response. The lower panel shows the computed discharge of Seal River, and the solid circles (●) give the measured discharge.

Model Results and Applications

To complete the model development, we need to evaluate the friction factor, k. In this mode, we must match several parameters: the lake level measurements at the BLM field camp, both the amplitude and the phase shift from the tidal forcing, and the discharge of Seal River, varying a single parameter. By trial and error we found that a value for k of 0.0029 s^{-1} provided an acceptable fit for all of the measured parameters. Figure 11 shows the results of a model simulation using the measurements of 2003, $Qg = -1550$ $m^3 s^{-1}$, and the measured lake and river geometry. The middle panel gives the tidal forcing derived from the first four tidal constituents from the U.S. National Oceanic and Atmospheric Administration (NOAA) tidal constituents at Yakutat. Mean sea level is 0.0 m. Although Yakutat is a considerable distance from Seal River, the tide rapidly propagates around the Gulf of Alaska so that there is only a small difference in phase. For example, the time difference between tides for Yakutat to the east and Wingham Island to the west is only 3 or 4 min.

In the top panel of Figure 11 the straight black line at 1.95 m is the mean lake level above mean sea level. This corresponds to the particular solution, equation 5. The sinusoid in the top panel shows the response of the lake to the tidal fluctuations. The model attenuation or ratio of the lake tidal amplitude to the open ocean tidal amplitude is 0.064, which compares well with the measured attenuation of 0.07. The model gives a phase lag for the lake of 240 min, which is in reasonable agreement with the observed lag of 210 min. In the lower panel of Figure 11 the sinusoidal line is the calculated discharge of Seal River, and the ● symbols are the values measured by the ADCP. Negative values correspond to flow out of Vitus Lake into the Gulf of Alaska. The agreement between the measured and modeled discharge is very good. Other values for k did not reproduce the observations. Therefore, given the good agreement between the modeled and measured values for a variety of independent parameters, we feel that our model realistically captures the complexity of the Bering Glacier–Vitus Lake–Gulf of Alaska system. The model thus can be used to simulate the response of the system to a variety of possible future scenarios and to determine the conditions necessary for seawater to travel up Seal River into Vitus Lake, an occurrence that has intrigued Bering investigators since the first CTD measurements were made in 2001.

The validated model not only allows us to understand current conditions as described above, but also provides a way to determine the discharge from the Bering Lobe by lake level measurements. The particular solution, equation 5, is the time independent balance between the friction of the flow in Seal River and the pressure gradient that results from the elevation of the mean lake elevation above mean sea level. For the calibration case, Figure 11, the mean lake level is ~2 m above mean tidal level in the Gulf of Alaska, the zero line in the upper panel of this figure. Hence the model relates the mean lake level to the model parameters and the amount of water entering the lake from the Bering Lobe. As such, the lake is acting as a gauge for the fresh water discharge from the entire Bering Lobe. Continuous measurements of the lake level, such as those shown in the middle panel of Figure 1, calibrated by ADCP measurements in Seal River, could be used to determine the discharge into the Gulf of Alaska. By varying the flux of water from the Bering Lobe into Vitus Lake, we found that the mean lake level increases or decreases by 1.25 m when the mean discharge increases or decreases by 1000 $m^3 s^{-1}$. In the winter, when there is little or no flow into the lake, the mean lake level would be near mean sea level, thus providing a datum to reference the lake level measurements during the melt season.

This model also allows us to examine possible future conditions for the Bering Glacier System by systematically varying the model parameters that represent different scenarios. These scenarios include winter cooling, continued climate change or increased melt and an expanding lake, changes in river length, glacier surge events, and Pacific Ocean storm surges. In the following discussion the model results are compared with current conditions, as shown in Figure 11. For each scenario we first discuss the response of the lake level and then the corresponding response by the flow in Seal River.

We used the model to investigate the impact of increasing or decreasing the flow into Vitus Lake (corresponding to climate warming or cooling) on the lake level and the flow in Seal River. Figure 12 shows the results of increasing or decreasing Qg to −2050 m^3/s or −1050 m^3/s. The middle panel shows the tidal forcing, and the solid lines in the upper and lower panels are the baseline case shown in Figure 11. The top panel of Figure 12 shows the response of the lake to these changes. For increased flow, the lake level rises (long dashed-dotted line), and for less flow the lake level lowers (dashed line). A change of 1000 $m^3 s^{-1}$ results in a mean lake level change of 1.25 m. However, the tidally driven fluctuations remain the same. The lower panel of Figure 12 shows the response of the flow in Seal River. An increase in flow increases the net flow in the river, as it must, and the tidal

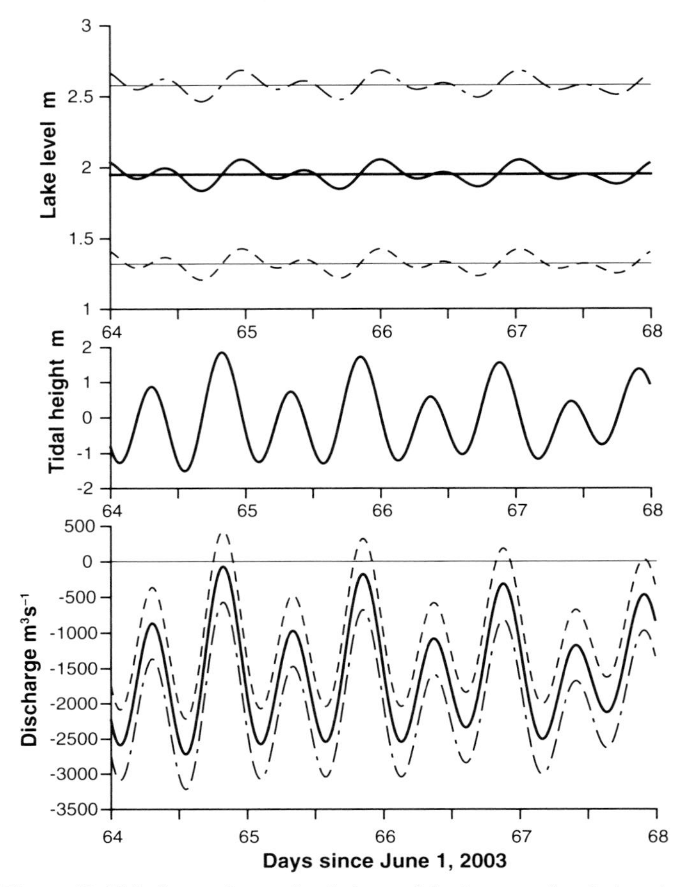

Figure 12. This figure shows simulations of the impact of variations in discharge on Vitus Lake. The middle panel is the tidal height. The heavy black lines in both the top and bottom panels are the calibration case. The dashed line and the long-short dashed line are the results from the model with the discharge *(Qg)* set to −2050 and −1050 $m^3 s^{-1}$, respectively.

perturbation is reduced. For reduced Qg, the opposite occurs, as shown by the dashed line. In fact, the flow in Seal River shows weak reversals near high tide.

To further investigate the effects of decreasing fresh water flow from the glacier and the associated increased possibility of seawater entering Vitus Lake, we ran the model with Qg varying from −1400 to 0 $m^3 s^{-1}$. For each value of Qg, we numerically integrated the flow in the river over the period of up-river velocity to determine the distance a water parcel at the seaward end of the river would travel toward the lake over a tidal cycle. The results are shown in Figure 13. The distance traveled increases as Qg becomes less negative and approaches zero. However, even for zero flow, the case for winter conditions, the distance traveled during a semidiurnal tidal cycle (6 h, 12.5 min) is only 6 km, which is 2.5 km less than the 8.5-km-long Seal River. Hence the system, in its present geometry, will not allow seawater to enter Vitus Lake under typical tidal forcing alone.

We also investigated the conditions that would result if Seal River were significantly shortened. Presently, Seal River runs

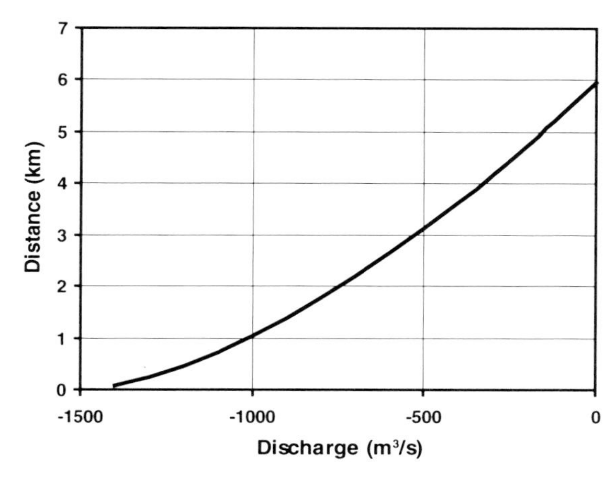

Figure 13. Distance traveled by a water parcel in Seal River during a flood, as a function of fresh water discharge from the Bering Lobe, is shown in this figure. Negative discharge values are the result of the coordinate system used by the model.

parallel to the shore behind a 5-km-long sand bar; however, maps of the area from the past century (Molnia and Post, this volume) show that the length of the longshore barrier bar is highly variable, responding to sediment supply, storms, and flow in Seal River. To simulate this occurrence we ran the model with present runoff conditions but with the river shortened by 5 km, from 8.5 km to 3.5 km. As shown in Figure 14, the resulting impacts are large; the mean lake level drops by 1.2 m, and the peak-to-peak tidal signal in the lake increases from 0.22 m to 0.53 m. The tidal fluctuations of flow in Seal River change dramatically; the peak-to-peak tidal component varies from 2000 to –4000 $m^3 s^{-1}$, which corresponds to peak velocities of 3.6 to –7.2 km hr^{-1}. Such large velocities will likely allow significant amounts of seawater to enter the lake and enhance the vertical mixing, which will change the current hydrographic structure of the lake, all with unknown consequences on the lacustrine ecosystems in the lake.

Vitus Lake is expanding, and it is likely to continue to do so until the Bering Glacier surges again or until it retreats to the point at which the grounding line for the glacier is at sea level. To understand the possible consequences in this scenario, we ran a model simulation in which the lake size was doubled from 1.08e8 m^2 to 2.16e8 m^2. The simulation, presented in Figure 15, shows that the lake level and its fluctuations remain nearly the same, but the tidal fluctuations in the Seal River would increase with considerable reversal at higher tides. Conversely, Vitus Lake could be greatly reduced in area as a result of a glacier surge (Molnia and Post, this volume). To simulate this scenario, we ran the model with the 1995 lake area. Figure 15 shows the resulting conditions. The level response to the tide remains nearly the same, but the tidal fluctuations of the flow in the river are greatly reduced. The fact that the lake level and its fluctuations remain nearly unchanged for either a much larger or smaller lake is because the model does not explicitly include the circulation in the lake. Also, the model assumes that changes in elevation at

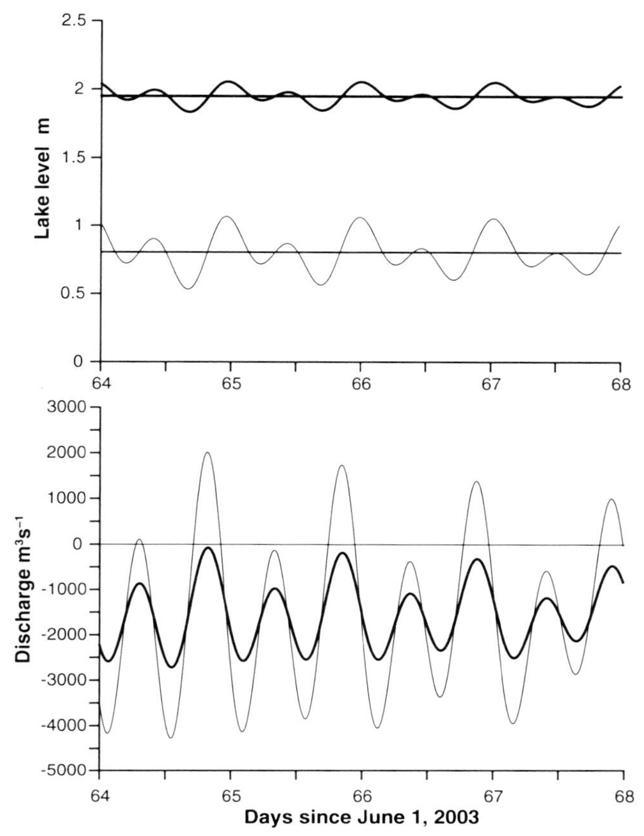

Figure 14. These plots show model results for the case in which the length of Seal River is reduced to 3.5 km, where the heavy line is the flow for the model calibration case.

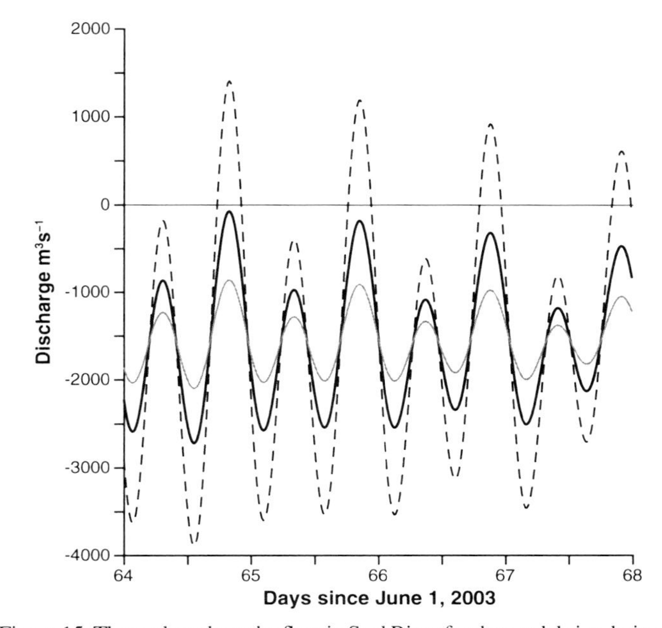

Figure 15. These plots show the flow in Seal River for the model simulations of doubling and halving the size of Vitus Lake (dashed line and thin gray line, respectively). The heavy line is the flow for the model calibration case.

the lake exit propagate rapidly throughout the lake such that the adjustment time scale of the lake is much shorter that the tidal fluctuations. The model does show large changes in the tidally driven flow in Seal River for each case, as expected.

Anecdotal reports from bush pilots in the area indicate that storm surges may raise sea level high enough to flood the entire Seal River region between the Gulf of Alaska and Vitus Lake. This would certainly allow seawater to enter the lake. The CTD survey for 2006 indicates that deep renewal may have occurred, and there were reports in the fall of 2005 of a strong surge resulting from the remnants of a typhoon coming ashore.

To simulate the effects of a storm surge, we used a step function as the ocean forcing in equation 4 rather than sinusoidal functions. In this case, we used Laplace transforms to obtain an analytic solution, which is a combination of exponentials. Figure 16 shows the response of the level of Vitus Lake to a unit surge that lasts for 5 d. The lake level response time is ~4 d, and a 1 m surge would bring in 4 psu km^3 of salt. Therefore, a storm surge of 2 m lasting for 3 or 4 d is sufficient to inject enough salt into Vitus Lake to account for the observed changes. Hence, storm surges appear to be the main cause for transporting water from the Gulf of Alaska into Vitus Lake.

CONCLUDING REMARKS

Since the last surge of the Bering Glacier in 1995, Vitus Lake has rapidly expanded in both area and volume. Despite these rapid changes the water chemistry of the lake remains relatively stable, and the water chemistry appears to be driven by subglacial discharge. Vitus Lake provides a unique outdoor laboratory for studying an evolving oligotrophic system.

The episodic exchange of water between the Gulf of Alaska and Vitus Lake, and the influx of melt water from the Bering Lobe of the Bering Glacier System, have produced a complex hydrographic structure in the lake. Despite intense stratification, oxygen depletion in the deep basin of Vitus Lake is not occurring, which is a result of low amounts of organic matter and vertical mixing. The latter is a result of fresh subglacial flow entering the lake at depth, the natural convection generated by melting ice along the terminus, and the melting of deep draft icebergs. CTD and DO surveys from 2001 to 2007 indicate that deep water exchange between the Gulf of Alaska and Vitus Lake occurred in the winters of 2005–2006 and 2006–2007.

We developed a model for the exchange of water between the gulf and the lake that includes the interaction of lake level; lake area; glacier melt; rainfall; river discharge; river length, width, and depth; and tides. The model shows that the level of Vitus Lake can be used to "gauge" the discharge. Winter lake level measurements would provide the no flow datum.

The model was used to simulate a variety of possible scenarios, including warming or cooling, changes in river length, glacier surge events, and storm surges. The most likely case is for accelerated glacier melt, continued terminus retreat, and the expansion of Vitus Lake. For this case the model shows that the lake level will rise, and the tidal fluctuations will become less significant.

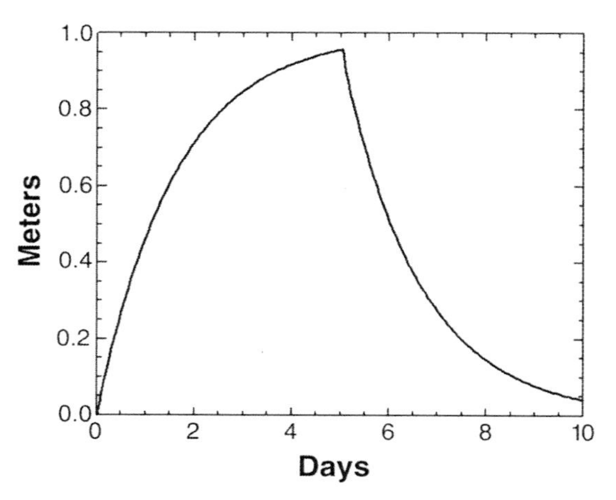

Figure 16. This plot gives the model results for the case of a storm surge. The plot shows that a storm surge of 2 m lasting for 3 or 4 d is sufficient to inject enough salt into Vitus Lake to account for the observed salinity and temperature changes.

ACKNOWLEDGMENTS

We gratefully acknowledge and thank John Payne, Scott Guyer, Chris Noyles, and Nathan Rathbun for their able assistance in carrying out all aspects of this complex field program as well as the staff of the University of Michigan's Marine Hydrodynamics Laboratories and the Michigan Tech Research Institute for creating innovative data collection platforms for extreme environments.

APPENDIX: HYDROLOGIC PROCESS OF BERING GLACIER AND VITUS LAKE, ALASKA: HORIBA AND ALWAS WATER QUALITY MEASUREMENT 2004-2007

This Appendix presents the extended set of water quality data for Vitus Lake. Water quality data were collected using a Horiba U-22 XD multi-parameter probe and a free-floating, robotic ALWAS buoy. Data are presented as graphs and summary statistics, for the Horiba and ALWAS instruments, respectively. In general, the data shows relatively consistent water quality values over the measurement time period at the sample sites (see Tables A1 and A2; Figs. A1, A2, and A3).

Horiba profile data were collected in 2004, 2005, 2006, and 2007, and ALWAS data were collected concurrently with the Horiba data in 2006 and 2007. Horiba measured the following water quality parameters: pH, dissolved oxygen (DO), conductivity, total dissolved solids (TDS), temperature, and turbidity. The ALWAS buoy measured the near-surface water and was configured to measure depth, temperature, conductivity, salinity, TDS, DO, pH, oxidation-reduction potential, turbidity, chlorophyll-a. The ALWAS buoys also provided the full GPS position and navigation suite including not only position but also course and speed over ground.

Hand held GPS receivers were used to navigate to the established water quality sampling sites (see Figs. A4 and A5). Horiba measurements were made at the surface, 1 m, 5 m, and 10 m below the surface, and then every 10 m to the bottom of the lake or until we reached the end of the 100 m cable. The ALWAS buoy was allowed to freely float at the surface while the profile measurements were being made.

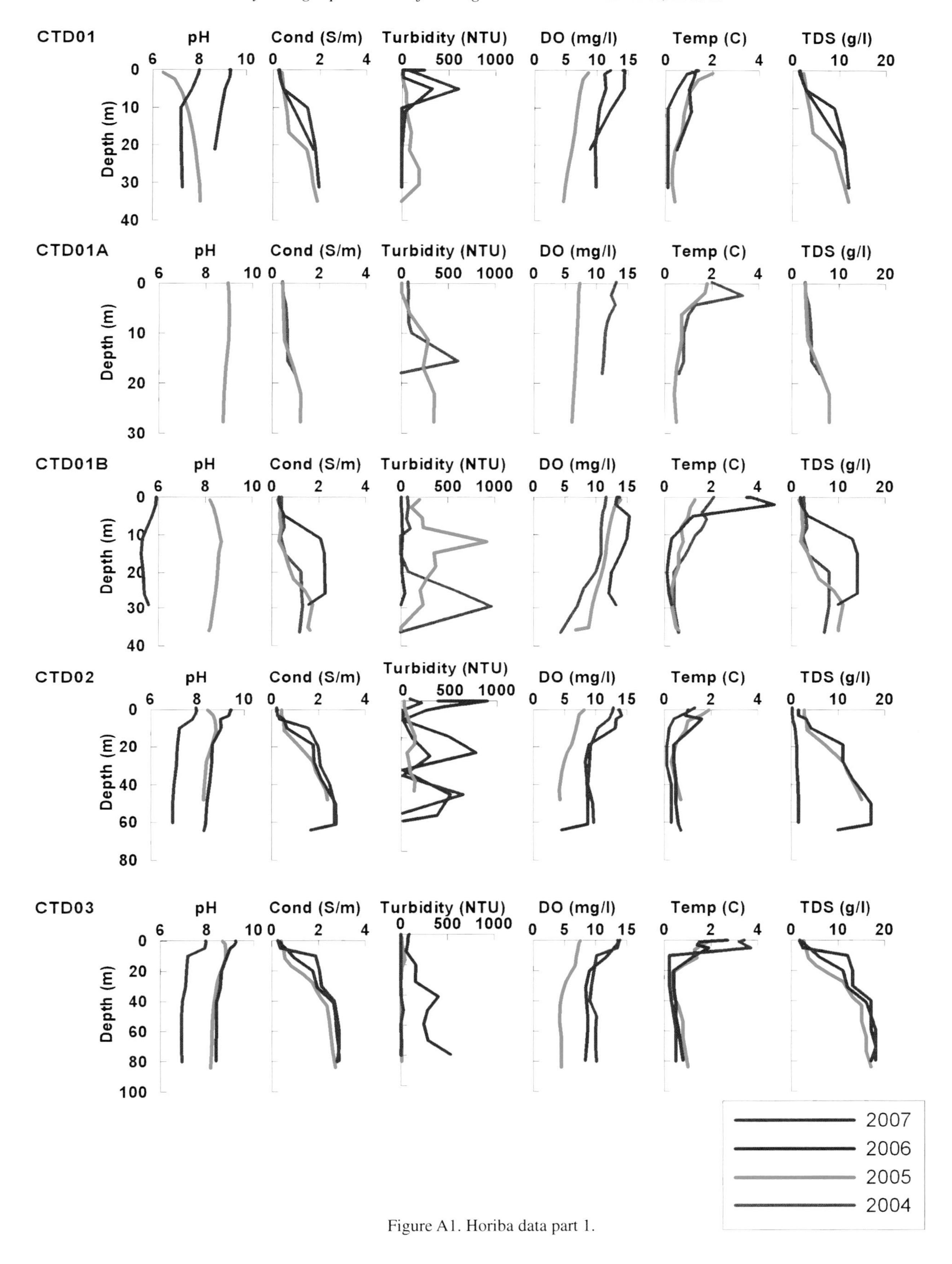

Figure A1. Horiba data part 1.

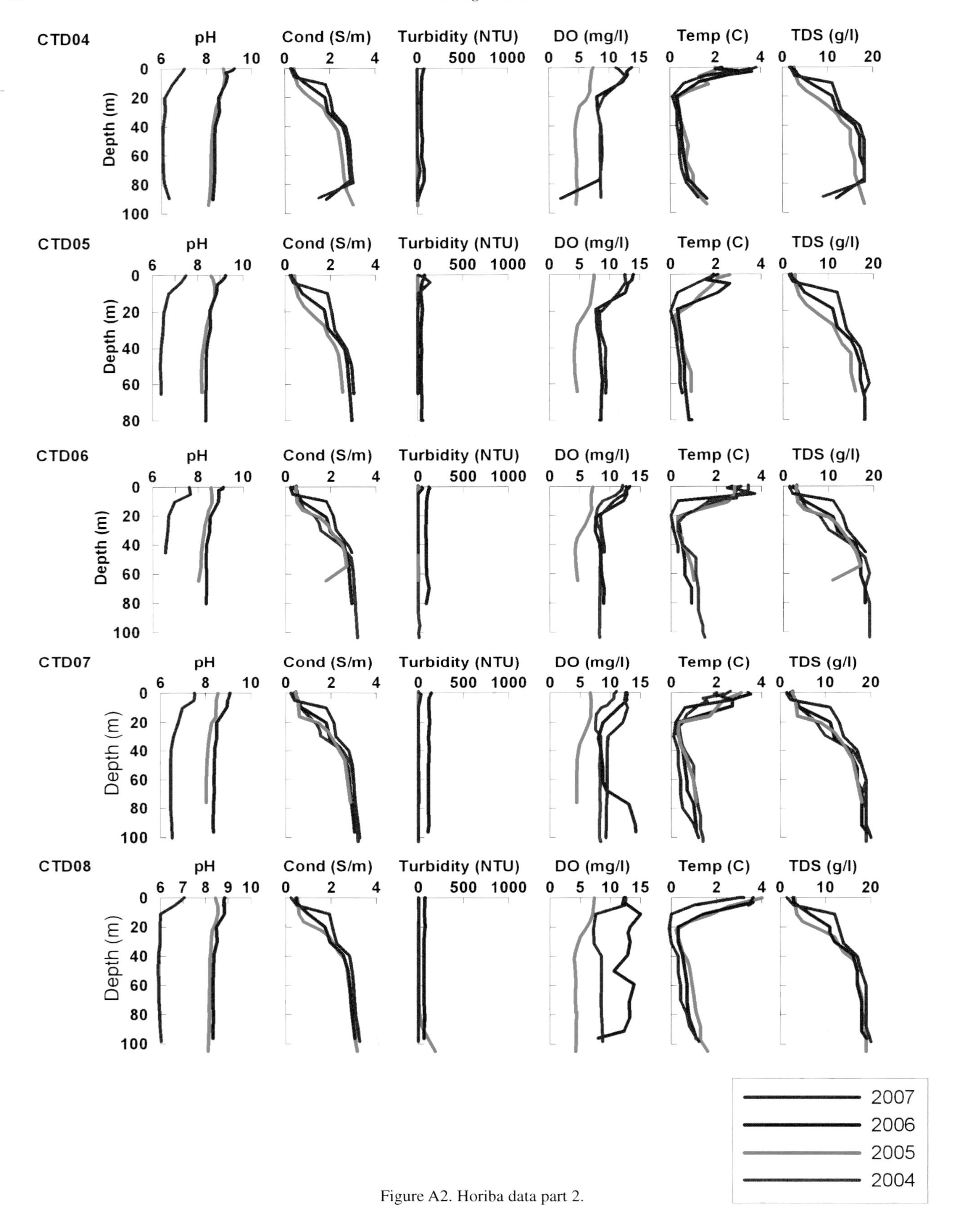

Figure A2. Horiba data part 2.

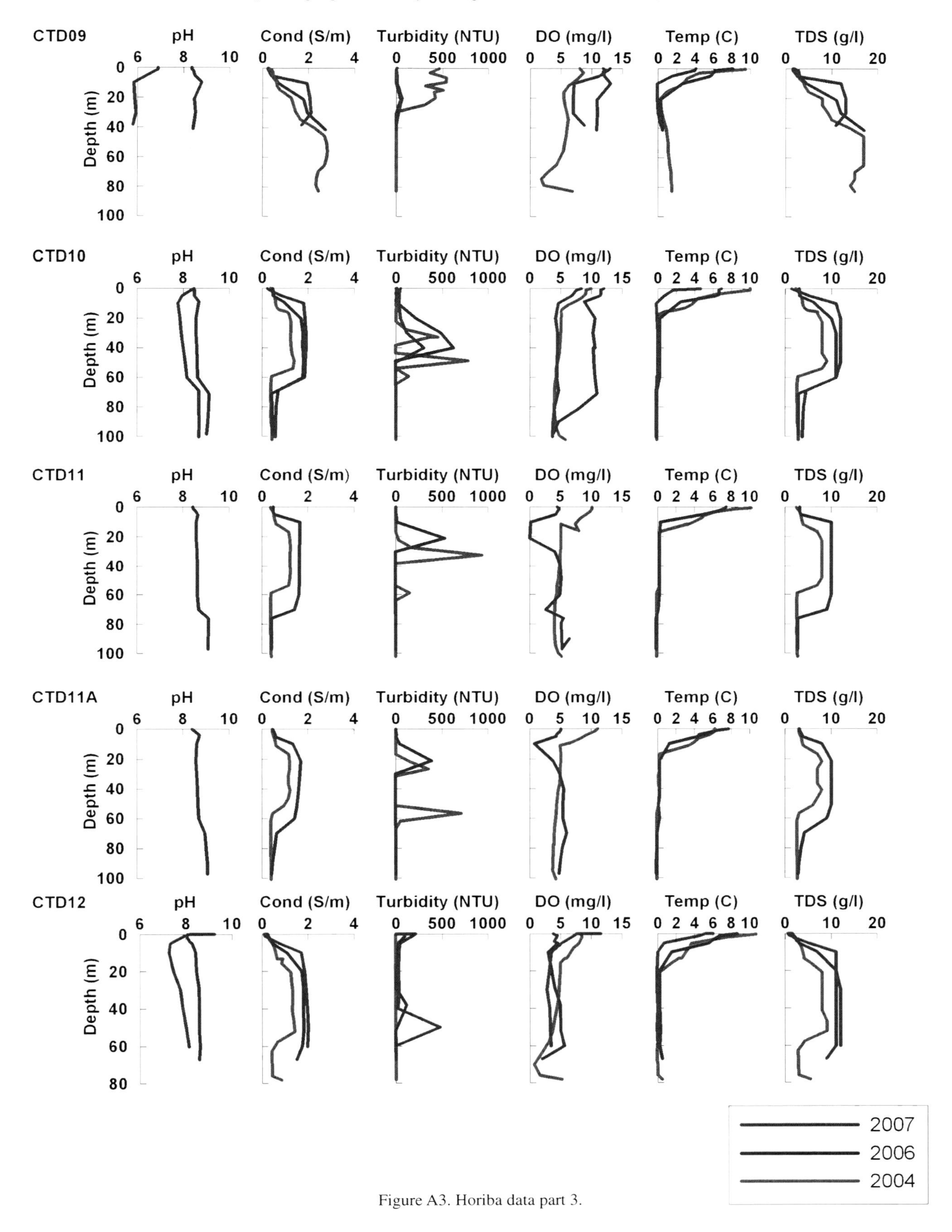

Figure A3. Horiba data part 3.

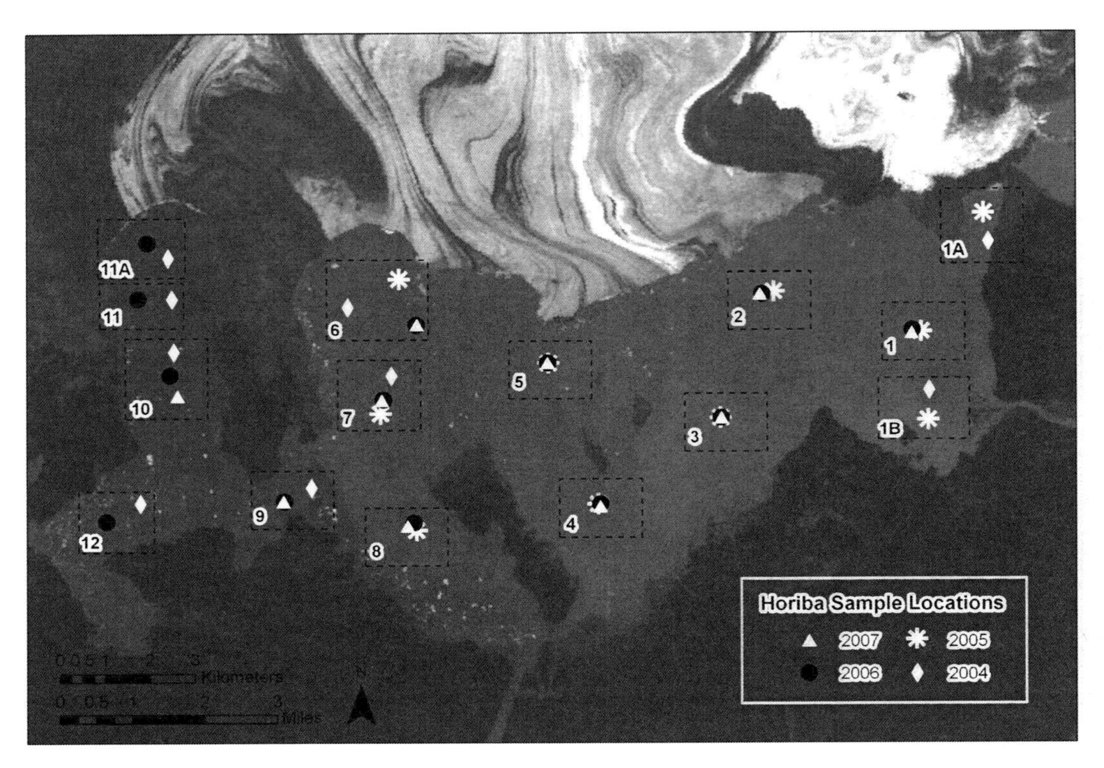

Figure A4. Horiba sample location map.

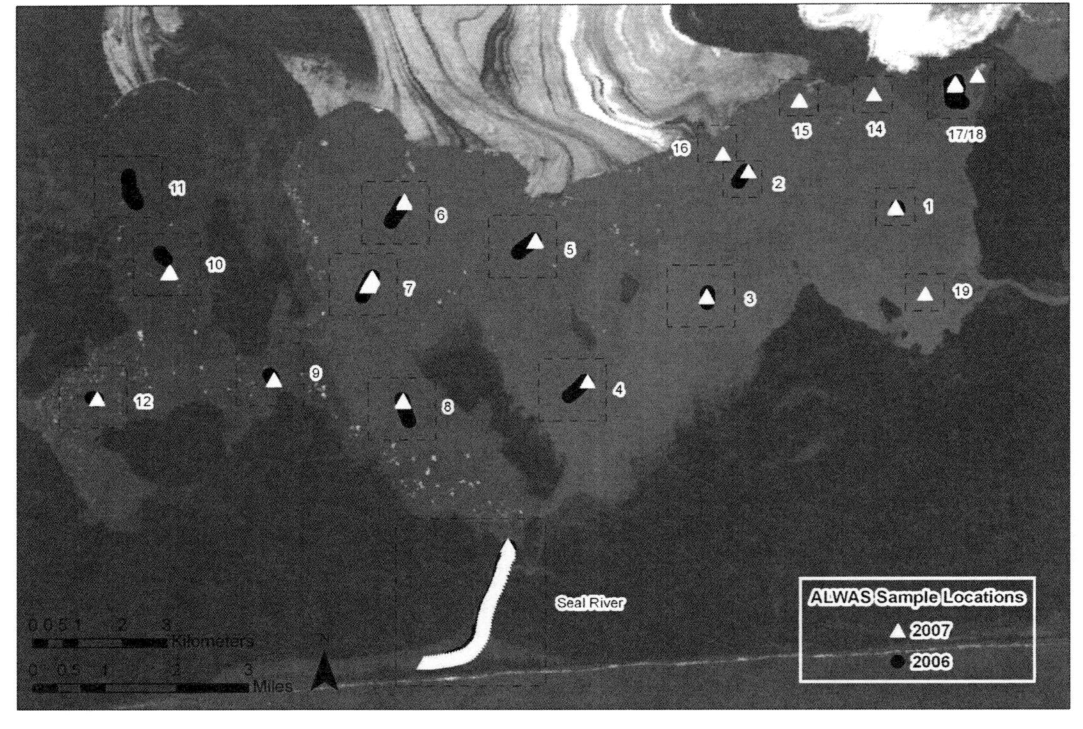

Figure A5. ALWAS sample location map.

TABLE A1. 2006 ALWAS SUMMARY STATISTICS

Temperature (°C)				
Site	Min	Max	Mean	SD
1	1.4	1.5	1.4	0.0
2	1.0	1.2	1.1	0.1
3	2.6	2.8	2.7	0.1
4	2.2	2.6	2.4	0.1
5	1.9	2.1	2.0	0.1
6	2.7	2.8	2.8	0.0
7	3.5	3.7	3.6	0.1
8	3.7	3.8	3.7	0.0
9	7.6	7.9	7.8	0.1
10	6.8	6.9	6.9	0.0
11	7.3	7.8	7.5	0.1
12	8.1	8.8	8.4	0.2
17/18	0.3	0.8	0.5	0.1
Seal	3.7	4.3	4.0	0.1
pH				
Site	Min	Max	Mean	SD
1	9.1	9.2	9.1	--
2	9.2	9.2	9.2	--
3	9.0	9.0	9.0	--
4	9.0	9.0	9.0	--
5	8.9	9.0	8.9	--
6	8.9	9.0	8.9	--
7	8.8	8.9	8.9	--
8	8.7	8.7	8.7	--
9	8.2	8.2	8.2	--
10	8.2	8.3	8.3	--
11	8.2	8.3	8.3	--
12	7.9	7.9	7.9	--
17/18	9.5	9.7	9.6	--
Seal	8.6	8.7	8.6	--
Dissolved Oxygen (mg/L)				
Site	Min	Max	Mean	SD
1	11.8	13.4	12.8	0.4
2	12.9	14.1	13.7	0.3
3	13.3	13.8	13.7	0.1
4	12.6	14.1	13.8	0.3
5	12.8	14.3	14.0	0.3
6	12.8	14.1	13.9	0.4
7	13.3	13.8	13.7	0.1
8	12.3	13.3	13.2	0.2
9	11.1	12.1	11.9	0.2
10	12.2	12.2	12.2	0.0
11	12.0	12.2	12.1	0.0
12	11.2	11.7	11.6	0.1
17/18	12.5	15.2	14.4	0.8
Seal	11.4	13.2	13.0	0.3
Conductivity (mS/cm)				
Site	Min	Max	Mean	SD
1	0.319	0.352	0.332	0.1
2	0.194	0.328	0.252	0.4
3	0.255	0.382	0.322	0.4
4	0.273	0.412	0.345	0.4
5	0.231	0.382	0.313	0.3
6	0.295	0.465	0.371	0.6
7	0.351	0.501	0.426	0.4

(*Continued*)

TABLE A1. 2006 ALWAS SUMMARY STATISTICS (*continued*)

Conductivity (mS/cm) (*Continued*)				
8	0.579	0.638	0.611	0.2
9	0.383	0.417	0.395	0.1
10	0.637	0.644	0.640	0.0
11	0.655	0.670	0.663	0.0
12	0.323	0.334	0.329	0.0
17/18	0.025	0.107	0.057	0.3
Seal	0.645	0.740	0.674	0.2
ORP (mV)				
Site	Min	Max	Mean	SD
1	150	156	153	2.0
2	163	169	164	1.6
3	150	174	159	7.3
4	145	166	152	5.8
5	124	144	128	4.5
6	136	152	140	4.2
7	148	170	154	5.3
8	158	178	166	5.9
9	118	127	122	2.8
10	141	162	146	5.4
11	142	167	147	5.3
12	143	160	150	5.4
17/18	27	108	82	26.0
Seal	140	150	143	3.3
Turbidity (NTU)				
Site	Min	Max	Mean	SD
1	2.9	418.0	74.0	110.8
2	110.4	230.2	149.5	25.1
3	7.5	22.5	12.0	5.2
4	2.2	21.1	6.7	5.3
5	9.7	31.2	13.7	4.8
6	2.7	31.0	6.7	6.0
7	2.5	20.1	5.7	5.2
8	2.5	19.8	4.2	3.4
9	17.1	19.3	18.1	0.8
10	2.6	3.4	3.0	0.2
11	1.7	3.0	2.2	0.3
12	45.7	58.3	51.9	3.4
17/18	995.5	999.2	997.0	1.0
Seal	2.4	27.3	11.4	7.7
Chlorophyll-a (Relative units)				
Site	Min	Max	Mean	SD
1	0.9	1.6	1.1	0.3
2	3.1	5.9	4.3	0.8
3	0.3	2.0	0.4	0.4
4	0.8	2.5	1.2	0.5
5	0.3	0.3	0.3	0.0
6	0.6	2.5	1.2	0.5
7	0.8	2.5	1.1	0.5
8	0.8	2.4	1.2	0.4
9	0.6	3.9	2.5	0.7
10	1.0	2.8	1.7	0.5
11	0.8	2.7	1.3	0.4
12	2.2	4.0	2.9	0.6
17/18	2.5	3.3	2.8	0.2
Seal	1.1	3.0	1.9	0.6

TABLE A2. 2007 ALWAS SUMMARY STATISTICS

Temperature (°C)

Site	Min	Max	Mean	SD
1	1.3	1.6	1.4	0.1
2	1.3	1.4	1.4	0.0
3	5.5	5.7	5.6	0.1
4	5.0	5.0	5.0	0.0
5	1.8	2.0	1.8	0.1
6	3.0	3.2	3.1	0.1
7	2.0	2.1	2.1	0.0
8	3.1	3.2	3.2	0.0
9	3.7	3.8	3.8	0.0
10	3.7	4.8	4.3	0.2
12	5.8	7.3	6.6	0.3
14	2.2	2.4	2.3	0.1
15	1.3	1.4	1.3	0.1
16	1.3	1.6	1.4	0.1
17	0.3	0.9	0.6	0.2
18	1.1	1.2	1.1	0.0
19	2.1	2.2	1.8	0.0
Seal	2.1	3.3	2.8	0.2

pH

Site	Min	Max	Mean	SD
1	7.9	9.0	8.6	--
2	9.1	9.1	9.1	--
3	8.8	8.9	8.8	--
4	8.8	8.8	8.8	--
5	8.8	9.0	9.0	--
6	8.9	8.9	8.9	--
7	9.0	9.0	9.0	--
8	8.8	8.9	8.9	--
9	8.7	8.8	8.7	--
10	7.6	8.4	8.3	--
12	7.5	7.7	7.6	--
14	8.8	9.0	8.9	--
15	9.0	9.1	9.1	--
16	9.1	9.1	9.1	--
17	9.1	9.3	9.2	--
18	9.1	9.2	9.1	--
19	8.9	9.0	7.8	--
Seal	8.4	8.6	8.5	--

Conductivity (mS/cm)

Site	Min	Max	Mean	SD
1	0.394	0.401	0.397	0.028
2	0.371	0.372	0.371	0.004
3	0.457	0.459	0.458	0.004
4	0.460	0.462	0.460	0.006
5	0.408	0.420	0.414	0.041
6	0.423	0.425	0.423	0.005
7	0.423	0.432	0.428	0.033
8	0.435	0.436	0.436	0.002
9	0.391	0.393	0.392	0.006
10	0.007	0.431	0.387	0.546
12	0.157	0.161	0.159	0.010
14	0.426	0.429	0.427	0.010
15	0.377	0.378	0.377	0.005
16	0.376	0.385	0.384	0.031
17	0.081	0.140	0.113	0.294
18	0.226	0.227	0.227	0.003
19	0.421	0.426	0.370	0.020
Seal	1.009	1.318	1.180	0.552

(*Continued*)

TABLE A2. 2007 ALWAS SUMMARY STATISTICS (*continued*)

ORP (mV)

Site	Min	Max	Mean	SD
1	176	186	182	3.5
2	153	158	155	2.2
3	156	159	158	0.9
4	158	160	159	0.8
5	155	157	155	0.8
6	154	157	155	1.1
7	151	158	154	1.8
8	158	160	159	0.6
9	158	162	160	1.1
10	137	203	190	9.5
12	217	232	222	3.0
14	163	170	167	2.9
15	157	161	159	1.1
16	154	159	156	1.7
17	135	156	148	7.2
18	152	158	156	1.9
19	144	153	132	3.3
Seal	127	147	133	4.8

Turbidity (NTU)

Site	Min	Max	Mean	SD
1	2.7	9.8	5.9	2.7
2	23.0	91.1	46.2	24.7
3	0.9	10.0	3.7	3.7
4	1.1	24.0	6.4	8.6
5	1.3	23.2	9.6	8.6
6	2.2	13.2	4.5	4.2
7	2.4	26.4	9.2	6.6
8	2.1	14.7	6.0	4.4
9	6.9	7.9	7.5	0.3
10	0.0	11.9	10.8	2.2
12	94.8	112.9	103.1	3.4
14	14.4	17.8	16.4	1.2
15	12.5	22.9	20.0	4.0
16	22.9	143.0	79.0	41.0
17	791.7	795.4	793.6	1.4
18	29.8	48.0	38.0	6.0
19	4.3	7.9	5.3	1.3
Seal	1.4	522.5	176.8	156.2

Chlorophyll-a (Relative units)

Site	Min	Max	Mean	SD
1	0.0	1.5	0.4	0.6
2	0.0	5.5	1.2	2.1
3	0.0	5.3	0.6	1.5
4	0.0	0.9	0.1	0.3
5	0.0	5.6	0.8	2.0
6	0.0	2.1	0.5	0.8
7	0.0	6.5	0.7	1.5
8	0.0	3.2	0.4	1.1
9	0.0	4.0	0.8	1.5
10	0.0	7.9	0.9	1.9
12	--	--	--	--
14	0.0	2.4	0.9	1.2
15	0.0	2.0	0.3	0.7
16	0.0	2.6	0.7	1.1
17	--	--	--	--
18	0.0	4.8	1.1	1.7
19	0.0	0.0	0.2	0.0
Seal	0.0	5.8	0.6	1.2

REFERENCES CITED

Alaska Climate Research Center, University of Alaska, Geophysical Institute; data available at http://climate.gi.alaska.edu/AKCityClimo/AK_Climate_Sum.html.

Arendt, A.A., Echelmeyer, K.A., Harrison, W.D., Lingle, C.S., and Valentine, V.B., 2002, Rapid wastage of Alaska glaciers and their contribution to rising sea level: Science, v. 297, p. 382–386, doi: 10.1126/science.1072497.

Fricker, H.A., Scambos, T., Bindschadler, R., and Padman, L., 2007, An active subglacial water system in West Antarctica mapped from space: Science, v. 315, p. 1544–1548, doi: 10.1126/science.1136897.

Hughes, W.F., and Brighton, J.A., 1967, Theory and Problems of Fluid Dynamics: New York, McGraw-Hill, 265 p.

Josberger, E.G., and Martin, S., 1981, A laboratory and theoretical study of the boundary layer adjacent to a vertical melting ice wall in salt water: Journal of Fluid Mechanics, v. 111, p. 439–473, doi: 10.1017/S0022112081002450.

Josberger, E.G., Shuchman, R.A., Meadows, G.A., Savage, S., and Payne, J., 2006, Hydrography and circulation of ice-marginal lakes at Bering Glacier, Alaska, U.S.A.: Arctic, Antarctic, and Alpine Research, v. 38, p. 547–560, doi: 10.1657/1523-0430(2006)38[547:HACOIL]2.0.CO;2.

Josberger, E.G., Bidlake, W.R., March, R.S., and Kennedy, B.W., 2007, Glacier mass-balance fluctuations in the Pacific Northwest and Alaska, USA: Annals of Glaciology, v. 46, p. 291–296 (papers presented at the International Symposium on Cryospheric Indicators of Global Climate Change, August 2006).

Krabill, W., Abdalati, W., Frederick, E., Manizade, S., Martin, C., Sonntag, J., Swift, R., Thomas, R., Wright, W., and Yungel, J., 2000, Greenland Ice Sheet: High-elevation balance and peripheral thinning: Science, v. 289, p. 428–430, doi: 10.1126/science.289.5478.428.

Meier, M.F., and Dyurgerov, M.B., 2002, How Alaska affects the world: Science, v. 297, p. 350–351, doi: 10.1126/science.1073591.

Merrand, Y., and Hallet, B., 1996, Water and sediment flux during the 1993–94 Bering Glacier surge, Alaska, USA: Annals of Glaciology, v. 22, p. 233–240.

Motyka, R.J., Hunter, L., Echelmeyer, K., and Connor, C., 2003, Submarine melting at the terminus of a temperate tidewater glacier, LeConte Glacier, Alaska: Annals of Glaciology, v. 36, p. 57–65, doi: 10.3189/172756403781816374.

Rignot, E., Box, J., Hanna, E., and Burgess, E., 2008, Mass balance of the Greenland Ice Sheet from 1958 to 2007: Geophysical Research Letters, v. 35, doi:10.10209/2008GL035417.

Royer, T.C., and Grosch, C.E., 2006, Ocean warming and freshening in the northern Gulf of Alaska: Geophysical Research Letters, v. 33, L16605, doi:10,1029/2006GL02676.

U.S. Geological Survey, Benchmark Glaciers Program; data available at http://ak.water.usgs.gov/glaciology/.

Warren, C.R., and Kirkbride, M.P., 2003, Calving speed and climatic sensitivity of New Zealand lake-calving glaciers: Annals of Glaciology, v. 36, p. 173–178, doi: 10.3189/172756403781816446.

Manuscript Accepted by the Society 02 June 2009

The Geological Society of America
Special Paper 462
2010

Botanical inventory of the Bering Glacier Region, Alaska

Marilyn Barker*
Department of Biological Sciences, University of Alaska Anchorage, 3101 Science Circle, Anchorage, Alaska 99508, USA

ABSTRACT

An inventory of the vascular flora of the Bering Glacier Region, south of the Bagley Ice Field, was conducted from 2000 to 2007. The area includes nunataks, mountains, and glacial forelands. The objectives of the inventory were to (1) assess the botanical biodiversity of the region, (2) identify rare taxa and areas of phytogeographic interest, and (3) provide data that would assist land managers with planning. The inventory has particular significance, as prior to this study the only specimens noted from this region were from the coastline. To date, 466 taxa have been identified, representing one quarter of the plant species of Alaska. Ninety-three of the species represent range extensions, 58 are gap fillers (species that fill distribution gaps), and 19 are rare species with a state rank of 3 or less on the Alaska Natural Heritage Program Vascular Plant Tracking List. Carbon Mountain, Robinson Mountains, and the Tashalich River sites contain 60% of the Heritage Tracking Plants. Three exotic weed species were found adjacent to Vitus Lake.

INTRODUCTION

The Bering Glacier Region covers an area of ~3000 km^2, with more than half of this area covered by glacial ice. The region is interesting floristically, as it is strategically adjacent to the Gulf of Alaska and includes a diverse array of ecosystems from coastal dunes to alpine mountain systems. Phytogeographically this region is a transition zone where the southeast flora mixes with northerly elements. Because of the logistical difficulties caused by the presence of the Bering Glacier, the region had been overlooked in previous botanical studies. Only a few specimens of coastal plants had been documented from this region until the present study. The area contains many diverse geomorphologic features, the largest being the glacial forelands. Prominent are several mountain features, such as the Grindle and Khitrov Hills, currently surrounded by ice, whereas the Robinson Mountains, Carbon Mountain, and Nichawak Mountain are at the periphery of current ice. There is evidence that many of these sites were covered by ice in previous advances of the glacier. We identified two potential glacial refugia: the Suckling Hills, which by their southwest coastal location may have escaped glaciation, and the slopes above Hanna Lake, which may have been given protection by the E-W–trending Robinson Mountains. Both of these potential refugia contain old forest components. Within the broad landscape are many other unique habitats: fresh water seeps, rich fens, salt water and fresh water marshes, poor fens, large lakes, and many smaller ponds as well as riparian sites. Each of these habitats supports a different plant community.

The objectives of this inventory were to (1) assess the botanical species richness of the region, (2) identify rare taxa and areas of phytogeographic interest, and (3) provide the botanical information necessary for proper land management.

METHODS

Botanical inventories were conducted from 2000 to 2007. The number of days in the field varied each year, from 2 to 10 d.

*afmhb@uaa.alaska.edu

Barker, M., 2010, Botanical inventory of the Bering Glacier Region, Alaska, *in* Shuchman, R.A., and Josberger, E.G., eds., Bering Glacier: Interdisciplinary Studies of Earth's Largest Temperate Surging Glacier: Geological Society of America Special Paper 462, p. 129–165, doi: 10.1130/2010.2462(07). For permission to copy, contact editing@geosociety.org.

(No work was done in 2006, as helicopter support was unavailable.) Field work was both weather and helicopter dependent.

Prior to the start of each field season, target sites were selected on the basis of different lithologies and landforms. Landsat images (2001) and U.S. Geological Survey (USGS) topographic maps (1959) were used to identify potentially unique habitats and localities for study. For the purpose of this review, study sites were grouped on the basis of landform and ecological affinities into seven broad categories: Early Colonization, Glacial Forelands, Temperate Forest, Subalpine, Alpine, Coastal Uplands, and Wetlands (Table 1; Fig. 1). The Wetlands category was further subdivided into Fens, Fresh Water Marsh, Salt Water Marsh, Ponds and Lakes, and Beaches. Ecosystems at each site were described, emphasizing associations between landforms, plant communities, and fauna. The dominant plant species of each community were noted, and all remaining species on the site listed. For more detailed information on the locations of these sites, see Appendix Table 1.

Once on site, the team of botanists walked through the area, noting plants seen and describing their associations. Collections were made for confirmation and documentation. Special attention was paid to the effects of microtopography, such as cutbanks, stream bank seeps, snow beds, and frost scars, because these provide additional niches for plants.

TABLE 1. BOTANICAL SITES WITH ELEVATIONS

Category	Site #	Site name	Elevation (m)
Colonization sites	1	Arrowhead Island	35
	2	Ancient forest	10
	3-4	Weeping Peat Island	28
	5-6	Berg Lake	142–152
Glacial forelands	7	Camp	10
	8	Cabin	7
	9	Midtimber Lake	3-15
	10	Tsiu landing strip	23
	11	Tsiu drainage	2
	66–73	Riparian systems	4-66
Temperate forest	12	Hanna Forest	28
	13	River Forest	47
	14	Tashalich Forest	6
	15	Kaliakh Forest	39
	16	Swan Lake Forest	29
Subalpine	17–18	Sunshine Point	451–477
Alpine	19–24	Grindle Hills	417–752
	25–35	Robinson Mountains	360–947
	36–37	Donald Ridge	360–644
	38–43	Khitrov Hills	385–924
	44	Nichawak Mountain	480–514
	45–46	Carbon Mountain	627–822
Coastal uplands	47–50	Suckling Hills	218–500
Wetlands			
Fens	51–60	East Bering fens	39–360
Fresh water marsh	61-63	Marshes	5-22
Salt water marsh	64-65	Oaklee River Marsh	3-4
Lakes and ponds	74-77	Lakes and ponds	5-66
Beaches	78–82	Gulf of Alaska beaches	1-2

It is important to note that the sites were not equally studied, and that sites were not randomly chosen within a given category, owing to logistical limitations such as weather and helicopter availability. Some of the sites were visited for only an hour or less, others all day, and still others with multiple visits allowing time for gathering quantitative data on plant communities. When possible, plotlines were established, and vegetation cover and frequency determined, using the 2 × 5 dm Daubenmire frame (Daubenmire, 1959).

Plants that were newly identified as present in the region, or for which doubt existed as to their identification, were collected. Specimens were pressed and cross-referenced to field notes. When rare plants (S1–S3 on the Alaska Natural Heritage Program Vascular Plant Tracking List) or potentially rare plants were encountered, the habitat was described, and the population of rare plants was assessed.

In the herbarium, keys and reference collections were used to accurately confirm specimen identification. Reference keys included *Flora of Alaska and Neighboring Territories* (Hultén, 1968), *Flora of North America (1993–2006)*, *Anderson's Flora of Alaska and Adjacent parts of Canada* (Welsh, 1974), *Alaska Trees and Shrubs* (Viereck and Little, 1972), and *Flora of the* Yukon Territory (Cody, 1996). Most names in the text were standardized to reflect taxa in the U.S. Department of Agriculture (USDA) Plants Database. Duplicate collections are at the University of Alaska, Fairbanks Museum of the North and at the University of Alaska, Anchorage herbaria.

RESULTS

Four hundred sixty-six taxa, belonging to 68 plant families, were found at the Bering Glacier. Forty-four of the families are dicotyledons. The remainder is divided among monocotyledons (11), gymnosperms (1), and ferns–fern allies (12). A complete list of all plants identified can be found in Appendix Table 2 as well as the location where found. Appendix Table 3 provides a cross-reference list of scientific and common plant names.

1. Colonization Sites: Arrowhead Island, Ancient Forest, Weeping Peat, Berg Lake

Glacial retreat exposes new land for plant invasion and colonization. The areas recently exposed consist of low elevation morainal-outwash complexes. They contain compact boulder till interfingered with irregular masses of stratified sands, silts, and gravels. Buried peats and/or stratified layers are common (A. Pasch, 2005, personal commun.). Mushrooms on the surface often indicate the location of buried organic material. These sites were overridden by the Bering Glacier during the Holocene glacial maximum, and have been exposed by retreat of the ice front in the 1990s.

When a glacier recedes it sets the stage for plant colonization by exposing new unvegetated areas. In several respects the debris left by a receding glacier is a favorable habitat for plant invasion.

Botanical Inventory of Bering Glacier: Site Locations

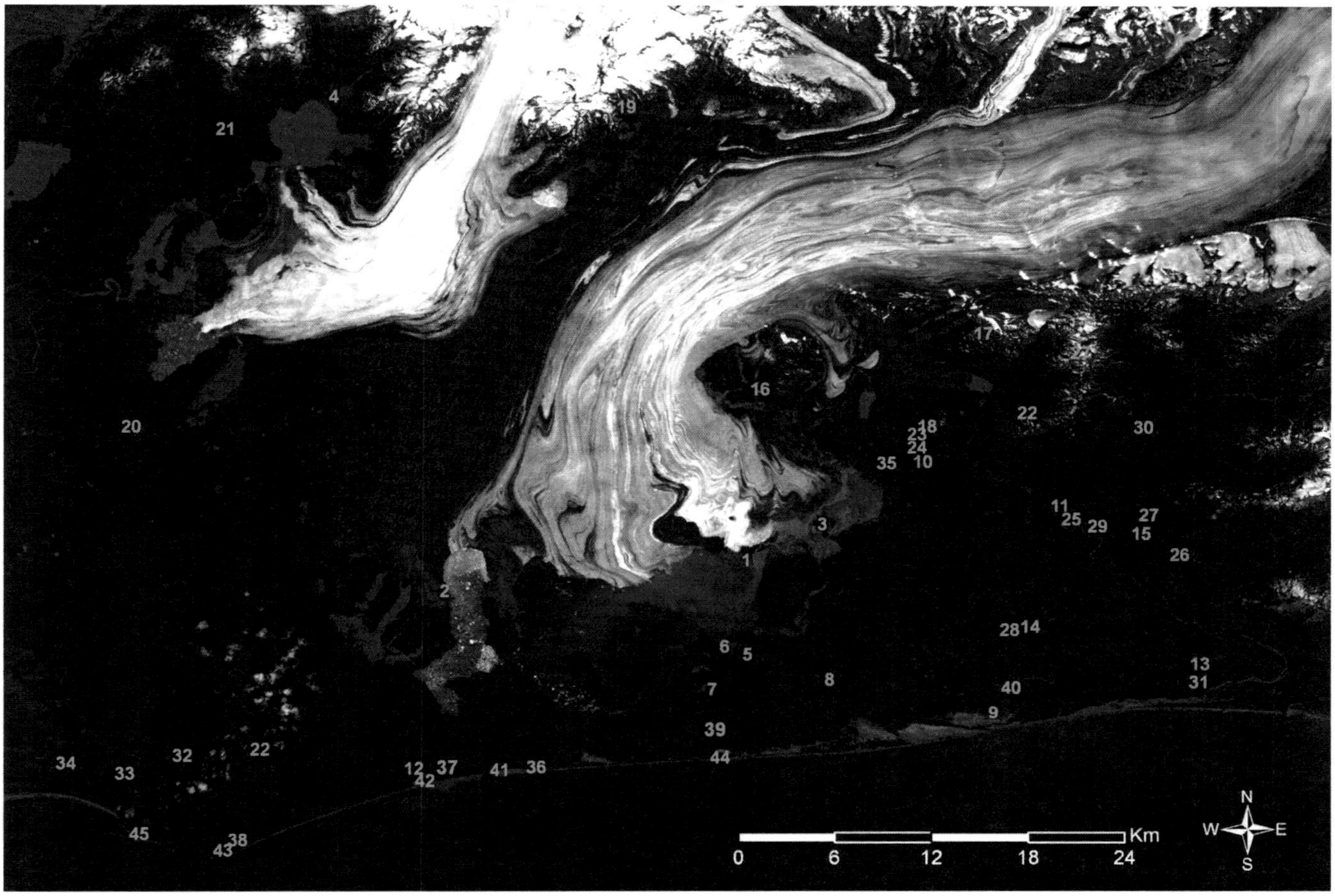

Site #	Site Name	Site #	Site Name	Site #	Site Name	Site #	Site Name	Site #	Site Name	Site #	Site Name
1	Arrowhead Island	9	Tsiu Drainage	17	Robinson Mountains	25	Kaliakh River Fen	33	Oaklee Freshwater Marsh	41	Seal Beach
2	Ancient Forest	10	Hanna Forest	18	Donald Ridge	26	Hope Fen	34	Oaklee Saltwater Marsh	42	Tashalich Beach
3	Weeping Peat	11	River Forest	19	Khitrov Hills	27	Odor Creek Fen	35	Hanna Creek	43	Kiklukh Beach
4	Berg Lake	12	Tashalich Forest	20	Nichawak Mountain	28	Swan Lake Fen	36	Seal River	44	Mid-Timber Beach
5	Camp	13	Kaliakh Forest	21	Carbon Mountain	29	Kaliakh/Kulthieth	37	Tashalich River	45	Cape Suckling
6	Cabin	14	Swan Lake Forest	22	Suckling Hills	30	Upper Kaliakh Fen	38	Kiklukh River		
7	Midtimber Lake Trail	15	Sunshine Point	23	Donald Ridge Fen	31	Kaliakh River Marsh	39	Mid-timber River		
8	Tsiu Landing strip	16	Grindle Hills	24	Hanna Sloping Fen	32	Suckling Lowland	40	Tsivat River		

Figure 1. Map of the Bering Glacier Region, indicating botanical site locations.

It consists of a mixture of mineral particles from coarse to fine, all kept moist by water from the melting ice. Colonization occurs in three steps: immigration, establishment, and multiplication (Daubenmire, 1968). When plants first invade an area, they have little relationship to each other and are widely separated. Later, when more plants settle in, they begin to interact with each other and begin to alter the environment. Over the course of time the habitat changes from one openly exposed to sun and wind to one so protected from the elements as to offer very different conditions. This leads to a progressive change in plant populations. These changes or stages are known as seres. Each seral stage has its own influence on modifying the soil and atmosphere as well as attracting animal species. Sequential seral stages are known as *plant succession*.

Plant succession in front of receding glaciers typically begins with colonization of mosses and lichens. Concurrently, seedlings of dominant species of later seral stages can be found. Over time, these early herbaceous communities are overtopped with woody plants, which in the Bering Region will develop into a Sitka spruce–Western hemlock forest. The time required from glacial retreat to a mature spruce-hemlock forest is 500–800 a.

Arrowhead Island, the Ancient Forest site, and Weeping Peat Island offer good examples of early stages of primary succession. These sites were exposed at different times by retreating ice in the 1990s. Invading species typically are prolific seed-spore producers, with the seeds-spores often adapted to wind dispersal. The plant diversity of these early communities is dependent upon the distance between the site and a seed-spore source. Typically species richness is poor on newly exposed surfaces. However, representative species from all seral stages can be found. Because of differential growth rates, different species take turns dominating the landscape. Fifteen species were found on Arrowhead Island, and 10 were found at the Ancient Forest. Weeping Peat Island has been exposed for a greater period of time and had 88 species.

Arrowhead Island

Most of this island emerged from the ice in 1999. The southernmost edge emerged ~10 a earlier as evidenced by 2–3-m-tall *Populus balsamifera* trees, estimated to be 8–10 a old, growing on the south shoreline. The surface consists of poorly sorted sands, silts, fine gravels, and coarse gravels, with a few large boulders. Plants were scattered, with an estimated 1% coverage. Common plants included horsetails *(Equisetum variegatum* and *arvense)* and *Chamerion latifolium* (river beauty). Woody seedlings of alders and poplars were present. Occasionally seedlings of conifers were found. Of note was a *Picea sitchensis* seedling 5 cm tall and a *Tsuga mertensiana* seedling measuring 10 cm.

Ancient Forest

The Ancient Forest site was exposed in 1998 as ice retreated. The retreat exposed large stumps of spruce trees, which were later radiocarbon dated at A.D. 800. In addition there were several peat blocks with radiocarbon dates of 2000 B.C. For information on the dating and size of the Ancient Forest itself, see Crossen and Lowell (this volume).

Figure 2. Colonization by *Equisetum arvense*, a plant disseminated by windborne spores.

The present day plants showed a coverage of 1%–2% when the site was examined in 2000. The most common plants were *Equisetum arvense* and willowherbs *(Epilobium hornemannii* and *Chamerion latifolium).* Woody seedlings were not noted in 2000. Figure 2 shows colonization of *Equisetum arvense*; the green tuft is the gametophyte phase, and the straight stem the sporophyte.

Weeping Peat Island

We visited Weeping Peat Island over a period of 3 a (1999–2001), with a final visit in 2007 (Fig. 3). Species richness increased by 20% during the 5 a hiatus. The first year was a brief reconnaissance visit. In 2000 the primary emphasis was on fossil invertebrate shells. In 2001, some vegetation measurements were taken. On Weeping Peat Island it is obvious that plant communities become more diverse in direct proportion to the length of time the land has been ice free. The ice retreat across the island

Figure 3. Weeping Peat Island in relation to the ice front (2005).

was well documented (J. Fleisher, 2006, personal commun.). Five zones, from barren, recently exposed land to densely vegetated thickets, were identified (Table 2).

Zone 1. Zone 1 represents recently exposed sites with vegetation covering <1% of the surface. *Alnus viridis* is the most common species, with an estimated 0.6% cover (Fig. 4). The mean height of *A. viridis* is 27 cm. The two other prominent species are *Equisetum variegatum* and the moss genus *Racomitrium*. Other plants found at this site include *Sagina nivalis, Chamerion latifolium, Populus balsamifera*, and *Salix* spp. Marking the end of this zone is a moraine deposited in 1999. A few additional species were found on the moraine: *Epilobium ciliatum, Arctagrostis latifolia, Juncus alpinoarticulatus, Festuca rubra, Stellaria* calycantha, and a *Picea sitchensis* seedling.

Zone 2. Although species diversity remained the same, the plants in zone 2 were considerably more developed. Alders averaged 71 cm in height. Only the herbaceous *Chamerion latifolium* was shorter than in zone 1. *Chamerion latifolium* disappears entirely in zone 3, emphasizing its need for bright sun. Plant cover increased to 2%–5%. The estimated age of this community was 2–3 a.

Zone 3. Zone 3 is a mosaic of dense shrubs (alders and willows 2–3 m tall with >150% coverage) alternating with open herb-moss areas with 40% cover. The cover averaged 75% for the area. Cover data often exceed 100% in plant communities because each species is given a cover estimate, and tall plants overlap shorter plants, so the same area on the ground can be covered by multiple layers. Of note were *Picea sitchensis* seedlings 10 cm tall and smaller seedlings of *Tsuga mertensiana*. Species richness increases dramatically as new habitats are created with increases in shade and soil richness. Ferns appear, along with other interesting species such as the ladies tress orchid *Spiranthes romanzoffiana* and pearly everlasting, *Anaphalis margaritacea*. However, plants dependent upon bright sun disappear with increasing shade.

Zone 4. This zone developed on glacial drift that was said to have been exposed in 1997 (Fleisher et al., this volume); the plants, however, indicate an older date. The canopy was dominated by *Alnus viridis* and *Populus balsamifera* at 4 m in height. The understory was diverse (20+ species), including *Orthilia secunda, Platanthera aquilonis*, and *Athyrium filix-femina*. The moss layer was predominantly *Rhytidiadelphus squarrosus*. Of note were *Picea sitchensis* trees up to 130 cm in height and with 9 to 13 nodes. Typically *Picea* produces one whorl of branches each year, so its age is easy to determine. *Tsuga mertensiana* was also present, though much shorter at 10 cm.

Zone 5. Zone 5 represents the oldest vegetation on the island. *Populus balsamifera* (4 m) is beginning to overtop the *Alnus* (3–4 m) thicket. *Picea sitchensis* is >2 m tall. Layers of dense vegetation form a deep shade. The understory includes *Cerastium fontanum, Orthilia secunda*, and *Moehringia lateriflora*, as well as more species of mosses.

The south end of Weeping Peat Island has a series of fen pools and a small lake. These additional habitats bring the total species richness of Weeping Peat Island to 88.

TABLE 2. COMPARATIVE DATA ON PLANT SUCCESSION AT WEEPING PEAT ISLAND

	Zone 1	Zone 2	Zone 3	Zone 4	Zone 5
Alnus viridis height (cm)	27	71	300	200–400	200–400
Populus balsamifera height (cm)	8	21.1	370	400	400+
Salix alaxensis height (cm)	4	11.3	250	150–250	
Salix barclayi height (cm)	3	9.6	240		
Salix sitchensis height (cm)	7	7.4	250		
Chamerion latifolium height (cm)	17	11.9			
Picea sitchensis height (cm)			10	75–130	200
Tsuga mertensiana height (cm)			+	10	
Date of moraine	1999	1998		1997	
Plant cover (%)	1	2–5	150+	220+	300+
Estimated age (years)	1–2	2–3	5–6	8–9	10

Figure 4. *Alnus viridis* colonizing recently exposed ground moraine

Berg Lake

Berg Lake is a classic ice dammed lake that has had a dynamic history during the past 100 a. Early USGS maps (ca. 1906) show that the Berg Lake basin was nearly filled with ice, damming the four valleys to the north of the ice front. When the ice retreated, the four lakes merged into one. Berg Lake is surrounded by a series of fans, moraines, and strandlines formed within the past century (Fig. 5). Several retreats of the Steller Lobe are documented by the strandlines. The exposed lake basin (of blue clay, loess, and stratified and unstratified silts, sands, and gravels) forms the surface for plant colonization. The ground surface has been reworked by wave action and multiple episodes of flooding, destroying and

Figure 5. West shore of Berg Lake, showing strandlines.

creating habitats for plants. This constant reworking of the surface has caused succession to repeatedly begin anew.

Berg Lake has relatively low species richness (75 species) owing to its recent geologic history of flooding and glacial recession. The vegetation adjacent to Berg Lake is sparse, consisting primarily of early seral species. One species was unique to this site, *Athyrium americanum*.

In 2001, species lists were compiled for each of four areas thought to represent different seral stages: the spit, the fan, the recessional moraine, and the strandline zone (Table 3).

The **spit** (1). The spit was the most recently colonized site, and, as expected, had the lowest species richness (16 species). Overall plant cover was estimated between 1% and 2%. The few alders present were <0.5 m tall. The poplars averaged 20 cm, and the willows, 3.5 cm. A lone *Picea sitchensis* seedling was 2 cm tall. Both *Poa alpina* and *Carex macrochaeta* were heavily grazed by geese; there were no seed heads.

The **fan** (2). The fan, which is older than the spit, contained 33 species, with the plant cover estimated to be 10%. *Chamerion latifolium* was the most common plant with a 3%–4% cover, followed by *Alnus viridis*, *Salix sitchensis*, and *Equisetum arvense*.

TABLE 3. COMPARISON OF FOUR SUCCESSIONAL SITES AT BERG LAKE

	Spit	Fan	Moraine	Strandline
Elevation (m)	142	142	143	212
Number of species	16	33	27	31
Percent cover	1–2	10	20–30	180
Height of *Alnus* (cm)*	50	96	147	200+
Height of *Picea* (cm)*	2		8	78

*Measurements are based on the mean height of 10 plants.

All plants were estimated at a 1%–2% cover. Height measurements were taken on four species: *Populus balsamifera* 11.5 cm, *Salix sitchensis* 77.7 cm, Salix *alaxensis* 114.8 cm, and *Alnus viridis* 96.8 cm.

The **recessional moraine** (3). Twenty-seven species were found on the recessional moraine. The plant cover was estimated between 15% and 20%. *Chamerion latifolium* and *Alnus viridis* were estimated each to have 5% cover, whereas *Salix sitchensis* and *Populus balsamifera* were estimated at 2% each. Plant heights averaged 147 cm for *Alnus viridis*, 54.4 cm for *Populus balsamifera*, and 7.9 cm for *Picea sitchensis*.

The **strandline zone** (4). The strandline represents a previous shoreline ~70 m above the current lake level. Thirty-one species of vascular plants were found here. Total plant cover was estimated at 180%. (Plant cover estimates often exceed 100% owing to multiple layers of plants overtopping each other.) *Alnus viridis* formed 90% of the cover, followed by *Heracleum maximum* at 30%, *Salix alaxensis* and *Populus balsamifera* each at 10%, *Chamerion latifolium* at 9%, and *Aruncus dioicus* at 4%. *Picea sitchensis* averaged 78 cm tall.

Animal visitation was evident on all early colonization sites. Evidence left behind included tracks and feces as well as evidence of grazing and browsing. The probable visitors were moose, wolves or coyotes, black and brown bear, geese, and sandhill cranes.

2. Glacial Forelands

The Bering Camp and Midtimber Lake trail sit on the Bering Glacier forelands (Fig. 6). The forelands are underlain by a neoglacial recessional moraine and an outwash plain complex of silts, sands, and gravels with blocks of clay and scattered boulders. The fresh morainal relief of 10–30 m is constantly

Figure 6. Glacial forelands, showing a mix of open gravels, ponds, and alder thickets.

being reworked within the region, destroying and creating habitats for plants.

The most prominent landmarks in the forelands are Vitus and Midtimber Lakes. Midtimber Lake represents an old river channel that at one time carried much of the runoff of the glacier. Today, the Seal River carries most of the glacial runoff. The region also is home to drainages of seven rivers: From east to west these rivers are Oaklee, Kiklukh, Tashalich, Seal, drainage of Midtimber Lake, Tsiu, Tsivat, and Kaliakh. Parts of this region have no streams or rivers but very good drainage, and these regions appear desertlike. Other parts have poor drainage and support a mosaic of wetland communities.

Much of the vegetation in the forelands represents various stages of plant succession relating to landscape age. Examples of mesic, xeric, and hydric successions can be seen. Mesic succession occurs on sites where water is not a limiting factor, with neither too much nor too little water. Xeric succession occurs where water is a limiting factor because of drainage and/or porosity of soil, e.g., bare rock surfaces or talus slopes. Third, hydric succession occurs where there is an overabundance of water, e.g., ponds and lakes.

Small rivers cut through the forelands, forming riparian zones along their shores. These zones support a diverse flora as they thread through a mosaic of communities from shrub swamps and wet meadows to cottonwood stands and evergreen forests. Riparian communities often reflect upstream communities, where chance seeds are carried by water and deposited along the shores. One hundred seventy-five species were found along the rivers. Only five were common to all rivers, *Achillea millefolium* var. *borealis*, *Epilobium ciliatum ssp. glandulosum*, *Fragaria chiloensis*, *Geum macrophyllum*, and *Heracleum maximum.* One species, *Poa laxiflora*, was found only on these waterways. *Poa laxiflora* is on the Heritage Tracking List and a range extension from SE Alaska. *Limosella aquatica* is also on the Heritage Tracking List, and the Tashalich and Tsiu River populations fill gaps in its sparsely documented distribution in Alaska.

There are 214 species of vascular plants in the foreland. Eleven are not found elsewhere; these include *Poa laxiflora*, *Carex utriculata*, *Carex lenticularis* var. *limnophila*, *Dodecatheon pulchellum*, *Galium trifidum* ssp. *columbianum*, *Juncus albescens*, *Platanthera obtusata*, *Rorippa curvisiliqua*, *Poa annua*, *Matricaria discoidea*, and *Taraxacum officinale* ssp. *officinale*. *Platanthera* and *Rorippa* are both range extensions; *Rorippa* and *Poa laxiflora* are also on the Heritage Tracking List, state ranking 1 and 3, respectively. Three of these, *Matricaria*, *Poa annua*, and *Taraxacum*, are considered aggressive exotics. *Poa annua* is a particularly aggressive exotic that transports seeds on aircraft wheels and boot treads; it is common on small airstrips throughout the state.

Much of the land cover is alder thicket *(Alnus viridis)* with smaller amounts of willow *(Salix sitchensis*, *S. alaxensis)* and elderberry *(Sambucus racemosa)*. The understory is *Athyrium filix-femina*, *Pyrola asarifolia*, *Calamagrostis canadensis*, and *Stellaria crispa.*

There are many ponds and small lakes. These are commonly edged with *Carex* spp., *Equisetum arvense*, and *Eleocharis palustris*. Aquatic plants included floating-leaved and submerged potamogetons. A discussion of aquatic plants can be found in the Wetlands section of this chapter. Some sites with impeded drainage developed into rich fens and other wetland types, thus increasing the overall diversity.

Evidence of many mammals was found, including short-tailed weasels, river otters, black and brown bear, and moose.

3. Temperate Forest

Forest sites are identified as plant communities with a closed tree canopy (canopy coverage of 70%–100%). Mature forests occupy older sites, such as the slopes of former nunataks and coastal-beach-ridge complexes east and west of the glacial advance of the last century. The terrain can be flat or steep (up to a 30°–40° slope). Soil development is shallow over glacial drift. The dominant trees are notably shallow rooted and prone to windfall. Seven different forest sites were visited, ranging in elevation from 6 m to 47 m.

Seventy-eight species were noted in the forest sites; 19 were woody species, 13 were ferns and fern allies, and the remaining 46 were herbs. Only three species were limited to the forest habitat: *Lonicera involucrata*, *Luzula divaricata*, and *Monotropa hypopitys* ssp. *lanuginosa.* Both *Lonicera* and *Monotropa* are significant range extensions from SE Alaska. Another important plant was *Ranunculus pacificus*, found ecotonally between the forest and wet herbaceous meadow. It is both a range extension from SE Alaska and is on the Heritage Tracking List of rare plants (S3, G3).

The dominant trees are *Picea sitchensis* and *Tsuga mertensiana*, with *P. sitchensis* predominating at low elevations. The trees form a canopy reaching heights of 30–50 m. Individual trees reach circumferences of 3.5–4.1 m. *Tsuga heterophyla*, also present, was not common. Beneath the canopy is a tall shrub layer (1–3 m) composed *of Oplopanax horridus*, *Rubus spectabilis*, *Viburnum edule*, and, less commonly, *Lonicera involucrata.* Shorter shrubs (to 1 m) were also present, including *Menziesia ferruginea* and *Vaccinium ovalifolium.* Beneath the shrubs was a rich herb layer dominated by the ferns *Dryopteris expansa*, *Athyrium filix-femina*, and *Blechnum spicant.* Other ground cover species present were *Coptis trifolia*, *Rubus pedatus*, *Streptopus amplexifolius*, and *Tiarella trifoliata.* Large pleurocarpous mosses complete the understory. The mosses included *Rhytidiadelphus loreus*, *Plagiothecium undulatum*, and *Hylocomium splendens*. Epiphytes were also present, including the fern *Polypodium glycyrrhiza*, the moss *Antitrichia curtipendula*, and many lichens, including *Sphaerophorus globosus* and *Lobaria pulmonaria (*Fig. 7).

4. Subalpine Sites: Sunshine Point

Subalpine sites are found near the boundary of the upper limit of tree growth and the alpine zone. The higher elevations merge into the treeless tundra vegetation, and the lower elevation sites interfinger with tall shrub or forest zones. Soils are well developed and typically support rich herb meadows. The appearance of subalpine vegetation is parklike, with bands of trees intermixing with herb meadows. The trees provide no more than 10% of the overall canopy cover. Two sites were visited along a southern ridge in the Robinson Mountains at elevations between 400 m

Figure 7. *Picea sitchensis*, towering over mosses and other small plants.

and 500 m. They are characterized by islands of heath vegetation and rich herb meadow interfingered with *Tsuga-Picea* forest fragments. *Empetrum nigrum* was the dominant heath plant. Several species dominated the herb meadow, including *Veratrum viride, Chamerion angustifolium, Aconitum delphinifolium, Valeriana sitchensis*, and *Senecio triangularis*. These plants reached heights of 0.5–1.5 m (Fig. 8).

Ninety-three species were found on the subalpine sites; of these, 24 were restricted to shade zones beneath the *Tsuga-Picea* forest fragments. Seven species were range connectors, and two species were short range extensions from across the Bagley Ice Field *(Lloydia serotina* and *Anemone narcissiflora)*.

5. Alpine Sites: Khitrov Hills, Grindle Hills, Robinson Mountains, Donald Ridge, Carbon Mountain, and Nichawak Mountain.

Alpine sites are those that rise above the upper limit of tree growth and are elevationally between the tree line and the snowline. The position of the tree line is correlated with the altitude where the temperature for the warmest month is at least 10 °C (Crawford, 1989). Tree seedlings that establish in this habitat are reduced to a krumholz life-form (Fig. 9). Herbaceous plants in this zone are characterized by short stature (1–20 cm) both to conserve heat and gain protection from the wind. Limitations to growth are imposed by a combination of available heat, light, wind, and moisture. A true alpine community has sparse vegetation spread over a talus or gravel substrate (Körner, 1999). Key indicator species of alpine sites in the Bering Glacier Region are *Loiseleuria procumbens, Arctostaphylos alpine*, and *Silene acaulis*.

All of the mountain sites are particularly interesting at Bering Glacier, because many of the summits have been completely surrounded by ice for decades and even centuries. Thus isolated, they can serve as refugia for a variety of life-forms. Two of these sites, the Grindle Hills and the Khitrov Hills, are still islands of land surrounded by a sea of ice. The remaining four sites are connected to land on at least one boundary. Three hundred two taxa were found in the alpine sites.

Figure 9. Krumholz (*Picea glauca*) on mixed heath tundra, Nichawak Mountain.

Figure 8. Lush subalpine meadow with *Veratrum viride* and *Chamerion angustifolium* as dominants, interfingering with forest.

The flora of these sites is the most diverse group in the Bering Region (Table 4). In part this is due to the variety of microhabitats. Microtopography is more important in the alpine communities for controlling heat, moisture, and wind protection than in other plant communities. On any given site there is a large range of ecologic factors. North- and south-facing slopes each have their own distinct communities. The south face is warmer and drier, and the north face is cooler and wetter. North faces often retain snow longer and hence have a better supply of water and a shorter growing season. Drainage is also important; i.e., exposed bare rock or fields of gravel do not retain water, creating a xeric environment. In contrast, sites near small streams, waterfalls, or seeps have no water limitations. Wind also plays an important factor in plant habitation, and mountain slopes offer varying degrees of protection from the wind.

Evidence of wildlife was found at all the mountain sites, including tracks and scat, and actual sightings of black bear, brown bear, mountain goat, and wolf.

TABLE 4. SPECIES RICHNESS AT MOUNTAIN SITES

Site	Species richness
Khitrov Hills	198
Grindle Hills	185
Robinson Mountains	232
Donald Ridge	148
Carbon Mountain	165
Nichawak Mountain	130

Khitrov Hills

The Khitrov Hills rise steeply from the ice, as nunataks, to heights of 1400 m. They separate the Steller Lobe from the main body of the Bering Glacier. Bedrock exposures are prominent and consist mostly of pale gray Tertiary sandstone. Embedded in the sandstone are bits of quartz, muscovite, and hornblende. The sandstone has faint laminar bedding and weathered low-angle cross-bedding, suggesting a calcareous cement. The outcrops are dominated by small spires, trenches, or pits, all with vertical sides (A. Pasch, 2005, personal commun.). In mid-August there were still many persistent snow patches. The Khitrov Hills merge southward into a substantial medial moraine between the Steller and Bering Lobes. Both coal deposits and leaf fossils were noted at the 2007 landing site.

Most of the vegetative cover is heath or herbaceous meadow. There were also scattered alders and spruce, particularly at lower elevations, and open scree slopes and rocky outcrops at higher elevations. The heath was dominated by *Harrimanella stelleriana*, *Phyllodoce glanduliflora*, and *Empetrum nigrum*. These subshrubs form a dense layer 10–20 cm tall. Other common heath species present were *Vaccinium uliginosum* and *Loiseleuria procumbens*. Lichens made up ~10% of the cover. Dominance varied in the herbaceous meadow between the taller *Lupinus nootkatensis*, *Chamerion angustifolium*, and shorter *Nephrophyllidium crista-galli* and *Geum calthifolium* communities. Other common herbs were *Erigeron peregrinis*, *Valeriana sitchensis*, *Arnica latifolia*, *Veratrum viride*, and *Athyrium filix-femina*.

In addition to heath and herbaceous meadow, the Khitrov Hills abound with unique microhabitats such as waterfalls, streams, rock crevices, pools, and wetlands. These variations promote species diversity.

The flora is composed of 198 species. Four plants were found only in the Khitrov Hills: *Anemone multifida*, *Antennaria friesiana* ssp. *neoalaskana*, *Ranunculus karelinii*, and *Saxifraga tricuspidata*. All four are range extensions. In addition, *Agoseris aurantiaca* var. *aurantiaca*, *Agrostis humilis*, and *Carex phaeocephala* are on the Heritage Tracking List.

Grindle Hills

The Grindle Hills are a small group of nunataks with summits ranging from 629 m to 833 m at the point where the Bering Glacier turns direction from west to south. The hills are underlain by Cretaceous sandstone, which was moderately deformed by uplift of the Robinson Mountains (J. Fleisher, 2007, personal commun.). The dominant vegetation is low heath (Fig. 10). At lower elevations, alder thickets are prominent. The Grindles have many topographic features that increase the variety of habitats, including streams, tarns, waterfalls, cliffs, and scree slopes. The bryoflora indicates a siliceous rock base (Slack, this volume).

Six visits were made to the Grindle Hills over the course of 6 a, some as brief as 10–15 min, others 6–8 h. One hundred eighty-five species were found in the Grindle Hills. Three plants were found only in the Grindle Hills: *Minuartia arctica*, *Packera ogotorukensis*, and *Senecio lugens*; all three are range extensions. The list also includes five Heritage Tracking Plants: *Carex phaeocephala*, *Romanzoffia unalaschcensis*, *Gentianella propinqua* ssp. *aleutica*, and *Agrostis humilis*.

Figure 10. Low heath dominated by *Phyllodoce glanduliflora* and *Vaccinium uliginosum*.

Robinson Mountains

The Robinson Mountains are a range of mountains trending east-west for ~100 km. McIntosh Peak is the high point of the Bering Glacier Region at 961 m. Only the western slopes of these mountains are considered a part of the Bering Glacier Region for this vegetation study.

The Robinson Mountains form a divide that apparently confined the ice flow of the Bering Glacier, forcing it to flow west for miles before turning south to the Gulf of Alaska. The surface is underlain by steeply dipping bedrock, some of which is dark gray basalt that weathers to an orange color. The dip is near vertical, and the strike E-W. The structure of the bedrock appears to control topographic features, such as a series of benches. Immature micaceous siltstone and shale clasts showed an abundance of slickensides and fault gouge. Higher elevations show permafrost features such as well-developed cryoplanation terraces, stone stripes, and solifluction lobes. Moraines on the north side of the ridge suggest that the margin of the Bering Glacier was once higher but never topped the divide (A. Pasch, 2005, personal commun.).

Ten visits were made to the Robinson Mountains over a period of 8 a, and 232 taxa were documented from these visits. Fourteen of these are found only in the Robinson Mountains, 11 of which are range extensions. Notable of these extensions are *Silene involucrata* from the Alaska Range and *Antennaria monocephala* ssp. *angustata* from the Alaska Range and Kluane National Park, Yukon Territory. Four species are on the Heritage Tracking List.

On the lower mountain slopes the vegetation forms complex mosaics of alternating heath and herbaceous meadow. The heaths are dominated by *Harrimanella stellariana*, *Phyllodoce glanduliflora*, and *Empetrum nigrum*. The herbaceous meadows are dominated by *Valeriana sitchensis*, *Arnica latifolia*, *Lupinus nootkatensis*, and *Aconitum delphinifolium*. Scattered in the area were *Picea* trees held in a krumholz life-form to 0.5 m.

Upper drier slopes have well-developed colonies of *Saxifraga oppositifolia*, *Silene acaulis*, and *Salix arctica*. *Oxytropis jordalii* thrives in extremely dry gravels. Figure 11 depicts an upper slope characterized by scattered dwarf willows and cushions of *Silene acaulis*.

Moist slopes around seeps are dominated by *Nephrophyllidium crista-galli* with an understory of *Hippuris montana*. Mossy areas close to the seeps show colonies of *Leptarrhena pyrolifolia*, *Pedicularis sudetica*, *P. oederi*, and *Primula eximia*. One particular seep had pure white plants of *Pedicularis sudetica*. The seeps are also home to large colonies of a small pale gentian, *Gentianella propinqua* ssp. *aleutica*, a Heritage Tracking Plant. One particularly nice waterfall was surrounded by a massive colony of *Epilobium luteum*, the yellow fireweed.

In places where drainage is impeded, both cottongrass fens and ponds were found. Common plants at this site were *Carex saxatalis*, *Carex macrochaeta*, *Carex lenticularis*, and *Eriophorum angustifolium*. See Slack (this volume) for commentary on the moss flora at this site.

Donald Ridge

Donald Ridge is a moderately high elevation area underlain by steeply dipping Tertiary fossiliferous sedimentary bedrock. It contains resistant sandstone beds that form E-W–trending

Figure 11. Upper slope of Robinson Mountains with cushions of *Silene acaulis*.

hogback ridges. In places the ridges exhibit an unusual form of weathering in which jumbles of frost-rived boulders lie at the top of voids created by solution along vertical joints. Linear valleys between are eroded in softer shales and siltstones (A. Pasch, 2005, personal commun.).

A total of 148 vascular plant species were identified and divided among several different plant communities. *Festuca brevissima* and *Salix myrtifolia* have been identified only from this site. *Festuca brevissima,* a range extension, normally is found in the Alaska Range and the Wrangell Mountains.

Eight plant associations were identified, controlled primarily by topography:

1. Ericaceous dwarf scrub meadow: The dominant community at the landing site was an ericaceous dwarf scrub meadow. Atypical was the dominance of *Harrimanella stelleriana* (rather than *Cassiope tetragona* or *C. mertensiana*, as in other parts of Alaska) with 70% cover. Other species with significant cover were *Luetkea pectinata* (14%) and *Phyllodoce glanduliflora* (11%). Bryophytes formed a carpet underneath. The average height of the community is 0.2 m. Important herbs were *Nephrophylidium crista-galli* (6%), *Hippuris montana* (5%), and *Carex macrochaeta* (4%).

2. Hogback ridges: The hogback ridges were covered with a thick soil and supported a rich flora of *Arnica lessingii*, *Conioselinum gmelinii*, *Geum calthifolium*, and the ferns *Athyrium filix-femina* and *Dryopteris expansa.*

3. South-sloping shale scree: South-facing scree slopes supported widely scattered plants including *Kumlienia cooleyae*, *Saxifraga ferruginea* (Fig. 12), *Minuartia biflora*, and *Arnica latifolia.*

4. Valleys between hogback ridges: Drainage from the snow pack formed wet swales that supported a rich flora of primulas and saxifrages.

5. Herbaceous meadows: Rich herbaceous meadows included *Aconitum delphinifolium*, *Anemone narcissiflora*, *Lupinus nootkatensis*, *Pedicularis verticillata*, and *Fritillaria camschatcensis.*

Figure 12. *Saxifraga ferruginea* on a south-facing scree slope.

6. Rocky knoll with *Rhododendron camtschaticum*: One clump of *Rhododendron camtschaticum* lay on a rocky knoll below camp. It was growing with *Diapensia lapponicum*, *Loiseleuria procumbens*, and the lichen *Sphaerophorus*. We also found *R. camtschaticum* 60 km to the east on Nichawak Mountain and the Suckling Hills.

7. Upland herbaceous slopes: Deer cabbage, *Nephrophyllidium crista-galli*, formed a thick carpet between other communities. *Nephrophyllidium* had 95% cover and a height of 0.2–0.4 m. *Hippuris montana* as well as moss made up the understory.

8. Fen, stream, and pool. A shelf at 360 m contained a fen and stream complex. The fen was dominated by *Eriophorum angustifolium*. Other common species found were *Dodecatheon jeffreyi* and *Carex pluriflora*. Sphagnum moss was also present. Aquatic plants in the pools included *Callitriche palustris* and *Sparganium hyperboreum*. The data from these small fens are discussed in more detail in the section on the wetlands.

Carbon Mountain

Carbon Mountain is a large mountain rising from the north shore of Berg Lake. It towers to 1570 m and spawns multiple glaciers and drainages. Two visits were made to Carbon Mountain, the first to the eastern flank, and the second to the western flank. Elevations at the landing sites were 627 m and 822 m, respectively. The high point reached during these forays was 1087 m.

The Carbon Mountain sites were mostly on barren, dry, windswept ridge crests with *Arctostaphylos alpina*, *Salix rotundifolia*, and *Silene acaulis*, typical alpine vegetation. On the sheltered ledges below the crest we found *Parnassia fimbriata*, *Polystichum lonchitis*, and *Heuchera glabra*. Away from the crest, as with other sites, is an alternating heath-herb meadow complex. The heath was dominated by *Phyllodoce glanduliflora*, *Empetrum nigrum*, and *Luetkea pectinata*. The herb meadow was dominated by *Sanguisorba canadensis* and *Carex macrochaeta*. Open gravel areas had scattered *Salix arctica* and *Lupinus nootkatensis* (Fig. 13).

One hundred sixty-five species were documented, including four plants from the Natural Heritage Tracking List. One of these, *Agoseris aurantaica*, is listed as an S1 (Alaska Natural Heritage Program, 2006). Carbon Mountain is one the richest sites in the Bering Glacier Region for Heritage plants as well as hosting six species not yet found elsewhere in the Bering Glacier Region.

Two other interesting finds were noted for Carbon Mountain. The first was a mudstone 5- × 8-cm fern frond fossil found on a scree slope 1056 m upslope. The second was the abundance of Herkimer diamonds (double-pointed quartz crystals) scattered among the scree gravels on both east and west slopes.

Nichawak Mountain

Nichawak Mountain is an isolated mountaintop approximately halfway between Carbon Mountain and the Suckling Hills on the western edge of the Bering Region. At the present time ice has retreated from all sides of the mountain. The

Figure 13. Steep west-facing slope supporting a community of *Heracleum maximum*, *Geranium erianthum*, *Lupinus nootkatensis*, and *Arnica latifolia*.

closest adjacent mountain is Mount Campbell, at 556 m. The landing site was near the tree line adjacent to several windswept *Picea* 1–4 m tall.

Heath plant communities dominated the landscape. They showed a different species dominance than the other heaths did. These heaths were dominated by *Vaccinium uliginosum*, *Empetrum nigrum*, and *Salix arctica* with scattered tufts of *Deschampsia beringensis*. *Phyllodoce glanduliflora* was absent, and *Harrimanella stelleriana*, though present, did not play an important role in these communities.

To the west, on a 30°–40° slope, we found a complex mixture of *Carex nigricans*, *Nephrophyllidium crista-galli*, and *Caltha leptosepala*. One hundred thirty plants were documented. *Artemisia furcata* was the only plant found at Nichawak and at no other Bering Glacier site. *A furcata* forms a range connection between the Alaska Range and southern Yukon Territory. Two Heritage Tracking plants were found as well: *Gentianella propinqua* ssp. *aleutica* and *Agrostis humilis*.

6. Coastal Uplands: Suckling Hills

The Suckling Hills are underlain by Tertiary bedrock consisting of dipping sedimentary sandstones, volcanics, and tephra units (Fig. 14). Clasts at the surface and on scree slopes appear to have been derived from weathering of frost rived bedrock. All are similar in lithology to the bedrock (A. Pasch, 2005, personal commun.). The microtopograpy is rilled rather than flat. It is significant that this area probably was ice free during the last ice advance.

The Suckling Hills sites ranged from 218 m to 500 m in elevation. The predominant vegetation cover in these hills is a lush herbaceous meadow; the drier ridge sites supported low heath vegetation. Altogether the flora has 184 species, three of which were found only in the Suckling Hills. These include *Juncus filiformis*, *Urtica dioica*, and *Vaccinium vitis-idaea*. *Juncus filiformis* is an extension from the Wrangell Mountains and Kenai Peninsula, and *Gentianella propinqua* ssp. *aleutica* was the only Heritage Tracking Plant.

The ridge tops are covered by a lush herbaceous meadow with scattered salmonberry patches. Graminoids make up a significant part of the ground cover (25%–75%). The average height of vegetation was 0.75 m. Common grasses included *Festuca rubra*, *Calamagrostis canadensis*, and *Deschampsia cespitosa*. Predominant herbs were *Chamerion angustifolium*, *Sanguisorba canadensis*, *Angelica lucida*, *Heracleum maximum*, *Aconitum delphinifolium*, and *Gentiana platypetala*. Locally *Athyrium filix-femina*, *Alnus viridis*, or *Picea sitchensis* were dominant.

Of particular interest was a shaly scree along the ridge in a saddle. This site was home to some notable plants, most of which were considerable range extensions. These included *Koenigia islandica* (connection between Canada and central Alaska), *Malaxis brachypoda* (connection between Kodiak Island and southeast Alaska), and *Lagotis glauca* (extension from the Aleutians). In addition a large population of *Gentianella propinqua* ssp. *aleutica* (connection between Kodiak Island and Glacier Bay), numbering over a hundred individual plants, was documented. *Gentianella* had previously been found in small numbers at other alpine sites (Grindle Hills, Nichawak Mountain, and Robinson Mountains).

Figure 14. Topography of the Suckling Hills, showing rich graminoid-herbaceous meadows; note the uneven ground surface.

7. Wetlands: Fens, Coastal Marsh, Fresh Water Marsh, Lakes, Ponds, and Beaches

Wetlands are characterized by soils inundated or saturated by surface water or groundwater. Three wetland criteria must be present: (1) water or evidence that water was present for a length of time, (2) wetland vegetation (plants that have adapted to survive in saturated soils), and (3) hydric soils. In Alaska, sites are characteristically flooded with 15 cm or more of water but may have no standing water late in the summer; even so, the soils remain saturated. There are many kinds of wetlands, with their characterization dependent on such factors as pH, water source, water amount, drainage, and presence of salts and minerals (Viereck et al., 1992).

Fens

Fen is a general term for a peat-forming ecosystem that may contain sphagnum moss (Figs. 15, 16). Fens are defined by having a mineral-rich source of flowing water. Fens have a neutral to slightly acid pH and are generally low in nitrogen. Poor fens can be acid and low in nutrients with plants similar to bogs (J. Glime, 2007, personal commun.). The vegetation is dominated by grasses and sedges along with many bryophytes.

Ten fens were visited, all in the eastern Bering Glacier Region. All were sloping fens ranging from a 5% to a 15% slope. Most of the fens were at lower elevations between 40 and 160 m. Two fens were higher at 279 m and 360 m. Most of the fens had small pools or small streams dissecting them. The fens we observed would be classed as mesotrophic fens, i.e., fens that are dominated by sedges rather than by sphagnum moss. Typical dominant plants were *Eriophorum chamissonis*, *Trichophorum cespitosum*, *Carex aquatilis* var. *dives*, and *Carex pauciflora*. Other important herbs included *Dodecatheon jeffreyi*, *Platanthera* spp., and *Gentiana douglasiana*. Important woody plants included *Andromeda polifolia* and *Vaccinium uliginosum*.

Notable are three species of carnivorous plants, *Drosera rotundifolia*, *D. anglica*, and *Pinguicula vulgaris*. Carnivorous plants are often associated with nitrogen-poor fens and bogs. These plants trap and digest insects as an alternate source of nitrogen.

We found a total of 158 species in the fen habitats. The combination of pH and drainage allows fens to support plants not found in other habitats; hence 14 species are restricted to the fens: *Microseris borealis*, *Carex laeviculmis*, *C. livida*, *C. echinata* ssp. *phylomanica*, *C. magellanica*, *Drosera anglica*, *Juncus ensifolius*, *Lycopodium lagopus*, *Malaxis paludosa*, *Malus fusca*, *Vaccinium oxycoccos*, *Scheuchzeria palustsris*, *Selaginella selaginoides*, and *Thelypteris quelpaertensis*. *Malus fusca* is a crab apple tree found only in fens. Mature trees were twenty feet tall. At these sites are young to mature fruit-bearing trees. *Malaxis paludosa*, a range connection, is on the Heritage Tracking List.

Fens are a high use area for both black and brown bears, as evidenced by actual sightings, interlacing trail systems, and scat.

Fresh Water Marsh

Marshes are vegetated flat plains dissected with streams (Fig. 17). Plant communities that develop on marshes are herb

Figure 15. Sloping mesotrophic lowland fen on upper Kulthieth River.

Figure 16. Upland cottongrass fen in the Robinson Mountains.

Figure 17. Fresh water marsh east of the Suckling Hills, showing drainages and animal trails. Willows and alders occupy the edges of drainages.

dominated and have few woody plants. The soils are usually high in humus content and may be inundated for long periods of time, if not continuously.

Four different fresh water marshes were visited. All were associated with river drainages. The first was between forest fingers near the Tashalich River and was dominated by broad-leaved herbs. *Menyanthes trifoliata* and *Angelica genuflexa* provided most of the plant cover. The second and third fresh water marshes were in the Oaklee River drainage (Fig. 17). These were dominated by graminoids, mostly *Calamagrostis canadensis*, *Deschampsia beringensis*, and *Carex aquatilis* var. *dives*. Other interesting plants were *Lysimachia thyrsiflora*, *Hierochloe odorata*, and *Myrica gale*. *Myrica gale* is particularly notable because of its nitrogen-fixing capabilities. The fourth fresh water marsh was near the Tsiu River drainage. This broad area included a large lake and broad, open mudflats. The Tsiu was a particularly rich area for spike rushes, including two species found only at this site: *Eleocharis acicularis* and *E. quinqueflora*. The latter is an S1 Tracking List plant.

One hundred thirty-seven plant species were found in the fresh water marshes. Sixteen species were found only in these fresh water marshes, and four Heritage Tracking List Plants: *Ceratophyllum demersum*, *Eleocharis quinqueflora*, *Limosella aquatica*, and *Ranunculus pacificus*.

Salt Water Marsh

Salt water marshes are low elevation ecosystems adjacent to the ocean with little topographic relief. Typically these communities are dissected by tidal channels. Because of its closeness to the ocean, this region is subject to tidal inundation on a daily and/or seasonal basis. Typically coastal marshes are dominated by graminoid plants, including *Carex* spp. and *Triglochin maritima*. Only one coastal marsh was sampled. It was on the Oaklee River east of the Suckling Hills (Fig. 18).

The vegetation in a coastal marsh shows transitions from the land to the ocean on the basis of salt tolerances of the plants. Oaklee salt marsh is dominated by dense swards of *Carex lyngbyei* and *Deschampsia beringensis*. Thirty-nine species were noted at the site. Eleven species were restricted to this habitat. Many of these are salt tolerant species such as *Carex mackenzii*, *C. glareosa*, *C. lyngbyei*, *Chrysanthemum arcticum*, *Eleocharis kamtschatica*, *Lomatogonium rotatum*, *Stellaria humifusa*, and *Triglochin maritima*. It should be noted that all the *Lomatogonium* (Fig. 19) flowers were white, an unusual color form for the naturally blue-flowered plant.

Lakes and Ponds

Inland bodies of water vary in area from small ditches one can easily step across to vast expanses of water that one cannot see across. Aside from size, they vary in other ways. Some are vernal puddles that dry out completely in the summer months, others can retain constant water levels for centuries. Some are deep enough to be thermally stratified, and some have uniform temperatures throughout. And finally, some can be penetrated by light, whereas others extend below the depth of effective light penetration. In the Bering Region, water bodies come in all sizes,

Figure 18. Coastal marsh near the mouth of the Oaklee River. Differences in color indicate different plant assemblages. The deepest green is *Carex lyngbyei.*

Figure 19. White-flowered *Lomatogonium rotatum*.

but they all support plants. These plants are adapted to aquatic habitats and fall into five categories (Daubenmire, 1974):

1. Emergent anchored plants—plants that are rooted to the substratum but stand erect above the water level *(Carex, Hippuris).*
2. Floating-leaved anchored plants—plants that are rooted to the substratum, but their leaves lie on the surface of the water body *(Potamogeton, Sparganium, Nuphar).*
3. Submerged anchored plants—plants that grow entirely underwater and are attached to the lake bottoms *(Isoetes, Myriophyllum, Potamogeton).*
4. Suspended plants—plants that lack attachment to the soil but do not penetrate the surface of the water *(Utricularia, Ceratophyllum).*
5. Floating plants—plants that are in contact with water and air, but not soil *(Lemna).*

All five categories are represented in the Bering Glacier Region; however, for purposes of this chapter, emergent anchored plants will not be considered, as they were discussed in other wetland categories. Additionally some plants are amphibious, being equally at home in an aquatic habitat or on land. Two notable amphibious plants in the Bering Glacier Region are *Polygonum amphibium* and *Ranunculus trichophyllus*.

All three species of *Utricularia* in Alaska were found in shallow ponds. These suspended water plants are particularly interesting because their leaves have small bladder traps that actively trap small aquatic organisms to use as a source of nitrogen.

Most aquatic plants have broad distributions, owing to the ease of seed transport by waterfowl. It is not unusual that of the

26 aquatic species, there are only 8 not in range; 3 are connections, and 3 only extended from across the Bagley Ice Field, leaving only two real range extensions: *Ceratophyllum demersum* is known from the Denali region, and *Potamogeton foliosus* is a range extension from the Alaska Range.

Beaches

The ocean beaches form a harsh habitat for plants. They are the boundary between ocean and land, and hence are mostly barren. During spring high tides and storm tides they are subject to inundation by salt water. Beach substrate is sand with little organic matter. The unstable sand is easily moved by wind, so often dunes are associated with beaches. Other microhabitats that support some unique plants are formed by large driftwood logs (Fig. 20).

Plants growing adjacent to the ocean must be salt and desiccation tolerant. Twelve species have been found in the beach habitat. The dominant plant on the upper beach was *Leymus mollis*. It is associated with *Honckenya peploides* and *Lathyrus japonicus*. Five other species are found only on the upper beach and include *Mertensia maritima*, *Galium aparine*, *Poa macrantha*, *Glenia littoralis*, and *Cakile edentula*. Both *Poa macrantha* and *Glenia littoralis* are range connections and are on the Heritage Tracking Plant List, S1 and S3, respectively.

DISCUSSION AND CONCLUSION

Plant species in the Bering Glacier Region have diverse origins and histories. They include circumpolar species with broad distributions as well as floristic elements from the Aleutian-Asian connection, the SE coastal ranges, and the Cordilleran Range of the interior. This discussion will focus on overall findings, HeritageTracking Plants, range connections and extensions, and management considerations.

Figure 20. Pacific beach with beach greens, *Honckenya peploides*, among drift logs.

Overall Findings

The botanical inventory of the Bering Glacier Region consists of 68 families, 213 genera, 445 species, and 21 subspecies, for a total of 466 taxa (fully listed in Appendix Table 2). Five of these plants are carnivorous, and one is parasitic. The families with the best representation are Cyperaceae with 47 taxa, Poaceae with 45 taxa, and Asteraceae with 35 taxa. Twenty families are monotypic in the Bering Glacier Region. Most of these taxa have been collected and specimens placed either in the University of Alaska, Anchorage (UAA) or in the University of Alaska, Fairbanks (UA) Museum herbaria. At the end of the 2005 field season the UA Museum of the North collection consisted of 1099 specimens from 55 localities. A complete listing of the UA Museum Bering Glacier Region specimens with all collection data can be found on the Arctos Database Web site (see listing in References).

No one species was found at all 17 sites. The most cosmopolitan species was *Equisetum arvense*, which was found at 15 sites, all sites except the lake and subalpine environments. *Equisetum* is transported by windblown spores and colonizes any bare substrate. Once established, its rhizomes go deep (up to 2 m), so it is not easily eliminated once established. *Alnus viridis* and *Salix barclayi* were found at 14 of the sites. The importance of these tall shrubs as members of the plant community is often underrated, particularly by those attempting to cross an alder-willow thicket. The roots of *Alnus* support nitrogen-fixing bacteria that convert nitrogen gas into a form usable by other organisms; *Salix* is important for moose browse.

Six species were found at 13 sites: *Achillea millefolium*, *Athyrium filix-femina*, *Calamgrostis canadensis*, *Picea sitchensis*, *Rubus spectabilis*, and *Salix sitchensis*. These plants all have broad ecological tolerances. Table 5 compares the 17 ecological sites for species richness. This table underscores the need to visit as many sites as possible to gain a clear picture of the flora.

It is notable that the site with the lowest number of taxa contained the highest percentage of plants unique to a site. The sites with the most unique collections of species were characterized by an overriding ecological feature; e.g., both beaches (50% unique) and salt water marshes (28% unique) have saline soils. Four sites had no unique taxa: two of these were early colonization sites where unique taxa would not be expected. The third site was the subalpine site at Sunshine Point. Sunshine Point is a lower elevation site in the Robinson Mountains; hence most of these plants overlapped the Robinson Mountains survey. The fourth site, Lakes and Ponds, had many unique plants, but these were also listed as part of the geographic category where they were found. For example, Tashalich Lake plants were also included with those of the Tashalich Fresh Water Wetland.

Three of the herbaceous species are aggressive exotics: *Taraxacum officinale*, *Matricaria discoidea*, and *Poa annua*. All three were found adjacent to Vitus Lake; only one of these, *Poa annua*, was found in abundance. In 2007 we attempted to find the *Matricaria* location again; no *Matricaria* were found. We

TABLE 5. COMPARATIVE SPECIES RICHNESS

Category of sites	Site #	# taxa	% total flora	Unique taxa	% unique/site
Early colonization sites	1–2	22	4.7	0	0
Berg and Weeping Peat	4–6	116	24.9	0	0
Glacial forelands	7–11, 66–73	214	46	11	5.1
Temperate forest	12–16	78	16.7	3	3.8
Subalpine	17–18	93	20	0	0
All alpine sites	19–46	302	65.1	23	7.6
Grindle Hills	19–24	185	39.7	3	1.6
Robinson Mountains	25–35	232	49.8	14	6
Donald Ridge	36–37	148	31.8	2	1.3
Khitrov Hills	38–43	198	42.5	4	2
Nichawak Mountain	44	130	27.9	1	0.7
Carbon Mountain	45–46	165	35.2	6	3.6
Suckling Hills	47–50	184	39.5	3	1.6
Fens	51–60	158	33.9	13	8.2
Fresh water marsh	61–63	137	29.6	16	11.6
Salt water marsh	64–65	39	8.3	11	28.2
Lakes and ponds	74–77	27	5.8	0	0
Beaches	78–82	12	2.5	6	50

also relocated the *Taraxacum* site and destroyed 60+ plants. *Poa annua* is still abundant.

This project has been complementary to a floristic inventory carried out by the U.S. National Park Service north of the Bagley Ice Field. These two studies have filled in a major void in the knowledge of the flora of Alaska. A recent report of Wrangell–Saint Elias Park inventories showed 832 species (887 taxa) (Cook and Roland, 2002). This park is adjacent to the Bering Region, perhaps suggesting that many more plants are yet to be found.

Heritage Plants

Twenty species encountered in the Bering Glacier Region are on the Alaska Natural Heritage Program Vascular Plant Tracking List (listed in the References) (Table 6). An additional three Bering Glacier species were recently removed from the list: *Platanthera chorisiana*, *Carex lenticularis* var. *dolia*, and *Minuartia biflora*. These plants are still rare but are no longer imperiled. The Heritage Tracking List assigns each taxon a numeric rank of relative imperilment on the basis of standard rank factors applied at state (S) and global (G) levels. A ranking of 1 indicates that a species is critically imperiled because of extreme rarity or because of some factor(s) making it especially vulnerable to extinction. In this category, individual organisms number less than a thousand. A ranking of 2 indicates that a species is imperiled (as 1) but that there are at least 2000–10,000 individuals. A ranking of 3 indicates that a species is vulnerable because it is very rare and uncommon throughout its range, or is found only in a restricted range. A ranking of 4 indicates that the species status is apparently secure; the plant is uncommon but not rare. A ranking of 5 indicates that a species is secure; the species is common, widespread, and abundant. The Heritage Program in Alaska focuses on S1–3 plants.

The heritage plants were found at many diverse sites; however, the Robinson Mountains and Carbon Mountain are the richest sites for rare plants. Seven of the heritage plants were found there along with 20 taxa that were found nowhere else in the region. The second richest site was the Tashalich River–Cape Suckling coastal area. There were six Heritage Plants and 17 taxa not found elsewhere. Together, these three sites were home to more than half of the rare plants.

TABLE 6. HERITAGE TRACKING PLANTS CURRENTLY IN THE BERING REGION

	Scientific name	State ranking	Global ranking
1	*Agoseris aurantiaca*	S1	G5
2	*Agrostis thurberiana = A. humilis*	S2	G5
3	*Botrychium alaskense*	S2	G2–3
4	*Carex phaeocephala*	S3	G4
5	*Ceratophyllum demersum*	S1	G5
6	*Eleocharis kamtschatica*	S2–3	G4
7	*Eleocharis quinqueflora*	S1	G5
8	*Gentianella propinqua* ssp. *aleutica*	S2–4	G5
9	*Glehnia littoralis* ssp. *leiocarpa*	S3	G5
10	*Isoetes occidentalis*	S1–2	G4–5
11	*Limosella aquatica*	S3	G5
12	*Lonicera involucrata*	S2	G4–5
13	*Malaxis paludosa*	S3	G4
14	*Papaver alboroseum*	S3	G3–4
15	*Poa laxiflora*	S2–3	G3–4
16	*Poa macrantha*	S1	G5
17	*Potentilla drummondii*	S2	G5
18	*Ranunculus pacificus*	S3	G3
19	*Rorippa curvisiliqua*	S1	G5
20	*Romanzoffia unalaschcensis*	S3	G3

Following is a brief description and the general location for each of the heritage plants.

1. *Agoseris aurantiaca* (S1, G5). *Agoseris aurantiaca*, commonly known as mountain dandelion, is a herb that grows 1–4 dm tall and looks superficially like a dandelion. It can be

found on high slopes and meadows. Though not rare globally, there are few known Alaska populations. In the Bering Glacier Region, two populations of *Agoseris* were found: a single site on a bench high on the slopes of Carbon Mountain and two sites along a northeast ridgeline of the Khitrov Hills.

2. *Agrostis thurberiana* = *A. humilis* (S2, G5). *Agrostis thurberiana*, alpine bentgrass, is a slender 1–3-dm-tall grass found in wet meadows, mostly at alpine sites. It has a disjunct statewide distribution, from the outmost Aleutian Islands, skipping to the Kenai Peninsula, and another skip to southeastern Alaska. We found this plant at five sites: Grindle Hills, Khitrov Hills, Robinson Mountains, Nichawak Mountain, and Donald Ridge.

3. *Botrychium alaskense* (S2–3, G2–3). *Botrychium alaskense,* Western moonwort, is a diminutive eusporangiate fern in which the pinnae are bipinnate. *B. alaskense* was first discovered near Fairbanks in 1999 (Wagner and Grant, 2002). Since then it has been documented for Kodiak Island and the Alaska Peninsula. Our Bering Glacier collection represents a major range extension for this plant.

4. *Carex phaeocephala* (S3, G4). *Carex phaeocephala,* the dunhead sedge, has sessile lateral spikes in a tight group. Its habitat is alpine rocky ridges and slopes, with its primary distribution in southern Alaska, southern Yukon Territory, and south to California. In the Bering Glacier Region, this sedge was found on alpine slopes in the Grindle Hills, Robinson Mountains, Khitrov Hills, and Carbon Mountain.

5. *Ceratophyllum demersum* (S1, G5). *Ceratophyllum demersum* is commonly known as coon's tail. Plants are aquatic, entirely submersed in water, and lack roots. Stiff leaves are serrated and whorled at the nodes, crowded toward the tip, giving the appearance of a "coon tail." *Ceratophyllum* has a broad global distribution but is rare in Alaska, known only from half a dozen sites in the interior of the state. It is most often found in hard-water lakes.

6. *Eleocharis kamtschatica* (S2–3, G4). *Eleocharis kamtschatica*, the Kamchatka spikerush, is a 10-cm-tall graminoid plant growing in brackish soils along the coast. It is identified from other spikerushes by the large size of the tubercle, a solitary basal scale, and its coastal habitat.

7. *Eleocharis quinqueflora* (S1, G5). *Eleocharis quinqueflora*, few-flowered spikerush, is similar in size to *E. kamtschatica* but has a very small, barely distinguishable tubercle and prefers fresh water marshes. It has a Cordilleran distribution, following the mountains into central Alaska. Its near coastal appearance at the Bering Glacier is a range extension.

8. *Gentianella propinqua* ssp. *aleutica* (S2–4, G5). *Gentianella propinqua* ssp. *aleutica*, the Aleutian gentian, is a small (up to 10 cm) pale gentian with yellowish flowers (Fig. 21). It is not particularly rare at the Bering Glacier, having been found in the Robinson Mountains, Grindle Hills, Suckling Hills, and Nichawak Mountain. It is a range connection between the Aleutian Islands and one population known from Skagway.

9. *Glehnia littoralis* ssp. *leiocarpa* (S3, G5). *Glehnia littoralis* ssp. *leiocarpa*, commonly known as silvertop (Fig. 22),

Figure 21. *Gentianella propinqua* ssp. *aleutica* growing with *Salix arctica.*

Figure 22. *Glehnia littoralis* ssp. *leiocarpa* growing at Cape Suckling.

is found scattered on seashores along the west coast of North America. It has a herbaceous stem growing from a woody taproot; typically the petioles of the leaves are buried in the sand. In the Bering Glacier Region, this plant was found at the mouth of the Tsiu River and on the beach at Cape Suckling. Other known sites in Alaska are Kodiak Island and Lynn Canal.

10. *Isoetes occidentalis* (S1–2, G4–5). *Isoetes occidentalis* is commonly known as the western quillwort. It is a small aquatic plant consisting of a number of awl-shaped leaves from a somewhat swollen base. Disjunct populations are found coastally in Alaska, from the southeast to the Aleutian Islands.

11. *Limosella aquatica* (S3, G5). *Limosella aquatica* is a small aquatic plant commonly known as mudwort. It is found along mud banks and shallow water in widely disjunct localities

Figure 23. *Limosella aquatica* growing in the muddy shores of the Tashalich River.

Figure 24. *Papaver alboroseum* on a steep talus slope.

in Alaska, including the Seward Peninsula, Bristol Bay, the Aleutian Islands, and central Yukon Territory. It was found at two coastal river sites, Tashalich River (Fig. 23) and Tsiu River.

12. *Lonicera involucrata* (S2, G4–5). Commonly known as the twinberry honeysuckle, *Lonicera involucrata* is an understory shrub 1–3 m tall. Until it was found in the Bering Glacier Region, known locations were disjunct between Skagway and Ketchikan in SE Alaska. It was found only in the forest near the Tashalich River.

13. *Malaxis paludosa* (S3, G4). *Malaxis paludosa*, bog adder's tongue, is an inconspicuous (5–10 cm) orchid found in muskegs and bogs. The Bering Glacier plants are a range connection between disjunct populations in Cook Inlet and Skagway and Ketchikan.

14. *Papaver alboroseum* (S3, G3–4). *Papaver alboroseum*, commonly known as the pink poppy, is a small poppy that grows from 0.6 to 1.2 dm tall. It prefers gravelly soils and rocky outcrops. It occurs in discrete populations scattered in south-central and interior Alaska. In the Bering Glacier it was found on an alpine scree slope at Carbon Mountain (Fig. 24).

15. *Poa laxiflora* (S2–3, G3–4). *Poa laxiflora*, the loose-flower bluegrass, grows 0.8–1.2 m tall. It is found in lowland coastal sites in southeastern Alaska. At the Bering Glacier it was found at two coastal riparian sites, the Seal River and the Tashalich River.

16. *Poa macrantha* (S1, G5). *Poa macrantha*, commonly known as the seashore bluegrass, is a range extension from British Columbia and southeastern Alaska. It is a handsome grass, growing 50–75 cm tall. In the Bering Glacier Region it was found only on the sandy beach at Cape Suckling (Fig. 25).

17. *Potentilla drummondii* (S2, G5). *Potentilla drummondii*, Drummond's cinquefoil, is a small yellow rose found in alpine gravels and scree slopes. This flower is known in Alaska only in south-central Alaska and east to the Bering Glacier

Figure 25. *Poa macrantha* at Cape Suckling.

Region, where it was found only on a south-facing scree slope at Carbon Mountain.

18. *Ranunculus pacificus* (S3, G3). *Ranunculus pacificus*, the Pacific buttercup, is a perennial herb 2–5 dm tall. It prefers moist sites in meadows and woodlands. It is found in southeastern Alaska and is endemic to the Pacific Northwest. It was found at three sites, the Tashalich forest site, the glacier forelands, and the Oaklee River drainage. All of the populations were small.

19. *Rorippa curvisiliqua* (S1, G5). *Rorippa curvilisiqua*, commonly known as the curvepod yellowcress, is an annual (or possibly a biennial) plant similar to the common *Rorippa palustris*, except that the capsules are curved. It typically grows 20–40 cm tall with conspicuous yellow flowers. Only one solitary plant was found, so *R. curvisiliqua* was not collected. Apparently its seeds travel, clinging to waterfowl, and grow wherever deposited; hence plants are often solitary. As the plant requires 120 frost-free days with a minimum temperature of 52°, it is unlikely to become established in the Bering Glacier Region, according to the Plants Database of the U.S. Department of Agriculture (see listing in the References).

20. *Romanzoffia unalaschcensis* (S3, G3). *Romanzoffia unalaschcensis*, the Aleutian mist-maid, is a small, white flowered herb 5–20 cm tall, though usually shorter in the Bering Glacier Region. It can be found coastally from Prince William Sound, westward along the Alaska Peninsula, and to the Aleutian Islands. At the Bering Glacier this plant was found at Berg Lake, Grindle Hills, Robinson Mountains, and Donald Ridge, together constituting an eastward extension of its range (Fig. 26).

Figure 26. *Romanzoffia unalaschensis* in the Grindle Hills is distinguished from the common *R. sitchensis* by viscid pubescent leaves.

Range Connections and Extensions

To date, 466 taxa have been documented, representing approximately one-quarter of the Alaska flora (Hultén,1968). These plants include range extensions, range fillers, and range connections. Three hundred fifteen of the plant finds were within their known expected range. Ninety-three species are range extensions, plant populations found beyond their known range, as indicated by Hultén (1968). Forty-three of these extensions were short, just over the Bagley Ice Field. This indicates that the Bagley Ice Field is a major barrier to plant dispersal. The remainder of the extensions was from interior Alaska, the Aleutians, the Yukon Territory, or southeastern Alaska (Table 7). This table does not include those merely across the Bagley Ice Field or those just beyond. In addition, 58 species fill distribution gaps between disjunct populations. For example, *Adiantum aleuticum* has populations in the Aleutians, Kenai Peninsula, and southeastern Alaska.

Management Considerations

Public lands are mandated for multiple use. This mandate requires knowledge of ecological resources so that uses can be balanced and resource conflicts can be reconciled. A baseline

TABLE 7. LIST OF MAJOR RANGE EXTENSIONS

Range	Species found at the Bering Glacier
West of Copper River	*Calamagrostis deschampsioides*
Aleutians	*Saxifraga punctata* var. *insularis, Romanzoffia unalaschcensis*
Southeastern Alaska	*Poa laxiflora, Monotropa hypopitys, Lonicera involucrata, Ranunculus pacificus, Listera caurina, Rorippa curvisiliqua*
Alaska Range	*Primula eximia, Antennaria friesiana* ssp. *alaskana, Minuartia arctica, Ranunculus karelinii, R. gmelinii, Oxytropis jordalii, Carex magellanica, Ceratophyllum demersum, Cicuta virosa, Eleocharis quinqueflora, Festuca brevissima, Juncus biglumis, Potamogeton foliosus, Ranunculus occidentalis* var. *nelsoni, Selaginella sibirica, Silene involucrata, Stellaria crassifolia, Botrychium alaskense*
North of the Bagley Ice Field	*Arctagrostis latifolia, Arctostaphylos alpina, Carex maritima, C. membranacea, Diapensia lapponica, Dodecatheon frigidum, Draba lactea, Eleocharis acicularis, Erigeron humilis, Gentiana prostrata, Lagotis glauca, Linnaea borealis, Packera ogotorukensis, Pedicularis lanata, P. sudetica, Platanthera obtusata, Potentilla nana, Silene acaulis* var. *subacaulescens*
Yukon Territory	*Agoseris aurantiaca, Antennaria friesiana* ssp. *neoalaskana, Antennaria monocephala* ssp. *angustata*
Cordilleran Mountains	*Juncus filiformis, Festuca saximontana*

Note: This list does not include extensions merely across the Bagley Ice Field or those just beyond.

inventory of these important resources, including botanical surveys, is essential to planning and decision making. The East Alaska Resource Management Plan (RMP), which covers the Bering Glacier Region, states as a goal to manage vegetation so as to restore forest and riparian health, improve productivity and biological diversity, and provide for fish and wildlife habitat.

The Bering Glacier Region was designated as a Research Natural Area in 2007. This implies several management considerations in order to preserve the pristine environment. The vegetation at Bering Glacier shows all stages of community development from colonization on land recently exposed by the retreating glacier to climax forest communities. All of the stages are important. Each of the seral stages has significance for birds, mammals, and other wildlife. Many animals prefer to live at the boundary of two ecosystems, thereby diversifying the food supply and availability of shelters.

The goal should be to maintain a full range of natural plant communities with constituent species native to this part of south-central Alaska. The biggest threat to the plant communities is the introduction of exotic species. Of the three species already noted for the area, only one spreads by airborne seeds *(Taraxacum officinale)*, whereas the other two *(Poa annua* and *Matricaria discoidea)* require animal or human transport. It is recommended that the U.S. Bureau of Land Management (BLM) establish an aggressive program to prevent the introduction and spread of exotic species. It is also recommended that a program be put into place to remove exotic species. This will require training to ensure that only the exotic species are removed. For example, one of the rarest plants at the Bering Glacier is commonly known as the mountain dandelion *(Agoseris aurantiaca)*, and superficially it looks like the common dandelion, *Taraxacum officinale*. As this example illustrates, a poorly implemented exotic species program could do more harm than good.

The Bering Glacier Region has a wealth of natural beauty and is now becoming a destination for some ecotourism. Unfortunately, as people visit the Bering Region they can inadvertently carry with them seeds hidden on boot treads, folded into camping tarps and tents, or hitchhiking on airplane wheels. Exotic species that are able to establish themselves in the new environment then compete with native species and cause changes in the ecosystem. Some can become a major ecological threat by dominating their new community.

Two regions within the Bering region stand above others for their concentration of rare and unusual species. The first of these are the alpine sites, specifically Carbon Mountain, Khitrov Hills, and Robinson Mountains. They have seven of the heritage Tracking Plants and several range extensions, including, at Carbon Mountain: *Gentiana prostrata*, *Salix rotundifolia*, and *Selaginella sibirica*; at Khitrov Hills: *Saxifraga tricuspidata* and *Ranunculus karelinii*; and at Robinson Mountains: *Papaver radicatum*, *Dodecatheon frigidum*, and *Linnaea borealis.*

The second region includes Cape Suckling from the Tashalich River to the Oaklee River. This region is home to six more of the Heritage Tracking Plants as well as being the only site where we found *Carex macrocephala*, *Luzula divaricata*, *Eleocharis macrostachya*, and *Lomatogonium rotatum*. The unique combination of species in these two areas qualifies them for special conservation concern.

Few areas in the world as pristine as the Bering Glacier Region still exist. As such, it is critical to protect this unusual ecological system from human disturbance. The inventory of vascular flora found in diverse habitats surrounding the Bering Glacier is essential for prudent management of this globally significant region. This and other biological and geological surveys provide crucial information for resolving multiple use conflicts that may arise. This survey fills in a major geographical gap in our knowledge of plant biogeography in southern Alaska and serves as a baseline to track future habitat changes. Plant survey work should continue until all ecosystems in the region have been thoroughly documented. This survey also provides information critical for future ecological studies.

ADDENDUM

One additional field season occurred after the writing of this chapter. Due to weather conditions, most of the last field data was collected near the Gulf of Alaska. Twelve additional plants were added to the species list: *Atriplex gmelinii*, *Carex limosa*, *Juncus supiniformis*, *Malaxis brachypoda*, *Pinguicula villosa*, *Poa trivialis*, *Rhynchospora alba*, *Rosa rugosa*, *Rubus chamaemorus*, *Viola adunca*, and *Zanichellia palustris*. Only *Carex limosa* was a range extension, five species were range connectors: *Rhynchospora alba*, *Juncus supiniformis*, *Poa trivialis*, *Sisrinchium litorale*, *Viola adunca*, and *Zanichellia palustris*. Significantly two species were introduced exotics: *Rosa rugosa* and *Poa trivialis*. *Zanichellia palustsris* was the only addition to the Natural Heritage Tracking List and was evaluated at S3G3. The status of *Malaxis brachypoda* is under evaluation.

APPENDIX 1: SITE LOCATIONS AND YEAR VISITED

Category	Site #	Site	Lat-deg	Ladecmin	Long-deg	Lodecmin	Datum	Elev (m)	Year
Colonization sites	1	Arrowhead Island	60	10.238	143	17.41	Unk	35	2001
	2	Ancient Forest	60	8.21	143	37.07	Unk	10	2001–2
	3	Weeping Peat	60	11.37	143	12.27	Unk	28	2000–1
	3	Weeping Peat	60	11.446	143	11.922	NAD83	13	2007
	4	Berg Lake	60	25.504	143	50.67	Unk	142–143	2001
	4	Berg Lake	60	24.474	143	51.33	WGS84	152	2002
Glacial forelands	5	Camp	60	7.226	143	16.998	Unk	10	2000–5
	6	Cabin	60	7.4	143	18.6	NAD83	7	2002–4
		Kaliakh	60	5.662	142	55.057	NAD83	1	2008
	7	Midtimber Lake Trail	60	6.277	143	20.82	NAD27	5–12	2003
	8	Tsiu Landing strip	60	6.456	143	12.095	NAD27	23	2005
	9	Tsiu Drainage	60	5.588	142	59.212	NAD83	2	2007
Temperate forest	10	Hanna Forest	60	13.731	143	6.598	Unk	28	2001
	11	River Forest	60	13.558	142	59.41	NAD27	47	2003
	12	Tashalich Forest	60	2.23	143	38.91	NAD27	6	2003
	13	Kaliakh Forest	60	13.55	142	59.31	NAD27	39	2005
	14	Swan Lake Forest	60	8.254	142	58.412	NAD83	29	2007
		Cape Suckling Forest	59	59.776	143	54.681	NAD83	14	2008
Subalpine	15	Sunshine Point	60	10.775	142	50.389	NAD27	477	2005
	15	Sunshine Point	60	11.444	142	48.579	NAD27	451	2005
Alpine	16	Grindle Hills	60	16.67	143	18.97	Unk	610	2000
	16	Grindle Hills	60	16.304	143	16.809	Unk	417–752	2001
	16	Grindle Hills	60	16.813	143	16.511	WGS84	443	2002
	16	Grindle Hills	60	16.7	143	19.006	NAD27	484	2003
	16	Grindle Hills	60	15.79	143	8.67	NAD27	611	2003
	16	Grindle Hills	60	16.34	143	16.66	NAD27	743	2005
	17	Robinson Mts.	60	18.59	143	2.41	Unk	615	2000
	17	Robinson Mts.	60	18.282	143	1.155	Unk	620	2001
	17	Robinson Mts.	60	15.768	142	58.898	WGS84	583	2002
	17	Robinson Mts.	60	18.82	142	57.033	WGS84	947	2002
	17	Robinson Mts.	60	18.52	143	2.049	NAD27	616	2003
	17	Robinson Mts.	60	18.16	143	0.17	NAD27	633	2003
	17	Robinson Mts.	60	18.61	143	1.81	NAD27	712	2004
	17	Robinson Mts.	60	17.95	142	55.46	NAD83	926	2004
	17	Robinson Mts.	60	18.516	143	1.164	NAD27	736	2005
	17	Robinson Mts.	60	18.522	143	1.701	NAD 83	653	2007
	17	Robinson Mts.	60	18.444	143	0.732	NAD 84	711	2007
		Robinson Mts	60	15.470	142	59.359	NAD83	705	2008
	18	Donald Ridge	60	15.471	143	4.902	WGS84	644	2002
	18	Donald Ridge	60	15.59	143	5.64	NAD27	360	2003
	19	Khitrov Hills	60	25	143	28	Unk	473–700	2000
	19	Khitrov Hills	60	26.575	143	25.168	Unk	736	2001
	19	Khitrov Hills	60	26.6	143	25.2	Unk	924	2001
	19	Khitrov Hills	60	20.996	143	32.125	WGS84	385	2002
	19	Khitrov Hills	60	24.14	143	30.8	NAD27	659	2003
	19	Khitrov Hills	60	26.708	143	15.09	NAD83	1102	2007
	20	Nichawak Mt.	60	13.25	143	59.6	NAD83	480–514	2005
	21	Carbon Mountain	60	23.902	143	42.47	NAD27	627	2004
	21	Carbon Mountain	60	26.428	143	53.389	NAD27	822	2005

(*continued*)

APPENDIX 1: SITE LOCATIONS AND YEAR VISITED (*continued*)

Category	Site #	Site	Lat-deg	Ladecmin	Long-deg	Lodecmin	Datum	Elev (m)	Year
Coastal uplands	22	Suckling Hills	60	5.62	143	45.68	WGS84	400–500	2002
	22	Suckling Hills	60	6.287	143	45.181	NAD27	245–254	2003
	22	Suckling Hills	60	2.427	143	50.458	NAD27	218–243	2003
	22	Suckling Hills	60	4.001	143	46.768	NAD27	364	2004
		Cape Suckling Cliffs	59	59.796	143	56.041	NAD83	5	2008
Fens	23	Lower Hanna Fen	60	13.731	143	6.598	Unk	150	2001
	23	Donald Ridge Fen	60	15.59	143	5.64	NAD27	360	2003
	24	Hanna Sloping Fen	60	13.99	143	5.89	NAD27	141	2003
	25	Kaliakh River Fen	60	13.55	142	59.41	NAD27	42	2003
	25	Kaliakh River Fen	60	13.55	142	59.31	NAD27	39	2004
	26	Hope Fen	60	10.875	142	45.568	NAD27	126	2005
	27	Odor Creek Fen	60	12.595	142	48.14	NAD27	300	2005
	28	Swan Lake Fen	60	8.148	142	56.683	NAD83	13	2007
	29	K_K Fen	60	11.399	142	52.875	NAD83	65	2007
	30	Upper K Fen	60	17.035	142	46.878	NAD83	147	2007
		Cape Fen	59	59.776	143	54.681	NAD83	12	2008
		E. Suckling Fen	60	0.749	143	54.207	NAD83	68	2008
Fresh water marsh	31	Kaliakh R. Marsh	60	6.06	142	54.715	NAD27	22	2005
	32	Suckling Lowland	60	1.215	143	57.002	NAD27	5	2005
	33	Oaklee FW Marsh	60	2.233	144	4.273	NAD83	5	2007
		North Tashalich Lagoon	60	2.884	143	37.638	NAD83	6	2008
Salt marsh	34	Oaklee R.Marsh	60	2.588	144	1.912	NAD27	3	2005
	34	Oaklee R.Marsh	60	2.238	144	3.621	NAD83	4	2007
Riparian	35	Hanna Creek	60	14.08	143	7.6	Unk	60	2001
	35	Hanna Lake Shore	60	14.262	143	7.218	NAD27	60	2003
	36	Seal River	60	3.226	143	28.144	WGS84	7	2002
	37	Tashalich River	60	2.22	143	38.9	NAD27	6	2003
	37	Tashalich River	60	2.515	143	37.076	NAD83	6	2007
	38	Kiklukh River	59	59.308	143	52.272	NAD27	4	2004
	39	Mid-timber River	60	3.51	143	18.79	NAD27	4	2005
	40	Tsivat River	60	5.866	142	54.747	NAD27	20	2005
Ponds and lakes		Hanna Lake SW	60	14.32	143	7.5	NAD27	66	2004
		Swan Lake	60	8.148	142	56.683	NAD83	13	2007
		Tashalich Lake	60	2.364	143	37.381	NAD84	5	2007
		Tsiu Lake	60	5.349	142	59.956	NAD83	10	2007
Beaches	41	Seal Beach	60	3.226	143	28.144	WGS84	1	2002
	42	Tashalich Beach	59	59.3	143	52.27	NAD27	1	2003
	43	Kiklukh Beach	60	3.37	143	18.79	NAD27	1	2004
	44	Mid-Timber Beach	60	5.709	142	54.747	NAD27	1	2005
	45	Cape Suckling	60	0.125	143	58.07	NAD83	1	2007

Note: 2008 sites are not included in the chapter text. Ladecmin—latitude in decimal minutes; lodecmin—longitude in decimal minutes; NAD27—North American Datum 1927; NAD83—North American Data 1983; WGS84—Word Geodetic System 1984.

APPENDIX 2: TAXA, RANGE, SITE, AND HERITAGE TRACKING LIST STATUS

									Alpine sites												
New in 2008	Plants Database Name	Range[1]	Status	Early colonization	Berg Lake, Weeping peat	Glacial Forelands	Temperate Forest	Subalpine Park	Grindle Hills	Robinson Mts	Donald Ridge	Khitrov Hills	Nichawak Mountain	Carbon Mt	Suckling Hills	Fens	Fr Marsh	Salt Marsh	Lakes, Ponds, ect	Beach	CONSTANCY
	SITES			1-2	3-4	6-9,35-40	10-14	15	16	17	18	19	20	21	22	23-30	31-33	34	§	41-45	
	Achillea millefolium var. borealis	R			1	1		1	1	1	1	1	1	1	1		1	1		1	13
	Aconitum delphiniifolium	R						1	1	1	1	1	1	1	1	1					9
	Actaea rubra ssp. arguta	R			1	1	1					1	1		1		1				7
	Adiantum aleuticum	C			1	1									1						3
	Agoseris aurantiaca var. aurantiaca	E	S1,G5									1		1							2
	Agrostis aequivalvis	C				1		1		1		1	1		1	1					7
	Agrostis alaskana#	R			1					1		1				1	1				5
	Agrostis exarata	R		1	1	1			1			1	1		1	1	1				9
	Agrostis humilis	C	S2,G5					1	1	1		1	1			1					6
	Agrostis mertensii	R			1	1			1	1	1	1	1	1	1	1					10
	Agrostis stolonifera	Cs				1											1				2
	Alnus viridis ssp. sinuata	R		1	1	1	1	1	1	1	1	1	1	1	1	1	1				14
	Alopecurus aequalis var. aequalis	R				1		1		1											3
	Alopecurus alpinus	C								1					1						2
	Anaphalis margaritacea	C			1	1						1		1							4
	Andromeda polifolia var. polifolia	R			1											1					2
	Anemone multifida	B										1									1
	Anemone narcissiflora var. monantha	B						1	1	1	1	1	1	1	1	1					9
	Anemone richardsonii	R								1		1									2
	Angelica genuflexa	R													1	1	1				3
	Angelica lucida	R				1	1	1	1	1	1	1	1		1	1	1				11
	Antennaria alpina	C								1	1	1									3
	Antennaria friesiana ssp. alaskana	E								1		1									2
	Antennaria friesiana ssp. neoalaskana	E										1									1
	Antennaria monocephala ssp. angustata	E								1											1
	Antennaria monocephala ssp. monocephala	R							1			1		1							3
	Antennaria rosea ssp. confinis*	R							1	1											2
	Antennaria rosea ssp. pulvinata	B							1	1		1		1							4
	Aquilegia formosa	R			1	1			1	1		1			1						6
	Arabis eschscholtziana	R			1	1									1		1			1	5
	Arabis kamchatica	R			1	1			1	1		1	1	1	1						8
	Arctagrostis latifolia	E							1			1			1	1					4
	Arctophila fulva	R				1										1	1				3
	Arctostaphylos alpina	E							1	1		1		1							4
	Arctostaphylos uva-ursi	B												1							1
	Argentina egedii ssp. egedii	R				1											1	1			3
	Arnica amplexicaulis	C							1							1					2
	Arnica latifolia	R				1		1	1	1	1	1	1	1	1						9
	Arnica lessingii	R							1	1	1	1		1	1						6
	Artemisia arctica ssp. arctica	R						1	1	1	1	1	1	1	1						8
	Artemisia furcata	Cs											1								1
	Artemisia tilesii ssp. elatior	B			1	1				1					1						4
	Aruncus dioicus ssp. vulgaris	R			1	1	1	1	1	1	1	1		1	1	1					11
	Astragalus alpinus var. alpinus	B			1					1				1							3
	Athyrium americanum	R			1																1
	Athyrium filix-femina ssp. cyclosorum	R			1	1	1	1	1	1	1	1	1	1	1	1	1				13
*	Atriplex gmelinii	R																		1	1
	Barbarea orthoceras	R				1	1		1	1						1					5
	Blechnum spicant	R				1	1	1		1						1					5
	Botrychium alaskense	E	S2,G2-3							1											1
	Botrychium lanceolatum	R				1				1	1	1	1	1							6
	Botrychium lunaria	R				1			1			1		1							4
	Botrychium minganense	R				1								1							2
	Botrychium pinnatum	R								1		1	1	1							4
	Cakile edentula ssp. edentula	R																		1	1
	Calamagrostis canadensis ssp. canadensis	R			1	1	1	1	1	1	1	1	1	1	1	1	1				13
	Calamagrostis canadensis ssp. langsdorffii	R															1				1
	Calamagrostis deschampsoides	E															1	1			2
	Calamagrostis nutkaensis	R				1	1	1					1		1	1					6
	Calamagrostis stricta ssp. inexpansa	R				1									1	1	1	1	1		6

(continued)

APPENDIX 2: TAXA, RANGE, SITE, AND HERITAGE TRACKING LIST STATUS (*continued*)

									Alpine sites													
New in 2008	Plants Database Name	Range[1]	Status	Early colonization	Berg Lake, Weeping peat	Glacial Forelands	Temperate Forest	Subalpine Park	Grindle Hills	Robinson Mts	Donald Ridge	Khitrov Hills	Nichawak Mountain	Carbon Mt	Suckling Hills	Fens	Fr Marsh	Salt Marsh	Lakes, Ponds, ect	Beach	CONSTANCY
	Callitriche palustris	R				1					1	1			1	1	1				6
	Caltha leptosepala ssp. leptosepala	R						1	1	1	1	1	1	1	1						8
	Caltha palustris var. palustris	R				1	1					1			1	1	1				6
	Campanula lasiocarpa	R							1	1	1	1	1	1	1						7
	Campanula rotundifolia	R			1	1			1	1	1	1		1	1						8
	Cardamine bellidifolia var. bellidifolia	B							1	1	1	1		1	1						6
	Cardamine oligosperma var. kamtschatica	R			1	1	1		1	1	1	1	1	1	1		1				11
	Carex anthoxanthea	R				1		1	1	1	1	1	1	1	1	1					10
	Carex aquatilis var. aquatilis	R			1	1						1				1					4
	Carex aquatilis var. dives	R				1									1	1	1				4
	Carex bicolor	B			1	1											1				3
	Carex canescens	R													1	1	1				3
	Carex circinata	R							1	1	1	1	1		1						6
	Carex diandra	B															1				1
	Carex echinata ssp. phyllomanica	R														1					1
	Carex garberi	B			1	1															2
	Carex glareosa	C																1			1
	Carex gmelinii	R				1											1				2
	Carex gynocrates	R															1				1
	Carex lachenalii	B							1	1	1	1	1								5
	Carex laeviculmis	R														1					1
	Carex lenticularis var. dolia	R			1	1			1	1					1	1					6
	Carex lenticularis var. limnophila	C				1															1
	Carex lenticularis var. lipocarpa	R		1	1	1			1			1					1				6
*	Carex limosa	B														1					1
	Carex livida	R														1					1
	Carex lyngbyei	R				1										1	1	1			4
	Carex mackenzii	R																1			1
	Carex macrocephala	R				1														1	2
	Carex macrochaeta	R			1			1	1	1	1	1	1	1	1						9
	Carex magellanica	E														1					1
	Carex maritima	E			1	1											1				3
	Carex membranacea	E								1											1
	Carex mertensii	R			1	1	1		1	1				1							6
	Carex nigricans	R							1	1		1	1	1		1					6
	Carex pachystachya	R				1						1					1				3
	Carex pauciflora	R							1							1					2
	Carex phaeocephala	C	S3,G4						1	1		1		1							4
	Carex pluriflora	R			1										1	1	1				4
	Carex pyrenaica ssp. micropoda	C						1	1	1	1	1	1	1							7
	Carex saxatilis	R				1				1					1	1	1				5
	Carex scirpoidea	B								1				1							2
	Carex spectabilis	R								1											1
	Carex utriculata	R				1															1
	Carex viridula ssp. viridula	B				1											1				2
	Cassiope lycopodioides	R							1	1	1			1							4
	Castilleja parviflora var. parviflora	R			1			1	1	1	1	1	1	1							8
	Castilleja unalaschcensis	R			1	1		1	1	1	1	1	1	1	1		1				11
	Cerastium beeringianum ssp. beeringianum	R			1	1			1	1	1	1	1	1	1						9
	Cerastium fontanum	R			1	1															2
	Ceratophyllum demersum	E	S1,G5														1		1		2
	Chamerion angustifolium	R			1	1		1	1	1	1	1	1	1	1	1	1				12
	Chamerion latifolium	R		1	1	1			1	1	1	1		1	1						9
	Chrysanthemum arcticum	R																1			1
	Cicuta douglasii	R				1	1									1	1			1	5
	Cicuta virosa	E															1				1
	Cinna latifolia	R				1					1				1						3
	Circaea alpina	R				1	1	1		1			1				1				6
	Claytonia sibirica var. sibirica	R				1						1	1		1						4
	Comarum palustre	R				1	1			1					1	1	1				6
	Conioselinum gmelinii	R				1		1	1	1	1	1	1	1	1						9

(*continued*)

APPENDIX 2: TAXA, RANGE, SITE, AND HERITAGE TRACKING LIST STATUS (*continued*)

									Alpine sites												
New in 2008	Plants Database Name	Range[†]	Status	Early colonization	Berg Lake, Weeping peat	Glacial Forelands	Temperate Forest	Subalpine Park	Grindle Hills	Robinson Mts	Donald Ridge	Khitrov Hills	Nichawak Mountain	Carbon Mt	Suckling Hills	Fens	Fr Marsh	Salt Marsh	Lakes, Ponds, ect	Beach	CONSTANCY
	Coptis asplenifolia	R					1									1					2
	Coptis trifolia	R					1					1			1	1	1				5
	Cornus canadensis	R					1	1	1		1	1	1			1					7
	Cornus suecica	R							1	1	1	1	1	1	1	1					8
	Cryptogramma acrostichoides	R							1	1	1	1		1	1						6
	Cystopteris fragilis	R			1	1			1	1	1	1			1						7
	Dactylorhiza viridis	R							1	1		1		1							4
	Deschampsia beringensis	R				1		1	1	1	1	1			1	1	1	1			10
	Deschampsia cespitosa	R							1	1	1	1	1		1						6
	Diapensia lapponica ssp. obovata	E							1	1	1	1									4
	Dodecatheon frigidum	E								1											1
	Dodecatheon jeffreyi ssp. jefferyi	C						1	1	1		1			1	1					6
	Dodecatheon pulchellum	R				1									1						2
	Draba lactea	E								1			1								2
	Draba stenoloba var. stenoloba	R							1	1	1	1	1	1							6
	Drosera anglica	R														1					1
	Drosera rotundifolia	R				1										1	1				3
	Dryas integrifolia	B								1											1
	Dryas octopetala	B								1											1
	Dryopteris expansa	R			1	1	1	1	1	1	1	1	1	1	1	1	1				13
	Eleocharis acicularis	E															1				1
	Eleocharis kamtschatica	Cs	S2-3,G4															1			1
	Eleocharis macrostachya	R			1	1											1				3
	Eleocharis palustris var. palustris	R			1	1											1	1			4
	Eleocharis quinqueflora	E	S1,G5														1				1
	Elliottia pyrolaeflora	C							1					1							2
	Elymus hirsutus	C					1	1							1	1	1				5
	Empetrum nigrum	R			1	1			1	1	1	1	1	1	1	1					10
	Epilobium anagallidifolium	R			1	1			1	1	1	1			1						7
	Epilobium ciliatum ssp. ciliatum	B		1	1	1			1	1			1				1				7
	Epilobium ciliatum ssp. glandulosum	R		1	1	1							1		1	1	1				7
	Epilobium hornemannii ssp. behringianum	R			1	1					1	1		1	1						6
	Epilobium hornemannii ssp. hornemannii	R		1	1	1			1	1	1	1			1		1				9
	Epilobium lactiflorum	R										1				1					2
	Epilobium leptocarpum	R			1	1		1		1			1	1	1						7
	Epilobium luteum	R			1	1			1	1	1	1			1						7
	Epilobium palustre	C			1	1	1								1	1	1		1		7
	Equisetum arvense	R		1	1	1	1		1	1	1	1	1	1	1	1	1	1		1	15
	Equisetum fluviatile	R			1	1						1				1	1		1		6
	Equisetum palustre	R				1										1	1				3
	Equisetum variegatum	R		1	1	1				1							1				5
	Erigeron humilis	E								1		1		1							3
	Erigeron peregrinus ssp. peregrinus	R							1	1	1	1	1	1	1	1					8
	Eriophorum angustifolium	R			1	1			1	1		1			1	1					7
	Eriophorum chamissonis	R			1	1				1						1	1				5
	Eriophorum scheuchzeri	R				1			1	1											3
	Euphrasia mollis	R		1	1	1			1	1		1		1			1				8
	Eurybia sibirica	B			1	1				1					1						4
	Festuca brachyphylla	R			1				1	1	1	1	1	1	1						8
	Festuca brevissima	E									1										1
	Festuca rubra	R				1			1			1			1		1	1		1	7
	Festuca saximontana var. saximontana	E								1											1
	Fragaria chiloensis	R		1	1	1									1		1				5
	Fritillaria camschatcensis	R				1		1	1	1	1	1	1	1	1	1					10
	Galium aparine	R				1										1	1			1	4
	Galium trifidum ssp. columbianum	R				1															1
	Galium trifidum ssp. trifidum	B				1	1										1				3
	Galium triflorum	R				1	1								1		1				4
	Gentiana douglasiana	R													1	1					2
	Gentiana platypetala	R						1	1	1	1	1	1	1	1						8
	Gentiana prostrata	E												1							1

(*continued*)

APPENDIX 2: TAXA, RANGE, SITE, AND HERITAGE TRACKING LIST STATUS (*continued*)

									Alpine sites												
New in 2008	Plants Database Name	Range†	Status	Early colonization	Berg Lake, Weeping peat	Glacial Forelands	Temperate Forest	Subalpine Park	Grindle Hills	Robinson Mts	Donald Ridge	Khitrov Hills	Nichawak Mountain	Carbon Mt	Suckling Hills	Fens	Fr Marsh	Salt Marsh	Lakes, Ponds, ect	Beach	CONSTANCY
	Gentianella amarella ssp. acuta	R				1						1		1	1		1				5
	Gentianella propinqua ssp. aleutica	E	S2-4,G5						1	1			1		1						4
	Gentianella propinqua ssp. arctophila	B				1			1	1	1				1						5
	Geranium erianthum	R					1	1	1	1	1	1	1	1	1	1					10
	Geum calthifolium	R			1	1		1	1	1	1	1	1	1	1	1					11
	Geum macrophyllum	R			1	1	1	1				1	1		1		1				8
	Glehnia littoralis ssp. leiocarpa	C	S3,G5																	1	1
	Gymnocarpium dryopteris	R				1	1	1	1	1	1	1	1	1	1	1	1				12
	Harrimanella stelleriana	R						1	1	1	1	1	1	1	1	1					9
	Hedysarum alpinum	B			1		1			1											3
	Heracleum maximum	R			1	1		1	1	1	1	1	1	1	1		1				11
	Heuchera glabra	R			1	1		1	1	1	1	1	1	1	1	1					11
	Hieracium triste	R						1	1	1	1	1	1	1	1						8
	Hierochloe alpina	R							1	1	1	1	1	1							6
	Hierochloe hirta ssp. arctica	R				1				1					1		1	1			5
	Hippuris montana	R						1	1	1	1	1	1	1	1	1					9
	Hippuris tetraphylla	R																1	1		2
	Hippuris vulgaris	R				1	1									1	1		1		5
	Honckenya peploides ssp. major	R				1														1	2
	Hordeum brachyantherum ssp. brachyantherum	R			1	1									1		1	1			5
	Huperzia chinensis	C				1	1	1		1					1						5
	Huperzia haleakalae	R				1	1	1	1	1	1	1	1	1	1	1					11
	Iris setosa ssp. setosa	R				1									1	1	1	1			5
	Isoetes occidentalis	Cs	S1-2, G4-5													1			1		2
	Juncus albescens	R				1															1
	Juncus alpinoarticulatus	R		1	1	1											1				4
	Juncus ambiguus	R															1				1
	Juncus arcticus	R		1	1	1				1					1		1	1			7
	Juncus biglumis	E				1				1											2
	Juncus bufonius	R			1	1											1	1			4
	Juncus castaneus	R			1	1				1											3
	Juncus drummondii	C								1	1	1		1							4
	Juncus ensifolius	C														1					1
	Juncus falcatus ssp.sitchensis	C			1	1		1									1	1			5
	Juncus filiformis	E													1						1
	Juncus mertensianus	R							1	1	1	1		1		1					6
*	Juncus supiniformis	C														1					1
	Koenigia islandica	C				1				1					1						3
	Kumlienia cooleyae	R								1	1										2
	Lagotis glauca	E											1		1						2
	Lathryus palustris	R														1	1	1			3
	Lathyrus japonicus	R				1														1	2
	Lemna minor	B				1											1		1		3
	Leptarrhena pyrolifolia	R			1			1	1	1	1	1		1	1	1					9
	Leymus mollis	R				1									1		1			1	4
	Ligusticum scoticum ssp. hultenii	R				1									1		1	1			4
	Limosella aquatica	Cs	S3,G5			1											1				2
	Linnaea borealis	E					1			1											2
	Listera caurina	E				1	1	1					1								4
	Listera cordata	R				1	1	1		1	1	1		1	1	1					9
	Lloydia serotina	B							1	1		1		1							4
	Loiseleuria procumbens	C							1	1	1	1	1	1	1						7
	Lomatogonium rotatum	B																1			1
	Lonicera involucrata var. involucrata	E	S2,G4-5			1	1														2
	Luetkea pectinata	R						1	1	1	1	1	1	1							7
	Lupinus nootkatensis	R			1	1			1	1	1	1	1	1	1		1				10
	Lupinus polyphyllus	C															1				1
	Luzula arcuata ssp. unalaschcensis	R				1			1	1	1	1	1	1	1						8
	Luzula divaricata	R				1	1														2
	Luzula multiflora ssp. frigida	E				1			1	1			1	1		1					6

(*continued*)

APPENDIX 2: TAXA, RANGE, SITE, AND HERITAGE TRACKING LIST STATUS (*continued*)

									Alpine sites												
New in 2008	Plants Database Name	Range[†]	Status	Early colonization	Berg Lake, Weeping peat	Glacial Forelands	Temperate Forest	Subalpine Park	Grindle Hills	Robinson Mts	Donald Ridge	Khitrov Hills	Nichawak Mountain	Carbon Mt	Suckling Hills	Fens	Fr Marsh	Salt Marsh	Lakes, Ponds, ect	Beach	CONSTANCY
	Luzula multiflora ssp. multiflora var. kobayasii	R			1	1		1				1	1		1		1				7
	Luzula parviflora	R				1		1	1	1	1	1	1	1	1		1				10
	Luzula spicata	R							1	1		1		1							4
	Luzula wahlenbergii	R							1	1		1	1	1	1						6
	Lycopodium alpinum	R				1		1	1	1		1	1	1							7
	Lycopodium annotinum	R					1	1	1		1	1	1	1	1	1					9
	Lycopodium clavatum ssp. clavatum	R					1	1	1	1	1	1	1	1	1	1					10
	Lycopodium lagopus	B								1						1					2
	Lycopodium sitchense	R						1	1	1	1	1		1		1					7
	Lysichiton americanus	R				1	1								1	1	1				5
	Lysimachia thyrsiflora	Cs														1	1				2
	Maianthemum dilatatum	R				1	1								1	1	1				5
	Malaxis brachypoda	C				1	1								1		1				4
*	Malaxis diphyllus	?				1															1
	Malaxis paludosa	C	S3,G4													1					1
	Malus fusca	C					1									1					2
	Matricaria discoidea	R				1															1
	Menyanthes trifoliata	R				1									1	1	1		1		5
	Menziesia ferruginea	R					1	1	1	1	1					1					6
	Mertensia maritima	R																		1	1
	Microseris borealis	C														1					1
	Mimulus gutattus	R			1	1			1		1			1	1		1				7
	Minuartia arctica	E							1												1
	Minuartia biflora	B								1	1	1		1							4
	Minuartia rubella	B							1	1			1	1							4
	Mitella pentandra	C							1		1	1		1	1	1					6
	Moehringia lateriflora	R			1	1	1		1	1		1		1	1						8
	Moneses uniflora	R				1	1	1							1	1					5
	Monotropa hypopitys	E					1														1
	Montia chammissoi	R															1				1
	Montia fontana ssp. fontana	R													1		1				2
	Myrica gale	R				1										1	1	1			4
	Myriophyllum sibiricum	B				1										1	1		1		4
	Nephrophyllidium crista-galli	C							1	1	1	1	1	1	1	1					8
	Nuphar lutea ssp. polysepala	R														1			1		2
	Oenanthe sarmentosa	C															1				1
	Oplopanax horridus	R				1	1	1	1		1		1		1	1	1				9
	Orthilia secunda	R			1	1	1		1	1			1	1							7
	Osmorhiza purpurea	R				1	1	1	1	1	1	1	1	1	1		1				11
	Oxyria dygyna	R			1				1	1	1	1		1							6
	Oxytropis jordalii *	?								1		1									2
	Packera ogotorukensis	E							1												1
	Papaver alboroseum	Cs	S3,G3-4											1							1
	Papaver radicatum ssp.alaskanum	R								1											1
	Parnassia fimbriata var. fimbriata	R								1	1	1		1		1					5
	Parnassia kotzebuei	R			1	1			1		1	1									5
	Parnassia palustris	R			1	1			1	1		1		1	1		1	1			9
	Pedicularis capitata	B							1	1	1										3
	Pedicularis lanata	E							1	1			1								3
	Pedicularis oederi	C								1	1	1	1	1	1						6
	Pedicularis parviflora ssp. parviflora	R													1	1	1				3
	Pedicularis sudetica ssp. interior	E							1	1				1							3
	Pedicularis verticillata	R						1	1	1	1	1	1	1	1		1				9
	Petasites frigidus var. frigidus	R			1	1		1	1	1	1	1	1	1	1	1					11
	Phegopteris connectilis	R					1	1	1	1	1		1		1						7
	Phleum alpinum	R			1	1		1	1	1	1	1	1	1	1		1				11
	Phyllodoce glanduliflora	R				1	1	1	1	1	1	1	1	1	1	1					11
	Picea sitchensis	R		1	1	1	1	1	1	1	1	1		1	1	1	1				13
*	Pinguicula villosa	R														1					1
	Pinguicula vulgaris	R													1	1	1				3
	Plantago macrocarpa	R														1	1	1			3

(*continued*)

APPENDIX 2: TAXA, RANGE, SITE, AND HERITAGE TRACKING LIST STATUS (*continued*)

									Alpine sites												
New in 2008	Plants Database Name	Range[1]	Status	Early colonization	Berg Lake, Weeping peat	Glacial Forelands	Temperate Forest	Subalpine Park	Grindle Hills	Robinson Mts	Donald Ridge	Khitrov Hills	Nichawak Mountain	Carbon Mt	Suckling Hills	Fens	Fr Marsh	Salt Marsh	Lakes, Ponds, ect	Beach	CONSTANCY
	Plantago maritima	R																1			1
	Platanthera aquilonis	R			1	1	1		1	1		1			1	1	1				9
	Platanthera chorisiana	C						1			1					1					3
	Platanthera dilatata	R		1	1	1	1	1		1		1			1	1	1				10
*	Platanthera huronensis	R				1															1
	Platanthera obtusata	E				1	1														2
	Platanthera stricta	R			1		1	1	1	1	1	1	1	1	1	1					11
	Poa alpina	R			1	1			1	1	1	1		1							7
	Poa annua	R				1															1
	Poa arctica ssp. arctica	R				1			1	1	1			1							5
	Poa arctica ssp. lanata	R				1				1											2
	Poa eminens	R																1		1	2
	Poa laxiflora	E	S2,G3			1															1
	Poa macrantha	C	S1,G5			1														1	2
	Poa paucispicula	R								1	1	1	1	1							5
	Poa pratensis ssp. alpigena	R							1	1		1									3
	Poa pratensis ssp. colpodea	R				1			1			1									3
	Poa stenantha	R			1					1		1			1						4
*	Poa trivialis	C				1															1
	Polemonium acutiflorum	R							1	1	1		1		1	1					6
	Polemonium pulcherrimum	B								1		1	1	1	1						5
	Polygonum amphibium	R															1		1		2
	Polygonum fowleri	R				1											1			1	3
	Polygonum viviparum	R							1	1	1	1	1	1	1		1				8
	Polypodium glycyrrhiza	R				1	1								1	1	1				5
	Polystichum braunii	C				1	1		1												3
	Polystichum lonchitis	R			1	1	1			1		1		1		1					7
	Populus balsamifera ssp. trichocarpa	R		1	1	1			1	1							1				6
	Potamogeton alpinus	R			1	1											1		1		4
	Potamogeton epihydris	C														1			1		2
	Potamogeton foliosus	E				1										1	1		1		4
	Potamogeton friesii	Cs				1													1		2
	Potamogeton gramineus	R			1	1										1			1		4
	Potamogeton natans	R														1			1		2
	Potamogeton praelongus	Cs														1			1		2
	Potamogeton pusillus	R														1			1		2
	Potamogeton richardsonii	R				1											1		1		3
	Potentilla drummondi ssp. drummondi	Cs	S2,G5											1							1
	Potentilla nana	E								1		1		1							3
	Potentilla uniflora	B								1		1		1							3
	Potentilla villosa	R							1	1	1	1	1	1	1						7
	Prenanthes alata	R			1	1	1	1	1	1	1		1	1	1						10
	Primula cuneifolia ssp. saxifragifolia	C							1	1	1										3
	Primula egaliksensis	C															1				1
	Primula eximia	E							1	1	1										3
	Prunella vulgaris	C			1	1								1	1						4
	Puccinellia nutkaensis	R				1												1	1		3
	Puccinellia phryganodes	C																1			1
	Pyrola asarifolia	R			1	1	1		1	1				1			1				7
	Pyrola minor	R				1			1	1				1		1					5
	Ranunculus cymbalaria	Cs				1												1		1	3
	Ranunculus eschscholtzii	R							1	1	1	1		1							5
	Ranunculus flammula var. filiformis	R															1				1
	Ranunculus gmelinii	E															1				1
	Ranunculus karelinii	E										1									1
	Ranunculus occidentalis var. brevistylis	R				1				1		1		1	1						5
	Ranunculus occidentalis var. nelsonii	E													1		1				2
	Ranunculus pacificus	E	S3,G3			1	1										1				3
	Ranunculus pygmaeus	B								1											1
	Ranunculus trichophyllus var. trichophyllus	R			1	1											1		1		4
	Ranunculus uncinatus	R			1	1	1								1		1				5

(*continued*)

APPENDIX 2: TAXA, RANGE, SITE, AND HERITAGE TRACKING LIST STATUS (*continued*)

									Alpine sites													
New in 2008	Plants Database Name	Range[1]	Status	Early colonization	Berg Lake, Weeping peat	Glacial Forelands	Temperate Forest	Subalpine Park	Grindle Hills	Robinson Mts	Donald Ridge	Khitrov Hills	Nichawak Mountain	Carbon Mt	Suckling Hills	Fens	Fr Marsh	Salt Marsh	Lakes, Ponds, ect	Beach	CONSTANCY
	Rhinanthus minor ssp. groenlandicus	R				1									1		1	1			4
	Rhodiola integrifolia ssp. integrifolia	R							1	1	1	1	1	1							6
	Rhododendron camtschaticum ssp. camtschaticum	C									1		1		1						3
*	Rhynchospora alba	C														1					1
	Ribes bracteosum	R				1	1	1								1					4
	Ribes laxiflorum	R				1	1		1		1	1				1					6
	Romanzoffia sitchensis	R							1	1				1							3
	Romanzoffia unalaschcensis	E	S3,G3		1				1	1	1										4
	Rorippa curvisiliqua	E	S1,G5			1															1
	Rorippa palustris	R			1	1										1	1				4
*	Rosa rugosa	R				1															1
	Rubus arcticus ssp. stellatus	R				1	1	1	1	1		1	1	1	1	1	1				11
*	Rubus chamaemorus	R														1					1
	Rubus pedatus	R				1	1	1	1	1	1	1	1	1	1	1					11
	Rubus spectabilis	R			1	1	1	1	1	1	1	1	1	1	1	1	1				13
	Rumex aquaticus var. fenestratus	R				1				1					1	1	1	1		1	7
	Rumex salicifolius var. transitorius	R				1											1				2
	Sagina maxima ssp. crassicaulis	R				1											1	1			3
	Sagina nivalis	R		1	1	1			1	1		1	1	1	1						9
	Sagina saginoides	R			1	1				1		1		1							5
	Salix alaxensis	R		1	1	1	1		1	1		1					1				8
	Salix arctica	R			1	1			1	1	1	1	1	1	1	1					10
	Salix barclayi	R		1	1	1	1	1	1	1	1	1		1	1	1	1	1			14
	Salix commutata	R				1				1			1			1					4
	Salix glauca ssp. glauca var. villosa	B			1	1			1	1											4
	Salix myrtillifolia	B									1										1
	Salix polaris	B							1	1		1			1						4
	Salix reticulata	R							1	1				1							3
	Salix rotundifolia	B												1							1
	Salix sitchensis	R		1	1	1	1	1	1	1	1	1	1		1	1	1				13
	Salix stolonifera	R							1	1	1	1	1		1						6
	Sambucus racemosa var. racemosa	R				1	1	1		1	1	1	1		1		1				9
	Sanguisorba canadensis	R			1	1	1	1	1	1	1	1	1	1	1	1					12
	Saxifraga bronchialis	R							1	1		1	1	1							5
	Saxifraga caespitosa	B							1	1			1	1	1						5
	Saxifraga ferruginea	R							1	1	1				1						4
	Saxifraga lyallii ssp. hultenii	R				1			1	1	1	1		1		1					7
	Saxifraga mertensiana	R							1	1	1	1									4
	Saxifraga nelsoniana ssp. carlottae	R						1	1	1		1									4
	Saxifraga nelsoniana ssp. insularis	E							1	1	1										3
	Saxifraga nelsoniana ssp. pacifica	R			1					1											2
	Saxifraga nelsoniana ssp. nelsoniana	E						1	1	1	1		1	1	1	1					8
	Saxifraga nivalis	R								1											1
	Saxifraga oppositifolia	R								1		1		1							3
	Saxifraga rivularis	B							1	1		1		1							4
	Saxifraga tricuspidata	B										1									1
	Scheuchzeria palustris	Cs														1					1
	Scirpus microcarpus	R				1	1	1							1						4
	Selaginella selaginoides	R														1					1
	Selaginella sibirica	E												1							1
	Senecio lugens	B							1												1
	Senecio pseudoarnica	R																		1	1
	Senecio triangularis	C				1		1	1	1	1	1	1	1	1	1					10
	Sibbaldia procumbens	R			1				1	1	1	1	1	1	1	1					9
	Silene acaulis var.acaulis	R							1	1	1	1	1	1	1						7
	Silene acaulis var. subacaulescens	E								1											1
	Silene involucrata	E								1											1
*	Sisyrinchium littoralae	R													1						1
	Solidago canadensis var. lepida	R													1	1	1				3
	Solidago multiradiata	R				1			1	1	1	1	1	1							7

(*continued*)

APPENDIX 2: TAXA, RANGE, SITE, AND HERITAGE TRACKING LIST STATUS (*continued*)

									Alpine sites												
New in 2008	Plants Database Name	Range†	Status	Early colonization	Berg Lake, Weeping peat	Glacial Forelands	Temperate Forest	Subalpine Park	Grindle Hills	Robinson Mts	Donald Ridge	Khitrov Hills	Nichawak Mountain	Carbon Mt	Suckling Hills	Fens	Fr Marsh	Salt Marsh	Lakes, Ponds, ect	Beach	CONSTANCY
	Sorbus sitchensis	R					1	1	1	1	1	1	1	1		1					9
	Sparganium angustifolium	R			1	1									1	1	1		1		6
	Sparganium hyperboreum	R														1	1		1		3
	Spiranthes romanzoffiana	R			1	1				1						1	1				5
	Stellaria borealis ssp. borealis	R			1	1			1	1							1				5
	Stellaria calycantha	R			1	1			1	1	1	1	1	1	1						9
	Stellaria crassifolia	E															1				1
	Stellaria crispa	R			1	1	1	1	1	1	1	1			1		1				10
	Stellaria humifusa	R																1			1
	Stellaria longifolia	B		1	1	1										1					4
	Stellaria longipes	R							1	1	1	1		1		1	1	1			8
	Streptopus amplexifolius	R				1	1	1	1	1	1	1	1		1	1	1				11
	Stuckenia filiformis	R				1											1	1	1		4
	Swertia perennis	R							1				1		1	1					4
	Symphyotrichum subspicatum var. subspicatum	R				1						1		1	1	1	1	1			7
	Taraxacum officinale ssp. ceratophorum	R								1				1							2
	Taraxacum officinale ssp. officinale	R				1															1
	Taraxacum phymatocarpum	B								1	1	1									3
	Tellima grandiflora	R			1	1									1						3
	Thelypteris quelpaertensis	R								1						1					2
	Tiarella trifoliata	R				1	1	1	1	1	1	1		1	1	1	1				11
	Tofieldia coccinea	R							1	1		1	1	1	1						6
	Torreyochloa pallida var. pauciflora	R				1									1	1					3
	Triantha occidentalis ssp. brevistyla	R				1										1	1				3
	Trichophorum cespitosum	R							1	1		1				1					4
	Trientalis europaea ssp. arctica	R				1	1	1	1	1	1	1	1	1	1	1	1				12
	Triglochin maritima	R																1			1
	Triglochin palustris	R			1	1											1		1		4
	Trisetum canescens	R				1	1	1					1		1	1					6
	Trisetum spicatum	R			1	1			1	1	1	1	1	1	1	1					10
	Tsuga heterophylla	R			1		1									1					3
	Tsuga mertensiana	R		1		1	1	1	1	1	1	1	1	1		1					11
	Urtica dioica ssp. gracilis	R													1						1
	Utricularia intermedia	B														1	1		1		3
	Utricularia macrorhiza	R															1		1		2
	Utricularia minor	B															1		1		2
	Vaccinium alaskaense	R					1	1		1	1	1	1		1	1					8
	Vaccinium cespitosum	R									1	1		1		1					4
	Vaccinium cespitosum var.paludicola#	C										1		1							2
	Vaccinium ovalifolium	R				1	1	1	1	1	1	1	1	1	1	1					11
	Vaccinium oxycoccos	R														1					1
	Vaccinium uliginosum	R							1	1	1	1	1	1	1	1					8
	Vaccinium vitis-idaea	R													1	1					2
	Vahlodea atropurpurea	R						1	1	1	1	1	1	1	1	1					9
	Valeriana sitchensis	R						1	1	1	1	1	1	1	1						8
	Veratrum viride	C				1	1	1	1	1	1	1	1	1	1	1					11
	Veronica americana	R				1										1	1				3
	Veronica wormskjoldii ssp. wormskjoldii	R					1		1	1	1	1	1	1	1						8
	Veronica wormskjoldii ssp.stelleri	C							1	1		1		1	1						5
	Viburnum edule	R				1	1									1					3
*	Viola adunca	C				1															1
	Viola epipsila ssp. repens	R				1	1		1			1			1	1	1				7
	Viola glabella	R				1	1	1			1	1	1		1	1	1				9
	Viola langsdorffii	R				1	1	1	1	1	1	1	1	1	1	1	1				12
*	Zannichellia palustris	Cs	S3,G3			1													1		2
				22	116	234	90	93	192	240	148	198	130	165	194	176	155	39	31	19	

*Plants added in the 2008 survey.
†Range categories: R—in the expected range; C—connection; Cs—scattered statewide distribution; B—short extension over the Bagley Ice Field; E—extension.
§Lakes and ponds are contained within other units, i.e. Mid-timber Lake is in the forelands.
#Name from Hulten, not from Plants Database.

APPENDIX 3. CROSS-LIST OF SCIENTIFIC AND COMMON PLANT NAMES

	Generic and specific names	Common names	Family
	Ferns & Fern Allies		
1	Adiantum aleuticum	Maidenhair fern	Adiantaceae
2	Dryopteris expansa	Wood fern	Aspidaceae
3	Gymnocarpium dryopteris	Oak fern	Aspidaceae
4	Polystichum braunii	Braun's hollyfern	Aspidaceae
5	Polystichum lonchitis	Northern hollyfern	Aspidaceae
6	Athyrium americanum	Alpine ladyfern	Athyriaceae
7	Athyrium filix-femina ssp. cyclosorum	Lady fern	Athyriaceae
8	Cystopteris fragilis	Fragile fern	Athyriaceae
9	Blechnum spicant	Deer fern	Blechnaceae
10	Cryptogramma acrostichoides	Parsley fern	Cryptogammaceae
11	Equisetum arvense	Field horsetail	Equisetaceae
12	Equisetum fluviatile	Water horsetail	Equisetaceae
13	Equisetum palustre	Marsh horsetail	Equisetaceae
14	Equisetum variegatum	Variegated scouringrush	Equisetaceae
15	Isoetes occidentalis	Western quillwort	Isoetaceae
16	Huperzia chinensis	Chinese clubmoss	Lycopodiaceae
17	Huperzia haleakalae	Pacific clubmoss	Lycopodiaceae
18	Lycopodium alpinum	Alpine clubmoss	Lycopodiaceae
19	Lycopodium annotinum	Stiff clubmoss	Lycopodiaceae
20	Lycopodium clavatum ssp. clavatum	Running clubmoss	Lycopodiaceae
21	Lycopodium lagopus	One cone clubmoss	Lycopodiaceae
22	Lycopodium sitchense	Sitka clubmoss	Lycopodiaceae
23	Botrychium alaskense	Western moonwort	Ophioglossaceae
24	Botrychium lanceolatum	Lanceleaf grapefern	Ophioglossaceae
25	Botrychium lunaria	Common moonwort	Ophioglossaceae
26	Botrychium minganense	Mingan moonwort	Ophioglossaceae
27	Botrychium pinnatum	Northern moonwort	Ophioglossaceae
28	Polypodium glycyrrhiza	Licorice fern	Polypodiaceae
29	Selaginella selaginoides	Northern spikemoss	Selaginellaceae
30	Selaginella sibirica	Siberiain spikemoss	Selaginellaceae
31	Phegopteris connectilis	Fox face fern	Thelypteridaceae
32	Thelypteris quelpaertensis	Queen's veil	Thelypteridaceae
	Gymnosperms		
1	Picea sitchensis	Sitka spruce	Pinaceae
2	Tsuga heterophylla	Western hemlock	Pinaceae
3	Tsuga mertensiana	Mountain hemlock	Pinaceae
	Monocotyledons		
1	Carex anthoxanthea	Grassyslope arctic sedge	Cyperaceae
2	Carex aquatilis var. aquatilis	Water sedge	Cyperaceae
3	Carex aquatilis var. dives	Sitka sedge	Cyperaceae
4	Carex bicolor	Two color sedge	Cyperaceae
5	Carex canescens	Silvery sedge	Cyperaceae
6	Carex circinata	Coiled sedge	Cyperaceae
7	Carex diandra	Lesser panicled sedge	Cyperaceae
8	Carex echinata ssp. phyllomanica	Star sedge	Cyperaceae
9	Carex garberi	Elk sedge	Cyperaceae
10	Carex glareosa	Lesser saltmarsh sedge	Cyperaceae
11	Carex gmelinii	Gmelin's sedge	Cyperaceae
12	Carex gynocrates	Northern bog sedge	Cyperaceae
13	Carex lachenalii	Twotipped sedge	Cyperaceae
14	Carex laeviculmis	Smoothcone sedge	Cyperaceae
15	Carex lenticularis var. dolia	Enander's sedge	Cyperaceae
16	Carex lenticularis var. limnophila	Lakeshore sedge	Cyperaceae
17	Carex lenticularis var. lipocarpa	Kellogg's sedge	Cyperaceae
18	*Carex limosa	Mud sedge	Cyperaceae
19	Carex livida	Livid sedge	Cyperaceae
20	Carex lyngbyei	Lyngbye's sedge	Cyperaceae
21	Carex mackenzii	Mackenzies sedge	Cyperaceae
22	Carex macrocephala	Largehead sedge	Cyperaceae
23	Carex macrochaeta	Longawn sedge	Cyperaceae
24	Carex magellanica	Boreal bog sedge	Cyperaceae
25	Carex maritima	Curved sedge	Cyperaceae
26	Carex membranacea	Fragile sedge	Cyperaceae
27	Carex mertensii	Mertens' sedge	Cyperaceae
28	Carex nigricans	Black alpine sedge	Cyperaceae
29	Carex pachystachya	Chamisso sedge	Cyperaceae
30	Carex pauciflora	Fewflower sedge	Cyperaceae
31	Carex phaeocephala	Dunhead sedge	Cyperaceae
32	Carex pluriflora	Manyflower sedge	Cyperaceae

(continued)

APPENDIX 3. CROSS-LIST OF SCIENTIFIC AND COMMON PLANT NAMES (continued)

	Generic and specific names	Common names	Family
	Monocotyledons (continued)		
33	Carex pyrenaica ssp. micropoda	Pyrenean sedge	Cyperaceae
34	Carex saxatilis	Rock sedge	Cyperaceae
35	Carex scirpoidea	Bachelor sedge	Cyperaceae
36	Carex spectabilis	Showy sedge	Cyperaceae
37	Carex utriculata	Northwest territory sedge	Cyperaceae
38	Carex viridula ssp. viridula	Little green sedge	Cyperaceae
39	Eleocharis acicularis	Needle spikerush	Cyperaceae
40	Eleocharis kamtschatica	Kamchatka spikerush	Cyperaceae
41	Eleocharis macrostachya	Pale spikerush	Cyperaceae
42	Eleocharis palustris var. palustris	Common spikerush	Cyperaceae
43	Eleocharis quinqueflora	Few flowered spikerush	Cyperaceae
44	Eriophorum angustifolium	Tall cottongrass	Cyperaceae
45	Eriophorum chamissonis	Chamisso's cottongrass	Cyperaceae
46	Eriophorum scheuchzeri	White cottongrass	Cyperaceae
47	*Rhynchospora alba	White beak sedge	Cyperaceae
48	Scirpus microcarpus	Panicled bulrush	Cyperaceae
49	Trichophorum cespitosum	tufted bulrush	Cyperaceae
50	Iris setosa ssp. setosa	Blue flag/Wild Iris	Iridaceae
51	*Sisyrinchium littoralae	Blue-eyed grass	Iridaceae
52	Juncus albescens	Northern white rush	Juncaceae
53	Juncus alpinoarticulatus	Northern green rush	Juncaceae
54	Juncus ambiguus	Seasice rush	Juncaceae
55	Juncus arcticus	Mountain rush	Juncaceae
56	Juncus biglumis	Twoflowered rush	Juncaceae
57	Juncus bufonius	Toad rush	Juncaceae
58	Juncus castaneus	Chestnut rush	Juncaceae
59	Juncus drummondii	Drummons's rush	Juncaceae
60	Juncus ensifolius	Swordleaf rush	Juncaceae
61	Juncus falcatus ssp.sitchensis	Falcate rush	Juncaceae
62	Juncus filiformis	Thread rush	Juncaceae
63	Juncus mertensianus	Merten's rush	Juncaceae
64	*Juncus supiniformis	Hairy leaf rush	Juncaceae
65	Luzula arcuata ssp. unalaschcensis	Curved woodrush	Juncaceae
66	Luzula divaricata	Forked woodrush	Juncaceae
67	Luzula multiflora ssp. frigida	Common woodrush	Juncaceae
68	Luzula multiflora ssp. multiflora var. kobayasii	Common woodrush	Juncaceae
69	Luzula parviflora	Small flowered woodrush	Juncaceae
70	Luzula spicata	Spiked woodrush	Juncaceae
71	Luzula wahlenbergii	Wahlenberg's woodrush	Juncaceae
72	Triglochin maritima	Seaside arrowgrass	Juncaginaceae
73	Triglochin palustris	Marsh arrowgrass	Juncaginaceae
74	Lemna minor	Common duckweed	Lemnaceae
75	Fritillaria camschatcensis	Chocolate Lily	Liliaceae
76	Lloydia serotina	Alp lily	Liliaceae
77	Maianthemum dilatatum	False lily-of-the-valley	Liliaceae
78	Streptopus amplexifolius	Wild cucumber	Liliaceae
79	Tofieldia coccinea	Northern asphodel	Liliaceae
80	Triantha occidentalis ssp. brevistyla	Sticky tofieldia	Liliaceae
81	Veratrum viride	Corn lily/False hellebore	Liliaceae
82	Dactylorhiza viridis	Frog orchid	Orchidaceae
83	Listera caurina	Northwest twayblade orchid	Orchidaceae
84	Listera cordata	Twayblade orchid	Orchidaceae
85	*Malaxis brachypoda	White adder's-mouth orchid	Orchidaceae
86	Malaxis diphyllus	Aleutian adder's-mouth orchid	Orchidaceae
87	Malaxis paludosa	Bog adder's tongue	Orchidaceae
88	Platanthera aquilonis	Northern Green Orchid	Orchidaceae
89	Platanthera chorisiana	Chimisso's orchid	Orchidaceae
90	Platanthera dilatata	Bog candle/Scen bottle	Orchidaceae
91	*Platanthera huronensis	Huron green orchid	Orchidaceae
92	Platanthera obtusata	Blunt leaf orchid	Orchidaceae
93	Platanthera stricta	Slender bog orchid	Orchidaceae
94	Spiranthes romanzoffiana	Lady's tress orchid	Orchidaceae
95	Agrostis aequivalvis	Arctic bentgrass	Poaceae
96	Agrostis alaskana *	Alaska bentgrass	Poaceae
97	Agrostis exarata	Sipke bentgrass	Poaceae
98	Agrostis humilis	Alpine bentgrass	Poaceae
99	Agrostis mertensii	Northern bentgrass	Poaceae
100	Agrostis stolonifera	Creeping bentgrass	Poaceae
101	Alopecurus aequalis var. aequalis	Shortawn foxtail	Poaceae

(continued)

APPENDIX 3. CROSS-LIST OF SCIENTIFIC AND COMMON PLANT NAMES (*continued*)

	Generic and specific names	Common names	Family
	Monocotyledons (*continued*)		
102	Alopecurus alpinus	Boreal alopecurus	Poaceae
103	Arctagrostis latifolia	Polargrass	Poaceae
104	Arctophila fulva	Pendant grass	Poaceae
105	Calamagrostis canadensis canadensis	Bluejoint grass	Poaceae
106	Calamagrostis canadensis langsdorffii	Bluejoint grass	Poaceae
107	Calamagrostis deschamsoides	Circumpolar reedgrass	Poaceae
108	Calamagrostis nutkaensis	Pacific reedgrass	Poaceae
109	Calamagrostis stricta ssp.inexpansa	Northern reedgrass	Poaceae
110	Cinna latifolia	Drooping woodreed	Poaceae
111	Deschampsia beringensis	Bering tufted hairgrass	Poaceae
112	Deschampsia cespitosa	Tufted hairgrass	Poaceae
113	Elymus hirsutus	Northern ryegrass	Poaceae
114	Festuca brachyphylla	Colorado fescue	Poaceae
115	Festuca brevissima	Alaska fescue	Poaceae
116	Festuca rubra	Red fescue	Poaceae
117	Festuca saximontana var. saximontana	Rocky mountain fescue	Poaceae
118	Hierochloe alpina	Alpine holeygrass	Poaceae
119	Hierochloe hirta ssp. arctica	Vanilla grass	Poaceae
120	Hordeum brachyantherum ssp. brachyantherum	Meadow barely	Poaceae
121	Leymus mollis	Beach rye	Poaceae
122	Phleum alpinum	Aline timothy	Poaceae
123	Poa alpina	Alpine bluegrass	Poaceae
124	Poa annua	Annual bluegrass	Poaceae
125	Poa arctica ssp. arctica	Arctic bluegrass	Poaceae
126	Poa arctica ssp. lanata	Arctic bluegrass	Poaceae
127	Poa eminens	Speargrass	Poaceae
128	Poa laxiflora	Looseflower bluegrass	Poaceae
129	Poa macrantha	Seashore bluegrass	Poaceae
130	Poa paucispicula	Alaska bluegrass	Poaceae
131	Poa pratensis ssp. alpigena	Kentucky bluegrass	Poaceae
132	Poa pratensis ssp. colpodea	Kentucky bluegrass	Poaceae
133	Poa stenantha	Northern bluegrass	Poaceae
134	*Poa trivialis	Rough Bluegrass	Poaceae
135	Puccinellia nutkaensis	Nootka alkaligrass	Poaceae
136	Puccinellia phryganodes	Creeping alkaligrass	Poaceae
137	Torreyochloa pallida var. pauciflora	Pale false mannagrass	Poaceae
138	Trisetum canescens	Tall trisetum	Poaceae
139	Trisetum spicatum	Spike trisetum	Poaceae
140	Vahlodea atropurpurea	Mountain hairgrass	Poaceae
141	Potamogeton alpinus	Alpine pondweed	Potamogetonaceae
142	Potamogeton epihydris	Ribbon leafed pondweed	Potamogetonaceae
143	Potamogeton foliosus	Leafy pondweed	Potamogetonaceae
144	Potamogeton friesii	Fries' pondweed	Potamogetonaceae
145	Potamogeton gramineus	Variable leaf pondweed	Potamogetonaceae
146	Potamogeton natans	Floating pondweed	Potamogetonaceae
147	Potamogeton praelongus	White stem pondweed	Potamogetonaceae
148	Potamogeton pusillus	Small pondweed	Potamogetonaceae
149	Potamogeton richardsonii	Richardson's pondweed	Potamogetonaceae
150	Stuckenia filiformis	Fineleaf pondweed	Potamogetonaceae
151	Scheuchzeria palustris	Rannoch-rush	Scheuchzeriaceae
152	Sparganium angustifolium	Narrowleaf bur-reed	Sparganiaceae
153	Sparganium hyperboreum	Northern bur-reed	Sparganiaceae
154	*Zannichellia palustsris	Horned pond week	Zannichelliaceae
	Dicotyletons		
1	Angelica genuflexa	Kneeling angelica	Apiaceae
2	Angelica lucida	Wild celery	Apiaceae
3	Cicuta douglasii	Poison water hemlock	Apiaceae
4	Cicuta virosa	Mackenzies water hemlock	Apiaceae
5	Conioselinum gmelinii	Chinese parsley	Apiaceae
6	Glehnia littoralis ssp. leiocarpa	Silvertop	Apiaceae
7	Heracleum maximum	Cow parsnip	Apiaceae
8	Ligusticum scoticum ssp. hultenii	Hulten's sea lovage	Apiaceae
9	Oenanthe sarmentosa	Water parsley	Apiaceae
10	Osmorhiza purpurea	Sweet cicely	Apiaceae
11	Lysichiton americanus	Skunk cabbage	Araceae
12	Oplopanax horridus	Devil's walking stick	Araliaceae
13	Achillea millefolium var. borealis	Yarrow	Asteraceae
14	Agoseris aurantiaca var. aurantiaca	Mountain dandelion	Asteraceae
15	Anaphalis margaritacea	Pearly everlasting	Asteraceae

(*continued*)

APPENDIX 3. CROSS-LIST OF SCIENTIFIC AND COMMON PLANT NAMES (*continued*)

	Generic and specific names	Common names	Family
	Dicotyletons (*continued*)		
16	Antennaria alpina	Alpine pussy-toes	Asteraceae
17	Antennaria friesiana ssp. alaskana	Fries pussy-toes	Asteraceae
18	Antennaria friesiana ssp. neoalaskana	Fries pussy-toes	Asteraceae
19	Antennaria monocephala ssp. angustata	Pygmy pussy-toes	Asteraceae
20	Antennaria monocephala ssp. monocephala	Pygmy pussy-toes	Asteraceae
21	Antennaria rosea ssp. confinis	Rosy pussy-toes	Asteraceae
22	Antennaria rosea ssp. pulvinata	Pink pusspy-toes	Asteraceae
23	Arnica amplexicaulis	Clasping arnica	Asteraceae
24	Arnica latifolia	Broadleaf arnica	Asteraceae
25	Arnica lessingii	Nodding arnica	Asteraceae
26	Artemisia arctica ssp. arctica	Boreal sagebrush	Asteraceae
27	Artemisia furcata	Forked wormwood	Asteraceae
28	Artemisia tilesii ssp. elatior	Tilesius wormwood	Asteraceae
29	Chrysanthemum arcticum	Arctic daisy	Asteraceae
30	Erigeron humilis	Arctic-alpine fleabane	Asteraceae
31	Erigeron peregrinus ssp. peregrinus	Subalpine fleabane	Asteraceae
32	Eurybia sibirica	Siberian aster	Asteraceae
33	Hieracium triste	Woolly hawkweed	Asteraceae
34	Matricaria discoidea	Pineapple weed	Asteraceae
35	Microseris borealis	Northern microseris	Asteraceae
36	Packera ogotorukensis	Ogotoruk Creek ragwort	Asteraceae
37	Petasites frigidus var. frigidus	Arctic coltsfoot	Asteraceae
38	Prenanthes alata	Rattlesnake root	Asteraceae
39	Senecio lugens	Blacktip ragwort	Asteraceae
40	Senecio pseudoarnica	Seaside ragwort	Asteraceae
41	Senecio triangularis	Arrowleaf ragwort	Asteraceae
42	Solidago canadensis var. lepida	Canada goldenrod	Asteraceae
43	Solidago multiradiata	Rocky Mt.Goldenrod	Asteraceae
44	Symphyotrichum subspicatum var. subspicatum	Douglas aster	Asteraceae
45	Taraxacum officinale ssp. ceratophorum	Common dandelion	Asteraceae
46	Taraxacum officinale ssp. officinale	Common dandelion	Asteraceae
47	Taraxacum phymatocarpum	Northern dandelion	Asteraceae
48	Alnus viridis ssp. sinuata	Green Alder	Betulaceae
49	Mertensia maritima	Oysterleaf	Boraginaceae
50	Arabis eschscholtziana	Hairy rockcress	Brassicaceae
51	Arabis kamchatica	Kanchatka rockcress	Brassicaceae
52	Barbarea orthoceras	Wintercress	Brassicaceae
53	Cakile edentula ssp. edentula	American searocket	Brassicaceae
54	Cardamine bellidifolia var. bellidifolia	Alpine bittercress	Brassicaceae
55	Cardamine oligosperma var. kamtschatica	Umbel bittercress	Brassicaceae
56	Draba lactea	Milky draba	Brassicaceae
57	Draba stenoloba var. stenoloba	Alaska draba	Brassicaceae
58	Rorippa curvisiliqua	Curvepod Yellowcress	Brassicaceae
59	Rorippa palustris	Yellowcress	Brassicaceae
60	Callitriche palustris	Vernal water-starwort	Callitrichaceae
61	Campanula lasiocarpa	Mountain hairbell	Campanulaceae
62	Campanula rotundifolia	Scottish hairbells	Campanulaceae
63	Linnaea borealis	Twin flower	Caprifoliaceae
64	Lonicera involucrata var. involucrata	Twinberry honeysuckle	Caprifoliaceae
65	Sambucus racemosa var. racemosa	Red elderberry	Caprifoliaceae
66	Cerastium beeringianum ssp. beeringianum	Bering chickweed	Caryophyllaceae
67	Cerastium fontanum	Common mouse-ear chickweed	Caryophyllaceae
68	Honckenya peploides ssp. major	Beach greens	Caryophyllaceae
69	Minuartia arctica	Artic sandwort	Caryophyllaceae
70	Minuartia biflora	Mountain sandwort	Caryophyllaceae
71	Minuartia rubella	Beautiful sandwort	Caryophyllaceae
72	Moehringia lateriflora	King river sandwort	Caryophyllaceae
73	Sagina maxima ssp. crassicaulis	Sticky stem pearlwort	Caryophyllaceae
74	Sagina nivalis	Snow pearlwort	Caryophyllaceae
75	Sagina saginoides	Arctic pearlwort	Caryophyllaceae
76	Silene acaulis var. subacaulescens	Moss campion	Caryophyllaceae
77	Silene acaulis var.acaulis	Moss campion	Caryophyllaceae
78	Silene involucrata	Arctic catchfly	Caryophyllaceae
79	Stellaria borealis ssp. borealis	Boreal starwort	Caryophyllaceae
80	Stellaria calycantha	Northern starwort	Caryophyllaceae
81	Stellaria crassifolia	Fleshy starwort	Caryophyllaceae
82	Stellaria crispa	Curled starwort	Caryophyllaceae

(*continued*)

APPENDIX 3. CROSS-LIST OF SCIENTIFIC AND COMMON PLANT NAMES (*continued*)

	Generic and specific names	Common names	Family
	Dicotyletons (*continued*)		
83	Stellaria humifusa	Saltmarsh starwort	Caryophyllaceae
84	Stellaria longifolia	Longleaf starwort	Caryophyllaceae
85	Stellaria longipes	Longstalk starwort	Caryophyllaceae
86	Ceratophyllum demersum	Coon's tail	Ceratophyllaceae
87	*Atriplex gmelinii	Gmelini's saltbrush	Chenopodiaceae
88	Cornus canadensis	Canadian bunchberry	Cornaceae
89	Cornus suecica	Alpine bunchberry	Cornaceae
90	Rhodiola integrifolia ssp. integrifolia	King's Crown	Crassulaceae
91	Diapensia lapponica ssp. obovata	Lapland diapensia	Diapensiaceae
92	Drosera anglica	English sundew	Droseraceae
93	Drosera rotundifolia	Roundleaf sundew	Droseraceae
94	Empetrum nigrum	Crowberry	Empetraceae
95	Andromeda polifolia var. polifolia	Bog rosemary	Ericaceae
96	Arctostaphylos alpina	Alpine bearberry	Ericaceae
97	Arctostaphylos uva-ursi	Bear berry	Ericaceae
98	Cassiope lycopodioides	Clubmoss mountain heather	Ericaceae
99	Elliottia pyrolaeflora	Copper bush	Ericaceae
100	Harrimanella stelleriana	Mountain heath	Ericaceae
101	Loiseleuria procumbens	Alpine azalea	Ericaceae
102	Menziesia ferruginea	Fool's huckleberry	Ericaceae
103	Phyllodoce glanduliflora	Aleutian heather	Ericaceae
104	Rhododendron camtschaticum ssp. camtschaticum	Kamchatka rhododendron	Ericaceae
105	Vaccinium alaskaense	Alaska blueberry	Ericaceae
106	Vaccinium cespitosum	Dwarf bilberry	Ericaceae
107	Vaccinium cespitosum var.paludicola*	Cespitose blueberry	Ericaceae
108	Vaccinium ovalifolium	Early blueberry	Ericaceae
109	Vaccinium oxycoccos	Bog cranberry	Ericaceae
110	Vaccinium uliginosum	Bog blueberry	Ericaceae
111	Vaccinium vitis-idaea	Lingonberry	Ericaceae
112	Astragalus alpinus var. alpinus	Alpine milkvetch	Fabaceae
113	Hedysarum alpinum	Eskimo potato	Fabaceae
114	Lathryus palustris	Marsh pea	Fabaceae
115	Lathyrus japonicus	Beach pea	Fabaceae
116	Lupinus nootkatensis	Nootka lupine	Fabaceae
117	Lupinus polyphyllus	Bigleaf lupine	Fabaceae
118	Oxytropis jordalii	Jordol's locoweed	Fabaceae
119	Gentiana douglasiana	Swamp gentian	Gentianaceae
120	Gentiana platypetala	Broad petal Gentian	Gentianaceae
121	Gentiana prostrata	Pygmy gentian	Gentianaceae
122	Gentianella amarella ssp. acuta	Autumn dwarf gentian	Gentianaceae
123	Gentianella propinqua ssp. aleutica	Aleutian gentian	Gentianaceae
124	Gentianella propinqua ssp. propinqua	Dwarf gentian	Gentianaceae
125	Lomatogonium rotatum	Marsh felwort	Gentianaceae
126	Menyanthes trifoliata	Bog bean	Gentianaceae
127	Nephrophyllidium crista-galli	Deer cabbage	Gentianaceae
128	Swertia perennis	Star gentian	Gentianaceae
129	Geranium erianthum	Wild geranium	Geraniaceae
130	Ribes bracteosum	Stink currant	Grossulariaceae
131	Ribes laxiflorum	Trailing black currant	Grossulariaceae
132	Hippuris montana	Mountain mare's tail	Haloragaceae
133	Hippuris tetraphylla	4-leaf mare's tail	Haloragaceae
134	Hippuris vulgaris	Common mare's tail	Haloragaceae
135	Myriophyllum sibiricum	Shortspike watermilfoil	Haloragaceae
136	Romanzoffia sitchensis	Alaska mistmaiden	Hydrophyllaceae
137	Romanzoffia unalaschcensis	Alaska mistmaiden	Hydrophyllaceae
138	Prunella vulgaris	Selfheal	Lamiaceae
139	*Pinguicula villosa	Hairy butterwort	Lentibulariaceae
140	Pinguicula vulgaris	Common butterwort	Lentibulariaceae
141	Utricularia intermedia	Flat-leaf bladderwort	Lentibulariaceae
142	Utricularia macrorhiza	Common bladderwort	Lentibulariaceae
143	Utricularia minor	Lesser bladderwort	Lentibulariaceae
144	Myrica gale	Sweet gale	Myricaceae
145	Nuphar lutea ssp. polysepala	Yellow pond lily	Nymphaeaceae
146	Epilobium anagallidifolium	Pimpernel willowherb	Onagraceae
147	Epilobium ciliatum ssp. ciliatum	Fringed willowherb	Onagraceae
148	Epilobium ciliatum ssp. glandulosum	Fringed willowherb	Onagraceae
149	Epilobium hornemannii ssp. behringianum	Hornemann's willowherb	Onagraceae
150	Epilobium hornemannii ssp. hornemannii	Hornemann's willowherb	Onagraceae

(*continued*)

APPENDIX 3. CROSS-LIST OF SCIENTIFIC AND COMMON PLANT NAMES (*continued*)

	Generic and specific names	Common names	Family
	Dicotyletons (*continued*)		
151	Epilobium lactiflorum	Milkflower willowherb	Onagraceae
152	Epilobium leptocarpum	Slenderfruit willowherb	Onagraceae
153	Epilobium luteum	Yellow fireweed	Onagraceae
154	Epilobium palustre	Marsh willowherb	Onagraceae
155	Chamerion angustifolium	Fireweed	Onagracee
156	Chamerion latifolium	River beauty	Onagracee
157	Circaea alpina	Enchanger's nightshade	Onagracee
158	Boschniakia rossica	Northern groundcone	Orobanchaceae
159	Papaver alboroseum	Pink poppy	Papaveraceae
160	Papaver radicatum ssp.alaskanum	Rooted poppy	Papaveraceae
161	Plantago macrocarpa	Seashore plantain	Plantaginaceae
162	Plantago maritima	Goose tongue	Plantaginaceae
163	Polemonium acutiflorum	Tall jacob's ladder	Polemoniaceae
164	Polemonium pulcherrimum	Jacob's ladder	Polemoniaceae
165	Koenigia islandica	Island purslane	Polygonaceae
166	Oxyria dygyna	Mountain sorrel	Polygonaceae
167	Polygonum amphibium	Water smartweed	Polygonaceae
168	Polygonum fowleri	Fowler knotweed	Polygonaceae
169	Polygonum viviparum	Alpine bistort	Polygonaceae
170	Rumex aquaticus var. fenestratus	Western dock	Polygonaceae
171	Rumex salicifolius var. transitorius	Toothed willow dock	Polygonaceae
172	Claytonia sibirica var. sibirica	Siberian springbeauty	Portulaceae
173	Montia chammissoi	Water minerslettuce	Portulaceae
174	Montia fontana ssp. fontana	Water blinks	Portulaceae
175	Dodecatheon frigidum	Northern shooting star	Primulaceae
176	Dodecatheon jeffreyi ssp. jefferyi	Sierra shooting star	Primulaceae
177	Dodecatheon pulchellum	Darkthroat shooting star	Primulaceae
178	Lysimachia thyrsiflora	Tufted loosestrife	Primulaceae
179	Primula cuneifolia ssp. saxifragifolia	Pixie-eye primrose	Primulaceae
180	Primula egaliksensis	Greenland Primrose	Primulaceae
181	Primula eximia	Chukchi Primrose	Primulaceae
182	Trientalis europaea ssp. arctica	Starflower	Primulaceae
183	Moneses uniflora	Froggy's nightlamp	Pyrolaceae
184	Monotropa hypopitys	Pinesap	Pyrolaceae
185	Orthilia secunda	Side bells	Pyrolaceae
186	Pyrola asarifolia	Pink pyrola	Pyrolaceae
187	Pyrola minor	Snowline wintergreen	Pyrolaceae
188	Aconitum delphiniifolium	Monk's hood	Ranunculaceae
189	Actaea rubra ssp. arguta	Baneberry	Ranunculaceae
190	Anemone multifida	Pacific anemone	Ranunculaceae
191	Anemone narcissiflora var. monantha	Narcissis-flowered anemone	Ranunculaceae
192	Anemone richardsonii	Richardson's anemone	Ranunculaceae
193	Aquilegia formosa	Wild columbine	Ranunculaceae
194	Caltha leptosepala ssp. leptosepala	Mountain marigold	Ranunculaceae
195	Caltha palustris var. palustris	Marsh marigold	Ranunculaceae
196	Coptis asplenifolia	Fernleaf goldthread	Ranunculaceae
197	Coptis trifolia	Threeleaf goldthread	Ranunculaceae
198	Kumlienia cooleyae	Cooley's buttercup	Ranunculaceae
199	Ranunculus cymbalaria	Alkali buttercup	Ranunculaceae
200	Ranunculus eschscholtzii	Eschscoltz's buttercup	Ranunculaceae
201	Ranunculus flammula var. filiformis	Greater creeping spearwort	Ranunculaceae
202	Ranunculus gmelinii	Gmelin's butteracup	Ranunculaceae
203	Ranunculus karelinii	Ice cold buttercup	Ranunculaceae
204	Ranunculus occidentalis var. brevistylis	Western buttercup	Ranunculaceae
205	Ranunculus occidentalis var. nelsonii	Nelson's buttercup	Ranunculaceae
206	Ranunculus pacificus	Pacific buttercup	Ranunculaceae
207	Ranunculus pygmaeus	Pygmy buttercup	Ranunculaceae
208	Ranunculus trichophyllus var. trichophyllus	Whitewater crowfoot	Ranunculaceae
209	Ranunculus uncinatus	Earle's buttercup	Ranunculaceae
210	Argentina egedii ssp. egedii	Silverweed	Rosaceae
211	Aruncus dioicus ssp. vulgaris	Goatsbeard	Rosaceae
212	Comarum palustre	Marsh 5-finger	Rosaceae
213	Dryas integrifolia	Arctic avens	Rosaceae
214	Dryas octopetala	Mountain avens	Rosaceae
215	Fragaria chiloensis	Wild strawberry	Rosaceae
216	Geum calthifolium	Caltha-leaved avens	Rosaceae
217	Geum macrophyllum	Largeleaf avens	Rosaceae
218	Luetkea pectinata	Partridgefood	Rosaceae
219	Malus fusca	Oregon Crabapple	Rosaceae
220	Potentilla drummondi ssp. drummondi	Drummond's cinquefoil	Rosaceae
221	Potentilla nana	Arctic cinquefoil	Rosaceae

(*continued*)

APPENDIX 3. CROSS-LIST OF SCIENTIFIC AND COMMON PLANT NAMES (*continued*)

	Generic and specific names	Common names	Family
	Dicotyletons (*continued*)		
222	Potentilla uniflora	Oneflower cinquefoil	Rosaceae
223	Potentilla villosa	Villous cinquefoil	Rosaceae
224	*Rosa rugosa	Rugosa rose	Rosaceae
225	Rubus arcticus ssp. stellatus	Nagoon berry	Rosaceae
226	*Rubus chamaemorus	Cloud berry	Rosaceae
227	Rubus pedatus	Trailing raspberry	Rosaceae
228	Rubus spectabilis	Salmonberry	Rosaceae
229	Sanguisorba canadensis	Canadian burnet	Rosaceae
230	Sibbaldia procumbens	Creeping sibbaldia	Rosaceae
231	Sorbus sitchensis	Sitka mountain ash	Rosaceae
232	Galium aparine	Stickywilly	Rubiaceae
233	Galium trifidum ssp. columbianum	Threepetal bedstraw	Rubiaceae
234	Galium trifidum ssp. trifidum	Threepetal bedstraw	Rubiaceae
235	Galium triflorum	Fragrant bedstraw	Rubiaceae
236	Populus balsamifera ssp. trichocarpa	Cottonwood	Salicaceae
237	Salix alaxensis	Feltleaf willow	Salicaceae
238	Salix arctica	Arctic willow	Salicaceae
239	Salix barclayi	Barclay's willow	Salicaceae
240	Salix commutata	Undergreen willow	Salicaceae
241	Salix glauca ssp. glauca var. villosa	Grayleaf willow	Salicaceae
242	Salix myrtillifolia	Blueberry willow	Salicaceae
243	Salix polaris	Polar willow	Salicaceae
244	Salix reticulata	Net-leaf willow	Salicaceae
245	Salix rotundifolia	Mouse-ear willow	Salicaceae
246	Salix sitchensis	Sitka willow	Salicaceae
247	Salix stolonifera	Sprouting leaf willow	Salicaceae
248	Heuchera glabra	Alum root	Saxifragaceae
249	Leptarrhena pyrolifolia	Leather-leaf saxifrage	Saxifragaceae
250	Mitella pentandra	Miterwort	Saxifragaceae
251	Parnassia fimbriata var. fimbriata	Fringed grass of Parnassus	Saxifragaceae
252	Parnassia kotzebuei	Kotzebue's grass of Parnassus	Saxifragaceae
253	Parnassia palustris	Grass of Parnassus	Saxifragaceae
254	Saxifraga bronchialis	Yellow dot saxifrage	Saxifragaceae
255	Saxifraga caespitosa	Tufted alpine saxifrage	Saxifragaceae
256	Saxifraga ferruginea	Russethair saxifrage	Saxifragaceae
257	Saxifraga lyallii ssp. hultenii	Redstem saxifrage	Saxifragaceae
258	Saxifraga mertensiana	Wood saxifrage	Saxifragaceae
259	Saxifraga nelsoniana ssp.carlottae	Heart-leaf saxifrage	Saxifragaceae
260	Saxifraga nelsoniana ssp.insularis	Heart-leaf saxifrage	Saxifragaceae
261	Saxifraga nelsoniana ssp.pacifica	Pacific saxifrage	Saxifragaceae
262	Saxifraga nelsoniana ssp.nelsoniana	Heart-leaf saxifrage	Saxifragaceae
263	Saxifraga nivalis	Snow saxifrage	Saxifragaceae
264	Saxifraga oppositifolia	Purple mountain saxifrage	Saxifragaceae
265	Saxifraga rivularis	Weak saxifrage	Saxifragaceae
266	Saxifraga tricuspidata	Prickley saxifrage	Saxifragaceae
267	Tellima grandiflora	Big flower tellima	Saxifragaceae
268	Tiarella trifoliata	Foam flower	Saxifragaceae
269	Castilleja parviflora var. parviflora	Smallflower Indian paintbrush	Scrophulariaceae
270	Castilleja unalaschcensis	Alaska Indian paintbrush	Scrophulariaceae
271	Euphrasia mollis	Eyebright	Scrophulariaceae
272	Lagotis glauca	Weasel snout	Scrophulariaceae
273	Limosella aquatica	Water mudwort	Scrophulariaceae
274	Mimulus gutattus	Monkey flower	Scrophulariaceae
275	Pedicularis capitata	Capitate lousewort	Scrophulariaceae
276	Pedicularis lanata	Woolly lousewort	Scrophulariaceae
277	Pedicularis oederi	Oeders lousewort	Scrophulariaceae
278	Pedicularis parviflora ssp. parviflora	Smallflower lousewort	Scrophulariaceae
279	Pedicularis sudetica ssp. interior	Walrus flower	Scrophulariaceae
280	Pedicularis verticillata	Verticilate lousewort	Scrophulariaceae
281	Rhinanthus minor ssp. groenlandicus	Rattlebox	Scrophulariaceae
282	Veronica americana	American speedwell	Scrophulariaceae
283	Veronica wormskjoldii ssp. wormskjoldii	American alpine speedwell	Scrophulariaceae
284	Veronica wormskjoldii ssp.stelleri	Steller's alpine speedwell	Scrophulariaceae
285	Viburnum edule	High-bush cranberry	Scrophulariaceae
286	Urtica dioica ssp. gracilis	Stinging nettle	Urticaceae
287	Valeriana sitchensis	Sitka valerian	Valerianaceae
288	*Viola adunca	Bird violet	Violaceae
289	Viola epipsila ssp. repens	Dwarf marsh violet	Violaceae
290	Viola glabella	Pioneer violet	Violaceae
291	Viola langsdorffii	Aleutian violet	Violaceae

*Plants added in 2008.

ACKNOWLEDGMENTS

Funding for this project has come from many sources, but three deserve special mention: the U.S. Bureau of Land Management (BLM) for supplying logistics, National Fish and Wildlife Foundation (NFWF) for a generous grant for the 2007–2008 field seasons, and the Alaska Native Plant Society for small grants throughout the life of the project. In addition, many people deserve special thanks for this project; at the top of the list is John Payne (BLM) for his vision of the Bering Glacier as a research area. Special thanks go to the team of botanists who have worked on the project through the years. These include Al Batten, Verna Pratt, Garry Davies, Tony Reznicek, and Nancy Slack. Al Batten also acted as a liaison to the databases at the University of Alaska, Fairbanks Museum and was helpful when I had questions. Additional thanks go to all the other researchers who kept an eye out for unique plants while completing their own studies. Still more thanks go to Christopher Noyles, Nathan Rathbun, and Scott Guyer for their excellent management of the BLM camp. Also, I would like to send a thank-you to Anne Pasch, Al Batten, Janice Glime, and Frank von Hippel for their efforts in editing this document. The list cannot end before thanking the helicopter pilots, and Fishing and Flying, for their extraordinary efforts in getting us to and from camp and into some challenging landing sites.

REFERENCES CITED

Alaska Natural Heritage Program, data available on the World Wide Web, updated April 2007, at URL http://aknhp.uaa.alaska.edu/pdfs/botany/RPWG_06pdf.

Arctos: A Collaborative Database System, data available on the World Wide Web at URL http://arctos.database.museum/home.cfm.

Cody, W.J., 1996, Flora of the Yukon Territory: Ottawa, Canada, National Research Council of Canada, NRC Research Press, 643 p.

Cook, M.B., and Roland, C.A., 2002, Notable vascular plants from Alaska in Wrangell–St. Elias National Park and Preserve with comments on the floristics: Canadian Field Naturalist, v. 116, p. 192–304.

Crawford, R.M.M., 1989, Studies in Plant Survival: Ecological Case Histories of Plant Adaptation to Adversity: Cambridge, Massachusetts, Blackwell Science, 296 p.

Daubenmire, R., 1959, A canopy coverage method of vegetation analysis: Northwest Science, v. 33, p. 43–64.

Daubenmire, R., 1968, Plant Communities: A Textbook of Plant Synecology: New York, Harper & Row, 294 p.

Daubenmire, R., 1974, Plants and Environment: A Textbook of Autecology: New York, Wiley & Sons, 422 p.

Flora of North America, 1993–2006: New York, Oxford University Press (several volumes).

Hultén, E., 1968, Flora of Alaska and Neighboring Territories: Stanford, California, Stanford University Press, 1008 p.

Körner, C., 1999, Alpine Plant Life: Functional Plant Ecology of High Mountain Ecosystems: New York, Springer, 333 p.

U.S. Department of Agriculture, Plants Database: http://plants.usda.gov/.

Viereck, L.A, and Little, E., Jr., 1972, Alaska trees and shrubs, second edition: Snowy Owl Books, University, 265 p.

Viereck, L.A., Dyrness, C.T., Batten, A.R., and Wenzlick, K.J., 1992, The Alaska Vegetation Classification: Technical Report PNW-GTR-286, 278 p.

Wagner, W.H., and Grant, J.R., 2002, Botrychium alaskene, a New Moonwort from the interior of Alaska: American Fern Journal, v. 92, p. 164–170, doi: 10.1640/0002-8444(2002)092[0164:BAANMF]2.0.CO;2.

Welsh, S.L., 1974, Anderson's Flora of Alaska and Adjacent Parts of Canada: Provo, Utah, Brigham Young University Press, 724 p.

MANUSCRIPT ACCEPTED BY THE SOCIETY 02 JUNE 2009

The Geological Society of America
Special Paper 462
2010

Biogeography and ecological succession in freshwater fish assemblages of the Bering Glacier Region, Alaska

Heidi L. Weigner*
Frank A. von Hippel*
Department of Biological Sciences, University of Alaska Anchorage, 3211 Providence Drive, Anchorage, Alaska 99508, USA

ABSTRACT

An inventory of fish species was conducted in the Bering Glacier Region, Alaska, in 2002–2006. Ten species were collected: surf smelt, coho salmon, sockeye salmon, rainbow trout, Dolly Varden, threespine stickleback, prickly sculpin, slimy sculpin, Pacific staghorn sculpin, and starry flounder. All are either marine in origin or tolerant of salt water; consistent with this, fishes in the watershed tolerate a wide range of water qualities in fresh water. Stickleback, prickly sculpin, slimy sculpin, coho salmon, and Dolly Varden are found most commonly either because they are early colonizing species or they are able to out-compete early colonizers. Species that readily assume residence in fresh water were found equally often in isolated and connected lakes and streams, whereas diadromous and marine species were found primarily in lakes and streams with outlets. Except for Dolly Varden, species were more likely to be found in nonglacial than in glacial lakes and streams. Greater species richness was associated with the presence of aquatic vegetation and algae, both of which provide structural complexity and indicate more abundant nutrient levels in otherwise oligotrophic waters. With the exception of Vitus Lake, which is tidally influenced, fish species richness was low. Older lakes and streams support more species than younger aquatic habitats, presumably owing to greater time for colonization and the formation of habitat complexity. This ever-changing aquatic landscape has been colonized by typical fish species in the region, but these colonists have evolved in atypical ways, including populations of dwarf Dolly Varden and stickleback species pairs.

INTRODUCTION

Understanding the biogeography of a region is paramount to population and community studies. Knowledge of the history of individual taxa allows us to anticipate the role each will exhibit in a new community (Wilson, 1961). New lakes and streams formed in the wake of receding glaciers provide a diverse landscape for colonization and evolution of oceanic fishes in fresh water. These young, reticulated freshwater landscapes are subject to numerous disturbance factors. Habitat types include proglacial (ice contact at glacial terminus) and periglacial (ice contact at sides of glacier) lakes, kettle ponds, glacial and snow meltwater streams, oxbow lakes, tidally influenced lakes and streams, and older, mature lakes, streams, and estuaries. The combination of various biotic and abiotic features of each type of freshwater system plays an important and complex role in the biogeography of fishes (Johnson et al., 1977; Dunham et al., 2003).

The freshwater biogeography of newly deglaciated terrain is poorly understood. In Glacier Bay National Park, Alaska, Milner (1987, 1994; Milner and Bailey, 1989) investigated the

*Emails: Weigner: anhlw@uaa.alaska.edu; von Hippel: affvh@uaa.alaska.edu.

Weigner, H.L., and von Hippel, F.A., 2010, Biogeography and ecological succession in freshwater fish assemblages of the Bering Glacier Region, Alaska, *in* Shuchman, R.A., and Josberger, E.G., eds., Bering Glacier: Interdisciplinary Studies of Earth's Largest Temperate Surging Glacier: Geological Society of America Special Paper 462, p. 167–180, doi: 10.1130/2010.2462(08). For permission to copy, contact editing@geosociety.org.

colonization and development of freshwater stream communities, including colonization by salmonids, in relation to stream age and the temporal succession of aquatic macroinvertebrate communities. These studies focused on succession from newly formed glacier meltwater streams to mature clear-water streams with a relatively stable hydrology. The biogeography of freshwater fishes in newly deglaciated terrain with a complex system of streams, rivers, lakes, kettle ponds, and wetlands is less well understood.

Most biogeographic studies of freshwater fishes relative to glacial landscapes have been reconstructions of historical distributions during the Pleistocene glaciations on the basis of geological evidence (McPhail and Lindsey, 1970, 1986; Lindsey, 1975; Lindsey and McPhail, 1986; McPhail, 2007). Currently, reconstructions of historic distributions are performed using phylogenetic analyses (e.g., Bernatchez and Wilson, 1998; Johnson and Taylor, 2004; Fraser and Bernatchez, 2005; Gagnon and Angers, 2006; Ray et al., 2006). The ongoing recession of the Bering Glacier gives us the opportunity for studying fish colonization in real time in a complex, newly formed freshwater landscape.

The purposes of this study are to document associations between fish species and abiotic and biotic factors in the Bering Glacier Region and to document the distribution of fish species in specific locations. We examine associations among fish species and their associations with substrate type, water appearance, water quality, aquatic vegetation, and algae. We investigate which fishes are found in lakes versus streams, which colonize glacial lakes and streams, and which exist in isolated waters versus those connected to the Gulf of Alaska. We also discuss fish species richness in the context of the approximate year of a site's most recent deglaciation. We assemble all of the variables and build a predictive model to determine the occurrence of fish species based on these variables. By examining a suite of biotic and abiotic parameters, we investigate fish colonization and fish succession in aquatic habitats newly formed around the terminus of North America's largest glacier.

Finally, we report on the distribution of dwarf Dolly Varden *(Salvelinus malma)* and species pairs of threespine stickleback (*Gasterosteus aculeatus* species complex; von Hippel & Weigner, 2004) and discuss management implications.

Study Area

Bering Glacier (Fig. 1) is the largest glacier in the world outside of Greenland and Antarctica. It drains ~5200 km^2 of south-central Alaska (Merrand and Hallet, 1996; Jaeger and Nittrouer, 1999) and is ~200 km long (Herzfeld and Mayer, 1997). It is the temperate zone's largest surging glacier (Wiles et al., 1999), with six surges reported since 1900 (Molnia et al., 1996; Jaeger and Nittrouer, 1999). Surges typically occur every 20–30 a (Molnia and Post, 1995; Muller and Fleisher, 1995). During a surge the

Figure 1. Landsat imagery of the Bering Glacier Region, showing all 80 sampling sites. Some sites were sampled more than once, but each site is indicated only once on the map. Landmarks named are discussed in the text. Image was captured 10 September 2001.

terminus of the glacier overruns the glacial forelands and causes major changes to the young ecosystem (Herzfeld and Mayer, 1997). The most recent surge, which occurred in 1993–1995, caused rapid advance in the central region of the terminus of the Bering Lobe, including a 1500 m advance in 17 d (Herzfeld and Mayer, 1997). Meanwhile, the western side of Vitus Lake around Tashalich Arm underwent a retreat of 25 m/d (Herzfeld and Mayer, 1997). The dynamics of surges cause different areas of the terminus to move at different rates throughout the surge (Herzfeld and Mayer, 1997), which has been the pattern for the past 1500 a (Molnia and Post, 1995). Therefore, different parts of the surrounding ecosystem are disturbed at different rates.

The termini of the Bering and Stellar Lobes form a chain of proglacial lakes separated from the Gulf of Alaska by a narrow strip of land (Fig. 1; Molnia and Post, 1995; Herzfeld and Mayer, 1997). The largest of these proglacial lakes is Vitus Lake, which is ~25 km long, 10 km wide and up to 150 m deep, though it is constantly changing (Brouwers and Forester, 1993; Molnia and Post, 1995; Josberger et al., 2006). Since the end of the 1950s, Vitus Lake has drained through the Seal River into the Gulf of Alaska. The Seal River is 5 km long and 5–10 m deep; at high tide its surface is below sea level (Molnia and Post, 1995; Merrand and Hallet, 1996). Before the 1950s, Bering Glacier was drained simultaneously by the Tsivat, Tsiu, Midtimber, Seal, Tashalich, Kiklichk, Kosacuts, and Bering Rivers (Molnia and Post, 1995). Now only the Bering, Kosacuts, and Seal Rivers provide primary drainage.

Most sampling was conducted in the forelands of the Bering Glacier between the Gulf of Alaska and the current terminus (Table 1; Fig. 1). The terminus of the glacier has been receding from its Neoglacial maximum since ~1900 (Molnia and Post, 1995; Herzfeld and Mayer, 1997), though some of the lakes and streams were overridden by the glacier as recently as the latest surge in 1993–1995. The area sampled is a heterogeneous landscape of proglacial and periglacial oligotrophic lakes, glacial and clear meltwater streams, kettle ponds, and tidally influenced lakes, streams, and wetlands. There is no history of fish stocking in the area, and humans have caused minimal disturbance.

METHODS

Fish were trapped using unbaited 1/8 in. mesh minnow traps, unbaited semi-oval traps, dip nets, and seine nets. Six to 10 traps were set for ~24 h at each site. Some species of fish may have been missed at some sites because of sampling methods that were biased toward small individuals at the margins of aquatic habitat.

Fish were killed with an overdose of MS-222 anesthetic and preserved in 95% ETOH for DNA analysis or fixed in 10% buffered formalin and then preserved in 70% ETOH for morphological analysis. Fish were visually identified (when necessary under a Leica MZ6 dissecting microscope) following the methods of Morrow (1980), Pollard et al. (1997), and Mecklenburg et al. (2002). Anadromous and resident freshwater threespine stickleback were distinguished from each other on the basis of methods described in von Hippel and Weigner (2004). Because Dolly Varden were not in breeding coloration, dwarfism was assumed on the basis of small size and the appearance of dark parr marks coincident with the presence of spots (Klemetsen and Grotnes, 1980; Parker and Johnson, 1991). In normal-size Dolly Varden, parr marks disappear when the spots appear. Normal-size resident freshwater Dolly Varden are between 10 and 25 cm fork length (reviewed by McPhail, 2007).

Sampling was conducted from early May until early September 2002–2006. We sampled for fish 221 times in 80 lakes (including kettle ponds and former rivers that evolved into lakes) and streams. All traps where placed within 3 m of the shore and between 0.3 and 2 m deep. Field sites were reached by foot, all-terrain vehicle (ATV), helicopter, and inflatable boat with an outboard engine. We did not sample bodies of water at random but on the basis of accessibility and in a few cases, previous reports from the Alaska Department of Fish and Game (1999).

Water quality measurements were made using a YSI-85 oxygen, conductivity, salinity, and temperature meter and a Hanna pH meter. pH readings for 1 June–2 July 2004 were excluded from analysis owing to a faulty meter. Surface water temperature was verified in the field using an alcohol thermometer. Salinity was verified in the field using a salinity refractometer. Latitude and longitude coordinates were collected using a Garmin Vista global positioning system (GPS) receiver, in decimal degrees and in datum WGS 1984.

Other observations, including substrate type, water appearance, presence of aquatic vegetation, and presence of macroscopic algae, were made in the field, and isolation of lake or stream was recorded. Substrate type was divided into six categories: silt-clay-mud, sand (<0.25 cm), gravel (0.25–5 cm), cobble (5–25 cm), boulders (>25 cm), and bedrock. Water appearance was divided into five categories: clear, muddy-silty, scummy-foamy, cloudy, or oily sheen; these categories were subsequently reduced either to clear or to silty–not clear. Isolation is defined as lack of connection to the Gulf of Alaska by surface waters. Many sites sampled are seasonally connected, based on rainfall and snowmelt. However, if they were not connected at the time of sampling, as determined visually from the air and walking the perimeter of the site, then they were classified as isolated, and these sites are likely to be isolated most of the time. Spawning salmon were visually identified by observation on the spawning grounds when possible.

Data were analyzed using SPSS version 11.5. Although some lakes and streams were sampled multiple times, each lake or stream is treated as a single observation in analyses. Chi-square analyses were used to test for significance of associations between pairs of fish species, occurrence in lakes versus streams, occurrence in isolated versus connected lakes and streams, occurrence in glacial versus nonglacial lakes and streams (a glacial lake is defined as in glacial contact, and a glacial stream as fed by glacial meltwater), fish associations with aquatic vegetation, fish associations with algae, and fish associations with water appearance. Substrate type was not mutually exclusive and therefore could not

TABLE 1. LOCATION OF ALL SITES SAMPLED, NUMBER OF TIMES EACH SITE WAS SAMPLED, AND FISHES FOUND AT EACH SITE

Site name	Times sampled	Location		Fish species										
		Lat (N)	Long (W)	Hp	Ok	On	Om	Sm	GaA	GaR	Ca	Cc	La	Ps
Bentwood Lake	2	60.18	143.26											
Berg Lake Pond	1	60.4244	143.76715					X						
Berg Lake Stream 1	3	60.41	143.86					X						
Berg Lake Stream 2	1	60.42326	143.84619					X						
Berg Lake Stream 3	2	60.43	143.79					X						
Berg Lake Stream 4	6	60.427	143.67					X						
Berg Lake Stream 5	2	60.412	143.73					X						
Lake on Bering River	6	60.36	143.973		X				X	X	X	X		
Stream near Bering River	1	60.2756	143.97131		X			X			X			
Creek 1	12	60.103	143.36		X				X	X	X	X	X	X
Creek 2	9	60.09	143.35		X				X	X	X	X		X
Hana Lake	1	60.23716	143.12201								X			
Khitrov Lake	1	60.3951	143.47476					X						
Stream in Khitrov Hills	1	60.39976	143.48659					X						
Kiklichk River	3	60.017	143.803		X			X		X	X			
Kosakuts River	4	60.27	143.015		X			X				X		
Midtimber Lake	9	60.07	143.33						X	X	X	X		
Stream in Suckling Hills	1	60.0345	143.83661					X						
Stream below Suckling Hills	1	60.02073	143.82176		X		X	X						
Tashalich Lake	6	60.039	143.63		X				X	X			X	
Tsiu Lake	2	60.19115	143.20921								X	X		
Tsiu River	10	60.08	143.05		X	X		X	X	X				X
East Tsivat Lake	1	60.13512	142.97578							X				
Tsivat Lake	2	60.2315	143.13					X			X			
East Fork Tsivat River	1	60.13509	142.98289		X		X			X				
Tsivat River	3	60.065	143.25						X	X	X	X		
West Fork Tsivat River	1	60.16788	142.98744		X	X		X				X		
Unnamed Lake	5	60.102	143.55		X				X	X	X			
Stream into Vitus Lake	1	60.1498	143.26956		X			X			X	X		
Vitus Lake	59	60.1	143.3	X	X			X	X	X	X	X	X	X
Unnamed	1	60.08626	143.59593							X	X			
Unnamed	1	60.12875	143.5657								X			
Unnamed	2	60.0933	143.54							X			X	X
Unnamed	1	60.07311	143.44769										X	
Unnamed	1	60.07116	143.4408		X					X			X	X
Unnamed	3	60.11	143.475							X	X	X		
Unnamed	1	60.1161	143.47565							X	X			
Unnamed	1	60.11666	143.47559							X	X			
Unnamed	1	60.07656	143.39952							X	X			
Unnamed	1	60.17439	142.88318		X					X	X			
Unnamed	1	60.28817	143.27142											
Unnamed	1	60.28145	143.26598											
Unnamed	1	60.27804	143.27007											
Unnamed	1	60.2735	143.11859											
Unnamed	2	60.274	143.12											
Unnamed	2	60.23	143.12		X		X				X			
Unnamed	1	60.19212	143.20744											
Unnamed	1	60.17346	143.17477		X					X		X		
Unnamed	1	60.15723	143.21997				X			X		X		
Unnamed	2	60.1702	143.286		X							X		
Unnamed	1	60.15536	143.27312		X									
Unnamed	1	60.12015	143.27409								X			
Unnamed	1	60.1204	143.28148											
Unnamed	1	60.12083	143.27992											
Unnamed	2	60.11587	143.29259		X					X	X			
Unnamed	2	60.11794	143.30305		X					X	X			
Unnamed	2	60.1185	143.304		X					X	X			

(continued)

TABLE 1. LOCATION OF ALL SITES SAMPLED, NUMBER OF TIMES EACH SITE WAS SAMPLED, AND FISHES FOUND AT EACH SITE (*continued*)

		Location		Fish species										
Site name	Times sampled	Lat (N)	Long (W)	Hp	Ok	On	Om	Sm	GaA	GaR	Ca	Cc	La	Ps
Unnamed	2	60.1172	143.304							X	X			
Unnamed	1	60.11991	143.30804							X	X			
Unnamed	2	60.12018	143.303		X					X				
Unnamed	1	60.1217	143.30627											
Unnamed	1	60.12245	143.31848		X									
Unnamed	1	60.12247	143.32005											
Unnamed	1	60.11792	143.33102							X	X			
Unnamed	1	60.42302	143.8459											
Unnamed	1	60.14207	143.72112							X	X			
Unnamed	1	60.14292	143.658											
Unnamed	1	60.14577	143.6366								X			
Unnamed	1	60.1189	143.67191							X	X			
Unnamed	1	60.10201	143.6897							X	X			
Unnamed	1	60.10155	143.68843							X	X			
Unnamed	1	60.08538	143.59091							X	X			
Unnamed	1	60.455	144.2083							X				
Unnamed	1	60.27523	143.97087											
Unnamed	1	60.2760	143.97153								X			
Unnamed	1	60.39596	143.476											
Unnamed	1	60.39689	143.47847											
Unnamed	1	60.3978	143.47951											
Unnamed	1	60.06223	143.31793							X	X			
Unnamed	1	60.08607	143.59273							X	X			

Note: Site name is given if the lake or stream is named. Differences in the specificity of the GPS coordinates depend on the size of the lake or stream and the number of times sampled. Hp—surf smelt; Ok—coho salmon; On—sockeye salmon; Om— rainbow trout; Sm—Dolly Varden; GaA—anadromous threespine stickleback; GaR—resident freshwater threespine stickleback; Ca—prickly sculpin; Cc—slimy sculpin; La—Pacific staghorn sculpin; Ps—starry flounder.

be analyzed statistically. Associations for many fish species could not be analyzed statistically because of small sample sizes.

Approximate ages of lakes and streams were determined from geographic information system (GIS) layers that were built from aerial photographs and historical data from the region. GIS layers were constructed by ALTARUM. Approximate ages were divided into six categories on the basis of when the lake or stream was deglaciated. Several of these sites were overrun by glacial surges, and the year of the most recent retreat of the glacier was considered as the year of formation. Data for the age of deglaciation were available only for lakes and streams in the Bering Glacier forelands. Lakes and streams on the perimeter were not included in the age analysis. If any part of the site sampled was not glaciated in 1900, the whole site was categorized as older than 1900. The categories were divided into before 1900, 1900–1967, 1968–1978, 1979–1981, 1982–1990, and after 1995. The categories were picked on the basis of available GIS data. The 1993–1995 glacial surge covered all sampled sites that were formed after 1990, and hence there is no age category between 1991 and 1995; these sites formed again after the ice retreated subsequent to 1995. These age categories were then given age ranks from 1 (oldest) to 6 (youngest) and subjected to Spearman's correlation against rank of the number of fish species present in the site.

Binary logistic regressions were performed to build predictive models for the presence or absence of fish species based upon biotic and abiotic parameters. Continuous water quality variables were recoded and binned into groups to normalize their distributions. Salinity was divided into fresh water (0–0.5 ppt), brackish water (0.6–30.0 ppt), and marine water (>30.1 ppt). However, marine water was not present at any study site. Each variable was tested to determine if it alone had a significant effect on the presence or absence of a species. Nonsignificant terms and terms that were highly correlative with each other (>.80) were removed from the model. If a term needed to be removed owing to a correlation with another term, the model was tested separately with each correlated variable to determine which had the best fit. Two-way interaction terms were tested between the significant main effects (Landau and Everitt, 2004), but because none were significant they are not reported.

RESULTS

Fish were captured in 79% (n = 63) of the 80 lakes and streams and in 90% of the 221 sampling trials. We collected 10 fish species (Table 1), with resident freshwater threespine stickleback and prickly sculpin *(Cottus asper)* occurring in the greatest

TABLE 2. LOCATIONS, SPECIES PAIR CATEGORY (LAKE OR STREAM), AND HABITAT DETAILS OF THE SPECIES PAIRS OF ANADROMOUS AND RESIDENT FRESHWATER THREESPINE STICKLEBACK

Location	Latitude (N)	Longitude (W)	Species pair category	Habitat details
Bering River Lake site	60.35771	143.97263	Lake	Small lake adjacent to Bering River
Creek 1	60.10482	143.35008	Stream	Stream draining into Vitus Lake
Creek 2	60.09770	143.35117	Stream	Stream draining into Vitus Lake
Midtimber Lake	60.08	143.35	Lake	Lake formerly a river
Tashalich Lake	60.03905	143.62566	Lake	Lake formerly a river
Tsiu River	60.08	143.06	Stream	River and wetlands
Tsivat River	60.065	143.247	Lake	Lake formerly a river
Unnamed Lake	60.10195	143. 55113	Lake	Small lake adjacent to Vitus Lake
Vitus Lake	60.1	143.3	Lake	Proglacial lake

Note: Differences in the specificity of the GPS coordinates depend on the size of the lake or stream and the number of times sampled.

number of lakes and streams (n = 37 sites for each, 46% of all sites sampled). The resident freshwater stickleback was the most abundant species wherever it occurred. Surf smelt *(Hypomesus pretiosus)* were collected only at three locations in Vitus Lake, where they were collected dead on the beach. They are most likely found sporadically in the region. Coho salmon *(Oncorhynchus kisutch)* were found in 25 lakes and streams (31% of sites sampled). Sockeye salmon *(Oncorhynchus nerka)* were rarely observed in the region (just two streams). Fry were not trapped, though spawning adults were observed in both sites. Rainbow trout *(Oncorhynchus mykiss)* were also rarely found in the region (just four lakes and streams). Dolly Varden were trapped in 18 lakes and streams (23% of sites). Anadromous threespine stickleback were trapped in nine lakes and streams (11% of sites). All nine sites that contained anadromous stickleback also contained resident freshwater stickleback, giving rise to sympatric species pairs (Table 2). Slimy sculpin *(Cottus cognatus)* were captured in 14 water bodies (17.5% of sites). Pacific staghorn sculpin *(Leptocottus armatus)* and starry flounder *(Platichthys stellatus)* were captured in six sites (7.5% of sites).

Associations between Fishes

We examined associations between fish species found in sympatry. With 10 species of fish and 2 varieties of stickleback, there are 55 possible pairwise associations, of which 47 were not encountered often enough to analyze statistically. Five of the eight association analyses were not significant (coho salmon and Dolly Varden, coho salmon and prickly sculpin, Dolly Varden and prickly sculpin, resident freshwater stickleback and slimy sculpin, and prickly sculpin and slimy sculpin). Resident freshwater stickleback have significant positive associations with coho salmon ($\chi^2 = 4.6$, df = 1, $p = .032$) and prickly sculpin ($\chi^2 = 19.8$, df = 1, $p <.001$), and a significant negative association with Dolly Varden ($\chi^2 = 8.2$, df = 1, $p = .004$).

Fish were captured alone at 19 sites. The species most commonly found alone was Dolly Varden. Resident freshwater species tended to occur sympatrically with few other fish species, while anadromous and marine species, in that order, were more often found coexisting with a larger number of other fish species (Fig. 2).

Fish in Lakes or Streams

The presence of fish was not significantly associated with whether a site was a lake (n = 51) or stream (n = 29; $\chi^2 = 1.5$, df = 1, p = .219). However, some species were trapped more

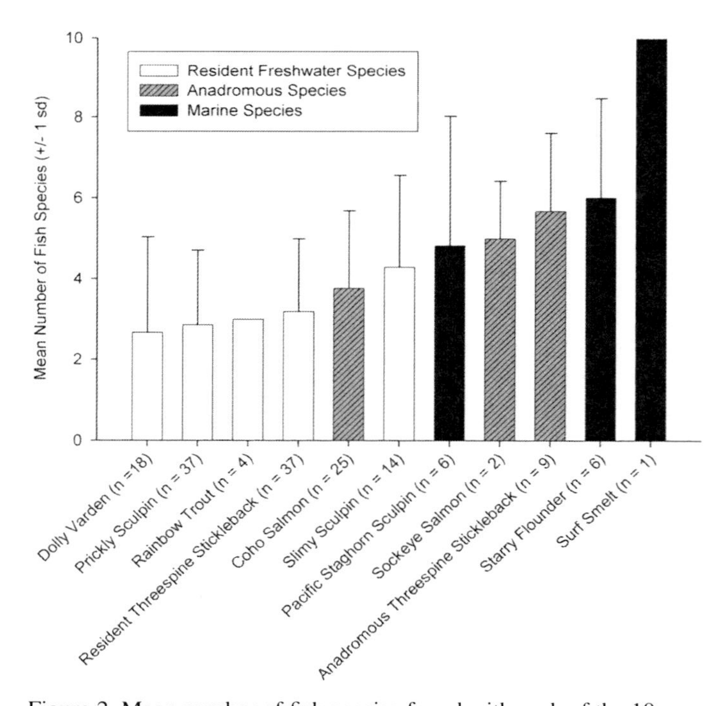

Figure 2. Mean number of fish species found with each of the 10 species trapped in freshwater sites. Resident and anadromous threespine stickleback are counted as a single species in terms of number of species in a site. Prickly sculpin may adopt either a resident freshwater or catadromous life history.

commonly in one habitat type than another. Coho salmon (χ^2 = 8.9, df = 1, p = .003) and Dolly Varden (χ^2 = 17.3, df = 1, p <.001) were trapped more often in streams, whereas sockeye salmon and rainbow trout were found only in streams (though the samples were too small to analyze statistically). Resident freshwater stickleback (χ^2 = 4.2, df = 1, p = .040) and prickly sculpin (χ^2 = 8.9, df = 1, p = .003) were trapped more often in lakes. Slimy sculpin were trapped equally often in lakes and streams (χ^2 = 1.4, df = 1, p = .239). Remaining species were not found often enough to analyze statistically.

Fish in Isolated Lakes and Streams

Fish were trapped in 37 (71%) of 52 lakes and streams that were isolated from the Gulf of Alaska and in 26 (93%) of 28 lakes and streams connected to the Gulf of Alaska. Therefore, fish were more commonly found in connected lakes and streams (χ^2 = 5.1, df = 1, p = .024).

Although most species were found more frequently in connected lakes and streams, this trend was only significant for coho salmon (χ^2 = 26.9, df = 1, p <.001). Dolly Varden (χ^2 = 2.3, df = 1, p = .130), resident freshwater stickleback (χ^2 = 0.2, df = 1, p = .622), and prickly sculpin (χ^2 = 0.2, df = 1, p = .622) had no significant difference in the probability of being trapped in isolated or connected lakes and streams. Sample sizes were too small to analyze probabilities for the remaining species, though of these, only anadromous stickleback were found in isolated lakes and streams. For anadromous stickleback to occur in isolated sites, the sites must have established seasonal connections that were not evident at the time of sampling.

Fish in Glacial Lakes and Streams

Twenty-six (33%) of the lakes and streams sampled were glacial. Fish were found in 14 of them (54%) compared with 91% of nonglacial lakes and streams. Therefore, we trapped fish in a higher proportion of nonglacial lakes and streams (χ^2 = 14.3, df = 1, p <.001).

With the exception of Dolly Varden, which were found more often in glacial lakes and streams (χ^2 = 8.7, df = 1, p = .003), all species are found more commonly in nonglacial lakes and streams. Dolly Varden were trapped in 11 different glacial lakes and streams, or 42% of the total. Coho salmon were found in only 3 (12%) out of the 26 glacial lakes and streams trapped (χ^2 = 7.0, df = 1, p = .008). Rainbow trout and sockeye salmon were not found in any glacial lakes or streams. Surf smelt, anadromous and resident freshwater stickleback, Pacific staghorn sculpin, and starry flounder were all found in only one glacial lake, Vitus Lake. Resident freshwater stickleback were trapped significantly more often in nonglacial lakes and streams (χ^2 = 27.9, df = 1, p <.001). Prickly sculpin are also seldom found in glacial lakes and streams compared with nonglacial lakes and streams (χ^2 = 14.8, df = 1, p <.001). Remaining species were not captured often enough to be analyzed statistically.

Fish Associations with Habitat Parameters

Fish were more commonly found in lakes and streams with aquatic vegetation. Out of the 37 lakes and streams with aquatic vegetation, 35 (95%) also had fish present, whereas out of 43 lakes and streams without aquatic vegetation, only 28 (64%) also had fish present (χ^2 = 10.3, df = 1, p <.001). Resident freshwater stickleback (χ^2 = 16.0, df = 1, p <.001) and prickly sculpin (χ^2 = 12.6, df = 1, p <.001) have statistically significant associations with the presence of aquatic vegetation. Remaining species were not captured frequently enough to be analyzed statistically, except for Dolly Varden, which showed no significant association.

Fish are also usually found in lakes and streams with algae. Out of the 49 lakes and streams with algae, 43 (88%) also had fish present, whereas out of 31 lakes and streams without algae, only 20 (65%) also had fish present (χ^2 = 6.1, df = 1, p = .013). All fishes are found more often in lakes and streams with algae than without. There were significant associations between the presence of algae and coho salmon (χ^2 = 7.9, df = 1, p = .005), resident freshwater stickleback (χ^2 = 6.0, df = 1, p = .014), prickly sculpin (χ^2 = 4.0, df = 1, p = .046), and slimy sculpin (χ^2 = 4.3, df = 1, p = .039). Dolly Varden had a nonsignificant association with algae. Remaining species were not captured frequently enough to be analyzed statistically.

Substrate types were not mutually exclusive. Any combination of substrate could occur in a lake or stream, with the exception of bedrock, which was not present at any site. Substrate type did not appear to affect whether fish were present or absent, nor which species of fish were found at which sites (Table 3).

Fish were found more often in clear water than in muddy-silty, scummy-foamy, cloudy, or oily water (Table 4). When water appearance was categorized as clear or not clear, fish were found more often in clear water sites (χ^2 = 15.5, df = 1, p <.001). Resident freshwater stickleback, however, were the only species to show a significant association with clear water (χ^2 = 10.1, df = 1, p <.001). Coho salmon, Dolly Varden, prickly sculpin, and

TABLE 3. PRESENCE AND ABSENCE OF FISH BY SUBSTRATE TYPE

	Substrate type by site (%)				
	Silt, clay, mud	Sand	Gravel	Cobble	Boulder
Fish present	22.1	22.1	21.6	20.1	28.0
Fish absent	17.5	22.8	22.8	21.1	15.8

Note: Substrate types are not mutually exclusive.

TABLE 4. PRESENCE AND ABSENCE OF FISH BY WATER APPEARANCE

	Water appearance by site (%)				
	Clear	Muddy/silty	Scummy/foamy	Cloudy	Oily sheen
Fish present	68.8	22.1	5.2	2.6	1.3
Fish absent	33.3	55.6	0	11.1	0

Note: Water appearance categories are mutually exclusive.

slimy sculpin had no associations with clear water when analyzed statistically. Remaining species were not captured enough to analyze statistically.

Water Quality

Pacific staghorn sculpin, prickly sculpin, slimy sculpin, starry flounder, coho salmon, Dolly Varden, and anadromous and resident stickleback were collected sufficiently often to estimate water quality tolerances within the range of sites tested. Most fish species were trapped in a wide range of temperatures (Fig. 3A), salinities (Fig. 3B), pH values (Fig. 3C), conductivities (Fig. 3D), and dissolved oxygen values (Fig. 3E) that closely matched the distribution from all 221 sampling events. The species trapped most commonly in high salinities were Pacific staghorn sculpin, starry flounder, and slimy sculpin. The mean pH of water bodies in the region is 8.4, which is well above the pH of a typical freshwater habitat (Dodds, 2002).

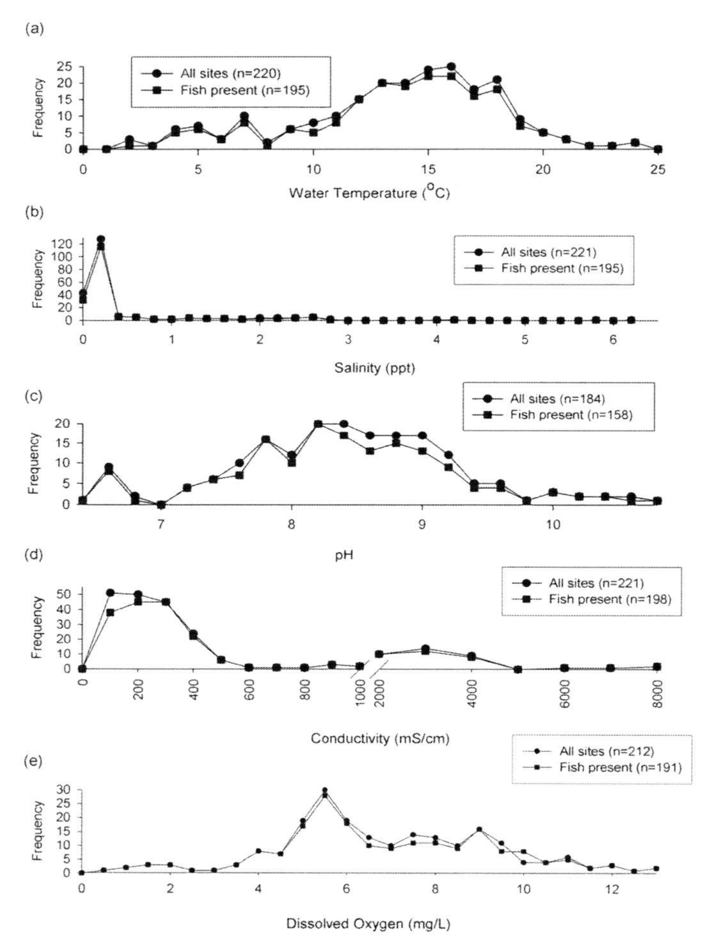

Figure 3. Fish species were found in a wide range of water qualities. (A) Frequency of sampling of sites of different water temperatures (°C) and frequency of any fish present in those sites. (B) Frequency of sampling of sites of different salinities (ppt) and frequency of any fish present in those sites. (C) Frequency of sampling of sites of different pH values and frequency of any fish present in those sites. (D) Frequency of sampling of sites of different conductivities (µS/cm) and frequency of any fish present in those sites. (E) Frequency of sampling of sites of different dissolved oxygen values (mg/L) and frequency of any fish present in those sites.

Approximate Age of Lakes and Streams

Age of the lakes and streams was divided into six categories, based on the timing of deglaciation. The older a lake or stream, the more likely it was to have fish, and the more fish species we captured (Fig. 4). Currently, few fish species are found in sympatry; the mean number of fish species per lake or stream that deglaciated in 1978 or before is 3.2, whereas the mean number per lake or stream formed in 1981 or after is 1.4. Species found in the youngest sites were prickly sculpin (n = 3), slimy sculpin (n = 2), and resident freshwater stickleback (n = 2).

Logistic Regression Models

The presence of fish is best predicted by clear water (Wald = 8.37, p = .004), a nonglacial water body (Wald = 4.98, p = .026), and water bodies connected to the Gulf of Alaska (Wald = 5.65, p = .017). The presence of resident freshwater stickleback is best predicted by the presence of prickly sculpin (Wald = 9.02, p = .003), the presence of coho salmon (Wald = 5.68, p = .017), the absence of Dolly Varden (Wald = 8.87, p = .003), the presence of aquatic vegetation (Wald = 3.99, p = .046), and the presence of clear water (Wald = 7.81, p = .005). The presence of coho salmon is best predicted by water bodies connected to the Gulf of Alaska (Wald = 20.37, p <.001) and being found in nonglacial

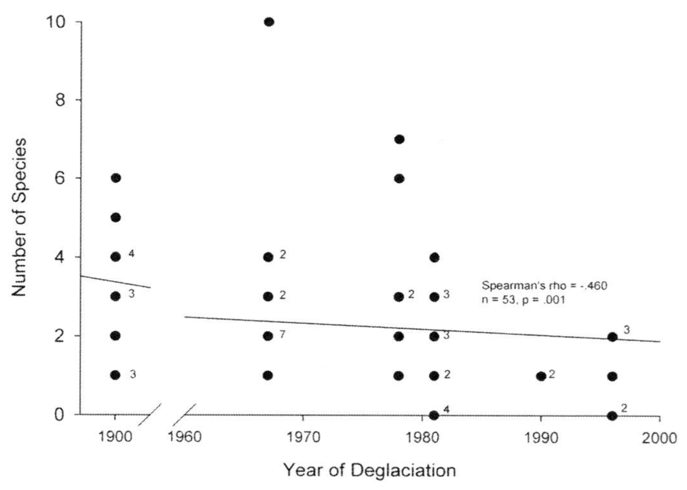

Figure 4. Number of fish species found in lakes and streams of different ages. Numbers next to data points indicate multiple sites with the same number of fish species. A regression line is fitted to the data, indicating a significant trend toward fewer species found in younger sites.

water bodies (Wald = 5.78, $p = .016$). The presence of Dolly Varden is best predicted by stream habitat (Wald = 11.22, $p = .001$) and a glacial water body (Wald = 6.77, $p = .009$). The presence of prickly sculpin is best predicted by lake habitat (Wald = 5.34, $p = .021$) and the presence of resident stickleback (Wald = 5.57, $p = .018$). The presence of slimy sculpin is best predicted by connection to the Gulf of Alaska (Wald = 8.44, $p = .004$).

Distribution of Dwarf Dolly Varden

Nearly all Dolly Varden trapped in the region appeared to be dwarfed (Table 5). Out of 18 lakes and streams where Dolly Varden were found, 9 are isolated (mostly Berg Lake area and Khitrov Lake area), and 11 are glacial.

DISCUSSION

Associations between Fishes

Fish communities are governed by predator-prey relations and competition for food and habitat (Evans et al., 1987). Age and stability of the watershed are important for the maturation of fish communities from loose, random assemblages to more stable communities (Evans et al., 1987). Freshwater habitats around the Bering Glacier are both young and unstable, as they undergo periodic glacial surges and glacial outburst floods. As the ice-free landscape around the Bering Glacier continues to change and expand with deglaciation, more stable aquatic communities would be expected to form (Evans et al., 1987).

Most species were rarely found in the region (Table 1), presumably owing to the youth of the landscape (most sites formed after 1900, and many formed in the last few decades; Fig. 4). However, we found a significant association between resident freshwater stickleback and coho salmon. Juvenile coho salmon prey on smaller stickleback and their eggs (Reimchen, 1994). A significant association also exists between resident freshwater stickleback and prickly sculpin, most likely because adult prickly sculpin prey upon stickleback (Moodie, 1972a; McPhail, 2007; Messler et al., 2007), and they share many habitat requirements. A negative association was found between Dolly Varden and resident freshwater stickleback. Although Dolly Varden prey on stickleback (McPhail, 2007), the two species are primarily found in different habitats. Dolly Varden were found in glacial streams, whereas resident stickleback were found mostly in nonglacial lakes and streams.

Fish in Isolated Lakes and Streams

Four of the salmonid species either were found significantly more often in streams than lakes or were found only in streams. In contrast, resident freshwater stickleback and prickly sculpin were found significantly more often in lakes. Lakes and streams can be isolated from the Gulf of Alaska or connected. Connection to the Gulf of Alaska is essential for anadromous life histories, such as those of most salmonids.

Since anadromous fish in the Bering Glacier Region spawn in rapidly changing rivers and ephemeral streams, they can often become trapped in oxbow lakes, pools, or former rivers

TABLE 5. LOCATIONS AND HABITAT DETAILS OF DWARF DOLLY VARDEN SITES

Location	Latitude (N)	Longitude (W)	Habitat type	pH
Berg Lake area	60.41	143.86	Isolated glacial stream	8.65
Berg Lake area	60.42326	143.84619	Isolated glacial stream	8.20
Berg Lake area	60.43	143.79	Isolated glacial stream	7.90
Berg Lake area	60.427	143.67	Isolated glacial stream	9.58
Berg Lake area	60.42440	143.76715	Isolated glacial pond	6.50
Berg Lake area	60.412	143.73	Isolated glacial stream	7.35
Stream near Bering River	60.27560	143.97131	Nonglacial stream draining into glacial lake	–
Khitrov Lake	60.39510	143.47476	Isolated glacial lake	7.50
Khitrov stream	60.39976	143.48659	Isolated glacial stream	9.10
Kiklichk River	60.017	143.803	Nonglacial river	7.75
Kosakuts River	60.27	143.015	Glacial river	8.09
Stream below Suckling Hills	60.02073	143.82176	Nonglacial stream	8.90
Stream in Suckling Hills	60.03450	143.83661	Isolated nonglacial stream	7.70
Tsiu River	60.08	143.05	Nonglacial river	8.38
Tsivat Lake	60.2315	143.13	Glacial lake	8.22
West Fork Tsivat River	60.16788	142.98744	Nonglacial stream	7.85
Stream into Vitus Lake	60.14980	143.26956	Nonglacial stream	6.60
Vitus Lake	60.1	143.3	Glacial lake	8.17

Note: Differences in the specificity of the GPS coordinates depend on the size of the lake or stream and the number of times sampled. All populations appear to be dwarfed with the possible exception of the Dolly Varden in the Tsiu River. These fish were not trapped; however, they were visually identified when angled, and placement into the proper life history category was not made.

that have become lakes. Some streams in the Bering Glacier Watershed are only connected once a year or every few years to the Gulf of Alaska, while others remain connected until the shifting ice front and outburst floods alter the hydrology sufficiently to change drainage patterns. Fish species that can assume a resident freshwater life history and reproduce in isolated freshwater habitat, such as Dolly Varden, threespine stickleback, and prickly sculpin, readily adapt to the changing aquatic landscape. This likely explains why these species were found equally often in isolated and connected lakes and streams. However, coho salmon are strictly anadromous (Fry, 1973), and hence they were found significantly more often in connected than in isolated lakes and streams. Those spawned in isolated habitats may survive if the habitat is reconnected to the sea during times of high water, a pattern that is common in the region. However, if an outlet to the sea fails to reconnect, these fish will retain the morphology of juveniles and fail to undergo sexual maturation (Fry, 1973).

Ephemeral hydrological networks provide unequal opportunities to different fish species to colonize different habitats, including isolated lakes and streams (Hershey et al., 2006). Movements of the glacier alter hydrological channels and thus allow some fishes to colonize lakes that have become unavailable to others (Hershey et al., 2006). As the stabilization of hydrological networks progresses with deglaciation, an increasing number of lakes and streams will likely become isolated from the lack of seasonal connections from glacial meltwater and the establishment of regular stream channels that no longer avoid ice dams by altering between drainages. This should further sort the species so that only those able to evolve into resident freshwater forms are found in isolated fresh waters.

Fish in Glacial Lakes and Streams

Glacial waters have low temperature, nutrient levels, and food resources, and high levels of suspended sediments (Gislason et al., 1999). Waters with a high suspended-sediment load present poor habitat for most fish species. Suspended sediments hinder visibility and thereby may interfere with reproduction and food acquisition. Sedimentation also leads to infilling of gravels and therefore poor oxygenation of fish embryos, such as those of salmonids, and developing young in the substrate. The lack of relationship between fish species and substrate size (Table 3) is consistent with findings in streams in Glacier Bay National Park, Alaska (Milner, 1987). Some fish species, such as salmon, require course substrates, such as large gravel and small cobbles for proper construction of their redds and development of their embryos. However, multiple substrate types were typically present at a site, and much of the sediment is coarse glacial outwash. As expected, fish were captured more often in clear water (Table 4), and fish are more likely to be found in nonglacial lakes and streams.

Dolly Varden were the only species that was more likely to be found in glacial lakes and streams. Previous studies in Glacier Bay National Park showed that Dolly Varden are able to thrive in the harsh environment of glacial streams and lakes because they require less habitat complexity than other fishes, such as coho salmon (Milner and Bailey, 1989); furthermore, Dolly Varden are the first of the salmonids to colonize new stream habitat (Milner, 1994).

Surf smelt, anadromous and resident threespine stickleback, Pacific staghorn sculpin, and starry flounder were found in one glacial lake, Vitus Lake. Vitus Lake is unique within the study area because it is both large and tidally influenced, which likely facilitate its colonization by anadromous or marine fish and diversify the water chemistry to such a degree that many microhabitats are available in the lake, from fresh water at stream inlets to salinities up to 18 ppt at depths of 60 m (Josberger et al., this volume). These factors likely explain the relatively high fish species richness in Vitus Lake (8 of 10 fish species from the region, including both forms of threespine stickleback; Table 1).With the exception of Vitus Lake, fish species diversity in glacial lakes is typically low (Table 1).

Fish Associations with Habitat Parameters

Habitat complexity promotes fish community development by providing diversified niches with a richer prey base and refuge from predators. The Bering Glacier Watershed is mostly young without much structural complexity to its aquatic habitats. As expected, the waters with aquatic vegetation and algae (and therefore some structural complexity and niche variation) were most likely to contain fish. Species such as sculpins, which are ambush predators, tend to occur in complex habitats that provide cover (Wootton, 1998; McPhail, 2007). Resident freshwater threespine stickleback with a benthic life history also tend to occur in shallow water with cover provided by aquatic vegetation and woody debris (Hagen 1967; Messler et al., 2007). Furthermore, stickleback males prefer to build their nest where it is concealed by aquatic vegetation to minimize risk of predation and to increase the quality of parental care by reducing courtship intrusions, nest raids, and territorial encounters (Hagen, 1967; Black and Wootton, 1970; Jenni, 1972; Moodie, 1972b; Kynard, 1978, 1979; Sargent and Gebler, 1980; Sargent, 1982; FitzGerald, 1983; Lachance and FitzGerald, 1992).

As the glacier continues to recede, the watershed will likely evolve from ice contact, unstable lakes and streams that are overrun by glacial surges and outburst floods every 20–30 a to mature lakes and streams with a relatively stable hydrology. This, in turn, should lead to a more complex habitat characterized by aquatic vegetation, algae, woody debris, and a diversified prey base. Local fish species richness (alpha diversity) will likely increase as a consequence. Our results are consistent with this pattern. The older a lake or stream, the more likely it is to have fish, and the more fish species it has (Fig. 4). In addition to having more time for habitat complexity to form, older lakes and streams have had more time for fish colonization. As lakes and streams age, their fish communities grow beyond the first few early colonizing species to include species more dependent on already established ecological interactions.

Water Quality

Fish can often tolerate brief changes in water quality by modifying their behavior. Long-term changes to water quality, however, require adjustments to biochemical and physiological processes (Wootton, 1998).

A broad range of surface-water temperatures was found in the region (Fig. 3A), which is consistent with a diversity of habitat types from massive proglacial lakes to small isolated ponds. Water temperature is one of the most important factors in determining the distribution of fishes owing to its influence on dissolved oxygen and rates of physiological and biochemical processes (Wootton, 1998). Deep lakes possess thermoclines, which a fish might need to swim through on a daily basis, and shallow water will often fluctuate several degrees over a 24 h period (Wootton, 1998). Not surprisingly, Bering Glacier fishes, most of which are early colonizing species and therefore adaptable to new environments, are found in a wide range of water temperatures characteristic of the habitat diversity (Fig. 3A).

Fish species in the Bering Glacier Region also have broad salinity tolerances (Fig. 3B), as is common in northern watersheds colonized after the melting of Pleistocene ice sheets (Pielou, 1991). This is expected, as the newly deglaciated watershed was colonized by marine species and salt-tolerant anadromous and resident freshwater species that pioneer as colonists in newly formed freshwater habitat.

Three marine species were collected (surf smelt, Pacific staghorn sculpin, and starry flounder). Even though these species are classified as marine, they are also sometimes found in brackish or fresh water (Gunter, 1942; McAllister, 1963; Morrow, 1980). Surf smelt were found only in Vitus Lake, which they reached by swimming up the Seal River, presumably with the incoming tide. Pacific staghorn sculpin and starry flounder are both found at the upper ends of tidally influenced lakes and streams (Gunter, 1942). Prickly sculpin, which can assume either a resident freshwater or a catadromous life history (McAllister and Lindsey, 1961), are found in freshwater lakes and streams as well as in estuaries (Brown et al., 1995). Taken together, our results indicate that resident freshwater fishes tend to be found in freshwater sites with few other fish species, while anadromous and marine fishes are found in freshwater sites with an intermediate and a high number of other fish species, respectively (Fig. 2). This may be because marine fishes can osmoregulate in lower salinities to a greater degree than freshwater fishes can osmoregulate in higher salinities (Gunter, 1942), leading to a more restricted habitat range for resident freshwater species.

Most lakes and streams sampled are less than ~100 a old and have a mean pH of 8.4, which is on the alkaline side of typical fresh water. Such elevated pH levels may be typical in glacially influenced freshwater systems owing to the effect of glacial sediments. Similar levels are found in Glacier Bay National Park (Engstrom et al., 2000). The tolerance of the fishes to elevated pH (Fig. 3C) is likely due to their marine origin. The normal range of pH for the marine environment is 7.5–8.5 (Knutzen, 1981), whereas the pH of fresh water usually ranges from 6.0 to 8.0 (Dodds, 2002). Even within Vitus Lake the pH varies considerably. Josberger et al. (2006, this volume) found the pH in two vertical profiles in Vitus Lake to be between 5 and 6, though these profiles were collected well away from our sampling locations. They found the pH along a vertical profile in Berg Lake to be between 8.5 and 9, which is more consistent with our results.

Conductivity levels varied widely in the region (Fig. 3D) owing to the tidal influence from the Gulf of Alaska. Higher conductivity levels in fresh water indicate a greater availability of dissolved ions, which translates into lower physiological expense for fishes to acquire needed ions.

Dissolved oxygen is essential for proper functioning of fish (Wootton, 1998). Salmonids are not tolerant of poorly oxygenated water, which results in greater mortality of eggs and alevins and poor growth of parr. Some other fishes are more tolerant of a wide range of dissolved oxygen (Wootton, 1998). The dissolved oxygen concentration of some shallow water sites with fish were too low for sustained fish survival (Fig. 3E); presumably, fish swim out of these areas when their oxygen demand increases.

Predictive Models

Logistic regression models support inferences from bivariate comparisons. The presence of fish is best predicted by clear, nonglacial water that is connected to the Gulf of Alaska. Models examining determinants of the presence of individual fish species are consistent with their habitat preferences. For example, Dolly Varden presence is best predicted by a habitat being a glacial stream.

Early Colonizing Species

Remarkably, 79% of all lakes and streams sampled had at least one species of fish. This is undoubtedly an underestimate of the true frequency of lakes and streams with fish, as our trapping efforts likely missed fish at some sites. These lakes and streams are colonized from the ocean by marine or diadromous species or from glacial refugia by freshwater species. Therefore, the potential pool of colonist species is relatively small.

Early colonizing species are those found by themselves (unless these displaced the original colonists) and those found in the youngest sites. The two species most commonly found by themselves were Dolly Varden (n = 9 sites) and prickly sculpin (n = 5 sites; Table 1). Species found in the youngest sites were prickly sculpin, slimy sculpin, and resident freshwater stickleback. If, as these data indicate, Dolly Varden, prickly sculpin, slimy sculpin, and resident freshwater stickleback are early colonizing species, then they should often be found in sites with relatively few other fish species. This is the case, with the exception of slimy sculpin, which are found in sites with an intermediate number of other fish species (Fig. 2). Our data further suggest that early colonizing species of fish can be expected to be found in most lakes and streams in a deglaciating landscape, including newly formed sites.

The most commonly occurring species in the Bering Glacier Watershed, resident freshwater stickleback and prickly sculpin, followed by coho salmon, Dolly Varden, and slimy sculpin (Table 1), may have a competitive advantage over subsequent colonists by virtue of their adaptability and head start. Their colonization sets the stage for more complex ecological interactions involving fish as predators, prey, and competitors. Anadromous species that colonize newly formed lakes and streams (such as salmonids and stickleback) are also an important source of marine-derived nutrients to an oligotrophic system, which supports the complexity of succession in both freshwater and surrounding riparian communities (Hicks et al., 2005). Early colonizing anadromous fishes are therefore of key importance to the successional processes that facilitate colonization by other species.

Stickleback Species Pairs

Coexistence of reproductively isolated anadromous and resident freshwater sticklebacks in streams occurs frequently throughout the threespine stickleback's range (McPhail, 1994; McKinnon and Rundle, 2002; Boughman, 2007). However, such pairs have been identified only in a few lakes: Lake Azabachije in Kamchatka (Ziuganov et al., 1987), Lake Sana in the Kuril Islands (discussed in Mori, 1990), lakes Harutori, Akkeshi, and Hyotan on Hokkaido Island (Mori, 1990; Higuchi et al., 1996), Mud Lake, Alaska (Karve et al., 2008), and coastal lakes formed subsequent to the Great Alaska Earthquake of 1964 on Middleton Island, Alaska (Gelmond et al., 2009). In the Bering Glacier Region we discovered three stream pairs and six lake pairs of anadromous and resident freshwater threespine stickleback species (Table 2; von Hippel and Weigner, 2004). One of the species pairs is in proglacial Vitus Lake, which is the first report of a stickleback species pair of any kind in a proglacial lake. The stickleback species pairs in the Bering Glacier Watershed, and particularly the Vitus Lake pair, represent an important component of the biodiversity of the Bering Glacier forelands.

Dwarf Dolly Varden

The populations of dwarf Dolly Varden in the Berg Lake and Khitrov Lake areas are isolated by massive waterfalls and glacial ice, respectively. Furthermore, these and other Dolly Varden populations are unusual and deserve further study (Table 5). Little is known about their life history and population structure. It is not understood if these populations are dwarfed because of environmental constraints, such as poor nutrition, high density, and elevated pH (Table 5), and/or because of genetic factors, such as a mutation in the insulin-like growth factor 1 allele or allele fixation resulting from the founder effect, genetic drift, inbreeding, or selection. Dwarf populations of Dolly Varden also occur in the Matanuska-Susitna Valley (M.A. Bell, 2003, pers. comm.) and the Bristol Bay region (McPhail and Lindsey, 1970) of Alaska, but these areas deglaciated thousands of years ago, whereas many of the Bering Glacier sites are less than 100 a old.

Because the Berg Lake and Khitrov Lake fish live above dispersal barriers, a phylogenetic analysis may provide interesting insights into how Dolly Varden colonized such isolated habitat. Possibilities include delivery by piscivorous birds, which bring live prey to their young (M.S. Christy, 2003, pers. comm.), or dispersal through glacial conduits. Such unusual methods of colonization would likely be reflected in low genetic diversity from a population bottleneck. Phylogenetic relationships with surrounding populations may yield the source population(s) of the colonists. Furthermore, the question arises as to whether Dolly Varden became dwarfed in these young sites or if they colonized these sites while already dwarfed.

Management Implications

The stickleback species pair in Vitus Lake deserves protection as the only known species pair that exists in a proglacial lake anywhere in the world. The greatest threats to unusual and important stickleback populations in Alaska are exotic fishes (such as northern pike, *Esox lucius*), stocked salmonids, human impacts on water quality, water withdrawals for human use, and climate change leading to habitat loss (von Hippel, 2008). None of these are current threats in Vitus Lake, but in the event that potentially damaging activities are proposed in the future, such as fish stocking, management decisions should be informed by the presence of this unique species pair. Similarly, every effort should be made to prevent the introduction of exotic aquatic species, including fish. Invasive aquatic species would likely harm many of the freshwater fish species native to the Bering Glacier Watershed.

The last surge and outburst floods of 1993–1995 destroyed the salmon runs in Vitus Lake. Coho salmon have returned to Vitus Lake (Table 1), though the magnitude of the run is unknown. The Tsiu and Kiklichk Rivers remain popular coho salmon sportfishing rivers, and we trapped coho parr in both. We also found coho parr in the Tsivat and Kosakuts Rivers and Tashalich Lake, as well as a small lake adjacent to the Bering River and 19 other locations (Table 1). Adult sockeye salmon have been observed in the Tsiu River and the East and West forks of the Tsivat River (Table 1), and there have been reports of adult pink salmon in the Kiklichk River. It is not known if other runs of Pacific salmon are present in the Bering Glacier Watershed. A thorough sampling of the streams in late summer and fall should be undertaken to determine which salmon species are present and the strength of their runs. Sampling in the spring when salmon emerge from their redds would provide additional information on population dynamics. For example, pink and chum salmon go to sea immediately after emergence from redds (McPhail and Lindsey, 1970), and hence they may have been missed in our sampling. The habitats of current and potential salmon streams need to be protected from pollution events or other causes of habitat degradation.

The dynamics of the Bering Glacier have created a complex of proglacial and periglacial lakes, kettle ponds, glacial and snow meltwater streams, oxbow lakes, tidally influenced lakes and streams, and older, mature lakes, streams, and estuaries. This ever-changing aquatic landscape has been colonized by typical fish species in the region, but some of these colonists have evolved in atypical ways. Berg Lake, which drains the Bering Glacier independently of Vitus Lake, underwent an outburst flood in 1994 that lowered its surface by >100 m in three days (Payne et al., 1997). Berg Lake is also seemingly impossible to colonize owing to the massive waterfalls that drain the lake, and yet its small tributaries contain dwarf Dolly Varden. Dwarf Dolly Varden are also found in the isolated Khitrov Lake area, and stickleback species pairs are found in at least nine locations. As we learn more about the fishes of the Bering Glacier Watershed, we may find additional, unusual populations. It is clearly a watershed of paramount importance to understanding colonization and succession of aquatic organisms in newly deglaciated terrain. It warrants its status as a Research Natural Area with protections in place to safeguard the natural succession of its aquatic habitats.

ACKNOWLEDGMENTS

The U.S. Bureau of Land Management (BLM) provided significant funding and logistical support for this project. John Payne, Scott Guyer, and Chris Noyles of the BLM made our field work possible. Dedicated field assistance was provided by Xavier Sotelo, Eric Sjoden, Lauren Smayda, and Julie Warren. John Tucker ably flew us to remote lakes and streams. Numerous helpful comments on the manuscript were provided by Mike Bell and an anonymous reviewer. Fish were collected under Alaska Department of Fish and Game permit nos. SF-2002-105, SF-2003-017, SF-2004-012, SF-2004-014, SF-2005-020, SF-2005-021, SF-2006-017, and SF-2006-018. All research protocols were approved by the University of Alaska Anchorage Institutional Animal Care and Use Committee.

REFERENCES CITED

Alaska Department of Fish and Game, Division of Habitat and Restoration, 1999, Yakataga State Game Management Plan: Douglas, Alaska, 127 p.

Bernatchez, L., and Wilson, C.C., 1998, Comparative phylogeography of Nearctic and Palearctic fishes: Molecular Ecology, v. 7, p. 431–452, doi: 10.1046/j.1365-294x.1998.00319.x.

Black, R., and Wootton, R.J., 1970, Dispersion in a natural population of three-spined sticklebacks: Canadian Journal of Zoology, v. 48, p. 1133–1135, doi: 10.1139/z70-197.

Boughman, J.W., 2007, Condition-dependent expression of red colour differs between stickleback species: Journal of Evolutionary Biology, v. 20, p. 1577–1590, doi: 10.1111/j.1420-9101.2007.01324.x.

Brouwers, E.M., and Forester, R.M., 1993, Ostracode assemblages from modern bottom sediments of Vitus Lake, Bering Piedmont Glacier, Southeast Alaska: U.S. Geological Survey Bulletin 2068, p. 228–235.

Brown, L.R., Matern, S.A., and Moyle, P.B., 1995, Comparative ecology of prickly sculpin, *Cottus asper*, and Coastrange sculpin, *C. aleuticus*, in the Eel River, California: Environmental Biology of Fishes, v. 42, p. 329–343, doi: 10.1007/BF00001462.

Dodds, W.K., 2002, Freshwater Ecology Concepts and Environmental Applications: San Diego, California, Academic Press, 569 p.

Dunham, J., Rieman, B., and Chandler, G., 2003, Influences of temperature and environmental variables on the distribution of bull trout within streams at the southern margin of its range: North American Journal of Fisheries Management, v. 23, p. 894–904, doi: 10.1577/M02-028.

Engstrom, D.R., Fritz, S.C., Almendinger, J.E., and Juggins, S., 2000, Chemical and biological trends during lake evolution in recently deglaciated terrain: Nature, v. 408, p. 161–165, doi: 10.1038/35041500.

Evans, D.O., Henderson, B.A., Bax, N.J., Marshall, T.R., Oglesby, R.T., and Christie, W.J., 1987, Concepts and methods of community ecology applied to freshwater fisheries management: Canadian Journal of Fisheries and Aquatic Sciences, v. 44, p. 448–470, doi: 10.1139/f87-347.

FitzGerald, G.J., 1983, The reproductive ecology and behaviour of three sympatric sticklebacks (*Gasterosteidae*) in a saltmarsh: Biology of Behaviour, v. 8, p. 67–79.

Fraser, D.J., and Bernatchez, L., 2005, Allopatric origins of sympatric brook charr populations: colonization history and admixture: Molecular Ecology, v. 14, p. 1497–1509, doi: 10.1111/j.1365-294X.2005.02523.x.

Fry, D.H., 1973, Anadromous Fishes of California: State of California, Department of Fish and Game, 112 p.

Gagnon, M.C., and Angers, B., 2006, The determinant role of temporary proglacial drainages on the genetic structure of fishes: Molecular Ecology, v. 15, p. 1051–1065, doi: 10.1111/j.1365-294X.2005.02828.x.

Gelmond, O., von Hippel, F.A., and Christy, M.S., 2009, Rapid ecological speciation in three-spined stickleback from Middleton Island, Alaska: The roles of selection and geographic isolation: Journal of Fish Biology, v. 75, p. 2037–2051.

Gislason, D., Ferguson, M.M., Skulason, S., and Snorrason, S.S., 1999, Rapid and coupled phenotypic and genetic divergence in Icelandic Arctic char (*Salvelinus alpinus*): Canadian Journal of Fisheries and Aquatic Sciences, v. 56, p. 2229–2234, doi: 10.1139/cjfas-56-12-2229.

Gunter, G., 1942, A list of the fishes of the mainland of North and Middle America recorded from both freshwater and sea water: American Midland Naturalist, v. 28, p. 305–326, doi: 10.2307/2420818.

Hagen, D.W., 1967, Isolating mechanisms in three-spined sticklebacks (*Gasterosteus*): Journal of the Fisheries Research Board of Canada, v. 24, p. 1637–1692.

Hershey, A.E., Beaty, S., Fortino, K., Keyse, M., Mou, P.P., O'Brien, W.J., Ulseth, A.J., Gettel, G.A., Lienesch, P.W., Luecke, C., McDonald, M.E., Mayer, C.H., Miller, M.C., Richards, C., Schuldt, J.A., and Whalen, S.C., 2006, Effect of landscape factors on fish distribution in arctic Alaskan lakes: Freshwater Biology, v. 51, p. 39–55, doi: 10.1111/j.1365-2427.2005.01474.x.

Herzfeld, U.C., and Mayer, H., 1997, Surge of Bering Glacier and Bagley Ice Field, Alaska: An update to August 1995 and an interpretation of brittle-deformation patterns: Journal of Glaciology, v. 43, p. 427–436.

Hicks, B.J., Wipfli, M.S., Lang, D.W., and Lang, M.E., 2005, Marine-derived nitrogen and carbon in freshwater-riparian food webs of the Copper River Delta, southcentral Alaska: Oecologia, v. 144, p. 558–569, doi: 10.1007/s00442-005-0035-2.

Higuchi, M., Goto, A., and Yamazaki, F., 1996, Genetic structure of threespine stickleback, *Gasterosteus aculeatus*, in Lake Harutori, Japan, with reference to coexisting anadromous and freshwater forms: Ichthyological Research, v. 43, p. 349–358, doi: 10.1007/BF02347634.

Jaeger, J.M., and Nittrouer, C.A., 1999, Marine record of surge-induced outburst floods from the Bering Glacier, Alaska: Geology, v. 27, p. 847–850, doi: 10.1130/0091-7613(1999)027<0847:MROSIO>2.3.CO;2.

Jenni, D.A., 1972, Effects of conspecifics and vegetation on nest site selection in *Gasterosteus aculeatus* L: Behaviour, v. 42, p. 97–118, doi: 10.1163/156853972X00121.

Johnson, L.S., and Taylor, E.B., 2004, The distribution of divergent mitochondrial DNA lineages of threespine stickleback (*Gasterosteus aculeatus*) in the northeastern Pacific Basin: Post-glacial dispersal and lake accessibility: Journal of Biogeography, v. 31, p. 1073–1083, doi: 10.1111/j.1365-2699.2004.01078.x.

Johnson, M.G., Leach, J.H., Minns, C.J., and Oliver, C.H., 1977, Limnological characteristics of Ontario lakes in relation to associations of walleye (*Stizostedion vitreum vitreum*), northern pike (*Esox lucius*), lake trout (*Salvelinus namaycush*), and smallmouth bass (*Micropterus dolomieui*): Journal of the Fisheries Research Board of Canada, v. 34, p. 1592–1601.

Josberger, E.G., Shuchman, R.A., Meadows, G.A., Savage, S., and Payne, J., 2006, Hydrography and circulation of ice-marginal lakes at Bering

Glacier, Alaska, USA: Arctic, Antarctic, and Alpine Research, v. 38, p. 547–560, doi: 10.1657/1523-0430(2006)38[547:HACOIL]2.0.CO;2.

Karve, A., von Hippel, F.A., and Bell, M.A., 2008, Isolation between sympatric anadromous and resident threespine stickleback species in Mud Lake, Alaska: Environmental Biology of Fishes, v. 81, p. 287–296, doi: 10.1007/s10641-007-9200-2.

Klemetsen, A., and Grotnes, P., 1980, Coexistence and immigration of two sympatric Arctic charr, *in* Balon, E.K., ed., Charrs: Salmonid Fishes of the Genus *Salvelinus*: Hingham, Massachusetts, Kluwer Boston, p. 757–763.

Knutzen, J., 1981, Effects of decreased pH on marine organisms: Marine Pollution Bulletin, v. 12, p. 25–29, doi: 10.1016/0025-326X(81)90136-3.

Kynard, B.E., 1978, Breeding behavior of a lacustrine population of threespine sticklebacks (*Gasterosteus aculeatus* L.): Behaviour, v. 67, p. 178–207, doi: 10.1163/156853978X00323.

Kynard, B.E., 1979, Nest habitat preference of low plate morphs in threespine sticklebacks (*Gasterosteus aculeatus*): Copeia, v. 1979, p. 525–528, doi: 10.2307/1443234.

Lachance, S., and FitzGerald, G.J., 1992, Parental care tactics of three-spined sticklebacks living in a harsh environment: Behavioral Ecology, v. 3, p. 360–366, doi: 10.1093/beheco/3.4.360.

Landau, S., and Everitt, B.S., 2004, A Handbook of Statistical Analyses Using SPSS: New York, Chapman and Hall, 368 p.

Lindsey, C.C., 1975, Proglacial lakes and fish dispersal in southwestern Yukon Territory: Verhandlungen der Internationalen Vereinigung für Theoretische und Angewandte Limnologie, v. 19, p. 2364–2370.

Lindsey, C.C., and McPhail, J.D., 1986, Zoogeography of the fishes of the Yukon and Mackenzie Basins, *in* Hocutt, C.H., and Wiley, E.O., eds., The Zoogeography of North American Freshwater Fishes: New York, Wiley & Sons, p. 639–674.

McAllister, D.E., 1963, A Revision of the Smelt Family, Osmeridae: Bulletin of the National Museum of Canada, v. 191, 53 p.

McAllister, D.E., and Lindsey, C.C., 1961, Systematics of the Freshwater Sculpins (*Cottus*) of British Columbia: Bulletin of the National Museum of Canada, v. 172, p. 66–89.

McKinnon, J.S., and Rundle, H.D., 2002, Speciation in nature: The threespine stickleback model systems: Trends in Ecology & Evolution, v. 17, p. 480–488, doi: 10.1016/S0169-5347(02)02579-X.

McPhail, J.D., 1994, Speciation and the evolution of reproductive isolation in the stickleback (*Gasterosteus*) of south-western British Columbia, *in* Bell, M.A., and Foster, S.A., eds., The Evolutionary Biology of the Threespine Stickleback: New York, Oxford University Press, p. 399–437.

McPhail, J.D., 2007, The Freshwater Fishes of British Columbia: Edmonton, University of Alberta Press, 620 p.

McPhail, J.D., and Lindsey, C.C., 1970, Freshwater Fishes of Northwestern Canada and Alaska: Fisheries Research Board of Canada, Bulletin 173, Ottawa, Queen's Printer for Canada, 381 p.

McPhail, J.D., and Lindsey, C.C., 1986, Zoogeography of the freshwater fishes of Cascadia (the Columbia System and rivers north to the Stikine), *in* Hocutt, C.H., and Wiley, E.O., eds., The Zoogeography of North American Freshwater Fishes: New York, Wiley & Sons, p. 615–637.

Mecklenburg, C.W., Mecklenburg, T.A., and Thorsteinson, L.K., 2002, Fishes of Alaska: Bethesda, Maryland, American Fisheries Society, 1037 p.

Merrand, Y., and Hallet, B., 1996, Water and sediment discharge from a large surging glacier: Bering Glacier, Alaska, U.S.A., summer 1994: Annals of Glaciology, v. 2, p. 233–240.

Messler, A., Wund, M.A., Baker, J.A., and Foster, S.A., 2007, The effects of relaxed and reversed selection by predators on the antipredator behavior of the threespine stickleback, *Gasterosteus aculeatus*: Ethology, v. 113, p. 953–963.

Milner, A.M., 1987, Colonization and ecological development of new streams in Glacier Bay National Park, Alaska: Freshwater Biology, v. 18, p. 53–70, doi: 10.1111/j.1365-2427.1987.tb01295.x.

Milner, A.M., 1994, Colonization and succession of invertebrate communities in a new stream in Glacier Bay National Park, Alaska: Freshwater Biology, v. 32, p. 387–400, doi: 10.1111/j.1365-2427.1994.tb01134.x.

Milner, A.M., and Bailey, R.G., 1989, Salmonid colonization of new streams in Glacier Bay National Park, Alaska: Aquaculture and Fisheries Management, v. 20, p. 179–192.

Molnia, B.F., and Post, A., 1995, Holocene history of Bering Glacier, Alaska: A prelude to the 1993–1994 surge: Physical Geography, v. 16, p. 87–117.

Molnia, B.F., Post, A., and Carlson, P.R., 1996, 20th-century glacial-marine sedimentation in Vitus Lake, Bering Glacier, Alaska, U.S.A.: Annals of Glaciology, v. 22, p. 205–210.

Moodie, G.E.E., 1972a, Predation, natural selection and adaptation in an unusual threespine stickleback: Heredity, v. 28, p. 155–167, doi: 10.1038/hdy.1972.21.

Moodie, G.E.E., 1972b, Morphology, life history and ecology of an unusual stickleback (*Gasterosteus aculeatus*) in the Queen Charlotte Islands, Canada: Canadian Journal of Zoology, v. 50, p. 721–732, doi: 10.1139/z72-099.

Mori, S., 1990, Two morphological types in the reproductive stock of three-spined stickleback, *Gasterosteus aculeatus*, in Lake Harutori, Hokkaido Island: Environmental Biology of Fishes, v. 27, p. 21–31, doi: 10.1007/BF00004901.

Morrow, J.E., 1980, The Freshwater Fishes of Alaska: Anchorage, Alaska Northwest Publishing, 248 p.

Muller, E.H., and Fleisher, P.J., 1995, Surging history and potential for renewed retreat: Bering Glacier, Alaska, U.S.A.: Arctic and Alpine Research, v. 27, p. 81–88, doi: 10.2307/1552070.

Parker, H.H., and Johnson, L., 1991, Population structure, ecological segregation and reproduction in non-anadromous Arctic charr, *Salvelinus alpinus* (L), in four unexploited lakes in the Canadian Arctic: Journal of Fish Biology, v. 38, p. 123–147, doi: 10.1111/j.1095-8649.1991.tb03098.x.

Payne, J.F., Coffeen, M., Macleod, R.D., Kempka, R., Reid, F.A., and Molnia, B.F., 1997, Monitoring change in the Bering Glacier Region, Alaska, using Landsat TM and ERS-1 imagery: Arctic Research of the United States, v. 11, p. 75–79.

Pielou, E.C., 1991, After the Ice Age: Chicago, University of Chicago Press, 366 p.

Pollard, W.R., Hartman, G.F., Groot, C., and Edgell, P., 1997, Field Identification of Coastal Juvenile Salmonids: Madeira Park, Canada, Harbour Publishing, 32 p.

Ray, J.M., Wood, R.M., and Simons, A.M., 2006, Phylogeography and post-glacial colonization patterns of the rainbow darter, *Etheostoma caeruleum* (Teleostei: *Percidae*): Journal of Biogeography, v. 33, p. 1550–1558, doi: 10.1111/j.1365-2699.2006.01540.x.

Reimchen, T.E., 1994, Predators and morphological evolution in threespine stickleback, *in* Bell, M.A., and Foster, S.A., eds., Evolutionary Biology of Threespine Stickleback: New York, Oxford University Press, p. 240–273.

Sargent, R.C., 1982, Territory quality, male quality, courtship intrusions, and female nest-choice in the threespine stickleback, *Gasterosteus aculeatus*: Animal Behaviour, v. 30, p. 364–374, doi: 10.1016/S0003-3472(82)80047-X.

Sargent, R.C., and Gebler, J.B., 1980, Effects of nest site concealment on hatching success, reproductive success, and paternal behavior of the threespine stickleback, *Gasterosteus aculeatus*: Behavioral Ecology and Sociobiology, v. 7, p. 137–142, doi: 10.1007/BF00299519.

von Hippel, F.A., 2008, Conservation of threespine and ninespine stickleback radiations in the Cook Inlet Watershed, Alaska: Behaviour, v. 145, p. 693–724.

von Hippel, F.A., and Weigner, H., 2004, Sympatric anadromous-resident pairs of threespine stickleback species in young lakes and streams at Bering Glacier, Alaska: Behaviour, v. 141, p. 1441–1464, doi: 10.1163/1568539042948259.

Wiles, G.C., Post, A., Muller, E.H., and Molnia, B.F., 1999, Dendrochronology and late Holocene history of Bering Piedmont Glacier, Alaska: Quaternary Research, v. 52, p. 185–195, doi: 10.1006/qres.1999.2054.

Wilson, E.O., 1961, The nature of the taxon cycle in the Melanesian ant fauna: American Naturalist, v. 882, p. 169–193.

Wootten, R.J., 1998, Ecology of Teleost Fishes (2nd edition): Norwell, Massachusetts, Kluwer Academic Publishers, 392 p.

Ziuganov, V.V., Golovatjuk, G.J., Savvaitova, K.A., and Bugaev, V.F., 1987, Genetically isolated sympatric forms of threespine stickleback, *Gasterosteus aculeatus*, in Lake Azabachije (Kamchatka-peninsula, USSR): Environmental Biology of Fishes, v. 18, p. 241–247, doi: 10.1007/BF00004877.

Manuscript Accepted by the Society 02 June 2009

Printed in the USA

The Geological Society of America
Special Paper 462
2010

Harbor seal (Phoca vitulina richardii) *use of the Bering Glacier habitat: Implications for management*

Danielle M. Savarese*
Jennifer M. Burns[†]
University of Alaska, Anchorage, Department of Biological Sciences, 3211 Providence Drive, Anchorage, Alaska 99508, USA

ABSTRACT

Harbor seal *(Phoca vitulina richardii)* use of a haulout in Vitus Lake at the terminus of the Bering Glacier (60° 5′ N, 143° 30′ W) was characterized by conducting 69 aerial surveys over the 2001–2003 period. Harbor seals were observed hauling out only on low, flat icebergs in Vitus Lake, predominantly (80% ± 3%) at the head of Seal River. There was a marked increase in seal abundance in the fall (from <200 to >900 seals), and as seal abundance increased, the average number of seals per iceberg remained relatively constant (10 ± 1 seals/iceberg), while the average number of icebergs occupied increased (from 12 to 48 icebergs in 2002). In 2003 the number of icebergs used was lower, and therefore seal density per berg was higher, than in 2002, suggesting that the availability of suitable ice influences grouping behavior in harbor seals. Diet and genetic studies suggest that harbor seals move into Vitus Lake from stocks in both Southeast Alaska and Prince William Sound in order to forage on local salmon runs. These findings have implications for ongoing efforts to determine appropriate stock structure for management decisions and suggest that the frequency and extent of seasonal movements must be considered when conducting regional population monitoring surveys. As human activity in the Bering Glacier area increases, monitoring and educational efforts should be initiated in order to ensure that harbor seals remain a functioning element of the ecosystem.

INTRODUCTION

At the terminus of the Bering Glacier lies Vitus Lake, an estuarine lake that is ~20 km long and 10 km wide. The large icebergs that calve (i.e., separate) from the Bering Glacier often ground within Vitus Lake, where they rest until melting enough to float and eventually they are carried out to the Gulf of Alaska via the Seal River. The icebergs tend to ground in the shallowest areas of the lake, particularly at the head of the Seal River. The size and shape of these icebergs vary widely, from flat platforms with little or no topography, to towering masses >50 ft high, to elaborate ice arches, and to small pieces of brash ice (Fig. 1). Only the flat, more stable icebergs that lie within Vitus Lake are used by harbor seals for resting on.

Harbor seals *(Phoca vitulina richardii)* are small pinnipeds (adults, 148–191 cm in length, weighing 85–115 kg; Boness et al., 2002) that are coastally distributed throughout the subarctic and temperate waters of the North Atlantic and North Pacific Oceans. In Alaska their range extends along the southeastern panhandle, across the Gulf of Alaska, throughout the Aleutian Island chain, and into the Bering Sea (Angliss and Outlaw, 2006). Throughout their range, harbor seals typically spend a part of

*Present address: LGL Alaska Research Associates, Inc., 1101 E. 76th Avenue, Suite B, Anchorage, Alaska 99518, USA; dsavarese@lgl.com.
[†]Corresponding author; afjmb4@uaa.alaska.edu.

Savarese, D.M., and Burns, J.M., 2010, Harbor seal *(Phoca vitulina richardii)* use of the Bering Glacier habitat: Implications for management, *in* Shuchman, R.A., and Josberger, E.G., eds., Bering Glacier: Interdisciplinary Studies of Earth's Largest Temperate Surging Glacier: Geological Society of America Special Paper 462, p. 181–191, doi: 10.1130/2010.2462(09). For permission to copy, contact editing@geosociety.org.

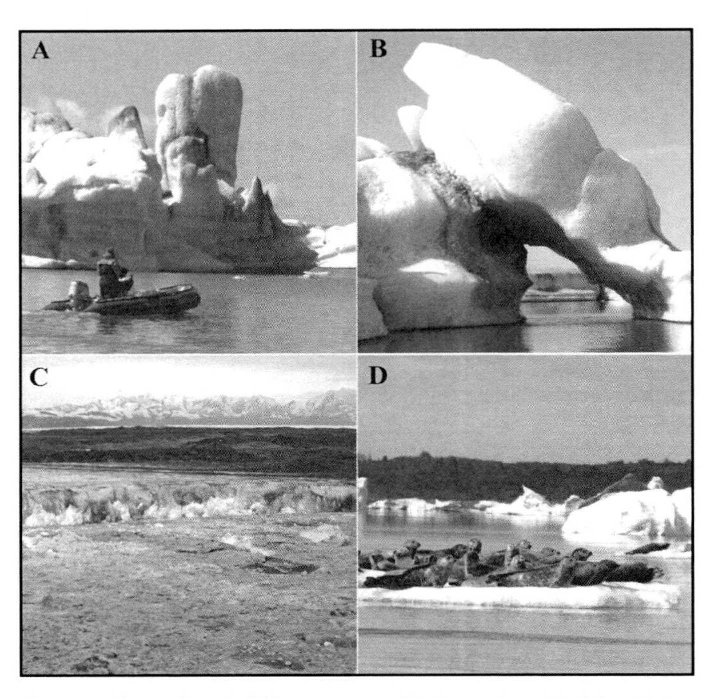

Figure 1. Examples of different types of icebergs found in Vitus Lake: (A) tall and less stable, (B) ice arches, (C) brash ice in Tashalich Arm, and (D) low-lying and stable, preferred by harbor seals.

every day resting on sand or gravel beaches, rocks, or glacial ice, a behavior termed *hauling out* (Thompson et al., 1989; Hoelzel, 2002). Although it is not known why seals select particular locations for hauling out, most areas selected are relatively undisturbed by human activity and relatively inaccessible to terrestrial predators such as bears, coyotes, and foxes (Riedman, 1990). Seals are most frequently seen ashore during midday when solar radiation is highest and/or during those portions of the tidal cycle when preferred substrate is available (Frost et al., 1999; Boveng et al., 2003; Simpkins et al., 2003). However, tidal influences on haulout behavior are typically absent in areas where seals utilize glacial ice (Hoover, 1983). In addition to the influence of time of day, haulout behavior also changes seasonally. Seals typically spend a greater proportion of time ashore during the early summer pupping period so that females can nurse their pups, and during the August molt period so that they can warm their skin and accelerate the molting process (Watts, 1992; Boveng et al., 2003; Simpkins et al., 2003).

Understanding the diel and seasonal shifts in the proportion of time harbor seals spend ashore is important, because accurate estimates of population size and trajectory are critical to making effective management decisions. However, because not all animals haul out at any given time, and seals cannot be counted when they are in the water, harbor seal surveys are typically conducted during those times of the year when the largest proportion of the population is hauled out (Boveng et al., 2003). The number of animals not observed is estimated on the basis of correction factors that account for weather, tidal state, time of day, and day of the year (Huber et al., 2001; Simpkins et al., 2003). Inherent in this design is the assumption that animals do not move between survey sites during the period of the survey (which can last for days or weeks, depending on the area to be covered), and that harbor seals do not show differential movements between survey areas between years. Because harbor seals are widely distributed within Alaska, their entire range has historically been surveyed only once every 5 years. However, to supplement rangewide counts, annual surveys are flown over a subset of the range each year (Frost et al., 1999). Trends at these "index" sites are assumed to mirror those of the population as a whole (National Marine Fisheries Service [NMFS] et al., 2003). Typically, index surveys are conducted during the August annual molt, and take 1–2 wk to complete. The closest index count sites to the Bering Glacier are in Prince William Sound.

In Alaska, harbor seal populations are managed jointly by the NMFS and the Alaska Native Harbor Seal Commission (ANHSC) under guidelines set out in the Marine Mammal Protection Act (MMPA, Sec. 119), and the co-management action plan. The MMPA mandates that federal agencies manage human impacts on marine mammal populations to ensure that marine mammals do not "diminish beyond the point at which they cease to be a significant functioning element in the ecosystem of which they are a part . . . ." (MMPA, Sec. 2(2)), while the harbor seal co-management action plan respects the tribal regulation of Native subsistence harvests and cultural use. Because the basic demographic unit that is managed is the local stock, rather than the total number of animals across the entire range, accurate identification of stock structure is critical.

Currently, the harbor seal population within Alaska is subdivided into three broad geographic stocks: Southeast Alaska, Gulf of Alaska, and Bering Sea (Fig. 2; Angliss and Outlaw, 2006) with seals at the Bering Glacier falling within the boundaries of the Southeast Alaska stock. These stock boundaries were created on the basis of geographic barriers, distance between haulouts, and observations that adult harbor seals rarely make long distance movements and display high site fidelity, especially during the breeding season (Pitcher and McAllister, 1981; Lowry et al., 2001). Because these stocks have been assumed to be demographically isolated, population trends are modeled separately for each stock.

More than a decade has passed since the three geographic stocks were identified, and new ecological and molecular studies suggest that Alaska harbor seal stock structure should be reevaluated. For example, within both the Gulf of Alaska and the Southeast Alaska stocks some local populations are increasing while others are decreasing, suggesting a greater degree of isolation than was previously appreciated (Angliss and Outlaw, 2006). Similarly, genetic studies have suggested that perhaps as many as a dozen local stocks are warranted (O'Corry-Crowe et al., 2003). In contrast, research on harbor seal movement patterns has demonstrated that seals sometimes move quite long distances (up to 500 km, Lowry et al., 2001). Interestingly, although glaciologists

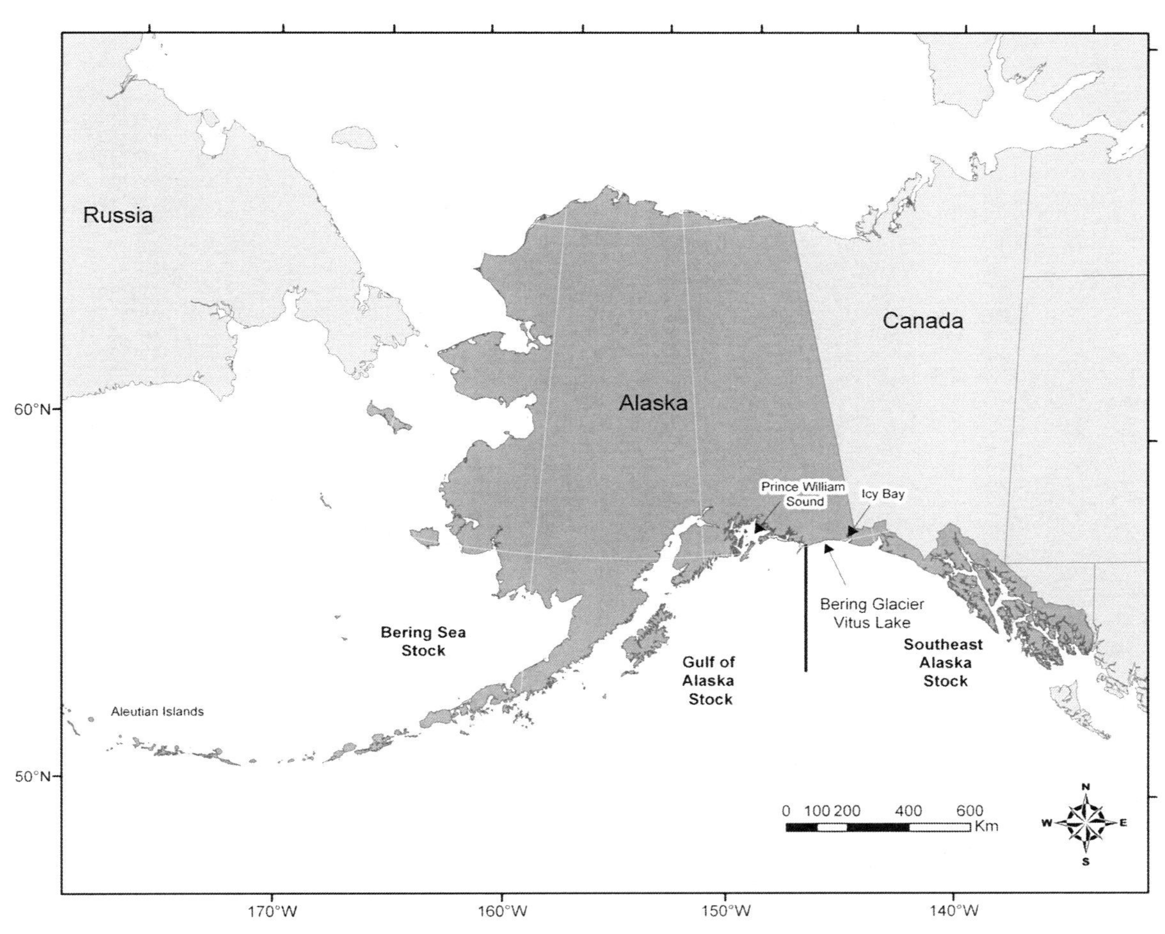

Figure 2. Range of the harbor seal in Alaska and the boundaries of the three current management stocks.

have observed harbor seals at the Bering Glacier since the 1970s, management agencies were unaware of this haulout; thus seals in the area were not included in population surveys until 2002. Because the Bering Glacier lies near the border between the Gulf of Alaska and Southeast Alaska management stocks, information on the behavior and region of origin of the seals that use the area is critical, as mixing between stocks at this site may present an opportunity for increased gene flow.

Understanding the factors that influence seal abundance and use of the Bering Glacier site is also important, because glacial ice haulouts are used by large numbers of seals, but the reasons for, and influences on, seal behavior at these sites are poorly understood (Calambokidis et al., 1987; Mathews and Kelly, 1996; Mathews and Pendleton, 2006; Bengtson et al., 2007). Given the large seasonal variation in seal numbers often seen at glacial haulouts (Mathews and Kelly, 1996), the potential impacts of global warming on glacial ice dynamics, and the growing impact of human visitation to these sites, it is becoming increasingly important that managers better understand the behavioral ecology of seals at glacial sites. Therefore, the objectives of this study were to (1) characterize harbor seal use of the habitat, (2) determine the number of harbor seals utilizing the Bering Glacier haulout, (3) investigate the environmental factors influencing their use of the site, and (4) develop management recommendations that could be used by relevant federal, state, and local agencies to ensure that harbor seals remain a functioning element of the Bering Glacier ecosystem.

METHODS

A total of 69 aerial surveys were flown between May and November 2001–2003 to localize and characterize harbor seal use of the habitat at the terminus of Bering Glacier. Surveys were flown at an altitude of ~230 m using a Fairchild-Hiller helicopter. All seals hauled out were photographed using a Canon D60 digital SLR camera with a 75–300 mm image-stabilizing zoom lens (Savarese, 2004). Surveys covered the entire Vitus Lake habitat as well as the shores of Seal River and the outer coastline at the mouth of Seal River. For the purpose of identifying seal localities within Vitus Lake, the lake was subdivided into three areas: head of Seal River, central Vitus Lake, and Tashalich Arm (Fig. 3). The numbers of seals in these three areas were tracked separately because these areas differed ecologically. Icebergs at the head of Seal River were closest to marine foraging areas; the central Vitus Lake icebergs were closer to the Bering Glacier, where fresh-water runoff and upwelling associated with submerged moraines may have concentrated prey, and the brash ice in Tashalich Arm, which was farthest from the open ocean. To determine if seals preferred certain areas, differences in the relative proportion of seals with respect to area, year, and month were tested for, using the nonparametric Kruskal-Wallace test. Significant differences were assumed if $p < 0.05$.

Determining the number of seals using the Bering Glacier ecosystem was accomplished by counting all seals in all photographs taken during each survey. The seasonal trend in harbor seal abundance and the impact of environmental factors (weather, time of day, and tidal state) on the number of seals hauled out was statistically modeled. In addition, photographs were studied to determine if pups were present, and photogrammetry was used to investigate whether the age class of the harbor seals hauling out at the Bering Glacier changed seasonally. This work was previously reported in Savarese (2004).

To assess how seal density changed as harbor seal abundance at Bering Glacier increased, we recorded the number of seals on each iceberg and the total number of icebergs used by the seals. Because we were unable to determine whether unoccupied icebergs were suitable for haulout, this index is not a true measure of habitat availability. For seals at the head of the Seal River, where abundance was highest, we used general linear models to test the effect of month, year, and the month-by-year interaction, on the average number of seals per iceberg, and the total number of icebergs used in each survey conducted between June and October of 2002 and 2003. All count data were log-transformed prior to analysis.

To investigate the relatedness of Bering Glacier harbor seals to those at other haulouts in Alaska, seals in Vitus Lake were captured using monofilament gill nets. Seals were manually restrained, and tissue samples for genetic analysis were collected via flipper punch. Tissue samples were stored in ethanol prior to analysis by the Southwest Fisheries Science Center, NMFS in San Diego, California. To assess genetic diversity and genetic differentiation, a 588-base pair region of the mitochondrial genome was amplified by polymerase chain reaction and sequenced (O'Corry-Crowe et al., 2003). Additional details on the capture and handling protocols can be found in Savarese (2004).

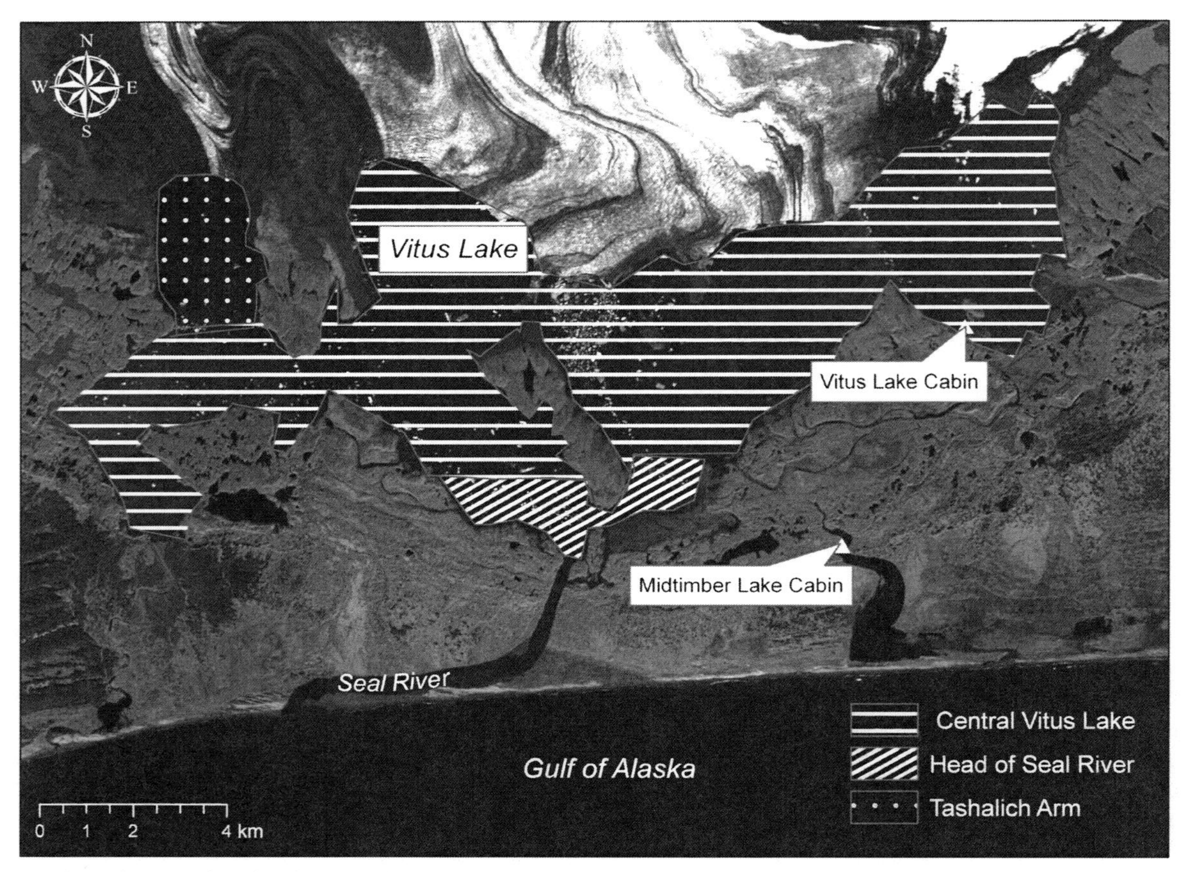

Figure 3. Map of Bering Glacier–Vitus Lake, outlining the three areas where harbor seals have been documented.

To determine if seals were foraging in Vitus Lake, and to determine the types of prey selected, scat samples were collected opportunistically from the icebergs in Vitus Lake. All prey hard parts were identified to the lowest possible taxon by Pacific Identifications, Vancouver, Canada, and seasonal shifts in diet were tested for using Chi-square analysis. For full details on this diet analysis, see Savarese (2004).

RESULTS

Distribution within Bering Glacier Ecosystem

Harbor seals were not observed hauling out on terrestrial substrates at Bering Glacier, either on the shores of Vitus Lake, the extensive sandy banks of the Seal River, or the surrounding coastal Gulf of Alaska beaches. Harbor seals were found hauling out only on icebergs within Vitus Lake. The seals that tended to haul out were on icebergs that were flat and low-lying as opposed to icebergs with substantial relief probably because these were the most stable. Seal distribution within the lake varied significantly by area, but not by year or month (Kruskall Wallis $p < 0.001$, Fig. 4). In both 2002 and 2003 the vast majority (>80%) of seals hauled out on icebergs at the head of the Seal River. In contrast, only 20% of the seals hauled out were in central Vitus Lake, and fewer than 5% were in Tashalich Arm, despite the abundance of low-profile ice there. The absence of a significant monthly effect indicated that no significant redistribution of seals took place within the lake as the total number of seals in the area increased.

Within the area at the head of the Seal River, where most seals were found, the group size ranged from 1 to 251 on a given iceberg, and the mean density was 10 ± 1 seals per iceberg. There was no monthly effect on the average number of seals per iceberg, but there was a significant annual effect ($F_{1,19} = 7.97$, $p = 0.011$), with average seal density higher in 2003 than in 2002 (Fig. 5A). There was no significant interaction between month and year. In contrast, there were significant monthly ($F_{4,19} = 6.175$, $p = 0.002$) and annual ($F_{1,19} = 5.180$, $p = 0.035$) effects on the total number of icebergs used by seals, but again no interactive effects (Fig. 5B). Bonferroni post-hoc tests revealed that seals used the greatest number of icebergs in September and the fewest in June, and that on average more icebergs were used in 2002 than in 2003. In combination, these findings suggest that the higher density of seals per iceberg in 2003 was due to the lower number of suitable icebergs in the area. Because we could not determine if unoccupied icebergs were actually suitable for haulout, we have no measure of suitable unoccupied habitat. Field notes indicated that some apparently suitable icebergs in the area were not occupied. Therefore, these results may have been influenced by seal social and aggregation behavior, or other ecological factors that we did not record.

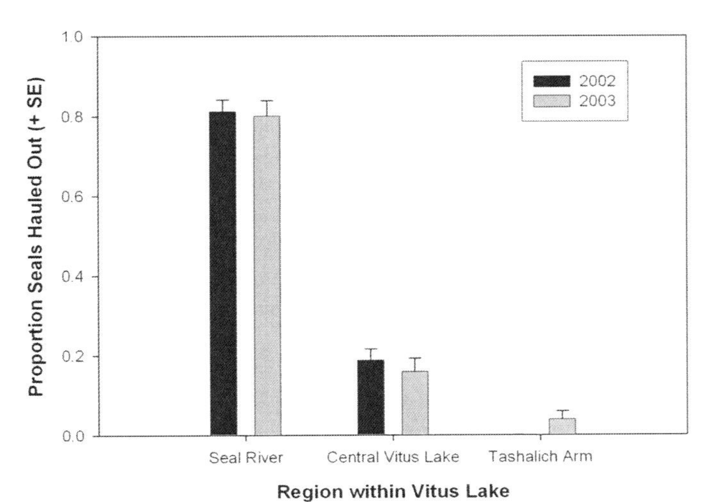

Figure 4. Mean (+S.E.) proportion of seals hauled out in each region of Vitus Lake during aerial surveys in 2002 and 2003.

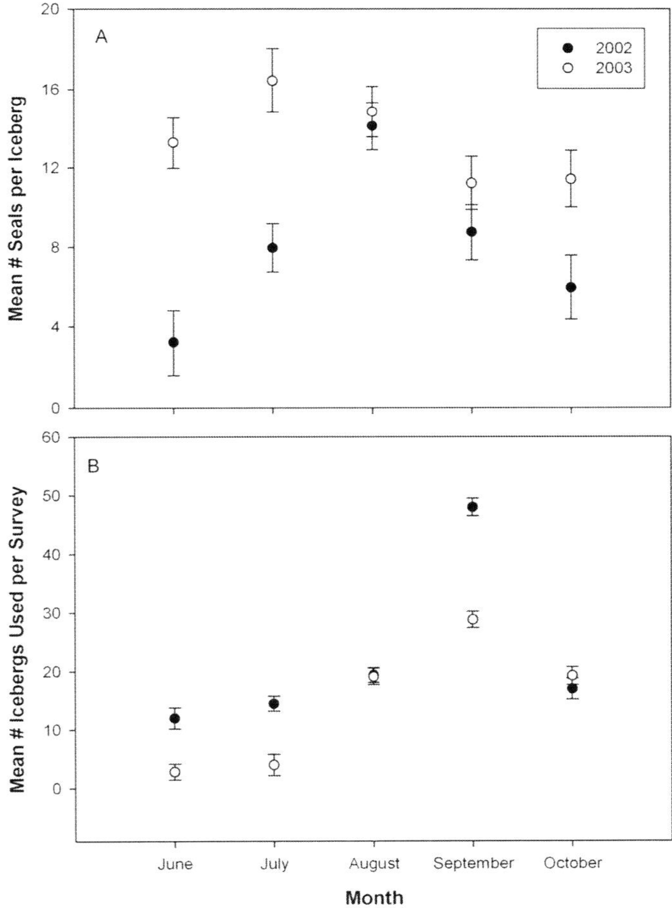

Figure 5. Mean (±S.E.) number of icebergs utilized (A) and seals per iceberg (B) during each survey in the region at the head of Seal River by month and year.

Seasonal Trends in Abundance

Harbor seals were observed in Vitus Lake on each of the 69 surveys flown between May and November 2001–2003, and ~150 seals were observed hauled out on two aerial overflights in February 2003 (G. Ranney, Fishing and Flying, Cordova, Alaska, 2003, personal commun.). In combination, these observations, coupled with our modeling data, suggest that harbor seals are present at the site year-round (Savarese, 2004), although additional winter surveys are necessary to confirm this. However, the number of seals was not stable throughout the year. Seals were most abundant in mid-September, with an estimated 930 seals (95% CI: 795–1089; Savarese, 2004) hauled out under locally ideal conditions on 14 September (Fig. 6). Fewer animals hauled out under rainy conditions. The seasonal increase was not due to reproduction, because pups were never observed at the site, nor was it due solely to an influx of juveniles, because there was no significant change in the mean size of the seals hauled out during the summer and fall (Savarese, 2004).

Genetics

A total of 18 skin samples for genetic analysis were collected from harbor seals at Bering Glacier during 2002–2003. Mitochondrial DNA haplotypes of the harbor seals at Bering Glacier matched those of both Southeast Alaska and Prince William Sound (part of the Gulf of Alaska stock) (G.M. O'Corry-Crowe, 2004, personal commun.). Genetic diversity measured at the site was high, with 17 unique haplotypes identified among 18 samples.

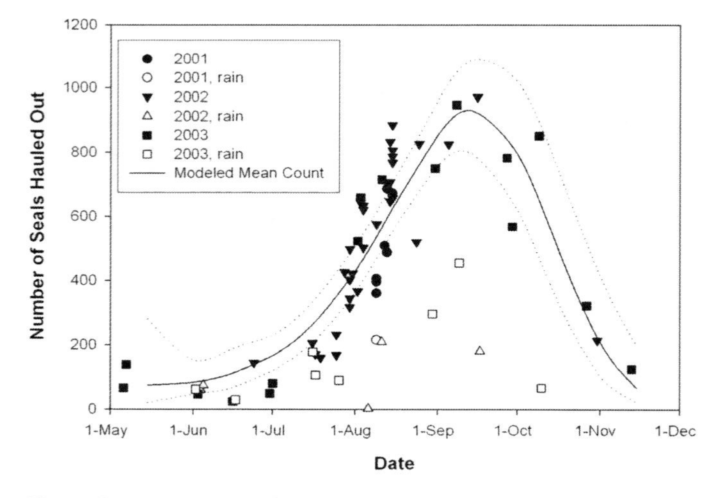

Figure 6. Mean number of harbor seals hauled out in Vitus Lake under locally ideal conditions, as modeled using nonlinear regression with a Poisson distribution. Dotted lines show 95% CI around the mean, and the raw count data are represented by solid point markers. Open markers indicate counts conducted during rain.

DISCUSSION

The research we conducted on harbor seals at the Bering Glacier clearly demonstrated that the glacial ice in Vitus Lake is an important haulout substrate for harbor seals in the area and that the site is a seasonally important haulout for nonreproductive seals. The results of genetic testing indicate that seals may come to the Bering Glacier from both the Gulf of Alaska and Southeast Alaska management stocks. These factors must be considered when developing management plans for seals in the area.

During the course of this study, harbor seals were never observed hauling out along the Gulf of Alaska coastline, along the banks of the Seal River, or on any shoreline in Vitus Lake. Instead, seals preferentially utilized the low-profile icebergs that calved from the Bering Glacier. The seals' preference for icebergs is not surprising, given the high density of bears in the surrounding area (Griese, 1991). Terrestrial haulouts are likely associated with increased risk of predation as well as an increased proportion of time spent in vigilance activities (DaSilva and Terhune, 1988). Most of the icebergs utilized by seals were large enough for multiple animals to haul out on, and the smaller pieces of brash ice that tended to accumulate in Tashalich Arm were not used to any significant extent. This was not because seals cannot use brash ice for haulout; at many areas in Alaska where glaciers calve directly into marine fjords the majority of the icebergs are very small, and seals are commonly found using brash ice as haulout substrate (Calambokidis et al., 1987; Cesarone and Withrow, 1999; Jansen et al., 2006; Bengtson et al., 2007).

The preference for the larger pieces of ice near the head of Seal River may reflect both habitat and energetic considerations. The current that drains Vitus Lake via the Seal River carries the small pieces of ice out to sea (Chapter 6, this volume), and the only icebergs that remain in the lake near the river mouth are large and/or grounded. Harbor seals may prefer to congregate on these larger icebergs because the amount of time an individual animal spends being vigilant decreases, and time resting increases, as group size increases (DaSilva and Terhune, 1988). However, seals may also preferentially use these larger icebergs because they provide safe haulout sites close to foraging grounds. Scat samples collected in Vitus Lake did not contain remains from any fresh-water species, suggesting that seals do not forage extensively on native lake fauna but instead target marine species (Savarese, 2004). Therefore, seals may congregate near Seal River rather than traveling farther into the lake to reduce time and energy spent in travel between foraging and resting areas. If so, changes in the abundance and characteristics of the ice in the lake may alter harbor seal haulout patterns.

Calving rates and outflow volumes can significantly impact the amount and type of ice available in the lake for seals to haul out on. We observed annual differences in the amount and distribution of ice in the lake. For example, when this study was initiated in 2001, harbor seals were found on large icebergs that were distributed along the length of Beringia Novaya Island (central Vitus Lake). However, by 2002 there were far fewer icebergs within the lake, and seals appeared to be more concentrated near the head of Seal River. The decline in icebergs within the lake and at the head of Seal River continued into 2003, when the mean number of icebergs used by seals was lowest. The coincident increase in the average number of seals per iceberg suggests that there was some degree of substrate limitation over time. Whereas changes in ice abundance and distribution within the lake may influence where seals are found and how they are aggregated, even large declines in the availability of suitable ice may not prevent seals from using the region for haulout. Throughout Alaska, harbor seals are commonly found on beaches and small islands (Pitcher and McAllister, 1981; Boveng et al., 2003; Simpkins et al., 2003; Bengtson et al., 2007), and though we did not observe seals utilizing terrestrial haulouts near Vitus Lake, glaciologists working at the Bering Glacier in the 1970s did note harbor seals hauled out on the sandy banks of the Seal River at that time (B. Molnia, 2002, personal commun.).

In addition to heterogeneity in the localities where seals are found, there is also significant seasonal variability in harbor seal abundance. While there are fewer than 200 harbor seals at the Bering Glacier haulout between November and June, more than 800 animals can be seen hauled out in late summer (Savarese, 2004). This large variation in numbers is typical of glacial and terrestrial haulout sites throughout Alaska, where numbers often peak in late summer during the molt period (Mathews and Kelly, 1996; Bengtson et al., 2007). In contrast, there was little variation in the number of animals hauled out within each day, indicating that, as at most glacial haulouts, there was little diel or tidal effect on haulout patterns (although see Hoover, 1983). However, weather did have a large effect, with significantly fewer animals seen under rainy conditions (Calambokidis et al., 1987; Watts, 1992; Boveng et al., 2003; Simpkins et al., 2003; Mathews and Pendleton, 2006). Since not all animals haul out at the same time, even under locally optimal conditions, correction factors must be applied to the number of seals counted in order to determine the total number of seals in the area (Huber et al., 2001; Simpkins et al., 2003, Boveng et al., 2003). There is substantial variability in the correction factors derived for different areas, seasons, haulout substrates, years, and even age classes, and we were unable to determine correction factors specific to the Bering Glacier haulout. However, if we apply the correction factors developed for harbor seals in different regions of Alaska (1.198 for seals using terrestrial haulouts, Simpkins et al., 2003; 1.92 for seals using glacial haulouts, Cesarone and Withrow, 1999) to the peak number of seals estimated to be hauled out under locally ideal conditions (930 on a sunny 14 September), we estimate that 1114–1786 harbor seals are present at the Bering Glacier in mid-September. This estimate should be treated cautiously because we know that this is not a closed population. For example, if more seals moved into the area after 14 September, but hauled out less frequently, these figures would underestimate total use. Conversely, they would overestimate use if most of the seals in Vitus Lake were foraging locally and hauling out regularly.

To assess the relative importance of the Bering Glacier haulout to harbor seals in Alaska, it would be best to place these numbers into the context of the surrounding population size, and the reasons for the seals' presence in the area. Although the absence of pups at Bering Glacier indicates that the site is not currently important from a reproductive perspective, new breeding areas are often first utilized by nonreproductive animals, suggesting that it may become so at some point in the future (Oxman, 1995). However, harbor seal numbers at Vitus Lake reached their peak only 2 wk after the date of peak molt for adult seals at Tugidak Island, Alaska, and molt dates are known to vary with latitude and between age and sex classes (Temte et al., 1991; Daniel et al., 2003). Therefore, Vitus lake may be a seasonally important site for molting seals. The NMFS estimates that 2%–4% of Alaska harbor seals that hauled out on glacial ice during the molt period can be found at the Bering Glacier, and this number increases to 5.3%–7% if the area is restricted to the region of south-central Alaska between Prince William Sound and Yakutat Bay (J. Jansen, 2004, personal commun.).

The Bering Glacier may also be seasonally important to regional harbor seal populations because it provides a safe haulout substrate that is in close proximity to seasonally abundant prey resources such as the salmonids that return to the Bering and Tsiu Rivers (Savarese, 2004). Salmon returning to river systems to spawn are an important prey resource for many pinnipeds throughout the Pacific Northwest (Brown and Mate, 1983; Bigg et al., 1990; Middlemas et al., 2006; Brown and Terhune, 2003). In both 2002 and 2003 the peak in harbor seal abundance coincided with peak regional salmon abundance (Ashe et al., 2005; Savarese, 2004), and as the numbers of seals at the Bering Glacier increased during the late summer, the proportion of salmon in their diet also increased. Thus, while scat analyses indicate that seals at the Bering Glacier, like those throughout the state, forage on eulachon, capelin, pollock, herring, flatfishes, cod, and salmon (Pitcher, 1980; Iverson et al., 1997), it may be the salmon that attract large numbers to the region in late fall.

If seasonally abundant prey is drawing harbor seals to the Bering Glacier haulout, then changes in the timing of their prey's return could affect the timing of peak harbor seal abundance, with implications for population management. In Alaska, trends in harbor seal population size have traditionally been determined by a combination of annual surveys of index sites, with complete enumeration at all haulout sites occurring only every 5 a (Boveng et al., 2003). If natural fluctuations in the timing of salmon runs, such as the 4 wk variation in the peak of coho returns documented at the nearby Bering River (Rick Merizon, ADF&G Cordova office, 2004, personal commun.), cause harbor seal abundance at an index site to vary, and this is not accounted for during the selection of survey dates or the analysis of survey results, accurate characterization of a population trend will be delayed until the next complete enumeration. Similarly, if progressive shifts in the timing of salmon runs occur in response to climate change (Houghton et al., 2001;

Juanes et al., 2004) or other factors, but are not detected, managers may assume that changes in local seal abundance reflect population trends rather than a behavioral response to changes in resource availability. Because this effect can be reduced by including multiple sites across a broad geographic area, and using a series of counts over multiple years to characterize trends, the influence of seasonally abundant prey resources on the movements of harbor seals should be considered carefully when population monitoring programs are designed (Savarese, 2004).

The influx of a large number of seals into the Bering Glacier area, possibly in response to seasonally available prey resources, also has implications for population management decisions that are based on stock structure. Currently harbor seals in Alaska are divided into three geographic stocks: Bering Sea, Gulf of Alaska, and Southeast Alaska, and there has been little consideration of seal movement between these areas (Westlake and O'Corry-Crowe, 2002; Angliss and Outlaw, 2006). However, our genetic findings support the theory that harbor seals at the Bering Glacier come from a combination of haulouts in Prince William Sound (>100 km to the west, Gulf of Alaska Stock) and Southeast Alaska (>100 km to the east), indicating that seals from two areas currently managed as separate stocks may mix at the Bering Glacier. This finding is timely, because the degree to which populations are demographically isolated is central to the ongoing effort to redefine Alaska harbor seal stocks into smaller units. This effort is based on recent genetic studies, tracking data, census counts that indicate heterogeneous population trends within current geographic stocks, and links between trends and genetic diversity (Lowry et al., 2001; Westlake and O'Corry-Crowe, 2002; O'Corry-Crowe et al., 2003). The NMFS and their co-management partner, the ANHSC, have discussed amending stock structure to include the Bering Glacier haulout in a stock that encompasses Prince William Sound to the west and Icy and Yakutat Bays to the east, as is supported by our data (NMFS–ANHSC proposal, October 2005).

If such a new management stock is recognized, then the impact of the periodic surges on seals in the area should be considered. Over the last century the Bering Glacier has undergone five surges, and each time the dramatic shifts in hydrological properties and increases in the amount of ice present in Vitus Lake have had large impacts on the biological communities in the lake and surrounding areas (Molnia and Post, 1995). For example, during the 1993–1995 surge the increased local turbidity, outflow volumes, and altered drainage conditions stopped sockeye salmon from returning to the Seal River (B. Molnia and J. Payne, 2001, personal commun.). While juvenile salmon have recently been trapped in Vitus Lake (Chapter 8, this volume) it is unclear whether the adult salmon run has been reestablished since the surge ended and turbidity decreased. In addition, harbor seals were absent from the lake during the last surge, and their presence was not noted again until 1997 (B. Molnia and J. Payne, 2001, personal commun.). It is not entirely clear what factors determine the magnitude, duration, and periodicity of glacial surges (Chapter 15, this volume), but warming conditions are known to be altering glacial dynamics in many systems (Houghton et al., 2001). If harbor seals typically abandon the site during surges, then the magnitude and periodicity of surges will affect harbor seal distribution within the region. Information on the geographic extent and duration of movements away from the Bering Glacier is necessary for accurate population monitoring, and for appropriately characterizing the potential for gene flow between regions.

IMPLICATIONS FOR MANAGEMENT OF HARBOR SEALS AT BERING GLACIER

Whereas natural perturbations such as glacial surges cannot be mitigated by federal agencies, several anthropogenic factors have the potential to impact harbor seal use of the Bering Glacier habitat, and these factors should be considered for management action. Anthropogenic threats include disturbance from aircraft, vessel traffic, habitat degradation, and competition-interaction with commercial fisheries. Potential human-caused impacts on the seals at Bering Glacier should be considered, and possibly mitigated, by the U.S. Bureau of Land Management (BLM), which is responsible for issuing permits for individuals and companies to conduct activities in the area.

Over the coming decades, one of the larger impacts that humans may have on harbor seals in the Bering Glacier area may be the increased number of interactions that will result from increases in visitor traffic. Interest in local ecotourism and kayak tours of the area has been growing in recent years (BLM Glennallen office, 2004, personal commun.), and the state of Alaska erected one public use cabin on Vitus Lake in 2002, and, as part of a plan to develop a series of interconnected remote cabins throughout south-central Alaska, connected it to a second nearby cabin on Midtimber Lake (see Fig. 3). As the only way to access the region is by small plane, increases in the number of people recreating in the area would be accompanied by an increase in air traffic, which would have the potential to disturb hauled out harbor seals (Johnson, 1977; Suryan and Harvey, 1999; Jansen et al., 2006). In addition, harbor seals can be disturbed by boat traffic, and they react strongly to kayakers, likely because a kayak's quiet approach does not provide sufficient warning (Lewis and Mathews, 2000; Harris et al., 2003). While the most common reaction of harbor seals to disturbance is increased vigilance, animals will abandon the haulout if the disturbance is too threatening (Suryan and Harvey, 1999). Continued flushing of seals into the water alters their natural activity budgets, and may be especially stressful during reproductive and molting periods, when animals increase their haulout frequency (Thompson et al., 1989; Daniel et al., 2003). Prolonged, frequent disturbances have caused seals to abandon haulout areas (Newby, 1973; Paulbitski, 1975).

Mitigation measures that may reduce the impact of recreational activities on the local harbor seal population could take several forms. Given the relatively low number of current visitors, perhaps the simplest would be an educational approach. Existing federal regulations (the Marine Mammal Protection Act) prohibit people from disturbing seals through close approach (defined as <100 m or any distance at which a behavioral reaction is noticed). However, the public and guides are often unfamiliar with these regulations. Increased outreach and educational efforts, such as posting informational flyers at the public use cabins, and training local tour operators and guides about methods that can enhance viewing opportunities while reducing disturbance, may increase public awareness of, and compliance with, federal regulations. Such efforts have dramatically reduced visitor disturbance of harbor seals within Ailik Bay, without the need to restrict the number of visitors or the viewing areas or seasons (Hoover-Miller et al., 2006).

Given that public access to the Bering Glacier is currently limited because of its remote location, the area could offer a unique environment in which to study harbor seal behavior in a glacial ecosystem without the potentially confounding effects of human disturbance, such that occurs in most other glacial environments by cruise ships, fishers, and tourists (Calambokidis et al., 1987; Suryan and Harvey, 1999; Lewis and Mathews, 2000; Jansen et al., 2006). Additional information is needed on the diel and seasonal haulout patterns of harbor seals in glacial ice environments in order to develop accurate correction factors for population size estimates and to better understand how the large-scale movements documented into and away from these sites affect gene flow (Bengtson et al., 2007). Observing harbor seal behavioral patterns under undisturbed conditions could also provide baseline information for testing hypotheses about how glacial surges and retreats may influence population dynamics and stock structure at the Bering Glacier, as well as at other glacial sites throughout the state.

Development activities in the area may threaten harbor seal populations. For example, mining of the hard rock minerals in the mountains above the terminus of the Bering Glacier could (if developed) impact animals directly through increased disturbance rates or indirectly through impacts on habitat quality, salmon runs, or water-borne contaminant levels. Harbor seals can accumulate toxins in their tissues, and increased heavy metal loads are associated with reproductive and neurological defects (O'Shea, 1999). In addition, harbor seals are vulnerable to oil exposure, which can be acutely toxic or cause chronic effects from ongoing low level exposure to remnant oil in the ecosystem and food chain (Frost et al., 1994). Therefore, it might be prudent for BLM to work with the Department of Environmental Conservation, the Alaska Department of Fish and Game Habitat and Restoration Division, and local oil spill agencies to designate Vitus Lake as a site of special concern or MESA (Most Ecologically Sensitive Area) so that it receives priority for boom placement in the event of an oil spill.

Finally, since harbor seals apparently move into the Bering Glacier area in order to forage on seasonally available salmonid prey, their use of these fish stocks should be considered when managing local fish stocks. Unfortunately, little relevant salmon data are currently collected in the Bering Glacier region. Commercial fishers can interact with pinnipeds through direct and indirect competition for prey, as well as through direct take in fishing gear (Northridge and Hofman, 1999; Angliss and Outlaw, 2006). While harbor seal bycatch levels are low relative to current stock size (31 and 33 harbor seals taken from the Southeast Alaska and Gulf of Alaska stocks, respectively; Angliss and Outlaw, 2006), changes in fishing practices or animal behavior may increase seals' vulnerability, whereas shifts in seal population size or stock structure may alter the impact of current take on population trajectories.

In conclusion, this study shows that harbor seals preferentially use the low, stable icebergs within Vitus Lake to haul out, but the importance of the site increases significantly in late summer. This increase may result from seals moving into the area from distant areas and distinct geographic stocks to take advantage of seasonally abundant salmonid prey resources. The large seasonal shifts in abundance and location documented here have implications for the design of population monitoring programs and conservation efforts. Whereas climate change and glacial surges have the potential to alter harbor seal use of the Bering Glacier haulout over time, these are natural effects that cannot be directly managed. Increased human activities in and around the Bering Glacier Region, however, have the potential to impact harbor seal use of the site, and careful consideration should be given to educational outreach and management actions that can minimize potential disturbances so that harbor seals remain a functioning element of the Bering Glacier ecosystem.

ACKNOWLEDGMENTS

The authors wish to sincerely thank all those involved with this project. We especially thank Frank von Hippel for his assistance with study design, and John Payne, Scott Guyer, and Chris Noyles of the Bureau of Land Management, who provided financial, logistical, and emotional support. We also wish to thank several field assistants, including Leslie Sarten, Shawn Harper, Monica Bando, Julie Koplan, Cheryl Clark, Lance Clark, Chris Burns, Steve Trumble, and Danielle Greaves. Special thanks go to John Tucker of Wilderness Helicopters for flying our aerial surveys. We also thank Greg O'Corry-Crowe of the National Marine Fisheries Service for providing advice on the management implications of the genetic data. The manuscript was significantly improved on the basis of comments from Anne Hoover-Miller and Marilyn Barker. Finally, we express our sincere gratitude to our extended Bering Glacier family, including all of our research partners, Gail Ranney, Larry and Ben Hancock of Fishing and Flying, and Steve Ranney and the staff at the Orca Lodge in Cordova, Alaska. This research was conducted under Marine Mammal Scientific Research Permit no. 1003-1665-01 and was approved by the University of Alaska Institutional Animal Care & Use Committee. Funding for this project was provided by a grant from the U.S. Department of the Interior to J. Burns.

REFERENCES CITED

Angliss, R.P., and Outlaw, R.B., 2006, Draft Alaska marine mammal stock assessments, 2006: National Marine Fisheries Service, Alaska Fisheries Science Center Report, 246 p.

Ashe, D., Gray, D., Lewis, B., Merizon, R., and Moffitt, S., 2005, Prince William Sound Management Area 2003 Annual Finfish Management Report 05-54: Anchorage, Alaska Department of Fish and Game, 191 p.

Bengtson, J.L., Phillips, A.V., Mathews, E.A., and Simpkins, M.A., 2007, Comparison of survey methods for estimating abundance of harbor seals (*Phoca vitulina*) in glacial fjords: Washington, D.C., Fishery Bulletin, v. 105, p. 348–355.

Bigg, M.A., Ellis, G.M., Cottrell, P.E., and Milette, L.L., 1990, Predation by harbour seals and sea lions on adult salmon in Comox Harbour and Cowichan Bay, British Columbia: Report 1769, Pacific Biological Station, Department of Fisheries and Oceans, Nanaimo, British Columbia, 31 p.

Boness, D.J., Clapham, P.J., and Mesnick, S.L., 2002, Life history and reproductive strategies, *in* Hoelzel, A.R., Marine Mammal Biology: An Evolutionary Approach: Oxford, UK, Blackwell Publishing, p. 278–324.

Boveng, P., Bengtson, J.L., Withrow, D., Cesarone, J., Simpkins, M.A., Frost, K.J., and Burns, J.J., 2003, The abundance of harbor seals in the Gulf of Alaska: Marine Mammal Science, v. 19, p. 111–127, doi: 10.1111/j.1748-7692.2003.tb01096.x.

Brown, C.L., and Terhune, J.M., 2003, Harbor seal (*Phoca vitulina* Linnaeus) abundance and fish migration in the Saint John Harbour: Northeastern Naturalist, v. 10, p. 131–140.

Brown, R.F., and Mate, B.R., 1983, Abundance, movements, and feeding habits of harbor seals (*Phoca vitulina*) at Netarts and Tillamook Bays, Oregon: Washington, D.C., Fishery Bulletin, v. 81, p. 291–301.

Calambokidis, J., Taylor, B.L., Carter, S.D., Steiger, G.H., Dawson, P.K., and Antrim, L.D., 1987, Distribution and haul-out behavior of harbor seals in Glacier Bay, Alaska: Canadian Journal of Zoology, v. 65, p. 1391–1396, doi: 10.1139/z87-219.

Daniel, R.G., Jemison, L.A., Pendleton, G.W., and Crowley, S.M., 2003, Molting phenology of harbor seals on Tugidak Island, Alaska: Marine Mammal Science, v. 19, p. 128–140, doi: 10.1111/j.1748-7692.2003.tb01097.x.

DaSilva, J., and Terhune, J.M., 1988, Harbour seal grouping as an anti-predator strategy: Animal Behaviour, v. 36, p. 1309–1316, doi: 10.1016/S0003-3472(88)80199-4.

Frost, K.J., Lowry, L.F., Sinclair, E.H., Ver Hoef, J., and McAllister, D.C., 1994, Petroleum hydrocarbons in tissues of harbor seals from Prince William Sound and the Gulf of Alaska, *in* Loughlin, T.R., Marine Mammals and the Exxon Valdez: San Diego, Academic Press, p. 331–358.

Frost, K.J., Lowry, L.F., and Ver Hoef, J., 1999, Monitoring the trend of harbor seals in Prince William Sound, Alaska, after the Exxon Valdez oil spill: Marine Mammal Science, v. 15, p. 494–506, doi: 10.1111/j.1748-7692.1999.tb00815.x.

Griese, H.J., 1991, Unit 6 brown bear survey-inventory performance report, *in* Abbott, S.M., Federal Aid in Wildlife Restoration Progress Report, Project V-23-4, Study 4.0: Annual Performance Report of Survey-Inventory Activities, pt. 5, Brown Bear, v. 22, p. 33–47.

Harris, D.E., Lelli, B., and Gupta, S., 2003, Long term observations of a harbor seal haul-out site in a protected cove in Casco Bay, Gulf of Maine: Northeastern Naturalist, v. 10, p. 141–148.

Hoelzel, A.R., 2002, Marine mammal biology: An evolutionary approach: Oxford, UK, Blackwell Publishing, 432 p.

Hoover, A.A., 1983, Behavior and ecology of harbor seals (*Phoca vitulina richardsi*) inhabiting glacial ice in Aialik Bay, Alaska [M.S. thesis]: University of Alaska, Fairbanks, 133 p.

Hoover-Miller, A., Jezierski, C., Conlon, S., and Atkinson, S., 2006, Harbor seal population dynamics and responses to visitors in Ailik Bay, Alaska: 2005 report to the Ocean Alaska Sciences and Learning Center and the National Park Service, April 2006 (unpublished), 44 p. Available from the Alaska SeaLife Center, P.O. Box 1329, Seward, Alaska 99664, USA.

Houghton, J.T., Ding, Y., Griggs, D.J., Noguer, M., and van der Linden, P.J., Dai, X., Maskell, K., and Johnson, C.A., 2001, Climate Change 2001: The Scientific Basis: Cambridge, UK, Cambridge University Press, 881 p.

Huber, H.R., Jeffries, S., Brown, R.F., DeLong, R.L., and VanBlaricom, G., 2001, Correcting aerial survey counts of harbor seals (*Phoca vitulina richardsi*) in Washington and Oregon: Marine Mammal Science, v. 17, p. 276–293, doi: 10.1111/j.1748-7692.2001.tb01271.x.

Iverson, S.J., Frost, K.J., and Lowry, L.F., 1997, Fatty acid signatures reveal fine scale structure of foraging distribution of harbor seals and their prey in Prince William Sound, Alaska: Marine Ecology Progress Series, v. 151, p. 255–271, doi: 10.3354/meps151255.

Jansen, J.K., Bengtson, J.L., Boveng, P.L., Dahle, S.P., and Ver Hoef, J., 2006, Disturbance of harbor seals by cruise ships in Disenchantment Bay, Alaska: An investigation at three spatial and temporal scales: Seattle, Washington, Alaska Fisheries Science Center, Processed Report 2006–02, National Marine Fisheries Service, 75 p.

Johnson, B.W., 1977, The effects of human disturbance on a population of harbor seals, *in* Environmental Assessment of the Alaskan Continental Slope, U.S. Department of Commerce, NOAA/OCSEAP Annual Report, v. 1, 708 p. [NTIS PB-280934/1].

Juanes, F., Gephard, S., and Beland, K.F., 2004, Long-term changes in migration timing of adult Atlantic salmon (*Salmo salar*) at the southern edge of the species distribution: Canadian Journal of Fisheries and Aquatic Sciences, v. 61, p. 2392–2400, doi: 10.1139/f04-207.

Lewis, T.M., and Mathews, E.A., 2000, Effects of human visitors on the behavior of harbor seals (*Phoca vitulina richardsi*) at McBride Glacier Fjord, Glacier Bay National Park: Final Report to Glacier Bay National Park, Resource Management Division, Gustavus, Alaska, 22 p.

Lowry, L.F., Frost, K.J., Ver Hoef, J., and DeLong, R.L., 2001, Movements of satellite-tagged subadult and adult harbor seals in Prince William Sound, Alaska: Marine Mammal Science, v. 17, p. 835–861, doi: 10.1111/j.1748-7692.2001.tb01301.x.

Marine Mammal Protection Act of 1972 (16 U.S.C. 1361-1407, P.L. 92-522, October 21, 1972, 86 Stat. 1027) as amended by P.L. 94-265, April 13, 1976, 90 Stat. 360; P.L. 95-316, July 10, 1978, 92 Stat. 380; P.L. 97-58, October 9, 1981, 95 Stat. 979; P.L. 98-364, July 17, 1984, 98 Stat. 440; P.L. 99-659, November 14, 1986, 100 Stat. 3706; P.L. 100-711, November 23, 1988, 102 Stat. 4755; P.L. 101-627, November 28, 1990, 100 Stat. 4465; P.L. 102-567, October 29, 1992, 106 Stat. 4284; P.L. 103-238, 3, April 30, 1994, 108 Stat. 532; P.L. 105-18, June 12, 1997, 111 Stat. 187; and P.L. 105-42, August 15, 1997, 111 Stat. 1125.

Mathews, E.A., and Kelly, B.P., 1996, Extreme temporal variation in harbor seal (*Phoca vitulina richardsi*) numbers in Glacier Bay, a glacial fjord in southeast Alaska: Marine Mammal Science, v. 12, p. 483–488, doi: 10.1111/j.1748-7692.1996.tb00603.x.

Mathews, E.A., and Pendleton, G.W., 2006, Declines in harbor seal (*Phoca vitulina*) numbers in Glacier Bay National Park, Alaska, 1992–2002: Marine Mammal Science, v. 22, p. 167–189.

Middlemas, S.J., Barton, T.R., Armstrong, J.D., and Thompson, P.M., 2006, Functional and aggregative responses of harbour seals to changes in salmonid abundance: Proceedings of the Royal Society of London, v. 273, p. 193–198, doi: 10.1098/rspb.2005.3215.

Molnia, B.F., and Post, A., 1995, Holocene history of the Bering Glacier, Alaska: A prelude to the 1993–1994 surge: Physical Geography, v. 16, p. 87–117.

National Marine Fisheries Service, Alaska Department of Fish and Game, Alaska SeaLife Center, and Alaska Native Harbor Seal Commission, 2003, Alaska harbor seal research plan 2003: http://www.fakr.noaa.gov/protectedresources/seals/harbor.htm, 87 p.

Newby, T.C., 1973, Observations on the breeding behavior of the harbor seal in the state of Washington: Journal of Mammalogy, v. 54, p. 540–543, doi: 10.2307/1379151.

Northridge, S.P., and Hofman, R.J., 1999, Marine mammal interactions with fisheries, *in* Twiss, J.R., and Reeves, R.R., eds., Conservation and management of marine mammals: Washington, D.C., Smithsonian Institution Press, p. 99–119.

O'Corry-Crowe, G.M., Martien, K.K., and Taylor, B.L., 2003, The analysis of population genetic structure in Alaskan harbor seals, *Phoca vitulina*, as a framework for the identification of management stocks: Southwest Fisheries Science Center, National Marine Fisheries Service, Administrative Report LJ-03–08, 66 p.

O'Shea, T.J., 1999, Environmental contaminants and marine mammals, *in* Reynolds J.E., III, and Rommel, S., eds., Biology of Marine Mammals: Washington D.C., Smithsonian Institution Press, p. 485–539.

Oxman, D.S., 1995, Seasonal abundance, movements, and food habits of harbor seals (*Phoca vitulina richardsi*) in Elkhorn Slough, California [M.S. thesis]: Stanislaus, California State University, 126 p.

Paulbitski, P.A., 1975, The seals of Strawberry Spit: Pacific Discovery, v. 28, p. 12–15.

Pitcher, K.W., 1980, Food of the harbor seal, *Phoca vitulina richardsi*, in the Gulf of Alaska: Fishery Bulletin (Washington, D.C.), v. 78, p. 544–549.

Pitcher, K.W., and McAllister, D.C., 1981, Movements and haulout behavior of radio-tagged harbor seals, *Phoca vitulina*: Canadian Field Naturalist, v. 95, p. 292–297.

Riedman, M., 1990, The Pinnipeds. Seals, Sea Lions, and Walruses: Berkeley, University of California Press, 439 p.

Savarese, D.M., 2004, Seasonal trends in harbor seal abundance at the terminus of the Bering Glacier in Southcentral Alaska [M.S. thesis]: University of Alaska, Anchorage, 70 p.

Simpkins, M.A., Withrow, D.E., Cesarone, J.C., and Boveng, P.L., 2003, Stability in the proportion of harbor seals hauled out under locally ideal conditions: Marine Mammal Science, v. 19, p. 791–805, doi: 10.1111/j.1748-7692.2003.tb01130.x.

Suryan, R.M., and Harvey, J.T., 1999, Variability in reactions of Pacific harbor seals, *Phoca vitulina richardsi*, to disturbance: Fishery Bulletin (Washington, D.C.), v. 97, p. 332–339.

Temte, J.L., Bigg, M.A., and Wiig, O., 1991, Clines revisited: The timing of pupping in the harbour seal (*Phoca vitulina*): Journal of Zoology, v. 224, p. 617–632, doi: 10.1111/j.1469-7998.1991.tb03790.x.

Thompson, P.M., Fedak, M.A., McConnell, B.J., and Nicholas, K.S., 1989, Seasonal and sex-related variation in the activity patterns of common seals (*Phoca vitulina*): Journal of Applied Ecology, v. 26, p. 521–535, doi: 10.2307/2404078.

Watts, P., 1992, Thermal constraints on hauling out by harbor seals *Phoca vitulina*: Journal of Zoology, v. 70, p. 553–560.

Westlake, R.L., and O'Corry-Crowe, G.M., 2002, Macrogeographic structure and patterns of genetic diversity in harbour seals (*Phoca vitulina*) from Alaska to Japan: Journal of Mammalogy, v. 83, p. 1111–1126, doi: 10.1644/1545-1542(2002)083<1111:MSAPOG>2.0.CO;2.

Manuscript Accepted by the Society 02 June 2009

Printed in the USA

The Geological Society of America
Special Paper 462
2010

The 1993–1995 surge and foreland modification, Bering Glacier, Alaska

P. Jay Fleisher*
State University of New York–Oneonta, Earth Sciences, Oneonta, New York 13820, USA

Palmer K. Bailey
U.S. Army Corps of Engineers (retired), 64710 Knob Hill Road, Anchor Point, Alaska 99556, USA

Eric M. Natel
Eastman Kodak, Research and Development, Rochester, New York 14650, USA

Ernie H. Muller[†]
Syracuse University, Department of Earth Sciences, Syracuse, New York 13244, USA

Don H. Cadwell[†]
New York State Geological Survey, Albany, New York 12230, USA

Andrew Russell
University of Newcastle upon Tyne, School of Geography, Newcastle, NE17RU, UK

ABSTRACT

A 25–30 yr surge cycle anticipated by Post (1972) was confirmed by the 1993–1995 surge, although the advance culminated more than a kilometer short of the 1965–1967 surge limit. During the initial 6 mo. of the 1993–1995 surge the eastern terminus of the Bering Glacier Piedmont Lobe advanced 1.0–1.5 km at a rate that varied between 1.0–7.4 m/d, and thickened by an estimated 125–150 m. One year after the surge began an outburst of pressured subglacial water temporarily interrupted basal sliding and slowed ice front advance. Within days gravel and blocks of ice transported and deposited by that flood partially filled an ice-contact lake, forming a 1.5 km² sandur. During the next few months a second outburst nearly dissected a foreland island with the resulting construction of two additional sandar, each nearly 1 km². Both outburst sites coincided with a subglacial conduit system that has persisted for decades and survived two surges. When the surge resumed, advance was intermittent and slower. A prominent push moraine marks the limit of ice advance on the eastern sector.

Although basal sliding across a saturated substrate was a major contributor to surge-related changes along the eastern sector, the most profound foreland altera-

*Email: fleishpj@oneonta.edu.
[†]Posthumously.

Fleisher, P.J., Bailey, P.K., Natel, E.M., Muller, E.H., Cadwell, D.H., and Russell, A., 2010, The 1993–1995 surge and foreland modification, Bering Glacier, Alaska, *in* Shuchman, R.A., and Josberger, E.G., eds., Bering Glacier: Interdisciplinary Studies of Earth's Largest Temperate Surging Glacier: Geological Society of America Special Paper 462, p. 193–216, doi: 10.1130/2010.2462(10). For permission to copy, contact editing@geosociety.org.

tion was the result of outburst-related erosion, deposition, and drainage modification associated with outburst floods. The dominant modification of overridden terrain was subglacial hydraulic scouring of sub-kilometer scale basins, 15–20 m deep, and outburst-related proglacial sandur development. Only after a decade of retreat was it possible to assess the limited direct effects of overriding ice, which were confined to deposition of a sub-meter-thick deformation till, decameter-scale flutes, and drumlinized topography accompanied by truncation of subglacial strata.

INTRODUCTION

The Bering Piedmont Lobe is a broad, warm-based glacier with an ice-marginal peripheral drainage system, all similar to the Laurentide Ice Sheet as it retreated from New York State ~14 ka. Field study in this modern glacial analogue was initially focused on ice-contact lake morphology and sedimentation and Neoglacial stratigraphy, but abruptly shifted from normal processes to surge events. The surge began in the spring of 1993 with the development of a kinematic wave that moved down the trunk glacier from the Bagley Ice Field to a position near the head of the Piedmont Lobe in early June, arriving on the eastern sector in July. We monitored ice front activity from eastern foreland islands where detailed, ground-based observations were possible. Icebergs compressed along the heavily crevassed southern margin made it virtually impossible to distinguish them from the actual ice front.

Initially, attention focused on surge-related changes as the glacier moved onto eastern sector islands and into lake basins (Fig. 1) (Fleisher et al., 1994a; Muller et al., 1993). We measured the rate of advance of the land-based ice front, but our focus again changed when outbursts of pressurized subglacial water incised the glacial foreland while depositing extensive sandar and modifying the peripheral drainage system. At the end of the 2 yr surge we had assessed the impact of the surge and outburst floods by comparing them with previously mapped, pre-surge terrain. The direct effects of ice advance were surprisingly limited; the dominant influence was from sub- and proglacial hydraulic erosion and flood-related deposition. Since 1998 we have focused on the rates of retreat and downwasting, which ultimately led to the return of the ice front to near its pre-surge position, thus bringing to a close a single surge cycle.

This study incorporates all events that occurred during the surge, from the beginning in early summer of 1993 through the culmination of glacial advance marked by the construction of a terminal push moraine during the fall of 1995. Furthermore, we monitored retreat during the decade that followed and assessed the effects of the surge and associated outburst floods on foreland terrain.

REGIONAL LOCATION AND GEOGRAPHIC SETTING

Bering Glacier, the largest known surging glacier in North America and the largest of nine major glaciers that originate in the Bagley Ice Field, flows 100 km westward from the Canadian border to transect the Chugach Range. Descending to coastal

Figure 1. Bering Piedmont Lobe surging onto Weeping Peat Island (3 January 1994).

lowlands at the foot of the Robinson Mountains, it coalesces with Steller Glacier. Together they constitute the largest known glacier system in North America (5200 km^2) with a Piedmont Lobe second only in size to that of Malaspina Glacier 150 km to the east. The Grindle Hills confine the Bering trunk glacier on its eastern flank at the head of the Piedmont Lobe. To the north, Kaliakh glacier tongue separates the Grindle Hills from Override Ridge, a landform reference where thickening ice covering the ridge signals the development of surge conditions (Fig. 2).

Much of the eastern margin of the Piedmont Lobe fronts in Tsivat, Tsiu, and Vitus Lakes. Together they form a peripheral chain of lakes that drain to the Gulf of Alaska by way of Seal River. Weeping Peat Island separates Tsivat and Tsiu Lake basins, and Bentwood Island lies between Tsiu and Vitus Lakes. In 1989, retreat from the 1965–1967 surge limit opened Bentwood Narrows, thus separating the glacier from Bentwood Island. Although ice still rested on the stoss side of Peat Falls Island, the glacier front had completely retreated from Weeping Peat Island prior to the surge. All of these foreland features, except the easternmost Bentwood Island, were covered by ice during the 1965–1967 surge, whereas the limit of advancing ice during the 1993–1995 surge fell short of this by more than a kilometer.

METHODS OF TERRAIN ASSESSMENT

U.S. Geological Survey (USGS) topographic maps of this region are decades out of date. They illustrate ice coverage conditions that place all islands and lake basins affected by the 1993–1995 surge concealed beneath the eastern Piedmont Lobe. Consequently, a basic topographic map was generated to supplement a 1992 false-color satellite image generated by the U.S. Bureau of Land Management (BLM) and Ducks Unlimited (Fig. 2). A

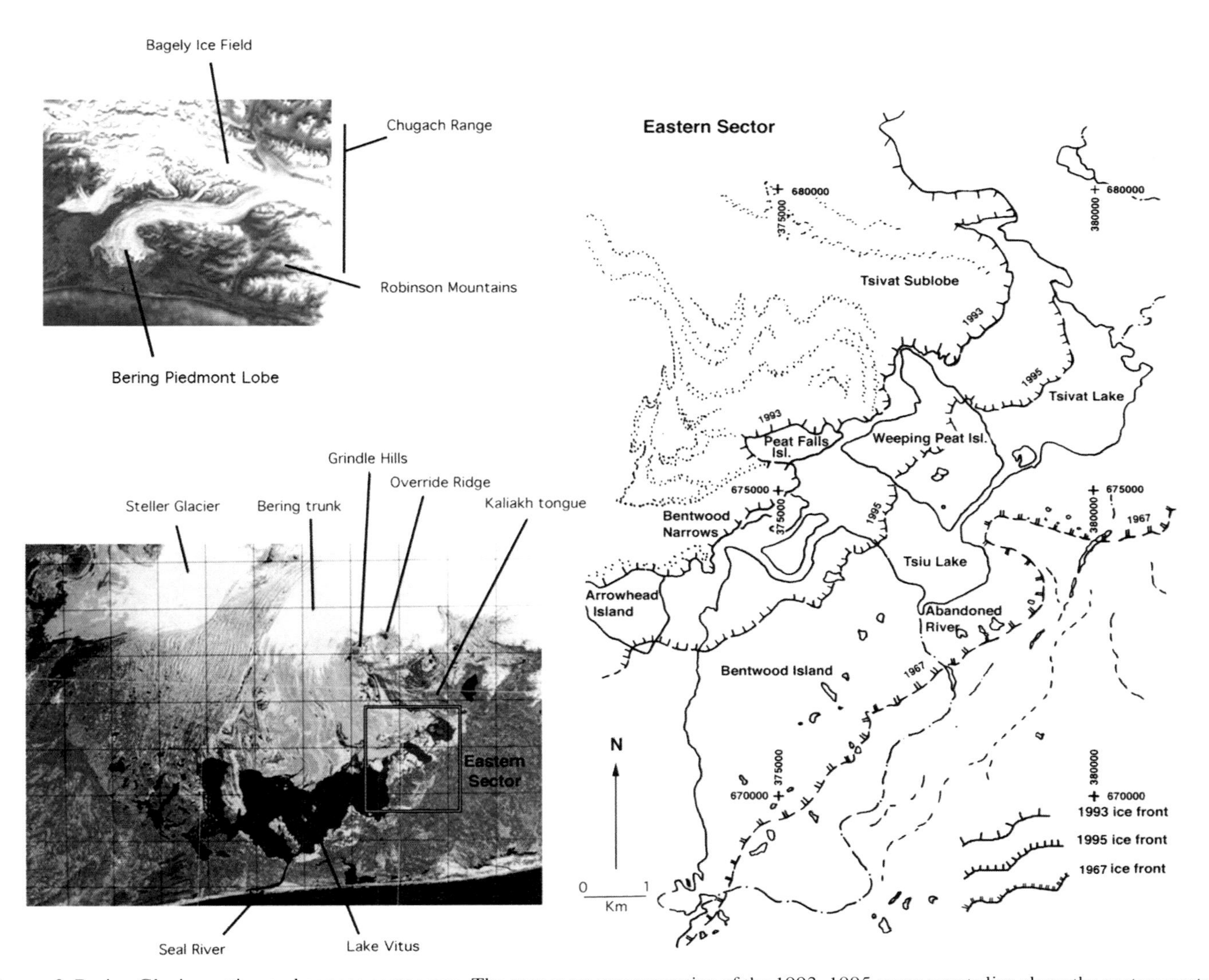

Figures 2. Bering Glacier setting and eastern sector map. The area most representative of the 1993–1995 surge events lies along the eastern sector of the Bering Piedmont Lobe.

superimposed, 5 km global positioning system (GPS) reference grid and latitude-longitude lines define the general position and configuration of the ice front and the peripheral drainage system. USGS aerial photographs capture the mid-1960s surge at its maximum limit and its progressive retreat during the decades that followed. These images depict the uncovering of landforms during gradual retreat through the 1970s and 1980s, and topographic details related to the 1993–1995 surge and associated outburst floods. A series of seven ground-based photo stations and several aerial views depict same-scene changes during advance and retreat phases of the surge. These provide an excellent frame of reference for the sequence of events and magnitude of change. GPS technology facilitated mapping that linked USGS aerial photography, satellite imagery, and the 15 min Bering A-5 quadrangle. The pre- and post-surge topography of island landforms prominent on Weeping Peat Island are known from conventional alidade-stadia surveys. Mapping from 1992 through 2007 includes control of ice-front positions, progressive retreat from the 1995 surge limit, and subglacial landforms uncovered during retreat. Multiple topographic cross sections were surveyed to define the position, relief and morphology of sandar surfaces, abandoned meltwater channels, moraines, scarps, and basins.

A GPS-controlled benchmark to which all surveyed elevations are tied is a 2 m boulder in a stable position on Weeping Peat Island. Satellite telemetry (compliments of Yann Merrand, University of Washington) determined the position of a steel pin placed within the top of this boulder to be at North 60°11′25.48″, West 143°12′21.50″, and 20.51 m above sea level (a.s.l.). A series of alidade-stadia land surveys links more than two dozen other established benchmarks, thus generating a benchmark network (Gerhard et al., 1996, and unpublished data).

Annual bathymetric surveys of Tsivat and Tsiu Lakes and peripheral embayments were initiated in 1990, 4 yr before surge activity began, and were conducted each successive year through 2006, with the exception of surge years 1994–1995, when limnic turbidity was too high for sonar signal transmission. Collection of data in Tsivat Lake ceased at the end of the 1997 field season, when the entire basin filled with fine sediment issuing from fountaining vents along the ice front. Changes in water depth and basin morphology derived from bathymetric data were used to calculate rates of sedimentation.

PRE-SURGE FORELAND

Foreland Islands

The general eastern sector island topography is characterized by prominent, kilometer-scale streamlined hills depicting a direction of overriding ice from northwest to southeast. The relief of this drumlinoid topography is low (<15 m), and the intervening swales may contain elongate, closed basins holding shallow ponds. Dry gullies of all sizes are cut into bluffs on the flanks of all the islands. Most are rather small, but some of the larger gullies are typically 100–200 m wide at their mouths, have steep longitudinal gradients, and convey intermittent discharge during extended periods of rain.

Elongate, low-relief streamlined flutes of different sizes and extent commonly mantle virtually all surfaces, including drumlins. Most flutes originate at meter- to submeter-size surface boulders and extend in the direction of ice movement for a few tens of meters or less. They are seldom more than a few decimeters in relief and consist of relatively noncohesive diamicton. Features resembling miniature eskers are draped across flutes and tend to favor stoss slopes. Seldom more than 40–50 cm high, they are best seen in oblique light. They appear in clusters with trends that subtly favor the direction of ice movement but approach a random orientation. They tend to be less conspicuous than flutes, owing to lower relief and arbitrary orientation, and consist of moderately sorted to poorly sorted silt-rich pebbly gravel. Only rarely do they contain well-sorted pebbles and cobbles.

A curious aspect of all the islands is the steeply sloping bluffs that flank virtually all island margins. Yet neither ice-contact nor proglacial, subaerial streams were ever present to incise these bluffs, which suggests an alternative erosional process related to subglacial meltwater movement during tunnel valley formation (Wright, 1973; Brennand and Shaw, 1994; Patterson, 1994; Piotrowski, 1994; Clayton et al., 1999).

Ice-Contact Lake-Basin Morphology

Flanking Weeping Peat Island on the north and south are the basins of Tsivat and Tsiu Lakes, respectively. Although in close proximity as ice-contact basins, bathymetric surveys indicate they are morphologically different. The Tsivat Lake basin developed in the crescent-shaped void produced during retreat of the stagnant Tsivat Sublobe. It reaches pre-surge depths of 40–50 m along a 3 km ice front indented by fracture-controlled canals. Remnants of detached, grounded, and submerged ice blocks, held fast within rapidly accumulating lake mud, form a highly irregular basin floor of odd-shaped mounds, deep holes, and troughs (Franz and Fleisher, 1992; Fleisher et al., 1993; Fleisher et al., 2003). In contrast, the glacier appears draped across the trough of Tsiu Lake, where the ice front is oriented perpendicular to the elongate lake basin. Steep basin walls descend from adjacent island bluffs to a broad, flat basin floor that gradually increases in depth from 10 to 15 m at its eastern end to 60–65 m along an actively calving, kilometer-wide western ice front. This shape, orientation, and morphology suggest that the Tsiu Lake trough continues beneath the glacier.

Prior to the surge a narrow, ice-floored embayment of Tsiu Lake, which separated the retreating ice front from the western shore of Weeping Peat Island, was 20–30 m wide and only 3–4 m deep, except for one 40–50 m reach in which water depths plunged within a confined cavity 20–25 m deep. Curiously, it was from within this same cavity that massive fragmented ice blocks would on occasion suddenly rise above the water surface to stand an estimated 5–10 m in relief. Furthermore, meter-size rafts of frazil ice persisted here, thus indicating the venting of supercooled water.

Pre-Surge Limnic Thermal and Turbidity Properties

During the summers of 1988 through 1993, subaqueous vents were common along the Tsivat ice front, where rising plumes of highly turbid, supercooled water carried clusters of frazil ice (Fleisher et al., 1993). So pervasive were these conditions in July 1991 that a sediment trap placed at a near bottom depth of 19 m, ~50 m from the ice front, was within a few days buoyed to the surface by the growth of large frazil ice crystals. In contrast, frazil ice was never observed forming in Tsiu Lake. These conditions indicate that most of the supercooled water was discharged directly into Tsivat Lake from a subglacial conduit system that favored this location.

Fleisher et al. (1996a, 1993) report that during the 3 yr period preceding the surge, supercooled conditions prevailed in Tsivat Lake, with temperatures ranging from –0.1 to –0.7 °C throughout the water column below a 3–6 m thermocline. In contrast, Tsiu Lake water remained at 0.0 to + 0.4 °C at depths >2 m, did not support the formation of frazil ice, and was void of ice front vents. The common occurrence of highly turbid, supercooled water in Tsivat Lake indicates that significant amounts of subglacial meltwater were discharged directly from a conduit system that vented exclusively to the Tsivat basin at several places along the ice front.

The amount of suspended sediment (turbidity) for Tsivat and Tsiu Lakes was determined from water samples taken every summer in both lakes. Samples were pumped through 47 micron filters, which were then dried and weighed to yield grams per liter. From this it was determined that pre-surge values ranged between 1.5 g/L near the surface and 2.7 g/L at depth (Fleisher et al., 1996a).

SURGE CYCLICITY AT BERING GLACIER

Surge cyclicity at Bering Glacier, first recognized by Austin Post (1972), is well represented by historic accounts, aerial photography, and direct observation (Muller and Fleisher, 1995; Fleisher et al., 1995). The earliest reference to what may have involved surge activity is implied in the ship's log of George Vancouver in 1794, including offshore sketches that depict a steep glacier front broken by orthogonal crevasses, creating a pattern similar to city blocks (Lamb, 1984). A historic record provided by personal observations of Belcher in 1837, published in 1843 (Pierce and Winslow, 1979); Bradford Washburn's 1938 photographs (prior to a major surge that culminated ~1940); and U.S. Air Force photography in 1946 illustrate anomalous thickening that placed the leading edge of the Tsivat Sublobe near its Holocene terminal moraine at the head of the eastern outwash fan. Molnia and Post (1995) provide a thorough synopsis of Holocene events.

By 1950, Bering Glacier was again well into a recessional mode of active retreat. USGS aerial photographs document yet another limited surge event between 1957 and 1960, when the ice margin advanced 1–5 km along a 35 km ice front (Post, 1972). Intensive crevassing as a result of rapid extending flow was noted far into the Bagley Ice Field, 50 km up-ice of the terminus. The volume of ice displaced exceeded that observed in any previous surge (Meier and Post, 1969). Renewed surging between 1965 and 1967 resulted in further advances of as much as 4 km, again bringing the eastern sector ice margin to within a few kilometers of its Holocene maximum. This apparent surge cyclicity justified high expectations when, in 1980–1981, the ice mounted and covered much of Override Ridge, but the weak kinematic wave dissipated before reaching the ice front of the Piedmont Lobe. Post (1972) hypothesized that prominently distributed kinked debris bands on the Bering Piedmont Lobe are evidence of an inherent 25–30 yr surge cycle. By the early 1990s Post suggested that a surge was overdue and could be anticipated within the next few years (Austin Post, 1990, personal commun.). The first direct evidence of outburst flooding midway through a surge is well represented on USGS aerial photographs (2 September 1966). The resulting sandur remains a prominent landform between the Tsivat and Tsiu basins, east of Weeping Peat Island, where ground subsidence over buried ice blocks continued into the late 1990s.

SYNOPTIC CHRONOLOGY OF SURGE EVENTS

Phase One

The first phase of the surge began in late spring 1993 with the formation of a surface bulge above the snowline on the upper Bering trunk in the vicinity of the Bagley Ice Field. By early June the bulge had progressed to the upper Piedmont Lobe, where orthogonal crevasses bound snow-capped seracs hundreds of meters across. Within weeks the bulge spread at an estimated rate of 200 m/d to the central piedmont area, transforming a gently sloping, crevasse-free, low-relief surface into extensively crevassed domes separated by basins and troughs (Fig. 3). Rows of en echelon crevasses developed along the flanks of shifting zones of more active flow. By mid-summer many segments of the eastern ice front began to advance and thicken (Fleisher et al., 1993). Similar activity was also apparent along the massive southern front in Lake Vitus, where rapid advance compressed large icebergs generated by widespread calving (Molnia, 1993).

In mid-October the highly fractured and broken ice front advanced across and onto foreland islands at rates of <1–2 m/d (Fig. 4). By late October the glacier crossed and obstructed a river channel (Bentwood Narrows), connecting the eastern peripheral lake system with Lake Vitus along the southern ice front. Consequently, upstream lakes rose 14 m to reoccupy a river (Abandoned River) abandoned during retreat from the mid-1960s surge limit.

High Magnitude Outburst Floods

The first of three high-discharge outburst floods occurred the following summer on 27 July 1994, when pressurized subglacial water suddenly burst from the ice front along the north side of Weeping Peat Island. Coarse gravel and house-size blocks of ice were initially propelled outward several hundred meters, forming a chaotic deltaic sandur that grew to 1.5 km^2 into Tsivat Lake. Loss of conduit ceiling support led

Figure 3. Crevassed domes, troughs, and rows of en echelon crevasses.

to collapse, thus forming Icewall Canyon (Fig. 5). The northwestern segment of Weeping Peat Island directly adjacent to the outburst conduit was affected most by initial erosion. An area of 0.75 km^2 that stood 25–30 above Tsivat Lake, consisting of 18.8 million m^3, was completely removed and served as source material for the sandur. Logs and tree stumps up to 40–60 cm in diameter, and peat mats 10–20 cm thick derived from a subglacial source, were incorporated in the sandur. Bathymetric surveys indicate that a minimum of 74.8 million m^3 of sediment was discharged during the initial outburst and from vents that remained active during the following 2 yr. Therefore, the difference (56 million m^3 or 75%) must have come from beneath the glacier (Fleisher et al., 2003).

Within hours of the initial outburst into Tsivat basin, continued failure of the subglacial conduit ceiling and walls quickly formed a 250-m-wide, 70-m-deep ice-walled canyon that grew headward (northwestward) along the northern side of Weeping Peat Island. An initial rate of headward growth (150–200 m/d) brought the canyon head to a position directly over an open vent 600 m behind the ice front. Vent activity then shifted to favor continued canyon growth for an additional 400 m parallel to the ice front. Headward growth ended in vertical canyon walls above a second open vent beneath a narrow ice bridge separating the canyon headwall from a collapsed window on the glacier surface, which had enlarged to hundreds of meters in diameter. This sequence of headward migration from one vent to another produced an angular-shaped canyon that extended up-glacier from the ice front (600 m), then turned parallel to the ice front (400 m), effectively separating a rectangular (300 m by 200 m) segment of the terminus from further surge activity, thus stranding it on Weeping Peat Island.

Discharge persisted for 2–3 d (Austin Post, 1994, personal commun.) then gradually diminished to normal flood conditions visible on USGS aerial photos taken 7 September 1994 (Fig. 5) before finally ending within the weeks that followed. On 18 October, after discharge had ceased, the icy floor of the canyon was observed covered with a discontinuous mantle of coarse gravel and a cluster of stranded, house-size ice blocks. Curiously, the blocks were not simply large chunks of lag glacial ice but rather consisted of an ice breccia cemented by frazil ice, thus indicating fragmentation by collapse followed by partial refreezing in a supercooled matrix.

Sometime between 7 September 1994 (date of USGS aerial photos) and field observations on 18–21 October 1994, subglacial flood discharge shifted to an outlet centered on Weeping Peat Island (Fig. 5). Although lesser in magnitude than the initial outburst, discharge was sufficiently forceful to breach a low confining ridge, cross a mid-island lowland, and move floodwater and 2–3 m ice blocks down a preexisting gully into Tsiu Lake. Extensive scouring of the terrain and subsequent gravel deposition formed a new sandur (100–175 m wide) that prograded ~100 m into Tsiu Lake. Simultaneously, floodwaters coursed along the broken ice front, undermining ice blocks too large to be moved, scouring a flight of terraces, and depositing a massive sandur (Fig. 6). During the year that followed, the Weeping Peat Island ice front leaked meltwater from a series of persistent channels that breached the actively forming push moraine.

Evidence for yet a third outburst was concealed beneath the glacier and remained undetected until retreat, nearly a decade after the surge ended, uncovering an assemblage of landforms directly behind the push moraine bordering the Riverhead Sandur. Unlike the previous outbursts, pressurized subglacial meltwater remained confined beneath the glacier within a conduit oriented parallel to the ice front. Water movement was away from the mid-island outburst site and toward Tsiu Lake basin. Consequently, the subglacial island surface was hydraulically scoured, forming a closed basin from which material was transported laterally toward Tsiu Lake and deposited beneath the ice at the island margin.

Phase Two

The second phase of the surge commenced immediately following the flood across Weeping Peat Island. Ice front movement

June 4, 1993

October 20, 1993

November 21, 1993

Figure 4. Time-lapse photo sequence. This three-image sequence shows initial changes observed from a photo station on Weeping Peat Island during the onset of surging in 1993. All images are of the identical scene as viewed from a photographic survey station on Weeping Peat Island 800 m from the pre-surge ice front, with a northwest view. A reference boulder is circled on each image. The initial 4 June 1993 image shows a heavily crevassed surface bulge on the main Bering trunk as it emerges from behind the western Grindle Hills 12 km away. The Bering surface slopes gently toward the debris-covered terminus, reflected in an ice-contact water body. The second image on 20 October repeats the scene four and a half months later. This view depicts an ice front advance of ~200 m and the Piedmont Lobe thickened by 50–75 m, based on triangulation calculations related to partially obscured views on the horizon. The third image on 21 November one month later shows that the glacier thickened rapidly during the initial phase of the surge. The terminus consists of ice blocks at the angle of repose after advancing an additional 200 m and thickened a total of 125–150 m. This photographic station was overridden four months later in March 1994.

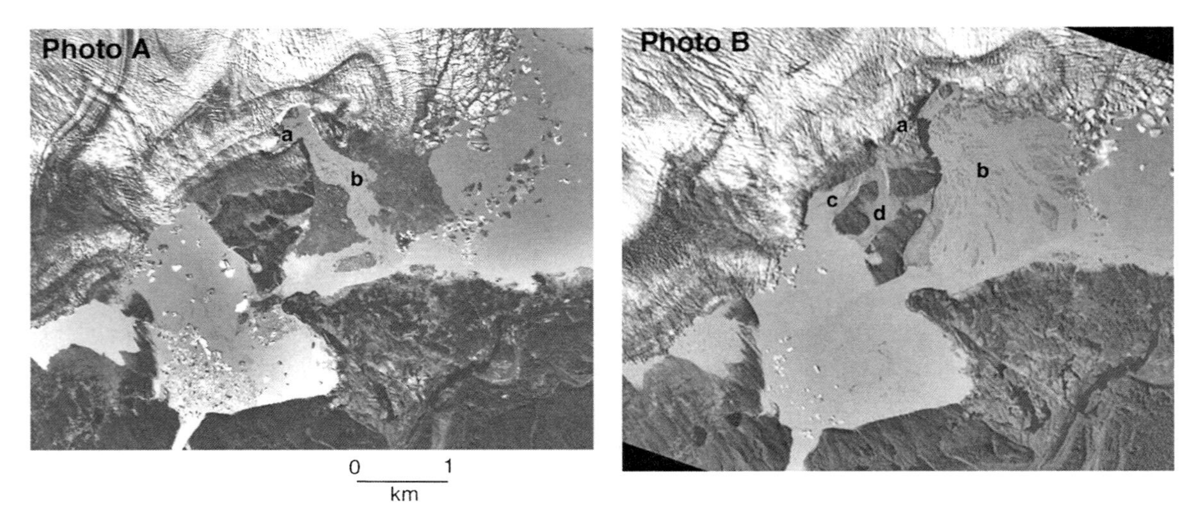

Figure 5. USGS aerial photographs (94V4–110 and 95V2–210), illustrating initial impacts of outburst floods. (A) 7 September 1994: First outburst formed Icewall Canyon (a) and deposited Icewall Sandur (b) on the north flank of Weeping Peat Island. (B) 27 September 1995: Icewall Canyon is closed (a) and an enlarged Icewall Sandur (b), plus the effects of a second outburst centered on Weeping Peat Island that formed Riverhead Sandur (c) and Splitlake Sandar (d).

Figure 6. Leaky ice front along Riverhead Sandur. Oblique aerial view of leaky Riverhead Sandur and ice front, and stranded ice blocks transported by outburst flood. Note ice front indentations, indicating areas of higher discharge.

along the eastern sector was sporadic and discontinuous, fluctuating from dormancy to a slow 1–2 m/d. This sluggish and intermittent movement gradually pressed the ice front a few tens of meters forward, which was enough to close Icewall Canyon in 1995 and construct a prominent push moraine (Fig. 5). Although the push moraine was breached by several leaky meltwater channels that remained active into the 1996 summer, little accumulation is evident at breach sites, and the moraine retained a relief of 8–10 m above the adjacent sandur, thus marking the 1993–1995 surge limit on Weeping Peat Island.

RATES OF ADVANCE

Measurements of ice front advance at four ground-based survey stations on Weeping Peat Island and two on Bentwood Island commenced in September 1993 (Fleisher et al., 1995) and continued throughout the surge until late July 1995. Supplementing these are measurements from a series of USGS aerial photographs that spans the entire surge event, as well as post-surge images by the BLM.

Indications of ice front advance on the eastern sector appeared first on the stoss side of Peat Falls Island, where aerial photographs dated 10 July 1993 show newly formed crevasses in a thickening ice front. Ice front retreat continued everywhere else along the entire eastern foreland into early September, except on Peat Falls Island, where in late September the ice front was advancing at a rate of 2.5 m/d.

Ice front advance was in progress at all eastern sector islands by mid-October and accelerated into mid-November. During this time, rates varied from 5.5 and 5.9 m/d on Weeping Peat Island to 7.1 m/d on Arrowhead Island and 7.4 m/d on Peat Falls Island.

Rates of advance increased significantly at every reference station throughout the fall of 1993, with the greatest contrast on Arrowhead Island, where earlier rapid retreat from a developing stoss-side ice-contact lake ceased and advance began at a rate of 6.7 m/d. Similar changes were observed in November on Bentwood Island, where advancing ice closed Bentwood Narrows. Measurements from all stations fluctuated on a weekly basis, with some temporarily slowing while others increased. The 3 mo. average for the entire eastern sector during the 1993 fall season was 4.0 m/d. From November 1993 through March 1994, rates of ice front advance varied from 1.0 to 7.4 m/d, fluctuating from place to place, with greatest rates approaching 10 m/d during the 1994 winter months at Bentwood Narrows (Fleisher et al., 1994b, 1995; Rosenfeld et al., 1994).

Rates for spring and early summer 1994 were derived from a combination of USGS aerial photographs and 35 mm oblique images obtained during reconnaissance flights in March and early June. Spring rates declined on Weeping Peat Island, but not uniformly, slowing from 6.6 to 1.5 m/d and from 4.7 to 4.0 m/d at two survey stations that were not yet overridden. Conversely, rates increased slightly on Bentwood Island from 2.8 to 3.0 m/d. June and July measurements at ground-based stations during the 1994 summer were made at intervals of 1–4 d during an 18-d period commencing on 4 June 1994. These, plus mid-summer measurements on July 20, immediately prior to the initial outburst flood, depicted a general trend for eastern sector ice front activity.

At one of the Bentwood stations, rates of advance increased steadily from 1.47 to 2.55 m/d through mid-June, then declined gradually to 1.23 m/d by late June. A single July measurement yielded a 28-d average rate of 1.02 m/d. The second Bentwood station also showed a slowing advance in mid-June from a maximum of 2.20 m/d to a minimum of 1.56 m/d, followed by a decline to a 0.42 m/d average in July. Of the four survey stations on Weeping Peat Island, the most detailed record shows rates that gradually increased from 0.19 m/d in early June to a high of 2.97 m/d in mid-June, then decreased to 1.48 m/d by late June. This 16-d rise and decline matches the trend on Bentwood Island.

This progressive increase, followed by a gradual decline, is representative of activity at other survey stations. The same trends were observed at other stations, but the timing differed from station to station. The most uniform rates were from central Weeping Peat Island, where 1.08 m/d in early June increased to 1.70 m/d by mid-June, then declined slightly to 1.36 to 1.40 m/d in late June. Of all stations monitored, the one centered on Weeping Peat Island showed the slowest surge activity, with rates below 0.41 m/d during the entire 1994 summer (Fleisher, et al., 1995). This is thought to be due to juxtaposition with Icewall Canyon, which served to separate part of the ice front from active flow.

The next opportunity to make field measurements came in mid-October, 3 mo. after the initial outburst flood of 27 July 1994. By October the ice front lacked a newly broken appearance, was significantly less steep, and was in retreat on Bentwood Island, where it occupied a position 13 m from a small push moraine. Similarly, retreat on Weeping Peat Island placed the ice front 30 m from the push moraine. Projecting the late July average rate of advance from the last measured position places the ice against these moraines, which coincides within days to the initial outburst event.

SUBGLACIAL CONDUIT AND VENT SYSTEM

Conduit System

Landforms and drainageways that emerged during decades of retreat from the 1965–1967 surge maximum in the eastern sector show that the primary location of meltwater escape is from a subglacial conduit system between the main Piedmont Lobe and the Tsivat Sublobe (Fleisher et al., 1998). This well-established system is represented on the glacier surface by a linear sag that extends several kilometers up-glacier between the northwestern side of Weeping Peat Island and the glacier draped over Peat Falls Island. USGS aerial photos (2 September 1966) of the ice front during the 1965–1967 surge show flood conditions in progress along the same trend, including the development of an ice-walled canyon. A semi-continuous record of USGS aerial photos between the 1965–1967 and 1993–1995 surges shows that subglacial discharge persisted along this trend during the entire retreat phase (Fleisher et al., 1998). Indeed, the initial outburst flood in July 1994 originated from an ice front position along this trend. Although bedrock is not exposed anywhere along the proximal eastern foreland, Fleisher et al. (1998) question the ability of an unconsolidated substrate to persist in the same location after enduring decades of continuous high-magnitude discharge. From this they speculate on a subsurface bedrock influence.

Discharge measurements of open-channel surface flow into and out of Tsivat Lake established that non-glacier-sourced water accounts for <2.5% of the total water balance (Fleisher et al., 1998). This indicates that virtually all of the water entering Tsivat Lake is from glacial sources, which are mainly ice front vents. Flow from Tsivat Lake to Tsiu Lake consistently conveys significantly less discharge than the Tsiu Outlet channel (Fleisher et al., 2003), which indicates that Tsiu Lake must have an additional inflow source. Considering the lack of foreland sources, and the relatively small input and clean nature of a supraglacial source, the only possible alternate source for the ubiquitous turbid water must be an active subglacial conduit, although no ice front vents are apparent.

Vent System

The location of upwelling vents and the chronology of their activity are summarized in Figure 7. With the onset of surge activity, vent areas became poorly defined, and much of the ice front on Weeping Peat and Bentwood Islands leaked turbid meltwater at many low-level discharge sites. Such was the case on the Riverhead ice front, where most meltwater was released from three

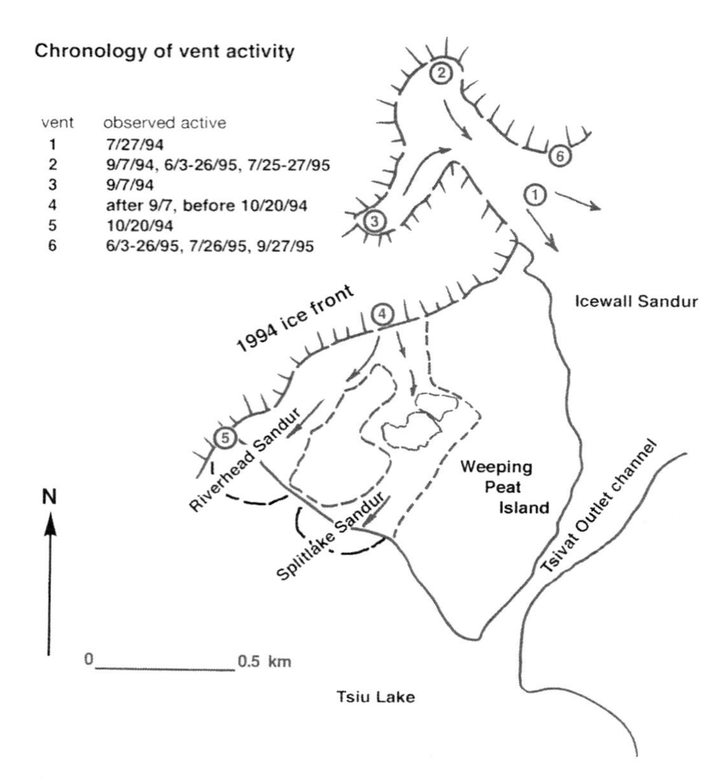

Figure 7. Vent locations and chronology of activity. Numerical sequence of vent activity is in chronological order, beginning with the 27 July 1994 outburst flood forming Icewall Canyon (1) and ending with vents migrating northward along the Tsivat ice front (6).

low-discharge channels that breached the push moraine even as it was forming. The initial July 1994 outburst flood was fed by a primary vent along the well-established subglacial Tsivat Basin Conduit System (Fleisher et al., 1998; Fleisher et al., 1996b). Collapse of the conduit ceiling during the outburst flood led to the formation of an ice-walled canyon. After the canyon grew headward from the ice front toward this vent, discharge shifted laterally to the south, and the canyon grew in that direction, thus initiating a dogleg turn shown in Figure 5A. By mid-October the main discharge had shifted once again, this time to the second outburst site centered on Weeping Peat Island, resulting in the formation of Splitlake and Riverhead Sandar (Fig. 5B). One year later, toward the end of the surge, vent activity shifted back to the primary site above the main conduit system on the north side of Weeping Peat Island (Fig. 7). During early post-surge years, vents began to shift back to their pre-surge locations as they progressively migrated northward along the Tsivat ice front. By 2005, glacier thinning and retreat from Weeping Peat Island was accompanied by renewed activity from fountaining vents that provided discharge to an ice-contact stream developing on the northwestern side of Weeping Peat Island. In 2007, vent-fed discharge was sufficiently great to form standing waves in a 30–50-m-wide channel that grew daily by bank failure.

RETREAT AND SURFACE MELTING

Retreat

Annual GPS mapping of ice front retreat along the entire eastern sector during post-surge years is illustrated in Figure 8. The database from which rates of retreat are derived required island traverses on foot and boat traverses on ice-contact water bodies. Factors other than weather and climate that influence rate of retreat on land include orientation of the surface from which the ice retreats (stoss or lee-side slopes) and whether or not an ice-contact stream flows along the glacier snout. The generalization derived from these conditions is that retreat is slowest on gently inclined lee slopes where the ice front is relatively dry. Conversely, retreat is most rapid from wet, stoss slopes that support ice-contact, low discharge runoff. The rate of retreat from ice-contact lakes is primarily controlled by factors that influence iceberg calving, such as more rapid calving of thin, highly fractured ice in deep water (Post, 1997). Measured rates of annual retreat on the island segments of the foreland range between 50 and 75 m/a. Calving retreat in ice-contact lakes varies significantly from a few hundred to several hundred meters per year.

Surface Melting

Diurnal Rates of Melting

Daily rates of ice surface lowering (downwasting) were measured at three readily accessible sites ~1 km up-glacier from the terminus. Rate was determined by measuring the progress exposure of 4-m-long PVC pipes placed vertically within the glacier surface at each site. Diurnal rates of ice surface lowering were monitored by repeatedly measuring the amount of exposed pipe every few days, and the pipes were periodically reset during a 3-wk period in June. This method typically yielded rates between 5.5 and 8.5 cm/d, well within a multiyear database of 4.4–10.9 cm/d.

Variations in rates of melting can be directly correlated with ambient ice surface temperature and intensity and duration of precipitation, but wind direction is also an important factor. Wind from an ice-free direction caused increased melting, whereas wind movement off the glacier had the opposite effect (Dworak and Fleisher, 2003).

Annual Rates of Melting

The amount of annual ice surface lowering is derived from multiyear ice surface surveys extending 1 km up-glacier from the terminus. A 9 yr record of survey results is presented in Figure 9. This information is combined with GPS mapping surveys to establish a database for ice front positions, thus yielding a dual approach to measuring rate of retreat.

These profiles indicate that the annual amount of surface lowering of relatively clean ice and ice with widely dispersed surface debris averages ~10 m/a. The lowest segments of the profiles show irregularities that coincide with a push moraine at the

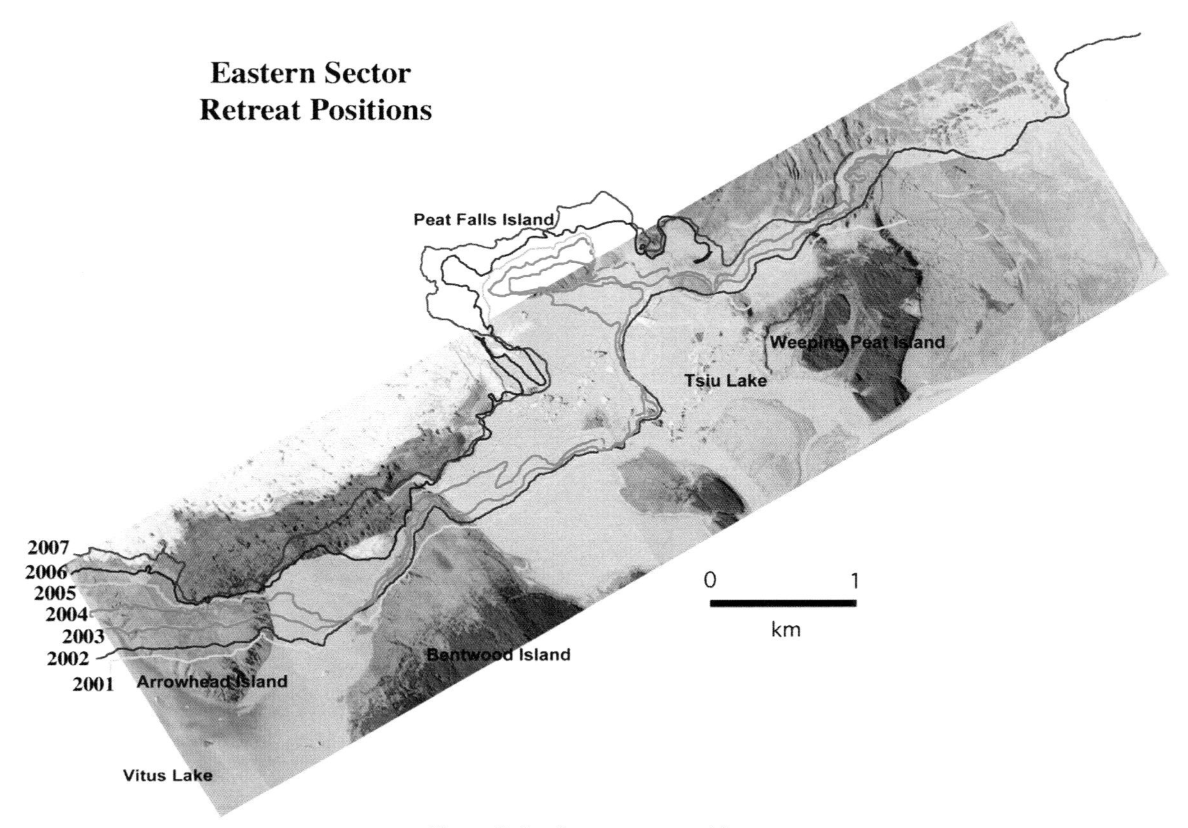

Figure 8. Ice front retreat positions.

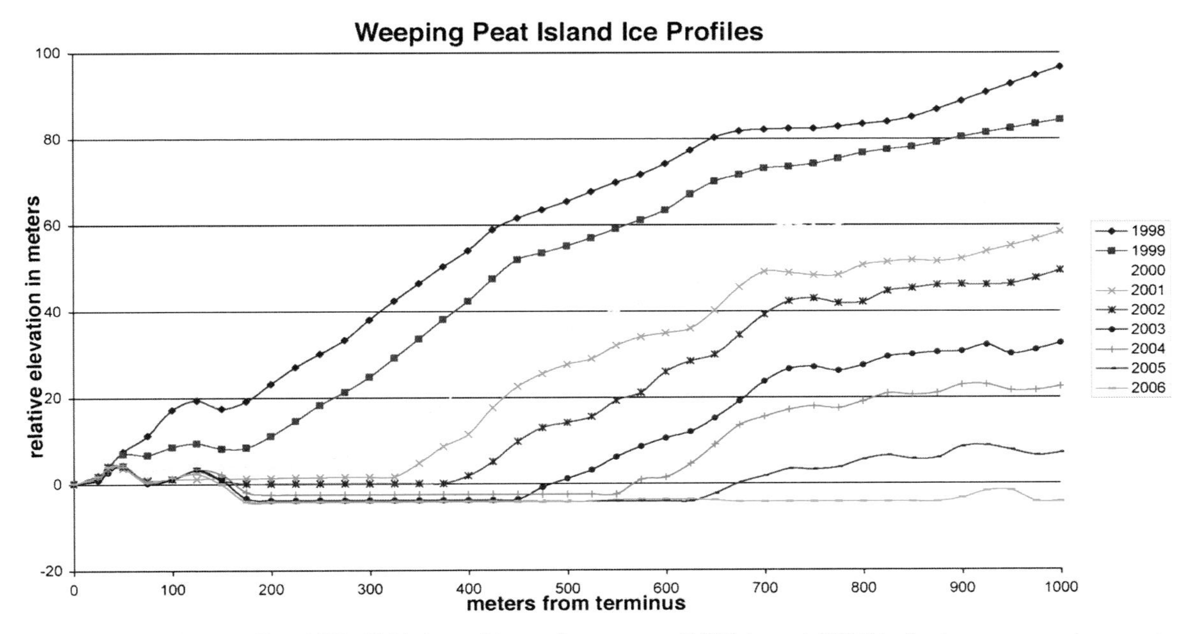

Figure 9. Ice surface profiles, 1998–2006. Annual ice surface surveys (1998 through 2006) indicate an average downwasting rate of ~10 m/a for the area within 1 km of Weeping Peat Island, and a rate of retreat that initially was retarded by the cover of a debris band, and then generally varied from 50 to 75 m/a. The flat segments of the profiles correspond to an ice-contact pond.

surge limit and an ice-marginal debris band. The flat segments represent a small ice-contact lake that developed during retreat, eventually as an embayment of Tsiu Lake.

A rate of annual downwasting on a debris band was derived from surface surveys that extend from a foreland benchmark several hundred meters up-glacier to a location on a debris band. Two consecutive annual surveys indicate that debris band downwasting takes place at a rate of ~8 m/a and that half of this occurs during June and July (Fleisher et al., 2006a).

SURGE-RELATED MODIFICATION OF ICE-CONTACT LAKE BASINS

Initial Effects on Lake Basins

During the full 2 yr span of the surge, the glacier advanced 2 km into Tsiu Lake, whereas surge energy was delayed to the more peripheral Tsivat Sublobe, thus resulting in a limited total advance of only 1 km. The closing of Bentwood Narrows and the subsequent 14 m rise in upstream water bodies caused lowland areas to be inundated. This was the case on Bentwood Island, where Bentwood Lake formed as a separate water body that ultimately became an embayment of Tsiu Lake during the ice front retreat. An embayment on Weeping Peat Island left high and dry by the 1989 fall of lake level, and many other low shoreline areas, were once again inundated. Tsiu Lake rose sufficiently to reach the threshold of Abandoned River, which was then reoccupied as the outlet for Tsiu Lake.

One year after the surge began the outburst flood from Icewall Canyon filled the southern one-third of Tsivat basin with 74.8 million m^3 of ice blocks and gravel (Fleisher et al., 2003), thus forming Icewall Sandur (Fig. 5). Included were large segments of trees, tree stumps, mattes of peat, and abundant fragments of littoral shells (Pasch et al., this volume). Within that portion of Tsivat basin unaffected by outburst deposition, previously existing, pre-surge, bottom irregularities with relief of 10–25 m at depths of 15–30 m were only partially masked by accumulating silt and sand. Two years after the surge ended in June 1997, suspended sediment and discharge from a series of ice front vents near the center of the Tsivat Sublobe abruptly increased, leading to the active growth of a delta in 16–18m of water. Daily increments of rapid progradation could be monitored as fast-moving supercooled water crossing the delta top, which then slowed at the leading edge, allowing frazil ice rafts to form and accumulate. When next observed in June 1998, this delta had grown to fill the entire basin with sand and silt, thus transforming the previously existing ice-contact lake into a broad sand plain across which braided channels meandered. With the loss of Tsivat basin as a sink for sedimentation, much of the suspended sediment (sand and silt) was bypassed into Tsiu Lake, thus accelerating the growth of a delta that progressively grew across Tsiu Lake basin toward the lake outlet (Fig. 5). Eventually, distributary flow shifted delta growth toward the unoccupied part of the basin, thus causing progradation toward the retreating ice front (Fleisher et al., 1998).

The 1995 limit of ice advance on Weeping Peat Island is represented by a prominent push moraine along the western flank of Riverhead Sandur. Immediately following the surge, a bathymetric survey showed that this moraine could be traced across the floor of Tsiu Lake where it connects with a similar feature on Bentwood Island (Fleisher et al., 2003). However, post-surge lacustrine sedimentation, combined with Tsiu Lake delta growth, eventually eliminated the expression of this moraine on the lake floor. No such moraine was found in Tsivat basin, where the ice front advance was limited to plowing into and compressing ice islands and grounded bergs. Retarded ice front advance, combined with rapid sedimentation related to highly turbid subglacial vent discharge, precluded moraine formation and/or masked its expression. Sedimentation remained stable for a full decade after the surge as accumulation rates returned to pre-surge values.

Water Temperature and Turbidity

Although ice-front advance during the first year of the surge brought about the most obvious changes, there were also subtle yet significant changes in the physical properties of the ice-contact lake water. Billowing plumbs of turbid water along the Tsivat ice front, common prior to the surge and indicative of discharge from submerged subglacial vents, was noticeably subdued during the surge, as was the associated occurrence of frazil ice rafts. Furthermore, supercooled water conditions within Tsivat Lake, common prior to the surge, no longer existed, whereas the temperature of Tsiu Lake simultaneously declined, and local pockets of sub-zero conditions were recorded. A subtle, yet recognizable thermal shift was in progress as the main conduit system that had consistently sustained supercooled conditions in Tsivat Lake temporarily redirected subglacial water into Tsiu basin.

Concurrently, the amount of suspended sediment within both lakes increased. Prior to the surge, values ranged between 1.5 g/L to 2.7 g/L throughout the water column, whereas during the first year of the surge turbidity spiked by five- to sixfold (Fleisher et al., 1996a; Fleisher et al., 1998; Merrand and Hallet, 1996). This is attributed to increased access to subglacial sediment from a broader distribution of water at the base of the glacier, which is considered essential to facilitate surge activity (Humphrey, et al., 1986).

Lake Level Fluctuations

Rare meter-scale changes are known to occur, such as a 6 h rise of 1 m in June 1991 owing to a rapid influx from an upglacier source of unconfirmed origin. In addition, a strandline etched on the flanks of the 1995 push moraine records a 6.5 m rise in Tsiu Lake above the normal mid-summer position owing to a severe coastal storm in late September 1995. Atypical warm weather in June–July 2004 and 2005 led to an approximate 2 m rise in Tsiu Lake and related water bodies. Conversely, annual fall and winter reductions in meltwater typically cause a lowering of 3.5–4.0 m.

Two substantial fluctuations in lake level occurred during the span of a single surge cycle. The first was in 1989, when retreat from Bentwood Narrows opened a lower, ice-contact outlet, and lakes upstream dropped 17 m (referred to colloquially as a *breakout*). This was followed 14 yr later by the 1993 surge that led to closure of the same outlet, causing an associated rise of 14 m. Although closure at both times (pre-1989 and post-1993) can be attributed to ice blocking the same outlet, the associated changes in lake level were not the same. Perhaps glacial ice related to the initial blockage was higher than the next, or remnant ice played a role in modifying the height of the alternate outlet channel (i.e., Abandoned River). In either case the drop in water level significantly altered the foreland landscape.

Relatively stable lake levels prevailed after outlet closure in 1993, with only meter-scale fluctuations influenced by seasonal changes in available meltwater. In August 2006, downwasting at Bentwood Narrows brought the glacier surface low enough to allow the Bentwood embayment of Tsiu Lake (e.g., Bentwood Lake) to encroach onto the glacier, where it entered a supraglacial cavity leading to englacial passageways. Water movement on and through the glacier led to enlargement of all passageways during a month-long, progressive 14 m decline in upstream water bodies, bringing them to near equilibrium with Lake Vitus. This was sufficient to vacate the channel of Abandoned River, thus leaving it high and dry once again (Fleisher et al., 2006a). The 2006 breakout and accompanying drop in base level caused incision of previously submerged, surge-related landforms, thus exposing a more thorough expression of the depositional effects of the surge.

OVERRIDDEN TERRAIN

Foreland Till

The surface deposits of foreland islands at Bering Glacier are similar to those reported in the ice-marginal areas of other large, lobate, retreating ice masses, some of which have surged. The foreland surface of some Iceland glaciers, namely Briedamerkurjökull (Price, 1969), Sidujökull (Kozarski and Szupryczynski, 1973), and Skeidararjökull (Galon, 1973; Klimek, 1973; Maizels, 1995; Russell et al., 2006) all have a close likeness to the Bering foreland landforms and deposits. At Bering Glacier the thickness of the 1993–1995 till ranges from as little as 10–15 cm to the less common 0.5–1.0 m, except where locally zoned till of questionable origin may reach several meters (Fleisher et al., 2006b). Resting directly upon well sorted and stratified sand and gravel outwash, the till contains rounded pebbles held firmly in a sandy matrix with a trace of silt and is fairly firm, resembling lodgment till. Flutes consisting of the same till are very common and are commonly attached to the downglacier side of submeter lodged boulders. At the time of deposition, however, the overriding ice lacked basal debris, yet a surge-related till is ubiquitous, which suggests that the till was formed from preexisting drift that was mobilized and redeposited, thus qualifying as a deformation till (Fleisher et al., 2006b).

Retreat from the strongly drumlinized surface of Bentwood Island exposed two sub-kilometer shallow (1–2 m) ponds occupying the elongate swales between drumlins. These features and adjacent gullies with decameter relief existed prior to the 1993–1995 surge, and upon retreat emerged from beneath the ice with the same form, orientation, and dimensions as before the surge, thus seemingly unaltered by the overriding ice. Similarly, other prominent gullies on Bentwood and Peat Falls Islands have also survived the surge without significant modification, which raises the question: How much erosional effect did the overriding ice have on the unconsolidated substrate and landforms? Gully wall exposures along all island flanks provide multiple cross-sectional views of the new surge-related deformation till mantling truncated beds of the undeformed outwash substrate, thus indicating limited shallow erosional effects of overriding ice during drumlin formation (Fleisher et al., 2004).

A uniquely different effect of overriding ice is demonstrated by deformed trees in the substrate of Weeping Peat Island within a consistently well-exposed bluff bordering the Tsivat Outlet channel (Fleisher et al., 2006b; Fleisher et al., 2002). Here, a 12-m-thick sequence of outwash gravel is interbedded with four separate, meter-scale lacustrine sand beds, each of which accumulated in a shallow, ephemeral lake that inundated a foreland outwash surface that had been sufficiently stable to accommodate various plants, including small alder and young spruce trees. Thus, the sand beds cover centimeter-scale basal peat, mantled by a thin (millimeter scale) veneer of deformed clay that grades upward into silt and sand. The trees remain in the living position, their base buried in sand and rooted in gravel below. Of particular significance is that all trees <20 cm in diameter show evidence of deformation. Some trunks are bent near their base, whereas others are offset laterally along horizontal zones of shearing confined to the base of each sand bed. Although the trees remain firmly rooted in gravel, they are bowed, sheared, and displaced 15–20 cm in the down-glacier direction. Equally significant is the depth at which this deformation occurred—3–7 m below the overridden surface.

Push Moraine

Marking the limit of surging advance is a push moraine of different dimensions and material from place to place. As the advancing glacier pressed forward onto the islands, a zone of saturated substrate became increasingly obvious at the base of the glacier. Overburden pressure squeezed this material to the leading edge of the ice, where it was constantly overridden by the sliding ice above. This semi-continuous, subglacial cushion of saturated substrate at the sole of clean, basal ice contained a tangle of uprooted and flattened alder. Occasional hesitations in ice front advance accommodated the development of a small, meter-scale mound, thus forming a push moraine, which would be overridden with renewed advance. As the glacier reached the surge limit, the push moraine grew in size and lateral extent, finally reaching a seldom exceeded 2–3 m relief. One exception is a prominent segment of the moraine along the western flank of the Riverhead Sandur (thus named the *Riverhead Push Moraine*) where relief reaches 8–10 m for a distance of several hundred meters from the shore of Tsiu Lake to the Riverhead outburst site midway across Weeping Peat Island. Here the moraine consists of sandur gravel that was literally pushed into position. The moraine is breached in several places by meter-scale meltwater channels that correspond to the leaky segments of the ice front (Fig. 6). A bathymetric survey of Tsiu Lake during the initial retreat phase showed a continuation of the moraine across the bottom of the lake to a shoreline position corresponding to a similar, although smaller, push moraine on Bentwood Island (Fleisher et al., 2003).

At the culmination of the surge, a prominent debris band occupied a position parallel to and immediately up-glacier from the Riverhead Push Moraine. Consequently, upon retreat the supraglacial material of the debris band retarded downwasting, thus causing the area behind the moraine to retain an ice core. With gradual melting of remnant ice, a conspicuous 80–100-m-wide belt of collapse topography developed behind the moraine. The highly angular surficial, ablation debris here ranges in size from very fine, subpebble clasts up to boulders several meters in diameter. A longitudinal topographic survey along the trend of this terrain shows that the normal hummocky relief is interrupted by three uniformly spaced crossing channels, leading to gaps in the moraine. Wood fragments, tree stumps, and splintered segments of tree trunks appear most frequently as lag within these crossing channels.

EROSION AND DEPOSITION RELATED TO OUTBURST FLOODS

Nova and Echo Basins

Retreat and downwasting from the 1995 surge limit exposed an assemblage of landforms that did not exist prior to the surge. Included are two sub-kilometer basins (Nova and Echo Basins) and their bordering bluffs, an esker between the basins, and englacial and subglacial outburst deposits. Of primary significance is the location of these features relative to outburst sites (Fig. 10).

Upon initial exposure from beneath the retreating ice in 1999, Nova Basin and Echo Basin emerged fully formed and independent of each other. They first appeared as ice-contact ponds, but as retreat progressed an inlet from Tsiu Lake developed, permitting boat access. Bathymetric surveys in 2002 showed that adjacent bluffs were subaerial extensions of the basin walls. At no time during the retreat phase did subaerial, ice-marginal, or proglacial streams flow adjacent to or around these basins. This suggests that Nova and Echo Basins, and their associated bluffs, emerged fully formed from beneath the glacier.

The extent to which the surface of Weeping Peat Island was scoured to form Nova Basin was calculated by comparing pre- and post-surge island surface elevations. A pre-surge (June 1991) topographic survey crossing the terrain into which Nova Basin formed used the surface of Tsiu Lake as a datum (Fig. 11). Ice front advance during the surge closed the Tsiu Lake outlet to Lake Vitus at Bentwood Narrows in November 1993, thus forcing Tsiu Lake and all upstream water bodies to rise 14 m. A GPS-controlled bathymetric survey of Nova Basin (June 2003) is tied to the post-surge lake surface datum. The post-surge 2003 bathymetric 3-D diagram depicts a maximum basin depth of 20 m. Superimposing the bathymetry on the pre-surge topographic profile and adjusting for surge-related lake level changes indicate that the bottom of Nova Basin is 14 m lower than the pre-surge island surface. This indicates that a scouring mechanism was active beneath the glacier. Juxtaposition with outburst sites suggests that scouring occurred during outburst events.

A southern extension of the linear band of collapsed topography (the Nova Kame Belt) between Nova Basin and the Riverhead Push Moraine forms a broad, lobate deposit that protrudes 100 m into Tsiu Lake beyond the pre-surge shoreline (Boatland) (Fig. 10). This hummocky landform, consisting of unsorted gravel littered with angular boulders resembling supraglacial debris, is capped by a discontinuous veneer of lake silt. A faint linear topographic fabric aligned with the direction of ice movement suggests that it was overridden. Its position relative to the push moraine indicates formation beneath the glacier, and its silt cap reflects a seasonally higher lake level.

Outburst Esker

Ice front retreat from the Riverhead Push Moraine (1996–1998) initially revealed a broad-crested, elongate mound, oriented with an apparent east-west trend, separating Echo Basin to the north from Nova Basin to the south (Fig. 10). This feature is situated in the source area for the Riverhead outburst and is aligned with the ice surface window noted during the surge, which implies a connection to an englacial and/or subglacial conduit system. First to be exposed at the eastern end of the mound was an unstable, subsiding surface mantled entirely with an anomalous matrix-free, well-sorted and well-rounded, clast-supported cobble gravel distributed across the full width of the mound (40–50 m). Conspicuous in appearance, this feature bears no resemblance to any stratigraphic units known to exist within Weeping Peat Island prior to the surge (Fleisher et al., 1998). Continued retreat and further exposure uncovered an ice-cored northern slope disturbed in part by meter-scale, en echelon slump scarps on the upper bluff descending into Echo Basin. In contrast, the opposite southern slope adjacent to Nova Basin appeared ice free. The linear symmetry of the broad crest contains subtle flutes oriented parallel to the dominant ice flow direction. When finally fully exposed by retreat in 2004, the mound was clearly recognized to have a form unlike the pre-surge landscape and consists of material atypical of Weeping Peat outwash. Although it lacks the classic sinuous form of an esker, alignment with the Riverhead outburst site combined with dimensions consistent with an outburst feeder tunnel (200 m long, 75–100 m wide, and 26 m in relief above

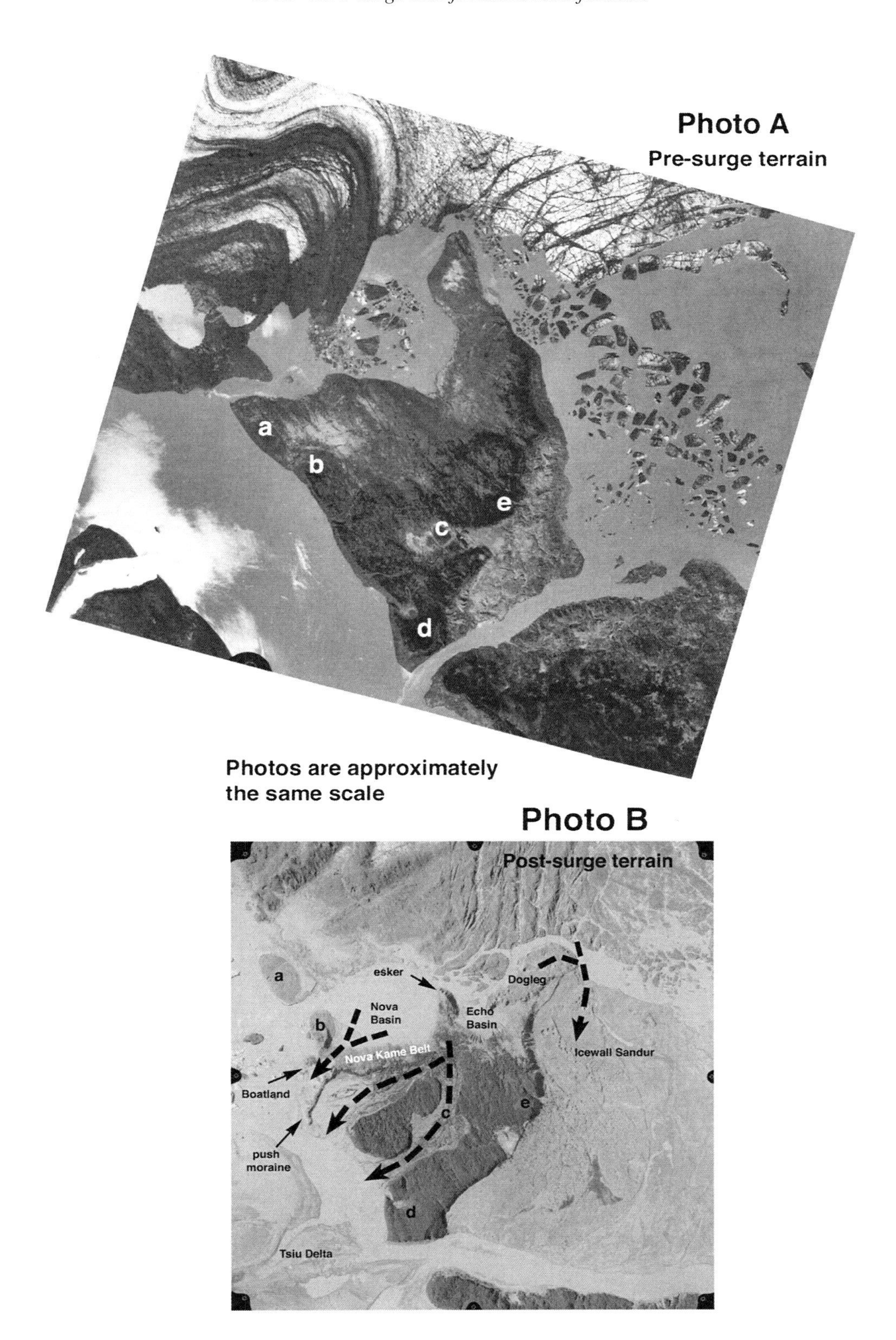

Figure 10. Reference images for pre- and post-surge landforms on Weeping Peat Island. (A) Pre-surge index to post-surge landforms (USGS aerial photograph 93V2–189, 10 July 1993). (B) Prominent outburst landforms (BLM aerial photograph R9, FL4, FR 1053, 25 September 2006). Arrows start at outburst sites and indicate direction of water movement.

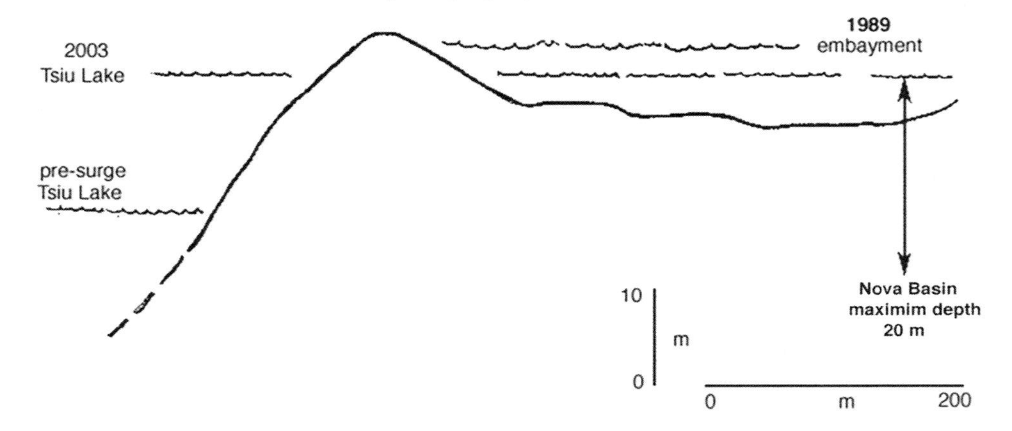

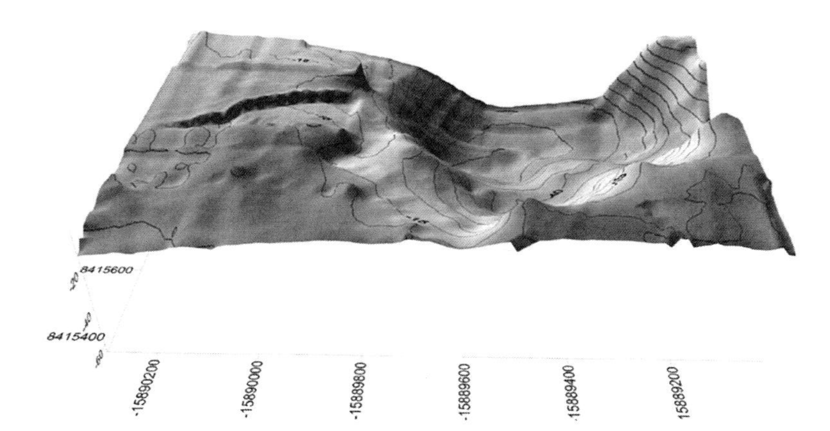

Figure 11. Pre-surge topographic cross-section of Weeping Peat Island and 2003 bathymetric 3-D diagram of Nova Basin. The topographic crosssection passes through an area of Weeping Peat Island into which Nova Basin was eroded. The pre-surge land surface is linked to the water level of Tsiu Lake, which rose when the advancing ice closed Bentwood Narrows. A 2003 bathymetric survey of Nova Basin indicates a basin depth of 20 m that projects a minimum of 14 m below the pre-surge surface. The 2003 bathymetric 3-D diagram illustrates the morphology of Nova Basin.

the adjacent water bodies), this feature is considered to be an esker deposited during the Riverhead outburst flood. However, the up-glacier end does not continue into the glacier. Instead, an initial prominent ice-contact bluff was exposed during retreat, yet no surface stream was ever present to incise the bluff. Continued retreat revealed a subglacial channel connection between Nova and Echo Basins at the western end of the esker.

Exposures flanking the esker contain an internal stratigraphic architecture typical of high discharge flood events (Russell et al., 2006) consisting of three units, all dipping 6° in the up-glacier direction (Fleisher, 2007). The basal unit contains crudely sorted, subtle rhythmic beds of pebble-cobble gravel containing isolated lenses of well-sorted coarse sand as well as lenses of clast-supported pebble gravel. Although the lower contact is not exposed, the upper contact is a massive, undulating, scoured, cut-and-fill channel that encompasses the entire width of the esker. This channel is lined with clusters of meter-size lag boulders capped by 0.5 m of matrix-free, clast supported cobbles and boulder gravel with distinct graded bedding that fines upward into pebble-size clasts. It is this graded gravel that rises eastward where it intersects the esker surface that forms the anomalous matrix-free cobble gravel that mantles the eastern end of the esker.

The uppermost unit is composed of subhorizontal, 15–20-cm-thick, rhythmically bedded pebble gravel alternating with coarse pebbly sand. Meter-wide lenses of coarse, unstratified sand appear randomly throughout this unit. Wood fragments and segments of broken tree trunks (up to 20 cm diameter) are common. The tree trunks are oriented parallel to the esker trend and inclined gently up-glacier.

Echo Basin

Echo Basin, on the north side of the esker, lies in an area that corresponds to the outburst vent at the head of Icewall Canyon. For this reason the formation of the basin is linked to a scouring mechanism active during the latter part of the July 1994 outburst flood. A bathymetric survey in 2003 showed basin depths that progressively increased toward the ice front, reaching a maximum of 20 m. As with Nova Basin, at no time during the retreat did an ice-margin stream flow in the vicinity of Echo Basin. Because the basin and adjacent bluffs emerged fully formed from beneath the retreating ice, the formation of both is attributed to subglacial scouring.

Associated with Echo Basin is an ice-cored deposit that originated as a veneer of coarse gravel on the floor of Icewall Canyon deposited during the July 1994 outburst flood event (Dogleg in Fig. 10). The canyon floor, including these deposits, was overridden when the canyon closed during the second phase of the surge. The leading edge of overriding ice carried a prominent debris band, just as at the Riverhead Push Moraine. In the years that followed, downwasting mantled the ice-cored gravel with a veneer of supraglacial debris, thus adding to an insulating mantle that continues to retard melting of remnant ice.

DISCUSSION

Effects of Overriding Ice

One might expect that basal sliding across a saturated unconsolidated substrate would result in significant erosional modification of the pre-surge terrain. However, field evidence does not support this assumption. Although the overriding ice essentially lacked basal debris, a sub-meter veneer of till was ubiquitously deposited. This supports the notion that the new till was not derived from the glacier proper, but rather consists of deformable bed material of preexisting drift that was mobilized and redeposited as a deformation till (Alley, 1992; Muller, 1983; Boulton and Hindmarsh, 1987; Benn and Evans, 1996). A prime example of deformation till in the process of formation is the observed mechanism for push moraine formation. Contrary to the implication of a plowing effect, the push moraine was in a constant state of formation as the saturated substrate was squeezed forward from beneath the ice, then overridden, only to be extruded and overridden again. A tangle of deformed alder within the push moraine implies significant mobilization. However, the large push moraine on the western flank of Riverhead Sandur does not consist entirely of deformation till. To the contrary, the southern end consists mostly of coarse sandur gravel, which would resist mobilization when saturated. Instead of being deformed, the sandur gravel was indeed pushed into place. The deflection of the moraine in the down-glacier direction at the lakeshore indicates a greater rate of advance over lake silts than island gravel, which would contribute to a plowing effect.

Macro- and micro-structural analyses of the deformation till support an origin related to subglacial deformation. Fleisher et al. (200ob) report clast fabric parallel to ice flow, halo structures and rotational grains, undulating fissility planes, scavenger structures, and clay layers that exhibit strong birefringence, all indicative of pervasive deformation (Menzies and Maltman, 1992; van der Meer, 1993). The common occurrence of flutes extending from the lee side of meter-size boulders, combined with fabric evidence of aligned clasts within the pervasively deformed till, suggests that flutes form as a result of mobilized substrate filling grooves at the sole of the glacier (Boulton, 1976; Paul and Evans, 1974).

Of related interest are two anomalous, sub-kilometer-size, shallow (1–2 m) ponds, occupying elongate swales on the strongly drumlinized surface of Bentwood Island. These may be attributed to sediment advection accompanying the downward migration of the deforming layer in what Hart et al. (1990) referred to as excavational deformation (Benn and Evans, 1998; Hart, 1995). The ponded swales were prominent components of the landscape prior to the surge and reappeared when uncovered during retreat. In addition, well-developed gullies with decameter relief that existed prior to the surge near these ponds, and elsewhere, also reappeared in what looks to be an unaltered condition after retreat, thus surviving both the erosional and depositional effects of overriding ice surprisingly well. The flanks of drumlinized terrain exposed within the walls of these gullies show clear evidence for bedding truncation, indicating some degree of erosion (Fleisher et al., 2004).

Factors that influence whether erosion occurs, and to what degree or whether till is deposited, are related to the attitude of slope relative to the direction of ice movement, the grain size of the overridden terrain, degree of substrate saturation, magnitude of overburden pressure, basal load carried by the ice, and rate and duration of ice movement. The diverse orientation of gully walls to ice movement, and lack of gully modification, suggest that orientation had little if any influence on erosion or deposition. Considering the coarse nature of the gully outwash and its resistance to saturation, erosion by mobilization is not likely, which translates into no erosion. Lack of mobilization means lack of deposition when the ice carries no basal load. Therefore, grain size and resistance to saturation, leading to resistance mobilization, appears rather significant. This leaves overburden pressure, rate of flow, and duration of flow. At best, the ice thickness beneath the overridden foreland would have been <200 m, which by glacier standards is fairly thin. Whereas rate of flow was relatively rapid, certainly much faster than normal glacier movement, the duration was relatively short. Therefore, all the factors—grain size, thin ice, and short duration—appear to be the main reasons why the overriding ice had little influence on the foreland terrain.

In contrast to what appears to be the relatively passive nature of overriding ice is the deformation of trees several meters beneath the ice-ground interface, as demonstrated by the sheared and bent trunks of alder and spruce trees within sand beds of the outwash substrate (Fleisher et al., 2006b). An explanation for this apparent inconsistency lies in the details of the host stratigraphy associated with tree deformation. All deformation is confined to very thin zones of shearing at the base of lacustrine sand beds, where sub-centimeter-scale clay grades upward into silt and ultimately sand. When saturated by infiltrating meltwater to depths of meters beneath the glacier, the diminished shear strength of wet clay facilitated lateral offset by directional stress applied by the overriding ice (Bindschadler, 1983). Additionally, the shear strength of sediments, which is reduced by increased pore-water pressure and lacks cohesion, is directly proportional to the vertical effective normal stress. In a sediment sequence in which layers of relatively impermeable material interrupts the isotropic distribution of pore-water pressure, the shear strengths of different strata differ significantly (Boulton et al., 2001). Thus the trees were deformed and truncated along clay seams at depth, whereas the well-drained, interstratified gravel resisted the same forces. At the same time, surface molding was active within 1 m of the ice interface, as indicated by a veneer of deformation till and flute formation.

Outburst Floods

The initial 1994 outburst and resulting formation of Icewall Canyon are linked to the Tsivat Basin Conduit System (Fleisher et al., 1998). The process of continuous canyon growth from one vent position to another suggests that the canyon formed by collapse along the course of an unroofed subglacial conduit. The common occurrence of house-size blocks of brecciated ice cemented by frazil ice on the canyon floor implies the occurrence of supercooled water that is attributed to meltwater rising along an adverse slope within an overdeepening beneath the Tsivat Sublobe (Evenson et al., 1999a; Evenson et al., 1999b; Alley et al., 1999). The canyon grew headward, first in an up-glacier trend toward one conduit vent, and then shifted to trend parallel to the ice front toward a second vent at the ultimate head of Icewall Canyon. An ice bridge ~150 m wide at the head of the canyon separated it from a 200-m-wide collapse window on the glacier surface directly above the ultimate position of Echo Basin.

Changes in the lacustrine thermal regime during the surge are also related to this conduit system. Rising plumes of supercooled water along the Tsivat ice front prior to the surge were noted by vertically moving currents of highly turbid water associated with the occurrence of frazil ice flocs that rose to the surface at many places. These are interpreted to be expressions of subglacial conduit vents discharging directly into the lake at some depth beneath the lake surface (Fleisher et al., 1993). A thermal shift occurred during the first summer of the surge (1994) as the temperature of Tsivat Lake changed a few tenths of a degree rising into values above zero Celsius. Simultaneously, the temperature of Tsiu Lake became cooler, thus suggesting a shift of water movement in the conduit from Tsivat to Tsiu basin (Fleisher et al., 1998). Post-surge (1996) cooling in Tsivat Lake signaled the return of supercooled discharge into that basin. The temporal sequence of subglacial vent activity (Fig. 7) along the ice front demonstrates that pre-surge conduit discharge did indeed shift temporarily to Tsiu basin during the surge. This indicates that the well-established conduit system consisted of multiple passageways through which water was directed to both sides of Weeping Peat Island during a surge. Subglacial discharge returned to pre-surge areas along the Tsivat ice front during post-surge years, with primary flow from the most active vents at the northwestern tip of Weeping Peat Island, in the same area as the pre-surge vents.

Limnic turbidity monitored prior to, during, and after the surge, indicates that suspended sediment load within both ice-contact lakes significantly increased during the surge. The conspicuous increase (5–6 times) of suspended sediment load in both lakes during times of flood discharge indicates greater sediment flux, which is attributed to a shift from divergent water movement at the base of the glacier during the surge to consolidated conduit discharge during the outburst phase that interrupted the surge (Fleisher et al., 1998). This supports implications of subglacial water movement interpreted from thermal data. Furthermore, a bedrock influence on the location of the conduit system is inferred. Outburst floods fed by pressurized subglacial discharge favored the same interlobate position during two consecutive, separate surges that took place decades apart (1965–1967 and 1993–1995). The massive scale and extent of erosional and depositional landforms associated with surge-related flood events indicate the inability of the sand and gravel substrate to resist such forces. Furthermore, the same ice front discharge locations remained active during the entire intervening retreat phase, which spanned nearly three decades. Therefore, the persistent position of the Tsivat Basin Conduit System supports a subglacial bedrock influence on water movement (Fleisher et al., 1998).

An estimate of flood discharge was derived from field data, a video shot during the outburst (Jamie Rouch, 1994, personal commun.), and an empirical approach (Shedd and Fleisher, 2005). Field work conducted within weeks of the outburst provided the scale of the active outburst channel width, the dimensions of ice blocks transported during the flood, water density based on turbidity, the gradient of the resulting sandur, and the roughness of the surface across which the floodwaters moved. An application of the Darcy-Weisbach equation to these observations (Clague, 1973; Clague and Matthews, 1973) yielded an approximate minimum paleo-discharge value on the order of ~3200 cms for the Icewall outburst flood.

Based on the degree of erosion and volume of sedimentation transported, the Riverhead outburst was less dynamic, yet it caused greater modification of Weeping Peat Island by virtue of its location. (Mitteager and Fleisher, 1996). Originating from a vent that was directly down-glacier from the collapse window on the glacier, a close genetic link with the Tsivat Basin Conduit System is implied. USGS aerial photographs dated 7 September 1994 lack evidence of this outburst (Fig. 5A), yet evidence of this outburst was observed in mid-October 1994. This places the outburst event within a four to five week window of time between early September and mid-October.

An estimate of outburst discharge is based on the stationary position of a 2.3 m boulder that resisted flood conditions, remaining in exactly the same pre-flood position and orientation (Fig. 12). Measurements of channel dimensions prior to the outburst, combined with the outburst sandur gradient and depth of gravel that nearly buried the boulder, yielded discharge values ranging from a low of 665 cms to a high of 1,100 cms (Clague, 1973; Shaw and Fleisher, 1996; Fleisher et al., 2000a; Shedd and Fleisher, 2005).

Figure 12. Splitlake boulder. The position and orientation of Splitlake boulder remained essentially unaltered by the passage of floodwaters during the Riverhead outburst.

Outburst Sites and Subglacial Hydraulic Activity

Sandar deposits contain evidence supporting forceful hydraulic erosion of subglacial material during outburst events. Tree trunks, meter-scale stumps, thick mats of peat, and fragments of littoral bivalves, all exhumed from buried forests, organic zones, and marine sources at depth beneath the glacier were found littering the sandar (Fleisher et al., 2000a; Wiles et al., 1999). Additionally, hydrofracture fills formed by forceful injection were exposed and accessible after post-surge ablation had lowered the surface of the Tsivat Sublobe. These fills consisted of silt-laden, frazil ice bands arranged in sub-centimeter accretionary increments within steeply inclined fracture walls. Several fills contained shattered wood fragments, and a few held broken tree trunks up to 30 cm in diameter. Consistent with this were fractures containing pebble and cobble gravel with dispersed boulders that must have been injected upward from a subglacial source by forces exceeding the overburden pressure (Fleisher et al., 1996a).

The location of Echo Basin coincides with the final position of headward erosion in Icewall Canyon, where rising subglacial flow emanated from an active conduit vent. Neither the 20–25-m-deep closed depressions of Nova and Echo Basins, nor the associated erosional bluffs, existed prior to outburst flooding, suggesting formation by a mechanism accompanying the outbursts. The excavation of massive quantities of coarse gravel, full-grown trees, and thick peat mats from beneath the glacier would have required a scouring mechanism working under an exceedingly high hydraulic pressure (Nye, 1976). A minimum of 14 m of hydraulic scouring is shown in the topographic cross section of Figure 11, which depicts the 20-m-deep Nova Basin superimposed on the pre-surge topography. The close association of these erosional features with identifiable outburst sites is the critical link between process and form.

These features are interpreted to be the result of hydraulic scouring that took place within confined, subglacial tunnels through which pressurized water meandered laterally and vertically above and below the ice-island interface before exiting the confines of the glacier (Johnson, 1999). The exit sites for the Icewall Canyon and Riverhead outbursts are obvious, but the occurrence of a subglacial outburst is demonstrated by erosional and depositional landforms uncovered after retreat behind the Riverhead Push Moraine.

Indicators of significant subglacial water movement favoring flow toward Tsiu Lake include the juxtaposition of the truncated blunt up-glacier end of the esker, the scouring of Nova Basin and associated bluff incision, the band of hummocky collapsed topography (the Nova Kame Belt) behind the push moraine, and the lobate shoreline extension of ice-cored deposits into Tsiu Lake (Boatland) (Fig. 10). The position of the Nova Basin with respect to the relatively un-breached push moraine implies that the direction for hydraulic flow was confined behind the push moraine before turning laterally into Tsiu Lake basin, all of which took place beneath the ice. The subtle linear alignment of Boatland mounds parallel to the ice flow direction indicates that the terrain was briefly overridden. Had these deposits formed earlier, they would have been reworked and incorporated as part of the moraine. Therefore, this lobate extension of the Weeping Peat Island shoreline formed during the waning stages of the surge, synchronous with the final movement of the ice. This timing is in agreement with the thermal shift in Tsiu Lake caused by the influx of supercooled water redirected by the conduit system from Tsivat Basin. This supports the interpretation that calls for subglacial water flow beneath the ice through Nova Basin and into Tsiu Lake.

Esker Formation

Although this feature lacks the classic sinuous topographic expression of an esker (Gorrell and Shaw, 1991; Brennand et al., 1996; Brennand, 2000), it nonetheless qualifies as one on the basis of several compelling physical attributes (Shulmeister, 1989; Russell et al., 2000). In addition to bearing no resemblance to Weeping Peat Island stratigraphy (Fleisher et al., 1998), it occupies a position consistent with the source area of the Riverhead outburst and is aligned on trend with the ice surface window that existed a few hundred meters up-glacier. Furthermore, 10 m bluffs on three sides expose a 3-D view of ice-cored, sorted and stratified gravel and lag boulders of unique internal architecture. It consists of water-laid materials with conspicuous backset beds, an exceptionally large-scale channel scour-and-fill structure lined with meter-size lag boulders and clast-supported boulder and cobble graded bedding, all of which are part of a sequence that is inclined to rise eastward at a gentle dip of 6° . A similar deposit with identical shape, morphology, and internal architecture has been attributed to outburst tunnel discharge associated with the 1996 jökulhlaup at the ice front of Skeidarárjökull, Iceland, by Russell et al. (2006). The linear alignment of subtle flutes and grooves on the benchlike top suggests a limited duration of overriding ice. Indeed, post-outburst ice advance was diminished and remained sluggish thereafter. Field evidence suggests that this feature is an englacial cavity filling deposited by processes related to exceptionally high discharge. The lack of up-glacier continuity into the ice is consistent with truncation by the channel connecting Nova Basin with Echo Basin.

The Surge Phenomena

A glacial surge is a rapid, short-lived ice-front advance at rates that may exceed 100 times normal flow, accompanied by rapid thickening. Surges are followed by quiescent periods when the glacier returns to "normal" flow conditions. Common to all surging glaciers is a large-scale, rapid transfer of mass from above the equilibrium line altitude (ELA), where the ice surface subsides, down-glacier to where thickening accompanies ice front advance. Also typical is the release of stored water in high-discharge flood events (Harrison and Post, 2002) that temporarily stalls ice front advance (Gomez et al., 2000; Russell et al., 2006).

Each glacier demonstrates a surge cycle on the order of a decade to several decades (Benn and Evans, 1998; Post, 1972). Given that only 207 of approximately 100,000 glaciers in Alaska are known to surge (Post, 1969), and that decades may pass between surges, it is indeed rare to observe a surge in progress, and even more so to monitor the entire surge cycle.

When addressing the questions, "Why do glaciers surge?" and "Why does one glacier surge but not its neighbors?" most authors assign significance to the storage of englacial and subglacial water. Storage of pressurized basal water in linked cavities above a bedrock substrate has been proposed as a means by which water may be widely distributed, thus reducing basal friction through buoyancy and facilitating rapid sliding (Kamb, 1987; Raymond, 1987). Impeded transmission of englacial water has also been proposed (Lingle and Fatland, 2003). Water that enters the pore spaces of a low permeability, unconsolidated substrate may lead to loss of shear strength, thus facilitating substrate deformation at rates greater than ice deformation (Alley, 1989; Boulton and Hindmarsh, 1987).

Most authors address one or more of several recurring issues, such as (1) the conditions that may lead to instability (Clarke et al., 1984), including anomalously high annual snowfall (Tangborn, 1997; Tangborn, 2002); (2) the physical changes that precede a surge that may serve as precursors, such as steepening glacier gradient or glacier "quakes" (Robin, 1969; Meier and Post, 1969; Clarke et al., 1984; Kamb et al.,1985; Harrison and Post, 2002); (3) the failure of a "plumbing system" to convey englacial and subglacial water (Weertman, 1969; Kamb, 1987; Kamb et al., 1985), leading to the trapping and storage of water (Hooke, et al., 1990; Fountain and Walder, 1998; Lingle and Fatland, 2003); (4) theoretical treatment of ice deformation (Boulton and Hindmarsh, 1987; Hart et al., 1990); (5) measurement of substrate deformation (Truffer et al., 2001); and (6) the mechanism for abrupt water escape, as during an outburst flood (Clarke, 1982; Humphrey et al., 1986; Molnia et al., 1994; Russell et al., 2006; Weertman, 1969).

The following summary of characteristics observed during the Bering surge are consistent with observations and reports elsewhere, thus reinforcing the significance of these attributes and events in the surge process:

1. The discharge of subglacial supercooled water from vents active along the Tsivat ice front was consistent with what had been observed for years prior to the surge, thus indicating that the normal plumbing system was working before surge activity began.

2. The initiation of the 1993–1995 surge 28 yr after a mid-1960s surge reinforced a cyclicity suggested at other surging glaciers.

3. Using data projected from a regional snowfall model, Tangborn postulated inordinately high snow accumulation during the 1992 winter 1 yr prior to the development of a kinematic wave up-glacier.

4. After the surge began, meltwater discharge from a well-established subglacial conduit system was interrupted in favor of the development of a "leaky" ice front.

5. Ice front advance was accompanied by ice thickening and basal sliding of clean ice (free of basal debris) across a saturated substrate.

6. The sudden discharge of pressurized subglacial water during several outburst floods temporarily interrupted ice front advance, after which ice movement was intermittent and slower.

7. The mechanism of push moraine formation, coupled with substrate mobilization during deposition of a deformation till, indicates that some of the glacier movement was due to substrate deformation.

8. The sudden release of pressurized water during outburst floods at sites linked to subglacial basin scouring indicates the rapid release of at least some water trapped within and beneath the glacier.

9. The shifting of ice front vents back to pre-surge positions indicates reoccupation of a pre-surge subglacial plumbing system along the trend of a well-established conduit system and the return to normal conditions.

SUMMARY

During the past half century the eastern terminus and foreland of Bering Piedmont Glacier has been the site of multiple surge events and associated outburst floods that impacted a peripheral drainage system. Following decades of retreat from a 1965–1967 surge maximum, the Bering once again surged in 1993, consistent with a 25–30 yr cyclicity. During the first phase of the surge the ice front advanced 1.5–2.0 km at rates from an early rate of 1.0 m/d to 7.4 m/d by mid-winter 1994, and averaged 2–3 m/d in June 1994.

The first and largest of three outburst floods temporarily interrupted active advance in July 1994 and deposited a 1.5 km^2 sandur into ice-contact Tsivat Lake. Sporadic ice front advance characterized phase two of the surge following a second outburst flood in the fall of 1994 that nearly dissected Weeping Peat Island, thus forming two additional sandar. Diminished ice movement throughout the 1995 summer, culminating in the construction of a push moraine, marked the maximum extent of the surge. The turbidity of supercooled water issuing from ice-front fountaining vents during the surge increased by 5–6 times, and the ice front took on a leaky appearance.

Several years of retreat exposed the subglacial effects of the surge on pre-surge terrain. Overriding ice remolded a drumlinized foreland, deposited a thin veneer of deformation till, and constructed a push moraine. These were minor in scale compared with the effects of outburst floods. Erosion by pressured water flowing through subglacial conduits scoured sub-kilometer-scale basins into pre-surge terrain in the vicinity of outburst sites, with accompanying esker and sandur deposition.

Significant changes in the level of peripheral ice-contact lakes were initiated in 1993 when advancing ice closed a strategically located outlet channel, causing upstream lakes to rise 14 m. Thirteen years later, glacial retreat and thinning accommodated a breakout that returned lakes to their pre-surge elevation. Rates of retreat from foreland islands ranged from 50 to 75 m/a, and surface melting has lowered the glacier 10 m/a within 1 km of the terminus.

ACKNOWLEDGMENTS

Funding for this research was provided by The State University of New York (SUNY) College at Oneonta and Earth Sciences Department, the College at Oneonta Foundation, the National Science Foundation, the National Geographic Society, the New York State Geological Survey, Museum and Science Service, the United University Professors, and most recently the International Earthwatch Institute. Graduate students Bill Morrow and Heidi Natel received funding through Geological Society of America Graduate Student Research Fellowships. Thirty-six SUNY Earth Sciences Department undergraduate students, whose work was partially supported by the College at Oneonta Research Committee, were significant contributors to this research, as were four cadets from the U.S. Military Academy at West Point.

The U.S. Army Corps of Engineers cooperated with field investigations during the surge years. We are grateful for the logistical support we received from the Prince William Sound Science Center, the Chugach National Forest Office, Wilderness Helicopter, and Cordova Air Service, with special recognition of pilot Patrick Kearney, all in Cordova, Alaska. Additional logistical support with field operations was received from the U.S. Bureau of Land Management (BLM), Anchorage Office. Christopher Noyles, BLM, provided timely imagery, and Jamie Roush, University of Alaska, Fairbanks, took the timely video of the 27 July 1994 outburst flood. Members of the research team who participated in various phases of this research, and whose input was particularly significant, include Austin Post, Charles Rosenfeld, Dorothy Peteet, Cal Huesser, and Brian Tormey.

REFERENCES CITED

Alley, R.B., 1989, Water-pressure coupling of sliding and bed deformation: II. Velocity depth profiles: Journal of Glaciology, v. 35, p. 119–129.

Alley, R.B., 1992, How can low-pressure channels and deforming tills coexist subglacially?: Journal of Glaciology, v. 38, p. 200–207.

Alley, R.B., Lawson, D.E., Evenson, E.B., Strasser, J.C., and Larson, J.G., 1999, Glaciohydraulic supercooling: A freeze-on mechanism to create stratified, debris-rich basal ice: II. Theory: Journal of Glaciology, v. 44, p. 563–569.

Benn, D.I., and Evans, D.J.A., 1996, The interpretation and classification of subglacially deformed materials: Quaternary Science Reviews, v. 15, p. 23–52, doi: 10.1016/0277-3791(95)00082-8.

Benn, D.I., and Evans, D.J.A., 1998, Glaciers and Glaciation: London, Edward Arnold, 734 p.

Bindschadler, R., 1983, The importance of pressurized subglacial water in separation and sliding at the glacier bed: Journal of Glaciology, v. 29, p. 3–19.

Boulton, G.S., 1976, The origin of glacially-fluted surfaces: Observations and theory: Journal of Glaciology, v. 17, p. 287–309.

Boulton, G.S., and Hindmarsh, R.C.A., 1987, Sediment deformation beneath glaciers; rheology and geological consequences: Journal of Geophysical Research, v. 92, p. 9059–9082, doi: 10.1029/JB092iB09p09059.

Boulton, G.S., Dobbie, K.E., and Zatsepin, S., 2001, Sediment deformation beneath glaciers and its coupling to the subglacial hydraulic system: Quaternary International, v. 86, p. 3–28, doi: 10.1016/S1040-6182(01)00048-9.

Brennand, T.A., 2000, Deglacial meltwater drainage and glaciodynamics: Inferences from Laurentide eskers, Canada: Geomorphology, v. 32, p. 263–293, doi: 10.1016/S0169-555X(99)00100-2.

Brennand, T.A., and Shaw, J., 1994, Tunnel channels and associated landforms, south central Ontario: Their implications for ice-sheet hydrology: Canadian Journal of Earth Sciences, v. 31, p. 505–522.

Brennand, T.A., Shaw, J., and Sharpe, D.A., 1996, Regional-scale meltwater erosion and deposition patterns, northern Quebec, Canada: Annals of Glaciology, v. 22, p. 85–92.

Clague, J.J., 1973, Sedimentology and paleohydrology of Late Wisconsinan outwash, Rocky Mountain Trench, Southeastern British Columbia, *in* Jopling, A.V., and McDonald, B.C., eds., Glaciofluvial and Glaciolacustrine Sedimentation: Society of Economic Paleontologists and Mineralogists Special Publication 23, p. 22–100.

Clague, J.J., and Mathews, W.H., 1973, The magnitude of jökulhlaups: Journal of Glaciology, v. 12, p. 501–504.

Clarke, G.J.C., 1982, Glacier outburst floods from "Hazard Lake", Yukon Territory, and the problem of flood magnitude prediction: Journal of Glaciology, v. 28, p. 3–21.

Clarke, C.J.C., Collins, S.G., and Thompson, D.E., 1984, Flow, thermal structure and subglacial conditions of a surge-type glacier: Canadian Journal of Earth Sciences, v. 21, p. 232–240.

Clayton, L., Attig, J.W., and Mickelson, D.M., 1999, Tunnel channels formed in Wisconsin during the last glaciation, *in* Mickelson, D.M., and Attig, J.W., eds., Glacial Processes Past and Present: Geological Society of America Special Paper 337: p. 69–82.

Dworak, R.J., and Fleisher, P.J., 2003, The Effect of Daily Weather on Ablation, Bering Glacier, Alaska: State University of New York, College at Oneonta, Earth Sciences Department, Open File Report, 13 p.

Evenson, E.B., Lawson, D.E., Larson, G.J., Alley, R.B., and Fleisher, P.J., 1999a, Observation of glaciohydraulic supercooling at the Bering and Malispina Glaciers, Southeastern Alaska: Geological Society of America Abstracts with Programs, v. 31, no. 7, p. A258.

Evenson, E.B., Lawson, D.E., Strasser, J.C., Larson, G.J., Alley, R.B., Ensminger, S.L., and Stevenson, W.E., 1999b, Field evidence for the recognition of glaciohydraulic supercooling, *in* Mickelson, D.M., and Attig, J.W., eds., Glacial Processes Past and Present: Geological Society of America Special Paper 337, p. 23–35.

Fleisher, P.J., 2007, An outburst esker, eastern sector, Bering Glacier, Alaska: Geological Society of America Abstracts with Programs, v. 39, no. 6, p. 117.

Fleisher, P.J., and Muller, E.H., 2000, Neoglacial History of Weeping Peat Island and Vicinity, Bering Glacier, Alaska: Albany, State University of New York, State Education Department, New York State Geological Survey, Open File Report 10kZ138, 16 p.

Fleisher, P.J., Gardner, J.A., and Franz, J.M., 1993, Bathymetry and sedimentary environment in proglacial lakes at the eastern Bering Piedmont Glacier of Alaska: Journal of Geological Education, v. 41, p. 267–274.

Fleisher, P.J., Muller, E.K., Cadwell, D.H., Rosenfeld, C.L., Thatcher, A., and Bailey, P.K., 1994a, Measured ice-front advance and other surge-related changes, Bering Glacier, Alaska: Eos (Transactions, American Geophysical Union), v. 75, p. 63.

Fleisher, P.J., Muller, E.K., Cadwell, D.H., Rosenfeld, C.L., Post, A., Bailey, P.K., and Bering Glacier Research Group (BERG), 1994b, Monitoring the 1993 Surge of the Bering Glacier, AK: Geological Society of America Abstracts with Programs, v. 26, no. 3, p. 17.

Fleisher, P.J., Muller, E.H., Cadwell, D.H., Rosenfeld, C.L., Bailey, P.K., Pelton, J.M., and Puglisi, M.A., 1995, The surging advance of Bering Glacier, AK; a progress report: Journal of Glaciology, v. 41, p. 207–213.

Fleisher, P.J., Muller, E.H., Bailey, P.K., and Cadwell, D.H., 1996a, Subglacial discharge and sediment load fluctuations during the 1993–95 surge of Bering Glacier, Alaska: Geological Society of America Abstracts with Programs, v. 28, no. 7, p. A110–A111.

Fleisher, P.J., Muller, E.H., Cadwell, D.H., Rosenfeld, C.L., and Bailey, P.K., 1996b, Subglacial conduit system endures repeated surges, Bering Glacier, Alaska: Geological Society of America Abstracts with Programs, v. 28, no. 3, p. 54.

Fleisher, P.J., Cadwell, D.H., and Muller, E.H., 1998, The Tsivat Basin Conduit System persists through two surges, Bering Piedmont Glacier, Alaska: Geological Society of America Bulletin, v. 110, p. 877–887, doi: 10.1130/0016-7606(1998)110<0877:TBCSPT>2.3.CO;2.

Fleisher, P.J., Shaw, L., Mitteager, W., Cadwell, D.H., Gerhard, D., Rosenfeld, C., Bailey, P.K., and Tormey, B.B., 2000a, Field Studies of Subglacial Outbursts and Ice Front Vent Migration, Eastern Sector, Bering Glacier, Alaska: Albany, State University of New York, State Education Department, State Geological Survey Open File Report 10kZ137, 23 p.

Fleisher, P.J., Lachniet, M.S., and Muller, E.H., 2000b, A Case for Mobilized Drift, Eastern Sector, Bering Glacier, Alaska: Albany, State University of New York, State Education Department, State Geological Survey Open File Report 10kZ139, 13 p.

Fleisher, P.J., Bailey, P.K., and Cadwell, D.H., 2003, A decade of sedimentation in ice contact, proglacial lakes, Bering Glacier, Alaska: Journal of Sedimentary Geology, v. 160, p. 309–324, doi: 10.1016/S0037-0738(03)00089-7.

Fleisher, P.J., Bailey, P.K., Natel, H.H., and Natel, E.M., 2004, The paradox of overriding ice: Landform continuity coupled with dynamic change, Bering Glacier, Alaska: Geological Society of America Abstracts with Programs, v. 36, no. 6, p. 172.

Fleisher, P.J., Bailey, P.K., and Natel, E.M., 2006a, Ice dam breakout at Bering Glacier, Alaska: Geological Society of America Abstracts with Programs, v. 38, no. 7, p. 236.

Fleisher, P.J., Lachniet, M.S., Muller, E.H., and Bailey, P.K., 2006b, Subglacial deformation of trees within overridden foreland strata, Bering Glacier, Alaska: Proceedings of the Annual Binghamton Geomorphology Symposium, 34th: Geomorphology, v. 75, p. 201–211.

Fountain, A.G., and Walder, J.S., 1998, Water flow through temperate glaciers: Reviews of Geophysics, v. 36, p. 299–328, doi: 10.1029/97RG03579.

Franz, J., and Fleisher, P.J., 1992, Origin of ice-contact lake basins, Eastern Bering piedmont lobe, AK: Geological Society of America Abstracts with Programs, v. 24, no. 1, p. 21.

Galon, R., 1973, Geomorphological and geological analysis of the proglacial area of Skeiðarárjökull: Geographia Polonica, v. 26, p. 15–56.

Gerhard, D.A., Cadwell, D.H., Fleisher, P.J., Rosenfeld, C.L., and Mitteager, W.A., 1996, Making topographic maps of recently-deglaciated terrain using GPS and GIS technologies: Geological Society of America Abstracts with Programs, v. 28, no. 3, p. 58.

Gomez, B., Smith, L.C., Magilligan, F.J., Mertes, F.A.K., and Smith, N.D., 2000, Glacier outburst floods and outwash plain development: Skeiðarársandur, Iceland: Terra Nova, v. 12, p. 126–131, doi: 10.1046/j.1365-3121.2000.00277.x.

Gorrell, G., and Shaw, J., 1991, Deposition in an esker, bead and fan complex, Lanark, Ontario: Sedimentary Geology, v. 72, p. 285–314, doi: 10.1016/0037-0738(91)90016-7.

Harrison, W.D., and Post, A.S., 2002, How much do we really know about glacier surging? International Symposium on Fast Glacier Flow, International Glaciological Society, Yakutat, Alaska, Abstract 56.

Hart, J.K., 1995, Subglacial erosion, deposition and deformation associated with deformable beds: Progress in Physical Geography, v. 19, p. 173–191, doi: 10.1177/030913339501900202.

Hart, J.K., Hindmarsh, R.C.A., and Boulton, G.S., 1990, Styles of subglacial glaciotectonic deformation within the context of the Anglian Ice Bed: Earth Surface Processes and Landforms, v. 15, p. 227–241, doi: 10.1002/esp.3290150305.

Hooke, R., Laumann, T., and Kohler, J., 1990, Subglacial water pressure and the shape of subglacial conduits: Journal of Glaciology, v. 36, p. 67–71.

Humphrey, N.F., Raymond, C.F., and Harrison, W.D., 1986, Discharges of turbid water during mini-surges of Variegated Glacier, Alaska, U.S.A.: Journal of Glaciology, v. 32, p. 195–207.

Johnson, M.D., 1999, Spooner Hills, northwest Wisconsin: High-relief hills carved by subglacial meltwater of the Superior Lobe, *in* Michelson, D.M., and Attig, J.W., eds., Glacial Processes: Past and Present: Geological Society of America Special Paper 337, p. 83–92.

Kamb, B., 1987, Glacier surge mechanism based on linked cavity configuration of the basal water conduit system: Journal of Geophysical Research, v. 92, p. 9083–9100, doi: 10.1029/JB092iB09p09083.

Kamb, B., Raymond, C.F., Harrison, W.D., Engelhardt, H., Echelmeyer, K.A., Humphrey, N., Brugman, M.M., and Pfeffer, T., 1985, Glacier Surge Mechanism: 1982–1983 surge of Variegated Glacier, Alaska: Science, v. 227, p. 469–479, doi: 10.1126/science.227.4686.469.

Klimek, K., 1973, Geomorphological and geological analysis of the proglacial area of the Skeiðarárjökull: Extreme eastern and western sections: Geographia Polonica, v. 26, p. 89–113.

Kozarski, S., and Szupryczynski, J., 1973, Glacial forms and deposits in the Sidujökull deglaciation area: Geographia Polonica, v. 26, p. 255–311.

Lamb, W.K., 1984, Volume 2 of a voyage of discovery to the north Pacific Ocean and round the world, 1791–1795: Hakluyt Society, issue 164, 2nd series, no. 163–166, 1752 p.

Lingle, C.S., and Fatland, D.R., 2003, Does englacial water storage drive temperate glacier surges?: Annals of Glaciology, v. 36, p. 14–20, doi: 10.3189/172756403781816464.

Maizels, J.K., 1995, Sediments and landforms of modern proglacial terrestrial environments, *in* Menzies, J., ed., Modern Glacial Environments: Processes, Dynamics and Sediments: Oxford, UK, Butterworth-Heinemann, p. 365–416.

Meier, M.F., and Post, A.S., 1969, What are glacier surges?: Canadian Journal of Earth Sciences, v. 6, p. 807–817.

Menzies, J., and Maltman, A.J., 1992, Microstructures in diamictons—Evidence of subglacial bed conditions: Geomorphology, v. 6, p. 27–40, doi: 10.1016/0169-555X(92)90045-P.

Merrand, Y.B., and Hallet, B., 1996, Water and sediment discharge from a large surging glacier: Bering Glacier, Alaska, U.S.A., summer 1994: Annals of Glaciology, v. 22, p. 233–240.

Mitteager, W., and Fleisher, P.J., 1996, Foreland modification by jökulhlaup discharge, Bering Glacier foreplain, Alaska: Geological Society of America Abstracts with Programs, v. 28, no. 3, p. 82–83.

Molnia, B.F., 1993, Major surge of the Bering Glacier: Eos (Transactions, American Geophysical Union), v. 74, p. 321.

Molnia, B.F., and Post, A., 1995, Holocene history of Bering Glacier: A prelude to the 1993–94 surge: Physical Geography, v. 16, p. 87–117.

Molnia, B.F., Post, A., and Fleisher, P.J., 1994, Abrupt glaciohydrologic change, Bering Glacier, AK: Eos (Transactions, American Geophysical Union), v. 75, p. 63.

Muller, E.H., 1983, Till genesis and the glacier sole, *in* Evenson, E.B., Schluchter, C., and Rabassa, J., eds., Till and Related Deposits: Rotterdam, A.A. Balkema, p. 19–22.

Muller, E.H., and Fleisher, P.J., 1995, Surge history and potential for renewed retreat, Bering Glacier, Alaska: Arctic and Alpine Research, v. 27, p. 81–88, doi: 10.2307/1552070.

Muller, E.H., Post, A., Fleisher, P.J., Cadwell, D.H., and Natel, E., 1993, Resurgence of the Bering Glacier: Geological Society of America Abstracts with Programs, v. 25, no. 6, p. A224.

Nye, J.F., 1976, Water flow in glacier jökulhlaups, tunnels and veins: Journal of Glaciology, v. 17, p. 181–188.

Pasch, A.D., Foster, N.R., and Irvine, G.V., 2010, this volume, Faunal analysis of late Pleistocene–early Holocene invertebrates provides evidence for paleoenvironments of a Gulf of Alaska shoreline inland of the present Bering Glacier margin, *in* Shuchman, R.A., and Josberger, E.G., eds., Bering Glacier: Interdisciplinary Studies of Earth's Largest Temperate Surging Glacier: Geological Society of America Special Paper 462, doi: 10.1130/2010.2462(13)

Patterson, C.J., 1994. Tunnel-valley fans of the St. Croix moraine, east-central Minnesota, U.S.A., *in* Warren, W.P., and Croot, D.G., eds., Formation and Deformation of Glacial Deposits: Rotterdam, A.A. Balkema, p. 59–87.

Paul, M.A., and Evans, H., 1974, Observations on the internal structure and origin of some flutes in glaciofluvial sediments, Blomstrandbreen, northwest Spitsbergen: Journal of Glaciology, v. 13, p. 393–400.

Pierce, R.A., and Winslow, J.H., eds., 1979: H.M.S. Sulfur on the Northwest and California Coasts, 1837 and 1839: The Accounts of Captain Edward Belcher and Midshipman Francis Guillemard Simpkinson: Kingston, Ontario, Canada, Limestone Press, 144 p.

Piotrowski, J.A., 1994, Tunnel-valley formation in north-western Germany—Geology, mechanisms of formation and subglacial bed conditions for the Bornhöved tunnel valley: Sedimentary Geology, v. 89, p. 107–141, doi: 10.1016/0037-0738(94)90086-8.

Post, A.S., 1969, Distribution of surging glaciers in western North America: Journal of Glaciology, v. 8, p. 229–240.

Post, A., 1972, Periodic surge origin of folded medial moraines on Bering Piedmont Glacier, Alaska: Journal of Glaciology, v. 11, p. 219–226.

Post, A., 1997, Passive and active iceberg producing glaciers, *in* Van der Veen, C.J., ed., Calving Glaciers: Columbus, Ohio State University, Byrd Polar Research Center, Report of a Workshop, 28 February–2 March 1997, BPRC Report 15, 194 p.

Price, R.J., 1969, Moraines, sandar, kames and eskers near Breiðamerkurjökull, Iceland: Transactions (Institute of British Geographers), v. 46, p. 17–43.

Raymond, C.F., 1987, How do glaciers surge? A review: Journal of Geophysical Research, v. 92, p. 9121–9134, doi: 10.1029/JB092iB09p09121.

Robin, G. de Q., 1969, Initiation of glacier surges: Canadian Journal of Earth Sciences, v. 6, p. 919–928.
Rosenfeld, C.L., Thatcher, A., Fleisher, P.J., and Bailey, P.K., 1994, GPS and video mapping of Alaska's Bering Glacier: GPS World, October issue, 26–27 p.
Russell, A.J., Tweed, F.S., and Knudsen, Ó., 2000, Flash flood at Sólheimajökull heralds the reawakening of an Icelandic subglacial volcano: Geology Today, May–June, p. 103–107.
Russell, A.J., Roberts, M.J., Fay, H., Marren, P.M., Cassidy, N.J., Tweed, F.S., and Harris, T., 2006, Icelandic jökulhlaup impacts; implications for ice-sheet hydrology, sediment transfer and geomorphology: Geomorphology, v. 75, p. 33–64.
Shaw, L.A., and Fleisher, P.J., 1996, Jökulhlaup paleohydrology, Weeping Peat Island, Bering Glacier, Alaska: Geological Society of America Abstracts with Programs, v. 28, no. 5, p. 111.
Shedd, B., and Fleisher, P.J., 2005, A Hydrologic Approximation of Paleo-Discharge during Glacial Outbursts, Bering Glacier, Alaska: An Empirical Approach: State University of New York, College at Oneonta, Earth Sciences Department, Open File Report, 12 p.
Shulmeister, J., 1989, Flood deposits in the Tweed Esker (southern Ontario, Canada): Sedimentary Geology, v. 65, p. 153–163, doi: 10.1016/0037-0738(89)90012-2.
Tangborn, W.V., 1997, Using low-altitude meteorological observations to calculate the mass balance of Alaska's Columbia Glacier related to calving and speed, *in* Van der Veen, C.J., Calving Glaciers: Columbus, Ohio State University, Byrd Polar Research Center, Report of a Workshop, 28 February–2 March 1997, BPRC Report 15, p. 141–161.
Tangborn, W.V., 2002, Connecting Winter Balance and Runoff to Surges of the Bering Glacier, Alaska: Data available on the World Wide Web at URL http://www.hymet.com/docs/yakuart4.pdf.
Truffer, M.K., Echelmeyer, K., and Harrison, W.D., 2001, Implications of till deformation on glacier dynamics: Journal of Glaciology, v. 47, p. 123–134, doi: 10.3189/172756501781832449.
van der Meer, J.J.M., 1993, Microscopic evidence of subglacial deformation: Quaternary Science Reviews, v. 12, p. 553–587, doi: 10.1016/0277-3791(93)90069-X.
Weertman, J., 1969, Water lubrication mechanism of glacier surges: Canadian Journal of Earth Sciences, v, 6, p. 929–942.
Wiles, G.C., Post, A., Muller, E.H., and Molnia, B.F., 1999, Dendrochronology and Late Holocene history of Bering Glacier, Alaska: Quaternary Research, v. 52, p. 185–195, doi: 10.1006/qres.1999.2054.
Wright, H.E., Jr., 1973, Tunnel valleys, glacial surges, and subglacial hydrology of the Superior Lobe, Minnesota, *in* Black, R.F., Goldthwait, R.P., and Williams, H.B., eds., The Wisconsinan Stage: Geological Society of America Memoir 36: p. 251–276.

Manuscript Accepted by the Society 02 June 2009

Printed in the USA

The Geological Society of America
Special Paper 462
2010

Structural geology and glacier dynamics, Bering and Steller Glaciers, Alaska

Ronald L. Bruhn
Department of Geology and Geophysics, University of Utah, 115 S, 1460 E, Room 383, Salt Lake City, Utah 84112-0102, USA

Richard R. Forster
Department of Geography, University of Utah, 260 S. Campus Drive, Room 270, Salt Lake City, Utah 84112-9155, USA

Andrew L.J. Ford
School of Conservation Sciences, Bournemouth University, Christchurch House, Talbot Campus, Poole, Dorset BH125BB, UK

Terry L. Pavlis
Department of Geology, University of Texas, 500 W. University Avenue, El Paso, Texas 79968, USA

Michael Vorkink
GeoStrata, LLC, 781 West, 14600 South, Bluffdale, Utah 84065, USA

ABSTRACT

The Bering and Steller Glaciers of southern Alaska provide the opportunity to investigate relationships between climate and tectonics in a glaciated mountain belt. The glaciers profoundly impact the climate, ecology, and landscape of the northeastern Gulf of Alaska margin. The glaciers flow among and over deformed and eroded rocks of the Yakutat microplate, where geological structures impart topographic variations in the landscape that strongly affect glacier dynamics. The Bering Glacier flows along a tectonic boundary within the microplate that separates two regions of different structural style and history. East of the glacier, erosion of folded and thrust-faulted sedimentary strata creates E-W ridges and valleys oriented at high angle to ice flow. Farther west, second-phase folds and faults are superimposed on these structures, creating mountain blocks with complex structural geometry. Where the second-phase limbs have an E-W structural grain the glaciers flow around broad headlands, and meltwater streams discharge southward through narrow canyons. N to NE trending fold limbs are streamlined by glacial scouring parallel to folded bedding, and the elongated mountains are separated by narrow ice- and water-filled troughs, or flat-floored sediment-filled valleys.

Measurements of ice motion and glacier surface topography are used in conjunction with geological mapping to constrain the location of the tectonic boundary beneath the Bering Glacier. The boundary is inferred to lie beneath the west-central terminus and extend up-glacier, passing west of the Grindle Hills and extending into the Khitrov Hills. The large volume of debris trapped in the Medial Moraine Band

Bruhn, R.L., Forster, R.R., Ford, A.L.J., Pavlis, T.L., and Vorkink, M., 2010, Structural geology and glacier dynamics, Bering and Steller Glaciers, Alaska, *in* Shuchman, R.A., and Josberger, E.G., eds., Bering Glacier: Interdisciplinary Studies of Earth's Largest Temperate Surging Glacier: Geological Society of America Special Paper 462, p. 217–233, doi: 10.1130/2010.2462(11). For permission to copy, contact editing@geosociety.org.

along the western edge of the Bering Glacier overlies a NNE-trending bedrock high formed by second-phase folding. The Bering and Steller Glaciers coalesce beneath the Medial Moraine Band, which then divides into several flows of faster and slower moving ice and debris. Thermokarst dominates the glacial structure on the lower part of the moraine band, where ice flow is 20 m/a or less. To the west, the Steller Glacier diverges into several lobes where it flows among remnants of second-phase folds.

The tectonic boundary beneath the Bering Glacier is inferred to be a concealed thrust or oblique-slip thrust fault that rises from the Aleutian megathrust or subduction zone, juxtaposing the second-phase folded terrain against and over the E-trending fold belt beneath the glacier. There is no surface expression of the tectonic boundary because of intense erosion and transport of rock debris by the glacier and meltwater rivers. Trunk river channels beneath the Bering Glacier are presumably affected by remnant structures at its base, where NNE-trending ridges and sediment filled troughs are juxtaposed against E-trending topography oriented at high angle to ice flow. This change in basal topography and structure presumably constricts the basal drainage network opposite a sharp bend in the Khitrov Hills, where surging initiated in 1993. Episodic freezing or deformation by ice flow in this part of the drainage network may create elevated fluid pressure that triggers episodic surging.

INTRODUCTION

Linear valleys and fjords that extend for tens and even hundreds of kilometers are hallmarks of glaciated mountain belts. These features commonly parallel regional metamorphic fabrics, and zones of intense jointing and faulting, where rock is mechanically anisotropic and weak parallel to ice motion (Gordon, 1981). Although the significance of glacial scouring along zones of weakness is widely recognized, much less is known about the interplay between active tectonics and glaciers. Glacier loading of the lithosphere certainly modifies tectonic stresses when ice loads wax and wane (Sauber and Molnia, 1997; Sauber et al., 2000; Stewart et al., 2000; Thorson, 2000; Sauber and Ruppert, 2008). There is also speculation that deep and over-steepened glacially scoured valleys concentrate tectonic stresses, potentially localizing both deformation and exhumation of rocks from depth (e.g., Pavlis et al., 1997; Bruhn et al., 2004; Berger et al., 2008). Because rates of glacial erosion and transport of rock debris commonly exceed rates of tectonic uplift (Harbor et al., 1988; Merrand and Hallet, 1996), topographic relief may be subdued even though deformation is vigorous (Pavlis et al., 1997).

The region surrounding the Bering and Steller Glaciers in southern Alaska is an excellent site to study the interplay between collisional tectonics and glaciation (Fig. 1). The Bering and Steller Glaciers originate in the Bagley Ice Valley, forming the largest temperate glacier system in North America. These glaciers profoundly impact the climate, ecology, and landscape of the northeastern Gulf of Alaska margin (Molnia, 1985). The dynamics of the Bering Glacier also merit special attention because of frequent and intense surging (Post, 1972; Molnia and Post, 1995; Muller and Fleisher, 1995; Merrand and Hallet, 1996).

The Yakutat microplate is colliding into and being partly subducted beneath North America at the northeastern end of the Aleutian megathrust, creating an active mountain belt with some of the highest topographic relief on Earth (Plafker, 1987; Plafker et al., 1994; Pavlis et al., 2004; Chapman et al., 2008). The middle and lower parts of the Bering Glacier flow along a tectonic boundary within the Yakutat microplate that juxtaposes regions of significantly different tectonic structure and topography (Plafker, 1987; Bruhn et al., 2004). Folds and faults of the central segment of the Saint Elias orogen strike predominantly W to WSW along the eastern edge of the glacier, but are refolded about plunging hinge lines farther west, where the structures in the microplate curve toward the strike of the Aleutian subduction zone (Fig. 2; Bruhn et al., 2004). The second-phase fold limbs contain complexly folded and faulted strata inherited from earlier deformation. These rocks are excavated by glacial scouring along zones of weakness caused by mechanical anisotropy of folded strata. Glacially carved and elongated N to NE trending mountain blocks and ridges develop where folded strata parallel directions of ice flow (Fig. 2). The topography is less streamlined in blocks where ice flows at high angle to the structural grain. These effects are obvious surrounding the Steller Glacier, which flows over and among bedrock ridges containing variably oriented rock structure created by second-phase folding (Fig. 2).

This study is part of a broader program of research into relationships between tectonics and climate in the Saint Elias orogen (Jaeger et al., 2001). Our research on the structural geology and ice dynamics of the Bering Glacier is motivated by interdisciplinary studies of relationships between active tectonics and climate (Jaeger et al., 2001; Bruhn et al., 2004), the interplay between glacier dynamics and seismicity (Stewart et al., 2000; Doser et al., 2007; Sauber and Ruppert, 2008), and the effect of climate change on ice dynamics, including mechanisms of glacier surge (Post, 1972; Roush et al., 2003). The Bering Glacier lies at a prominent structural boundary within the deforming Yakutat microplate and along the onshore projection of the Aleutian trench and megathrust (Bruhn et al., 2004). Our work on the

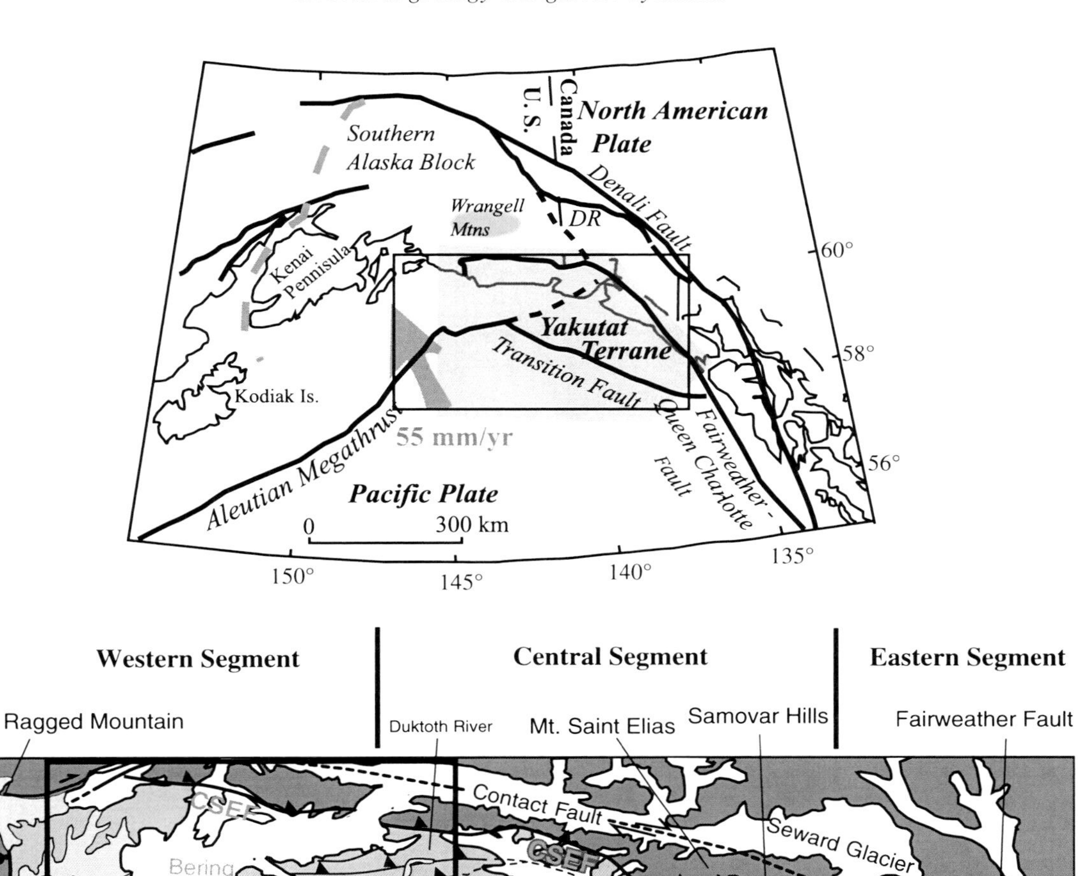

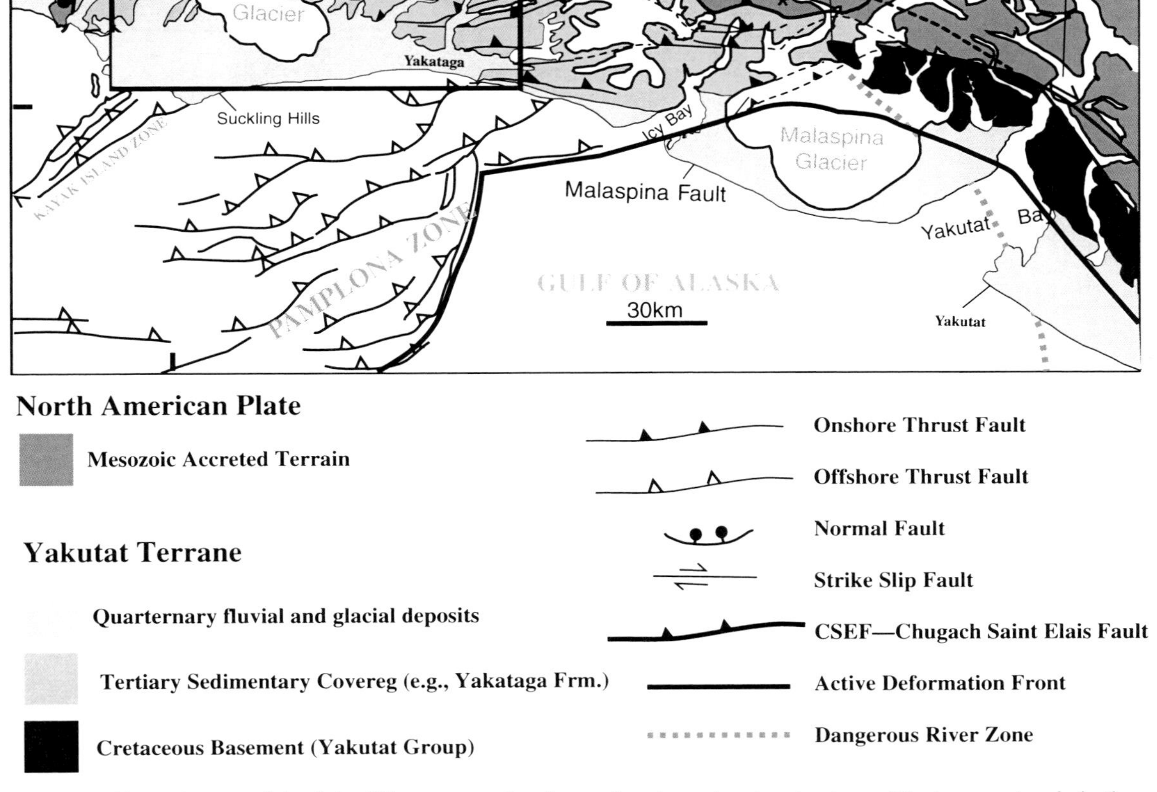

Figure 1. Tectonic map of the Saint Elias orogen, showing rock units and major structures. The large rectangle in the western part of the map encompasses the study area and location of Figure 2. The location of this tectonic map is shown as a gray box on the map of southern Alaska at the top of the figure. The large arrow indicates the rate and relative motion of the Pacific plate relative to North America (central Alaska). The figure is modified from Bruhn et al. (2004).

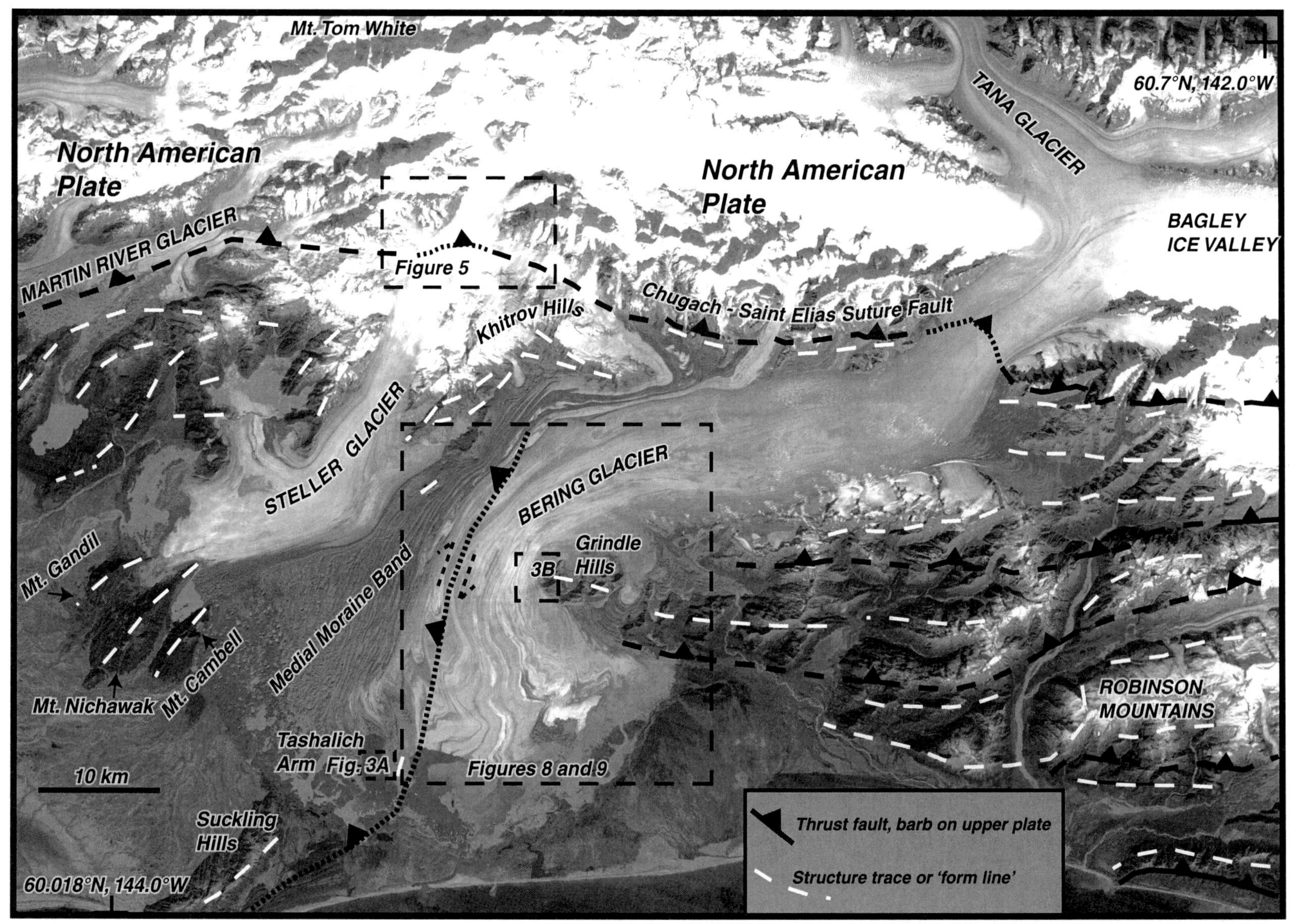

Figure 2. Generalized geography and structure map of the region surrounding the Bering and Steller Glaciers. The Chugach–Saint Elias thrust fault is the suture between the North American plate to the north, and Yakutat microplate to the south. Major east-trending thrust faults in the Robinson Mountains are sketched east of the Bering Glacier. White dashed lines are structural form lines that show in a generalized fashion the orientation of bedding strike, and fault and fold trends. Notice that bedding and structures strike mostly NE to E in the Robinson Mountains, but west of the Bering Glacier these structures are folded and faulted around by second-phase deformation, with large tracts of NNE to NE trending structure. The dotted thrust fault that extends from the eastern side of the Suckling Hills beneath the Bering Glacier to the structural bend or fold in the Khitrov Hills divides these two structural domains. This structure may accommodate oblique slip, with some right-lateral deformation as well as thrusting. The sharp structural bend or "groin" in the Khitrov Hills is located where the corner in the western wall of the Bering Glacier plunges downward beneath the ice. The Grindle Hills are an excellent site to see where E-trending structures extend beneath the eastern side of the Bering Glacier. Regions shown in figures 3, 5, 8, and 9 are indicated by labeled rectangles. The image was acquired by Landsat V during 1986.

structural geology of the region therefore has significant implications for both earthquake and tsunami potential in the Gulf of Alaska (Shennan et al., 2009). The primary motivation for the present work is to (1) better constrain the position of the tectonic boundary beneath the Bering Glacier, and (2) evaluate the influence of bedrock structure on glacial processes.

RESEARCH METHODS

Field and Remote Sensing Data

We integrate data from field observations of the Bering Glacier (including global positioning system, GPS surveying of ice motion) with remotely sensed data from optical (i.e., Landsat) and microwave (i.e., synthetic aperture radar [SAR]) sensors, and topographic data from the Shuttle Radar Topographic Mission (SRTM). A digital surface model (DSM) with a ground sample distance (GSD) of 30 m covering the lower Bering and Steller Glaciers was processed using SRTM data acquired by NASA in February 2000. This DSM was used for topographic analysis of the Bering and Steller Glaciers and terrain correction of ERS-1 SAR backscatter (σ^o) images. SAR data were processed from raw signal files to terrain corrected σ^o images using GAMMA software on the Department of Geography computer system at the University of Utah.

Sequences of images collected up to several years apart were used to track ice motion by image cross-correlation using IMCORR software (Scambos et al., 1992). We used panchromatic band images from LandSat 7 acquired 1 a apart, and ERS-1 SAR images acquired 5 a apart. The GSD of the Landsat 7 panchromatic images is 15 m, and the ERS-1 SAR GSD was 30 m. Preprocessing included co-registration of image pairs by matching ground control points, contrast enhancement by histogram adjustment, and speckle removal by adaptive filtering of SAR images.

Ice velocity was measured along three transects across the upper, middle, and lower parts of the Bering Glacier in August 2003 by repeated surveying of temporary monuments placed on the ice (Table 1). Trimble 4800 GPS receivers were deployed in fast-static differential recording mode with a base station at the U.S. Bureau of Land Management (BLM) Bering Glacier camp. GPS data were processed with Trimble GPS Office software and referenced to the WGS-84 datum. Survey precision is estimated to be within 5 cm horizontal, and <10 cm vertical.

Research Strategy

Glacier topography, structure, and flow velocity are influenced by the geometry and mechanical properties of a glacier's base (Nye, 1963; Fowler, 1982; Gudmundsson, 2003; Gudmundsson et al., 2003). This provides the opportunity to glean useful

TABLE 1. GPS SURVEY OF BERING GLACIER

Transect points	Northing (m)	Easting (m)	Elevation (m)	Northing (m)	Easting (m)	Elevation (m)	North displacement (m)	East displacement (m)	Vertical displacement (m)	Horizontal displacement (m)
Transect TA										
TA-1	6686710.17	370574.51	349.55	6686710.17	370574.49	349.46	0.00	–0.03	–0.10	0.027073973
TA-2	6687232.19	369982.60	348.36	6687232.12	369982.54	348.33	–0.06	–0.06	–0.03	0.087114866
TA-3	6688071.28	369235.88	358.78	6688071.17	369235.83	358.79	–0.12	–0.05	0.01	0.125717938
TA-4	6688852.33	368514.67	374.56	6688852.29	368514.71	374.51	–0.04	0.03	–0.05	0.055000001
TA-5	6689850.86	367999.52	382.68	6689850.87	367999.48	382.51	0.01	–0.04	–0.17	0.037107951
TA-6	6691026.25	367234.47	375.01	6691026.21	367234.45	375.09	–0.04	–0.02	0.08	0.046957428
TA-7	6692404.89	366548.04	380.67	6692404.92	366548.06	380.66	0.03	0.02	–0.01	0.037121422
TA-8	6693637.74	366050.63	375.98	6693637.75	366050.62	376.12	0.01	–0.01	0.14	0.013892444
Transect TB										
TB-1	6690285.63	381665.31	584.63	N.D.	N.D.	N.D.	N.D.	N.D.	N.D.	N.D.
TB -3	6692342.57	380197.70	553.59	6692342.65	380197.70	553.96	0.07	0.00	0.37	0.073027392
TB-4	6693476.68	379454.57	540.91	6693476.71	379454.47	540.56	0.03	–0.09	–0.35	0.095817535
TB-6	6695735.62	379367.83	559.98	6695735.57	379367.81	559.99	–0.05	–0.02	0.01	0.054781384
TB-8	6698496.98	378390.29	551.44	6698497.05	378390.24	551.35	0.07	–0.05	–0.09	0.087005747
TB-9	6700657.48	377797.06	524.70	6700657.50	377797.13	524.83	0.02	0.07	0.13	0.069641941
Transect TC										
TC-test	6703289.97	411510.03	967.22	6703288.73	411509.91	966.94	–1.24	–0.13	–0.28	1.242610157
TC-1	6703296.10	409925.61	937.65	6703294.56	409924.49	937.55	–1.54	–1.12	–0.10	1.905096323
TC-2	6703677.71	408519.64	940.47	6703676.12	408518.33	940.49	–1.59	–1.32	0.02	2.065374058
TC-3	6703897.90	407151.59	934.24	6703896.47	407150.17	934.32	–1.43	–1.42	0.08	2.019509346
TC-4	6704206.95	405789.43	915.26	6704205.61	405788.07	915.22	–1.34	–1.36	–0.04	1.910690974
TC-5	6704599.93	404301.11	906.15	6704598.93	404300.02	906.20	–1.01	–1.09	0.05	1.487581931

Note: Locations are given in Universal Transverse Mercator coordinates, Zone 7 North. Datum is WGS 84. Survey period: 2 August 2003 to 5 August 2003.

information concerning structural geology and tectonics where bedrock is buried by ice (Ford et al., 2003). Theoretical models relating surface perturbations to basal bedrock profile and "slipperiness" indicate that it is advantageous to monitor changes in the ice surface during surge, because the response of the ice surface to basal properties is amplified by increased sliding velocity (Gundmundsson, 2003; Gudmundsson et al., 2003). Changes in flow velocity and surface deformation above basal perturbations are expressed as low-pass filtered topographic features at the surface. These features will grow as velocity increases, then decay toward pre-surging amplitudes as velocity decreases. The Bering Glacier is ideal in this regard because it surges episodically, having averaged about once every twenty to thirty years during the twentieth century (Molnia and Post, 1995). The most recent surge in 1993–1995 occurred after deployment of the ERS-1 SAR, which provided, for the first time, the opportunity to repeatedly image the surface and monitor motion over the entire glacier even when obscured by cloud cover (Fatland and Lingle, 1998, 2002; Roush et al., 2003). An intensive campaign of field observations and measurements was also undertaken during and following surge (Molnia, 1993; Molnia and Post, 1995; Herzfeld, 1998).

OVERVIEW OF TECTONICS AND STRUCTURAL GEOLOGY

The Yakutat microplate was excised from the continental margin of southeastern Alaska, or possibly the Pacific Northwest, during the middle Tertiary and moved northward along the Fairweather transform fault (Plafker, 1987; Plafker et al., 1994). The microplate is currently colliding into southern Alaska at ~49 mm/a (Fletcher and Freymueller, 1999). Sedimentary rocks of the microplate are deformed in fold and fault belts both on land and offshore (Plafker, 1987). The suture between North America and the Yakutat microplate is a north-dipping thrust fault that crops out along the southern flank of the Saint Elias and eastern Chugach Mountains (Figs. 1, 2; Chugach–Saint Elias thrust fault). West of the Steller Glacier the Chugach–Saint Elias thrust is concealed beneath the Martin River Glacier and its outwash plain, but emerges farther west and south at Ragged Mountain, where the fault strikes almost N-S, ~90° counterclockwise relative to its strike farther east. Folding of the thrust fault is part of a second phase of deformation that extends throughout the orogen west of the Bering Glacier (Winkler and Plafker, 1993; Bruhn et al., 2004, 2006). The second-phase deformation is characterized by folding about plunging hinge lines and is superimposed on previously folded and thrust-faulted strata of the microplate.

The eastern edge of the second-phase folded terrain must lie beneath the Bering Glacier and its outwash along a line that projects from the coastline east of the Suckling Hills to the northern Khitrov Hills (Fig. 2; Miller, 1971; Plafker, 1987; Bruhn et al., 2004). At the terminus of the glacier the boundary is east of a bedrock-cored isthmus, where rocks crop out on the eastern shore of Tashalich Arm in Vitus Lake (Fig. 3A). Rocks exposed along the eastern shore of Tashalich Arm are strata of the Poole Creek Formation that strike N-S and dip west. The structural attitude is similar to that of bedding where the same formation is exposed in the Suckling Hills near the coast (Fig. 2).

Figure 3. (A) West-dipping and north-striking beds of Poole Creek Formation along the western shore of Tashalich Arm in Vitus Lake (see Fig. 2 for location). (B) Ice draped over bedrock promontory that extends beneath the Bering Glacier at the southwestern end of the Grindle Hills. A low-amplitude monocline is developed on the glacier surface above this ridge. The ridge is created by erosion of first-phase faults and folds along the eastern margin of the glacier. See small rectangles marked 3A and 3B in Figure 2 for locations of photos.

Farther up-glacier the next constraint on the location of the tectonic boundary is at the Grindle Hills, on the east side of the glacier (Fig. 2). Folds and thrust faults trend W to WSW in the Grindle Hills and are not affected by second-phase folding. South-facing monoclines are developed on the glacier's surface where bedrock ridges project beneath the ice at the western end of the Grindle Hills (Fig. 3B). The tectonic boundary must therefore lie west of these monoclines, which die out near the middle of the glacier.

The right-angle bend in the eastern margin of the Khitrov Hills is caused by second-phase folding and may lie near the northern termination of the major thrust fault that is inferred to lie beneath the lower Bering Glacier (Fig. 2; Bruhn et al., 2004). E-striking beds and fold hinges in the northern part of the mountain block curve abruptly into N to NE striking beds in the southern part because of superimposed folding about plunging hinge lines. Erosion of these reoriented structures created a bend or "groin" in the valley wall that plunges down beneath the Bering Glacier toward the initiation point of the 1993 surge, where ice first piled up against and flowed over a shallow bedrock ridge, according to Roush et al. (2003). Subsequently the surge front propagated down-glacier toward the terminus, and also up-glacier toward the Bagley Ice Valley (Fatland and Lingle, 1998, 2002; Roush et al., 2003).

The Medial Moraine Band lies above a shallow ice-covered ridge that extends southward from the Khitrov Hills. Radar echo sounding suggests that the ice thins in this area (Molnia and Post, 1995; Conway et al., 2009), an observation that is confirmed by the small bedrock nunatak that surfaces from the moraine band (Fig. 2). However, the valley wall must slope steeply eastward beneath the western margin of the glacier because flow foliation in the ice is nearly vertical, and moraine ridges are deformed into asymmetric Z-folds. These structures are indicative of shearing along a steeply dipping ice margin–valley wall contact zone (Fig. 4). Because the lower part of the glacier almost stagnates following surges, fold growth is nonuniform in time and occurs mostly during surges, when ice velocity and velocity gradients increase along the western edge of the glacier (Post, 1972). Folded moraine is subsequently rafted westward over the bedrock ridge, where it becomes incorporated into the slowly moving Medial Moraine Band.

FIELD OBSERVATIONS OF THE BERING GLACIER

We observed and measured structural features of the Bering Glacier during the GPS survey of ice motion in August 2003 (Fig. 4). Structural features included flow foliations, folded

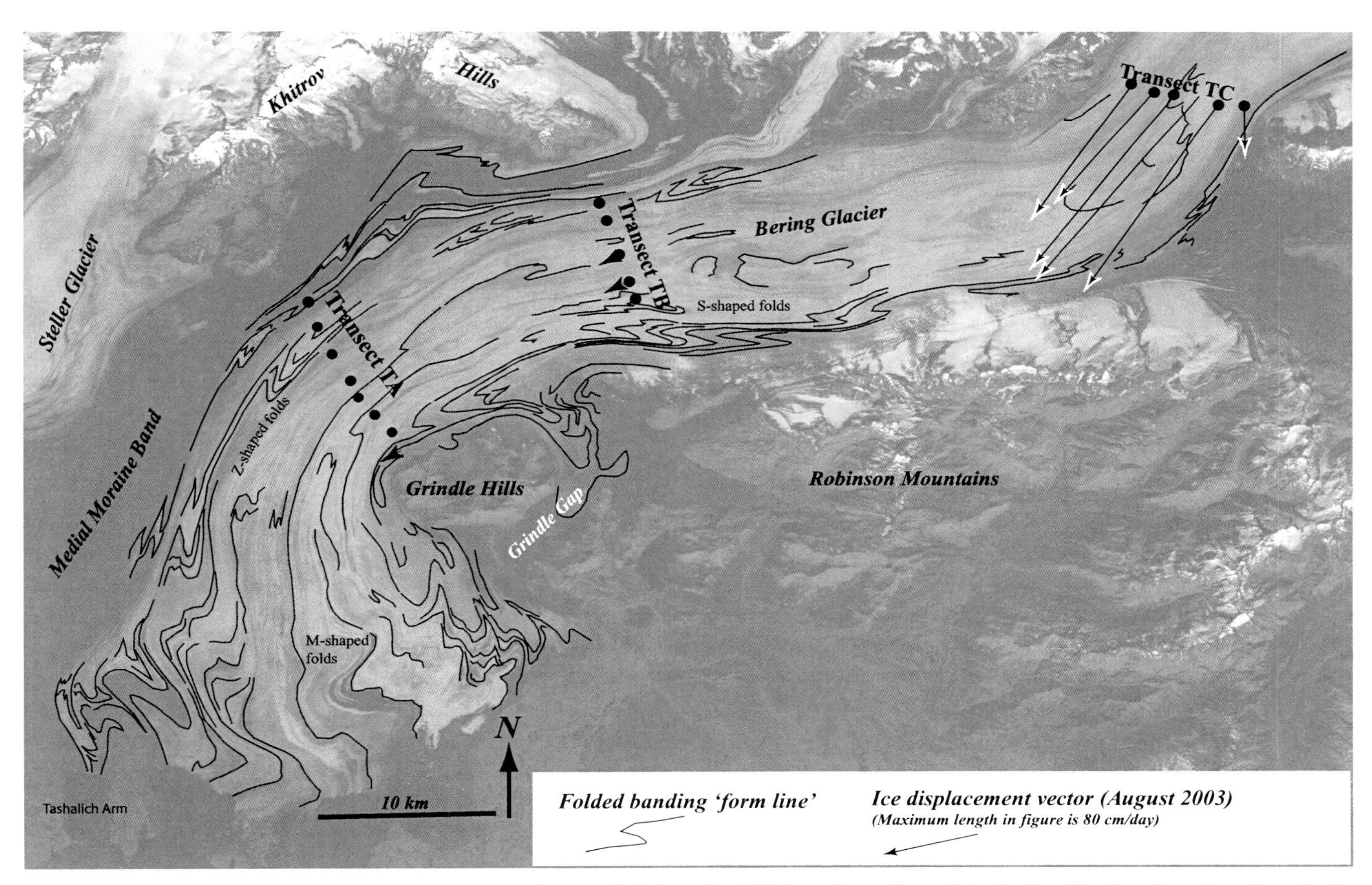

Figure 4. Structure form line map, showing symmetry (S, Z, and M shaped) of folds on the surface of the Bering Glacier when viewed vertically downward. Z-shaped folds dominate the western margin, S-shaped folds the eastern, and M-shaped folds the central terminus. S and Z folds reflect an important component of simple shearing where ice velocity decreases outward toward steep valley walls, especially during surging. Transects TA, TB, and TC show locations of GPS survey points with results presented in Table 1.

moraine, collapse pits, and the distribution of surface streams on the ice.

Flow foliations are complex in detail, but in general foliations dip steeply and strike subparallel to the longitudinal axis of the glacier. Foliation is defined by centimeter- to decimeter-wide subvertical bands of alternating debris laden and clean ice, which in aggregate form broader bands of light and dark reflectance on the glacier's surface that are detectable by aerial photography and satellite imagery. Intra-folio or "very tight to isoclinal" folds defined by remnant hinge zones and highly attenuated limbs are embedded within the foliation in some regions, and formed by extreme longitudinal extension and foliation-normal shortening strain.

The extreme crevassing that disrupted the glacier's surface during surging in 1993–1995 (Herzfeld, 1998) had largely disappeared by the summer of 1999, when we flew at low altitude over the Bering Glacier. Where devoid of snow the ice surface was marked by an intricate array of fracture traces, but most fractures were healed rather than open. This observation was further confirmed when we worked on the glacier during August 2003. At that time we also noted the distribution of surface water runoff on the glacier, because the snowline had retreated up into the Bagley Ice Valley (Fig. 2). Surface water percolated down-glacier in small streams past GPS transect TC just below the Bagley Ice Valley (Fig. 4) but disappeared into crevasses and moulins up-glacier from the Khitrov Hills (GPS transect TB). Below the Khitrov Hills meltwater flowed through englacial and basal conduits, while the ice surface was ablating rapidly.

Moraine ridges, similar in appearance to lateral moraines, develop by landsliding off trunk and tributary valley walls on both sides of the glacier. Decreasing flow velocity adjacent to valley walls creates shear strain that causes sinuous perturbations in moraine ridges to grow into large-amplitude asymmetrical folds with attenuated limbs and thick hinge zones (Post, 1972); these are "similar-style" folds in structural geology jargon. These folds exhibit different symmetry on either side of the glacier caused by lateral decrease in flow velocity toward steep valley walls. Z-shaped folds are concentrated along the western margin, and S-shaped folds along the eastern side of the glacier (Fig. 4). En-echelon crevasse arrays that developed in the ice marginal zone also indicate lateral shearing adjacent to the steep valley walls at depth.

Other fold styles include M-shaped or roughly symmetrical folds formed in the central terminus area, and highly attenuated folds with curved axial surfaces where the glacier flows around the western edge of the Grindle Hills (Fig. 4). Axial traces of M-shaped folds are subparallel to the ice front and form by longitudinal shortening strain where ice velocity decreases toward the terminus (Post, 1972; Roush et al., 2003). Isoclinal folds with curved axial traces are present where moraine is wrapped around bedrock ridges by diverging flow (Tashalich Arm) or where complex rotational strain is superimposed on S-shaped folds where ice flows around the western end of the Grindle Hills (Fig. 4). South-facing monoclines on the glacier surface indicate the presence of shallow bedrock extending beneath the ice from the Grindle Hills (Fig. 3B). In this area, strain axes must rotate spatially in response to (1) horizontal changes in directions and velocity of ice flow, and (2) vertical gradients in flow velocity created where ice flows over shallow bedrock at depth.

The surface of the Medial Moraine Band changes markedly toward the terminus from narrow moraine ridges to a broad thermokarst-pitted surface with vegetation cover near the terminus (Fig. 2). Cylindrical tunnels exposed in pit walls formed part of the englacial drainage system. Although many thermalkarst pits appear to be uniformly distributed over the surface, some are aligned in pit chains that presumably mark subsurface tunnels. We return to this issue when discussing the motion of ice in the Medial Moraine Band, and the upper threshold velocity for thermokarst development.

REMOTE SENSING OBSERVATIONS OF BERING AND STELLER GLACIERS

Surface Expression of Chugach–Saint Elias Fault

The Chugach–Saint Elias thrust fault is the first major structure encountered where ice flows southward from the Bagley Ice Valley (Fig. 2). Throughout the eastern Chugach and Saint Elias Mountains the thrust is marked by a pronounced south-facing escarpment formed by resistant metamorphosed sedimentary and volcanic rocks of the North American plate that are thrust southward over more easily eroded strata of the Yakutat microplate. Thus, the fault may affect both the basal topography and rheology of glacier beds because of topographic expression on the one hand, and a southward transition from stronger to weaker rocks on the other.

The position of the Chugach–Saint Elias thrust fault beneath the Steller Glacier is marked by a south-facing topographic step on the ice several kilometers north of where the concealed fault trace projects along the range front (Fig. 5). Below this step the glacier enters a trunk valley eroded into Tertiary strata of the Yakutat microplate. Ridges on either side of the trunk valley are underlain by gently plunging folds that are reoriented to a NNE trend by second-phase folding (Fig. 2).

Topographic evidence for the subglacier position of the Chugach–Saint Elias fault is more subdued and difficult to infer on the upper Bering Glacier, probably because of greater ice thickness and erosional retreat of the escarpment than on the smaller Steller Glacier. The fault contact may also be broken or folded by younger deformation beneath the glacier, but our geologic data are not yet sufficient to confirm this. Interferometric synthetic aperture radar (InSAR) observations of ice motion for 27–28 October 1995 show deceleration of flow directly down-glacier from where ice spills out of the Bagley Ice Valley (Fatland and Lingle, 2002). The transition from higher to lower velocity roughly coincides with the inferred position of the fault beneath the glacier, and this may reflect ice flow over the buried fault-line escarpment of the Chugach–Saint Elias fault. Below this point

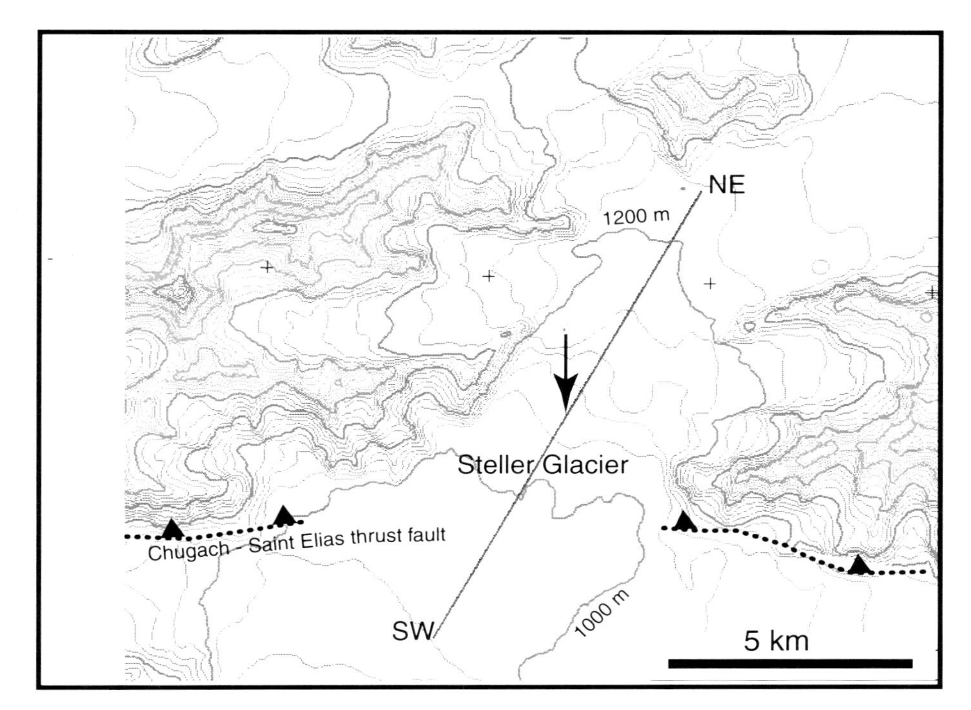

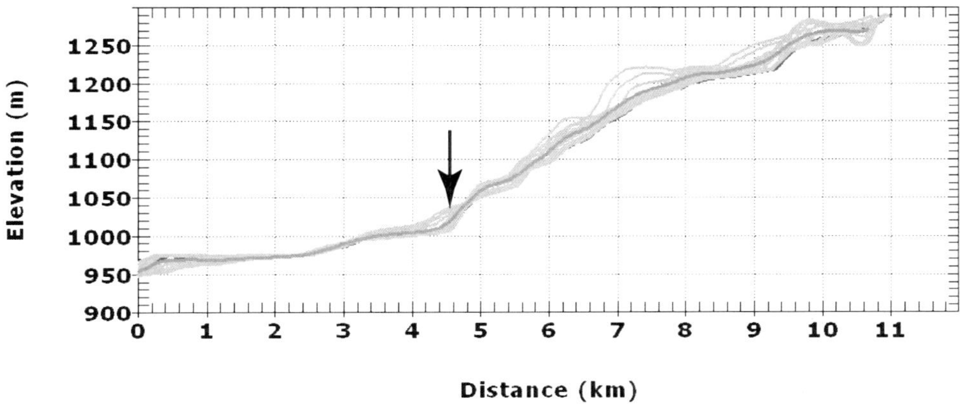

Figure 5. Topographic map of the upper Steller Glacier showing buried trace (dotted thrust fault symbol) of the Chugach–Saint Elias fault (Plafker, 1987). NE-SW line marks topographic profile shown in lower diagram. Vertical arrow marks abrupt step in surface slope that may represent surface expression of the glacially eroded fault-line escarpment beneath the glacier. Lower diagram shows 10 profiles stacked against each other along a 1-km-wide band centered on the profile line (NE-SW) on the map. Darkest profile is the average of the 10 stacked profiles. Contour interval is 50 m, and elevation data are from the National Elevation Data Set of the U.S. Geological Survey. See Figure 2 for location.

the ice flows through a long WSW oriented valley toward the sharp groin in the Khitrov Hills (Figs. 2, 4). Debris deposited on the ice from the southern flank of Waxell Ridge (eastern Chugach Mountains) and the northern part of the Khitrov Hills is transported along the western margin of the glacier and eventually incorporated into the Medial Moraine Band (Figs. 2, 4).

Steller Glacier Ice Motion

Displacements on the surface of the Steller Glacier are strongly influenced by bedrock topography that reflects the underlying geological structure. Figure 6 shows ice displacement measured by image cross-correlation between panchromatic band Landsat 7 ETM+ images acquired ~1 a apart (10 September 2001 and 29 September 2002). Displacements of ~300 m/a in the trunk valley decrease down-glacier onto the piedmont, where flow divides into several divergent tongues. These include (1) zone S-A, where ice turns abruptly east around the southeastern flank of the trunk valley and discharges northward into a narrow valley in the Khitrov Hills; (2) zone S-B, where ice flows slowly southward into a broad concave entrant in the Medial Moraine Band; (3) zone S-C, where ice discharges between the northern ends of Mounts Nichawak and Gandil; and (4) zone S-D, where ice flows east and then north to discharge into Berg Lake. Note the triangular-shaped band of moraine deformed against a rocky headland at the terminus between flow tongues S-C and S-D (Fig. 6). This headland is a remnant E-trending limb of a second-phase fold that blocks and diverts ice flow. The moraine presumably originated by landsliding onto the upper part of the glacier, and was subsequently carried over the piedmont. Flow zone S-D sometimes extends far enough west to block the narrow gorge leading south from Berg Lake, creating a temporary

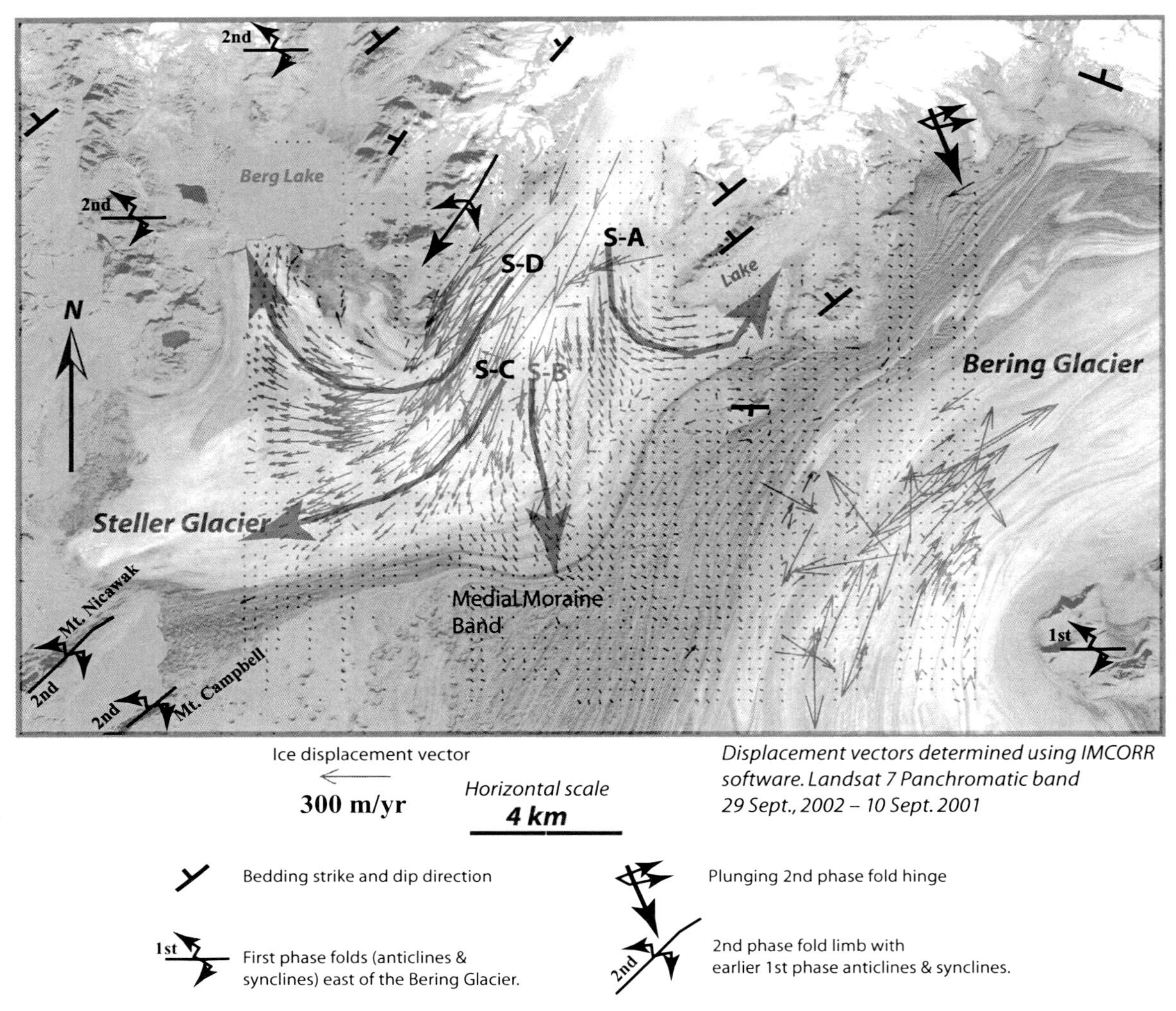

Figure 6. Displacement vectors showing ice motion on the Steller Glacier for the ~1 a interval between September 2001 and 2002. Small arrows denote displacement at grid points on the ice surface computed by cross-correlation of Landsat 7 panchromatic band images. Note that ice flow is divided into several tongues or zones (S-A through S-D) that diverge on the piedmont. Displacement vectors on Bering Glacier are caused mostly by lateral migration of dipping debris band contacts as the ice surface melts downward by ablation. Bedrock structural orientations are generalized from geological mapping by Miller (1971), Plafker (1987), Bruhn et al. (2004), and new work reported here.

dam that fails episodically and triggers outburst flooding of the lowlands to the south (Fig. 2).

Medial Moraine Band Ice Motion

Tracking motion within the Medial Moraine Band is a key to inferring the topography of bedrock at depth, and also for revealing dynamic interactions between the Bering and Steller Glaciers. We use two terrain corrected and coregistered ERS-1 SAR images acquired on 15 August 1998 and 27 September 2003, respectively, to map displacements of the Medial Moraine Band (Fig. 7). A time interval of 5 a is required to track motion over most of the moraine band because of slow displacement.

Displacement vectors (Fig. 7) define a more complex pattern of flow than we anticipated from just visual inspection of the images. There are four flow zones or tongues: M-A marks where ice from the Steller Glacier flows into the water-filled trough between Mount Campbell and Mount Nichawak at the terminus. Ice in flow zone S-B of the Steller Glacier enters the moraine band to form tongue M-B, which continues across the band to intersect SSW-flowing ice in zone M-D along the western edge of the Bering Glacier. The more rapidly flowing ice in zones M-A and M-B is partly deflected around the thermokarst topography (M-C), which dominates the lower part of the Medial Moraine Band, and is covered by vegetation at the terminus (Fig. 3B). Thermo-karst permanently pockmarks the surface where ice

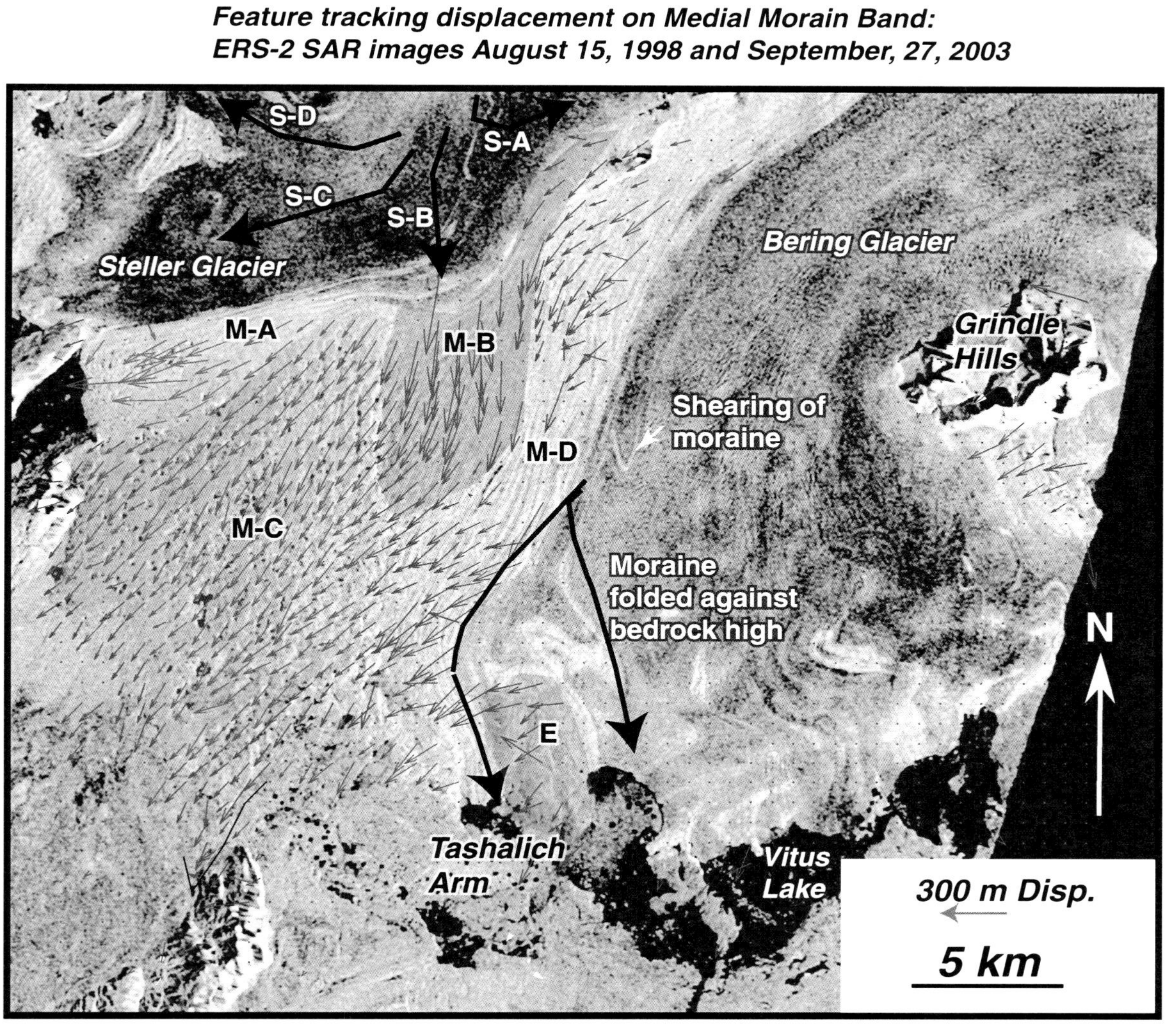

Feature tracking displacement on Medial Morain Band: ERS-2 SAR images August 15, 1998 and September, 27, 2003

Figure 7. Ice displacement for a 5 a interval on the Medial Moraine Band. Areas M-A–M-E are different regions of ice flow that are discussed in the text. Arrows S-A–S-D on Steller Glacier are flow regions defined in Figure 5. Thick black arrows on Bering Glacier show divergent flow at terminus around bedrock high defined by Roush et al. (2003) during surge in 1993. Base image is terrain corrected ERS-1 SAR backscatter image acquired on 15 August 1998.

motion is <~20 m/a. Region E marks complexly deformed ice and moraine that is folded or "wrapped around" a bedrock high near the terminus of the Bering Glacier. The bedrock high bounds a narrow channel through which ice discharged rapidly during the surge in 1993 (Roush et al., 2003).

Bering Glacier Ice Motion and Deformation

Mapping displacement of the Bering Glacier is challenging because flow accelerates from near stagnation to velocities >10 m/d during surge, and then slows markedly afterward (Post, 1972; Molnia and Post, 1995; Fatland and Lingle, 2002; Roush et al., 2003). The surface is also intensely fractured during surge, which destroys or severely modifies many features (Herzfeld, 1998). Here, we discuss changes in surface deformation and ice motion of the lower part of the glacier from 1992 through 2003, using Landsat and SAR images. We also rely on the results of Roush et al. (2003), who manually tracked surface motion using sequences of SAR scenes obtained during the surge in 1993. Image cross-correlation analysis of the two Landsat 7 scenes acquired between September 2001 and 2002 mostly failed on the Bering Glacier because annual displacements were too small in most areas. Where feature matches were found between scenes the displacement vectors formed no coherent pattern, and thus appear to be related to lateral migration of three-dimensional (3-D) structural features caused by ablation and lowering of the ice surface through dipping contacts (Fig. 6). This motivated us to measure ice motion along three transects in August 2003 using GPS surveying (Fig. 4).

Two ERS-1 SAR images of the surface of the lower Bering Glacier are shown in Figure 8. The top image (Fig. 8A) was acquired in 1992 prior to surging, and the lower image was acquired in August 1993, about two weeks after the surge front arrived at the terminus (Fig. 8B; Roush et al., 2003). The arrows on the 1993 image show the generalized flow directions determined by Roush et al. (2003) by manually tracking motion of surface features as the surge progressed. Surging was first detected as a newly formed undulation on the glacier's surface near the Khitrov Hills, where Roush et al. (2003) infer an ice-buried bedrock ridge or knob. Down-glacier from the surge initiation point, ice flow concentrated within a several-kilometer-wide channel bounded on the west by the Medial Moraine Band. Ice flow in this stream was diverted around the buried bedrock at Tashalich Arm, where ice and moraine were extensively fractured and folded around the upstream side of the ridge (Fig. 8B; Roush et al., 2003). The flow pattern was significantly different on the central and eastern part of the glacier, where ice flow fanned outward toward the south and southeast.

The radar texture of the glacier surface underwent a spectacular transformation during surging as the surface broke into a maze of crevasses (Herzfeld, 1998), and topography became amplified as rates of motion increased (Fatland and Lingle, 1998, 2002; Roush et al., 2003). Changes in backscatter brightness in Figure 8 are caused by a combination of processes that presumably include changes in surface roughness caused by crevassing, evolution of topography with kilometer and greater wavelength, and perhaps water content of the ice. Here we focus on kilometer and larger scale changes in radar texture.

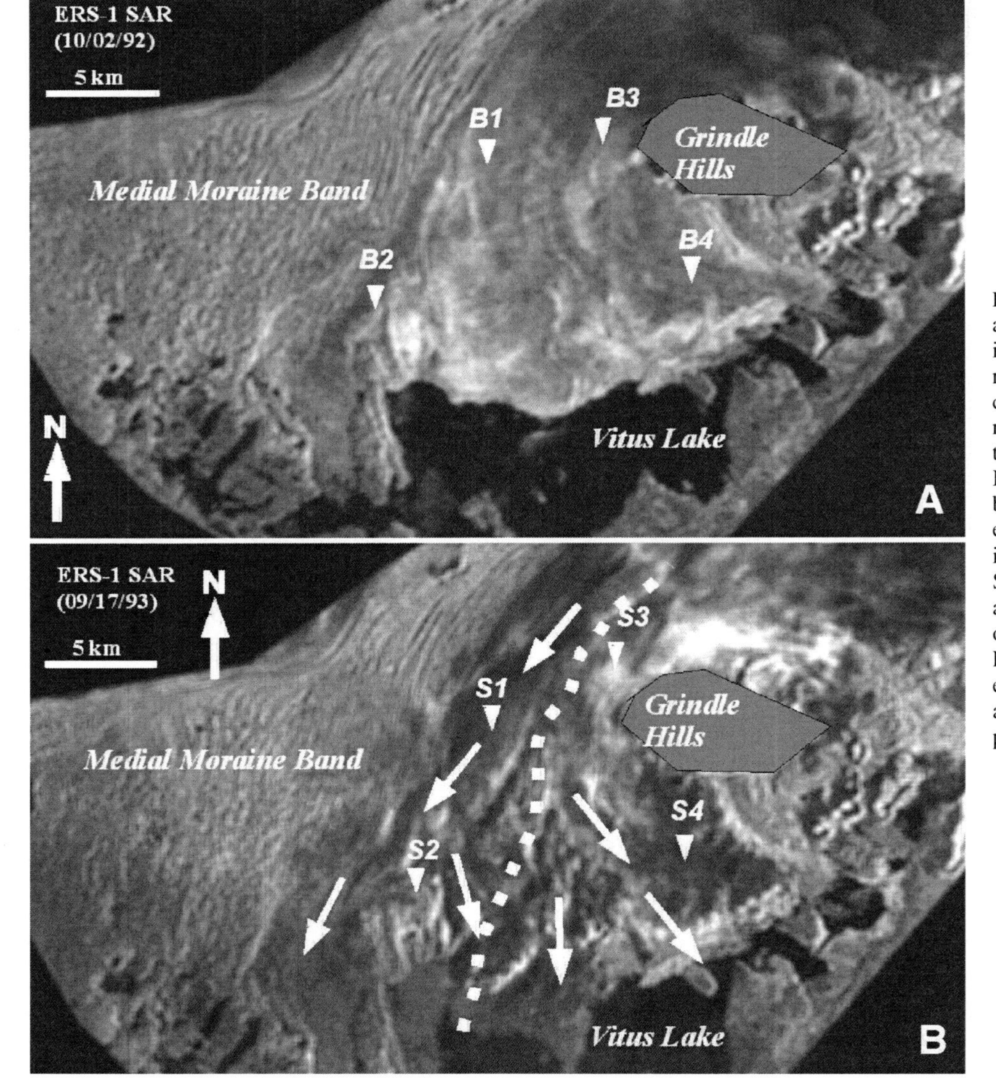

Figure 8. (A) Synthetic aperture radar image of Bering Glacier before onset of surge in October 1992. Note the relatively smooth radar texture on the surface of the glacier compared with that in panel B. B1–B4 are reference points before the onset of surging that are matched to points S1–S4 in panel B. (B) During surging the glacier surface became highly deformed. Notice the NNE-elongated topography (S1), piling up of ice against bedrock high (compare B2 and S2), and wrinkle undulations at S3. White arrows show flow directions during surge on lower glacier from feature tracking by Roush et al. (2003). Dashed heavy line is estimated position of the structural boundary or thrust fault beneath Bering Glacier projected to the glacier surface.

Compare the radar texture surrounding markers B1–4 on the 1992 image in Figure 8A with the texture or brightness variations at the same localities numbered S1–S4 during surge (Fig. 8B). NNE-elongated undulations tens of kilometers long grew along the western side of the glacier (B1 vs. S1) as surging progressed. Note also increased backscatter (brightness) where ice piled up against the bedrock ridge near the terminus in Tashalich Arm (B2 vs. S2). On the opposite side of the glacier, backscatter brightness also increased just up-glacier from the Grindle Hills (B3 vs. S3), presumably where ice flowed over and around bedrock cored by E-W–trending folds (Figs. 2, 3B). Note also that the radar texture is much different on the southeastern part of the glacier compared with the western part. The smooth radar texture of the eastern terminus (B4) evolved into short wavelength undulations (S4) of different size, shape, and orientation than the NNE-trending undulations (S1 and S2, Fig. 8B). The transition between these two regions of distinctly different radar texture and surface deformation lies above the projected eastern limit of second-phase folding.

Ice velocity and surface deformation decayed substantially following surging. By August of 2003, horizontal velocity was only a few centimeters per day along two transects surveyed across the glacier near the Khitrov Hills (Table 1; Transect TB, Fig. 4) and the Grindle Hills (TA, Fig. 4). Velocities were greater on Transect TC, between Barkley and Waxell Ridges on the upper part of the glacier (Fig. 4). The structure of the glacier's surface also changed in response to near-stagnation following surge. Most crevasses created during the 1993–1995 surge were filled or healed, and east of Tashalich Arm the terminus sloped gently into Vitus Lake, where water lapped over the ice surface.

Elevation data were acquired in February 2000 by the SRTM over the entire terminus region following surge (Fig. 9). Being at the extreme northerly extent of SRTM coverage (~60° N) the Bering Glacier was imaged during several passes of the Space Shuttle over consecutive days. The resulting raw data were averaged, which has the effect of suppressing the noise inherent in InSAR systems, to produce a very high quality DSM. The surface is marked by shallow troughs, some of which extend up-glacier >10 km from the terminus (Fig. 10). The topographic pattern is reminiscent of the large-scale dynamic changes in backscatter texture generated during surging (Fig. 8B). The NNE topographic grain of the western part of the glacier is parallel to and located in the same region as the elongated undulations that developed during surging (Fig. 8B). The topography on the eastern part of the terminus is also reminiscent of undulations that developed in that area during surging. Do these surface features reflect basal topography and/or trunk stream drainage patterns beneath the glacier? We cannot say for sure, but the dimensions and persistence suggest that they might. If so, then the surface topography of the glacier is directly affected by the underlying geological structure.

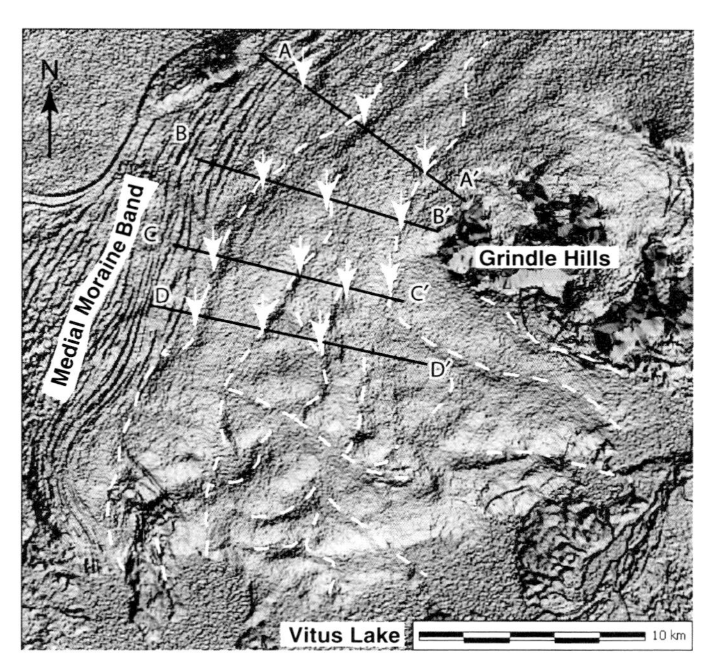

Figure 9. Digital Terrain Model (DTM) of lower Bering Glacier derived from Shuttle Radar Topographic Mission (SRTM). Sun illumination angle is from the SSE. Note elongated topographic troughs on the glacier surface that trend NNE on the west side, and more ESE on the east side of the glacial lobe. Troughs are marked by white dashed lines. Contour interval is 40 m. See Figure 2 for location of this figure.

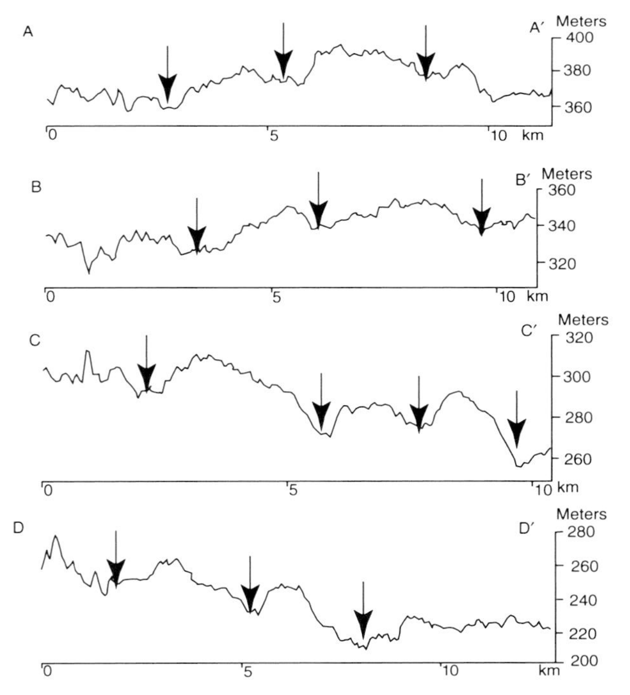

Figure 10. Topographic profiles across the lower Bering Glacier extracted from SRTM digital elevation data. Locations of profile lines (A-A′–D-D′) are shown in Figure 9. Note that troughs are up to 2 km wide and ~10 m to 20 m deep. Because the elevation data were acquired during the winter of 2001, the troughs are presumably partly filled by snow.

DISCUSSION

The tectonic setting and erosive power of the Bering and Steller Glaciers make them ideal for studying interactions between tectonics, topography, and glaciation. Our work establishes a strong empirical correlation between rock structure and observed spatial variations in the surface topography, displacement field, and structure of the glaciers. The Steller Glacier provides an obvious example, where differential erosion of second-phase structures creates topographic features that divert flow into several large tongues of ice. Where ice flows parallel to folded and faulted strata the mountain blocks are streamlined and elongated parallel to flow, but where ice flows across structures at high angle the mountains are more irregular in form, less streamlined by erosion, and contain deep gorges.

Where the Steller and Bering Glaciers coalesce in the Medial Moraine Band, ice flows within several entwined channels of faster and slower velocity that are presumably controlled by the topography of the underlying bedrock ridge, which forms the southern extension of the Khitrov Hills. Lastly, changes in ice surface topography and flow velocity during (Fig. 8B; Roush et al., 2003) and following the 1993–1995 surge (Fig. 9) of the Bering Glacier correlate in position and form with geometrical changes in rock structure and topography on either side of the inferred thrust fault boundary at the base of the glacier (Figs. 11, 12; Bruhn et al., 2004).

Correlating glacier surface features with basal structure and topography is little used, but is nonetheless a useful technique when investigating tectonics and geological structures in glaciated mountain belts (e.g., Ford et al., 2003). However, it is important to realize that there may not be a simple or direct spatial correlation between basal features and overlying deformation and topography on the glacier surface (Driedger and Kennard, 1986; Gundmundsson, 2003). Theoretical models predict that variations in surface velocity and deformation may shift spatially depending upon the nature of the basal irregularity (geometrical or rheological) and changes in the direction and speed of sliding. Mapping of deformed till and other deposits that were previously overrun at the terminus of the Bering Glacier shows that mechanical properties vary markedly (Fleisher et al., 2002), suggesting that changes in basal rheology may be as important as geometrical irregularities caused by the topographic and structural configuration of bedrock. However, at kilometer and greater scale, spatial variations in rheology will presumably mimic those of the underlying rock structure. For example, deformable water saturated sediment will thicken in basal troughs, and thin over bedrock highs, thus mimicking spatial variations in eroded bedrock. Basal stream systems will also mimic the directional grain of bedrock topography, an important concept when considering drainage of englacial water and potential changes in basal fluid pressure (Fig. 12).

The thrust faulted boundary beneath the Bering Glacier is an important tectonic feature and of regional significance (Fig. 11). Further work is required to determine if the fault remains active, and to refine its location and extent both offshore and also on land north of the Khitrov Hills. There is some additional evidence that the tectonic boundary extends well north of the Khitrov Hills. Doser et al. (2007) and Sauber and Ruppert (2008) identified a belt of enhanced seismicity in the vicinity of the Bering Glacier, and also found focal mechanisms that indicate thrust faulting on N to NE striking faults beneath the eastern Chugach Mountains in the region north of the Khitrov Hills and beneath the Bagley Ice Valley. Berger and Spotila (2006) and Berger et al. (2008) note that tectonic exhumation ages of rocks in the Khitrov Hills and Steller Glacier region are significantly younger than on the east side of the glacier, consistent with very young and perhaps still active uplift.

Two other studies use remote sensing of glaciers to infer geological structure in the Saint Elias orogen. Previously, several of our research group (Ford et al., 2003) used InSAR and

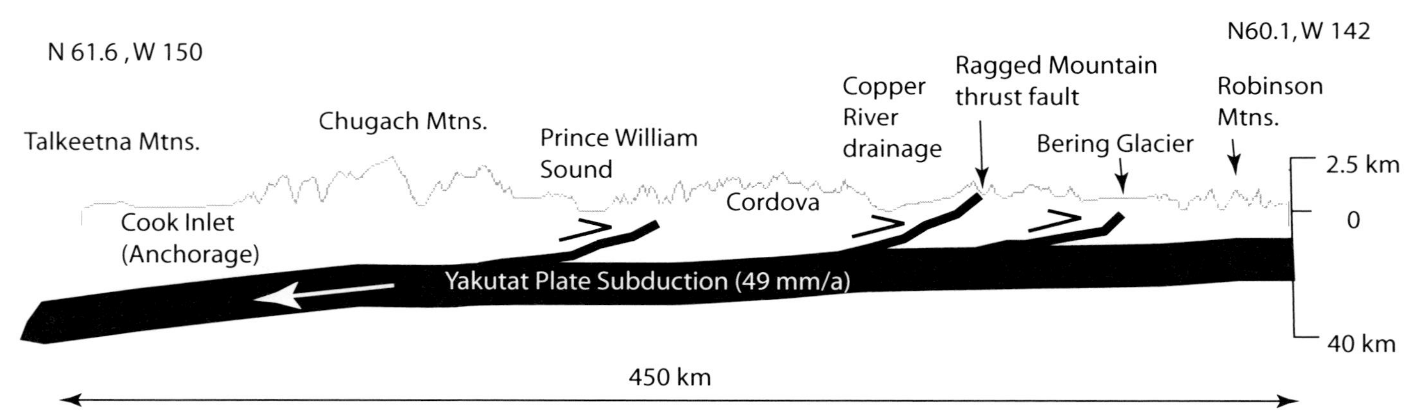

Figure 11. Schematic cross section from the Talkeetna Mountains in the west to the Robinson Mountains in the east, showing low-angle subduction of Yakutat microplate beneath southern Alaska. Note splay thrust faults emanating from the top of the subducting slab—one beneath Prince William Sound, the Ragged Mountain thrust (Bruhn et al., 2006), which may now be extending, and the inferred thrust buried beneath the Bering Glacier. This subduction zone last ruptured during the 1964 **M** 9.2 Great Alaskan Earthquake on 27 March 1964. The rupture caused co-seismic subsidence in Cook Inlet and uplift in Prince William Sound as far east as the western end of the Robinson Mountains. Large earthquakes in the fall of 1899 may have ruptured the subduction interface farther east, from the Robinson Mountains to Russell Fjord near Yakutat.

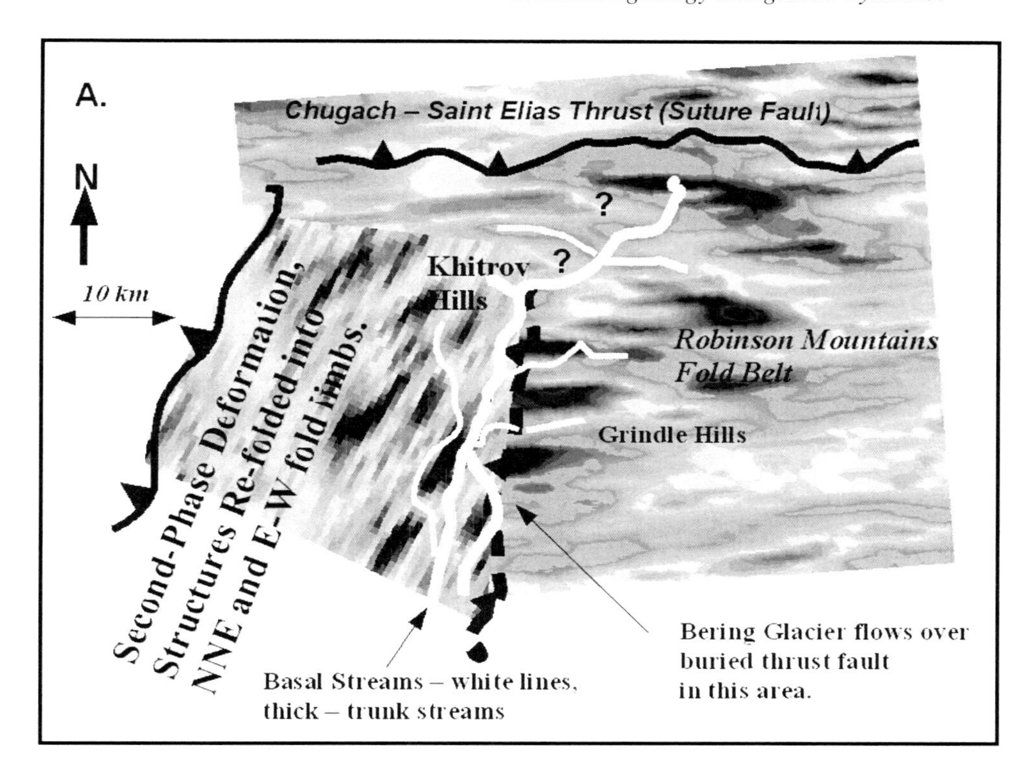

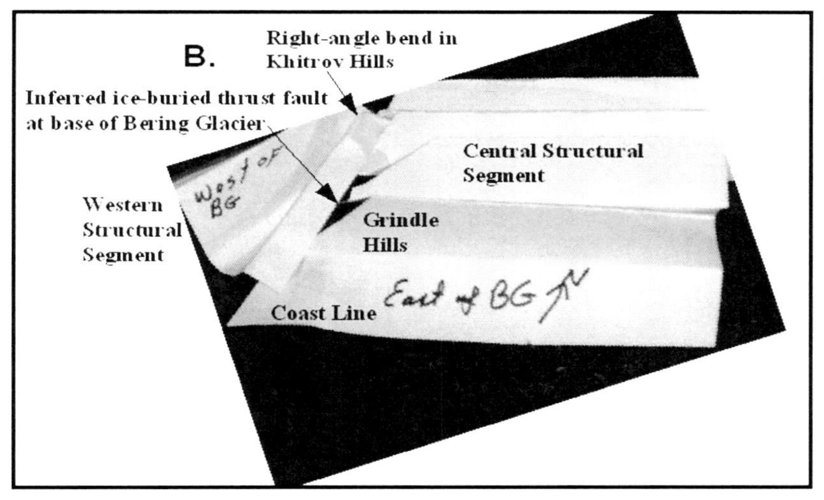

Figure 12. (A) Schematic diagram showing conceptual model of spatial changes in structure and topographic grain (orientations) where the Bering Glacier flows over a buried NNE striking thrust fault. Gray tones represent contour shapes of hills (darker) and valleys (lighter). This is a conceptual model and is not meant to be "accurate." (B) Paper model illustrating nature of folding and thrusting beneath the Bering Glacier and adjacent region. This figure is not to scale and is meant to convey only the generalized concept of sub-ice topography and structure.

topographic measurements to investigate the tectonics of the Seward Glacier basin, an ice filled depression at the NW end of the Fairweather transform fault along the old Contact fault system. Analysis of ice motion and topography on the Seward Glacier revealed a fault-bounded extensional basin at depth that is an important tectonic element in the orogen. Even earlier, side-looking airborne radar images of the Malaspina Glacier revealed elongated troughs on the surface that Molnia and Jones (1989) interpreted as having formed above fjord-like valleys at the base of the glacier. These features were also mapped by the SRTM mission in 2000, indicating that they are persistent features. The troughs are indeed interesting because they parallel an active thrust fault system that cuts across the Yakutat microplate offshore, projects on land beneath the western part of the Malaspina Glacier, and enters the Saint Elias Mountain front (Bruhn et al., 2004).

Deciphering the processes and timing of surges of temperate glaciers remains a fundamental problem in glaciology (e.g., Kamb, 1987; Fatland and Lingle, 2002). Although our work is focused on relationships between glaciers and tectonics, the results guide predictions about the morphology of the Bering's subglacial drainage network. Fortunately, these predictions can be tested in the future by radar sounding of ice thickness and basal topography, together with monitoring of ice surface and hydrological changes (e.g., Conway et al., 2009; Muskett et al., 2009). A conceptual model, or schematic, of the relationship between basal topography and stream patterns is shown in Figure 12. Key elements include the 90° bend in the valley wall at the Khitrov Hills, and west-plunging structural ridges that extend beneath the glacier from the Grindle Hills (Fig. 12A). The folded paper model in Figure 12B illustrates how the N-trending structures beneath the western part of the glacier are juxtaposed against the W-trending structures farther east. Of course, erosion by glacial and fluvial scouring, together with deposition of sediments, has modified the original topography significantly. Thick black lines represent trunk streams aligned along the N-trending topographic grain in the hanging wall (west side) of the inferred thrust fault. Thin lines are smaller feeder streams that channel flow into the trunk streams. In this conceptual model the area surrounding the groin (bend) in the Khitrov Hills is critical because trunk streams turn abruptly south and flow through the constriction formed by the 90° bend in the western valley wall, and the remnant W-trending ridges that plunge beneath the glacier from the Grindle Hills. Presumably, trunk stream conduits in this area are susceptible to partial closure by freezing, or by deforming ice where the glacier flows through the bend and constriction in the valley. This concept is supported by recent thermochronology work that suggests that the lower Bering Glaciers hydrologic system is isolated from its head water system in the Bagley Ice Valley (Headley and Enkelmann, 2009). This may explain why surging initiated near the Khirtov Hills in 1993, and then spread both up- and down-glacier accompanied by migration of over-pressured fluid as proposed by Fatland and Lingle (1998, 2002).

CONCLUSIONS

The Bering Glacier flows along a profound structural boundary in the Yakutat microplate, where the foreland fold and thrust belt of the central segment of the Saint Elias orogen becomes deformed by second-phase deformation west of the glacier. Analysis of glacier structure, surface topography, and patterns of ice displacement, when combined with geological data, constrain the position of the boundary. The structural boundary lies beneath the central part of the glacier at the terminus and extends beneath the ice at least as far north as the Khitrov Hills. The structural boundary is interpreted to be a thrust fault that dips westward to join the Aleutian megathrust at depth (Fig. 11).

The Steller Glacier flows among mountain blocks that are remnants of second-phase folds in which previous structures impart a strong mechanical anisotropy. Where ice flow is parallel to structural fabric the mountain blocks are elongated and streamlined in the directions of ice movement, and ridges are separated by sediment filled troughs. On the other, where ice flows across the structural grain of bedrock the mountain blocks are more irregular in shape, and deep gorges have been cut into the rock. The spatial distribution of these different shaped and oriented mountain blocks controls the flow of ice on the Steller Glacier, which splits into several tongues where the ice spreads over the piedmont. The Steller and Bering Glaciers coalesce in the Medial Moraine Band, a vast accumulation of moraine that is localized above a bedrock ridge that extends south from the Khitrov Hills. The orientation of this ridge is controlled by differential erosion along the N to NE trending structural grain imparted by second-phase deformation.

APPENDIX

We measured ice velocity along three transects crossing the glacier in August 2003, eight years after cessation of surging in 1995 (Fig. 4). Equipment included two Trimble 4800 GPS receivers operated in fast-track differential recording mode, with one receiver operated as a base station at the Bureau of Land Management camp on the shore of Vitus Lake, and the other as a roving receiver. Temporary survey monuments were constructed from plastic buckets with a bolt inserted through the lid for mounting the GPS antennae. The buckets were weighted down on the ice by filling them with rocks. Recording times at each station ranged from 8 to 15 min, depending upon satellite geometry and signal strength at the time of recording. The dates and times, UTM coordinates, and surface displacements measured by repeated surveying over periods of several days are given in Table 1.

ACKNOWLEDGMENTS

This work was supported by the U.S. National Science Foundation under grants 9725339 and 0408959, and by the U.S. National Aeronautics and Space Administration (NASA grants NAG-10136, NAG5-12552). We thank John Payne, Scott Guyer, and Christopher Noyles of the U.S. Bureau of Land Management for logistical support while in the field. Review of the draft manuscript and comments by Kristine Crossen and Jeanne Sauber are sincerely appreciated.

REFERENCES CITED

Berger, A.L., Spotila, J.A., Chapman, J.B., Pavlis, T.L., Enkelmann, E., Ruppert, N.A., and Buscher, J.T., 2008, Architecture, kinematics, and exhumation of a convergent orogenic wedge: A thermochronological investigation of tectonic-climatic interactions within the central St. Elias orogen, Alaska: Earth and Planetary Science Letters, v. 270, p. 13–24, doi: 10.1016/j.epsl.2008.02.034.

Bruhn, R.L., Pavlis, T.L., Plafker, G., and Serpa, L., 2004, Deformation during terrane accretion in the Saint Elias orogen, Alaska: Geological Society of America Bulletin, v. 116, p. 771–787, doi: 10.1130/B25182.1.

Bruhn, R.L., McCalpin, J., Pavlis, T., Gutierrez, F., Guerrero, J., Lucha, P., and Vorkink, M., 2006, Active tectonics of western Saint Elias Orogen, Alaska: Integration of LIDAR and field geology: Eos (Transactions, American Geophysical Union), v. 87.

Chapman, J.B., Pavlis, T.L., Gulick, S., Berger, A., Lowe, L., Spotila, J., Bruhn, R., Vorkink, M., Koons, P., Barker, A., Picornell, C., Ridgway, K., Hallet, B., Jaeger, J., and McCalpin, J., 2008, Neotectonics of the Yakutat collision: Changes in deformation driven by mass redistribution, *in* Freymueller, J.T., et al., eds., Active Tectonics and Seismic Potential in Alaska: American Geophysical Union, Geophysical Monograph Series, v. 179, p. 65–81, ISBN 978-0-87590-444-3.

Conway, H., Smith, B., Vaswani, P., Matsuoka, K., Rignot, E., and Claus, P., 2009, A low-frequency ice-penetrating radar system adapted for use from an airplane: Test results from Bering and Malaspina Glaciers, Alaska, USA: Annals of Glaciology, v. 51, p. 93–97.

Doser, D.I., Wiest, K.R., and Sauber, J.M., 2007, Seismicity of the Bering Glacier region and its relation to tectonic and glacial processes: Tectonophysics, v. 439, p. 119–127, doi: 10.1016/j.tecto.2007.04.005.

Driedger, C.L., and Kennard, P.M., 1986, Ice volumes on Cascade volcanoes: Mount Rainier, Mount Hood, Three Sisters, and Mount Shasta: U.S. Geological Survey Professional Paper 1365, http://www.cr.nps.gov/history/online_books/geology/publications/pp/1365/contents.htm.

Fatland, D.R., and Lingle, C.S., 1998, Analysis of the 1993–95 Bering Glacier (Alaska) surge using differential SAR interferometry: Journal of Glaciology, v. 4, p. 532–546.

Fatland, D.R., and Lingle, C.S., 2002, InSAR observations of the 1993–1995 Bering Glacier (Alaska, U.S.A.) surge and a surge hypothesis: Journal of Glaciology, v. 48, p. 439–451, doi: 10.3189/172756502781831296.

Fleisher, P.J., Lachniet, M.S., and Muller, E.H., 2002, Subglacial sediment deformation, Bering Glacier, Alaska: Annals of Glaciology, International Glaciology Society Fast Flow Symposium, Yakutat, Alaska, 2002, Abstract.

Fletcher, H.J., and Freymueller, J.T., 1999, New GPS constraints on the motion of the Yakutat block: Geophysical Research Letters, v. 26, p. 3029–3032, doi: 10.1029/1999GL005346.

Ford, A.L.J., Forster, R.R., and Bruhn, R.L., 2003, Ice surface velocity patterns on Seward Glacier, Alaska/Yukon, and their implications for regional tectonics in the Saint Elias Mountains: Annals of Glaciology, v. 36, p. 21–28, doi: 10.3189/172756403781816086.

Fowler, A.C., 1982, Waves on glaciers: Journal of Fluid Mechanics, v. 120, p. 283–321, doi: 10.1017/S0022112082002778.

Gordon, J.E., 1981, Ice-scoured topography and its relationships to bedrock structure and ice movement in parts of Northern Scotland and West Greenland: Geografiska Annaler, Ser. A, Physical Geography, v. 63, p. 55–65, doi: 10.2307/520564.

Gudmundsson, G.H., 2003, Transmission of basal variability to a glacier surface: Journal of Geophysical Research, v. 108, 2253, doi: 10.1029/2002JB002107.

Gudmundsson, G.H., Adalgeirsdottir, G., and Bjornsson, H., 2003, Observational verification of predicted increases in bedrock-to-surface amplitude transfer during a glacier surge: Annals of Glaciology, v. 36, p. 91–96, doi: 10.3189/172756403781816248.

Harbor, J.M., Hallet, B.B., and Raymond, C.G., 1988, A numerical model of landform development by glacial erosion: Nature, v. 333, p. 347–349.

Headley, R., and Enkelmann, E., 2009, Detrital thermochronology reveals efficiency of glacial erosion in the St. Elias Range, Alaska: Geological Society of American Abstracts with Programs, v. 41, no. 7, p. 305.

Herzfeld, U.C., 1998, The 1993–1995 Surge of Bering Glacier (Alaska)—A Photographic Documentation of Crevasse Patterns and Environmental Change: Trier, Germany, Universität Trier, Trierer Geographische Studien, 211 p.

Jaeger, J.M., Hallet, B., Pavlis, T., Sauber, J., Lawson, D., Milliman, J., Powell, R., Anderson, S.P., and Anderson, R., 2001, Orogenic and glacial research in pristine southern Alaska: Eos (Transactions, American Geophysical Union), v. 82, p. 213–216, doi: 10.1029/01EO00112.

Kamb, B., 1987, Glacier surge mechanism based on linked cavity configuration of the basal water conduit system: Journal of Geophysical Research, V. 92, no. B9, p. 9083–9100.

Merrand, J., and Hallet, B., 1996, Water and sediment discharge from a large surging glacier; Bering Glacier, Alaska, summer 1994, *in* Collins, D., et al., eds., Proceedings, International Symposium on Glacial Erosion and Sedimentation: Annals of Glaciology, v. 22, p. 233–240.

Miller, D., 1971, Geologic Map of the Yakataga District, Gulf of Alaska Tertiary Province, Alaska: U.S. Geological Survey Miscellaneous Investigations Map I-610, scale 1:125,000, 1 sheet.

Molnia, B.F., 1985, Processes on a glacier-dominated coast, Alaska: Zeitschrift für Geomorphologie, supplement-band, v. 57, p. 141–153.

Molnia, B.F., 1993, Major surge of the Bering Glacier: Eos (Transactions, American Geophysical Union), v. 74, p. 321, doi: 10.1029/93EO00582.

Molnia, B.F., and Jones, J.E., 1989, View through the ice: Are unusual airborne radar backscatter features from the surface of the Malaspina Glacier, Alaska, expressions of subglacial morphology?: Eos (Transactions, American Geophysical Union), v. 70, p. 701–710, doi: 10.1029/89EO00221.

Molnia, B.F., and Post, A., 1995, Holocene history of Bering Glacier, Alaska: A prelude to the 1993–94 surge: Physical Geography, v. 16, p. 78–117.

Muller, E.H., and Fleisher, P.J., 1995, Surging history and potential for renewed retreat: Bering Glacier, Alaska, U.S.A.: Arctic and Alpine Research, v. 27, p. 81–88, doi: 10.2307/1552070.

Muskett, R.R., Lingle, C.S., Sauber, J.M., Post, A.S., Tangborn, W.V., Rabus, B.T., and Echelmeyer, K.A., 2009, Airborne and spaceborne DEM- and laser altimetry-derived surface elevation and volume changes of the Bering Glacier system, Alaska, USA, and Yukon, Canada, 1972–2006: Journal of Glaciology, v. 55, no. 190, p. 316–326.

Nye, J.F., 1963, On the theory of the advance and retreat of glaciers: Geophysical Journal of the Royal Astronomical Society, v. 7, p. 431–456.

Pavlis, T.L., Hamburger, M.W., and Pavlis, G.L., 1997, Erosional processes as a control on the structural evolution of an actively deforming fold and thrust belt: An example from the Pamir-Tien Shan region, central Asia: Tectonics, v. 16, p. 810–822, doi: 10.1029/97TC01414.

Pavlis, T.L., Picornell, C., Serpa, L., Bruhn, R.L., and Plafker, G., 2004, Tectonic processes during oblique collision: Insights from the St. Elias orogen, northern North American Cordillera: Tectonics, v. 23, TC3001, doi: 10.1029/2003TC001557.

Plafker, G., 1987, Regional geology and petroleum potential of the northern Gulf of Alaska continental margin, *in* Scholl, D.W., et al., eds., Geology and Resource Potential of the Continental Margin of Western North America and Adjacent Ocean Basins: Circum-Pacific Council for Energy and Mineral Resources, Earth Science Series, v. 6, p. 229–268.

Plafker, G., Moore, J.C., and Winkler, G.R., 1994, Geology of the southern Alaska margin, *in* Plafker, G., and Berg, H.C., eds., The Geology of Alaska: Boulder, Colorado, Geological Society of America, Geology of North America, v. G-1, p. 389–449.

Post, A., 1972, Periodic surge origin of folded medial moraines on Bering piedmont glacier, Alaska: Journal of Glaciology, v. 11, p. 219–226.

Roush, J.J., Lingle, C.S., Guritz, R.M., Fatland, D.R., and Voronina, V.A., 2003, Surge-front propagation and velocities during the early-1993–95 surge of Bering Glacier, Alaska, U.S.A., from sequential SAR imagery: Annals of Glaciology, v. 36, p. 37–44, doi: 10.3189/172756403781816266.

Sauber, J.M., and Molnia, B.F., 1997, Ice mass moves the earth, *in* Shewe, P., and Stein, B.P., eds., Physics News in 1996: American Institute of Physics, p. 46–48.

Sauber, J.M., and Ruppert, N.A., 2008, Ice mass fluctuations and earthquake hazard in southern Alaska, *in* Freymueller, J.T., et al., eds., Active Tectonics and Seismic Potential in Alaska: American Geophysical Union, Geophysical Monograph Series, v. 179, p. 369–384, ISBN 978-0-87590-444-3.

Sauber, J.M., Plafker, G., Molnia, B.F., and Bryant, M.A., 2000, Crustal deformation associated with glacial fluctuations in the eastern Chugach Mountains, Alaska: Journal of Geophysical Research, v. 105, p. 8055–8078, doi: 10.1029/1999JB900433.

Scambos, T.A., Dutkiewicz, M.J., Wilson, J.C., and Bindschadler, R.A., 1992, Application of image cross-correlation to the measurement of glacier velocity using satellite image data: Remote Sensing of Environment, v. 42, p. 177–186, doi: 10.1016/0034-4257(92)90101-O.

Shennan, I., Bruhn, R.L., and Plafker, G., 2009, Multisegment earthquakes and tsunami potential of the Aleutian megathrust: Quaternary Science Reviews, v. 28, p. 7–13.

Stewart, I.S., Sauber, J.M., and Rose, J., 2000, Glacio-seismotectonics: Ice sheets, crustal deformation and seismicity: Quaternary Science Reviews, v. 19, p. 1367–1389, doi: 10.1016/S0277-3791(00)00094-9.

Thorson, R.M., 2000, Glacial tectonics: A deeper perspective: Quaternary Science Reviews, v. 19, p. 1391–1398, doi: 10.1016/S0277-3791(00)00068-8.

Winkler, G.R., and Plafker, G., 1993, Geologic Map and Cross Sections of the Cordova and Middleton Island Quadrangles, Southern Alaska: U.S. Geological Survey Miscellaneous Investigation Series Map I-1984, scale 1:250,000, 1 sheet and text.

Manuscript Accepted by the Society 02 June 2009

Printed in the USA

The Geological Society of America
Special Paper 462
2010

Holocene history revealed by post-surge retreat: Bering Glacier forelands, Alaska

Kristine J. Crossen*
Department of Geological Sciences, University of Alaska, 3211 Providence Drive, Anchorage, Alaska 99508, USA

Thomas V. Lowell*
Department of Geology, University of Cincinnati, Cincinnati, Ohio 45221-0013, USA

ABSTRACT

Since 1890, Bering Glacier has been retreating from its Little Ice Age position near the Gulf of Alaska. Although this retreat has been punctuated by spectacular surge events, areas previously under the ice have regularly become exposed following surges. Sedimentary deposits above the current level of Vitus Lake are well dated to the past 2200 a. Older deposits eroded from the floor of the previous fjord are deposited above lake level by ice advance, and these sediments and fossils allow us to reconstruct the past history of Bering Glacier.

Bering Glacier extended past the modern Pacific coastline during the Late Glacial Maximum (LGM), but retreated up its fjord by 13,000 a ago, leaving an embayed coastline inhabited by marine invertebrates until 5000 a ago. The shoreline was then uplifted, and terrestrial sandy outwash with intermittent peat bogs covered the landscape. Forests covered the area by 4000 a ago, with evidence of 800 a of continuous forest occupation (from 200 B.C. to A.D. 600) found at the Ancient Forest site. This forest was buried by rising lake levels, and between A.D. 600 and 1000, active gravel outwash deposition alternated with thin peats growing on more stable surfaces. Subsequently, active outwash deposits aggraded to 21 m above the current level of Vitus Lake as glacier outwash covered the forelands. A thin till capping the advance outwash is the only indicator of the Little Ice Age advance that occurred sometime after A.D. 1100. The sediment record does not hold evidence for multiple advances of Bering Glacier during the past 2200 a.

INTRODUCTION

In southern Alaska during the Late Glacial Maximum (LGM), massive glaciers flowed out of coastal mountains and extended across the continental shelf that was exposed when sea level lowered ~100 m (Carlson et al., 1982; Molnia, 1986). As climate warmed and the ice retreated, sea level rose, and coastal glaciers became tidewater glaciers calving into fjords along the Gulf of Alaska. Warming trends continued into the Holocene, beginning 10,000 yr before present (yr B.P.), and maximum temperatures were reached in many parts of Alaska ca. 9000 yr B.P. (the Birch Period). Following this time, cooling trends returned during the Neoglacial Period, and glaciers began cycles of advance and retreat responding to variations in temperature and precipitation for terrestrial glaciers, and to changes in water depth and sediment deposition for tidewater glaciers (Crossen, 1997).

*Emails: Crossen: afkjc@uaa.alaska.edu; Lowell: Thomas.Lowell@uc.edu.

Crossen, K.J., and Lowell, T.V., 2010, Holocene history revealed by post-surge retreat: Bering Glacier forelands, Alaska, *in* Shuchman, R.A., and Josberger, E.G., eds., Bering Glacier: Interdisciplinary Studies of Earth's Largest Temperate Surging Glacier: Geological Society of America Special Paper 462, p. 235–250, doi: 10.1130/2010.2462(12). For permission to copy, contact editing@geosociety.org.

Southern Alaskan glaciers are the focus of glacial research, because they have robust regimes that respond to climate change, and because their forelands commonly contain widespread organic materials. Forests colonize areas following glacial retreat, and if trees, soils, and peat are overrun by advancing ice and incorporated into glacial deposits, they allow dating of glacial events. In some of the most spectacular situations, buried forests become exhumed during subsequent cycles of ice retreat. Bering Glacier is the largest of the ice bodies along the coastal mountains, and researchers have used its deposits to help understand its past activity and implications for climate change.

Early explorers who described and mapped the area near Bering Glacier documented the ice terminus adjacent to the coast about A.D. 1900 (Seaton-Karr, 1887; Martin, 1905, 1907; Pierce and Winslow, 1979). This extended position of the ice correlates with worldwide glacial advances associated with the Little Ice Age (LIA), a cooling period from about A.D. 1350 to A.D. 1850 (ca. 650–150 yr B.P.). Subsequent workers have used buried trees and tills to suggest that earlier advances of Bering Glacier occurred ca. 1500 yr B.P. (A.D. 500) (Wiles et al., 1999), 1400 yr B.P. (A.D. 600) (Molnia and Post, 1995; Wiles et al., 1999), 800 yr B.P. (A.D. 1100) (Calkin, et al., 2001), and 700 yr B.P. (A.D. 1300) (Molnia and Post, 1995). Fleisher and Muller (Fleisher et al., 1998; Muller et al., 1991) suggest that multiple tills are interbedded within a complex sequence of stream outwash and lacustrine (lake bottom) sands. This would imply numerous advances over the past few thousand years. Wiles et al. (1999) suggested that an advance ca. 1500 yr B.P. (A.D. 500) is similar to that of other glaciers in southern Alaska.

Based on both the aforementioned research and the ages of trees growing on the terminal moraine (Wiles et al., 1999), most workers agree that Bering Glacier began its retreat from its LIA maximum in the late 1800s. However, this retreat has been punctuated by surges dated to about 1900, about 1920, 1938–1940, 1957–1960, 1965–1967, and, most recently, 1993–1995 (Post, 1972; Molnia, 1993; Fleisher et al., 1994; Fleisher et al., 1995; Muller and Fleisher, 1995). The surges are small, rapid short-term advances, lasting a few years, that punctuate the overall retreat. Following these surges, the glacier reverts to the retreat mode that is driven by post-LIA climate warming.

This research project began with the discovery of the Ancient Forest, a stand of trees uncovered by glacial retreat in 1998. Along the western edge of Vitus Lake, the ice in Tashalich Arm had retreated prior to 1993, but the 1993–1995 surge then buried the area with advancing ice (Molnia and Post, 1995). By 1998, the post-surge retreat exposed the forest for apparently the first time (Crossen et al., 2002). After 1998 the glacier continued to retreat northward up Tashalich Arm, and additional cliffs containing dead trees and peats became exposed, allowing dating of the organics contained within these deposits (Crossen, 2006a). Following these discoveries, we investigated all available sections around Vitus Lake with the exception of those on Arrowhead Island (on the far eastern part of the lake), as these were previously studied by J. Fleisher and his co-workers. The purpose of the project was to gather evidence for the sequence of past glacial advances and retreats, and to compare these data with those collected by earlier workers along the eastern ice margin. Using the dating of the organic materials and the interpretation of the interbedded sands and gravels, we wanted to reconstruct the Holocene history of the Bering Glacier forelands (Crossen, 2006b).

METHODS

Stratigraphic sections along the shoreline of Vitus Lake (Figs. 1 and 2) were located using aerial photography and aerial and boat surveys, and were then investigated on foot. Samples were excavated by hand tools from freshly exposed lakeshore sediments and from gullies cut by meltwater issuing from the ice.

Stratigraphic sections were measured using tapes, and elevations were established as levels above Vitus Lake (which averages 1–2 m above sea level). All locations were recorded using global positioning system (GPS) equipment. Positions of individual trees in the Ancient Forest trees were surveyed using a Trimble surveying station.

Organic samples were collected by hand and air-dried both in the field and in the Quaternary Geology Laboratory at the University of Alaska Anchorage. Samples were prepared and sometimes split before being sent for processing to the Beta Analytic Radiocarbon Dating Laboratory. Samples were dated using standard C-14 techniques and corrected for $^{13}C/^{12}C$ ratios. Table 1 shows dates as reported in both C-14 yr (yr B.P.) and calendar years (cal yr B.P.), with B.P. representing "before present" (the "present" being 1950). Calendar years were calibrated using standard techniques (Talma and Vogel, 1993; Stuiver and Pearson, 1993; Stuiver, 1998: Reimer et al., 2004). Wood samples were taken from whole trees, bark, or slabs cut from trunks, and peat was sampled from stratigraphic zones. The dating of these samples allowed us to identify when coastal forests, bogs, and other nonglacial environments existed in areas that were subsequently covered by ice and later exposed as Bering Glacier retreated.

USING STRATIGRAPHY TO INTERPRET PAST ENVIRONMENTS

Stratigraphic sections are used to interpret past environments through an understanding of how sediments are deposited. Till forms adjacent to the glacier as material is melted out of the base of the ice or from the ice face as the glacier retreats. This poorly sorted material contains everything from large automobile-size boulders to gravel, sand, and clay. This material is dumped off or melted out of the ice and left in place as the ice retreats.

In comparison, outwash material is deposited by braided streams as they flow from the melting ice toward a lake or the ocean. Streams carry the materials released from the ice and then sort the sand and gravel as they lose competence. Coarser grained gravels are deposited closer to the ice, and finer grained sands farther away (Boothroyd and Ashley, 1975). Moving water in the

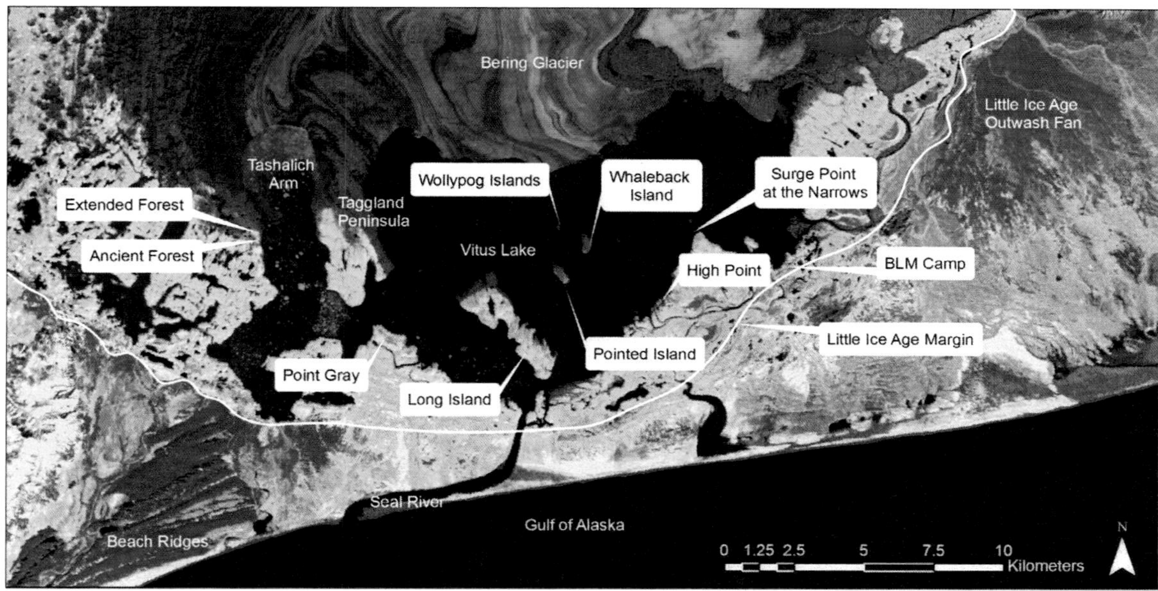

Figure 1. False color LANDSAT image of Bering Glacier Region. Place names and site locations in the Bering Glacier foreland. The late 1800 morainal position can be seen along the southern seaward margin of Vitus Lake. Along the southeast and southwest margins of this moraine, outwash plains prograde to the Gulf of Alaska and may be the modern analogues for the mid-Holocene environments adjacent to the advancing Bering Glacier. A series of raised beach ridges are seen along the coastline on either side of the Seal River. These features may have prograded across the fjord mouth in the mid-Holocene, changing Tashalich Arm from marine to fresh water. It is likely that these features were uplifted by earthquake events in the mid- to late Holocene.

streams can produce cross-bedding when the constantly changing streams cut and then fill their channels.

Farther from the ice, the finest grain sizes such as silt and clay will be carried by currents but are deposited only when quiet-water environments are encountered. Horizontal and parallel-bedded sands, silts, and clays commonly indicate settling in standing bodies of water. A common deposit in glacial lakes is a couplet of sand and clay. The sandy material is deposited during the summer melt season when turbid water flushes out of the glacial system. During the winter, glacial outflow decreases, and floating ice covers all or parts of the basin. In this low energy environment, a thin winter unit of fine-grained silt or clay drapes over the thicker sandy summer unit (Ashley, 1975; Shaw, 1975).

RESULTS

Ancient Forest Site

The Ancient Forest (Figs. 1 and 2) lies on the western side of Vitus Lake along the western shore of Tashalich Arm. Although this location was deglaciated before 1993, it was overrun during the 1993–1995 surge and was subsequently re-exposed in 1998. The site is on a small peninsula that protrudes from the west side of Tashalich Arm and was surrounded on three sides by ice when the forest was first discovered in 1998. (The lakeshore abutted only the south side of the peninsula.) The ice surged over the position of the forest and left an identifiable moraine on the plateau above the forest (Fig. 3). In 1999 the ice retreated and the peninsula became surrounded on three sides by lake water. A 22 m cliff composed of bedded sands and gravels extends above the Ancient Forest locale on its western side (Fig. 4).

When first uncovered, the Ancient Forest consisted of 39 large trees, 17 of which were rooted in the substrate with upright trunks. The remaining trees were tilted or horizontal, with a few in standing water. The largest tree measured 1.6 m in diameter, and several tree trunks contained 200–300 rings. The tops of three trunks were splintered toward the south in the direction of ice flow (Fig. 5).

Large blocks of peat were scattered around the forest. Several peat blocks were imbricated behind glacial boulders, indicating fluvial transport (Fig. 6); one continuous peat block ~20 m long may have been in situ. Spruce cones, spruce needles, bark, roots, and broken branches were identified on the surface and within the peat blocks (Fig. 7). This indicates a well-established forest at this location, similar to modern Alaskan coastal forests in areas adjacent to Bering Glacier.

The first clearly defined in situ area of peat was found along the north shore of the peninsula in 2005. This 8.2-m-long section was exposed only during lowest lake levels. Subhorizontal brown peat up to 1 m thick contained large wood fragments, branches, and needles similar to those of the Ancient Forest. The peat was underlain by bedded silts and fine sands, and overlain by sands.

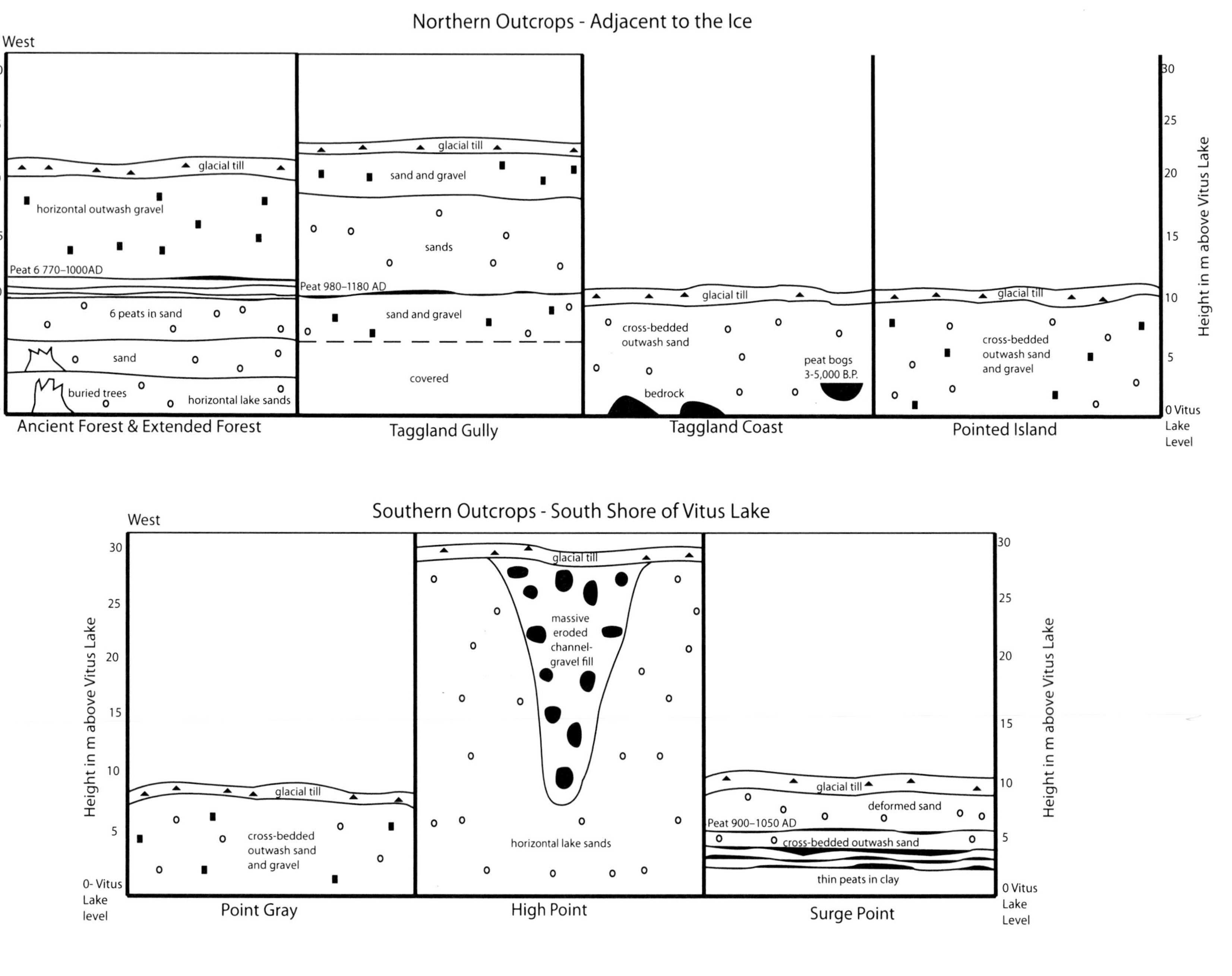

Figure 2. Stratigraphic sections along Vitus Lake bluffs. Refer to Figure 1 for map locations.

TABLE 1: RADIOCARBON DATES FROM BERING GLACIER FORELANDS

Site & sample	Lab number	Radiocarbon age ±1σ before present (BP)	Calibrated age 2σ BP	Calibrated date (BC–AD) 2σ	Material
Ancient Forest Site (AF)					
BG-98-03	Beta - 133377	2170 ± 60	Cal 2335–2000	50–385 AD	Basal peat—in situ - twigs - AF
BG-98-04	Beta - 133378	1590 ± 50	Cal 1565–1365	385–585 AD	Upper peat—in situ - wood sample - AF
BG-98-10	Beta - 133382	1220 ± 60	Cal 780–1280	670–970 AD	Outer rings—cut tree – AF
BG-C-04-04	Beta - 200061	1580 ± 60	Cal 1580–1330	370–620 AD	Upper peat—in situ - island - AF
BG-C-04-05	Beta - 200062	2080 ± 50	Cal 2150–1920	200 BC–30 AD	Basal peat—in situ - island - AF
Transported organics - AF					
BG-98-01	Beta - 133375	1450 ± 60	Cal 1485–1470 and 1430–1275	465–480 AD and 520–675 AD	Wood from ice surface—east of AF
BG-98-02	Beta - 133376	1800 ± 60	Cal 1870–1560	80–390 AD	Transported peat in stratigraphy—above AF
BG-98-06	Beta - 133379	1950 ± 70	Cal 2050–1720	100 BC–230 AD	Transported peat in gully—above AF
BG-98-07	Beta - 133380	3670 ± 60	Cal 3845–4155	1895–2205 BC	Transported wood—plateau above AF
BG-98-09	Beta - 133381	1580 ± 60	Cal 1335–1585	365–615 AD	Transported peat—adjacent to ice 1998
Extended Forest Site (EF)					
BG-C-04-01	Beta - 200059	1380 ± 60	Cal 1380–1180	570–770 AD	Peat 1—bark from vertical tree - in situ
BG C-04-02	Beta - 200060	1420 ± 60	Cal 1410–1260	540–690 AD	Peat 2—outer wood - vertical tree - in situ
BG-C-02-02	Beta - 200067	1370 ± 60	Cal 1360–1180	580–770 AD	Peat 3—root in peat - in situ
BG-C-02-01	Beta - 200066	1500 ± 50	Cal 1520–1300	430–650 AD	Peat 4—organic mat - in situ
BG-C-04-09	Beta - 200064	1200 ± 60	Cal 1270–970	680–980 AD	Peat 5—tree branch in peat - in situ
BG-C-04-07	Beta - 200063	1150 ± 50	Cal 1180–950	770–1000 AD	Peat 6—twigs in peat - in situ
Taggland					
BG-C-04-14	Beta - 200065	980 ± 50	Cal 970–780	980–1180 AD	Wood in peat layer—in situ
BG-06-01C	Beta - 228901	3090 ± 60	Cal 3440–3160	1490–1210 BC	Basal peat—in situ - south end of outcrop
BG-06-02C	Beta - 228902	3030 ± 60	Cal 3370–3060	1420–1120 BC	Basal peat—in situ - north end of outcrop
Surge Point–The Narrows					
BG-05-11C	Beta - 228900	1030 ± 50	Cal 1050–900 and 860–810	900–1050 AD and 860–810 AD	Upper peat—in situ - thin peat in sediments

The lack of till overlying the peat suggests that the Ancient Forest was not killed and buried by overriding ice but smothered by sands from a higher lake level.

Dating of the upper and lower layers of the continuous peat in the Ancient Forest and of the in situ peats adjacent to the Ancient Forest (Table 1) indicates that the thick peat at this locality accumulated between ~200 B.C. (2200 yr B.P.) and A.D. 600 (1400 yr B.P.). Both sections show similar dates for the basal and uppermost peat layers, and suggest an 800-a-long interval dominated by a terrestrial forest at this locality.

Extended Forest Site

The Extended Forest site lies ~50 m north of the Ancient Forest along the west side of Tashalich Arm (Fig. 1). In 1998 this area was covered by ice, but subsequent ice retreat exposed a 22-m-thick section of horizontally bedded sands and gravels (Fig. 2). Although the Ancient Forest beds are not exposed at this site, trees standing in lake water below lake level are probably associated with the Ancient Forest beds. The 22 m stratigraphic section (Figs. 8–10) reveals six additional forest and peat layers above lake level.

In the lowest part of the section, 0–4 m above the lake level (and presumably burying the Ancient Forest peat), is a 4-m-thick horizontally bedded sand. This unit is composed of 5–10-cm-thick sandy beds, each overlain by a 0.5–1 cm silt drape with sporadic soft sediment deformation (Fig. 11). These sediments lack diatoms and thus are unlikely to be marine (I. Shennan, 2003, personal commun.). These sands and silts presumably are lacustrine (lake) beds that buried the Ancient Forest when lake levels rose and rhythmic couplets were deposited by seasonal lake dynamics.

A peat layer (Peat 1) overlies these lacustrine beds. It is composed of a 50 cm organic silt exhibiting a reddish soil development that is overlain by a 10 cm silt layer. One large tree is clearly rooted in place within this red soil (Fig. 12). This layer contains numerous other trees in horizontal, vertical, and tilted positions. The four largest trees have ring counts of 104, 187, 198, and 204

Figure 3. Location of Ancient Forest along west side of Tashalich Arm. The small peninsula along the shoreline is covered by the Ancient Forest. The 1993–1995 moraine on the plateau above the Ancient Forest shows that ice covered the Ancient Forest during this surge event.

Figure 5. Largest tree in Ancient Forest site, with the top splintered in the down-ice direction, suggesting ice override. This tree may have been exposed prior to 1993 and overridden in the recent surge event. The size of the tree is similar to modern Sitka spruces (M. Barker, 2002, personal commun.).

Figure 4. Ancient Forest site in 1998. Seventeen rooted trees, 22 other trees and extensive peat beds were found along the shores of Tashalich Arm. A 22-m-high cliff extends above the Ancient Forest on the left, with the ice from Bering Glacier forming the dark cliff in the background.

Figure 6. Peat blocks imbricated (tilted) against boulders suggest that some blocks were transported by moving water below or adjacent to the ice. This raises the possibility that an outburst flood may have occurred at this site in addition to the well-documented outburst along the eastern margin of Bering Glacier. (Fleisher et al., this volume).

Figure 7. Excellent preservation of twigs, cones, and needles suggests that the peat originated from the floor of a coastal spruce forest (M. Barker, 2002, personal commun.).

Figure 8. Extended Forest site overview along the west side of Tashalich Arm. A 22 m cliff with six in situ peat and forest beds overlie the Ancient Forest. (Note three people for scale.) The large boulder in the center was deposited by ice that draped over the cliff and filled Tashalich Arm during the most recent surge event.

Figure 9. Six subhorizontal in situ peat beds at the Extended Forest site. The lower peats contain forest beds and large trees, whereas the upper peats are thinner.

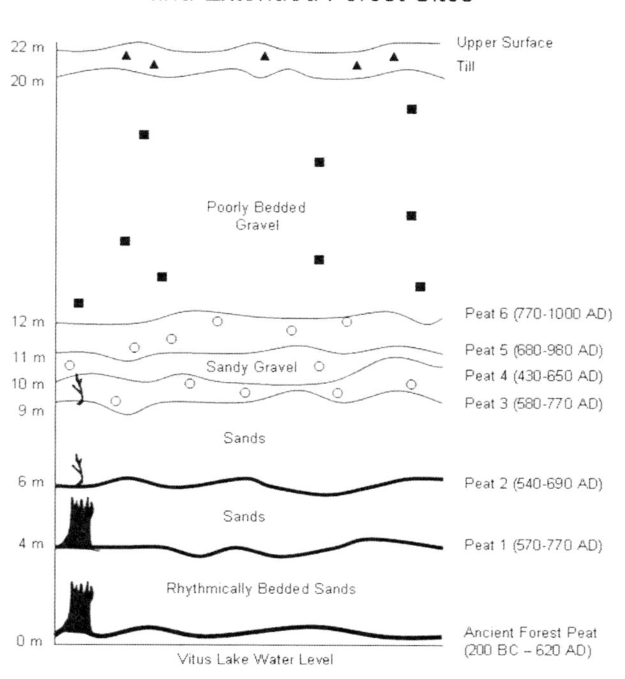

Figure 10. Statigraphic section showing sediments, peats, trees, and dates of materials in the Ancient Forest and Extended Forest sites along Tashalich Arm.

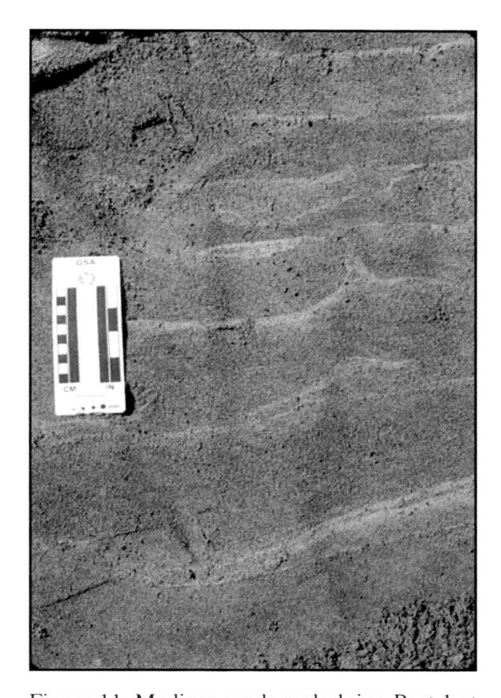

Figure 11. Medium sands underlying Peat 1 at the Extended Forest site. These same materials overlie the in situ Ancient Forest peats in the adjacent area. Fine drapes overlie the sand, indicating pulses of sediment deposited in the lake basin. Soft sediment deformation occurs along some bedding planes.

Figure 12. Peat 1, the lowest forest bed at the Extended Forest site. This large tree is rooted 4 m above lake level and 4 m above the Ancient Forest peats. Peat 1 shows the development of reddish soil below the tree trunk, which formed on the sandy material interpreted as lake deposits.

Figure 13. Peat 2 at the Extended Forest site. This 1-cm-thick peat dates to A.D. 540–690. It contains small trees bent over to the south. Note the sands and clays surrounding the wood.

rings, respectively. The rooted tree has its stem truncated at the level of overlying Peat 2, suggesting that it was buried and then truncated. Horizontally bedded sands 3.7 m thick bury both the tree and the red soil and underlie the next unit, Peat 2, which is a 1-cm-thick organic mat with small spruce trees with bent trunks oriented to the south (Fig. 13).

Peats 3 and 4 are exposed above lake level at 9.56 m and 10.16 m, respectively (Fig. 14). Both peats are only a few cm thick, with small alders rooted in place, and broken pieces of wood aligned along the peat surface. Peat 3 is buried by massive, poorly bedded sands alternating with well-bedded pebbly gravels.

Peats 5 and 6 are above lake level at 11.44 m and 12.24 m, respectively (Fig. 14). Both peats are thin mats up to 1 cm thick, mainly composed of small plants, roots, and twigs. Interbedded sediments between Peats 4 and 6 are horizontal sands and gravels, interpreted as glacial outwash deposits.

All the peat layers in this section were radiocarbon dated (Fig. 10). Peats 1 and 2 were dated using the bark from the large rooted trees, indicating that these trees died A.D. 570–770 and A.D. 540–690, respectively. Several trees at the Ancient Forest site that were two or more meters above the lake level appear to be associated with Peat 1 of the Extended Forest site. A date of A.D. 670–970 from the outer rings of one of these Ancient Forest trees overlaps with the dates from Peat 1 (Table 1). Peat 1

Figure 14. Peats 3–6 at the Extended Forest site. These uppermost peats are considerably thinner than Peat 1 and contain only small twigs and roots. These peats are interbedded with sands and gravels, suggesting that outwash sediments alternated with stable surfaces where peats could develop. Dates for the peats indicate 8 m of aggradation between A.D. 600 and 1000.

correlates with the Main Forest Bed along the eastern margins of Vitus Lake, as described by Wiles et al. (1999).

Peats 3, 5, and 6 were dated from wood fragments within the peats: roots from Peat 3, alder branches from Peat 5, and twigs from Peat 6. The organic mat was dated in Peat 4. The dates (Table 1) clearly indicate multiple periods of forests and peat growth at this locality from about A.D. 600 (1400 yr B.P.) to A.D. 1000 (1000 yr B.P.). Several of these peats appear to correlate with peat layers located at Tsui and Tsivat Lakes along the eastern margin of Bering Glacier (Wiles et al., 1999).

No additional peats are found above Peat 6, and massive and bedded sands and gravels make up the upper section from 12.24 m to 21.64 m above lake level. The uppermost 1–2 m of the section is composed of poorly sorted till with large angular boulders and cobbles deposited by overriding ice (Figs. 10 and 15).

Interpretations of the Ancient and Extended Forest Sites

The Ancient Forest and Extended Forest sites are instrumental in our understanding of the depositional environments of the Bering Glacier forelands between about 200 B.C. (1200 yr B.P.) and A.D. 1000 (1000 yr B.P.) (Fig. 10). The peats and buried trees indicate that a terrestrial environment occupied the area along the western edge of the active lobe of Bering Glacier from 200 B.C. until the present time. The Ancient Forest was established as a mature coastal forest lasting 800 a (from 200 B.C. to A.D. 600) and was then buried by lacustrine sediments. Between A.D. 600 and 800 (1400–1200 yr B.P.) a second mature forest was established over the lacustrine sediments, and a reddish soil was developed on the sandy substrate. This second forest was then buried by sedimentation ca. A.D. 800 (1200 yr B.P.). If both Peats 1 and 2 were buried by lacustrine sands, the ancient lake levels stood 6 m higher than at present. Both of the major forest beds along Tashalich Arm correlate with forest beds described along the eastern margins of Bering Glacier. This indicates that a widespread forest of large mature trees previously occupied localities that are now adjacent to the active margin of Bering Glacier, and that fluctuating lake levels buried these forests.

Figure 15. Uppermost 10 m of Extended Forest site. Subhorizontal gravels are devoid of peat deposits and indicate substantial aggradation by glacial outwash between A.D. 1000 and the LIA advance.

The sediments overlying these forest beds indicate conditions fluctuating between sediment aggradation and surface stability from A.D. 600 to 1000 (1400–1000 yr B.P.). The sediments overlying Peat 2 become increasingly coarse grained (Fig. 14), implying that the depositional environment changed from sandy lacustrine to gravelly outwash, probably as the ice advanced closer to this site.

The peats were also thinner, and the trees associated with these peats became smaller, above Peat 2. The alternating deposits of thin peats interbedded with gravels between Peats 3 and 6 imply that active outwash stream deposition alternated with periods of surface stabilization. Although the surface did not remain stable long enough for large forests to colonize, small alders and other plants became established.

The dating of these peats allows a calculation of the rate of outwash aggradation. From A.D. 600 to 1000 (1400–1000 yr B.P.), 8 m of outwash was deposited for an average aggradation rate of 2 cm/a. After A.D. 1000 (1000 yr B.P.), no peats were interbedded with outwash gravels (Figs. 10 and 15). This may be interpreted as 8 m of outwash being deposited so rapidly that no stable surface would be available for peat colonization, or alternatively that the peat surfaces were established but then eroded by subsequent active outwash activity. The uppermost outwash gravels contain transported and often imbricated peat blocks (Fig. 16). Three peat blocks dated to 100 B.C.–A.D. 230, A.D. 80–390, and A.D. 365–615 (Table 1) are the ages of the Ancient Forest peats and indicate that glacial ice and glacial outwash were eroding and transporting the Ancient Forest and other peats during multiple events after they had formed.

No tills are interbedded with the outwash deposits, and the uppermost till suggests that the glacier overrode the area only in the most recent times (Fig. 10). The thin surface till does not allow separation of the different glacial events such as the LIA advance and the numerous post-LIA surge events. The lack of multiple interbedded tills does suggest, however, that the LIA is the only major glacial advance in the past 2200 a at this location.

The three large trees with their trunks splintered in the down-ice direction (Fig. 5) suggest that ice override killed some of the trees and engulfed the forest. However, this is contradicted by the stratigraphic evidence that lake sands bury the trees up to 2 m above their bases, suggesting that the trees were killed by sediment inundation (Fig. 10). Additional evidence that may help settle this conundrum is the location of transported wood on the plateau above the Ancient Forest. The wood is found only in a limited area adjacent to the Ancient Forest, and dates on the transported wood and peat all fall within the range of the Ancient Forest dates (Table 1). B. Molnia (2007, personal commun.) reports that ice in Tashalich Arm had retreated north of the Ancient Forest site by 1993, and then advanced southward down Tashalich Arm during the 1993–1995 surge. Thus the

Figure 16. Imbricated peat block in upper gravel above the Ancient Forest site. This peat dates to A.D. 80–390 and formed at the same time as the Ancient Forest peat but was eroded and deposited above the Ancient Forest peat. This suggests the occurrence of additional Ancient Forest beds that have been reworked by glacial transport and outwash sedimentation.

Figure 17. Ancient Forest site in 1998. The dark cliff in the background is the ice of Bering Glacier. Note the buried tree in the foreground with the splintered top. This may have formed by ice overriding the partially exhumed tree during the most recent surge event. This may have been how all three splintered trees in the Ancient Forest formed.

conflicting evidence could be resolved if the Ancient Forest site had been exhumed prior to the 1993–1995 surge, leaving some of the tree trunks partially exposed. When the surging ice overran these trunks, it could have splintered their tops (Fig. 17). The same advance could have carried wood from the forest onto the plateau area adjacent to the forest as well. Subsequently, a flood event (likely an outburst flood, similar to the one along the eastern side of Vitus Lake—see Fleisher et al., this volume) could have exhumed the trees and forest beds, and deposited the imbricated peat blocks at the Ancient Forest site (Figs. 4–6).

Taggland Peninsula

A unique site lies on the west side of the Taggland Peninsula and along the east shore of Tashalich Arm directly across from the Ancient Forest site (Fig. 1). This site was previously investigated by B. Molnia, I. Shennan, and A. Pasch. Their findings of boring clams in situ within Tertiary sandstones and siltstones provide important information concerning the timing of the marine environment in Vitus Lake. Dates from this locale indicate that subtidal boring clams occupied this location from ca. 13,000–5000 yr B.P. (Shennan and Hamilton, this volume; Pasch et al., this volume).

Unconsolidated Holocene cross-bedded sands with minor gravel lenses overlie the consolidated Tertiary bedrock (Figs. 2, 18). This suggests that sandy outwash deposits prograded to at least 2 m above lake level here. At two localities, one north and the other south of the bedrock outcrop, peats are deposited above this sand, and all were overrun during the 1993–1995 surge. Prior to this surge, B. Molnia (Molnia and Post, 1995) collected the northern peat and dated it to ca. 4600 yr B.P. In 2006, following ice retreat from this area, the northern peat was not found in situ, but deformed fragments were found, presumably transported by the surge event. In the southern outcrop, although the surface peat had been disturbed by ice advance, the lower layers were intact and well rooted to the underlying cross-bedded sands. Two samples of basal peat (within 1 m of lake level) yielded dates of 3250 and 3340 calibrated (cal) yr B.P., respectively (Table 1). This indicates a changing environment at ca. 4000 yr B.P., when the rocky marine shoreline was transformed into a terrestrial sandy outwash plain interspersed with peat bogs. This may have been caused by 1–2 m of tectonic uplift, similar to that experienced here during the 1964 Alaska earthquake (Plafker and Rubin, 1967; Plafker, 1969).

In a gully behind the shoreline outcrop, 44 trees and numerous peat blocks were found, but none of the trees were rooted in place, suggesting that these organics were transported by glacier and stream processes. In the gully wall above the trees, a 13-m-long section of in situ peat was found (Fig. 2). The 1–2-cm-thick peat contained wood fragments that appeared to be alder. This peat was underlain by 3 m of coarse sandy gravel and overlain by 1–2 cm of fine sand and silt. An additional 15 m of medium tan and gray sand, overlain by 4 m of bedded sand and gravel, cap the section. The sand and gravel deposits both

Figure 18. Taggland site along the east side of Tashalich Arm. Tertiary sandstone is bored by marine clams (lower right) and is discussed in Pasch et al. (this volume) and Shennan and Hamilton (this volume). Cross-bedded sands interpreted as distal outwash overlie the bedrock and indicate the change from a marine to a terrestrial environment after 5000 yr B.P. These sands are cross-cut by fine-grained clastic dikes interpreted as having formed by injection of pressurized subglacial water during glacial surges.

Figure 19. Taggland site along the east side of Tashalich Arm. The 4-m-thick sandy outwash is overlain by 5 cm of glacial till. Additional clastic dikes cut the sandy outwash here.

above and below the peats are interpreted as glacial outwash. The peat dates to A.D. 980–1180 (Table 1) and indicates a time when the outwash surface was temporally stabilized. These dates overlap with those of Peat 6 at the Extended Forest site and at other sites along the eastern margin of Vitus Lake (Wiles et al., 1999), suggesting a widespread stabilization of the outwash surface at this time.

At all sites on the Taggland Peninsula, the uppermost meter of sediments is composed of thin till with striated rocks and angular supraglacial boulders (Figs. 2 and 19). This indicates ice override onto the advance outwash. The peat date above indicates that the ice advance must postdate A.D. 1100 (900 yr B.P.). The uppermost stratigraphic units on Taggland Peninsula (cross-bedded sands on the west side, and bedded sandy gravels on the east side) contain clastic dikes that extend from the base of the till into the subsurface. The fine-grained nature of the dikes implies that they were formed by pressurized subglacial water injected into fractured subglacial deposits (Figs. 18 and 19). Glacial surges are well known for producing massive amounts of englacial and subglacial water that could have formed these clastic dikes.

Vitus Lake Islands

The Mollypogs are two small islands in Vitus Lake northeast of Long Island (Fig. 1). These were completely overrun by ice during the 1993–1995 surge event and recently have become deglaciated. Both are low elevation islands that are covered during high lake levels. No stratigraphic sections, forest, or peat beds were identifed during examination of these islands.

Whaleback Island is one of the larger islands in Vitus Lake east of Long Island (Fig. 1). Most of the island was overrun by ice during the 1993–1995 surge event, but it was entirely deglaciated by 2005. The low surface gently rises to the south and culminates in a 10-m-high cliff along the south, southeast, and southwest sides of the island (Fig. 20). Along the southeastern coast a 5 m cliff exposes a 2.5 m sandy gravel overlain by a thin organic mat, which is overlain by 0.5 m of fines, 1 m of massive sand, 1 m of pebbly fines, and a till containing plucked and striated cobbles on the surface (Fig. 2).

Figure 20. Whaleback Island, showing fluted surfaces produced by the 1993–1995 surge. The 10 m cliffs along the south side of the island expose horizontal sand and gravels that are interpreted as glacial outwash deposits.

Pointed Island lies to the west of Whaleback Island, and adjacent to Long Island (Fig. 1). Like Whaleback Island, its slope rises gently from north to south, ending in 10-m-high bluffs on both the south and west sides of the island. Only the far northern part of this island was overrun during the surge event. The southern bluffs expose 10 m of both horizontal and cross-bedded sands and gravels (Figs. 2 and 21). These clearly represent glacial outwash deposits aggraded to 10 m above the current lake level. A thin gray till, up to 10 cm thick and containing large boulders, covers the outwash and suggests an ice advance that postdates many years of proglacial outwash deposition at this locality (Fig. 22).

Although both Pointed and Whaleback Islands currently lie in the center of the deepest parts of Vitus Lake (130 m deep; Molnia and Post, 1995), the stratigraphy indicates that this area was previously a large braided outwash channel aggraded to 10 m above the current lake level. The thin upper till indicates that ice advance is the most recent of geologic events.

Figure 21. Exposure at Pointed Island. Cross-bedded sand and gravels with transported wood are interpreted as outwash deposits. Note tree branch in foreground for scale.

Point Gray

This outcrop lies along the south side of Vitus Lake where Tashalich Arm joins the larger lake, east of the north end of Long Island (Fig. 1). Several sections are exposed for ~1 km along the lakeshore. The exposures are 4–8 m high and reveal ~0.5–1 m surface till underlain by 3–7 m of horizontal and cross-bedded sand and gravels (Fig. 2). These are interpreted to be outwash deposits that were overrun by ice. No trees or organic zones were exposed in the cliffs.

High Point

This locality is the highest bluff along the south shore of Vitus Lake, lying between the Seal River and Surge Point at The Narrows (Fig. 1). The 800-m-long section contains a 31-m-high cliff of 8–19-cm-thick rhythmically bedded horizontal sands overlain by 0.5–1-cm-thick silty drapes. These are interpreted as high elevation lacustrine beds (Fig. 2).

The sandy beds are eroded in several places by cut-and-fill structures that are up to 23 m deep and 20 m wide. These channel features are filled with massive, unsorted boulder deposits; coarse, angular gravels with sand lenses; and gravelly cross-bedded sands. These were presumably eroded by high-energy agents, such as debris flows or outwash channels adjacent to the ice. The uppermost 1 m exhibits a till with angular boulders and large transported tree trunks (Figs. 2 and 23). These deposits are interpreted as lake beds that were overrun and incised by proglacial or subglacial processes associated with ice advance.

Figure 22. Pointed Island: 10-m-thick subhorizontal sandy gravels are overlain by 10-cm-thick bouldery till at the surface. Neither Whaleback Island nor Pointed Island shows stratigraphic evidence of tills interbedded with gravels.

Surge Point at the Narrows

This section lies along the southeastern shoreline of Vitus Lake and marks the location where the 1993–1995 surge reached within a few hundred meters of the lakeshore. Drainage was then

Figure 23. High Point: highest elevation (31 m) along the south shore of Vitus Lake. Here the unsorted till overlies well-bedded sandy lake beds with clay drapes. This is the only place along the lakeshore where two tills are found in stratigraphic section, and both are within the uppermost 2 m of the section.

Figure 24. Surge Point at The Narrows. This locality along the southern Vitus Lake shoreline is where the ice nearly reached the shoreline during the 1993–1995 surge and where a narrow channel flowed between the ice margin and the shoreline cliff. The 11-m-high exposure contains 30 couplets overlain by 6 peats, the uppermost dating to A.D. 1010. The silty couplets are lake deposits that were subsequently vegetated by the peats as lake levels dropped or the area was uplifted. The peats are overlain by a 1–2-m-thick till, which indicates that the ice advance must postdate A.D. 1010.

restricted between the ice front and the shoreline (The Narrows), and a deep channel was eroded there (Fig. 1). A 500-m-long section exposes up to 11 m of horizontal layers (Fig. 24). The lowest section is composed of 4 m of rhythmically bedded sands with silty clay drapes containing 30 couplets. Five thin peat beds a few cm thick overlie the couplet unit. These peats are further overlain by 2 m of cross-bedded sands containing another 4 cm peat at the top. This upper peat is overlain by 4 m of deformed sands containing faults, folds, and clastic dikes that were probably deformed by overriding ice. This deformed sand is overlain by 1–2 m of silty till containing boulders (Fig. 2). The date of the uppermost peat of A.D. 950–1050 suggests that the ice advance postdates this time (Table 1).

These last two sections in adjacent areas along the shoreline of Vitus Lake (High Point and Surge Point) show that adjacent small basins clearly contain sediments originally deposited in quite disparate sedimentary environments, a common situation in glacial environments (Fig. 2).

Holocene History of Bering Glacier Foreland

The foreland surrounding Bering Glacier reveals numerous stratigraphic clues to the Holocene history of Bering Glacier. Although the buried forests and marine fossils released from melting ice (Pasch et al., this volume) are spectacular testimony to changing environments, the most significant finding is the youthful ages of the deposits exposed above lake level and the lack of multiple tills in the sections. In all of the exposed stratigraphic sections, glacial till is found only in the uppermost parts, whereas most of the deposits originated in nonglacial environments.

Based on submarine cores taken off Kayak Island, it appears that ice retreated from the continental shelf by 14,000 yr B.P. (T. Ager, 2007, personal commun.). Basal peat dated from the eastern foreland near Hanna Lake suggests that deglaciation in that area occurred by 11,000–12,000 yr B.P. (Fleisher et al., 1999). It seems likely that Bering Glacier retreated up its fjord as a tidewater glacier by at least 13,000 yr B.P., based on the age of the oldest in situ and glacially transported marine shells found in the forelands (Pasch et al., this volume). Shennan and Hamilton (this volume) suggest that sea level was up to 5 m higher at this time. A variety of marine organisms inhabited a shallow marine environment that existed along the edge of the fjords and islands similar to the scattered bays and inlets of Prince William Sound today (Fig. 25).

Peat deposits adjacent to the in situ boring clams indicate that a marine environment was replaced ca. 4600 yr B.P. with a terrestrial environment within a meter of the current elevation of the lake. Tectonic uplift of 1–2 m could have accomplished this transition (Hamilton et al., 2005). The mid-Holocene transition from a marine to a terrestrial environment could have been further accelerated by the progradation of baymouth bars across the Bering fjord. Raised beach ridges (currently undated) on either side of Seal River (Hayes et al., 1976) may have acted as a baymouth bar to cut off the mouth of the fjord and transform it into a fresh-water environment (Fig. 1).

Additional evidence suggests that forested environments existed in the forelands at other places by ca. yr 4000 B.P., as indicated by wood fragments transported in the ice (Table 1). The strongest evidence for forested environments at modern lake levels lies in the Ancient Forest beds dated from ca. 200 B.C. (2200 yr B.P.) to A.D. 600 (1400 yr B.P.). Large trees with >200

Figure 25. College Fjord in Prince William Sound is a modern analogue for the early to mid-Holocene environments at Bering Glacier. The ice would have receded toward the base of the mountains, and elongated fjords would have provided adequate habitat for marine coastal invertebrates. (See Pasch et al., this volume.)

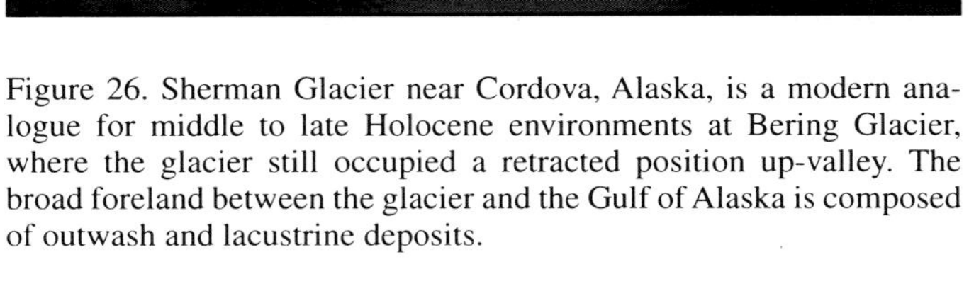

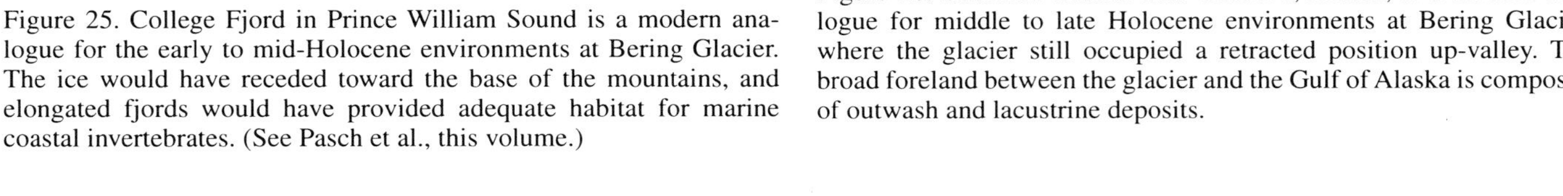

Figure 26. Sherman Glacier near Cordova, Alaska, is a modern analogue for middle to late Holocene environments at Bering Glacier, where the glacier still occupied a retracted position up-valley. The broad foreland between the glacier and the Gulf of Alaska is composed of outwash and lacustrine deposits.

rings surrounded by thick mats of forest peat containing cones, needles, and roots all suggest that a mature coastal forest occupied parts of the forelands for at least 800 a.

Rising lake levels after ca. A.D. 600 (1400 yr B.P.) buried the forest with as much as 4 m of sand, but another significant forest with trees over 200 a old was reestablished at the Extended Forest site and produced a reddish soil on the sandy surface. This forest likely correlates with the Main Forest bed along the eastern Bering Glacier margin (Wiles et al., 1999). This forest was also destroyed by sediment burial, as the surface aggraded to 10 m above the current lake level. Within the succeeding few hundred years, massive amounts of gravel were delivered to the forelands by braided streams flowing from the glacier. All of the lacustrine and fluvial deposits lie in subhorizontal beds across the foreland area and suggest that a continuous surface was aggrading to higher elevations by sediments delivered to this area. It seems likely that the previous rocky marine coastline was buried under an outwash plain. A modern analogue to this environment may be the coastal plains between Sherman and Sheridan Glaciers (adjacent to Cordova) today (Fig. 26). The outwash caused rapid aggradation of the surface, extending to 22 m above lake level at the Extended Forest site and to 10 m above lake level at Pointed and Whaleback Islands. This suggests expansion and possible advance of Bering Glacier up the valley from its current position sometime after ca. A.D. 800 (1200 yr B.P.). This may correlate with the A.D. first millennium advance of Tebenkov Glacier in western Prince William Sound (Crossen, 1997).

The thin till capping all stratigraphic sections around Vitus Lake is evidence for glacial advance across the forelands to the area of the current terminus. Dates on organics beneath the till show that Bering Glacier did not advance across the foreland until after ca. A.D. 1100 (900 yr B.P.). The juxtaposition of the horizontal stratigraphy interpreted as lacustrine and outwash deposits adjacent to deeply eroded channels in Vitus Lake today suggests that the ice advance produced considerable landscape erosion. This also supports the hypothesis that the LIA advance was the most substantial in the past 2000 a. This major event brought the ice margin within half a kilometer of the Gulf of Alaska and produced the substantial morainal ridges and outwash deposits now between Vitus Lake and the Gulf of Alaska (Fig. 1).

CONCLUSIONS

The Holocene history of Bering Glacier is written in the sedimentary deposits that were exposed as the ice retreated from its Little Ice Age maximum. Although Bering Glacier is famous for its multiple surge events, the sediments exposed in lake bluffs show no evidence of multiple glacial fluctuations during the past 2200 a. Instead, they suggest that the glacier advanced into the foreland quite recently, sometime after 900 a ago (A.D. 1100).

Prior to that LIA advance, nonglacial environments dominated the region. These earlier environments included a marine fjord between ~13,000 and 5000 a ago. About 5000 a ago the area changed from a marine to a terrestrial environment when peat bogs formed in multiple areas. This was likely due to an uplift event similar to the 1899 Yakutat earthquake (Tarr and Martin, 1912) or the 1964 Alaska earthquake (Stover and Coffman, 1993), or by beach ridges prograding across the mouth of the fjord, isolating it from marine conditions. By 4000 a ago a mature coastal forest covered the area, and forests with large trees and meter-thick peats existed from 2200 (200 B.C.) to 1400 (A.D. 600) a ago. This long-lived forest was then buried by lake sediments from a lake level at least 4 m higher than today. A second mature forest of large trees dated ca. A.D. 650 was again established, and a red soil horizon formed on the sandy sediments. This forest was buried by sandy gravels of an outwash plain. Interbedded peats and outwash gravels formed between ca. A.D. 600 (1400 yr B.P.) and A.D. 1000 (1000 yr B.P.) by active outwash deposition that was interrupted by short periods of surface stability where thin peat and small trees colonized the surface. The aggradation of the surface averaged 2 cm/a and built the outwash plain to an elevation 12 m above the current lake level. An additional 8 m of outwash devoid of peats was deposited above this unit as the ice advanced up-valley of its current terminus.

Only the upper 1–2 m of the section contains tills produced by glacier advance into the forelands sometime after A.D. 1100 (900 yr B.P.). Horizontal bedding of both the lake and outwash sediments suggests that the Bering Glacier foreland was completely filled by sediments, and the ice advance carved the deep embayments of modern Vitus Lake. The age of the upper till deposit constrains ice advance to the LIA, sometime after A.D. 1100 (900 yr B.P.). This advance was the most extensive since the end of the Ice Age (~13,000 a ago). The LIA advance produced the moraines that dam Vitus Lake, as well as the huge outwash plains that issue from these moraines (Fig. 1). The glacier terminus reached to within half a kilometer of the coast, and ice remained there until the end of the nineteenth century. Although the twentieth century was a time of dramatic surges of the glacier terminus, these surges have only punctuated 100 a of glacial retreat.

ACKNOWLEDGMENTS

The research at Bering Glacier would not have been possible without the support of the U.S. Bureau of Land Management, particularly of John Payne, who spear-headed the drive to make this facility available to scientists, and to Scott Guyer, Chris Noyles, and Nathan Rathbun, who keep the BLM camp running smoothly. The opportunity to interact with a variety of scientists at Bering Camp has made this an unusual and stimulating experience. University of Alaska Faculty Development grants also supported a portion of this research.

The geology crew included Anne Pasch, Ann Maglio, Jessequa Parker, and Marilyn Barker, as well as several other students, who have made this research enjoyable and rewarding. John Tucker and Gail Ranney acted as our trusted pilots for numerous field seasons. Ann Maglio and Danielle Savarese shared some of their photography for this chapter.

This manuscript has benefited from reviews by Anne Pasch, Marilyn Barker, Bob Shuchman, Gail Ashley, and Ron Bruhn.

REFERENCES CITED

Ashley, G.M., 1975, Rhythmic sedimentation in Glacial Lake Hitchcock, Massachusetts-Connecticut, *in* Jopling, A.V., and McDonald, B.C., eds., Glaciofluvial and Glaciolacustrine Sedimentation: Society of Economic Paleontologists and Mineralogists Special Publication 23, p. 304–320.

Boothroyd, J.C., and Ashley, G.M., 1975, Process, bar morphology and sedimentary structures on braided outwash fans, northeastern Gulf of Alaska, *in* Joplin, A.V., and McDonald, V.C., eds., Glaciofluvial and Glaciolacustrine Sedimentation: Society of Economic Paleontologists and Mineralogists Special Publication 23, p. 193–222.

Calkin, P.E., Wiles, G.C., and Barclay, D.J., 2001, Holocene coastal glaciation of Alaska: Quaternary Science Reviews, v. 20, p. 449–461, doi: 10.1016/S0277-3791(00)00105-0.

Carlson, P.R., Bruns, T.R., Molnia, B.F., and Schwab, W.C., 1982, Submarine valleys in the northeastern Gulf of Alaska: Evidence for glaciation on the continental shelf: Marine Geology, v. 47, p. 217–242, doi: 10.1016/0025-3227(82)90070-6.

Crossen, K.J., 1997, Neoglacial fluctuations of terrestrial, tidewater, and calving lacustrine glaciers, Blackstone-Spencer Ice Complex, Kenai Mountains, Alaska [Ph.D. thesis]: Seattle, University of Washington, 198 p.

Crossen, K.J., 2006a, Bering Glacier: Holocene history revealed by post-surge retreat: Alaska Geology, v. 37, p. 1–2.

Crossen, K.J., 2006b, Holocene history of Bering Glacier, Alaska: Geological Society of America Abstracts with Programs, v. 38, no. 5, p. 73.

Crossen, K.J., Pasch, A.D., and Barker, M.H., 2002, Bering Glacier retreat phase: Discoveries of overrun fiords and forests: Geological Society of America Abstracts with Programs, v. 34, no. 6, p. 477.

Fleisher, P.J., Muller, E.H., Cadwell, D.H., Rosenfeld, C.L., Bailey, P.K., Pelton, J.M., and Puglisi, P.A., 1994, Measured ice front advance and other surge-related changes, Bering Glacier, Alaska: Eos (Transactions, American Geophysical Union), v. 75, p. 63.

Fleisher, P.J., Muller, E.H., Cadwell, D.H., Rosenfeld, C.L., Gerhard, D., Shaw, L., and Mitteager, W., 1995, The surging advance of Bering Glacier, AK: A progress report: Journal of Glaciology, v. 41, p. 207–213.

Fleisher, P.J., Cadwell, D.H., and Muller, E.H., 1998, Tsivat Basin Conduit System persists through two surges, Bering Piedmont Glacier, Alaska: Geological Society of America Bulletin, v. 110, p. 877–887, doi: 10.1130/0016-7606(1998)110<0877:TBCSPT>2.3.CO;2.

Fleisher, P.J., Muller, E.H., Peteet, D.M., and Lachniet, M.S., 1999, Arctic enigma: Geotimes, v. 44, no. 1, p. 17–21.

Fleisher, P.J., Bailey, P.K., Natel, E.M., Muller, E.H., Cadwell, D.H., and Russell, A., 2010, this volume, The 1993–1995 surge and foreland modifi cation, Bering Glacier, Alaska, *in* Shuchman, R.A., and Josberger, E.G., eds., Bering Glacier: Interdisciplinary Studies of Earth's Largest Temperate Surging Glacier: Geological Society of America Special Paper 462, doi: 10.1130/2010.2462(10).

Hamilton, S., Shennan, I., Combellick, R., Mulholland, J., and Noble, C., 2005, Evidence for two great earthquakes at Anchorage, Alaska and implications for multiple great earthquakes through the Holocene: Quaternary Science Reviews, v. 24, p. 2050–2068.

Hayes, M.O., Ruby, C.H., Stephen, M.F., and Wilson, S.J., 1976, Geomorphology of the southern coast of Alaska: 15th Conference on Coastal Engineering, American Society of Civil Engineers, p. 1992–2008.

Martin, G.C., 1921, Petroleum in Alaska: U.S. Geological Survey Bulletin 719, 83 p.

Martin, G.C., 1907, Topographic Reconnaissance Map of the Pacific Coast from Yakutat to Prince William Sound, Alaska: U.S. Geological Survey Map, scale 1:1,200,000, 1 sheet.

Molnia, B.F., 1986, Late Wisconsin glacial history of the Alaska continental shelf, *in* Hamilton, T.D., Reed, K.M., and Thorson, R.M., eds., Glaciation in Alaska—The Geologic Record: Anchorage, Alaska Geological Society, p. 219–236.

Molnia, B.F., 1993, Major surge of the Bering Glacier: Eos (Transactions, American Geophysical Union), v. 74, p. 322, doi: 10.1029/93EO00582.

Molnia, B.F., and Post, A., 1995, Holocene history of Bering Glacier, Alaska: A prelude to the 1993–1994 surge: Physical Geography, v. 16, p. 87–117.

Muller, E.H., and Fleisher, P.J., 1995, Surging history and potential for renewed retreat: Bering Glacier, Alaska: Arctic and Alpine Research, v. 27, p. 81–88, doi: 10.2307/1552070.

Muller, E.H., Fleisher, P.J., Franzi, D.A., Stuckenrath, R., and Cadwell, D.H., 1991, Buried forest beds exposed by lowering of glacial Lake Tsui reveal late Holocene history of Bering Glacier, Alaska: Geological Society of America Abstracts with Programs, v. 23, no. 1, p. 106.

Pasch, A.D., Foster, N.R., and Irvine, G.V., 2010, this volume, Faunal analysis of late Pleistocene–early Holocene invertebrates provides evidence for paleoenvironments of a Gulf of Alaska shoreline inland of the present Bering Glacier margin, *in* Shuchman, R.A., and Josberger, E.G., eds., Bering Glacier: Interdisciplinary Studies of Earth's Largest Temperate Surging Glacier: Geological Society of America Special Paper 462, doi: 10.1130/2010.2462(13).

Pierce, R.A., and Winslow, J.H., eds., 1979, H.M.S. Sulphur on the Northwest and California Coasts, 1837 and 1839: The Accounts of Capt. Edward Belcher and Midshipman Francis G. Simpkinson: Hamilton, Limestone Press, 144 p.

Plafker, G., 1969, Tectonics of the March 27, 1964, Alaska Earthquake: U.S. Geological Survey Professional Paper 543-I, 74 p.

Plafker, G., and Rubin, M., 1967, Vertical displacements in south central Alaska during and prior to the great 1964 earthquake: Prague, Journal of Geosciences, v. 10, p. 1–7.

Post, A., 1972, Periodic surge origin of folded medial moraines on the Bering Piedmont Glacier, Alaska: Journal of Glaciology, no. 11, p. 219–226.

Reimer, P.J., Baillie, M.G., Bard, E., Bayless, A., Beck, J.W., Bertrand, C.J., Blackwell, P.G., Buck, C.E., Cutler, K.B., Damon, P.E., Edwards, R.L., Fairbanks, R.G., Friedrich, M., Guilderson, T.P., Hogg, A., Heugen, K.A., Kromer, B., McCormac, G., Manning, S., Ramsey, C.B., Reimer, R.W., Remmele, S., Southon, J.R., Stuiver, M., Talamo, S., Taylor, F.W., van der Plicht, J., and Weyhenmeyer, C.E., 2004, IntCal04 terrestrial radiocarbon age calibration, 0-26 cal kyr BP: Radiocarbon, v. 46, p. 1029–1058.

Seaton-Karr, H.W., 1887, Shores and Alps of Alaska: London, Simpson-Low, 248 p.

Shaw, J., 1975, Sedimentary successions in Pleistocene ice-marginal lakes, *in* Jopling, A.V., and McDonald, B.C., eds., Glaciofluvial and Glaciolacustrine Sedimentation: Society of Economic Paleontologists and Mineralogists Special Publication 23, p. 281–303.

Shennan, I., and Hamilton, S., 2010, this volume, Holocene sea-level changes and earthquakes around Bering Glacier, *in* Shuchman, R.A., and Josberger, E.G., eds., Bering Glacier: Interdisciplinary Studies of Earth's Largest Temperate Surging Glacier: Geological Society of America Special Paper 462, doi: 10.1130/2010.2462(14).

Stover, C.W., and Coffman, J.L., 1993, Seismicity of the United States, 1568–1989: U.S. Geological Survey Professional Paper 1527, 418 p.

Stuiver, M., 1998, INTCAL98 Radiocarbon age calibration: Radiocarbon, v. 40, p. 1041–1083.

Stuiver, M., and Pearson, G.W., 1993, High-precision bidecadal calibration of the radiocarbon timescale, AD 1950–500 BC and 2500–6000 BC: Radiocarbon, v. 35, p. 1–23.

Talma, A.S., and Vogel, J.C., 1993, Simplified approach to calibrating C14 dates: Radiocarbon, v. 35, p. 317–322.

Tarr, R.S., and Martin, L., 1912, The Earthquake of Yakutat Bay, Alaska in September, 1899: U.S. Geological Survey Professional Paper 69, 135 p.

Wiles, G.C., Post, A., Muller, E.H., and Molnia, B., 1999, Dendrochronology and late Holocene history of Bering Piedmont Glacier, Alaska: Quaternary Research, v. 52, p. 185–195, doi: 10.1006/qres.1999.2054.

Manuscript Accepted by the Society 02 June 2009

Printed in the USA

The Geological Society of America
Special Paper 462
2010

Faunal analysis of late Pleistocene–early Holocene invertebrates provides evidence for paleoenvironments of a Gulf of Alaska shoreline inland of the present Bering Glacier margin

Anne D. Pasch
Professor Emeritus, University of Alaska Anchorage, 7661 Wandering Drive, Anchorage, Alaska 99502, USA

Nora R. Foster
NRF Taxonomic Services, 2998 Gold Hill Road, Fairbanks, Alaska 99709, USA

Gail V. Irvine
U.S. Geological Survey, Alaska Science Center, 4210 University Drive, Anchorage, Alaska 99508, USA

ABSTRACT

A collection of marine invertebrates from Holocene glacial deposits of the Bering Glacier indicates that the Gulf of Alaska shoreline was several or more kilometers north of its present position during the late Pleistocene–early Holocene. Conventional radiocarbon dates of 29 bivalves from five localities range in age from 7590 ± 140 to 13,230 ± 25 ^{14}C yr B.P. These marine invertebrates provide a new tool not previously used to refine glacial chronology. Because the invertebrates were deposited by meltwater at the face of the receding ice, they must have been incorporated into the ice stream at points inland (north) of the existing glacier front and transported in portions of the Bering Glacier flowing southward. The logical position for an ancient shoreline would be the topographic break between the base of the coastal mountains and the present forelands. If that is correct, then the forelands were deposited within the past 7000 a. Known habits of the collected invertebrates were used to reconstruct the nature and biological composition of the ancient nearshore environment. Our results are consistent with other work, demonstrating that mollusks inhabiting nearshore environments in the Arctic are particularly useful in the reconstruction of paleocommunities and paleoenvironments.

The unusual preservation of delicate invertebrate skeletons during transport in glacial ice is difficult to explain. We have been unable to find a description of this type of entrainment, transport, and preservation elsewhere. Sediment preserved in the interior of some mollusk shells is a cohesive silt. If the invertebrates were encased in seafloor sediments and entrained into glacier ice, skeletons could have been protected by this sediment during transport and released intact when the silt blocks thawed during deposition of melt-out till. Melt-out till is sediment released by melting of stagnant or slowly moving debris-rich glacier ice. Fragments of shells are ubiquitous in all drift deposits (any type of sediment originating from glacial deposition) south of

Pasch, A.D., Foster, N.R., and Irvine, G.V., 2010, Faunal analysis of late Pleistocene–early Holocene invertebrates provides evidence for paleoenvironments of a Gulf of Alaska shoreline inland of the present Bering Glacier margin, *in* Shuchman, R.A., and Josberger, E.G., eds., Bering Glacier: Interdisciplinary Studies of Earth's Largest Temperate Surging Glacier: Geological Society of America Special Paper 462, p. 251–274, doi: 10.1130/2010.2462(13). For permission to copy, contact editing@geosociety.org.

the glacier front, but fragile intact skeletal remains were commonly found only at four localities, suggesting unique depositional conditions.

Intertidal and shallow subtidal species dominate the collection, indicating an origin close to a shoreline. A total of 110 species representing 6 invertebrate phyla were identified. Most species are mollusks (79%), but bryozoans (9%), arthropods (8%), polychaetes (2%), echinoderms (<1%), and a single protozoan (<1%) were also present. This biota includes species similar to those found in the modern Gulf of Alaska fauna. The identified species show a strong correlation with invertebrates described in studies of the contemporary Gulf of Alaska that include infaunal and epifaunal organisms from various depths and ecological habitats. Some invertebrate species indicate geographical shifts in distribution to the northwest and east.

INTRODUCTION

The Bering Glacier is the largest and longest glacier in North America and the largest temperate surging glacier on Earth (Molnia, 1995). It has surged six times within the last century. The latest surge (1993–1995) resulted in the U.S. Bureau of Land Management (BLM) developing a research program to investigate the dramatic changes produced in both the glacier and the surrounding area by the surge and subsequent rapid glacial retreat. The glacier, its related hydrologic systems, and the natural history of the adjacent areas have been the focus of research during the past decade.

In this study we show the existence of an unusual mechanism for the transport and deposition of fragile invertebrate skeletons by glacier ice, the existence of a late Pleistocene–early Holocene shoreline landward of the modern front of the Bering Glacier, significant differences in the fossil assemblages carried by specific moraines, and that these assemblages can be used to reconstruct paleocommunities and the conditions that existed adjacent to the ancient shoreline.

The age of this shoreline coincides with the period postulated for the migration of humans into North America using a coastal route (Jangala, this volume). An irregular landward shoreline would have been more hospitable for natives traveling in small watercraft than the current coastline, and the edible shellfish living there would have provided a food source.

Anne Pasch first visited the area in the fall of 1998 to observe a cluster of large tree trunks suddenly exposed by retreat of the 1993–1995 surge margin. The following summer a team from the University of Alaska Anchorage (Anne Pasch, Kristine Crossen, and Marilyn Barker) returned to the area to measure, map, and date the 39 trees that have become known as the Ancient Forest (Crossen, this volume).

The higher glacial drift plain adjacent to the Ancient Forest is covered by a veneer of melt-out till produced during the retreat of the 1993–1995 surge margin. There Crossen and Pasch discovered two *Macoma lipara* valves and a degraded limpet shell lying on the recently exposed drift surface. These marine shells appeared to be modern but were >13 km from the coast. The following year additional specimens were found on the melt-out till surface, and carbon dating of three shells showed them to be >7000 a old. Additional ancient specimens were collected during 8 subsequent field seasons from a total of 15 sites, and these specimens form the basis of this study.

AREA OF STUDY AND PHYSICAL SETTING

The study area, ~550 km^2 in size, lies between the ice margin of the active Bering Lobe and the Gulf of Alaska (Fig. 1). It is between 60° and 61° N latitude and 142° and 144° W longitude adjacent to the Bering Glacier, which is ~120 km southeast of Cordova, Alaska. The glacier descends from the Bagley Ice Field through a valley in the Chugach Mountains, joining the Steller Glacier to form a broad Piedmont Lobe with an area of ~910 km^2 (Molnia, 1995). It drains into the Gulf of Alaska between Cape Suckling to the west and the Tsiu River to the east. The Piedmont Lobe is made up of four major features: the active Steller Lobe to the west, the active Bering Lobe to the east, a large debris-covered medial moraine between them, and a terminal stagnant portion on the western margin. The stagnant lobe is covered with supraglacial melt-out till (rock debris derived from melting of the glacier surface). This thick layer of sediment supports stands of spruce trees, some of which are seen tipping over into actively forming kettles and kettle lakes. The foreland between the ice margin and the Gulf of Alaska consists of thick glacial deposits, reworked beach deposits, and a proglacial lake, Vitus Lake, which is 20 km long and up to 12 km wide. The lake continues to increase in size as the ice margin retreats.

The topography of the Bering Glacier is unusual in that parts lie below present sea level for a distance of ~55 km (Shuchman, this volume). In places the ice is nearly 200 m below sea level. This below–sea level trough extends 60 km up-glacier from the terminus (Molnia, 1995). The Piedmont Lobe ramps up on the foreland so that the terminus is several meters above sea level. Owing to the extraordinary thickness (800 m) and volume of the Piedmont Lobe, its flow path is not controlled by topography. The Piedmont Lobe has traversed Vitus Lake, which in profile shows parallel troughs and underwater ridges with a relief of >200 m (Shuchman, this volume). Therefore, the sediment incorporated into the basal portion of the Bering Glacier can be derived not only from bedrock

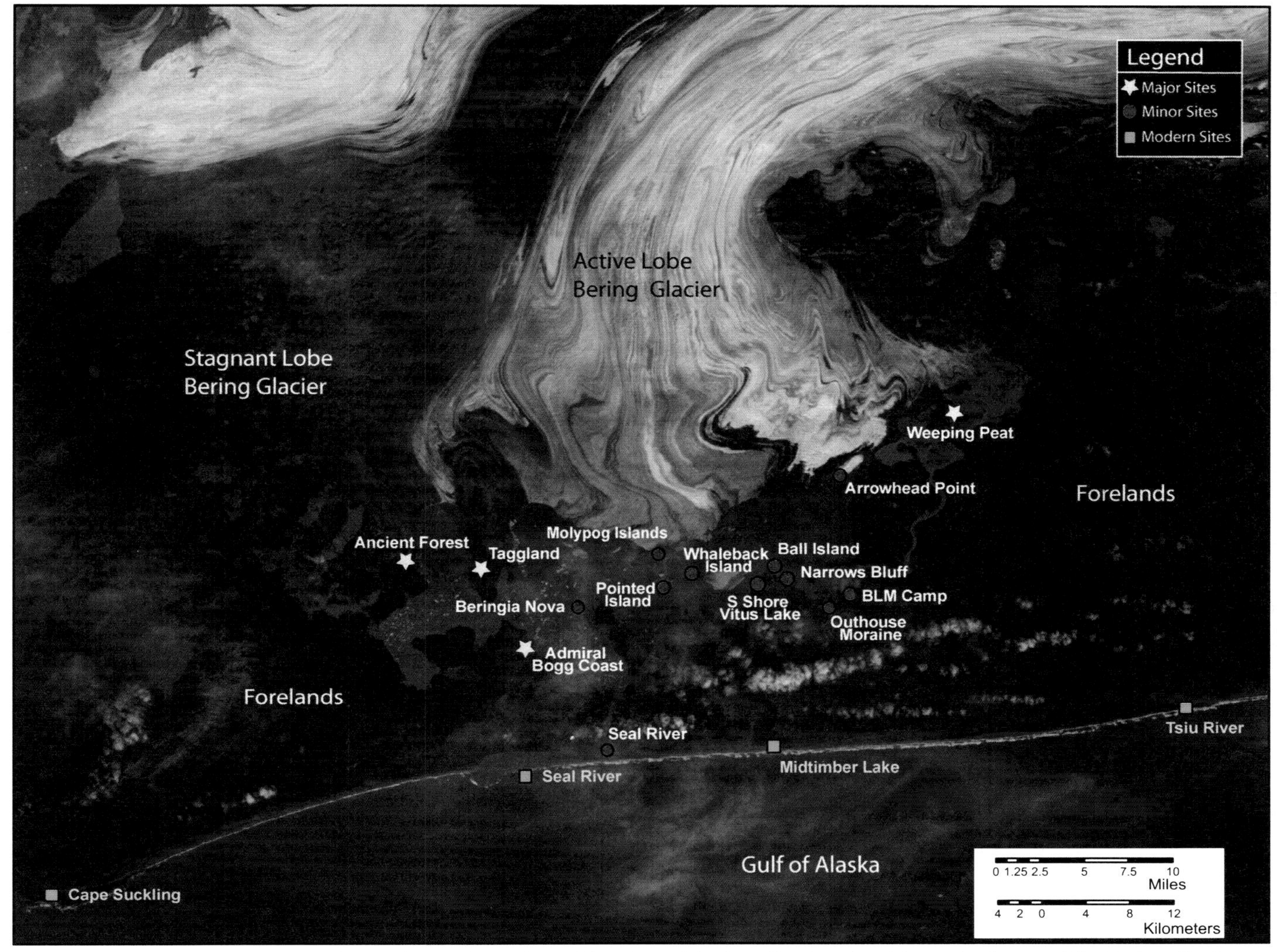

Figure 1. Study area at the margin of the active Bering Lobe and location of the 19 collection sites. True-color image of the Bering Glacier terminus acquired on 29 September 2002 by the Enhanced Thematic Mapper plus (ETM+) instrument aboard the USGS/NASA Landsat-7 satellite. NASA image by Robert Simmon, based on data provided by the Landsat 7 Science Team.

of the Chugach Mountains and the older glacial sediments of the foreland but also from a marine terrace now covered by ice. When released at the face of the ice, these sediments form a new veneer on the foreland, showing little deformation or transport. These deposits are characterized by extreme variations in size over short distances. This type of material provided the most specimens and characterized the most productive sites used in this study.

FIELD METHODS

Invertebrate material was collected over a 9 a period (1998–2006) during 9 to 14 d field seasons in late July–early August. Holocene glacial deposits adjacent to the active lobe were examined along the 30 km distance between Weeping Peat Island on the east and the Ancient Forest to the west. The 19 sites used in this study are shown in Figure 1, and taxa identified from the 15 ancient sites are listed in Appendix 1. Modern taxa identified from four beach sites along the current Gulf of Alaska shore are given in Appendix 2. Global positioning system (GPS) coordinates for the 15 ancient sites are provided in Appendix 3. Place names that are in common usage in the area are shown in Figure 1.

Nonquantitative sampling was used to acquire the greatest number of taxa, get a sense of their relative abundances, and obtain specimens that best represent each taxon. Sites with broad expanses of melt-out till surfaces were visited repeatedly, as they produced the most complete and best-preserved specimens. Other exposures adjacent to the ice front and in or adjacent to Vitus Lake were visited at least once. Shell material was present in every location examined either on the ground surface or in buried glacial deposits, where they weathered out of eroding vertical exposures. However, specimens in buried sands and gravels were fragmentary and not identifiable to species.

Transportation to sites in the forelands and the margin of the Gulf of Alaska was provided by helicopter. Avon boats with outboard motors were used to gain access to exposures in and around Vitus Lake. Under the supervision of Pasch, collecting was done mainly by student interns, other professionals, and even by helicopter pilots. Specimens were logged in by location with GPS coordinates, washed, and numbered for identification. Information on identifiable shells was entered into an Excel spreadsheet for analysis.

The most productive sites were those at the east and west margins of the active Bering Lobe: Weeping Peat and Ancient Forest, respectively (Table 1). The ice margin at the Ancient Forest site is grounded on a broad melt-out till plain that was easily accessible by helicopter. The greatest number of specimens was recovered here. Additional material is deposited each year as the ice front continues to retreat. The recently exposed surface is relatively stable, and few changes have been noted over an eight-year period other than the invasion of vegetation. The second most prolific site was Weeping Peat, on the east side, where material was retrieved from a pitted outwash plain adjacent to the retreating surge margin.

The age of the glacial deposits is a factor affecting the productivity of a locale, as the best collecting sites were on melt-out till that had been deposited in the previous year. Plant succession is rapid on these deposits. At the BLM campsite (Fig. 1) shrubs and conifers grew so high between 1998 and 2004 that they had to be cut down to facilitate detection of approaching bears. The camp is situated on a complex of glacial deposits left by the 1968 surge event, which, after 39 a, supports a dense cover of vegetation (mosses, herbaceous plants, and shrubs). Once the vegetation cover is established, a survey of ancient invertebrate material is impossible unless there are fresh erosional surfaces.

New areas are being exposed every year as the surge margin retreats, and they will undoubtedly be a source of additional material. Nevertheless, it is doubtful that the findings would greatly modify the conclusions reached to date.

SPECIMEN IDENTIFICATION, DATA ANALYSIS, AND RADIOCARBON DATING

Specimen Identification and Data Analysis

A database of all specimens was created, and species were identified to the lowest taxonomic level possible. Identifications are based on references to Alaska and North Pacific shelled fauna. Most were made by the authors, with assistance from Bruce L. Wing of the Alaska Fisheries Science Center, National Marine Fisheries Service, John Larson of the U.S. Minerals Management Service, and Roger J. Cuffey, Pennsylvania State University. Nomenclature is from Coan et al. (2000), Turgeon et al. (1998), Osburn (1952), and Kozloff (1996).

The modern Gulf of Alaska fauna is described in several publications. Hood and Zimmerman (1986) summarize relevant studies of the biology and oceanography of the Gulf of Alaska up to 1978. Feder and Jewett (1986) and Feder (2007, personal commun.) are the sources for feeding types and habitat preferences. The geographical distribution of most species found at the Bering Glacier is summarized in Foster (2000).

For analytical purposes, the database of shells and fragments was edited to select species that could be attributed to a specific depth, habit, and feeding type. For example, barnacles used in the analysis are suspension feeders and occur in an intertidal or subtidal setting. Similarly, polychaetes in the family Serpulidae were classified as intertidal-subtidal suspension feeders. Several different species of the bivalves *Macoma* and *Clinocardium* inhabit different depths, but for analysis, all *Macoma* are considered to be deposit feeders. *Clinocardium*, with short siphons extending only slightly above the sediment surface, feed on both suspended and detrital material and are classified as suspension-deposit feeders regardless of the depth they inhabit.

The taxa selected were described by depth preference as intertidal, intertidal-subtidal, subtidal, or shelf; and by habit, whether on rocks, at the sediment surface, or buried in sediment. Depths and habits reflect usage in Foster (2000). Feeding types (deposit feeder, suspension feeder, predator, scavenger, etc.) are based on usage in Feder and Jewett (1986).

Analysis of the fauna was limited to those from the four sites where the greatest numbers of specimens were found. Depth,

TABLE 1. ABUNDANCE OF SPECIMENS AND TAXA AT THE MOST PRODUCTIVE COLLECTION SITES COMPARED WITH ALL OTHER SITES COMBINED

	Ancient Forest	Weeping Peat	Taggland	Admiral Bogg	Other 11 sites combined	Total
No. of specimens	2192	1336	513	201	1988	6230
% of total	35%	22%	8%	3%	32%	100%
No. of entire valves	261	79	123	119	23	605
% of entire valves	43%	13%	20%	20%	4%	100%
No. of taxa	114	56	57	28	67	177
% of total no. of taxa	64%	32%	32%	15%	38%	

Note: Taxa include species as well as higher taxa.

habit, and feeding type of taxa from these four sites were compared to determine the environmental conditions represented by each of the four paleofaunas.

Dating of Shells

Radiocarbon ages were obtained from 29 bivalve specimens, with 9 analyses conducted by Beta Analytic and 20 analyses by the University of California, Irvine (UCI). The shells were entirely consumed in the Beta-Analytic analyses, whereas just small fragments of a shell were needed for the accelerator mass spectrometry (AMS) analyses performed by UCI. UCI reported radiocarbon concentrations given as fractions of the modern standard, $D^{14}C$, and conventional radiocarbon age, following the conventions of Stuiver and Polach (1977). All sample preparation backgrounds were subtracted, based on measurements of ^{14}C–free calcite. Results were corrected for isotopic fractionation according to the conventions of Stuiver and Polach (1977), with $\delta^{13}C$ values measured on prepared graphite using the AMS spectrometer. These can differ from $\delta^{13}C$ of the original material if fractionation occurred during sample graphitization or the AMS measurement.

We determined calibrated ages using the calibration program CALIB 5.0.2 (Stuiver and Reimer, 1993) and the Marine04.14C data set (Hughen et al., 2004). Radiocarbon ages are reported as conventional age (^{14}C yr B.P.) and calibrated age ranges (cal yr B.P.) (Table 2). Where there are multiple calibration curve intercepts, the separate ranges are presented along with their associated probabilities (that represent the relative area under the distribution). To account for the fact that upwelling ocean water is depleted in ^{14}C, we corrected for the local marine reservoir effect by using a delta R of 340 ± 50 yr, the value reported for

TABLE 2. CONVENTIONAL AND CALIBRATED RADIOCARBON AGES OF 29 BIVALVES

Location	Genus	Species	Conventional age (^{14}C yr BP)	Age uncertainty ±years*	Calibrated age range (cal yr BP)†	Probability	Lab code§
Admiral Bogg	*Macoma*	*lipara*	12,210	20	13,219–13,420		UCIAMS-26741
Admiral Bogg	*Macoma*	*lipara*	12,255	25	13,239–13,464		UCIAMS-26742
Admiral Bogg	*Macoma*	*middendorffi*	12,265	20	13,245–13,473		UCIAMS-26744
Admiral Bogg	*Macoma*	*brota*	13,060	25	14,020–14,629		UCIAMS-26743
Admiral Bogg	*Macoma*	*calcarea*	13,170	25	14,159–14,852		UCIAMS-26747
Admiral Bogg	*Macoma*	*calcarea*	13,215	25	14,206–14,914		UCIAMS-26746
Admiral Bogg	*Macoma*	*calcarea*	13,230	25	14,221–14,936		UCIAMS-26745
Admiral Bogg	*Macoma*	*middendorffi*	12,320	110	13,220–13,701		Beta-208865
Admiral Bogg	*Macoma*	*lipara*	13,050	70	13,963–14,737		Beta-208864
Ancient Forest	*Macoma*	*lipara*	9915	25	10,277–10,552		UCIAMS-26736
Ancient Forest	*Macoma*	*lipara*	9915	20	10,281–10,551		UCIAMS-26737
Ancient Forest	*Macoma*	*lipara*	10,390	20	10,802–10,861	0.047	UCIAMS-26740
					10,876–11,180	0.953	
Ancient Forest	*Macoma*	*lipara*	10,480	70	10,807–10,854	0.018	Beta-147769
					10,880–11,313	0.982	
Ancient Forest	*Macoma*	*lipara*	10,920	90	11,338–12,226		Beta-147767
Ancient Forest	*Macoma*	*lipara*	7590	140	7449–7995		Beta-147768
Ball Island	*Clinocardium*	*nuttalli* ?	10,240	70	10,594–11,094		Beta-173815
Seal River	*Saxidomus*	*gigantea*	12,965	30	13,895–14,249		UCIAMS-26731
Weeping Peat	*Saxidomus*	*gigantea*	8455	20	8427–8759		UCIAMS-26734
Weeping Peat	*Macoma*	*golikovi*	8510	15	8515–8873	0.996	UCIAMS-26749
					8877–8882	0.004	
Weeping Peat	*Saxidomus*	*gigantea*	8900	15	9037–9368		UCIAMS-26735
Weeping Peat	*Macoma*	*golikovi*	8900	15	9037–9368		UCIAMS-26748
Weeping Peat	*Saxidomus*	*gigantea*	8990	20	9160–9452		UCIAMS-26725
Weeping Peat	*Saxidomus*	*gigantea*	9010	15	9204–9465		UCIAMS-26733
Weeping Peat	*Saxidomus*	*gigantea*	9170	35	9379–9642		UCIAMS-26723
Weeping Peat	*Saxidomus*	*gigantea*	9275	15	9477–9757		UCIAMS-26724
Weeping Peat	*Saxidomus*	*gigantea*	9335	20	9526–9861		UCIAMS-26732
Weeping Peat	*Macoma*	*lipara*	9580	100	9665–10,300	0.998	Beta-162750
					10,305–10,313	0.002	
Weeping Peat	*Clinocardium*	*nuttalli*	9160	130	9128–9888		Beta-162751
Weeping Peat	*Saxidomus*	*gigantea*	9350	80	9486–10,037		Beta-162749

*The age uncertainty associated with the uncalibrated conventional radiocarbon age is ±1 sigma.
†The program Calib 5.0.2 (Stuiver and Reimer, 1993) was used to calculate calibrated age ranges, using the Marine04.14C data set (Hughen et al., 2004). Values presented are 2 sigma calibrated results, having a 95% probability of inclusion. Multiple calibration curve intercepts (and their error ranges) are presented, with their associated probabilities, which reflect the relative areas under the curve.
§UCIAMS—University of California, Irvine—accelerator mass spectrometry; Beta—Beta Analytic.

a pre-nuclear bomb specimen from Middleton Island, Alaska (McNeely et al., 2006), which is offshore and to the west of the Bering Glacier area. These calculations assume that the marine reservoir effect has not varied significantly over time; if this is not the case, then use of a constant local reservoir correction may produce inaccurate calibrations.

RESULTS

Description of Ancient Sites and Overview of Their Associated Invertebrates

The four most productive sites (Admiral Bogg Coast, Ancient Forest, Taggland, and Weeping Peat) accounted for 68% of the total number of specimens, with a range of 17–86 species per site. They provided 96% of the 605 entire valves found in the study area (Table 1). Nearly all specimens at these sites were collected from drift surfaces. At 3 sites, deposition within the past 8 a occurred adjacent to a rapidly retreating surge margin. However, the Admiral Bogg Coast site was located in a 1968 morainal complex.

Eleven less productive sites lie in the glacial drift along the margin of Vitus Lake and on small islands in the lake. These sites accounted for 32% of the total number of specimens collected and had 2–12 species per site. Only 23 entire shells were found at all of the less productive sites combined (Table 1). These sites include eroded moraines of the 1968 surge deposits and more recent melt-out till surfaces created during retreat of the 1993–1995 surge margin. Marine organisms were found at every site visited south of the ice front of the active Bering Lobe. Therefore, marine sediments must have been incorporated into the ice along its entire 30 km width.

Ancient Forest

This site was the most productive in terms of number of specimens (35% of the total) and number of taxa (64% of the total; Table 1). It lies in front of the ice margin and is easily accessible by helicopter. The melt-out till plain here, known informally as part of Herder Land, is a stable, gently undulating broad surface dotted with small ponds, low recessional moraines, flutes, and erratic boulders (Fig. 2). Although the distal portions of the 1993–1995 till plain are developing a thin plant cover, ancient

Figure 2. Overview of the Ancient Forest site, named after the nearby Ancient Forest. The 1993–1995 surge margin is marked by the low irregular ridge of the terminal moraine. Specimens were collected from the melt-out till plain lying between the 1993–1995 terminal moraine and the ice front of the active Bering Lobe.

invertebrates can still be found near the 1995 terminal moraine. Additional material is deposited each year as the ice front continues to retreat. The recently exposed surface is relatively stable; few changes have been noted over an eight-year period other than the invasion of vegetation.

The surface lies on a complex of older alluvial and lacustrine deposits with a minimum thickness of 25 m. The sequence is well exposed in cliffs along the west side of Tashalich Arm, where peat blocks and remains of buried spruce trees can be seen. See Crossen (this volume) for a more detailed description.

Taggland

The Taggland site lies due east of the Ancient Forest on the east margin of Tashalich Arm. It consists of a narrow headland between Tashalich Arm and a bay of Vitus Lake to the east, known as McMurdo Sound. The surface has a minimum relief of 20 m and consists of irregular morainal ridges, swales, and gullies that are dotted with erratic boulders and small relict ponds. Much of the surface consists of cobble and pebble pavements of lag gravel, suggesting a high-energy flow regime. The flat, smooth fluted surfaces characteristic of the Ancient Forest site are missing here. Whereas only 8% of the total number of specimens was found here, 20% of the total number of entire valves came from this site (Table 1). The count from Taggland is low, in part because this site was not visited as often as the Ancient Forest, and the ice-front retreat was slower, leaving less area exposed. Some of the most perfectly preserved gastropod specimens (naticids and *Colus*) were found here, but entire valves of *Macoma lipara*, so common at the Ancient Forest site, were not.

This site contains the only bedrock exposure in the study area. It is found on the east margin of Tashalich Arm near the level of Vitus Lake. Molnia (1995) reported the existence of boring *Penitella* clams in this exposure, which yielded a calibrated ^{14}C age of 4860 cal yr B.P. As these clams live from the mid-intertidal zone to a depth of 22 m (O'Clair and O'Clair, 1998), they are an indication of proximity to an earlier, more landward shoreline. The borings occur in a bedrock surface that has been exhumed from overlying glacial deposits (Fig. 3). In 2004 a drop in the level of Vitus Lake provided access to this surface, and specimens of *Penitella penita* and *Protothaca staminea* were recovered from bored holes in the bedrock (Poul Creek Formation of Oligocene age; Ron Bruhn, 2005, personal commun.). Shennan (this volume) reports four conventional ^{14}C dates from *P. penita* found in these boreholes ranging from 5310 ± 40 to 8800 ± 40 ^{14}C yr B.P. Seventy specimens of one foraminferan species, *Trichohyalus ornatissimus*, were recovered from silt packed between the valves of a *P. staminea* collected at the bottom of a *P. penita* borehole. Both adult and juvenile specimens of the foraminiferan were present in the silt sample. According to Bergen and O'Neil (1979), *Trichohyalus* dominates the foraminiferal fauna in the inner neritic zone, 18–29 m in depth. *P. staminea* is found from the intertidal area to depths of 18 m (Chew and Ma, 1987). As *P. penita* is found to depths of 22 m (O'Clair and O'Clair, 1998), the presence of these three species suggests that the Taggland bedrock surface was in the intertidal or subtidal environment to a depth of ~18 m. As the three species may not have co-occurred temporally, their presence may reflect differences in water depth over time.

Weeping Peat Island

The second most prolific site, with 22% of the total number of specimens and 32% of the total number of taxa, was Weeping

Figure 3. An articulated boring clam *(Penitella penita)* in its borehole in exposed bedrock at the Taggland site. Individuals from this recently exhumed site date from 5300 to 8800 ^{14}C yr B.P. (Shennan, this volume).

Peat Island (Table 1). It is an island on the eastern margin of the Bering Glacier Lobe (Fig. 1) that was overridden by several surges in the twentieth Century (Molnia, 1995). It is bounded by the Tsiu River on the east and the ice front on the west. A spectacular pitted outwash plain lies on the north side of the island, which has been modified by outwash streams issuing from the glacial front. A violent outburst flood occurred here in 1994 (John Payne, 2001, personal commun.). Streams disgorged from subsurface channels with such force that building-size chunks of ice and entire spruce trees were hurled onto the outwash plain. Subsequent melting of stranded ice blocks resulted in the formation of the numerous kettles with distinct outlines seen in 2004 (Fig. 4). The abrupt drop in velocity as the meltwater spread out across the outwash plain resulted in almost simultaneous deposition of coarse material. The ice blocks and finer gravel, sand, and silt were carried beyond to form a classic pitted outwash surface nearly a mile long. These conditions were optimal for the deposition of intact invertebrate shells. In 2001 the surface was entirely free of vegetation, and many delicate gastropods were recovered. The sandy surface is subject to strong winds that are filling in the kettles, creating small dunes, and abrading any remaining shell material. Thus, the surface degraded very rapidly, and by 2004 was no longer productive as a collecting site.

Admiral Bogg Coast

This site is situated on the west side of Vitus Lake, due west of Beringia Nova (Figs. 1, 5). The condition and number of shells found here were unexpected, because the drift surface associated with the 1968 surge moraine here has a well-developed cover of vegetation. The alder and willow shrubs exceed 2 m in height, and the herbaceous plant and moss cover is well established. However, a swale adjacent to the lake has been subject to waves from calving events with the vegetation partially scoured away, exposing part of the drift surface. This site had the highest density of shells: 56 fragments were counted in 1 m^2. The surface here is similar to that of the Ancient Forest, with a mosaic of gravel, lag gravel, and silty and sandy flat surfaces. The crest of the terminal moraine just north of this platform provides another rich collecting site. The section was not measured, but it is similar to others along the south shore of Vitus Lake, with several meters of stratified sands and gravels overlain by less than a meter of diamicton (unsorted material). This site was discovered in 2005 and was visited only once, but 3% (n = 201) of the total number of specimens was collected here. However, 59% of the shells collected here were entire valves; these represent 20% of the entire valves from all sites (Table 1). The shells remained intact after being exposed to the elements for 37 a (Fig. 6).

Seal River

The Seal River site (Figs. 1, 5) lies on the foreland drift plain adjacent to the beach along the Gulf of Alaska. It is not near the present ice front but lies just landward of the Gulf of Alaska beach south of Vitus Lake. Glacial drift here represents the ice maximum that occurred in the early part of the twentieth century (Molnia, 1995). The site can be seen in the 2001 satellite image as a channel cutting across the dunes from the beach west-northwest to the Seal River channel. This denuded surface was probably cut by overwash waves during a storm event that scoured away the vegetation. Marine invertebrates were found at this site during a plant survey. Although only 71 specimens

Figure 4. Overview of the Weeping Peat site. This sandy, pitted outwash plain is named for the adjacent island known as Weeping Peat. The pits are kettles formed during the 1994 outburst flood.

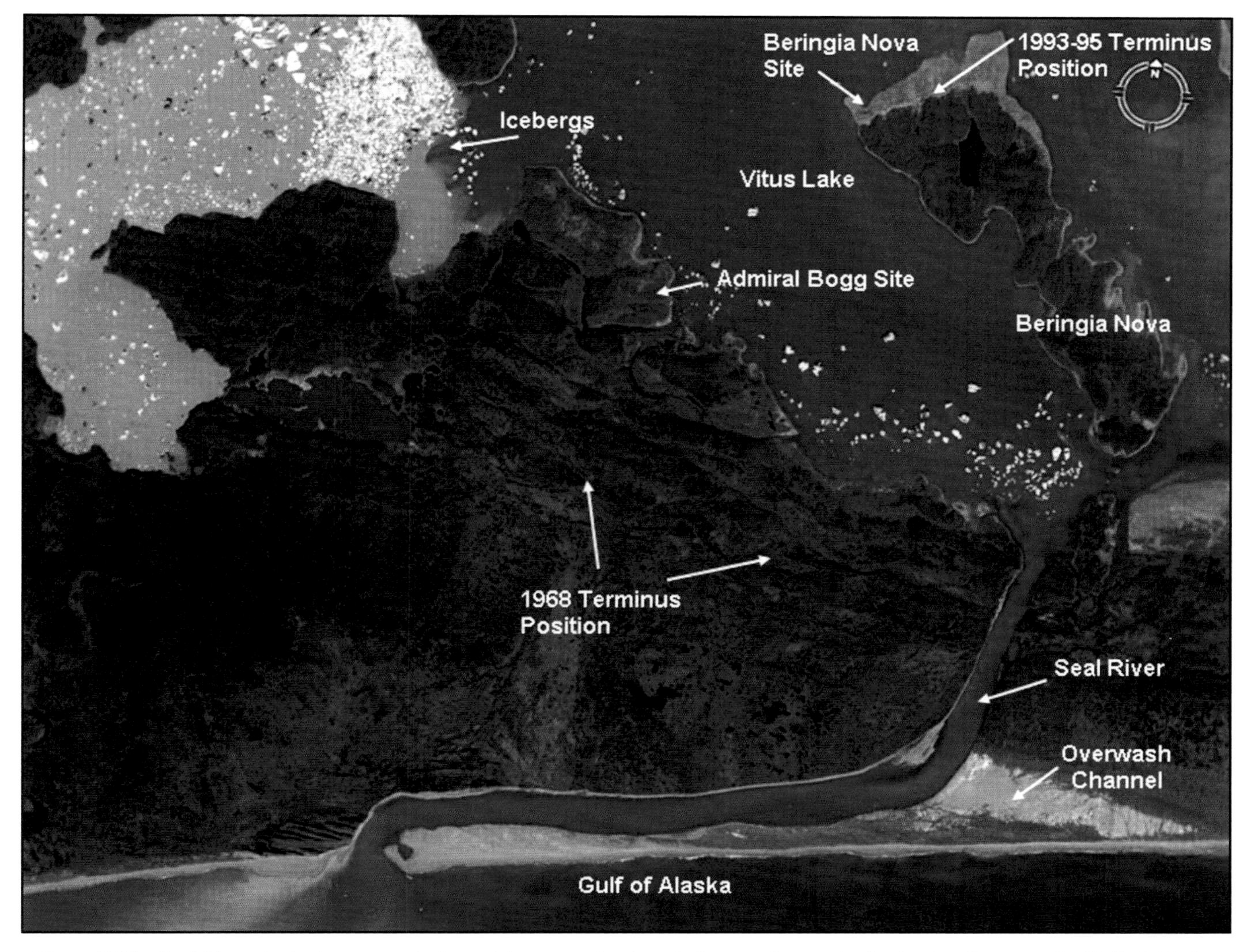

Figure 5. The Seal River site is the overwash channel (gray zone) stripped of vegetation on the forelands adjacent to the Gulf of Alaska. The Seal River drains Vitus Lake. The dense accumulation of icebergs is choking the entrance to Laurie Bay and Tashalich Arm (north of the photo). The Beringia Nova and Admiral Bogg Coast sites are also visible in this image.

(1% of the total number) were retrieved from this site; 13% of the specimens were entire. The conditions that prevailed to preserve these specimens, which were subject to deposition, burial, and retransport over the course of 100 a, are not understood.

Arrowhead Point

In aerial view, this is a triangular-shaped exposure of glacial drift extending south of the ice margin on the north shore of Vitus Lake (Fig. 1). It was almost completely overridden by the 1993–1995 surge, but glacial retreat has exposed a melt-out till surface. According to helicopter pilot John Tucker, it emerged in 1999. This platform consists of a thick sequence of older outwash and till that exceeds 35 m in thickness. The margins along the lake are cut by numerous gullies, providing good exposures of stratified sediments. Poorly sorted gravels with sandy stringers are common, as are zones of upright remnants of woody shrubs and peat. Three hundred fifteen shell fragments were collected here, but only seven taxa could be identified (Appendix 1). A few fragments were retrieved directly from basal ice.

Ball Island and Narrows Bluff

Ball Island and Narrows Bluff are sites that lie opposite a medial moraine (Fig. 1). Ball Island has a relief of <1 m and lies just offshore and north of Narrows Bluff on the south shore of Vitus Lake. Six species could be recognized from the shell fragments recovered from its wave-washed till surface (Appendix 1). The Narrows Bluff exposure is ~10 m thick, but only three genera were recovered from slope wash. Specimens from these sites are fragmentary and smaller than 5 mm. The exposure at Narrows Bluff consists of stratified silts, sands, and gravel with peat zones.

Figure 6. *Macoma* in situ on the surface of the 1968 glacial deposits at the Admiral Bogg Coast site.

BLM Campsite, Outhouse Moraine, and South Shore of Vitus Lake

Three other sites in the 1965–1968 surge margin drift, the BLM campsite, the Outhouse Moraine, and the South Shore of Vitus Lake (Fig. 1), were not very productive. Invertebrate material is ubiquitous but fragmentary at these sites, and was recovered from slope wash below the diamicton at the top of the bluff. The section consists of an upper diamicton 1–11 m thick that overlies 10–30 m of stratified sands and gravels of deltaic, alluvial, and lacustrine origin. The sand and gravel units are very clean and devoid of shell material.

The south shore of Vitus Lake is bordered by the undulating surface of the 1965–1968 surge margin moraines, with a relief of at least 19 m. A cover of herbaceous plants and alder-willow shrubs is well established on the surface. The eroding slopes facing Vitus Lake provide exposures where invertebrates can be found. The specimens, mostly bivalves, consisted of fragments 3–11 mm in size. They included three species and two genus-level identifications.

Whaleback, Pointed, and Wollypog Islands

These islands all lie in the central northern part of Vitus Lake (Fig. 1). The smaller ones (Pointed and Wollypog) were completely overridden by the 1993–1995 surge margin (Crossen, this volume). The stratigraphy of Whaleback and Pointed Islands is similar to that seen on the south shore with sorted sands and gravels overlain by a diamicton with a total thickness that exceeds 10 m. The Wollypogs (sometimes referred to as Molypogs) are two islands: the larger has a relief of ~15 m, whereas the smaller is <4 m above the level of Vitus Lake, with scoured surfaces from wave overwash. The surfaces are rather smooth with scattered erratics, cobbles, lag gravel, and a sparse vegetative cover. Invertebrate fragments were recovered from the surfaces and from the stratified sands and gravels.

Beringia Nova

This site (Figs. 1, 5) lies on the northwest tip of the island informally named Beringia Nova by Austin Post and Bruce Molnia. The northernmost tip of the island was covered by the 1993–1995 surge, and the topography was created by retreat of the surging ice. However, this site is geologically quite different from the others. Instead of a smooth surface, there are a series of "pyramids," most in the range of 2 × 2 × 2 m, some up to 5 m long, and a few larger "ridges" ~5 m high and 20 m long. They appear to be disjunct weathered blocks of cohesive sediment made up of contorted but distinct layers of silt and sand up to 10 cm thick (Fig. 7). Valves of *Nuculana pernula* were embedded in the silt. Also of interest were elongated masses of shell hash (~3 × 12 mm) that resembled the droppings of shellfish-eating marine birds. Both of these are evidence that the mounds are cohesive blocks of seafloor excavated by advancing ice. There is very deep water just north of Beringia Nova (Shuchman, this volume). The ice front traversed this deep section of the lake and then ramped up on Beringia Nova. The sudden decrease in velocity resulted in deposition of the bed load, now seen as eroding blocks of silt. On the basis of the fossils embedded in them, we interpret these blocks to be cohesive units of seafloor sediment deposited in conditions not existing elsewhere along the ice front. Here the volume of meltwater was not sufficient to dismantle these cohesive silt blocks. Elsewhere, ice fronts are retreating directly on the

Figure 7. Weathering mound of cohesive stratified silt on the island of Beringia Nova in Vitus Lake. The marine clam *Nuculana pernula* was found enclosed within such a mound.

melt-out till surfaces, and the volume of meltwater is sufficient to redistribute the fine sediment of the blocks as they thaw, leaving the invertebrate skeletons on the surface as residuals.

Comparison of Modern and Ancient Nearshore Gulf of Alaska Faunas

The distinct suites of shells and fragments found at the four richest sites invite comparison with contemporary fauna and habitats in the Gulf of Alaska. Thus, we surveyed the modern Gulf of Alaska shoreline bordering the study area for marine invertebrates. Diversity of the extant species suite is much lower than that of the ancient assemblage (33 versus 110 species), most likely because modern collecting was limited to the intertidal zone, and only three types of substrate (bedrock, sand, and a mix of sand and gravel) were present. Matches of modern to ancient specimens are reported by the lowest level of taxonomic identification possible. As shown in Table 3, 79% of the modern species match species of the ancient invertebrates, and 82% of the all modern taxa match ancient taxa. Most of those that did not match possessed fragile skeletons not likely to have been preserved in a beach environment.

Sandy beaches composed most of the shoreline in the study area where modern epifaunal invertebrates were collected. Attempts to retrieve infaunal organisms were unsuccessful. On these beaches energy levels and attrition rates are high, and fragile skeletons are not likely to be preserved over time. For example, Dungeness crab carapaces were common on the modern sandy

TABLE 3. EXTANT SPECIES FROM THE FOUR GULF OF ALASKA SHORELINE SITES AND THEIR MATCH TO ANCIENT TAXA AND SPECIES

Locality	Substrate	Match to ancient species	Match to ancient higher taxa	Number of species	Number of higher taxa	Number of specimens
Cape Suckling	Bedrock	20	6	27	6	349
Seal River	Sand	7	4	10	4	43
Midtimber Lake	Sand	2	2	4	3	51
Tsiu River	Sand-gravel	6	2	6	3	54
Total no. of specimens						497
Total (matching)		27	9			
Total modern species		34				
Total modern taxa			11			
Percentage match		79%	82%			

Note: Taxa and species are determined to the lowest taxonomic level possible. The number of taxa and species are not strictly additive in the columns because of overlap among sites.

beach bordering the Gulf of Alaska, whereas only one carapace was found in glacial drift.

A collection of epifaunal invertebrates from a bedrock marine terrace at Cape Suckling was similar to the ancient collection, as 20 of 27 modern species (74%) matched ancient species, and 6 of 6 (100%) higher taxa matched ancient taxa (Table 3). No modern muddy, low-energy marine environments existed in the study area, hence we have no modern invertebrates from this habitat type.

Results of Faunal Analysis

Our analysis is based on a total of 6214 invertebrate specimens collected from 15 localities over a 9 a period. All localities were situated in Holocene and Pleistocene glacial deposits bordering the active lobe of the Bering Glacier. Specimens encompass nine major taxon groups, as shown in Table 4. A miscellaneous group, including calcareous algae and boring organisms represented by holes in mollusk shells, was not used in the analysis. Bivalves are most numerous, with 70% of the total number of specimens; gastropods are the most diverse, with 44% of the total number of species. A total of 110 species, 70 genera, and 47 families were recognized.

All specimens consist of skeletal hard parts, primarily shells composed mainly of aragonite and/or calcite (Pojeta et al., 1987) that have the best chance of preservation in the fossil record (Clarkson, 1998). Valves or fragments of bivalves and gastropods were most abundant at every locality. Entire valves that maintained structures such as muscle scars, pallial lines, and growth lines were found at a few sites.

The retrieval of specimens, and their abundance, distribution, and diversity, were affected by several factors: original habitat of the species, original skeletal structure, degree of preservation, glacier movement, and mode of deposition. A benthic habit would promote incorporation into the basal portion of the advancing glacier as it scoured up parts of the seafloor. Epifaunal organisms (those that live all or part of their lives on the substrate) are always broken, suggesting that the tabulation of species is skewed by the mode of life. Large and robust skeletons of organisms that employed a relatively deep infaunal mode of life (living in burrows or buried in the substratum) were most likely to be preserved and detected; these are numerically dominant and include entire or nearly entire valves. The type of substrate (sand or silt) enclosing infaunal organisms would also affect the degree of preservation. For example, *Macoma lipara* and *Clinocardium nuttallii* were found in abundance. Both are fairly large (7–8 cm in length) and robust. However, entire valves of *M. lipara* are common, but *C. nuttalli* are always found as broken fragments of 3 cm or less across. The infaunal genus *Macoma* prefers a silty substrate that is cohesive and probably maintained a cohesive protective block around a shell. *C. nuttallii* lives close to or on the sediment surface in sandy material that is noncohesive. This sandy substrate incorporated into basal ice would not have maintained a cohesive protective block around the shell; thus *C. nuttallii* would be subject to the shear forces of moving ice, fragmenting the shell.

Fragments of bivalves are ubiquitous throughout the Holocene glacial deposits, as they can be found at every site either in situ, in the stratified outwash, or in the slopewash of eroding exposures. These fragments rarely exceed 1 cm in length. Entire valves of bivalves and gastropods were found only on drift surfaces. *Macoma lipara* valves, being large and robust and originally situated relatively deep in the substrate, were commonly found, but more surprising was the presence of the more delicate shells of other bivalves and gastropods. Chitons were recognized by skeletal plates. Bryozoans were identified from broken fragments of zoaria; all were <23 mm in length, and most were <10 mm. Echinoids were recognized from spine and test fragments ~1 × 5 mm in size. The evidence for annelid polychaetes (Serpulidae) consists of fragments of calcareous tubes <2.5 cm in size. One species of foraminifera was recovered from silty sediment removed from the interior of a *Protothaca staminea* found nestled in a burrow drilled into bedrock by the bivalve *Penitella penita*. Most of the arthropod parts found were calcareous barnacle plates. The most surprising finds consisted of three crabs: two leg fragments of tanner crab *(Chionoecetes bairdii)* and the carapace of a tiny Dungeness crab *(Cancer magister)* just 20 mm wide with the eye projections still in place (Fig. 8). How such delicate objects survived a 25 km ride in a glacier and subsequent deposition by meltwater is a difficult question to answer.

The entire shells exhibited such a high degree of preservation that the biota may have, in part, been alive when incorporated into glacier ice. The general appearance of most specimens is similar to that of extant shells found on the beach at the Gulf of Alaska. Some limpets retained color patterns. Even material found on the 1968 moraine looked very fresh. In this periglacial environment chemical weathering would be very slow. However, a few specimens did show evidence of weathering, such as roughness and chalky surfaces. Some bivalve shells contain drill holes commonly attributed to gastropod predators. The edges of fragments exhibited irregular surfaces showing little abrasion. This finding would suggest that most of the weathering observed took place in the original habitat.

Radiocarbon Dates

Conventional radiocarbon and calibrated ages for the 29 bivalves analyzed are presented in Table 2. Conventional ages ranged from 7590 ± 140 to 13,230 ± 25 ^{14}C yr B.P., whereas calibrated ages ranged from 7449 to 14,936 cal yr B.P. The oldest shells were found at Admiral Bogg Coast and Seal River, with conventional ages of ~12,200 to 13,300 ^{14}C yr B.P. (Fig. 9). A suite of younger shells was found at the Weeping Peat site (conventional ages from 8455 ± 20 to 9580 ± 100 ^{14}C yr B.P.). South of Weeping Peat at the Ball Island site, the one shell analyzed had a conventional age of 10,240 ± 70 ^{14}C yr B.P. The greatest variation in shell ages occurred at the Ancient Forest site, where

TABLE 4. TAXA FROM THE FOUR MOST PRODUCTIVE SITES USED IN THE PALEOECOLOGICAL ANALYSIS

Phylum: Class	Species	Phylum: Class	Species
Protozoa	*Trichohyalus ornatissimus*	Mollusca: Gastropoda	*Cryptonatica affinis*
			Diodora aspera
Annelida	*Paradexiospira vitrea*		*Euspira pallida*
	Serpula/Crucigera		*Fusitriton oregonensis*
			Kurtziella sp.
Mollusca: Bivalvia	*Astarte arctica*		*Lirabuccinum dirum*
	Astarte borealis		*Littorina sitkana*
	Astarte elliptica		*Lottia/Tectura*
	Astarte sp.		*Lottia pelta*
	Chlamys behringiana		*Margarites pupillus*
	Chlamys hastata		*Mohnia frielei* ?
	Chlamys rubida		*Neptunea lyrata*
	Chlamys sp.		*Neptunea phoenicia*
	Clinocardium ciliatum		*Neptunea pribiloffensis*
	Clinocardium fucanum		*Neptunea* sp.
	Clinocardium nuttallii		*Nucella* sp.
	Clinocardium sp.		*Nucella lamellosa*
	Cyclocardia ventricosa		*Nucella lima*
	Hiatella arctica		*Ocinebrina lurida*
	Humilaria kennerleyi		*Ocinebrina interfossa*
	Macoma balthica		*Oenopota excurvata*
	Macoma brota		*Oenopota levidensis*
	Macoma calcarea		*Oenopota rosea*
	Macoma golikovi		*Oenopota* sp.
	Macoma inquinata		*Oenopota turricula*
	Macoma lipara		*Olivella baetica*
	Macoma middendorffi		*Probela harpularia*
	Macoma sp.		*Puncturella galeata*
	Mactridae		*Puncturella multistriata*
	Mactromeris polynyma		*Puncturella* sp.
	Musculus sp.		*Tectura fenestrata*
	Mya truncata		*Tectura persona*
	Mytilus trossulus		*Tectura scutum*
	Nuculana pernula		*Trichotropis cancellata*
	Nuculana sp.		*Trichotropis insignis*
	Pandora wardiana		
	Panomya ampla		
	Panomya sp.	Mollusca: Polyplacophora	*Katharina tunicata*
	Patinopectin caurinus		
	Penitella penita	Arthropoda: Crustacea: Cirripedia	*Balanus crenatus*
	Pholadidae		*Balanus glandula*
	Pododesmus macroschisma		*Balanus rostratus*
	Protothaca staminea		Balanidae
	Protothaca tenerrima		*Semibalanus cariosus*
	Saxidomus gigantea		*Solidobalanus hesperius*
	Siliqua patula		
	Siliqua sp.	Arthropoda: Crustacea: Decapoda	*Cancer magister*
	Tellina sp.		*Chionoecetes bairdi*
	Tresus sp.		
	Yoldia sp.	Bryozoa	*Borgella tumulosa*
			Celleporaria smitti
Mollusca: Gastropoda	*Acmaea mitra*		*Celleporaria surcularis*
	Admete sp.		*Celleporaria* sp.
	Admete viridula		*Crisina* ? cf. *canariensis*
	Aforia circinata		*Electra arctica*
	Amphissa versicolor		*Heteropora cf. alaskensis*
	Beringius frielei		*Heteropora pelliculata*
	Boreotrophon multicostatus		*Heteropora* sp.
	Buccinum baeri		*Hippothoa hyalina*
	Buccinum plectrum		*Membranipora* sp.
	Colus halli		*Microporina articulata*
	Colus jordani		*Schizoporella smitti*
	Colus spitzbergensis		*Veleroa veleronis*
	Cryptobranchia alba		
	Cryptobranchia concentrica	Echinodermata: Echinoidea	*Strongylocentrotus* ?

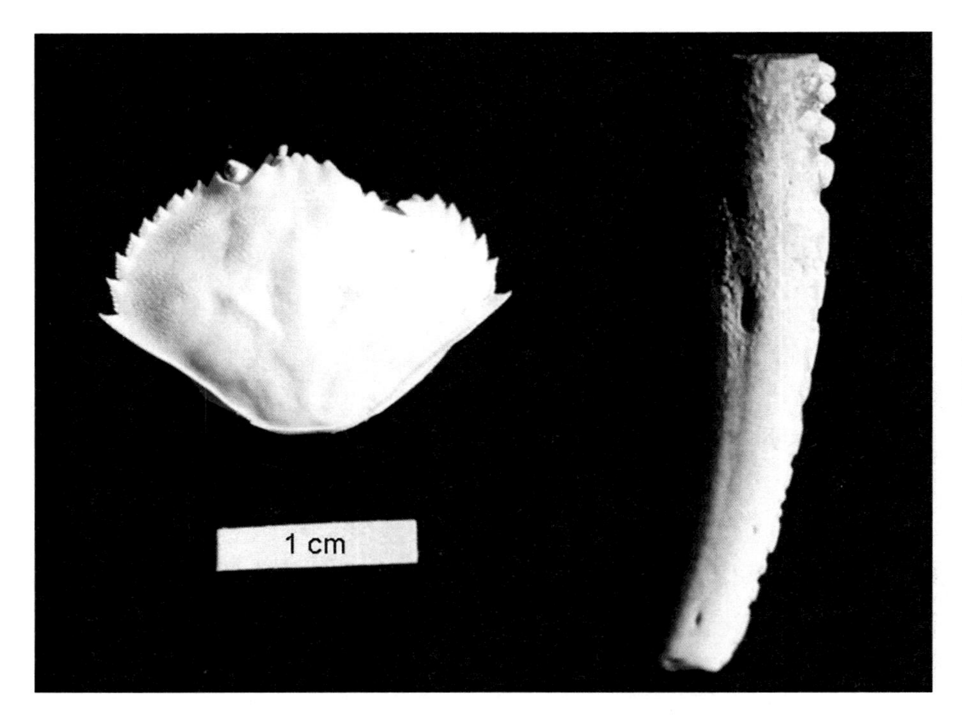

Figure 8. Carapace of *Cancer magister* and leg fragment of *Chionoecetes bairdi* found on the surface at the Ancient Forest site. Associated bivalve shells dated from 7590 to 10,920 ^{14}C yr B.P.

conventional shell ages ranged from 7590 ± 140 to 10,920 ± 90 ^{14}C yr B.P. (Table 2; Fig. 9).

INTERPRETATION

Paleoenvironments of the Ancient Shoreline

Mollusk remains in Quaternary assemblages are considered valid indicators of the original habitats (see review in Gordillo and Aitken, 2000). When the species assemblages from the four most productive sites are compared, using ecological attributes of the species (depth preference, habit, feeding type; Tables 5–7), it becomes apparent that the faunas represent different paleoenvironments. However, various factors could have affected the preservation and transport of specimens. The data could be skewed by factors that control the abundance and condition of the skeletal material, mode of life (e.g., whether infaunal or epifaunal), robustness of the skeleton, size, time of exposure to the elements (Gordillo and Aitken, 2000), and whether or not outburst flood conditions existed during the depositional process. Although our specimens are not in situ, we infer that their association with each other is probably the result of parts of the seafloor having been incorporated into glacier ice. These packages of sediment and biota would remain intact, following the parallel flow lines of the glacier as seen in parallel medial moraines. Organisms carried in moraines ending at these sites have taken different pathways, making it unlikely that organisms from one site could be "mixed" with those from another; note that the Ancient Forest site is ~30 km from Weeping Peat.

While the specimens from our four major sites represent transported species, we feel that an ecological analysis of the assemblages (Tables 5–7) is reasonable. Thus, even though conditions must have changed over the 5000–6000 a time span of the shoreline, comparison of taxa found at the four richest sites (Ancient Forest, Weeping Peat, Taggland, and Admiral Bogg Coast), indicate the occurrence of different paleoenvironments or habitats along the ancient shorelines.

Ancient Forest Habitat

The presence of skeletal fragments of the bryozoans *Celleporaria* spp. (Table 8), *Microporina* spp., and *Heteropora* spp. (Appendix 1) indicates an environment swept by currents, which favors such sessile suspension feeders. These bryozoans live on shells and stones on substrates of stones or silt at depths between 6 and 450 m (Kluge, 1975). In contrast, bryozoans are rare at Weeping Peat and Taggland and are absent from the Admiral Bogg Coast site. The presence of *Trichotropis*, *Puncturella*, Conidae, and calcareous tubes made by serpulid annelids at the Ancient Forest site is consistent with an environment where rock-boulder substrates are swept by currents carrying particulate organics and algal debris. Similar habitats along the contemporary Gulf of Alaska coast are found in Prince William Sound, lower Cook Inlet, and the Kodiak shores.

Over 45% of the fragments from the Ancient Forest are, however, deposit feeders (Table 7) and subtidal infauna (Tables 5, 6), including several *Macoma* species and small gastropods, suggesting the presence of an enriched soft substrate offshore of a rocky intertidal-nearshore setting. A similar

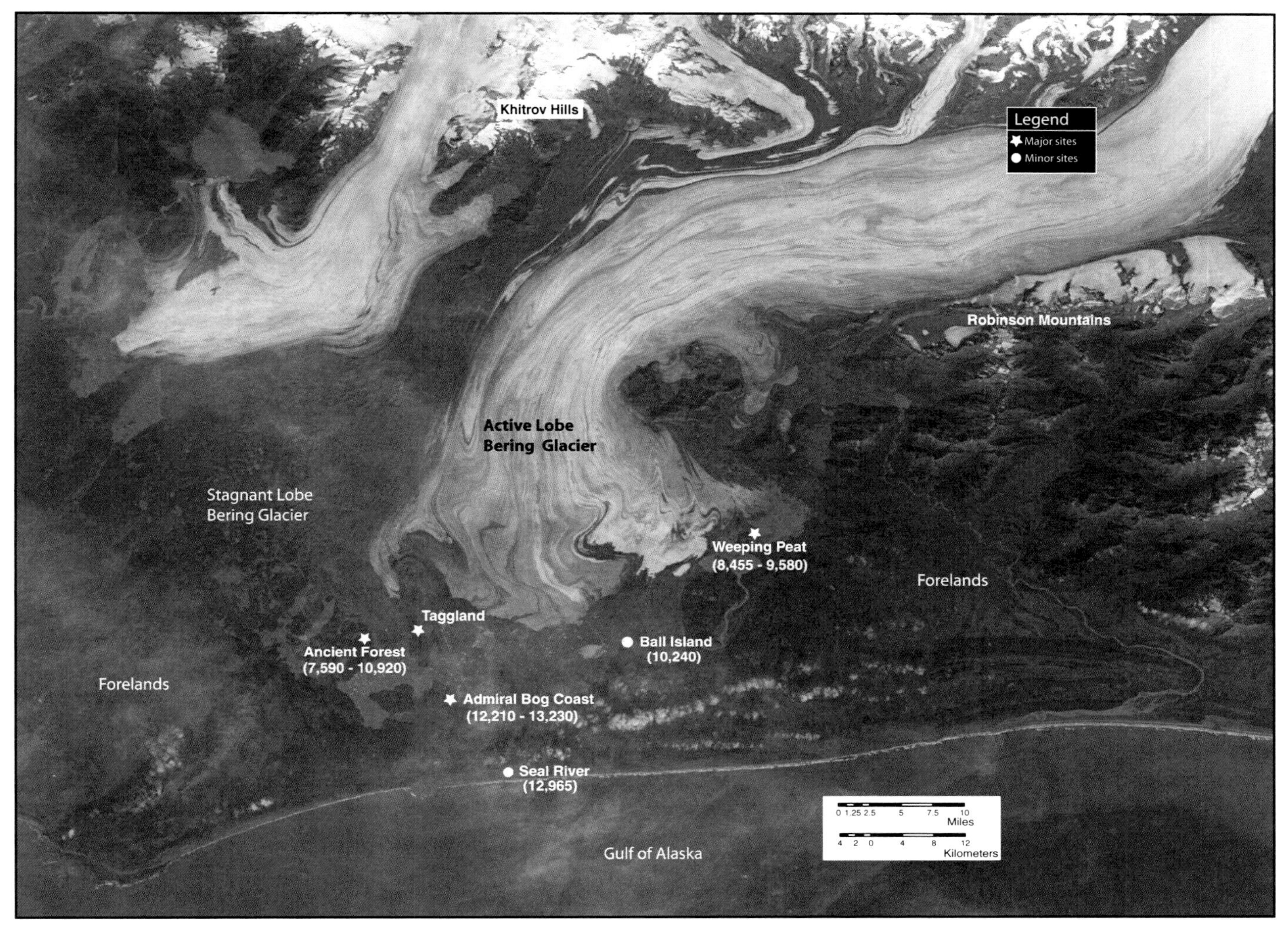

Figure 9. Ranges of conventional radiocarbon dates from shells found at five specific sites on the forelands.

TABLE 5. DIFFERENCES IN DEPTH PREFERENCES OF THE INVERTEBRATE FAUNAS FROM THE FOUR MOST PRODUCTIVE SITES

Depth	Admiral Bogg Coast (%)	Ancient Forest (%)	Taggland (%)	Weeping Peat (%)
Intertidal	9.1	26.9	17.1	42.4
Intertidal-subtidal	18.2	29.5	31.7	30.3
Subtidal	72.7	42.3	46.3	24.2
Not determined	0.0	1.3	4.9	3.1
Total	100	100	100	100

Note: Differences are expressed as the percentage of the total number of species found at each site.

TABLE 6. RELATIVE DIFFERENCES IN HABIT OR MODE OF LIVING OF THE INVERTEBRATE FAUNAS AT THE FOUR MOST PRODUCTIVE SITES

Habit	Admiral Bogg Coast (%)	Ancient Forest (%)	Taggland (%)	Weeping Peat (%)
Borer	0.0	0.0	2.4	0.0
Epifauna	31.8	38.5	41.5	33.3
Epifauna-rocks	4.5	32.0	17.1	30.3
Infauna	63.7	29.5	36.6	36.4
Not determined	0.0	0.0	2.4	0.0
Total	100	100	100	100

Note: Differences are expressed as the percentage of the total number of species found at each site.

TABLE 7. DIFFERENCES IN THE RELATIVE ABUNDANCE OF FEEDING TYPES OF THE INVERTEBRATE FAUNAS AT THE FOUR MOST PRODUCTIVE SITES

Feeding Type	Admiral Bogg Coast (%)	Ancient Forest (%)	Taggland (%)	Weeping Peat (%)
Deposit feeder	66.3	45.2	15.0	52.8
Grazer	0.6	4.0	14.6	1.2
Predator	3.0	4.7	7.8	3.2
Predator-scavenger	18.3	1.9	10.6	0.7
Scavenger	0.6	1.2	2.5	2.5
Deposit–suspension feeder	1.2	2.4	5.3	13.6
Suspension feeder	10.0	40.6	44.2	26.0
Total	100	100	100	100

Note: Differences are expressed as the percentage of the total number of species found at each site.

TABLE 8. RELATIVE ABUNDANCE OF THE DOMINANT TAXA (>10% AT ONE OR MORE SITES) AT THE FOUR MOST PRODUCTIVE SITES

Higher taxon	Name	Admiral Bogg Coast (%)	Ancient Forest (%)	Taggland (%)	Weeping Peat (%)
Barnacle	*Balanus glandula*		4		16
Bivalve	*Clinocardium* sp.		0.9	1.9	14.4
Bivalve	*Macoma calcarea*	20.3	0.3	5.8	
Bivalve	*Macoma golikovi*	18.1	0.8	1.3	8
Bivalve	*Macoma lipara*	19.2	10	0.6	0.2
Bivalve	*Macoma* sp.	7.7	33.1	5.5	
Bivalve	*Penitella penita*			10.7	
Bivalve	*Saxidomus gigantea*		0.6	0.3	19.2
Bryozoa	*Celleporaria* sp.		12.6	0.6	20
Gastropod (subtidal limpet)	*Cryptobranchia concentrica*	0.5	1.3	10.1	0.2

situation is described in lower Cook Inlet and Resurrection Bay (Feder and Jewett, 1986).

Weeping Peat Habitat

The Weeping Peat invertebrate suite, which includes large numbers of intertidal and intertidal subtidal species (Table 5), is composed mostly of the infaunal bivalves *Macoma* spp., *Saxidomus gigantea*, and *Clinocardium* spp. (Table 8). Over 50% of the specimens are intertidal and shallow subtidal deposit feeders: e.g., *Macoma golikovi* and *M. inquinata.* Suspension feeders constitute 26% of the fauna (Table 7). This site also includes the intertidal predatory gastropods *Nucella* spp., *Ocenibrina* spp., and *Cryptonatica.* This shelled fauna suggests a mixed substrate setting similar to intertidal habitats on Kodiak Island described by Nybakken (1969), which consisted of a substrate of boulders scattered on mud. The suspension feeders inhabited the tops of the boulders, whereas the deposit feeders occupied the muddy substrate in between them.

Admiral Bogg Habitat

Shells and fragments from the Admiral Bogg Coast represent infaunal and epifaunal subtidal species (Table 6), including the clams *Macoma calcarea*, *M. lipara*, and *M. golikovi* (Table 8), and the subtidal gastropods *Neptunea lyrata* and *Colus halli.* There were few low intertidal or rock substrate species at this site. Over 66% of the specimens (e.g., *Macoma*) represent deposit feeders (Table 7), a number that distinguishes it from the other sites. There were higher proportions (>18%) of the predators-scavengers such as *Colus* and *Neptunea* than at other sites (Table 7). The large proportion of deposit feeders suggests muddy sediments in a bay or fjord with a deep sill (Feder, 1979; Feder and Jewett, 1986).

Taggland Habitat

The boring clam *Penitella penita* and entire valves of the nestling clam *Hiatella arctica* are unique to the Taggland site (Table 8). Serpulid worm tubes are found only in the Ancient Forest and Taggland sites, suggesting a substrate of stiff clay or sedimentary rock at low intertidal or shallow subtidal depths in protected outer coasts (Ricketts et al., 1985). The dominance of suspension feeders (>44%) reflects the numbers of boring clams (*P. penita*), nestling clams (*Hiatella arctica*), barnacles (*Balanus cariosus*), and serpulids at this site (Table 7).The barnacle, *B. glandula,* common at the Ancient Forest and Weeping Peat sites, was not found here. *Penitella* and other boring species have been collected at a few places along the present Gulf of Alaska coast, including Bluff Point, west of Bishop's Beach near Homer, in Cook Inlet (Lees, 1978); however, studies of the intertidal region have not described this habitat or particular mix of species in Alaskan waters.

Extralimital Species

At least five species collected are not found in the contemporary Gulf of Alaska fauna. *Macoma middendorffi* is an Arctic–Western Pacific species, noted at Point Barrow, Bristol Bay, and west to Kamchatka. It is not found east of Bristol Bay. Similarly, *Chlamys behringiana* is an Arctic-Pacific species found in the same region and not east of Bristol Bay (Coan et al., 2000). The gastropods *Amphissa versicolor*, *Neptunea phoenecia*, *Ocinebrina lurida*, and the bivalve *Protothaca tenerrima*, are Eastern Pacific species with a more southerly distribution, known from southeast Alaska but not north of Juneau (Harbo, 1997; Coan et al., 2000; Foster, 2000).

The presence of extralimital species indicates changing conditions in what must have been and still is a dynamic environment. The Western Pacific species are associated with cooler conditions. Perhaps a weaker Alaska Coastal Current and favorable conditions in the benthic environment (of turbidity, water temperature, salinity, particulate organic carbon [POC] flux, and the absence of predators and competing species) allowed the ancient *Macoma* and *Chlamys* to live east of their current distributional range.

Similarly, a warmer ocean may have influenced the ability of the species now found south of Bering Glacier to live in the habitats that existed along the ancient shoreline. Cooler water, changes in salinity, increased turbidity, changing substrates, and altered trophic relationships associated with glacial conditions may have resulted in species shifts to the southeast.

Arguably the water temperature could not cool and warm at the same time. This contradiction could be explained if the extralimital species were separated by time. For example, the two species suggestive of cooler conditions, *M. middendorffi* (found at Admiral Bogg) and *C. behringiana* (found at Seal River), were collected from sites yielding older dates of 12,210–13,230 ^{14}C yr B.P. (Table 2). The Eastern Pacific gastropods, suggestive of warmer conditions, were all found at the Ancient Forest site with younger dates ranging from 7590 to 10,390 ^{14}C yr B.P. However, the single specimen of *P. tenerrima* found on the overwash channel at the Seal River site, which has one date of 12,965 ^{14}C yr B.P., does not fit this pattern.

Late Pleistocene–Early Holocene Glacial Chronology

During the late Cenozoic there have been dramatic fluctuations of glaciers in the Gulf of Alaska region. Hamilton (1994) cites advances of at least nine major ice sheets during the past 300,000 a with sea level fluctuations as much as 100 m. The coalescing ice sheets are not well dated because they advanced at least 100 km into the Gulf (Calkin et al., 2001). The known glacial sedimentary record is sparse, because it has been greatly altered by oceanic phenomena and subject to tectonic activity, changing mean sea level, and rapid erosion owing to the high relief of the region. Mountains within 55 km of the shoreline exceed 2500 m in elevation. Deposits of the last major ice shelf in the Gulf of Alaska are of early Pleistocene age (Hamilton, 1994). Estimates for the retreat of this last major episode are that it began ca. 12,000–15,000 ^{14}C yr B.P. (Hamilton, 1994; Wiles et al., 1999) or ca. 14,000–16,000 ^{14}C yr B.P. (Calkin et al., 2001). According to Hamilton (1994), Molnia (1995), and Wiles et al. (1999), the ice had completely withdrawn from the continental shelf by 10,000 ^{14}C yr B.P.

Our data indicate that the ice sheet was already in a retracted position by ca. 13,000 ^{14}C yr B.P. Marine invertebrates were established inland of the present ice front by that date. Marine conditions persisted in what are now the forelands until 5000 ^{14}C yr B.P. An inland position of a late Pleistocene shoreline is reasonable in that portions of the Bering Glacier lie nearly 200 m below sea level (Molnia, 1995; Shuchman, this volume). The base of its east-west–trending trough is below sea level as far as 55 km up-glacier. An encroaching ocean on this deglaciated landscape would have produced an irregular shoreline with bays and headlands and different types of substrates and energy levels, similar to that seen in Prince William Sound today. Marine conditions prevailed until ca. 5000–7000 ^{14}C yr B.P. At this time, bedrock now exposed at Taggland may have been an island or a topographic high ~22 m below mean sea level.

Subsequently the piedmont lobes of the Bering Glacier advanced into the intertidal zone with a sediment load that overwhelmed the marine environment. After ca. 5000 ^{14}C yr B.P., nonmarine conditions prevailed. Establishment of spruce forests between fluctuations of the ice front are well documented (Crossen, this volume).

CONCLUSIONS

The marine invertebrates collected provide evidence that a shoreline once existed north of the present ice margin. Conventional radiocarbon dates of shells ca. 7000–13,000 ^{14}C yr B.P. provide a range of dates for the existence of this marine environment. Therefore, the Bering-Steller Piedmont Glacier complex was in a retracted position at that time (late Pleistocene–early Holocene), and the present forelands did not exist until later. The rugged relief of the Chugach Mountains precludes the invasion of the Gulf of Alaska much more than 35–40 km inland at the base of the mountain front and up to 55 km into the deep gorge occupied by the valley portion of the Bering Glacier, where its base is currently below sea level.

At least four discrete invertebrate faunas (Admiral Bogg Coast, Ancient Forest, Taggland, and Weeping Peat) are recognized, and these correlate with modern Gulf of Alaska fauna that also inhabit a variety of environmental settings. Therefore, it is reasonable to conclude that the ancient shoreline was irregular, consisting of headlands and bays commonly found along glaciated terrain, and that it provided habitats with a variety of depths, energy levels, substrates, and food sources.

We are unsure of the exact mechanism by which glaciers can excavate and transport fragile invertebrate skeletal remains. We suspect that the organisms were entrained within blocks of a fine-grained protective substrate frozen by its proximity to the ice. Most of this material was sheared apart during glacier transport. However, a substrate of cohesive silt enclosing a complete shell or a crab carapace apparently had sufficient strength to maintain blocks up to several meters in length. Upon thawing at the ice front, the fine-grained material lost cohesiveness and was washed away by meltwater, releasing its "cargo" intact as residuals. Whatever the mechanism, we believe the depositional process is directly related to glacier surges that provide unique conditions for the release of fragile skeletal remains. Most of the identifiable specimens collected were found on glacial-surge drift deposits formed since 1995. Those from the Seal River and Admiral Bogg sites were deposited at the margin of the 1968 surge.

The assemblage of ancient species is similar to modern Gulf of Alaska fauna and may have some links to extant faunas now found both to the southeast and west, as indicated by the presence of extralimital species.

APPENDIX 1. LIST OF ANCIENT TAXA IDENTIFIED FROM 15 STUDY SITES

Genus	Species	AA	BA	CA	FA	OA	SR	TL	WA	AB	BN	SS	NB	WI	PI	MI	Total
Acmaea	*mitra*				3												3
Admete	*viridula*				4			1		1							6
Admete ?					1												1
Aforia	*circinata*				1												1
Amphissa	*versicolor*				3	1											4
Astarte	*arctica*									1							1
Astarte	*borealis*									3							3
Astarte	*elliptica*									1							1
Astarte	sp.				1												1
Balanus	*crenatus*							1									1
Balanus	*crenatus* ?											5					5
Balanus	*glandula*				141	3	2		96								242
Balanus	*rostratus*				18	5		8	8								39
Balanus	*rostratus* ?				10												10
Balanus	sp.				29	6								35	13	30	113
Balanus ?									1								1
Barnacle								1			7	21		5	2	59	95
Beringius	*frielei* ?				1												1
Bivalvia			41		33	9	14	27	20		48	86	9	47	32	21	387
Boreotrophon	*multicostatus*							1	2								3
Borer					4		1		9								14
Borgella	*tumulosa*				25												25
Bryozoa					4				1								5
Buccinum	*baeri* ?													1			1
Buccinum	*plectrum*				2												2
Calcareous algae ?														1			1
Cancer	*magister*				1												1
Carditid																2	2
Celleporaria	*smitti*				1												1
Celleporaria	*surcularis*				38				1								39
Celleporaria	sp.				167	5		1			1						174
Chionoecetes	*bairdi*				1			1									2
Chlamys	*behringiana*						4	2									6
Chlamys	*hastata*							2		3							5
Chlamys	*rubida*				1		1			1							3
Chlamys	sp.				3			2		1	1						7
Clinocardium	*ciliatum*				1												1
Clinocardium	*ciliatum* ?				2												2
Clinocardium	*fucanum*						1		25								26
Clinocardium	*fucanum* ?								2								2
Clinocardium	*nuttallii*	20	1		21	12		11	38	2	308						413
Clinocardium	sp.	6	5		15	8		6	79					1			120
Clinocardium ?									7								7
Columella					1			2	1	2	2						8
Colus	*halli*							7		15	8						30
Colus	*jordani*				8		5	6			1						20
Colus	*jordani* ?				2												2
Colus	*spitzbergensis*				1												1
Colus	*spitzbergensis*?								2								2
Crisina	*canariensis*?				2												2
Cyclocardia	*ventricosa*									1							1
Cryptobranchia	*alba*							6									6
Cryptobranchia	*concentrica*				20		1	31	1	1							54
Cryptobranchia	*concentrica* ?				2												2
Cryptonatica	*affinis*				13		2	7	2	3	1						28
Diodora	*aspera*				7						1					1	9
Diodora	sp.						1										1
Echinoidea					9			3	1						1		14
Electra	*arctica*			1													1
Euspira	*pallida*				1			2	1								4
Fusitriton	*oregonensis*				4		4	7	21	1	3	1				2	43
Fusitriton ?								1	3								4
Fusitriton/Neptunea							1									1	
Gastropoda					45	3		5	17						1	3	74

(*continued*)

APPENDIX 1. LIST OF ANCIENT TAXA IDENTIFIED FROM 15 STUDY SITES (*continued*)

Genus	Species	AA	BA	CA	FA	OA	SR	TL	WA	AB	BN	SS	NB	WI	PI	MI	Total
Gastropoda ?									2								2
Heteropora	*alaskensis*?				1												1
Heteropora	*pelliculata*				44												44
Heteropora	sp.				11												11
Hiatella	*arctica*							10						2		2	14
Hiatella	sp.												1				1
Hiatella ?								1									1
Hippothoa	*hyalina*			1													1
Humilaria	*kennerleyi*				1												1
Katharina	*tunicata*				1							1			1	3	6
Kurtziella	sp.				1												1
Kurtziella ?					2												2
Limpet					1												1
Lirabuccinum	*dirum*			1	1				4								6
Littorina	*sitkana*								1								1
Lottia	*pelta*				1												1
Lottia	*pelta* ?								1								1
Lottia/Tectura								1									1
Macoma	*balthica*								1		2						3
Macoma	*balthica*?							6									6
Macoma	*brota* ?				4					3							7
Macoma	*calcarea*				1		7	18		37							63
Macoma	*calcarea* ?				4												4
Macoma	*golikovi*				13		1	4	48	33							99
Macoma	*inquinata*				11				6								17
Macoma	*inquinata* ?								15								15
Macoma	*lipara*				161		3	2	1	35							202
Macoma	*lipara* ?	1			3	1											5
Macoma	*middendorffi*									3							3
Macoma	sp.				536	10		17	505		14						1082
Macoma ?					4		2		1								7
Macoma/Saxidomus		115		14	356	245		79	176	29	158	33	5	42	7		1259
Mactridae					1			3	8	2	20						34
Mactromeris	*polynyma*	1	2		7	2	4										16
Mactromeris	*polynyma* ?				1												1
Margarites	*pupillus*				1				6								7
Membranipora	sp.				1												1
Micropora ?					1												1
Microporina	*articulata*				64												64
Mohnia	*frielei*				3				2								5
Mohnia	*frielei* ?				2												2
Musculus ?					1												1
Mya	*truncata*				1				2	1							4
Mya/Panomya								4									4
Mytilus	*trossulus*										2					1	3
Mytillus	sp.							1	2								3
Mytilus ?									2							1	3
Natica ?										1							1
Naticid					2			3	2								7
Naticidae								7	2								9
Neptunea	*lyrata*				6			9	3	13	8						39
Neptunea	*lyrata* ?							1		3							4
Neptunea	*phoenicia*				1												1
Neptunea	*pribiloffensis*				1												1
Neptunea	sp.							10	1		1						12
Nucella	*lamellosa*				2				22								24
Nucella	*lamellosa* ?								2								2
Nucella	*lima*				1												1
Nucella	sp.				1												1
Nucella ?					2												2
Nuculana	*fossa*									1							1
Nuculana	*pernula*							1			7						8
Nuculana	*pernula* ?		1														1
Nuculana	sp.											2					2

(*continued*)

APPENDIX 1. LIST OF ANCIENT TAXA IDENTIFIED FROM 15 STUDY SITES (*continued*)

Genus	Species	AA	BA	CA	FA	OA	SR	TL	WA	AB	BN	SS	NB	WI	PI	MI	Total
Nuculanidae					3	3	1										7
Ocinebrina	*interfossa*				2												2
Ocinebrina	*lurida*				10												10
Oenopota	*excurvata*				1		1	2									4
Oenopota	*excurvata* ?								2								2
Oenopota	*levidensis*				1												1
Oenopota	*rosea*				2												2
Oenopota	*rosea* ?				2												2
Oenopota	*turricula*				14			1									15
Oenopota	sp.				10			1			1						12
Oenopota ?					3												3
Olivella	*baetica*				14												14
Pandora	*wardiana*				4												4
Panomya	*ampla*				4												4
Panomya	sp.									3	2						5
Paradexiospira	*vitrea*				1												1
Patinopectin	*caurinus* ?				1												1
Penitella	*penita*							33									33
Pholadidae					1												1
Pododesmus	*macroschisma*				7												7
Pododesmus	sp.														2		2
Probela	*harpularia*				1												1
Protothaca	*staminea*	2			18	14		16	26							3	79
Protothaca	*staminea* ?	1													1		2
Protothaca	*tenerrima*						1										1
Puncturella	sp.				1			3									4
Puncturella	*galeata*				1												1
Puncturella	*multistrata*							3									3
Saxidomus	*gigantea*	85				14			76								175
Saxidomus	*gigantea* ?		15		6		14		5								40
Saxidomus	sp.							1	34		6			6	9	2	58
Saxidomus ?		7			3										5		15
Schizoporella	*smitti*				1												1
Semibalanus	*cariosus*	6	1		33	1		10	25		1			2	1	56	136
Semibalanus	*cariosus* ?				3												3
Serpula/Crucigera					14		1	28								1	44
Siliqua	*patula*				17												17
Siliqua	sp.							1			2						3
Solidobalanus	*hesperis*	1															1
Solidobalanus	*hesperis* ?				1												1
Strongylocentrotus ?					2												2
Tectura	*fenestrata*				1												1
Tectura	*persona*				16				3								19
Tectura	*scutum*				1												1
Tellina ?															1		1
Tellinidae									10								10
Tresus	*capax* ?				2			1									3
Tresus	sp.															1	1
Trichohyalus	*ornatissimus*							70									70
Trichotropis	*cancellata*				77			15	1		1						94
Trichotropis	*cancellata* ?				2				1								3
Trichotropis	*insignis* ?				2												2
Unknown		53			2			9		1							65
Veleroa	*veleronis*				1												1
Venerid						1											1
Yoldia	*myalis* or *hyperbora*				1												1
Yoldia	sp.				3												3
Yoldia ?					1												1
Locale total		298	66	17	2192	343	71	513	1336	202	606	149	15	142	76	188	6214
Percentages		4.7	1	0.2	35.2	5.5	1.1	8.2	21.5	3.2	9.7	2.3	0.2	3	1.2	3	100

Note: Ancient specimens collected from 15 study sites were identified to the lowest taxonomic level possible. Taxonomic identifications are presented with the number collected from each site. Site names and abbreviations: AA—Arrowhead Point; BA—Ball Island; CA—BLM Camp; FA—Ancient Forest; OA—Outhouse Moraine; TL—Taggland; SR—Seal River; WA—Weeping Peat; AB—Admiral Bogg Coast; BN—Beringia Nova; SS—S. Shore Vitus Lake; NB—Narrows Bluff; WI—Whaleback Island; PI—Pointed Island; MI—Molypog Islands.

APPENDIX 2. COMPARISON OF COLLECTED MODERN GULF OF ALASKA TAXA WITH ANCIENT TAXA

Genus	Species	Present	No. Cape Suckling	No. Seal R.	No. Midtimber Lake	No. Tsiu R.
Acmaea	*mitra*	x	8			
Admete	*viridula*					
Admete ?						
Aforia	*circinata*					
Amphissa	*columbiana*	x	4			
Amphissa	*versicolor*					
Astarte	*borealis*					
Astarte	*elliptica*					
Astarte	sp.					
Balanus	*crenatus*					
Balanus	*glandula*	x	81	1		
Balanus	*rostratus*					
Balanus	sp.					
Balanus ?						
Bankia	***setacea***	**x**	**1**	**3**		
Balanidae						
Beringius	*frielei* ?					
Bivalve		x	1	18		9
Boreotrophon	*multicostatus*					
Borings		x	1			
Bryozoan		x	6	1	1	
Buccinum	*baeri* ?					
Buccinum	*plectrum*	x				1
Calcareous algae ?		x	x			
Cancer	*magister*	x	1	1	2	
Cardiitidae						
Chionoecetes	*bairdi*					
Chlamys	*behringiana*					
Chlamys	*hastata*					
Chlamys	*rubida*					
Chlamys	sp.					
Clinocardium	*ciliatum*	x				1
Clinocardium	*fucanum*					
Clinocardium	*nuttallii*					
Clinocardium	sp.					
Columella						
Colus	*halli*					
Colus	*jordani*	x		5		
Colus	*spitzbergensis* ?					
Cryptobranchia	*concentrica*					
Cryptonatica	*affinis*	x		1 ?		
Diodora	*aspera*					
Diodora	sp.					
Echinoid						
Entodesma	***navicula***	**x**	**3**			
Euspira	*pallida*					
Fusitriton	*oregonensis*	x	20			
Fusitriton/Neptunea						
Gastropod		x				1
Heteropora	*alaskensis* ?					
Heteropora	sp.					
Hiatella	*arctica*	x	1			
Hiatella	sp.					
Humilaria	*kennerleyi*					
Isodictya	**sp.**	**x**				**1**
Katharina	*tunicata*	x	3			
Kurtziella	sp.					
Lepas	***anatifera***	**x**	**24**	**5**	**27**	**5**
Limpet						
Lirabuccinum	*dirum*	x	27			
Littorina	*sitkana*	x	19			
Lottia	*pelta*					
Lottia/Tectura						
Macoma	*balthica*	x	17			

(continued)

APPENDIX 2. COMPARISON OF COLLECTED MODERN GULF OF ALASKA TAXA WITH ANCIENT TAXA *(continued)*

Genus	Species	Present	No. Cape Suckling	No. Seal R.	No. Midtimber Lake	No. Tsiu R.
Macoma	*brota ?*					
Macoma	*calcarea*					
Macoma	*expansa*	x				15
Macoma	*golikovi*					
Macoma	*inquinata*	x	4			
Macoma	*lipara*					
Macoma	*middendorffi*					
Macoma	sp.					
Macoma/Saxidomus						
Mactricidae						
Mactromeris	*polynyma*					
Margarites	***costalis***	**x**	**2**			
Margarites	*pupillus*					
Membranipora	sp.					
Micropora ?						
Microporina	*articulata*					
Mohnia	*frielei*					
Musculus ?						
Mya	***arenaria***	**x**	**8**			
Mya	***pseudoarenaria***	**x**	**1**			
Mya	*truncata*					
Mya/Panomya						
Mytilus	sp.					
Mytilus	*trossulus*	x	15		13	2
Natica ?						
Naticidae						
Neptunea	*lyrata*	x		1		1
Neptunea	*phoenicia*					
Neptunea	*pribiloffensis*					
Neptunea	sp.	x		1		
Neptunea ?		x	4			
Nucella	*lamellosa*					
Nucella	*lima*					
Nucella	sp.					
Nuculana	*fossa*					
Nuculana	*pernula*					
Nuculana	sp.					
Nuculanidae						
Ocenebrina	*interfossa*					
Ocenebrina	*lurida*					
Oenopota	*excurvata*					
Oenopota	*levidensis*					
Oenopota	*rosea*					
Oenopota	sp.					
Oenopota	*turricula*					
Olivella	*baetica*					
Pandora	*wardiana*					
Panomya	*ampla*					
Panomya	sp.					
Paradexiospira	*vitrea*					
Patinopectin	*caurinus* ?					
Penitella	*penita*	x	3			
Pholadidae						
Pododesmus	*macroschisma*	x	1	1		
Pododesmus	sp.					
Probela	*harpularia*					
Protothaca	*staminea*	x	13			
Protothaca	*tenerrima*					
Puncturella	*galeata*					
Puncturella	*multistriata*					
Saxidomus	*gigantea*	x	18			
Saxidomus	sp.					
Macoma/Saxidomus						
Schizobranchia	***insignis***	**x**	**1**	**2**		
Semibalanus	*cariosus*	x	22			
Serpula/Crucigera		x	1	1		
Siliqua	*patula*	x		1	6	16

(continued)

APPENDIX 2. COMPARISON OF COLLECTED MODERN GULF OF ALASKA TAXA WITH ANCIENT TAXA *(continued)*

Genus	Species	Present	No. Cape Suckling	No. Seal R.	No. Midtimber Lake	No. Tsiu R.
Siliqua	sp.					
Solidobalanus	*hesperius*					
Spirobis	**sp.**	**x**			**1**	
Strongylocentrotus ?	*droebachiesnesis*	x	1			
Tectura	*fenestrata*					
Tectura	*persona*	x	36			
Tectura	*scutum*	x	1			
Tellina ?						
Tellinidae		x			1	
Tresus	*capax* ?					
Tresus	sp.					
Trichohyalus	*ornatissimus*					
Trichotropis	*cancellata*					
Trichotropis	*insignis* ?					
Unknown		x	1	2		2
Veneridae						
Yoldia	*myalis* or *hyperbora*					
Yoldia	sp.					

Note: Modern taxa collected from four Gulf of Alaska shoreline sites are compared with the list of ancient taxa. Bold type indicates modern species not found in collections from the ancient sites.

APPENDIX 3. GPS LOCATIONS FOR 15 STUDY SITES

Site	Outcrop	Abbr.	Lat.	Long.	Elev. ± 5 m
Arrowhead Point	1993–95 surge moraine retreat surface	AA	60° 10.489′	143° 17.347′	37
Ball Island	1993–95 surge moraine	BA	60° 08.010′	143° 20.510′	1
BLM camp	1965–68 surge moraine	CA	60° 07.285′	143° 16.903′	16
Ancient Forest	1993–95 surge moraine retreat surface	FA	60° 09.447′	143° 39.141′	31
Outhouse Moraine	1965–68 surge moraine	OA	60° 08.487′	143° 37.250′	25
Taggland	1993–95 surge moraine retreat surface	TL	60° 07.594′	143° 32.275′	5
Seal River	1900? moraine, reworked	SR	60° 03.226′	143° 28.148′	7
Weeping Peat	1993–95 surge moraine retreat surface	WA	60° 11.747′	143° 11.577′	22
Admiral Bogg Coast	1968 surge moraine	AB	60° 06.112′	143° 31.766′	6
Beringia Nova	1993–95 surge moraine	BN	60° 06.910′	143° 29.593′	6?
S. shore Vitus Lake	1965–68 surge moraine	SS	60° 07.459′	143° 20.829′	6
Narrows Bluff	1965–68 surge moraine	NB	60° 07.876′	143° 20.413′	12
Whaleback Island	1993–95 surge moraine retreat surface	WI	60° 07.929′	143° 24.769′	0
Pointed Island	1993–95 surge moraine retreat surface	PI	60° 07.266′	143° 25.503′	0
Molypog Islands	1993–95 surge moraine retreat surface	MI	60° 08.208′	143° 25.607′	2

ACKNOWLEDGMENTS

We wish to thank the U.S. Bureau of Land Management for its support and maintenance of the Bering Glacier field camp, which made this project possible. We particularly thank John Payne for his vision and ability to put the research program together. We also thank Scott Guyer and Chris Noyles for their support and efforts in keeping the camp running efficiently. We are especially indebted to John Tucker, whose special skills as a helicopter pilot, knowledge of the landscape, and generous nature made every trip not only productive but enjoyable. We thank Bruce Molnia, Howard Feder, and Erica Madison for their helpful discussions and willingness to share their expertise. Special thanks go to Marilyn Barker and Kristine Crossen, who helped initiate this project and provided support and expertise every field season. We are also grateful for the many researchers, volunteers, staff members, and pilots who so willingly assisted us in the field. Our data set is the result of their enthusiastic collecting. We sincerely thank two anonymous reviewers for their thoughtful comments that greatly improved the manuscript. Any use of trade names is for descriptive purposes only and does not imply endorsement by the United States Government.

REFERENCES CITED

Bergen, F.W., and O'Neil, P., 1979, Distribution of Holocene foraminifera in the Gulf of Alaska: Journal of Paleontology, v. 53, p. 1267–1292.

Calkin, P.E., Wiles, G.C., and Barclay, D.J., 2001, Holocene coastal glaciation of Alaska: Quaternary Science Reviews, v. 20, p. 449–461, doi: 10.1016/S0277-3791(00)00105-0.

Chew, K.K., and Ma, A.P., 1987, Species profiles: Life histories and environmental requirements of coastal fishes and invertebrates (Pacific Northwest)—common littleneck clam: U.S. Fish and Wildlife Biological Report 82 (11.78), U.S. Army Corps of Engineers, TR EL-82-4, 22 p.

Clarkson, E.N.K., 1998, Invertebrate Palaeontology and Evolution (4th edition): Cambridge, Massachusetts, Blackwell Science, 452 p.

Coan, E.V., Scott, P.V., and Bernard, F.R., 2000, Bivalve Seashells of Western North America. Marine Bivalve Mollusks from Arctic Alaska to Baja California: Santa Barbara, California, Museum of Natural History Monographs (2), Studies in Biodiversity (2), 764 p.

Feder, H.M., 1979, Distribution and abundance of some epibenthic invertebrates of the northeastern Gulf of Alaska with notes on the feeding biology of selected species: Boulder, Colorado, Department of Commerce, National Oceanic and Atmospheric Administration, Environmental Assessment of the Alaska Continental Shelf, Final Reports of Principal Investigators, v. 4, Biological Studies, 74 p.

Feder, H.M., and Jewett, S.C., 1986, The subtidal benthos, *Chapter 12 of* Hood, D.W., and Zimmerman, S.T., eds., The Gulf of Alaska: Physical Environment and Biological Resources: Boulder, Colorado, U.S. Minerals Management Service Publication OCS Study MMS-86–0095, p. 347–398.

Foster, N.R., 2000, Biodiversity of Prince William Sound, *Chapter 10 of* Hines, A.H., and Ruiz, G.M., eds., Biological Invasions of Cold-Water Ecosystems: Ballast-Mediated Introductions in Port Valdez/Prince William Sound, Alaska: Valdez, Alaska, Final Project Report to Citizen's Advisory Council of Prince William Sound, 313 p.

Gordillo, S., and Aitken, A.E., 2000, Palaeoenvironmental interpretation of late Quaternary marine molluscan assemblages, Canadian Arctic Archipelago: Géographie physique et Quaternaire, v. 54, p. 301–315.

Hamilton, T.D., 1994, Late Cenozoic glaciation of Alaska, *in* Plafker, G., and Berg, H.C., eds., The Geology of Alaska: Boulder, Colorado, Geological Society of America, The Geology of North America, v. G-1, p. 813–844.

Harbo, R.M., 1997, Shells and Shellfish of the Pacific Northwest: Madeira Park, British Columbia, Canada, Harbour Publishing, 271 p.

Hood, D.W., and Zimmerman, S.T., 1986, The Gulf of Alaska: Physical Environment and Biological Resources: Boulder, Colorado, U.S. Minerals Management Service Publication OCS Study, MMS-86–0095, 655 p.

Hughen, K.A., Baillie, M.G.L., Bard, E., Bayliss, A., Beck, J.W., Bertrand, C.J.H., Blackwell, P.G., Buck, C.E., Burr, G.S., Cutler, K.B., Damon, P.E., Edwards, R.L., Fairbanks, R.G., Friedrich, M., Guilderson, P., Kromer, B., McCormac, F.G., Manning, S.W., Bronk Ramsey, C., Reimer, P.J., Reimer, R.W., Remmele, S., Southon, J.R., Stuiver, M., Talamo, S., Taylor, F.W., van der Plicht, J., and Weyhenmeyer, C.E., 2004, Marine04 Marine radiocarbon age calibration, 0–26 Cal kyr BP: Radiocarbon, v. 46, p. 1059–1086.

Kluge, G.A., 1975, Bryozoa of the northern seas of the U.S.S.R. (Mshanki Severnykh Morei S.S.S.R.), translated by B.R. Sharma from 1962 publication in Russian: New Delhi, Amerind Publishing, 711 p.

Kozloff, E.N., 1996, Marine Invertebrates of the Pacific Northwest: Seattle, University of Washington Press, 537 p.

Lees, D.C., 1978, Reconnaissance of the intertidal and shallow subtidal biotic, Lower Cook Inlet, Research Unit 417, *in* Environmental Assessment of the Alaska Continental Shelf: Boulder, Colorado, Final Reports of Principal Investigators, Biological Studies, v. 3, p. 179–506.

McNeely, R., Dyke, A.S., and Southon, J.R., 2006, Canadian Marine Reservoir Ages, Preliminary Data Assessment: Geological Survey of Canada Open File 5049, 3 p.

Molnia, B.F., 1995, Holocene history of Bering Glacier, Alaska: A prelude to the 1993–1994 surge: Physical Geography, v. 16, p. 87–117.

Nybakken, J.W., 1969, Pre-earthquake intertidal ecology of Three Saints Bay, Kodiak Island, Alaska: Fairbanks, University of Alaska, Biological Papers, v. 9, 115 p.

O'Clair, R.M., and O'Clair, C.E., 1998, Southeast Alaska's Rocky Shores: Animals: Auke Bay, Alaska, Plant Press, 563 p.

Osburn, R.C., 1952, Bryozoa of the Pacific Coast of America, Part 2, Cheilostoma: Ascophora: Allan Hancock Pacific Expeditions, v. 14, p. 271–611.

Pojeta, J., Jr., Runnegar, B., Peel, J.S., and Gordon, M., Jr., 1987, Phylum Mollusca, *Chapter 14 of* Boardman, R.S., Cheetham, A.H., and Rowell, A.J., eds., Fossil Invertebrates: Cambridge, Massachusetts, Blackwell Science, p. 270–435.

Ricketts, E.R., Calvin, J., and Hedgpeth, J.W., 1985, Between Pacific Tides (5th edition, revised by D.W. Phillips): Stanford, California, Stanford University Press, 652 p.

Stuiver, M., and Polach, H.A., 1977, Discussion: Reporting of ^{14}C data: Radiocarbon, v. 19, p. 355–363.

Stuiver, M., and Reimer, P.J., 1993, Extended ^{14}C database and revised CALIB radiocarbon calibration program: Radiocarbon, v. 35, p. 215–230.

Turgeon, D.D., Quinn, J.F., Jr., Bogan, A.E., Coan, E.V., Hochberg, F.G., Lyons, W.G., Mikkelsen, P.M., Neves, R.J., Roper, C.F.E., Rosenberg, G., Roth, B., Scheltema, A., Thompson, F.G., Vecchione, M., and Williams, J.D., 1998, Common and Scientific Names of Aquatic Invertebrates from the United States and Canada (2nd edition): Bethesda, Maryland, American Fisheries Society Special Publication 26, 536 p.

Wiles, G.G., Post, A., Muller, E.H., and Molnia, B.F., 1999, Dendrochronology and late Holocene history of Bering Piedmont Glacier, Alaska: Quaternary Research, v. 52, p. 185–195, doi: 10.1006/qres.1999.2054.

Manuscript Accepted by the Society 02 June 2009

The Geological Society of America
Special Paper 462
2010

Holocene sea-level changes and earthquakes around Bering Glacier

Ian Shennan
Sarah Hamilton
Sea Level Research Unit, Department of Geography, Durham University, Durham DH1 3LE, UK

ABSTRACT

The Bering Glacier Region reveals evidence of relative sea-level change or land uplift and subsidence resulting from predominantly glacio-isostatic and tectonic causes. From its glacial maximum extent on the continental shelf, Bering Glacier had retreated inland of the present coast and north of its present terminus by ca. 14,000 yr B.P. Relative sea level was above present ca. 9200 yr B.P. to at least 5000 yr B.P. before falling to below present. Analyses of sediment cores from the marsh east of Cape Suckling show a great earthquake ca. 900 cal (calendar) yr B.P. that caused coseismic land uplift and a tsunami. Relative sea level rose to above present by the early twentieth century, before coseismic land uplift, relative sea-level fall, in A.D. 1964. Correlation with sites from Cook Inlet to Icy Bay shows some critical differences between previous earthquake deformation cycles and A.D. 1964, related to the regional-scale tectonic setting.

CONTEXT AND AIMS

Bering Glacier lies in a particularly dynamic, yet poorly understood, region in terms of relative sea-level change. The challenge to understanding this area of science lies in the interplays between environmental processes that range in scale from global to local. During the Last Glacial Maximum, ~22,000 yr ago, Bering Glacier extended beyond the present coastline onto the continental shelf (Molnia and Post, 1995). This increased mass of ice caused deformation of the Earth's crust (Fig. 1A). As global climate warmed, great ice sheets then present across much of North America, Northern Europe, parts of Asia and South America, and Antarctica started to melt, causing global, or eustatic, sea-level rise (Fig. 1B). Patterns of sea-level rise vary from region to region in response to changing distributions of ice and water. Termed *relative sea-level change* for any particular location on the Earth's surface, the pattern depends on distance from the ice sheet, size of nearby ice sheets and glaciers, their rates of retreat, and the structure of the Earth's crust in the region (Fig. 1C).

The Bering Glacier Region adds particular scientific challenges and insights. It lies at the convergent plate margin between the Pacific and North American plates (Fig. 2). Episodic movement at the subduction zone interface, the Aleutian megathrust, and the Yakutat microplate produces vertical land-surface deformation that varies in magnitude depending upon the size of the earthquake and distance from the zone of plate slippage. Proximity to the subduction zone also means that the thickness of the lithosphere is much less than in nonseismic locations. A thin lithosphere is more sensitive to change, and therefore land uplift and subsidence may be more likely from short term advances and retreat of local glaciers (Fig. 1D).

We have little knowledge of how these different factors interact, as most current studies of relative sea level are from nonseismic localities, or seismic environments distant from glaciated areas. Bering Glacier lies between the major area of uplift during the great 1964 Alaska earthquake and the area uplifted by an earthquake to the east in 1899 (Fig. 2).

Most quantitative models used for explaining ice–Earth–sea-level interactions, such as those shown in Figure 1C, use information from nonseismic areas, away from major plate boundaries, such as eastern North America, Australia, Barbados, and Northwest Europe. Because Bering Glacier lies at the complex collision between the North American and Pacific

Shennan, I., and Hamilton, S., 2010, Holocene sea-level changes and earthquakes around Bering Glacier, *in* Shuchman, R.A., and Josberger, E.G., eds., Bering Glacier: Interdisciplinary Studies of Earth's Largest Temperate Surging Glacier: Geological Society of America Special Paper 462, p. 275–290, doi: 10.1130/2010.2462(14). For permission to copy, contact editing@geosociety.org.

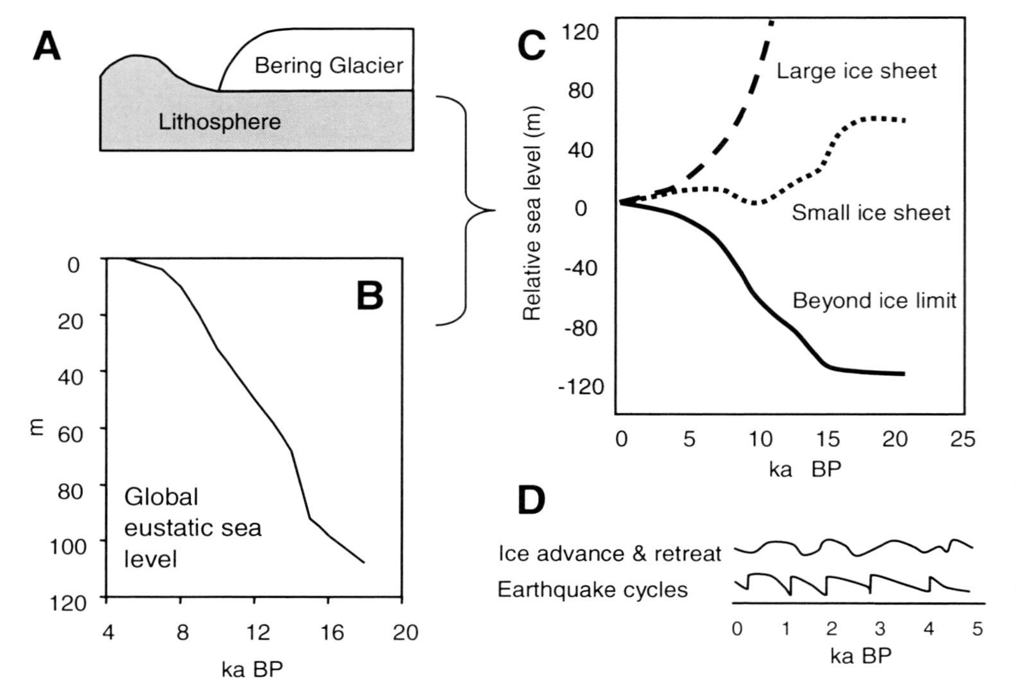

Figure 1. Schematic model of controls on relative sea level. Loading and unloading cycles of the lithosphere by ice sheet advance and retreat causes surface deformation within and beyond the ice limit (A). Land uplift and subsidence combine with global average sea-level change (B) (Peltier and Fairbanks, 2006), resulting from changes in all the ice sheets since the Last Glacial Maximum >20,000 a ago, to give relative sea-level change. Relative sea-level change varies from site to site, depending on the relative location of an individual site to ice sheets and their previous dimensions (C) (Peltier, 2002). Around Bering Glacier, proximity to an active plate boundary and a dynamic glacier adds further contributions to relative sea-level change (D) (Larsen et al., 2003; Larsen et al., 2005; Olsen et al., 1985; Shennan and Hamilton, 2006). In panel D the vertical scale is on the order of a meter or two, and the timescale, a few centuries.

plates, analysis of relative sea-level change will increase our understanding of tectonic processes and development of models to include these extra factors. Studies of relative sea-level change, together with quantitative models to predict the interactions between climate, ice sheets, sea level, and crustal uplift or subsidence, contribute to a wide range of interdisciplinary science that ranges from planetary interior, evolution and dynamics of the environment, to future environmental change from natural and anthropogenic causes.

Therefore, the primary aims of this chapter are to (1) illustrate how the morphology, sedimentology, and biostratigraphy of coastal deposits around Bering Glacier record relative (land-and) sea-level change since the Last Glacial Maximum; (2) show how relative sea-level change can record past great earthquakes, including possible correlation with late Holocene earthquakes in other parts of Alaska; (3) determine vertical displacements caused by glacier loading-unloading of the lithosphere; and (4) start to develop a paleogeographic model to explain the complex landscape of the glacier foreland and coastal zone.

TECTONIC SETTING

Bering Glacier originates in the Bagley Ice Field of the Wrangell–Saint Elias Mountain Range. Uplift and mountain building resulted from the Yakutat microplate colliding into and subducting beneath Alaska, changing from oblique convergence along the Fairweather–Queen Charlotte transform fault to subduction along the Aleutian megathrust (Fig. 2). The present rate of convergence of the Pacific plate with respect to southern Alaska is ~55 mm/a. Bering Glacier lay at the eastern limit of uplift during the 1964 $\mathbf{M} = 9.2$ Prince William Sound megathrust earthquake (Plafker, 1965, 1969) and the area between Yakutat and Cape Yakataga uplifted in 1899 during the two $\mathbf{M} \approx 8$ earthquakes (Bruhn et al., 2004). At present we have no detailed evidence of the pattern of deformation from Cape Suckling to Cape Yakataga during these two earthquakes.

Numerous earthquake-deformation-cycle models exist (Nelson et al., 1996; Shennan et al., 1999; Thatcher, 1984). These three describe periods of land-level change as reflected by relative sea-level change, associated with large plate boundary earthquakes and intervening periods (Fig. 3). Shennan et al. (1999) propose a four-stage earthquake deformation cycle model for sites in south-central Alaska undergoing coseismic subsidence (sudden relative sea-level rise). Rapid post-seismic uplift (relative sea-level fall) occurred in the decades after the earthquake, followed by centuries of slower inter-seismic uplift (relative sea-level fall), then pre-seismic relative sea-level rise immediately before the next earthquake. Subsequent work shows that this model is applicable to six great earthquakes in the past 3500 a, affecting sites around upper Cook Inlet (Hamilton and Shennan, 2005a; Hamilton et al., 2005; Hamilton and Shennan, 2005b; Shennan and Hamilton, 2006). For areas closer to the subduction zone (Fig. 3), observations (Plafker, 1965, 1969; Plafker et al., 1992) and a simple model would predict coseismic uplift (sudden relative sea-level fall), rapid post-seismic subsidence (relative sea-level rise) in the decades after the earthquake, and centuries of slower inter-seismic subsidence (relative sea-level rise). At present we have no data to suggest pre-seismic relative sea-level fall in the zone of coseismic uplift.

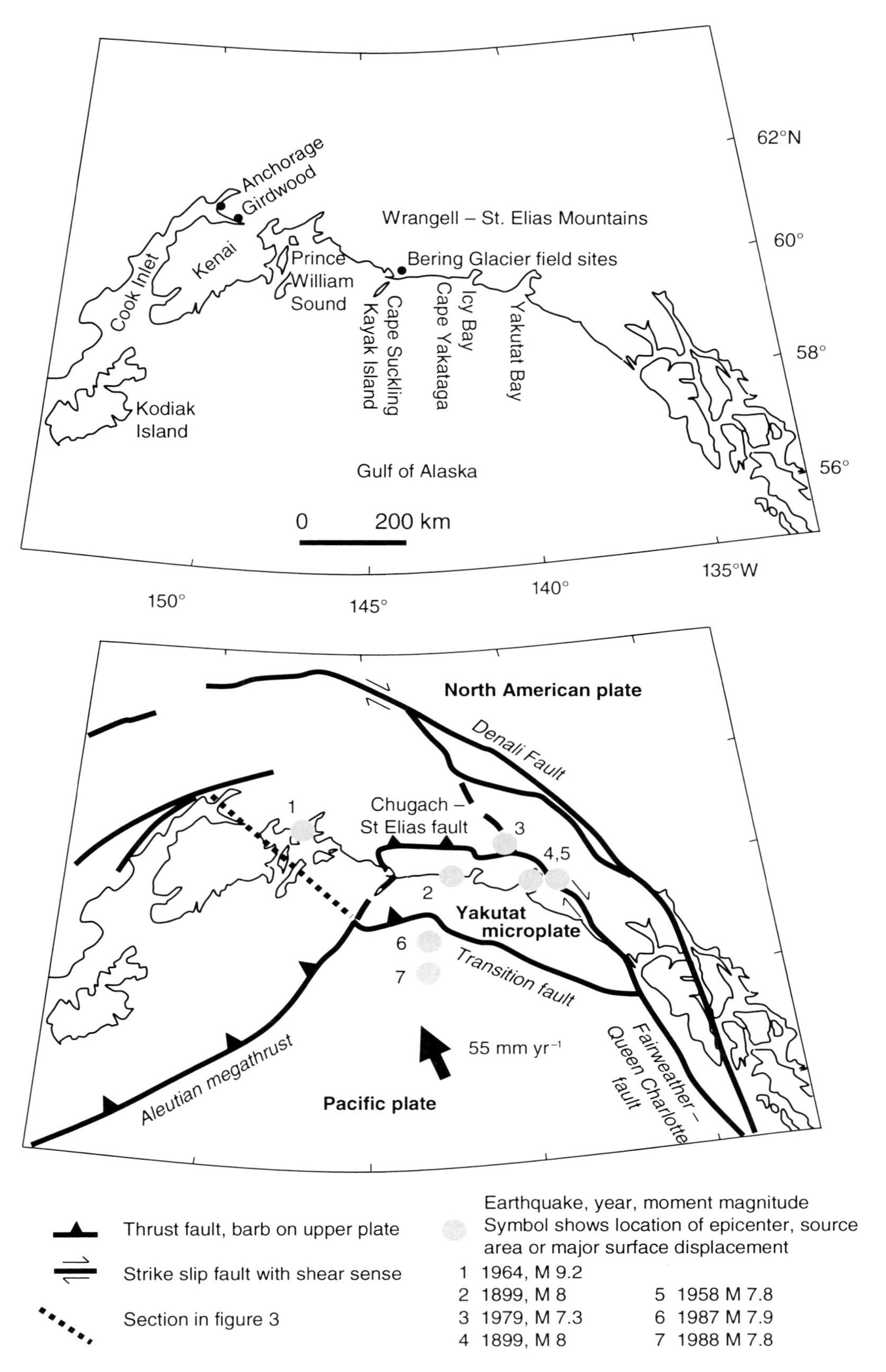

Figure 2. Site locations and tectonic setting (modified from Bruhn et al., 2004).

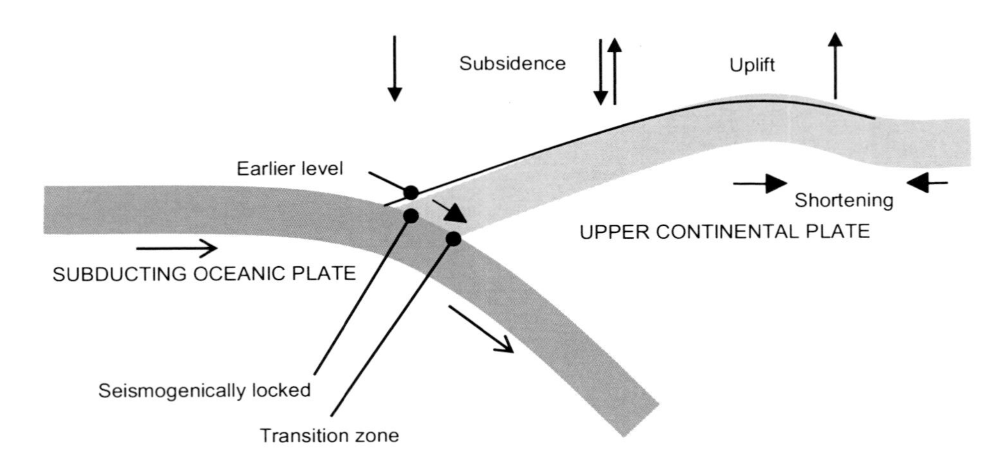

Figure 3. Schematic diagrams showing the pattern of (A) inter-seismic and (B) coseismic deformation associated with a subduction zone earthquake during an earthquake deformation cycle. Adapted from Nelson et al. (1996) to reflect the spatial pattern of coseismic deformation during the A.D. 1964 earthquake in Alaska along the section in Figure 2.

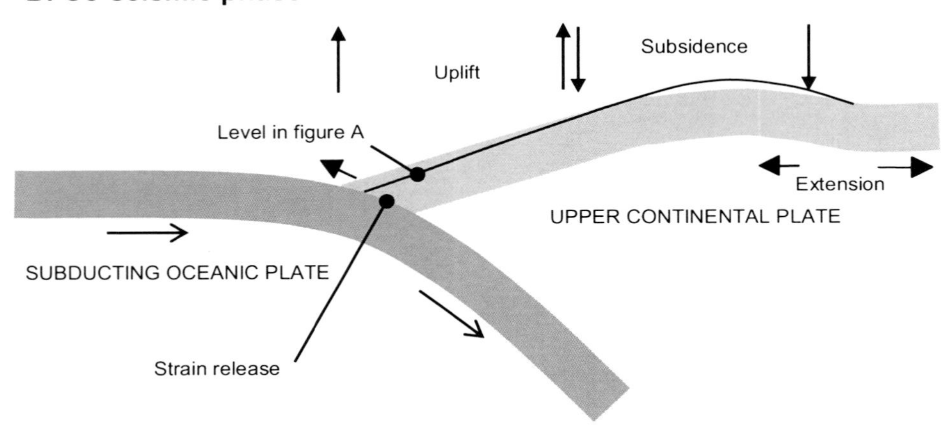

SITE SELECTION

Practical constraints on field sampling (helicopter time, weather, safe landing sites, access by boat from Bering Glacier field camp) and previous work, mainly for glaciological and paleo-vegetation studies, greatly influenced site selection, as published evidence of past relative sea level is limited.

Landforms, visible from satellite imagery and helicopter flights, which possibly reflect past sea level, include terraces on the surrounding uplands and large linear ridges, with infilled depressions between them, along the present coast. These ridges are absent around Seal River and Vitus Lake, possibly eroded by glacio-fluvial action during advance of Bering Glacier to its neoglacial maximum (Fig. 4). This is just south of the shore of Vitus Lake and was reached possibly as early as 960 yr B.P. around Seal River (Wiles et al., 1999), with other ages for the neoglacial limit between the mid-seventeenth century A.D. and the late nineteenth century A.D. (Molnia and Post, 1995; Wiles et al., 1999). Since A.D. 1900, Bering Glacier has undergone multiple surge and retreat cycles: ~1920, ~1938–1940, 1957–1960, 1965–1967, and 1993–1995 (Molnia and Post, 1995; Muller and Fleisher, 1995). Continued retreat since 1995 exposes new sediment sequences

Figure 4. Landsat image of Bering Glacier area, showing field locations (yellow circles) and approximate neoglacial limit of Bering Glacier (dashed line). Image courtesy of U.S. National Aeronautics and Space Administration (NASA), acquired under the Scientific Data Purchase Program.

each year. With field work restricted to between one week and two and a half weeks each year from 2002 to 2005, our field investigations fall into five categories: (1) elevated terracelike features, (2) exposures around Vitus Lake, (3) exposures along Seal River, (4) marsh sediments west of Cape Suckling, and (5) infilled depressions between tree-covered linear ridges along the coast (Fig. 4).

METHODS

Analysis of coastal sediments and the microfossils within them help to identify seismic from nonseismic relative sea-level changes (Long and Shennan, 1994; Nelson et al., 1996). Typically, in areas of coseismic subsidence, fresh-water peat rapidly submerges into the intertidal zone. This results in a peat-silt couplet, peat overlain by fine-grained clastic sediment, with a sharp stratigraphic boundary between the two (Atwater, 1987; Nelson et al., 1996). For coseismic uplift we would expect fine-grained clastic sediment overlain by peat, again with a sharp stratigraphic boundary between the two. Assessment of suddenness and amount of subsidence, lateral extent of peat-silt couplets with sharp upper contacts, evidence of tsunami deposits, and synchroneity of subsidence with other sites are critical in evaluating a possible coseismic cause (Nelson et al., 1996). In contrast, gradual changes in sediment type and the microfossils contained within them more likely indicate nonseismic relative sea-level change, such as those caused by eustatic sea-level change and isostatic land movements.

We follow well-established procedures (e.g., Hamilton and Shennan, 2005a) of coring and sampling exposures to record changes in sediment stratigraphy, followed by analyses of their fossil content. Ground elevations are estimated from hand-held global positioning system (GPS) receivers and U.S. Geological Survey (USGS) topographic maps, or relative to lake level or tide level, with error terms estimated for each site. Fossils include shells, visible plant remains (macrofossils), including wood (twigs and roots), leaves and stems of mosses and herbaceous plants, and diatom microfossils (visible only with a microscope). Our previous studies show that diatoms provide the most useful information to reconstruct sea-level changes, with the added advantage that exploratory samples can be prepared at base camp, giving instant information to maximize sampling time in the field.

Diatoms live in both marine and fresh-water environments. In broad terms, the order of diatom salinity classes (Table 1) should reflect change from, for example, tidal flat through salt marsh, to fresh-water marsh and bog. Marine (polyhalobous) and brackish (mesohalobous) groups usually dominate tidal flat environments, and fresh-water groups tolerant of different degrees of salinity (oligohalobous-halophile and oligohalobous-indifferent) become dominant through the transition from salt marsh to fresh-water marsh (e.g., Zong et al., 2003). Salt-intolerant species (halophobous) characterize the most landward communities, including those from acidic bogs above the level of highest tides.

Although this simple model works well for estuarine tidal marshes, and can be improved using quantitative techniques known as transfer functions (e.g., Gehrels et al., 2001; Hamilton and Shennan, 2005a), we also require information on the diatoms found in the transition from glaciolacustrine to marine environments, such as Vitus Lake to the mouth of Sea River, to gain insights into the different sedimentary environments of the Bering Glacier Region. Therefore, we collected surface diatom samples by cleaning away the modern surface film and collecting the top 1 cm of sediment to allow for seasonal variations in diatom blooms and effects of winter freezing. In subsequent sections we show only the summary diatom data into the classes described in Table 1. Overall we identified >175 different species in the modern samples and 125 species in fossil cores and sections.

TABLE 1. HALOBIAN CLASSIFICATION SCHEME OF DIATOMS

Classification	Salinity range (‰)	Environment
Polyhalobous	>30	Marine
Mesohalobous	0.2 to 30	Brackish
Oligohalobous—halophile	<0.2	Fresh water—stimulated at low salinity
Oligohalobous—indifferent	<0.2	Fresh water—tolerates low salinity
Halophobous	0	Salt-intolerant

Note: Classification scheme of Hemphill-Haley (1993).

We use AMS radiocarbon dating to establish the age of sediments, using either well-preserved in situ plant macrofossils (mainly herbaceous rootlets or stems) or in situ articulated bivalve molluscs (Table 2). We report ages as calendar yr B.P. (cal yr B.P., where "the present" is A.D. 1950) as the 95% range, rounded to 10 yr, calibrated using CALIB 5.0 (Stuiver and Reimer, 1993) with the terrestrial data set (file intcal.14c) and marine data set (file marine04.14c) for plant macrofossil samples and mollusc samples, respectively (Hughen et al., 2004; Reimer et al., 2004). We use a regional mean marine age correction of 460 ± 46 yr, obtained from the weighted mean of the 13 samples held in the CHRONO Marine Reservoir Database (http://calib.qub.ac.uk/marine/regioncalc.php 3 August 2006) between 55°–65° N and 130°–160° W.

MODERN DIATOM ASSEMBLAGES

Diatoms from modern surface sediment samples around Vitus Lake and Sea River show spatial zoning related to elevation and tidal influence (Fig. 5). Water salinity values, measured at the time of sampling, range from 0.01‰ to 2.6‰. Values >2.0‰ come from Seal River and Vitus Lake samples close to the entrance to Seal River. These fit well with the diatom summary classes, which reveal a marine influence greatest in Seal River. A lack of extensive saline tidal marshes around Vitus Lake, Seal River, and the adjacent coast means that we are presently limited to combining these modern distributions with those from Cook Inlet undertaken during other research projects (Hamilton and Shennan, 2005a) to interpret fossil

TABLE 2. RADIOCARBON DATES

Site & sample	Laboratory code	$^{13}C/^{12}C$ ratio (‰)	Radiocarbon age ±1σ B.P.	Calibrated age (2σ) B.P.	Material dated
Seal River – site 1					
SR/03/1-13cm	Beta - 186607	–28.6	300 ± 40	Cal BP 450 to 290	Herbaceous macrofossils, roots-stems
SR/03/1-33cm	Beta - 186608	–26.2	320 ± 40	Cal BP 470 to 300	Wood root, descending from organic layer at 13 cm
SR/03/1-38cm	Beta - 186609	–27.1	2680 ± 40	Cal BP 2850 to 2750	Herbaceous macrofossils, roots-stems
Seal River – site 4					
SR/03/4-5cm	Beta - 186610	pMC - 30	119.13 ± 0.44 pMC	Reported as a % of the modern reference standard, indicating the material was living within the last 50 yr	Herbaceous macrofossils, roots-stems
East shore Tashalich Arm					
TL 03/1	Beta - 191525	1.9	5310 ± 40	Cal BP 5740 to 5590	*Penitella pennita*, in situ 2m above lake water level
TL 03/3	Beta - 191526	1.1	8800 ± 40	Cal BP 9450 to 9080	*Penitella pennita*, in situ 0.1m below lake water level
TL 03/5	Beta - 191527	–2.3	5210 ± 40	Cal BP 5640 to 5480	*Penitella pennita*, in situ 3.5m above lake water level
TL 03/6	Beta - 191528	1.0	5130 ± 40	Cal BP 5580 to 5430	*Penitella pennita*, in situ 3.1m above lake water level
Marsh west of Cape Suckling – site 7					
CS/04/7 85cm	Beta - 203418	–28.1	130 ± 40	Cal BP 290 to 0	Herbaceous macrofossils, roots-stems
CS/04/7 108cm	Beta - 203419	–27.7	200 ± 40	Cal BP 310 to 260 and Cal BP 220 to 140 and Cal BP 30 to 0	Herbaceous macrofossils, roots-stems
CS/04/7 120cm	Beta - 203420	–26.8	290 ± 40	Cal BP 460 to 290	Herbaceous macrofossils, roots-stems
CS/04/7 397cm	Beta - 203421	–27.4	640 ± 40	Cal BP 670 to 540	Herbaceous macrofossils, roots-stems
Marsh west of Cape Suckling – site 2					
CS/05/2/224cm	Beta - 212214	–25.4	530 ± 40	Cal BP 630 to 600 and Cal BP 560 to 510	Herbaceous macrofossils, roots-stems
CS/05/2/266cm	Beta - 212212	–26.4	880 ± 40	Cal BP 920 to 700	Herbaceous macrofossils, leaves-stems
CS/05/2/268cm	Beta - 212211	–26.5	970 ± 40	Cal BP 950 to 780	Herbaceous macrofossils, roots-stems
CS/05/2/269cm	Beta - 212213	–26.9	920 ± 40	Cal BP 930 to 740	Herbaceous macrofossils, roots-stems
Ridges east of Tsiu River					
RET/05/1/55cm	Beta - 212215	–26.2	310 ± 40	Cal BP 480 to 290	Herbaceous macrofossils, rootlet-fragments
RET/03/1/3cm	Beta - 212216	–27.2	180 ± 40	Cal BP 300 to 240 and Cal BP 230 to 70 and Cal BP 40 to 0	Herbaceous macrofossils, root

diatom assemblages in Holocene sediments. We use these diatom assemblages in estimating paleo-marsh surface elevation relative to tide levels at the time of sediment accumulation and then infer late Holocene relative sea-level change around Bering Glacier. Future work around Bering Glacier should include the modern diatom flora of the extensive tidal marshes west of the Suckling Hills. This would allow more precise quantitative approaches for reconstructing sea-level change (Hamilton and Shennan, 2005a).

FOSSIL SITES

This section summarizes our investigations to reconstruct relative sea-level changes and evidence for earthquakes since the retreat of Bering Glacier from its maximum extent on the continental shelf about 20,000–22,000 yr ago.

Raised Terracelike Landforms and Sediments

In many deglaciated landscapes, relaxation of the Earth's crust following removal of a large mass of ice results in relative land uplift (Fig. 1), leaving evidence of marine or intertidal sediments and landforms above present sea level. We visited many areas on the hills around Bering Glacier (Fig. 4), concentrating on low-gradient terraces and muskeg across a wide range of elevations, ~15 to >450 m above present sea level (Fig. 6). Sediment cover on these low-gradient landforms ranges from a decimeter to more than 5 m. Samples from two terraces on

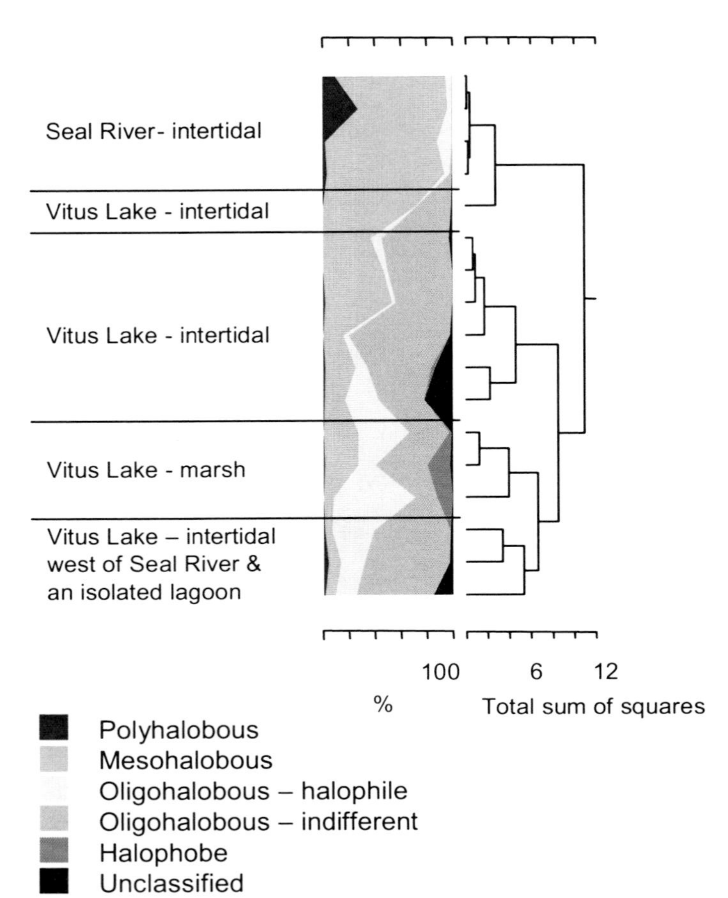

Figure 5. Summary of modern diatom samples from surface sediments around Vitus Lake, Seal River, and the adjacent coast, showing percentage sum of species in each halobian class (Table 1) and how the 17 samples cluster into 5 classes that reflect different sedimentary environments. Cluster analysis is based on percentage frequencies of all individual species counted, using constrained incremental sum of squares (Grimm, 1993).

the Suckling Hills show a thick cover of peat and mud overlying a basal coarse sand and gravel. Diatoms are absent in the sand and gravel, whereas the overlying mud and peat contain only fresh-water species. This typifies the evidence recovered so far. To date, we find no data to demonstrate that any of these terracelike features are raised shorelines, so we have no direct evidence of the elevation of relative sea level between the time of ice retreat and the oldest ages from the Tashalich Arm exposure (next section).

East Shore of Tashalich Arm

A small, <50 m by 30 m, exposure of Cretaceous rocks of the Yakutat Group on the east shore of Tashalich Arm (Fig. 7) is the only solid rock exposed around Vitus Lake and the foreland zone between the Grindle Hills and Suckling Hills (Fig. 4). The area lies within the neoglacial limit of Bering Glacier and was overrun by ice during the twentieth century surges, including 1993–1995. Following glacier retreat, shoreline and stream erosion of overlying cross-bedded sand capped by a thin diamicton exposed the rock outcrop, with more becoming visible each year during our annual field visits (2002–2005). It is not certain that this is the same exposure reported by Molnia and Post (1995) prior to the 1993–1995 surge. During field visits in 2002–2005 we found no trace of peat deposits just above the level of the outcrop (c.f. Molnia and Post, 1995), although ~50–100 m south of the outcrop we found contorted, transported peat interbedded with thin silt layers.

The rock outcrop extends from below lake level to ~4 m above, along an actively eroding stream channel through the overlying sand and diamicton. The outcrop has abundant burrows formed by the rock-boring *Penitella penita*, a low intertidal to shallow subtidal (to 22 m) species, common on soft rock and muddy shores around the Gulf of Alaska (Rehder, 1981). The fact that we now find these forms above lake level, itself ~2 m above high tide at the ocean, indicates relative sea level significantly above present at the time of their growth. Radiocarbon ages for a number of these shells (Table 1) indicate that this high sea level persisted between ~9000 and 5000 yr ago, with the coastline much farther to the north. The younger age is the same as that reported by Molnia and Post (1995).

Pasch (this volume and 2006) records many bivalves and other invertebrates from sediments exposed during the current retreat phase of Bering Glacier. These marine organisms must have been transported by the glacier from their original growth sites and deposited to the north. Radiocarbon ages range from ca. 8000 to more than 13,000 cal yr B.P. Although these samples confirm the presence of a marine embayment to the north of the Tashalich Arm, they do not give any further information on the elevation of sea level.

The Tashalich Arm exposure offers further scope for research, with more of it being exposed each year. We have found examples of secondary occupation of *Penitella penita* burrows, both by individuals of the same species and by *Prototheca staminea* (Fig. 7). These bivalves are infaunal, typically buried ~0.1 m in the gravel-sand-mud substrate, and occur from the mid-intertidal zone to 10 m depth (Pasch, 2006, personal commun.). Some of the sediments filling the bivalves contain foraminifers and diatoms. Analyses of these may give further detail on relative water depths and hence sea level. Also, we do not know if the *Prototheca* are the same age as the *Penitella*.

Seal River

Coastal erosion and channel migration produce extensive exposures of sediments along Seal River and its tributaries. Just downstream of the exit from Vitus Lake they reveal a transect through the end moraine formed at the neoglacial maximum extent of Bering Glacier (Fig. 4). From there to the ocean the bluffs along the north bank expose fluvioglacial and dune sequences, including interbedded peat layers. Of particular value to recon-

2003-07-31 Donald Ridge
60°15.829′N, 143°03.766′W, elevation ~462 m

2003-8-8 Khitrov Hills
60°24.152′N, 143°30.222′, elevation ~458 m

2004-8-10 Suckling Hills
60°00.109′N, 143°55.749′W, elevation ~55 m

2004-8-11 Suckling Hills
60°00.133′N, 143°57.040′W, elevation ~20 m

Figure 6. Examples of the types of sites investigated for evidence of raised shorelines (locations in Fig. 2) in the hill ranges surrounding Bering Glacier and the foreland. Sediment cover on these low-gradient landforms ranges from a decimeter to >5 m. Field assistants Elizabeth Jesclard and Garrett Schultz participated, courtesy of the High School Intern Program with Anchorage School District.

structing sea-level change are two sequences: site 1, within 200 m of the mouth of Seal River at the open ocean, and site 4, an overflow channel between Vitus Lake and Seal River (Fig. 4).

At site 1 a ~50-cm-thick sequence of finely laminated sand, mud, and organics lies within the mid- to upper intertidal zone on the north bank of Seal River (Fig. 8). During four field seasons it was visible only for one week, in 2003. At other times it lay buried beneath seasonal sediment deposited from river discharge and bluff collapse. It lies beneath interbedded coarse and fine sand layers, themselves beneath dunes presently vegetated by mature forest. The lowest part of the sample section (Fig. 8C) comprises fine laminae without organic material to date and no diatoms. The lowest organic layer, with small herbaceous roots and stems, has an age of 2860–2750 cal yr B.P. Diatoms are infrequent or absent up to a farther 5 cm above (Fig. 8D). They show that the laminated sediments accumulated in a predominantly fresh-water environment with some input of marine or brackish water. Up-section they show a slight increase in brackish diatoms to the base of a distinct organic zone at 13 cm (Fig. 8B, 8D). Small herbaceous roots and stems from this layer give an age of 470–290 cal yr B.P. Numerous wood roots extend down from this organic zone into the underlying sediment, and one was sampled, giving

Figure 7. Rock exposure on east shore of Tashalich Arm, 31 July 2003, location 60° 8.839′ N, 43° 35.181′ W, showing in situ shells of *Penitella penita* (middle) and secondary occupation by *Prototheca staminea* (bottom).

an age of 470–300 cal yr B.P. Brackish diatoms show an abrupt decline in abundance at the base of the organic layer, followed by a gradual increase up to the top of the section. We interpret this sequence as indicating relative sea level below present from at least ca. 2800 cal yr B.P to ca. 300 cal yr B.P because the diatom assemblages indicate only occasional input of marine or brackish water, probably around the limit of extreme high tide at the time of deposition rather than the present position in the mid- to upper intertidal zone. Below the organic layer at 13 cm the diatoms show little change—only a gradual but consistent increase in brackish influence up-section. This may indicate a slight relative sea-level rise. This episode ended with a sharp reversal, with the temporary exclusion of marine and brackish water and development of the distinct organic layer and growth of woody shrubs. Without a more extensive exposure of the sequence, we cannot be sure that this is an abrupt fall in sea level (or land uplift) or a local change in coastal morphology.

Site 4 shows a 1-cm-thick peat layer above the level of extreme tides (Fig. 8E, 8F). Diatoms occur only directly below and within the peat. They show a distinct decrease in brackish diatoms, suggesting relative sea-level fall (or land uplift). Given the radiocarbon result (Table 2), indicating an age within the past 50 a, we interpret this sequence as indicating slight land uplift during the A.D. 1964 great earthquake. From measurements of the upper limit of storm driftwood, Plafker (1969) indicated ~2–3 m uplift around Seal River.

Cape Suckling

West of Cape Suckling and the Suckling Hills lies an extensive area of tidal flat, marsh, and fresh-water wetlands in the lee of Okalee Spit and inland to the neoglacial limit of Bering Glacier (Fig. 4). A series of cores across the southern part of the marsh reveal multiple mud-peat couplets within the upper 3 m of sediment. Away from the spit all cores end within a gray silt unit, similar to the one seen at the base of the core shown in Figure 9, so deeper cores in the future will provide a longer record. Our current analyses rely on hand-drilled cores, limited to a maximum of ~5 m depth.

The present marsh shows an intricate pattern of tidal channels, low levees, backswamp, and open water. Their spatial pattern changes through time, and we require many cores to build up a clear picture of past environments. The sections of cores shown in Figure 9 summarize a consistent stratigraphy traced across hundreds of meters. These show three mud-peat couplets where slightly laminated mud underlies a peat or silt-peat unit. A sharp boundary at the base of a peat zone may indicate land uplift during an earthquake, raising an intertidal mudflat above the level of high tides. In contrast, a gradual transition suggests a slow change in relative sea level, perhaps as a consequence of glacio-isostatic land uplift.

Site 2 has the best record of the lower part of the sequence (Table 2; Fig. 9). Marine and brackish diatoms dominate gray laminated silt from 269 cm down to the base of the core (Fig. 9). These confirm sedimentation in a tidal flat environment. A <1-cm-thick coarse sand overlies the silt, with a sharp boundary between the two units. Peat, containing abundant *Sphagnum* moss and herbaceous roots, overlies the sand, and up-core becomes progressively silt-dominated within its top 10 cm, grading into

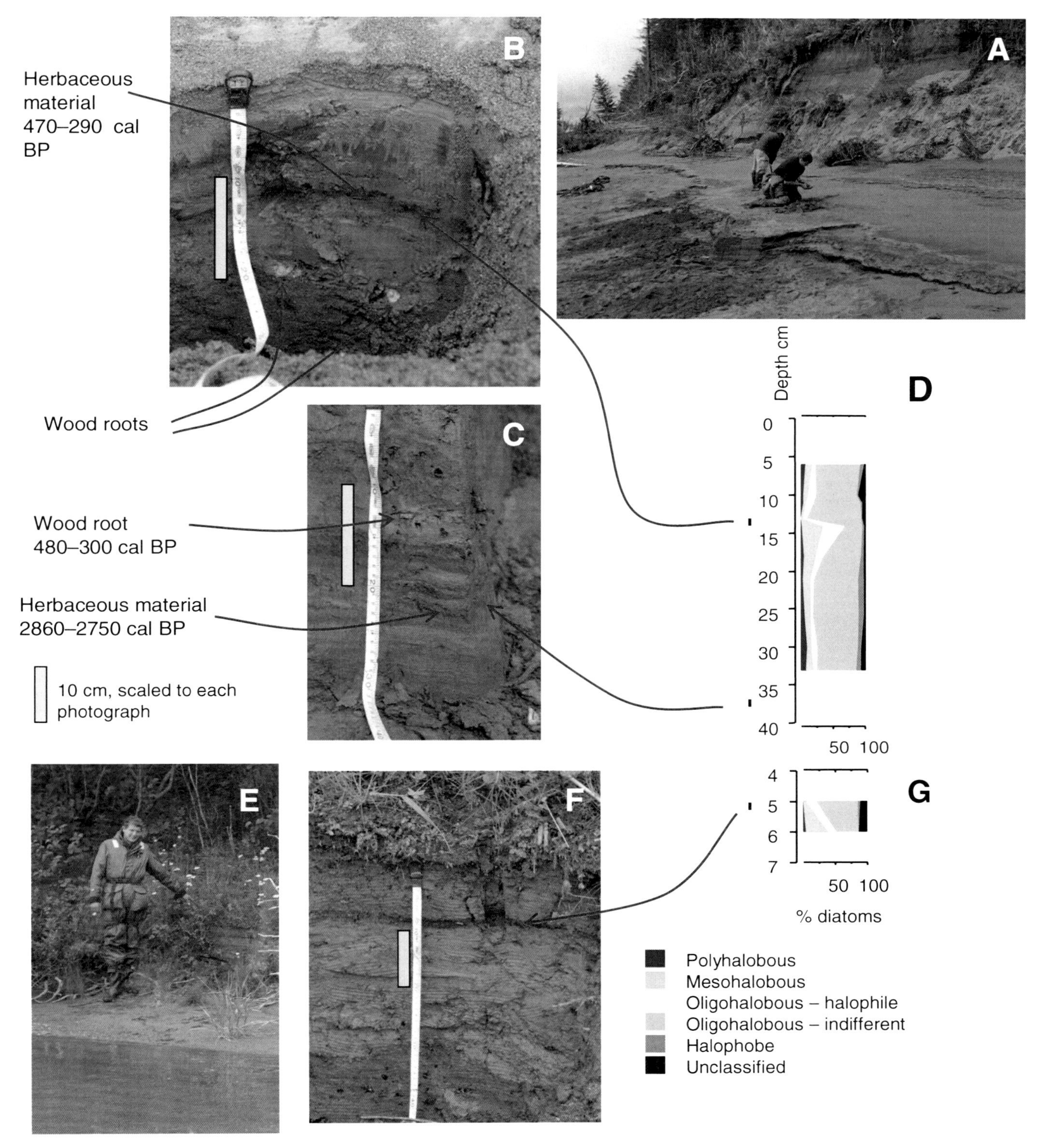

Figure 8. Seal River stratigraphy and summary diatom data. (A) sample site Seal River 1, close to the ocean (Fig. 4), 60° 3.010′ N, 143° 33.162′ W. Sampled during rising tide 3 August 2003. Composite section from shallow pit (B) and exposed face (C) ~1 m apart. Details of radiocarbon samples appear in Table 2. (D) Summary of diatom samples from composite section, showing percentage sum of species in each halobian class (Table 1). (E) Sample site Seal River 4, close to Vitus Lake (Fig. 4), 60° 4.304′ N, 143° 27.274′ W, sampled just after high tide 3 August 2003. Cleaned section (F) used for radiocarbon sample (Table 2) and diatom analysis (G). Field assistants Ryan Cooper and John Zurfelt participated, courtesy of the High School Intern Program with Anchorage School District.

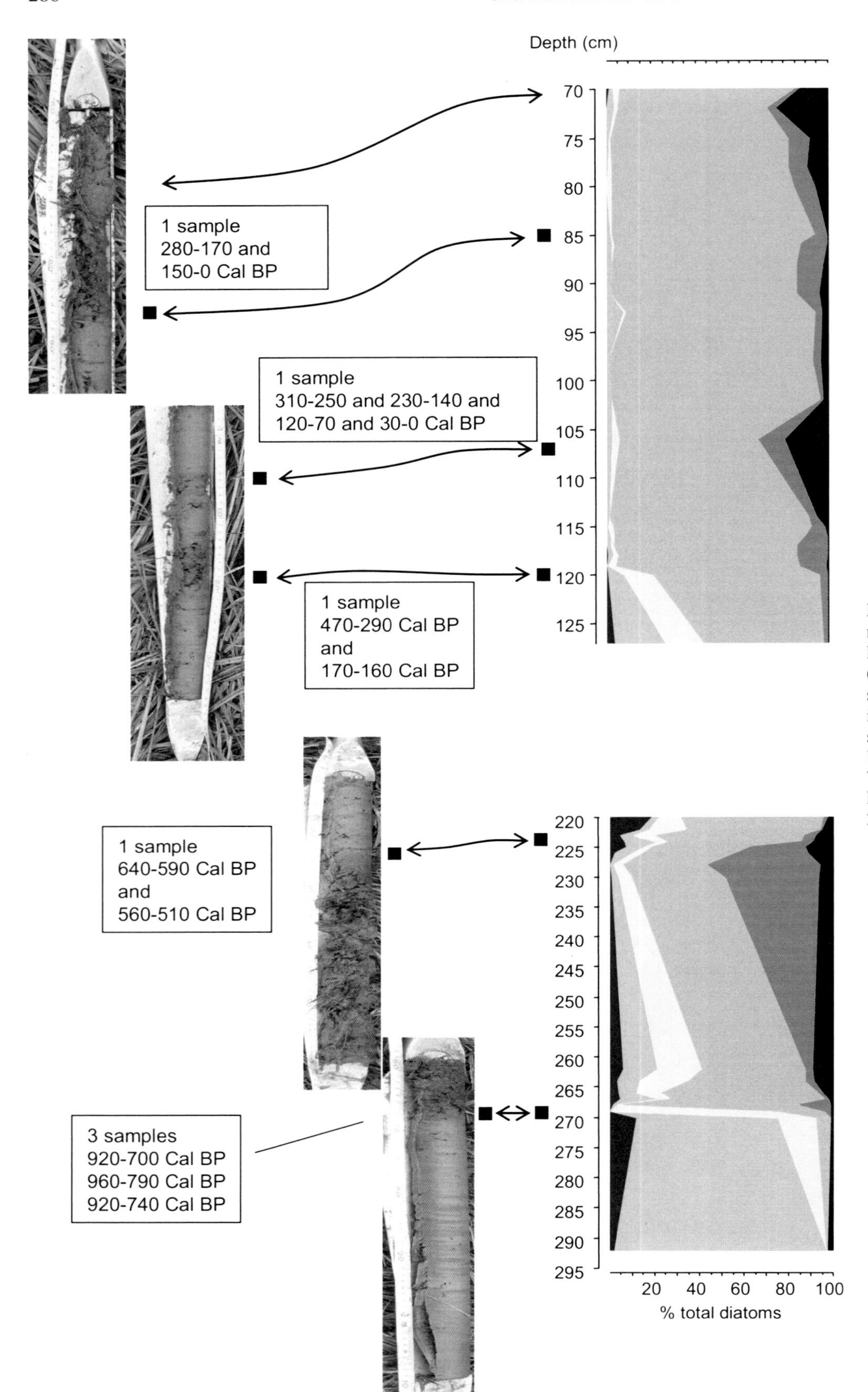

Figure 9. Core samples from beneath the marsh west of Cape Suckling: stratigraphy, radiocarbon ages (Table 2), and summary diatom data showing percentage sum of species in each halobian class (Table 1); same color key as Figures 5 and 8. Upper section, 70–126 cm, from site 7, 60° 1.285′ N, 143° 57.030′ W, and lower section, 220–295 cm, from site 2, 60° 1.197′ N, 143° 57.183′ W.

overlying laminated silt with herbaceous roots. Diatoms show the peat dominated by fresh-water and salt-intolerant diatoms, which decline toward the top of the unit. As silt content increases above 230 cm (Fig. 9), salt-intolerant diatoms decrease in abundance, and marine and brackish species increase.

We interpret the changes in lithology and diatoms across the base of the peat to be a result of rapid uplift during a great earthquake that raised the paleo-tidal flat above normal high tide level, allowing *Sphagnum* bog communities to develop. Coarse sand between the silt and peat units probably represents a tsunami caused by the earthquake. It is observed in a series of cores in the southern part of the marsh and so is unlikely to be an isolated channel flood deposit. We do not know whether the sand is derived from an open tidal flat or by washover from a sand spit, comparable to the present coastal morphology. In order to accurately date this earthquake we took three samples from in situ herbaceous or *Sphagnum* macrofossils. The sample from 269 cm contains herbaceous roots within the sand. All three samples give ages of ca. 950–700 cal yr B.P. This correlates with the penultimate great earthquake recorded in Cook Inlet (Shennan and Hamilton, 2006) and indicates a broadly similar pattern of co-seismic land movement to those observed in AD 1964 (see also the model in Fig. 3).

In contrast, gradual changes in lithology and diatom assemblages toward the top of the peat layer indicate gradual land subsidence–relative sea-level rise, with a return to tidal flat sedimentation by ca. 600–500 cal yr B.P. In the earthquake deformation cycle model (Fig. 3) this is consistent with inter-seismic relaxation, producing land subsidence.

Two more mud-peat couplets occur farther up-section, but in neither case was the lower boundary quite so sharp as the lowermost peat, nor was there a sand layer present. The diatom assemblages also reveal a different sequence of environmental change. At 120 cm core depth (Fig. 9), marine and brackish diatoms decrease, replaced by fresh-water and salt-intolerant species. The change is nowhere near as great as observed lower down, at ~269 cm. At this stage we could not confirm that this represents rapid co-seismic uplift, although the radiocarbon dates (470–290 cal yr B.P. and 170–160 cal yr B.P.) suggest possible correlation with a similar episode of slight land uplift at Seal River (Fig. 8). Diatoms through the upper couplet, boundary at 85 cm, show no evidence of relative sea-level change. Indeed both the silt and peat units formed in predominantly fresh-water environments (Fig. 9).

Ridges East of the Tsiu River

Prominent ridges, running approximately parallel to the present coast, are present east of Tsiu River and east of Cape Suckling (Fig. 4). On the basis of their morphology we support previous interpretations that they are fossil beach ridges (Muller and Fleisher, 1995) capped with windblown sediment and speculate that they were once more extensive, perhaps even coalescing before being eroded during Bering Glacier's advance to its neoglacial maximum. Assuming they are fossil beach ridges, we may expect the low-lying areas to include fossil lagoon sequences that may record both coastline evolution and relative sea-level change. These low-lying areas are very poorly drained, with extensive muskeg, making helicopter landing and hence sampling difficult. From numerous exploratory cores we have only recorded a thin surface peat, usually <50 cm, overlying massive gray silt or sand. We have recovered no diatoms from silt or sand units and only fresh-water species from any peat samples. Thus we cannot attribute infilling of these between-ridge depressions to lagoons connected to the sea. Lagoon sediments recording past sea levels still await discovery, but the silt and sand we recorded most likely come from fluvioglacial sedimentation, infilling between the coastal ridges. Therefore the radiocarbon ages we have from the base of the peat at two sites (Table 2; Fig. 4) clearly postdate formation of the ridges. They were in place at least before 480–300 cal yr B.P.

DISCUSSION

Radiocarbon ages on glacially transported bivalves (Pasch, this volume, and Pasch, 2006) show that the front of Bering Glacier had retreated from its Last Glacial Maximum position (ca. 22,000 cal yr B.P.) on the continental shelf to north of its present position before 13,000 cal yr B.P., possibly separated by the Khitrov Hills into two tidewater glaciers (Fig. 4). But they do not give any information on the elevation of sea level prior to 9000 cal yr B.P., the oldest age from the in situ bivalves at Tashalich Arm. Only future research will show whether sea level prior to this time was above or below present (Fig. 10). This is critical if we are to understand fully the relationships and interactions between deglaciation, climate, Earth rheology, crustal movements, and sea-level change.

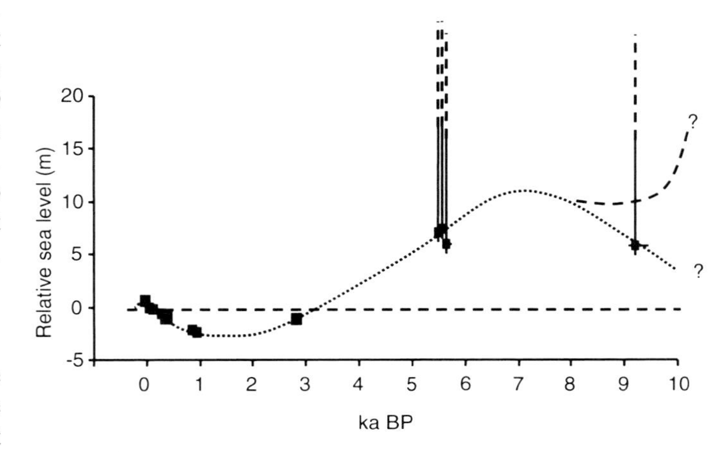

Figure 10. Summary of relative sea-level change. Error terms for age and elevation are only visible if greater than the solid symbol that represents each indicator of past sea level described in the text and summarized in Table 2; "B.P." refers to "before present," where "present" is A.D. 1950. Elevations relate to zero meters in A.D. 2005, or −55 B.P. Dotted and dashed lines represent two general trends discussed.

Relative sea level was above present from ca. 9000 to at least 5000 cal yr B.P. (Fig. 10). The broad depth range of *Penitella penita* (~20 m) means we cannot give a precise level. Further investigation and dating of the different mollusc species and analysis of foraminifers and diatoms within infilling sediment may give more precise reconstructions of sea-level change at Tashalich Arm during this period.

Between ca. 5000 and ca. 3000 cal yr B.P. (Fig. 10), relative sea-level fell to below present, indicated by the sequence at Seal River. We infer that this was a time of major changes in sedimentary environments and paleogeography. Forests were present in the western forelands by at least 4000 cal yr B.P. (this volume, and Crossen, 2006; Molnia and Post, 1995; Muller and Fleisher, 1995). In Tashalich Arm, interbedded peat and lacustrine sediments indicate lake levels lower than present ca. 2200–1400 cal yr B.P., rising to ~6 m above present by the latter time (this volume, and Crossen, 2006). However, we do not know the spatial extent of the lake in which these sediments were laid down. The paleogeography would have been very different to that today. Perhaps the coastal ridges we see today east of Tsiu River and east of Cape Suckling are remnants of a barrier system that extended across the whole Bering Glacier foreland. Fresh-water drainage could have been primarily around the north end of the Suckling Hills. We need much more evidence to reconstruct the paleogeographic evolution of the area, not least to explain high lake levels around Tashalich Arm at the same time as sea level below present at Seal River and in the marshes west of the Suckling Hills.

Sediments from Seal River and west of Cape Suckling indicate relative sea level below present from at least ca. 2800 cal yr B.P. until the twentieth century, with possible decimeter- to meter-scale changes superimposed on the general trend shown in Figure 10, resulting from cosesimic and inter-seismic land movements. At present we have good evidence of a great earthquake, with an accompanying tsunami, ~900 yr ago, producing land uplift of tidal flats west of Cape Suckling. Post-seismic and inter-seismic land subsidence led to gradual drowning of acidic bog, then tidal marsh and return to tidal flat within 300–400 yr. Evidence for a great earthquake ~900 yr ago, with coseismic land subsidence, is found at Girdwood and Anchorage (Shennan and Hamilton, 2006), suggesting a similar pattern of coseismic deformation as in A.D. 1964. Evidence for earthquake-related land movements ~300 yr ago from the sediments described above at Seal River and the marsh west of Cape Suckling marsh is inconclusive. Sediment lithology and diatoms at both places indicate a rapid change, but the magnitude of change is much less than observed for the ~900 yr ago earthquake and for any of the earlier ones recorded by sediment and diatom evidence around Cook Inlet (Shennan and Hamilton, 2006). We require more detailed modern diatom data sets from the Bering Glacier Region to calibrate these smaller changes.

Our record of uplift at Seal River, attributed to the A.D. 1964 earthquake, is corroborated by Plafker's (1969) measurements. He estimated ~3 m uplift at Seal River, decreasing to zero by Yakataga, based on elevations of the upper limit of storm-tide driftwood, recorded in 1965. We should note, however, that the first estimates of coseismic land uplift in 1964 show only 1 m at Cape Suckling (Plafker, 1965), and the 1965 published map was redrawn and published in Plafker et al. (1992). Assuming an error term of at least ±1 m for such indicators, the 1969 estimate is larger than the <1 m uplift indicated by the A.D. 2003 elevation of the peat layer at site 4 in Seal River. This may be indicative of post-seismic land subsidence over the past 40 yr and glacio-isostatic land movement.

Although the information gathered over the past few years advances our understanding of relative sea-level changes in the Bering Glacier Region, it also reinforces the need for more investigations of century to millennial-scale relative land- and sea-level changes and their spatial expression. We can highlight a number of unresolved issues that may illustrate the complex tectonics and glacio-isostatic setting of the area. We noted above the contrast between the elevations of lake sediments ca. 2200–1400 cal yr B.P. in Tashalich Arm and marsh sediments in Seal River. For an earlier period, pre–5000 cal yr B.P., Plafker and Rubin (1967) describe shells on an elevated terrace at Cape Suckling 6 m above the pre-1964 datum of a similar age to most of those described above from Tashalich Arm. They also report ages from rooted tree stumps ~3.7 to ~4.3 m below the pre-1964 datum at Cape Suckling ca. 400–800 cal yr B.P.; their growth and death are possibly related to coseismic uplift ca. 900 cal yr B.P. and subsequent inter-seismic land subsidence described above. These records contrast markedly with observations of long term uplift, at least 4000–8000 yr B.P. just 30 km WNW at Katalla (Plafker, 1969; Plafker and Rubin, 1967).

Finally, our investigations of marsh sediments west of Cape Suckling so far reveal no marsh emergence sequences that we can relate to A.D. 1964. This seems difficult to equate with estimates of 4–5 m emergence derived from barnacles and storm-tide driftwood at Cape Suckling (Plafker, 1969). We can only speculate that the sequences we sampled were shallow lagoonal prior to uplift, or there was significant compaction of marsh sediments from ground shaking. Another hypothesis is that local scale differences in coseismic movements exist in alignment along the western boundary of the Yakutat microplate through Kayak Island and the Suckling Hills to Bering Glacier (Fig. 2; Bruhn et al., 2004).

CONCLUSIONS

In summary, the Bering Glacier Region has proven potential to provide evidence of decadal to millennial-scale changes in relative sea level. Investigations to date show that Bering Glacier had retreated inland of the present coast and north of its present terminus by ca. 14,000 yr B.P. Relative sea level was above present ca. 9200 yr B.P. to at least 5000 yr B.P. before falling to below present. Buried mud-peat couplets in the marsh east of Cape Suckling show a great earthquake at ca. 900 cal yr B.P., including evidence of a tsunami. Relative sea level rose to above present by the early twentieth century, before coseismic land uplift, relative sea-level fall, in A.D. 1964.

This 10,000 yr pattern of relative sea-level change differs from what may be expected in comparison with model predictions for other seismic and nonseismic locations. Our observations reveal intricate interactions through time and over short spatial gradients related to ice sheet and glacier dynamics, glacio-isostasy, multiple earthquake cycles, and the tectonics of an active and complex plate boundary. Correlation with sites from Cook Inlet to Icy Bay (Fig. 1) shows some critical differences between previous earthquake deformation cycles and A.D. 1964, related to the regional-scale tectonic setting (Shennan, 2009; Shennan et al., 2009). But we require much more evidence from the Bering Glacier Region to fully understand all the intricate interactions and spatial gradients outlined above.

ACKNOWLEDGMENTS

We thank Caroline Gregory for contributing to the diatom analyses; Kris Crossen, Jay Fleisher, and Colin Murray-Wallace for insightful reviews; John Payne, Scott Guyer, Chris Noyles, Nathan Rathburn, and Mike Nolen of the Alaska office of the U.S. Bureau of Land Management for their vision and field logistics; Elizabeth Jesclard, Nicole Swenson, Ryan Cooper, Garrett Shultz, and John Zurfelt (field assistants, participation courtesy of the High School Intern Program with Anchorage School District); and Smiley Shields for logistical support in Alaska. This research was supported by the Department of Geography Research Development Fund, Durham University, UK, and the Bering Glacier Project of the Alaska office of the Bureau of Land Management.

REFERENCES CITED

Atwater, B.F., 1987, Evidence for great Holocene earthquakes along the outer coast of Washington State: Science, v. 236, p. 942–944, doi: 10.1126/science.236.4804.942.

Bruhn, R.L., Pavlis, T.L., Plafker, G., and Serpa, L., 2004, Deformation during terrane accretion in the Saint Elias orogen, Alaska: Geological Society of America Bulletin, v. 116, p. 771–787, doi: 10.1130/B25182.1.

Crossen, K., 2006, Holocene history of Bering Glacier, Alaska, *in* Bering Glacier to Glacier Bay—From Tectonics to Ice: Session in Honor of Austin Post, Paper 27–5, *in* North to Alaska: Geoscience, Technology, and Natural Resources: Anchorage, Alaska, Annual Meeting of the Cordilleran Section, Geological Society of America, 102nd; Annual Meeting of the Pacific Section, American Association of Petroleum Geologists, 81st; and Western Regional Meeting of the Alaska Section, Society of Petroleum Engineers; 8–10 May 2006.

Gehrels, W.R., Roe, H.M., and Charman, D.J., 2001, Foraminifera, testate amoebae and diatoms as sea-level indicators in UK saltmarshes: A quantitative multiproxy approach: Journal of Quaternary Science, v. 16, p. 201–220, doi: 10.1002/jqs.588.

Grimm, E., 1993, TILIA: A Pollen Program for Analysis and Display: Springfield, Illinois State Museum, http://www.museum.state.il.us/pub/grimm/.

Hamilton, S., and Shennan, I., 2005a, Late Holocene relative sea-level changes and the earthquake deformation cycle around upper Cook Inlet, Alaska: Quaternary Science Reviews, v. 24, p. 1479–1498, doi: 10.1016/j.quascirev.2004.11.003.

Hamilton, S.L., and Shennan, I., 2005b, Late Holocene great earthquakes and relative sea-level change at Kenai, southern Alaska: Journal of Quaternary Science, v. 20, p. 95–111, doi: 10.1002/jqs.903.

Hamilton, S., Shennan, I., Combellick, R., Mulholland, J., and Noble, C., 2005, Evidence for two great earthquakes at Anchorage, Alaska and implications for multiple great earthquakes through the Holocene: Quaternary Science Reviews, v. 24, p. 2050–2068, doi: 10.1016/j.quascirev.2004.07.027.

Hemphill-Haley, E., 1993, Taxonomy of Recent and Fossil (Holocene) Diatoms (Bacillariophyta) from Northern Willapa Bay, Washington: U.S. Geological Survey Open File Report 93–289, 151 p.

Hughen, K.A., Baillie, M.G.L., Bard, E., Beck, J.W., Bertrand, C.J.H., Blackwell, P.G., Buck, C.E., Burr, G.S., Cutler, K.B., Damon, P.E., Edwards, R.L., Fairbanks, R.G., Friedrich, M., Guilderson, T.P., Kromer, B., McCormac, G., Manning, S., Ramsey, C.B., Reimer, P.J., Reimer, R.W., Remmele, S., Southon, J.R., Stuiver, M., Talamo, S., Taylor, F.W., van der Plicht, J., and Weyhenmeyer, C.E., 2004, Marine04 marine radiocarbon age calibration, 0–26 cal kyr BP: Radiocarbon, v. 46, p. 1059–1086.

Larsen, C.F., Echelmeyer, K.A., Freymueller, J.T., and Motyka, R.J., 2003, Tide gauge records of uplift along the northern Pacific–North American plate boundary, 1937 to 2001: Journal of Geophysical Research, v. 108, 2216, doi: 10.1029/2001JB001685.

Larsen, C.F., Motyka, R.J., Freymueller, J.T., Echelmeyer, K.A., and Ivins, E.R., 2005, Rapid viscoelastic uplift in southeast Alaska caused by post–Little Ice Age glacial retreat: Earth and Planetary Science Letters, v. 237, p. 548–560, doi: 10.1016/j.epsl.2005.06.032.

Long, A.J., and Shennan, I., 1994, Sea-level changes in Washington and Oregon and the 'earthquake deformation cycle': Journal of Coastal Research, v. 10, p. 825–838.

Molnia, B.F., and Post, A., 1995, Holocene history of Bering Glacier, Alaska: A prelude to the 1993–1994 surge: Physical Geography, v. 16, p. 87–117.

Muller, E.H., and Fleisher, P.J., 1995, Surging history and potential for renewed retreat: Bering Glacier, Alaska, USA: Arctic and Alpine Research, v. 27, p. 81–88.

Nelson, A.R., Shennan, I., and Long, A.J., 1996, Identifying coseismic subsidence in tidal-wetland stratigraphic sequences at the Cascadia subduction zone of western North America: Journal of Geophysical Research, v. 101, p. 6115–6135, doi: 10.1029/95JB01051.

Olsen, C.R., Larsen, I.L., Lowry, P.D., Cutshall, N.H., Todd, J.F., Wong, G.T.F., and Casey, W.H., 1985, Atmospheric fluxes and marsh soil inventories of 7Be and 210Pb: Journal of Geophysical Research, v. 90, p. 10,487–10,495, doi: 10.1029/JD090iD06p10487.

Pasch, A.D., 2006, What are clams doing in the Bering Glacier?, *in* Bering Glacier to Glacier Bay—From Tectonics to Ice: Session in Honor of Austin Post, Paper 27–4, *in* North to Alaska: Geoscience, Technology, and Natural Resources: Anchorage, Alaska, Annual Meeting of the Cordilleran Section, Geological Society of America, 102nd; Annual Meeting of the Pacific Section, American Association of Petroleum Geologists, 81st; and Western Regional Meeting of the Alaska Section, Society of Petroleum Engineers; 8–10 May 2006.

Peltier, W.R., 2002, Global glacial isostatic adjustment: Palaeogeodetic and space-geodetic tests of the ICE-4G (VM2) model: Journal of Quaternary Science, v. 17, p. 491–510, doi: 10.1002/jqs.713.

Peltier, W.R., and Fairbanks, R.G., 2006, Global glacial ice volume and Last Glacial Maximum duration from an extended Barbados sea level record: Quaternary Science Reviews, v. 25, p. 3322–3337, doi: 10.1016/j.quascirev.2006.04.010.

Plafker, G., 1965, Tectonic deformation associated with the 1964 Alaska earthquake: Science, v. 148, p. 1675–1687, doi: 10.1126/science.148.3678.1675.

Plafker, G., 1969, Tectonics of the March 27, 1964, Alaska Earthquake: U.S. Geological Survey Professional Paper 543-I, 74 p.

Plafker, G., and Rubin, M., 1967, Vertical tectonic displacements in south central Alaska during and prior to the great 1964 earthquake: Prague, Journal of Geosciences, v. 10, p. 1–7.

Plafker, G., Lajoie, K.R., and Rubin, M., 1992, Determining recurrence intervals of great subduction zone earthquakes in Southern Alaska by radiocarbon dating, *in* Taylor, R.E., Long, A., and Kra, R.S., eds., Radiocarbon after Four Decades: An Interdisciplinary Perspective: New York, Springer-Verlag, p. 436–453.

Rehder, H.A., 1981, National Audubon Society Field Guide to North American Seashells: New York, Alfred A. Knopf, 894 p.

Reimer, P.J., Baillie, M.G.L., Bard, E., Bayliss, A., Beck, J.W., Bertrand, C.J.H., Blackwell, P.G., Buck, C.E., Burr, G.S., Cutler, K.B., Damon, P.E., Edwards, R.L., Fairbanks, R.G., Friedrich, M., Guilderson, T.P., Hogg, A.G., Hughen, K.A., Kromer, B., McCormac, G., Manning, S.,

Ramsey, C.B., Reimer, R.W., Remmele, S., Southon, J.R., Stuiver, M., Talamo, S., Taylor, F.W., van der Plicht, J., and Weyhenmeyer, C.E., 2004, IntCal04 terrestrial radiocarbon age calibration, 0–26 cal kyr BP: Radiocarbon, v. 46, p. 1029–1058.

Shennan, I., 2009, Late Quaternary sea-level changes and palaeoseismology of the Bering Glacier region, Alaska: Quaternary Science Reviews, v. 28, no. 17–18, p. 1762–1773, doi: 10.1016/j.quascirev.2009.02.066.

Shennan, I., and Hamilton, S., 2006, Coseismic and pre-seismic subsidence associated with great earthquakes in Alaska: Quaternary Science Reviews, v. 25, p. 1–8, doi: 10.1016/j.quascirev.2005.09.002.

Shennan, I., Scott, D.B., Rutherford, M.M., and Zong, Y., 1999, Microfossil analysis of sediments representing the 1964 earthquake, exposed at Girdwood Flats, Alaska, USA: Quaternary International, v. 60, p. 55–73, doi: 10.1016/S1040-6182(99)00007-5.

Shennan, I., Bruhn, R., and Plafker, G., 2009, Multi-segment earthquakes and tsunami potential of the Aleutian megathrust: Quaternary Science Reviews, v. 28, p. 7–13, doi: 10.1016/j.quascirev.2008.09.016.

Stuiver, M., and Reimer, P.J., 1993, Extended 14C database and revised CALIB 3.0 radiocarbon calibration program: Radiocarbon, v. 35, p. 215–230.

Thatcher, W., 1984, The earthquake deformation cycle, recurrence and the time predictable model: Journal of Geophysical Research, v. 89, p. 5674–5680, doi: 10.1029/JB089iB07p05674.

Wiles, G.C., Post, A., Muller, E.H., and Molnia, B.F., 1999, Dendrochronology and late Holocene history of Bering Piedmont Glacier, Alaska: Quaternary Research, v. 52, p. 185–195, doi: 10.1006/qres.1999.2054.

Zong, Y., Shennan, I., Combellick, R.A., Hamilton, S.L., and Rutherford, M.M., 2003, Microfossil evidence for land movements associated with the AD 1964 Alaska earthquake: The Holocene, v. 13, p. 7–20, doi: 10.1191/0959683603hl590rp.

Manuscript Accepted by the Society 02 June 2009

Printed in the USA

The Geological Society of America
Special Paper 462
2010

Surges of the Bering Glacier

Bruce F. Molnia
U.S. Geological Survey, 926A National Center, 12201 Sunrise Valley Drive, Reston, Virginia 20192, USA

Austin Post
U.S. Geological Survey (retired), 2014 Bradley Street, Dupont, Washington 98327-7712, USA

ABSTRACT

Bering Glacier, the largest glacier in continental North America, is a surging glacier. A surging or surge-type glacier is one that periodically discharges an ice reservoir by means of one or more sudden, brief, large-scale ice displacement(s). These displacements typically transfer ice from up-glacier accumulation areas to down-glacier ablation areas. Most surges occur with a periodicity of about a decade to more than a century. Typically during periods of ice displacement, flow rates increase dramatically, often as much as 10–100 or more times faster than normal. Most surges do not result in terminus displacements. However, surges of the Bering Glacier typically result in significant terminus thickening and displacement.

Retreat of Bering Glacier from its Little Ice Age maximum position began during the first decade of the twentieth century. At least five major surges have interrupted this ongoing retreat. The combination of these two processes, retreat in response to changing climate and surging, has resulted in a number of short term fluctuations in Bering Glacier's ice velocity, thickness, and terminus position. Another consequence of the periodic surge cycles has been multiple drawdowns of ice from the glacier's accumulation area. Major surges of the Bering Glacier occurred in ca. 1900, ca. 1920, ca. 1938–1940, 1957–1967, and 1993–1995. A smaller magnitude surge occurred in 2008–2009. Hence, during the twentieth and early twenty-first centuries, Bering Glacier surged approximately every 20 a. The surges that occurred during the second half of the twentieth century have been closely monitored. This chapter presents details about the recent surge behavior of Bering Glacier.

INTRODUCTION

The twentieth-century retreat of Bering Glacier has been interrupted by at least six major episodes of surging. Although rapid retreat and surging may seem incompatible, the results of these two processes have been large-scale fluctuations in the position of the terminus of Bering Glacier and in the significant drawdown of ice from the accumulation area of the glacier. The surges occurred in ca. 1900, ca. 1920, ca. 1938–1940, 1957–1967, and 1993–1995, with an approximate average of 20 a between surges (Post, 1972). A smaller magnitude surge occurred in 2008–2009.

The two surges that occurred during the second half of the twentieth century were closely monitored. Both surges resulted in a rapid multiple-kilometer advance of Bering Glacier's terminus and were accompanied by the transfer of a significant volume of ice into the expanding terminus area. A rapid retreat of the terminus, significantly enhanced by large-scale passive calving (disarticulation) into Vitus Lake, and rapid melting up-glacier from the terminus also followed each surge. The net result was a Bering Glacier that was much thinner than before surge onset. Intensive fracturing and calving accompanying the retreat phase resulted in far more ice being

Molnia, B.F., and Post, A., 2010, Surges of the Bering Glacier, *in* Shuchman, R.A., and Josberger, E.G., eds., Bering Glacier: Interdisciplinary Studies of Earth's Largest Temperate Surging Glacier: Geological Society of America Special Paper 462, p. 291–316, doi: 10.1130/2010.2462(15). For permission to copy, contact editing@geosociety.org.

removed from Bering Glacier than could have been through melting alone.

The two most recent major surges (1957–1967, 1993–1995) consisted of a pair of ice-displacement events separated by periods of stagnant ice. In the earlier surge, ice-displacement occurred from 1957 to 1960 and from 1965 to 1967, and ice stagnation occurred from late 1960 to early 1965. Maximum ice displacement exceeded 10 km (Post, 1972).

Post (1960, 1965, 1967b, 1969, 1972), Meier and Post (1969), and Post and LaChapelle (1971, 2000) have shown through direct observations of many surging glaciers that the folding (or contortion) of surface medial and lateral moraines is an expected result of surging. From analysis of vertical aerial photography, Post (1972) determined that individual chevron-folded medial moraines are the product of a separate surge event. He wrote (Post, 1972, p. 219):

> Vertical aerial photography taken before and after surges disclosed the direction and magnitude of ice flow in various parts of the piedmont lobe. The ice moved toward the terminus and expanded in a normal, radial pattern with no evidence of unusual shearing that would result in the formation of large folds. Many surging glaciers display repeated lateral displacements in their medial moraines which result from periodic surging of the main glacier past non-surging tributaries. Moraines of the Bering Glacier display small periodic irregularities of this nature. The large "accordion" folds in the moraines in the Bering Lobe are judged to be due to the combined effects of compressive flow and lateral or transverse expansion of those previously formed irregularities. The initially small pre-existing perturbations in the moraines are simply spread laterally and shortened radially into large folds as the ice spreads out.

At Bering Glacier, field observations and analysis of aerial photography and synthetic aperture radar (SAR) images indicate that at least 25 folded moraines are present in the Central Medial Moraine Band (CMMB). Since Bering Glacier has a 20 to 30 a cycle (Post, 1972), at least 500–750 a of periodic surges would be necessary to create the observed number of folded moraines (Fig. 1).

THE 1957–1960, 1965–1967, AND 1993–1995 SURGES OF BERING GLACIER

1957–1960 and 1965–1967 Surges

Following a surge that ended ~1940, Bering Glacier began to retreat. Retreat continued through 1957. This retreat resulted in the expansion of Vitus Lake to an approximate size of 14 × 5 km and an approximate area of 48 km^2. A 1958 aerial photograph collected by D.J. Miller showed a new surge developing in the eastern Bagley Ice Valley. There, the surface of Bering Glacier was lowered, and recent shearing of the glacier's margin had occurred at Waxell Ridge. Surge displacement of glacier ice probably began in 1957 and reached the terminus in 1959. Much of the margin of Bering Lobe advanced, and ice reoccupied much of Vitus Lake, greatly reducing its area. By

Figure 1. Summer 1938 view of the Central Medial Moraine Band of Bering Glacier, a north-looking oblique aerial photograph of most of the northern part of this moraine band. Individual folded medial moraine loops stand several meters above the bare ice surface of the glacier. Photograph no. 1894 by Bradford Washburn, Museum of Science (Boston).

1960 the terminus of Bering Lobe had advanced to a position from 0.5 to 3.0 km beyond its 1948 margin, reducing Vitus Lake to a maximum size of ~3.5 × 2.5 km with an approximate area of 3 km^2. Post (1972) reported maximum displacements of as much as 9 km that probably occurred during a 12 to 18 mo. period. By the fall of 1960, no further terminus activity was observed. The position of the terminus changed little through the spring of 1963 (Molnia and Post, 1995, Fig. 13), and retreat was under way by summer 1963.

Between 1965 and 1967 a second, smaller phase of surging reactivated much of the terminus and resulted in up to 1.0 km of additional terminus advance. Ice was displaced by as much as 4.0 km (Post, 1972). By 1967 the glacier margin was within 1–3 km of its early twentieth century maximum position (Fig. 2). Molnia and Post (1995) quantified the post-1967 rate of terminus retreat throughout the eastern margin of Bering Glacier and found that the terminus retreated drastically (maximum recession of as much as 10.7 km) between 1967 and the late August 1993 onset of terminus displacement associated with the 1993–1995 surge. Retreat did not occur at a constant rate, because individual years and areas showed considerable variability.

Molnia and Post (1995, p. 112, Table 1) presented annual average retreat rates along four approximately north-south transects extending from the 1967 ice limit to the August 1993 presurge ice margin (transect locations shown by Molnia and Post, 1995, p. 111, Fig. 15). For the entire terminus of the Bering Lobe, the average retreat rate was 0.43 km a^{-1}.

Annual retreat rates were calculated along each of the transects and ranged from 0.04 to 1.0 km a^{-1}. Exceptions were short-lived periods when rapid passive calving and disarticulation were occurring. Because ice-margin positions were plotted from available aerial photography, intervals between adjacent dated lines were not necessarily multiples of full years. Except for periods of rapid passive calving and disarticulation, the maximum measured retreat rates along each transect occurred post-1990.

The largest retreat—1.0 km—occurred along transect A in Tashalich Arm between June 1992 and March 1993. The total amount of terminus retreat along each transect was 8.80 km (transect A), 10.70 km (transect B), 8.85 km (transect C), and 7.00 km (transect D). In transects A and D, initial retreat rates were low (0.15 km a^{-1} or less). In transects B and C, there was an initial rapid loss of what was perhaps a floating, thin extension of ice as retreat began, followed by a much slower rate of terminus retreat.

In 1977 (transect B) and 1984 (transect D) (Molnia and Post, 1995, p. 111), two episodes of very rapid passive calving and disarticulation occurred; each instance resulted in >2.0 km of ice-margin retreat in a period of a few days or weeks. These rapid recessions occurred when the thinning, retreating glacier terminus decreased to a minimum thickness, after which buoyancy would not permit it to remain in contact with the bottom. As it floated, it rapidly disarticulated. Many large icebergs separated from the terminus, resulting in a very rapid recession of the margin. This produced substantial short-lived increases in the

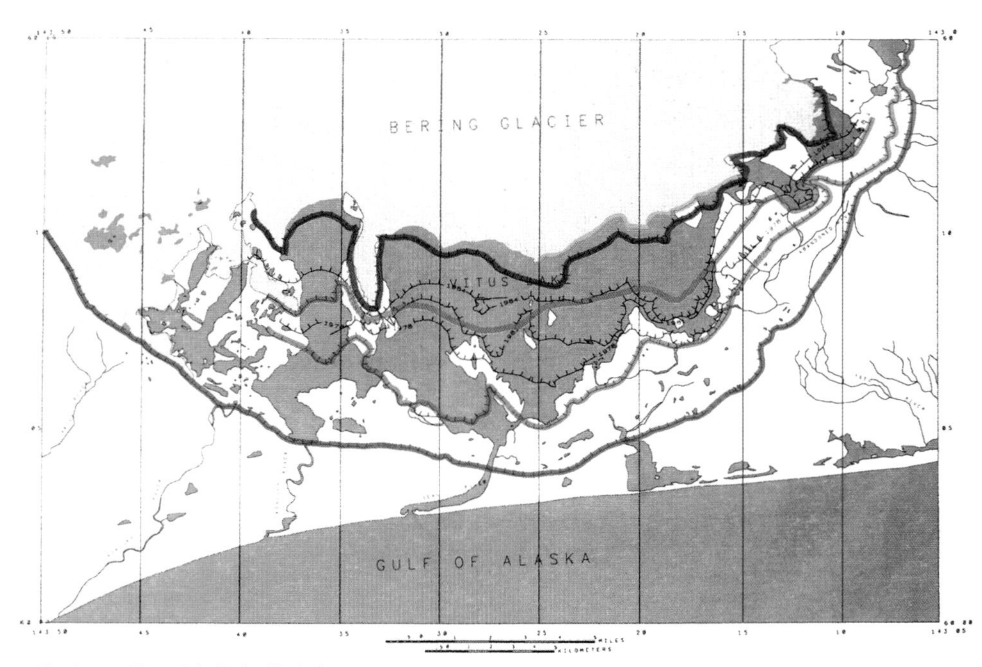

Terminus positions of the Bering Glacier in year:

Figure 2. Map showing positions of the terminus of Bering Glacier between 1900 and 1993. The 1993 position shown is the pre-surge (pre–28 August 1993) position of the terminus.

retreat rate. Before the summer of 1991, the retreating margin at the head of Tashalich Arm changed from one with a low rate of iceberg production to one with a much higher rate, and the fjord completely filled with floating calved icebergs. Correspondingly, the 1992–1993 retreat rate—1.0 km a^{-1}—was nearly twice as high as the next highest rate (1987–1990, 0.53 km a^{-1}) and more than six times higher than the initial rate of retreat (1967–1969, 0.15 km a^{-1}).

Sequential aerial photography provided additional insights into the characteristics of the post-1967 retreat, as a 17 August 1979 oblique aerial photograph by the first author shows (Fig. 3). For instance, rates of retreat in the ice-filled basins of Vitus Lake were much greater than they were on the adjacent islands and land areas. Similarly, as the resession progressed (Fig. 4), the surface gradient and elevation of Bering Glacier decreased significantly. On areas adjacent to Vitus Lake, large stagnant masses of glacier ice, remnants, covered the land surface. Some of these remnants were remobilized when they were contacted by advancing surge-displaced land-based ice (Fig. 5).

In addition to rapid retreat, Bering Glacier also thinned significantly during the period of recession. Between 1967 and 1993 the ablation area of Bering Lobe was virtually stagnant ice, and there was little detectable ice flow. The reduced flow from upglacier contributed to a significant thinning of the ice in the lower 80 km of the glacier. A comparison of ice-surface elevations at the terminus of Bering Glacier derived from the 1:63,360-scale U.S. Geological Survey (USGS) Bering Glacier A–7 topographic map (1984), which was based on 1972 aerial photography, with 1991 elevations derived from USGS vertical aerial photography, showed a thinning of ~150 m at the Bering Lobe terminus, an average surface lowering of ~7.9 m a^{-1}. Elsewhere, Molnia and Trabant (1992) compared 1990–1992 surface-elevation profiles made with a precision altimeter with elevations measured from the 1972 Bering Glacier topographic map (USGS, 1972). They determined that a thinning of between 85 and 180 m occurred along the centerline of Bering Lobe (a range of ~4.2–9 m a^{-1}. The locations of many of their precision altimeter survey points are shown in Figure 6. In some places, ice loss within the subsequent 20 a period represents ~20%–25% of the 1972 thickness of Bering Glacier at that location.

Molnia and Post (1995) reported that a comparison of 1946 and 1990 vertical aerial photographs shows that almost 200 m of thinning occurred between photographs along the margin of the Bering Glacier at an elevation of 280 m, south of the Grindle Hills. Molnia and Post (1995, p. 113, Fig. 16) presented Bering Glacier surface-elevation profiles for 1900, 1957, the 1960s, 1972, and 1991.

The 1993–1995 Surge

The 1993–1995 surge was anticipated and closely monitored. More than a year before the appearance of any evidence that a surge was beginning, the tongue of ice that filled the head of Tashalich Arm began to rapidly calve icebergs and retreat. The entire fjord was filled with floating ice within less than 3 mo. An analysis of sequential vertical aerial photographs led R.M. Krimmel (USGS, retired, Tacoma, Washington, 1996, personal commun.) to report that, during the 9-mo. period between 12 June 1992 and 16 March 1993, ice advanced toward Tashalich Arm at a velocity of 1.3 m d^{-1}, yet the terminus retreated ~300 m, more than 1 m d^{-1}. In the 4 mo. between 16 March 1993 and 10 July 1993, the velocity increased to 2–3 m d^{-1}. A similar

Figure 3. Oblique aerial photograph, 17 August 1979, showing the post-1967 surge retreat of the terminus of Bering Glacier. North-looking view of the southern part of Vitus Lake from offshore, east of the mouth of Seal River. In the 12 a after the end of the surge in 1967, the glacier had retreated ~3 km and thinned significantly. Bering Glacier has a very low gradient and almost no relief at its southern margin. The two large plumes of icebergs are the result of calving from rapidly disarticulating parts of the terminus of the glacier on either side of the island. Photograph by Bruce F. Molnia, U.S. Geological Survey.

Figure 4. Oblique aerial photograph, 20 July 1993, showing the post-1967 surge retreat of the east-central Piedmont Lobe region of Bering Glacier. Northwest-looking view along the terminus of Bering Glacier, which is rapidly calving and disarticulating along old crevasse planes. The height of the shoreline scarp at the northeast edge of Vitus Lake is <5 m. Because of their buoyancy, several calved blocks of ice have risen above the elevation of the surface of Bering Glacier. Photograph by Bruce F. Molnia, U.S. Geological Survey.

Figure 5. East-looking view, 29 July 1990, of part of a 300 × 600 m mass of debris-covered stagnant Bering Glacier ice near the head of the Taggland Peninsula. This remnant had a maximum height of ~30 m. When the rapidly advancing ice of the 1993 surge encountered this stagnant ice, it reconnected to the stagnant-ice mass and pushed it forward. Note the person for scale. Photograph by Bruce F. Molnia, U.S. Geological Survey.

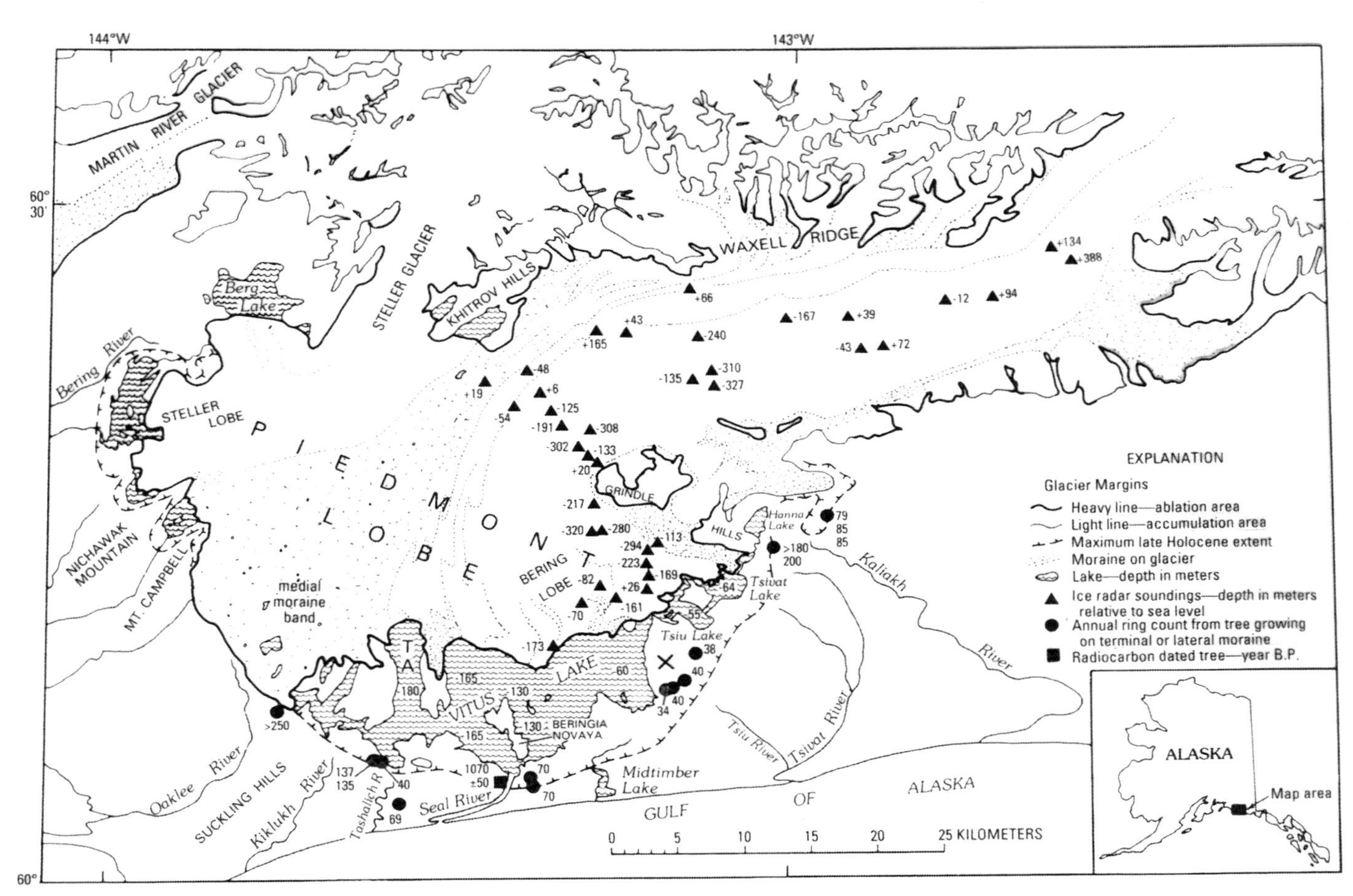

Figure 6. Map of the lower reaches of the Bering Glacier shows geographic features and locations from which data were collected. Triangles identify locations of ice-radar soundings and precision–altimeter surveys. The numbers adjacent to the triangles are basal ice depths relative to sea level expressed in meters. Numbers in Vitus, Tsiu, and Tsivat Lakes are maximum water depths in meters. Tsiu and Tsivat Lake depths are from Fleisher et al. (1993). Circles identify positions of trees growing on terminal and lateral moraines from which cores were obtained. The adjacent numbers are annual ring counts. The square adjacent to Seal River marks a radiocarbon-dated spruce tree recovered from the neoglacial terminal moraine. X marks the location of the Giant Log. TA is Tashalich Arm. The glacier margin and Vitus Lake are shown as they appeared in 1991. From Molnia and Post (1995, Fig. 2, p. 91).

rapid-calving process was characteristic of changes at several Bering Lobe ice-marginal areas when they were impacted by the arriving surge front, as shown on a 6 October 1993 oblique aerial photograph taken by the first author (Fig. 7).

In late August 1993, Vitus Lake had an area of ~70 km^2. Its maximum dimensions were ~20 km by 10 km. This significant increase in size was the result of more than two decades of late twentieth century recession. The 1993–1995 surge produced a maximum of ~10 km of advance of the terminus of Bering Lobe. Terminus-ice displacement occurred in two discrete phases: between late August 1993 and 17 September 1994, and between early May 1995 and mid-September 1995. A 7 mo. period without detectable ice displacement—from 17 September 1994 to early May 1995—separated the two advances. The surge resulted in a substantial increase in iceberg production; significant changes in the size, bathymetry, hydrology, and water chemistry of ice-marginal Vitus Lake; advance over vegetation (Fig. 8); and the complete or partial covering of all of the islands within Vitus Lake by advancing ice. In places, maximum ice-displacement rates and maximum surge-front-displacement velocities approached 100 m d^{-1} during the initial period of the surge.

The 1993–1994 Phase

In the central Bering Lobe the first visible evidence of the onset of the surge was the development of a large pressure ridge along the southwestern edge of the Grindle Hills, first observed during the spring of 1993, although the ridge probably had been building up for months before observation. During the early stages of the surge, this location served as the initiation point for the fracturing and shattering of the surface of Bering Glacier. Within weeks the up-glacier surface of Bering Lobe was

Figure 7. Northeast-looking oblique aerial photograph, 6 October 1993, of the intensively fractured, disintegrating terminus of the Bering Glacier. Fracturing extends many kilometers up-glacier. Several large surge-produced pressure ridges can be seen migrating toward the terminus. Photograph by Bruce F. Molnia, U.S. Geological Survey.

Figure 8. Oblique aerial photograph, 14 July 1994, of a lobe of Bering Glacier, Chugach Mountains, which was overriding an alder (*Alnus* sp.) forest as it advanced during the 1993–1995 surge. The trimline above the advancing terminus and the very young vegetation suggest that a thicker ice mass recently existed in this area. Photograph by Bruce F. Molnia, U.S. Geological Survey.

fractured for a distance of >50 km, and the thickening glacier was overtopping ridges, such as Override Ridge, in the lower Bagley Ice Valley (Fig. 9).

A comparison of three sets of mosaicked vertical aerial photographs of the terminus of Bering Glacier in Vitus Lake flown on 10 July 1993, 10 September 1993, and 25 February 1994 (Fig. 10) shows how various points along the glacier's terminus responded very differently to the surge process. In some areas the glacier's terminus rapidly advanced; in other areas it retreated. During the first 2 mo. of the surge the ice in part of the terminus adjacent to the eastern side of the peninsula that separates Vitus Lake from Tashalich Arm shattered and retreated rapidly following the arrival of the surge front. Simultaneously, to the east, Bering's terminus was rapidly advancing. By November the entire terminus was advancing rapidly into Vitus Lake.

A comparison of two sets of vertical aerial photographs acquired on 17 October 1993 and 16 May 1994 showed that the terminus adjacent to the eastern side of Beringia Novaya advanced ~7.78 km in this 211-d period. The average rate of terminus displacement was ~36.7 m d^{-1}. The actual displacement was higher because a significant volume of ice was continuously fracturing and calving from the advancing glacier margin. Cross-glacier fracturing proceeded at a much slower rate. But by July 1994 the CMMB showed fracturing and folding caused by both compressional and extensional stresses (Fig. 11). In essence, the fractures were a complex network of parallel and subparallel cracks and crevasses that opened in the previously near-stagnant CMMB.

R.M. Krimmel (USGS, retired, Tacoma, Washington, 1996, personal commun.) analyzed nine sets of sequential vertical aerial photographs acquired between 12 June 1992 and 7 September 1994 to derive many physical characteristics of the surging Bering Glacier (12 June 1992, 16 March 1993, 10 July 1993, 10 September 1993, 15–17 November 1993, 25 February 1994, 16 May 1994, 13 August 1994, 7 September 1994). He reported velocities lower than those cited above but noted that "the highest speed measured from the successive aerial photographs was 22 m d^{-1}. This speed was measured at two locations: in the west terminus area above the entrance to Tashalich Arm and in the area a few kilometers west of Grindle Hills. These speeds were the average speed for a several month period, and it is reasonable to expect that the speed was higher for portions of the measured period." Krimmel's plots for seven positions of the terminus of Bering Glacier, between 10 July 1993 and 7 September 1994, are shown in Figure 12.

Subsequent examination of sequential SAR imagery by Roush (1996) confirmed and enhanced the understanding of many of the events that occurred during the first 8 mo. of the surge. Roush found that the surge was in progress by 30 April 1993 and may have begun as early as 26 March 1993. Measurements showed that the surge front propagated down-glacier at a mean velocity of 90 m d^{-1} between 19 May and 25 August 1993, the migrating surge front first reaching the terminus of the glacier in Vitus Lake just before 24 August 1993. Subsequently the calving terminus advanced rapidly into Vitus Lake. On radar imagery the advancing surge front consisted of a distributed region of undulations and elongated bulges on the surface of Bering Glacier, having heights estimated from the SAR data of 40–110 m and widths varying from ~0.7 to 1.5 km. Roush (1996) compared 9 August 1993 and 18 October 1993 radar images and calculated that during this 71-d period the average advance rate in its central area was 19 m d^{-1} and that the mean rate of advance across the

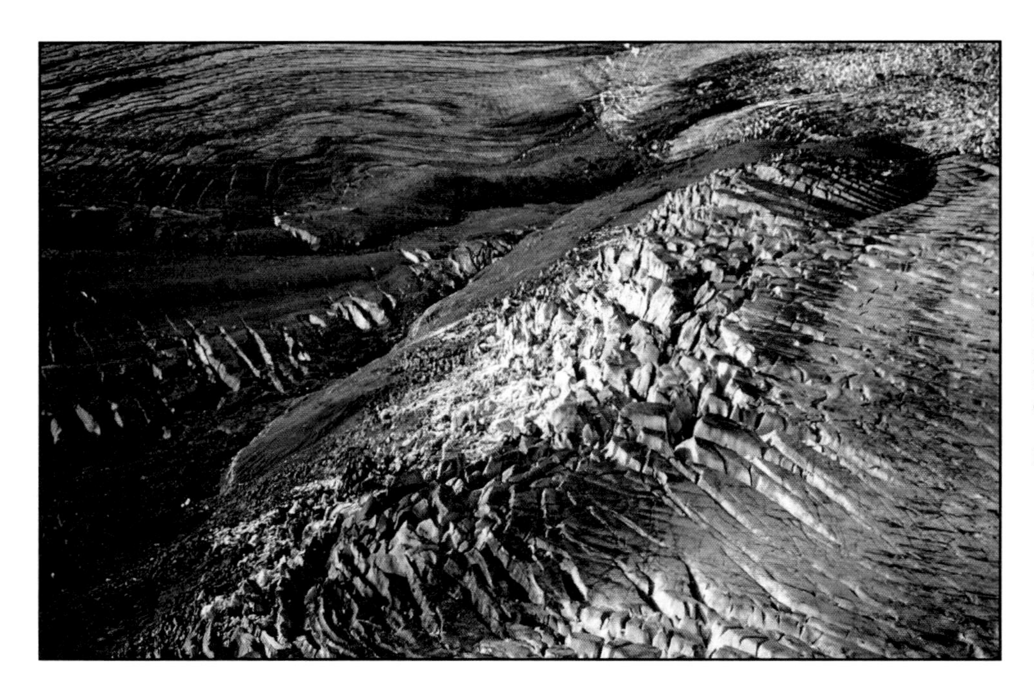

Figure 9. Oblique aerial photograph, 6 October 1993, shows surge-thickened ice at the edge of the Bagley Ice Valley expanding over an exposed bedrock ridge. The ridge, named Override Ridge by the second author, was also overridden by surge-thickened ice during the 1957–1960 surge. Photograph by Bruce F. Molnia, U.S. Geological Survey.

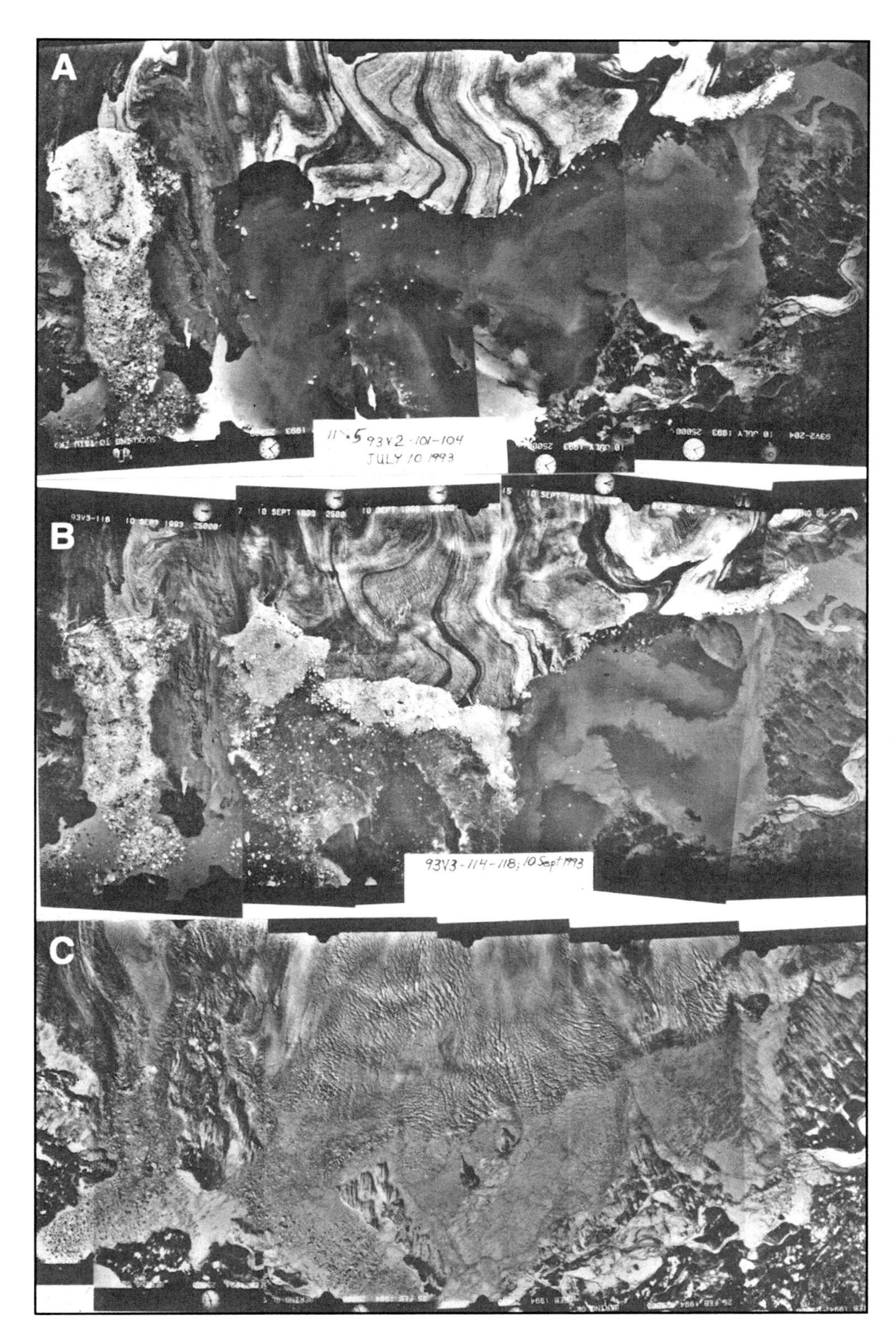

Figure 10. Three sets of mosaicked vertical aerial photographs of the terminus of Bering Glacier in Vitus Lake show rapid terminus changes during the first six months of the 1993–1995 surge. (A) Mosaic of 10 July 1993, showing the pre-surge terminus. (B) Mosaic of 10 September 1993, showing the response of the terminus to the arrival of the surge front. In addition to an increase in fracturing and crevassing, part of the terminus is retreating, while most of the terminus is actively advancing. (C) Mosaic of 25 February 1994, showing the dramatic advance of the entire terminus.

Figure 11. Photograph of several extensional ruptures located on the western side of the Central Medial Moraine Band of the Bering Glacier on 24 July 1994. Before the surge, this area was vegetation-covered stagnant ice. Photograph by Bruce F. Molnia, U.S. Geological Survey.

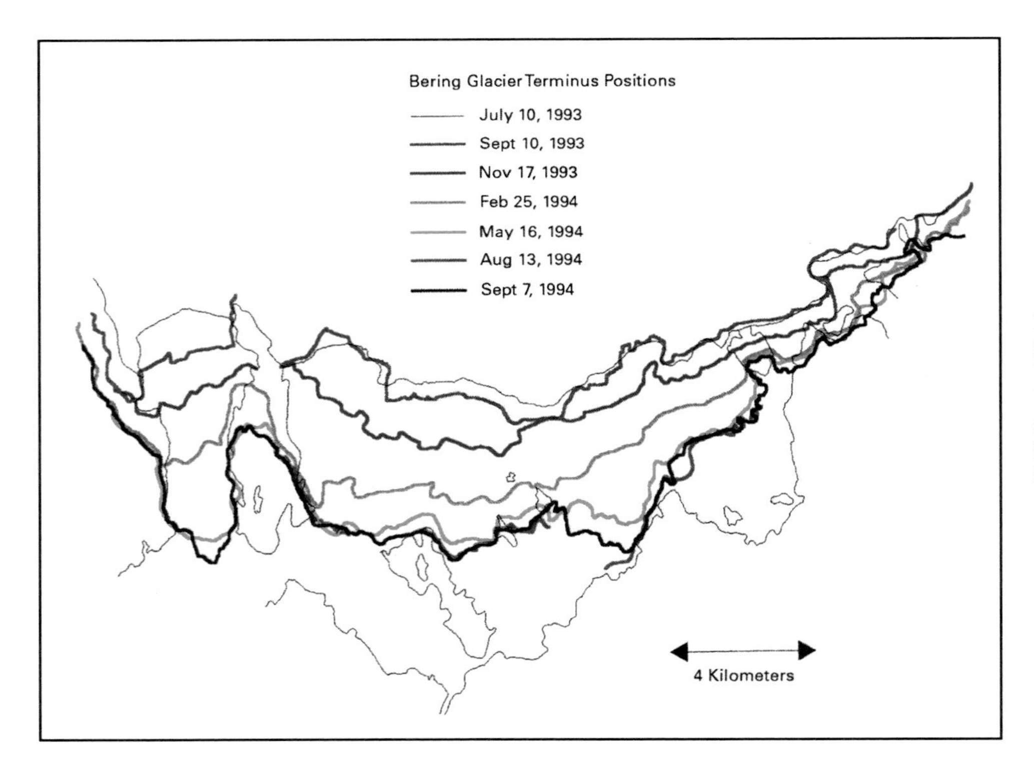

Figure 12. Terminus positions of Bering Glacier are shown between 10 July 1993 and 7 September 1994. The July 1993 shoreline of Vitus Lake is shown as a fine black line. From data compiled by Robert M. Krimmel, U.S. Geological Survey.

entire width of the terminus was 11 m d^{-1}. As previously noted, these rates pertain to the displacement of the ice margin and do not take into account the large volume of ice fractured and calved from the end of the advancing margin during the entire period of observation. Additionally, had Roush used 24 August 1993 as a start date, his average rates would have been >20% higher.

Studies by Fatland and Lingle (1998, 2002) analyzed the surge by using differential SAR interferometry (InSAR) on two pairs of ERS-1 radar images, one pair collected during a 3-d period in January 1992 before the onset of the surge, and the second during a 3-d period between 4 and 7 February 1994 at the peak of the surge. Their resulting high-resolution velocity data clearly show that the Bagley Ice Valley—southwest of Tana Glacier and northeast of Steller Glacier—underwent a 2.7-fold increase in velocity during the interval between image pairs, velocity having increased from 0.36 to 0.95 m d^{-1}. They also reported a drawdown of the surface of Bering Glacier of between 5 and 10 m and speculated that the velocity increase may have been caused by increased longitudinal stress gradients resulting from coupling to the surging main trunk of Bering Glacier. They also reported that InSAR was unsuccessful in determining velocities of the rapidly moving ice of Bering Glacier because there was no image correlation between sequential SAR images.

The advancing ice moved much more rapidly when it filled subglacier and sublacustrine fjord channels than it did when it advanced over land. Average land advance rates rarely exceeded 4 m d^{-1}. The advancing ice would frequently "bulldoze" large wedges of sediment in front of its advancing face. At many areas, advance was accompanied by upward thrusting of ice and sediment. Individual spatulate fingers of ice, ranging from a few meters to >100 m in width, built distinctive push moraines. Several of the islands in Vitus Lake were partially overtopped by ice during the first phase of the surge. The northern part of 1.7-km-long Tsitus (Arrowhead) Island, which had only become free of retreating ice during the summer of 1992, was quickly covered by advancing ice (Fig. 13A). By the end of the first phase of the surge, nearly 1 km of the northern end of the island was covered by advancing glacier ice (Fig. 13B). The same lobe of ice that overtopped the north end of Tsitus (Arrowhead) Island filled the channel east of the island that connected two eastern ice-marginal lakes with the main body of Vitus Lake. All of the drainage from the eastern part of Bering Glacier that previously flowed through this channel into Vitus Lake was diverted several kilometers to the south, where it flowed through an abandoned former ice-marginal channel adjacent to the 1967 ice-maximum position.

Advancing ice slowly covered the northern end of Pointed Island. However, a much more rapidly moving tongue of ice, at least 170 m thick, advanced past the island on its western side at a rate that exceeded 20 m d^{-1}. Before the first phase of the surge ended, two small islands, the Wallypogs, were completely covered, as was the northern two-thirds of Whaleback Island. The northwestern shoreline of Taxpayer's Bay, the area at the southeast side of Vitus Lake, was also overtopped by the advancing glacier.

R.M. Krimmel (USGS, retired, Tacoma, Washington, 1996 personal commun.) compared sequential vertical aerial photographs acquired on 16 May 1994, 13 August 1994, and 7 September 1994 to determine when this phase of the surge ended. He reported that ice velocities were high for the period between 16 May 1994 and 13 August 1994. Terminus changes were minor between 13 August 1994 and 7 September 1994. By mid-September 1994 the advance of Bering Glacier had ceased. Field observations by the first author suggest that detectable terminus displacement ceased on 14 September 1995. By then, Vitus Lake was so filled with the terminus of Bering Glacier and icebergs that almost no open water remained.

Surge-Produced Changes in Vitus Lake

Gray et al. (1994) documented changes in the physical and sedimentary characteristics of Vitus Lake that occurred during the first year of the surge by comparing pre-surge measurements made during the July 1992 retreat phase with peak surge measurements made in July 1994. In 1992 the lake had an area of ~160 km^2, and ~90% of its surface was free of icebergs. The lake was stratified into two distinct layers, the topmost of which extended from the surface to a depth of ~7 m. Generally, specific conductance values increased from ~2000 to ~5000 $\mu S\ cm^{-1}$ (microsiemens per centimeter), and water temperature decreased from ~5 °C to ~3 °C. Specific conductance values of ~17,000 $\mu S\ cm^{-1}$ at a depth of 10 m increased to a maximum of ~32,400 $\mu S\ cm^{-1}$ at a depth of 142 m. Most water temperature measurements made were <1 °C. The depth of the transition zone between the two layers corresponded to the depth of the thalweg (longitudinal outline of riverbed from source to mouth) of the Seal River. Additionally, the mean Secchi disk depth (measurements of water clarity) obtained from measurements made at localities >1 km from sediment-rich glacial water inflows was 2.3 m.

When Vitus Lake was measured in mid-July 1994, the surge had reduced its area to ~50 km^2, and more than 90% of its surface was covered by floating ice. Specific conductance values ranged from ~2000 to ~3200 $\mu S\ cm^{-1}$, and water temperature ranged from ~4 °C to ~0 °C. No evidence of stratification was found. Mean Secchi disk depth obtained from measurements made near the head of the Seal River was 0.4 m. Visual observations of water samples collected with a Van Dorn sampling device suggested that suspended sediment concentration increased with depth, unlike it did in 1992, when the lake was essentially sediment free. Lastly, in contrast to the 1992 sediment discharge rate of ~10 kg s^{-1}, the suspended sediment discharge in 1994 was ~10^4 kg s^{-1}.

The 1995 Phase

Oblique aerial photographs of the terminus region of Bering Glacier acquired in late November 1994 and again in late January 1995 showed no evidence of surge activity. Following a 7-mo. period beginning in September 1994, characterized by minor

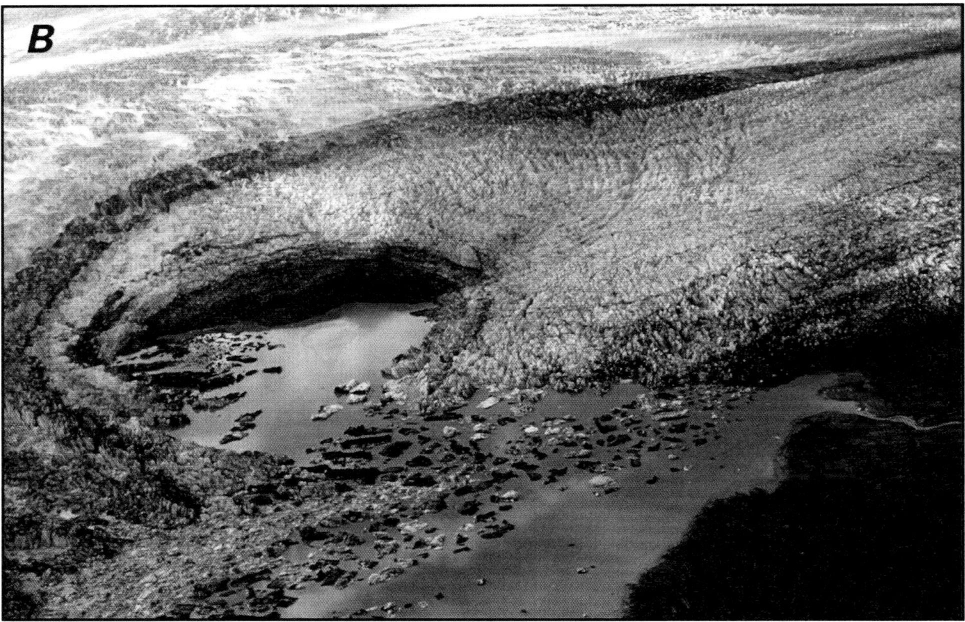

Figure 13. Two oblique aerial photographs of the Tsitus (Arrowhead) Island area that show changes that occurred during the first and second phases of the 1993–1995 surge of the Bering Glacier. (A) North-looking oblique aerial photograph, 7 October 1993, shows the advancing margin of the Bering Glacier just as it made contact with the north end of the island. The channel on the east side of the island, which connects to the eastern marginal lakes, is still open. Between the summers of 1992 and 1993 the north end of Tsitus (Arrowhead) Island had become ice free. (B) North-looking oblique aerial photograph, 9 July 1995, shows that the advancing margin of Bering Glacier had covered all but the southernmost 45 m of the island. Advancing ice masses in the channels on both sides of the island advanced toward each other but never joined. This photograph depicts the maximum ice advance and ice cover of Tsitus (Arrowhead) Island. Photographs by Bruce F. Molnia, U.S. Geological Survey.

retreat and near-stagnation, parts of the eastern Bering Glacier resumed surging. Several areas of the terminus advanced ~750 m in the 13-d period between 19 May and 1 June 1995 (57.7 m d^{-1}). The first evidence that new surge activity had begun was noted on 14 April 1995 by pilots Gayle and Steve Ranney of Cordova, Alaska. They observed that a section of the winter-ice cover of Vitus Lake was being compressed into a series of accordion-like folds. They also observed numerous deep fresh cracks and rifts as well as a number of blue-water lakes forming on the surface of Bering Glacier. The lakes, fractures, and rifts were characteristic features of the 1993–1994 phase of the surge.

USGS vertical aerial photography of the Bering Lobe terminus and Vitus Lake, acquired on 1 May 1995, confirmed the new fracturing and rifting as well as the numerous lakes. When compared with the late November 1994 vertical aerial photographs, the 1 May 1995 photographs showed that the terminus was advancing over the northern end of Beringia Novaya, the largest island in Vitus Lake, and over the north-central part of Pointed Island. The margin showed a significant increase in fracturing and iceberg production. Several icebergs were observed that were >0.5 km long. Oblique aerial photography on 19 May 1995 confirmed that the terminus was continuing to advance. On the southeastern shoreline of Vitus Lake, just west of Taxpayer's Bay, the advancing ice margin was redirecting the entire drainage from the eastern part of the Bering Glacier to a narrow ice-marginal channel. The result was several tens of meters of shoreline retreat and the development of a high bluff, as two oblique aerial photographs taken on 6 and 9 July 1995 by the first author show (Fig. 14).

The first author's 1 and 2 June 1995 visit to the glacier confirmed that not only was the surge affecting the eastern terminus region but also that it was affecting the northern part of the Bering Lobe and the southern part of the Bagley Ice Valley, as much as 30 km north of the terminus (Fig. 15). There, the winter 1994–95 snow surface was complexly fractured and rifted, and several large ice-surface (supraglacier) lakes reappeared at places where lakes had existed during the January to July 1994 period.

On 1 June 1995, only the southernmost 50 m of Pointed Island remained exposed, with ~750 m of the island having been covered during the 12 days since 19 May 1995. On Pointed Island, thousands of nesting birds were displaced or killed and their nests destroyed. Overtopping of the nests by the advancing ice margin was accompanied by ice blocks falling from the advancing terminus. On Tsitus (Arrowhead) Island, the total ice advance during this phase of the surge was <200 m. This phase of the surge was much shorter lived than the first, lasting ~5 mo.

Total terminus advance during the 1993–1995 surge was as significant as it was in the 1957–1967 surge, and advance velocities into proglacial Vitus Lake again greatly exceeded flow velocities over islands and adjacent land areas. This advance revitalized the stagnant mass of ice to the north in the Weeping Peat Island area, and the advancing ice reconnected with the toe of the glacier that had been detached by a jökulhlaup in December 1994 (see following section). Here the total amount of advance was ~250 m. As was the case to the west, ice advance ended by mid-September 1995.

Muskett et al. (2000) reported that, in June 1995, just after the onset of the second phase of the 1993–1995 surge, Bering Glacier was profiled by a geodetic airborne laser altimeter using methods described by Echelmeyer et al. (1996). These results were used to compare elevation and volume changes that occurred between the early 1970s and June 1995. They found that despite a significant volume of ice being transported to the Bering Lobe by the surge in 1993 and 1994, the surface of the glacier was "generally lower in 1995 than in 1972–1973" (Muskett et al., 2000, p. F404). They estimated that the total volume lost was 41 ± 10 km^3 of ice, which they presented as a corresponding area-average mass balance of –0.8 ± 0.2 m a^{-1} water equivalent. Elevation change was nonuniform. Surface lowering was greatest on the Bering Lobe near the terminus, where a maximum of 75–100 m of lowering had occurred. Although they did not quantify it, Muskett et al. (2000) noted that an area of the Bering Lobe 13 km upglacier from the terminus had undergone a small thickening. A maximum of 50 m of thickening was seen farther up-glacier above the equilibrium line. They concluded that "negative mass balance predominated over the massive downglacier transport of ice caused by the 1993–1995 surge" (Muskett et al., 2000, p. F404).

A comparison of 1950s topographic map data of the glacier with surface elevation data obtained during airborne profiling surveys in the middle and late 1990s showed that on an annual basis Bering Glacier thinned by 0.914 m a^{-1} and that its volume decreased by 1.78 km^3 a^{-1}. However, rates of change increased more than 300% between the middle 1990s and 1999. During this period, on an annual basis, Bering Glacier thinned by 3.077 m a^{-1}, and its volume decreased by 6.0 km^3 a^{-1} (K.A. Echelmeyer, W.D. Harrison, V.B. Valentine, and S.I. Zirnheld, University of Alaska, Fairbanks, March 2001, personal commun.).

The 1994–1995 Jökulhlaup

During the spring of 1994, a number of large blue-water ice-surface (supraglacier) lakes formed at several places on Bering Lobe and in the Bagley Ice Valley, as a 24 July 1994 oblique aerial photograph by the first author shows (Fig. 16). The largest of these was >1 km long. On 27 July 1994 a large sediment-laden jökulhlaup began at the terminus of Bering Lobe directly east of Weeping Peat Island. During the first few hours of the jökulhlaup, hundreds of large blocks of ice, many with maximum dimensions estimated to be >30 m, separated from the glacier face and were jetted into the deep lake adjacent to the ice margin. Within 24 h, a canyon more than 1 km long was cut into the ice margin as the point of water discharge retreated up-glacier. Initial peak discharges were estimated to be >10^5 m^3-s. Within days of the onset of flooding, many of the supraglacier lakes were drained.

In early September 1994 the retreating point of origin of the high-volume discharge reached the northern end of Weeping Peat Island. The discharge point then shifted direction nearly 90°. It continued to cut a widening west-trending channel through the

Figure 14. Two views show changes occurring on the southeast shoreline of Vitus Lake during July 1995. (A) Northeast-looking oblique aerial photograph, 9 July 1995, of the advancing ice margin that forced the entire drainage from the eastern part of the Bering Glacier to flow into a narrow channel. (B) Southeast-looking photograph, 6 July 1995, of the shoreline retreat and the development of a high bluff. Photographs by Bruce F. Molnia, U.S. Geological Survey.

Figure 15. Two oblique aerial photographs, showing complex fracturing and crevassing that occurred during the latter part of phase one and the early part of phase two of the 1993–1995 surge. (A) South-looking view, 9 May 1994, shows a 4-km-wide section of the Bagley Ice Valley, with several types of crevasses, including rectilinear crevasses, and the development of subparallel rifts. (B) North-looking view, 24 July 1994, across the central Piedmont Lobe shows part of a 5-km-long developing rift set in a wide section of rectilinear crevasses. The east-west–trending snow-filled crevasses represent fractures that formed during early fall 1993 and that were subsequently filled with winter 1993–1994 snow. The north-trending fractures and crevasses began forming during the spring of 1994. Photographs by Bruce F. Molnia, U.S. Geological Survey.

Figure 16. North-looking pre-jökulhlaup oblique aerial photograph, 24 July 1994, of a blue-water lake formed on the surface of Bering Glacier in the Bagley Ice Valley taken less than a week before the onset of the 28 July 1994 jökulhlaup. The photograph shows part of an ~350-m-diameter circular lake ~10 km west of Juniper Island. Photograph by Bruce F. Molnia, U.S. Geological Survey.

stagnant terminus of the glacier during the next few months and dissected and separated the toe from the remainder of the glacier.

During the first year of the flood, ~0.3 km^3 of sediment and ice were deposited into the ice-marginal lake system on the southeastern side of the glacier, and a channel was cut through the central area of Weeping Peat Island. The high-volume discharge continued for much of the next year and deposited a large outwash-fan delta, part of which formed in water as deep as 50 m. By late July 1995, following a lowering of the lake's level, the melting of ice blocks buried in the sediment created several dozen near-circular pit pond depressions (kettles) on the surface of the northern end of the outwash-fan delta adjacent to the ice margin. The largest had a diameter of ~20 m and a depth of 5 m. At several lakeshore and riverbank bluffs, blocks of ice buried in the sediment were exposed in profile.

By the spring of 1996 the volume of water discharging from the base of the glacier had decreased substantially, to near pre-jökulhlaup levels. As a result, the lake's surface elevation dropped several meters, and >6 km^2 of the outwash plain-fan delta was exposed. The stream discharging at the base of the glacier began to meander across the exposed outwash plain-fan delta. An inversion of topography occurred at many areas where the tops of ice blocks that were buried in flood sediment were exposed above the sediment plain. Some extended more than 3 m above the outwash plain–fan delta's surface (Fig. 17A, 17B). Many blocks were surrounded by moats, some water filled. Erosion and scour resulting from lateral stream-channel migration had removed at least 4 m of sand and gravel and left behind the newly exposed ice boulders. This erosion and downcutting were responsible for the reversal of topography.

All of the exposed ice boulders had melted away by 20 July 1996, and the surface of the outwash plain–fan delta was covered with several hundred kettle depressions (Fig. 17C). Of the 90 kettles the first author studied during a 2-wk field investigation in mid-August 1996, the largest was ~35 m in diameter and ~5 m deep. More than a dozen had ice exposed in their walls. Most of the studied kettles were continuing to enlarge through slumping and melting of buried ice. Another stream channel migration event was under way on 17 September 1996, the date of the last 1996 site visit. Between 26 August and 17 September 1996 the stream migration had completely removed ~40 of the easternmost kettles and exposed ice boulders at the surface of several of the former kettles on the eastern side of the westward-migrating stream. Many other areas showed evidence of surface slumping,

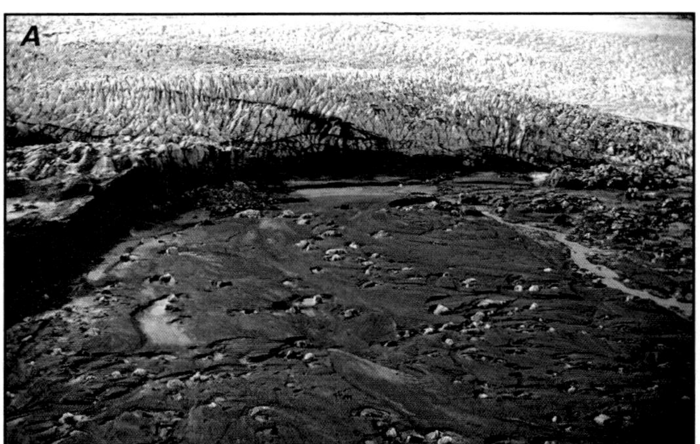

Figure 17. Three photographs showing the evolution of kettle ponds in the outwash-fan delta formed by sediment deposited during the 1994–1995 jökulhlaup. (A) Oblique aerial photograph, 30 April 1996, of the northern end of the outwash-fan delta shows two stages in the development of the kettle ponds. Meandering of the stream to the right of center has eroded the sediment cover from many large ice boulders to its east. Their rapid melting resulted in the formation of only a few small kettle ponds. To the west of the stream, where only the top part of the ice boulders is exposed, melting over the ensuing three months formed many large kettle ponds. (B) Ground photograph, 1 May 1996, of a 20-m-long sediment-banded ice boulder present in panel A. Three months later a 6-m-deep kettle pond had formed at this site. (C) Oblique aerial photograph, 20 July 1996, of the same area shown in panel A. Individual kettle ponds can be correlated with exposed ice blocks in the earlier photograph. Photographs by Bruce F. Molnia, U.S. Geological Survey.

an indication that melting of subsurface glacier ice was continuing. When the area was observed in August 2001, there was no evidence of continuing kettle growth, and various types of vegetation were becoming established in the area between the kettles.

It was initially thought that the July 1994 jökulhlaup was a surge-ending flood event, because ice velocities decreased following the onset of the jökulhlaup, and ice displacement ceased within 60 days. But it was not, because the ice resumed its advance following a 7-mo. pause. Perhaps the jökulhlaup was caused by a partial failure of the subglacier "barrier" or "dam" responsible for the initiation of the 1993–1994 surge. However, the July 1994 jökulhlaup was the only major jökulhlaup associated with the 1993–1995 surge of Bering Glacier.

Following the onset of the July 1994 jökulhlaup, a large suspended-sediment plume began to exit Vitus Lake and enter the Gulf of Alaska through Seal River, as a 7 September 1994 vertical aerial photograph shows (Fig. 18). It is likely that the plume originated from one or more subglacier channels on the eastern side of the Bering Lobe. Opening of the constriction that partially or even totally blocked flow through these sediment-dammed channels may have been a factor responsible for the onset of the 1993 surge. The discharge of sediment appears to be closely tied to the mechanics of Bering Glacier's surge cycle. From the 1970s through the middle1980s, sediment-laden water flowed from Vitus Lake through Seal River into the Gulf of Alaska. Sometime during 1985 the plume disappeared and remained absent until just after the beginning of the 1994 jökulhlaup. The sediment plume's reappearance at about the same time as the jökulhlaup's occurrence adds credence to the hypothesis that Bering Glacier's surges are caused by a buildup of subglacier water pressure and water volume resulting from blockage of Bering Glacier's subglacier channels with sediment and that the jökulhlaup resulted from the failure of the subglacier sediment dam.

The Seal River sediment plume is visible on nearly every Landsat image collected during the 1970s and early 1980s (Fig. 19A; see Post, 1976). During the 1970s and early 1980s the Seal River sediment plume was the second largest and second densest sediment plume entering the Gulf of Alaska. The largest—the Copper River plume—annually transports ~10^8 metric tons of sediment into the Gulf of Alaska (Reimnitz, 1966) (Fig. 19B). In addition to Landsat imagery, 1979 aerial photographs of the Seal River also show the sediment plume, but it is absent from 1990 and 1992 photographs. A 22 August 2003 Moderate Resolution Imaging Spectroradiometer (MODIS) image (Fig. 19C) covers approximately the same area encompassed by the 24 August 1978 Landsat 3 RBV image (Fig. 19A) and shows multiple plumes. A very large sediment plume is visible on Bradford Washburn's summer 1938 photograph of the mouth of the Seal River (Fig. 19D).

Figure 18. Vertical aerial photograph of Seal River, 7 September 1994, shows the large suspended sediment plume exiting Vitus Lake. At the time of this photograph the suspended sediment load exceeded 2 g L^{-1}. For much of the decade before the onset of the 1994 glacier outburst flood (jökulhlaup), the suspended sediment load dropped by more than an order of magnitude. USGS photograph no. 94–V4–85 by Robert M. Krimmel, U.S. Geological Survey.

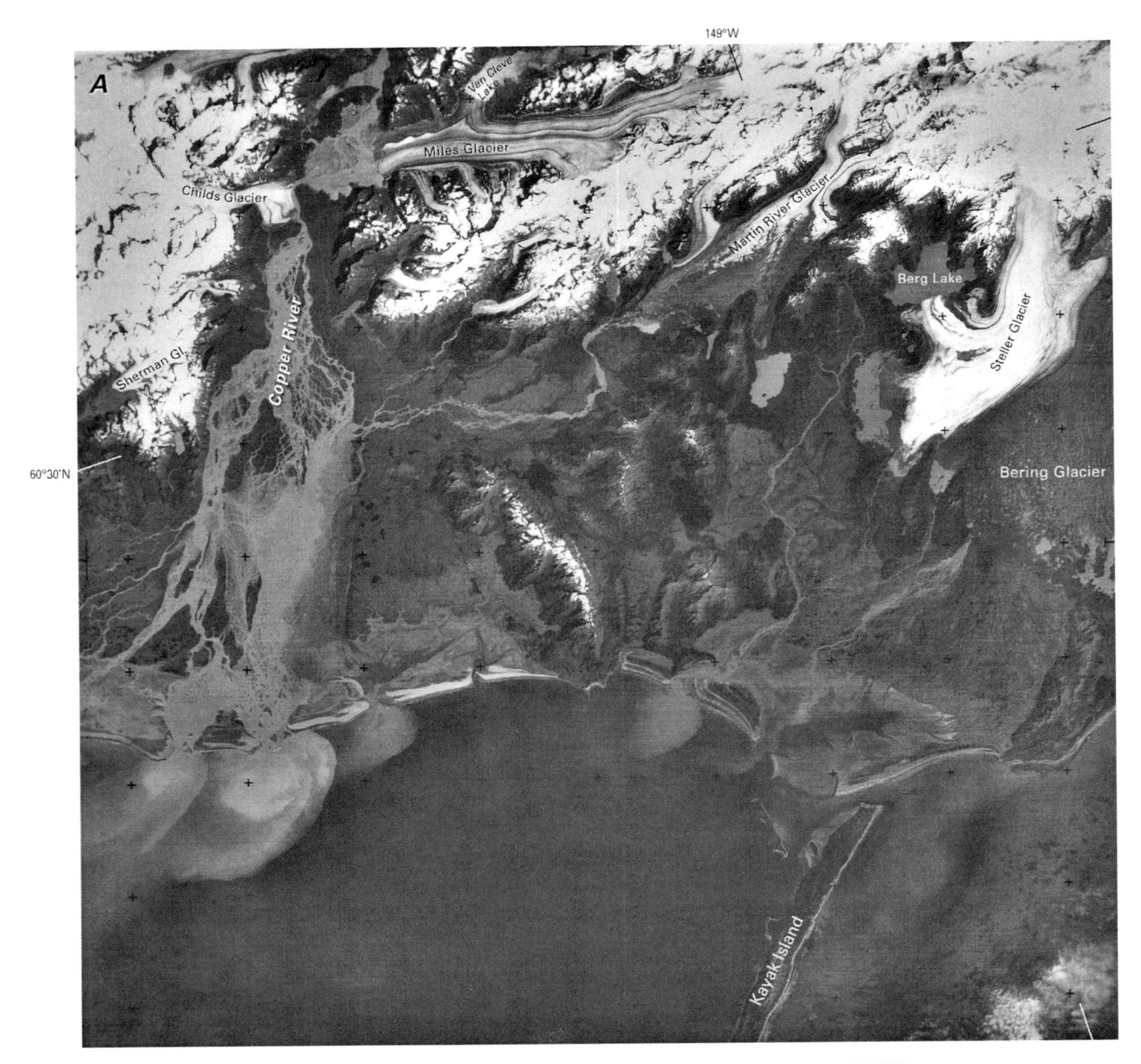

Figure 19 (*on this and following page*).

Figure 19. Two 1970s views, a 2003 view, and a 1938 view of the suspended sediment plume from the Seal River. The magnitude of the sediment plume before the middle 1980s suggests that little impediment existed to its flow through subglacial drainage channels of the Bering Glacier. The subsequent closure of the drainage channels, resulting in the middle 1980s' decrease in suspended sediment load, may have been a factor in the onset of the 1993 surge. The Seal River sediment plume, west of the terminus of Bering Glacier, separates into two components: part of the plume extends offshore, and part flows along the west side of Kayak Island, merging with the Copper River suspended sediment plume. In the 1970s and early 1980s the Seal River plume was the second largest sediment plume entering the Gulf of Alaska. For comparison, Reimnitz (1966) calculated that the Copper River plume, largest in Alaska, annually introduced ~10^{11} kg of suspended sediment into the Gulf of Alaska. (A) Annotated Landsat 3 RBV image of the Copper River Delta area and environs on 24 August 1978. Seal River is to the right, off the image, but part of its sediment plume is visible in the lower right corner. The Steller Lobe dams Berg Lake, which had jökulhlaups in 1984, 1986, and 1994. The Copper River is very turbid; its sediment plume extends far into the Gulf of Alaska. Landsat 3 RBV image (30172–20175–B; 24 August 1978; Path 71, Row 18) and caption courtesy of Robert M. Krimmel, U.S. Geological Survey, is from the EROS Data Center, Sioux Falls, South Dakota. (B) Photograph, 12 July 1976, taken from the rail of the R.V. Sea Sounder, ~3 m above the ocean surface, of the edge of the Seal River plume in the Gulf of Alaska. Photograph by Bruce F. Molnia, U.S. Geological Survey. (C) Moderate Resolution Imaging Spectroradiometer (MODIS) image, 22 August 2003, of the Gulf of Alaska coastal area from just east of Prince William Sound (left) to Grand Plateau Glacier (right).The same area shown in panel A is on the western third of the MODIS image. MODIS image is from the U.S. National Aeronautics and Space Administration. (D) Oblique aerial photograph, summer 1938, of the southernmost part of the terminus of Bering Glacier, Vitus Lake, Seal River, and the Gulf of Alaska. In this post-1920s surge photograph, a large sediment plume can be seen entering the Gulf of Alaska and flowing around Kayak Island. A large disarticulatiom event is also taking place at the glacier's terminus. Photograph no. 1000 by Bradford Washburn, Museum of Science (Boston).

Post-Surge Retreat of Bering Glacier

As was the case during the advance phase, when land-based advance rates were significantly lower than advance rates in adjacent lacustrine basins, the retreat rate of the terminus from the land areas was significantly less than that of the thicker tongues of ice that filled the deeper basins of Vitus Lake. Before the summer of 1996, little change was noted on the ice-covered islands within Vitus Lake or on the ice-covered part of the mainland east of Vitus Lake. By early summer 1996, the retreat of this land-based ice was under way at an average rate of <0.5 m d^{-1}. By 1997, multiple recessional moraines indicated hiatuses in the land-based retreat cycle at several places.

Following the end of terminus advance in mid-September 1995 the parts of Vitus Lake that were not filled by surge-advanced ice were filled with denselypacked icebergs. Almost no water was visible between the glacier's margin and the southern shore of Vitus Lake. Active calving, which was ongoing during the early rapid advance of the terminus, continued as surge advances waned. The glacier began to transition to a retreat-dominated regime (Fig. 20A), mostly through the production of many icebergs (Fig. 20B). Without the influx of surge-transported ice, rapid retreat of the terminus from the deeper basins of Vitus Lake ensued. Both Vitus Lake and Tashalich Arm quickly developed semicircular calving embayments (Fig. 21). Active calving in both areas continued through the summer of 1998. By late 1999, Beringia Novaya and Pointed Island were ice free; Whaleback Island was ice free by early 2001.

The retreat from Tsitus (Arrowhead) Island was much slower. Less than 50 m of retreat had occurred by 10 August 1998 (Fig. 22); <300 m of the southern end of the island had been exposed by August 2001. The ice that had covered Tsitus

Figure 20. Two oblique aerial photographs of the continued retreat of Bering Glacier from the north end of Beringia Novaya. (A) Photograph, 6 June 1997, shows the surge-maximum moraine and several recessional moraines, numerous developing kettles, and the retreating ice margin. (B) Photograph, 12 August 2001, taken more than 6 a following the cessation of surge motion within the terminus of the Bering Glacier. The boundary between vegetation and bare ground marks the 1995 maximum advance of the glacier. Continued thinning and retreat have resulted in the rapid disarticulation of the terminus of Bering Glacier. Photographs by Bruce F. Molnia, U.S. Geological Survey.

Figure 21. Oblique aerial photograph of the central terminus of Bering Glacier in Vitus Lake, Chugach Mountains, on 16 August 1998. The amphitheater-shaped embayment produced by the rapid calving retreat of the terminus is ~2 km in diameter. Photograph by Bruce F. Molnia, U.S. Geological Survey.

(Arrowhead) Island was the thickest of that for any island in Vitus Lake. The channel to the east that had connected to the eastern ice-marginal lakes was still filled by a large volume of surge-advanced ice. This connection was only reestablished in 2007.

Similarly, ice that had advanced over the mainland east of Vitus Lake still covered much of the land, as a 12 August 1998 oblique aerial photograph taken by the first author shows (Fig. 23). Before the summer of 2001, all of the ice that had advanced onto Weeping Peat Island had melted. The surface of the eastern part of the island displayed at least four recessional moraines.

By late summer 1997, the rate of calving decreased in the Vitus Lake embayment west of Beringia Novaya. Several concentric, arcuate crevasses developed parallel to the perimeter of the terminus in the calving embayment. By 26 September 1997, intense compressional forces rotated a single large mass of the glacier bounded by two subparallel crevasses and raised it more than 70 m above the surface of the adjacent ice (Fig. 24). After ~72 h, this massive 80 × 50 × 10 m pyramidal-shaped block of ice fractured and calved into Vitus Lake.

Active calving of the margin continued at a reduced rate from late 1997 through late 1998; however, the physical characteristics of the margin changed significantly. The height of the face of the glacier decreased by >50% between 1996 and late 1998. The height of the ice cliff along much of the margin in Vitus Lake continued to decrease in 1999 and 2000. In places the height was only 1–4 m above lake level.

By 1999, part of the terminus had retreated >5 km. By 2001, continuing retreat of the glacier had resulted in much of the terminus reoccupying positions similar to those held by the ice front in the pre-surge period from 1992 to 1993. In 2000, and again in 2001, as successions of large icebergs passively calved from the margin and drifted into Vitus Lake (Fig. 20B), parts of the terminus were observed retreating as much as 700 m in less than 24 h. By 12 August 2001, the surface of the glacier in the eastern terminus region closely resembled the pre-1993-surge surface of the glacier (Figs. 20B and 25).

In the central part of Bering Lobe, several kilometers behind the terminus, dramatic changes occurred during the post-surge period. When the surge ended in September 1995, Bering Lobe's surface was fractured with large rifts, irregular rectangular blocks of crevasse-bounded ice, and many seracs. Local relief was as much as 20 m in places (Fig. 15). Rapid melting was accelerated by the large surface area of the numerous surface features. By the end of summer 1997 the surface of Bering Lobe had little local relief. In August 2001, many areas of the generally flat surface of Bering Lobe showed scars from the intense deformation that the glacier had undergone during the surge. In addition, extensional fractures (Fig. 11) that had opened in the CMMB closed within 24 months.

Sauber et al. (2000) recognized that substantial quantities of ice were transported from the accumulation area of Bering Glacier to the terminus region during the 1993–1995 surge of the glacier. They noted substantial rates of near-instantaneous uplift in the reservoir region in response. These rates ranged from 18.2 ± 6.6 to 29.9 ± 5.7 mm a^{-1}. Sauber and Molnia (2000, 2004) also found that seismicity in the region of the dramatic thinning increased during the surge interval relative to the pre-surge period. Following the surge, during the 2 a post-surge period of 1998–2000, no earthquakes >**M** 2.5 occurred.

Josberger et al. (2001, 2006, and this volume) measured the bathymetry, temperature, and conductivity of Vitus Lake beginning in August 2001. They found that intense vertical convection in Vitus Lake was controlled by the salt content of the water and that there was strong saline stratification in the deeper parts of the lake. They concluded that thermal diffusion across the pycnocline (zone of change in water density as a function of depth)

Figure 22. Oblique aerial photograph of the deglacierized terrain following the post-surge retreat of Bering Glacier from the south end of Tsitus (Arrowhead) Island. East-looking view, 10 August 1998, of the southern end of Tsitus (Arrowhead) Island shows the retreating terminus of Bering Glacier. As was the case for Pointed Island, several recessional moraines mark the post-surge retreat. Many grooves and furrows mark the areas of maximum advance. Photograph by Bruce F. Molnia, U.S. Geological Survey.

Figure 23. North-looking oblique aerial photograph, 12 August 1998, of the post-surge retreat of Bering Glacier west of Weeping Peat Island. The most obvious features are the numerous grooves and furrows. Retreat during the 3 a period since the cessation of the surge amounts to ~500 m. Photograph by Bruce F. Molnia, U.S. Geological Survey.

Figure 24. Photograph of an uplifted pyramidal-shaped block of ice, ~80 × 50 × 10 m, produced by unusual compressional stresses in the calving embayment of Bering Glacier in Vitus Lake on 26 September 1997. Southwest-looking view across the Bering Glacier west of Beringia Novaya. The uplifted triangular block of ice was not observed while being rotated. However, its debris-covered former surface has been rotated toward the west. Note the single-engine aircraft flying behind the ice block. Photograph by Bruce F. Molnia, U.S. Geological Survey.

Figure 25. Oblique aerial photograph, 12 August 2001, of the nature of the terminus of Bering Glacier 6 a after the end of the 1995 surge. The north-looking view shows the low-relief terminus of Bering Glacier and the recently emerged Whaleback Island. With the exception of disarticulation of the terminus on the left side of the picture, calving is minimal. The Piedmont Lobe has a very low gradient, and the ice face ranges in height from <1 m to ~3 m . Photograph by Bruce F. Molnia, U.S. Geological Survey.

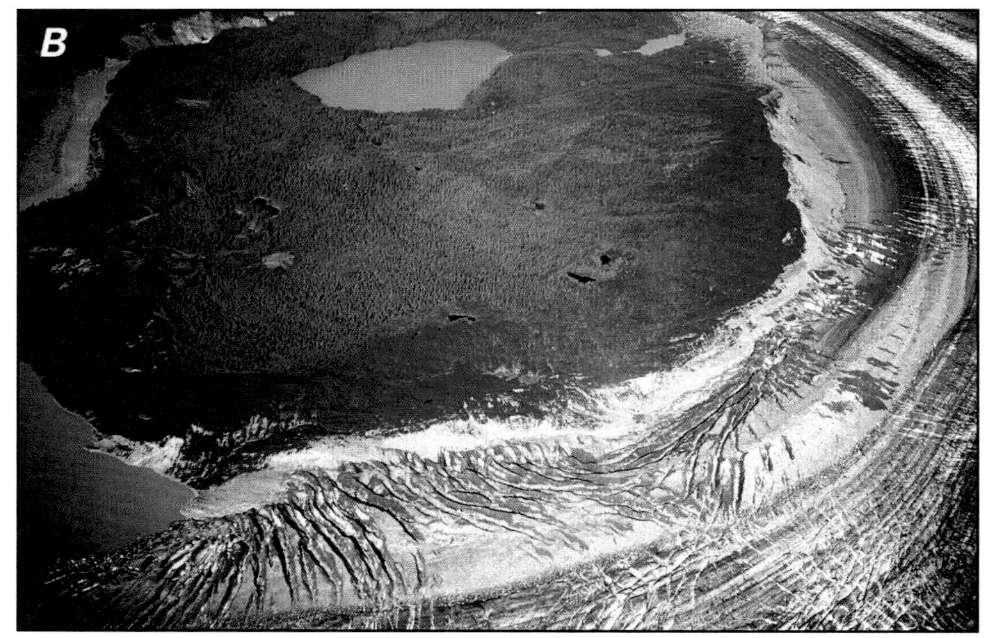

Figure 26. Two oblique aerial photographs of the effects of the surge on the ice-cored lateral moraine of the Steller Lobe, north of Nichawak Mountain. (A) Close-up north-looking view, 12 August 2000, of the distal end of the lateral moraine. In late spring 2000 the ice-cored moraine started to disintegrate. By late summer the Steller Lobe rapidly advanced into the adjacent lake. (B) By 12 August 2001 the surge was over. In a higher altitude northeast-looking view, most of the seracs and ice pinnacles visible in 2000 have melted. The rapid advance of the terminus has also ceased. Photographs by Bruce F. Molnia, U.S. Geological Survey.

may produce frazil ice growth, whereas melting of the glacier terminus produced convection at the margin of the lake.

A smaller magnitude surge of Bering Glacier occurred in the 2008–2009 period. When Chris Larsen of the University of Alaska, Fairbanks (2009, personal commun.) observed Bering Glacier in late summer 2009, he noted that the glacier's surface northeast of the Grindle Hills showed evidenced of surge-related fracturing. To determine the timing and extent of this surge, the first author examined photography of Bering Glacier that covered an area extending from the Central Valley Reach, approximately 15 km east of the Grindle Hills, on the east side, to the Central Medial Moraine Band on the west side, to the terminus of the Bering Lobe on the south side, to the Khitrov Hills on the north side. The photography was collected between May 2008 and February 2010. The examination revealed that the May 2008 photography showed no evidence of the surge, while the June and July 2009 photography showed complex patterns of surge-related surface fracturing that spanned much of the northern and western perimeter of the Grindle Hills and much of the Central Valley Reach, east of the Central Medial Moraine Band and south of the Khitrov Hills. Examination of the January 2010 photography showed that the surge fracturing extended both further west and south than in 2009. The western limit of fracturing was from 1 to 2 km east of the Central Medial Moraine Band, the northern limit was adjacent to the Khitrov Hills, while the southern limit was approximately 5 km north of the terminus. No difference in the extent and type of surge-related surface features was noted when January and February 2010 images were compared. Therefore, it appears that a surge of Bering Glacier, restricted to the Central Valley Reach and upper Bering Lobe recently occurred, originating sometime after May 2008. The surge was still underway in July 2009, but it ended prior by January 2010. Through January 2010, retreat of the Bering Lobe terminus continues.

Steller Glacier Activity

Steller Glacier did not actively surge during the 1993–1995 Bering Glacier surge. However, in 1994 it began to develop a series of large subparallel cracks along the southern margin of the Steller Lobe as a result of stresses that were transmitted through the CMMB. By 1998, part of the southwestern terminus of the Steller Lobe underwent a pulse that caused it to advance several hundred meters. By 2000, much of the surface of the Steller Lobe was fractured, and the main trunk of Steller Glacier and the northern lobe of its terminus, which enters (and dams) Berg Lake, began a mini-surge. This surge continued through the late summer of 2001. Since then the entire length of the Steller Glacier terminus has been retreating.

The ice-cored lateral moraine on the northwestern side of the part of the Steller Lobe that ends in an unnamed lake, north of Nichawak Mountain, became broken up in 2000 and rapidly advanced into the lake (Fig. 26). By fall 2001, the mini-surge was over. The southern part of the terminus of the Steller Lobe never showed any evidence of having been involved in this mini-surge.

SUMMARY

This chapter summarizes the twentieth and early twenty-first century surge behavior of Bering Glacier, the largest surging glacier in continental North America. Major surges of the Bering Glacier occurred in ca. 1900, ca. 1920, ca. 1938–1940, 1957–1967, and 1993–1995, or, on average, approximately every 20 a. A smaller magnitude surge occurred in 2008–2009. The focus of this presentation has been on a number of aspects of these last two surge cycles. Each of these surges (1957–1967 and 1993–1995) consisted of a pair of ice-displacement events separated by a period of ice stagnation. In the first of these surges, ice-displacement phases occurred from 1957 to 1960 and from 1965 to 1967, with maximum ice displacement exceeding 10 km. An ice stagnation phase occurred from late 1960 to early 1965. The 1993–1995 surge was observed in real time from onset to completion. Like the 1957–1967 surge it also produced a maximum of ~10 km of advance of the terminus of the Bering Lobe. Terminus-ice displacement also occurred in two discrete phases—between late August 1993 and 17 September 1994, and between early May 1995 and mid-September 1995. A 7-mo. period without detectable ice displacement—from 17 September 1994 to early May 1995—separated the two advances. Following each surge, Bering Glacier underwent a multi-decadal period of thinning and retreat.

REFERENCES CITED

Echelmeyer, K.A., Harrison, W.D., Larsen, C.F., Sapiano, J., Mitchell, J.E., DeMallie, J., Rabus, B., Adalgeirsdóttir, G., and Sombardier, L., 1996, Airborne surface profiling of glaciers: A case-study in Alaska: Journal of Glaciology, v. 42, no. 142, p. 538–547.

Fatland, D.R., and Lingle, C.S., 1998, Analysis of the 1993–95 Bering Glacier (Alaska) surge using differential SAR interferometry: Journal of Glaciology, v. 44, p. 532–546.

Fatland, D.R., and Lingle, C.S., 2002, InSAR observations of the 1993–95 Bering Glacier (Alaska, U.S.A.) surge and a surge hypothesis: Journal of Glaciology, v. 48, p. 439–451, doi: 10.3189/172756502781831296.

Fleisher, P.J., Franz, J.M., and Gardiner, J.A., 1993, Bathymetry and sedimentary environments in proglacial lakes at the eastern Bering Piedmont Glacier of Alaska: Journal of Geological Education, v. 41, p. 267–274.

Gray, J.R., Hart, R.J., and Molnia, B.F., 1994, 1994 changes in physical and sedimentary characteristics of proglacial Vitus Lake resulting from the surge of Bering Glacier, Alaska: Eos (Transactions, American Geophysical Union), 1994 Fall Meeting Supplement, v. 75, no. 44, p. 63.

Josberger, E.G., Meadows, G.A., Shuchman, R.A., and Payne, J.R., 2001, Sediment control of convection in glacier dammed lakes: Eos (Transactions, American Geophysical Union), v. 82, no. 47, Fall Meeting Supplement, Abstract IP52A–0743, p. F556.

Josberger, E.G., Shuchman, R.A., Meadows, G.A., Savage, S., and Payne, J.R., 2006, Hydrography and circulation of ice-marginal lakes at Bering Glacier, Alaska: Arctic, Antarctic, and Alpine Research, v. 38, p. 547–560, doi: 10.1657/1523-0430(2006)38[547:HACOIL]2.0.CO;2.

Meier, M.F., and Post, A., 1969, What are glacier surges?: Canadian Journal of Earth Sciences, v. 6, p. 807–817.

Molnia, B.F., and Post, A., 1995, Holocene history of Bering Glacier, Alaska: A prelude to the 1993–1994 surge: Physical Geography, v. 16, p. 87–117.

Molnia, B.F., and Trabant, D.C., 1992, Ice thickness measurements on Bering Glacier, Alaska and their relation to satellite and airborne SAR image patterns: Eos (Transactions, American Geophysical Union), v. 73, no. 43, Fall Meeting Supplement, p. 181.

Muskett, R.R., Lingle, C.S., Echelmeyer, K.A., and Valentine, V.B., 2000, Mass balance of Bering Glacier–Bagley Icefield during a surge cycle: Eos (Transactions, American Geophysical Union), v. 81, no. 48, Fall Meeting Supplement, p. F403–F404.

Post, A., 1960, The exceptional advances of the Muldrow, Black Rapids, and Susitna Glaciers: Journal of Geophysical Research, v. 65, p. 3703–3712.

Post, A., 1965, Alaskan glaciers—Recent observations in respect to the earthquake-advance theory: Science, v. 148, p. 366–368, doi: 10.1126/science.148.3668.366.

Post, A., 1967, Walsh Glacier surge, 1966 observation: Journal of Glaciology, v. 6, p. 763–765.

Post, A., 1969, Distribution of surging glaciers in western North America: Journal of Glaciology, v. 8, p. 229–240.

Post, A., 1972, Periodic surge origin of folded medial moraines on Bering Piedmont Glacier, Alaska: Journal of Glaciology, v. 11, p. 219–226.

Post, A., 1976, Environmental geology of the central Gulf of Alaska coast, *in* Williams, R.S., Jr., and Carter, W.D., eds., ERTS–1, a new window on our planet: U.S. Geological Survey Professional Paper 929, p. 117–119.

Post, A., and LaChapelle, E.R., 1971, Glacier Ice: Seattle, University of Washington Press, 110 p.

Post, A., and LaChapelle, E.R., 2000, Glacier Ice (revised edition): Seattle, University of Washington Press, in association with the International Glaciology Society, Cambridge, England, 145 p.

Reimnitz, E., 1966, Late Quaternary history and sedimentation of the Copper River Delta and vicinity, Alaska [Ph.D. thesis]: University of California, San Diego, 188 p.

Roush, J.J., 1996, The 1993–1994 surge of Bering Glacier, Alaska, observed with satellite synthetic aperture radar [M.S. thesis]: University of Alaska, Fairbanks, 101 p.

Sauber, J., and Molnia, B.F., 2000, Glacial fluctuations and the rate of deformation in south central coastal Alaska: Eos (Transactions, American Geophysical Union), v. 81, no. 48, Fall Meeting Supplement, p. F325.

Sauber, J., and Molnia, B.F., 2004, Glacier ice mass fluctuations and fault instability in tectonically active southern Alaska: Journal of Global and Planetary Change, v. 42, p. 279–293, doi: 10.1016/j.gloplacha.2003.11.012.

Sauber, J., Plafker, G., Molnia, B.F., and Bryant, M.A., 2000, Crustal deformation associated with glacial fluctuations in the eastern Chugach Mountains, Alaska: Journal of Geophysical Research, v. 105B, p. 8055–8077.

U.S. Geological Survey, 1972, Bering Glacier topographic map, scale 1:250,000, 1 sheet.

U.S. Geological Survey, 1984, Bering Glacier A–7 topographic map, scale 1:63,360, 1 sheet.

Manuscript Accepted by the Society 02 June 2009

Printed in the USA

The Geological Society of America
Special Paper 462
2010

Subarctic hunters to cold warriors: The human history of the Bering Glacier Region

John W. Jangala
Bureau of Land Management, Glennallen Field Office, P.O. Box 147, Glennallen, Alaska 99588-0147, USA

ABSTRACT

The coast of Alaska, generally, has a long history of human occupation and resource extraction. The Gulf of Alaska's coast was one leg of a long coastal migration route that was likely used by the first Native Americans entering North America more than 11,000 a ago, before an ice free corridor through the interior was open. Yet, although there is ample evidence of Native Alaskan occupations in the area at the time of their contact with the first European explorers in the mid to late 1700s, as well as subsequent habitation of the area by European and Euro-American fur traders, fishermen, and miners, there is no documented evidence for the prehistoric human use of this section of coastline. There are several possible reasons for this, including the harshness of the Gulf of Alaska's weather and seas. However, the main cause for this lack of information is the fact that there has been no systematic archaeological work in the area. Similarly, modern subsistence uses in the area are unstudied, and only anecdotal reports of current subsistence practices in the Bering Glacier vicinity are available for researchers.

INTRODUCTION

Human history in the Bering Glacier's vicinity is characterized by the use of the area's natural resources for subsistence and economic development, spanning between early hunters and gatherers, their descendents, European seal fur hunters, American miners and prospectors, and Alaska's Cold War veterans. Some of the ancestors of modern Native Americans likely passed through this region as they moved south along the Pacific Coast. However, no direct evidence of their passage along the Gulf of Alaska's shores has been identified and documented. Evidence for the prehistoric use of the area is limited only to the last couple of hundred years, mainly because little archaeological research has been conducted in the area. By the late 1700s and 1800s numerous European countries, and later the United States, began harvesting a variety of resources from the Alaskan coast, including otters, salmon, oil, and coal, changing a way of life for Native Alaskans. Today some rural Alaskans continue to live a subsistence lifestyle along the Gulf of Alaska, using a variety of resources near the Bering Glacier. However, even these modern uses of the area are poorly known and have not been well documented.

PREHISTORY OF THE GULF OF ALASKA

More than 10,000 a ago North American natives began traveling and living along the Gulf of Alaska's coast, harvesting marine mammals, fish, and plants for food, clothing, and tools. Archaeologists have long hypothesized that during the last Ice Age Alaska was the gateway for humans into North America from earlier populations in Asia. The earliest theories suggested that the first North Americans followed a dry, ice free corridor through interior Alaska to reach the rest of the continent (Antevs, 1935). However, this interior migration theory did not fit well with either the glacial history of the Alaskan and Canadian interior or the oldest archaeological evidence for humans in the Americas.

An ice free corridor through Alaska and Interior Canada, which appeared no earlier than 11,000 a ago, would not have provided enough time for people to have spread across much of

Jangala, J.W., 2010, Subarctic hunters to cold warriors: The human history of the Bering Glacier Region, *in* Shuchman, R.A., and Josberger, E.G., eds., Bering Glacier: Interdisciplinary Studies of Earth's Largest Temperate Surging Glacier: Geological Society of America Special Paper 462, p. 317–324, doi: 10.1130/2010.2462(16). For permission to copy, contact editing@geosociety.org.

North and South America (Dixon, 2001). A number of Clovis occupations in North America have been reliably dated as early as 11,200 yr B.P. (Haynes, 1992). In South America, ancient human evidence like Chile's Monte Verde site has been discovered, which is older than 12,500 yr B.P. and may be as old as 33,000 yr B.P. (Dillehay, 1997). These findings suggest that another route along the West Coast of North America may have brought the earliest ancestors of American Indians onto the continent. Here, the coast may have been habitable and relatively unblocked by glaciers by as early as 14,800 yr B.C. (Dixon, 1999). However, the problem with this hypothesis is its testability, as rising sea levels after the end of the Ice Age would have flooded many of the potential shoreline camps used by any early coastal migrants. Even so, some evidence of very early use of the Alaskan coastline has survived until today in Southeast Alaska at the Ground Hog Bay Site 2 and at the On Your Knees Cave site several hundred miles southeast of the Bering Glacier (Sites 1 and 2, Fig. 1).

These two sites provide the Alaska coast's earliest human occupation dates. The oldest human coastal camp was discovered at On Your Knees Cave on Prince of Wales Island (Site 1, Fig. 1). Here the oldest artifact found, dating from ca. 10,300 yr B.P., was a bone tool fashioned from a thin splinter of a land mammal's skeleton (Dixon, 2001). However, the human remains at the site, dated to ca. 9200 yr B.P., provide evidence that these people relied heavily on marine resources such as ringed seal, sea otter, and marine fish (Dixon, 2001; Kemp et al., 2007). Isotopes of carbon 13 from these human remains as well as those of marine animals are similar, suggesting that the early occupants of Alaska's coastline had a diet composed primarily of marine life. The oldest parts of Ground Hog Bay Site 2, on the tip of the Chilkat Peninsula, are approximately contemporary with the On Your Knees Cave site and date somewhere between 10,180 and 9130 yr B.P. (Ackerman, 1996). This site contains obsidian, a form of volcanic glass, from Sumez Island, in the Alexander Archipelago, as well as from Mount Edziza, near the upper Stikine River, which suggests that watercraft existed at least by that time to transport this tool-making glass across more than a hundred and fifty miles of inside passage. Ground Hog Bay Site 2 is also notable for the length of time it was occupied, about 5000–6000 a, and it was abandoned between 5000 and 4000 yr B.P. (Ackerman et al., 1979). Although archaeological research has been very limited in the Bering Glacier region itself, the distribution of these earliest of coastal sites shows that a strong maritime cultural tradition existed along Alaska's coast by more than 10,000 a ago.

Similarly, there are a large number of archaeological sites along Alaska's southern coast, including Ground Hog Bay Site 2, which span the period from ca. 7000 to ca. 1000 yr B.P. (Davis, 1990). However, a like number of archaeological sites from the same period have not been discovered closer to the Bering Glacier. There are several possible reasons for this. First and foremost is that almost no archaeological investigations have been conducted by federal, state or private land managers in the area, except on small project areas related to land use actions. A second important possibility is that the area was not used as heavily in prehistory as Southeast Alaska, because of the greater difficulty in using early watercraft in unprotected waters. The Gulf of Alaska's coast is well known for its seasonal storms and heavy surf, making its environment much rougher than the calmer Inside Passage of Southeast Alaska. Early maritime cultures using small craft for travel and subsistence may have preferred areas that were better protected from large ocean swells and unpredictable storms such as in the Inside Passage, or possibly the protected bays within Prince William Sound. A third reason is that re-advancing glaciers from 3000 to 5000 a ago, and to a lesser extent during the Little Ice Age from ca. 450 yr B.P. to 100 yr B.P., may have destroyed or overrun any previously deposited archaeological sites and precluded occupation of the Alaskan Gulf coast during that period (Cruikshank, 2001). During much of the Little Ice Age, glaciers extending from Yakutat and Dry Bays also may have provided a significant impediment for natives traveling along the shore in small watercraft. Oral histories handed down among the Eyak and Yakutat-Tlingit may recount the destruction of older villages in nearby Icy Bay almost two hundred years ago (De Laguna, 1958).

POST-CONTACT HISTORY OF THE BERING GLACIER REGION

Only one archaeological excavation has been conducted near the Bering Glacier in Yakutat Bay. The Old Town Site at Yakutat (Site 3, Fig. 1), which is dated to about A.D. 1750 A.D. or later, represents the time period just before or immediately after contact with the first Russians (De Laguna et al., 1964). However, De Laguna (1974) considers the excavation more of an ethnography, meaning the study of a culture, of the Yakutat-Tlingit in the eighteenth century. Essentially the excavation of the Old Town Site at Yakutat showed evidence of how the Yakutat-Tlingit lived in the recent past, while there was still a strong oral tradition and memories of that former village. Unfortunately, the site does not shed light on the more distant prehistory of the region. However, the Eyak and Tlingit who occupied the area during recent history have indicated that several important places existed along the coast from Yakutat Bay west to Controller Bay.

The area around the Bering Glacier was occupied by three cultural groups about the time that Europeans arrived in the region in the mid to late 1700s: the Chugach, the Eyak, and later the Tlingit. Originally, the Chugach, a Pacific Eskimo group, occupied much of the Prince William Sound area to about Cordova (Clark, 1984) and perhaps to the mouth of the Copper River (De Laguna, 1990). Although a village may have existed as far east as Cordova, their only travel east of the Copper River was related to raids or warfare with the neighboring Eyak (Birket-Smith and De Laguna, 1938).

The Eyak, an Athapaskan language speaking group, are the earliest known inhabitants of the Bering Glacier Region, with several reported villages and other named locations (De Laguna, 1990). They originally occupied the coast from about Cordova

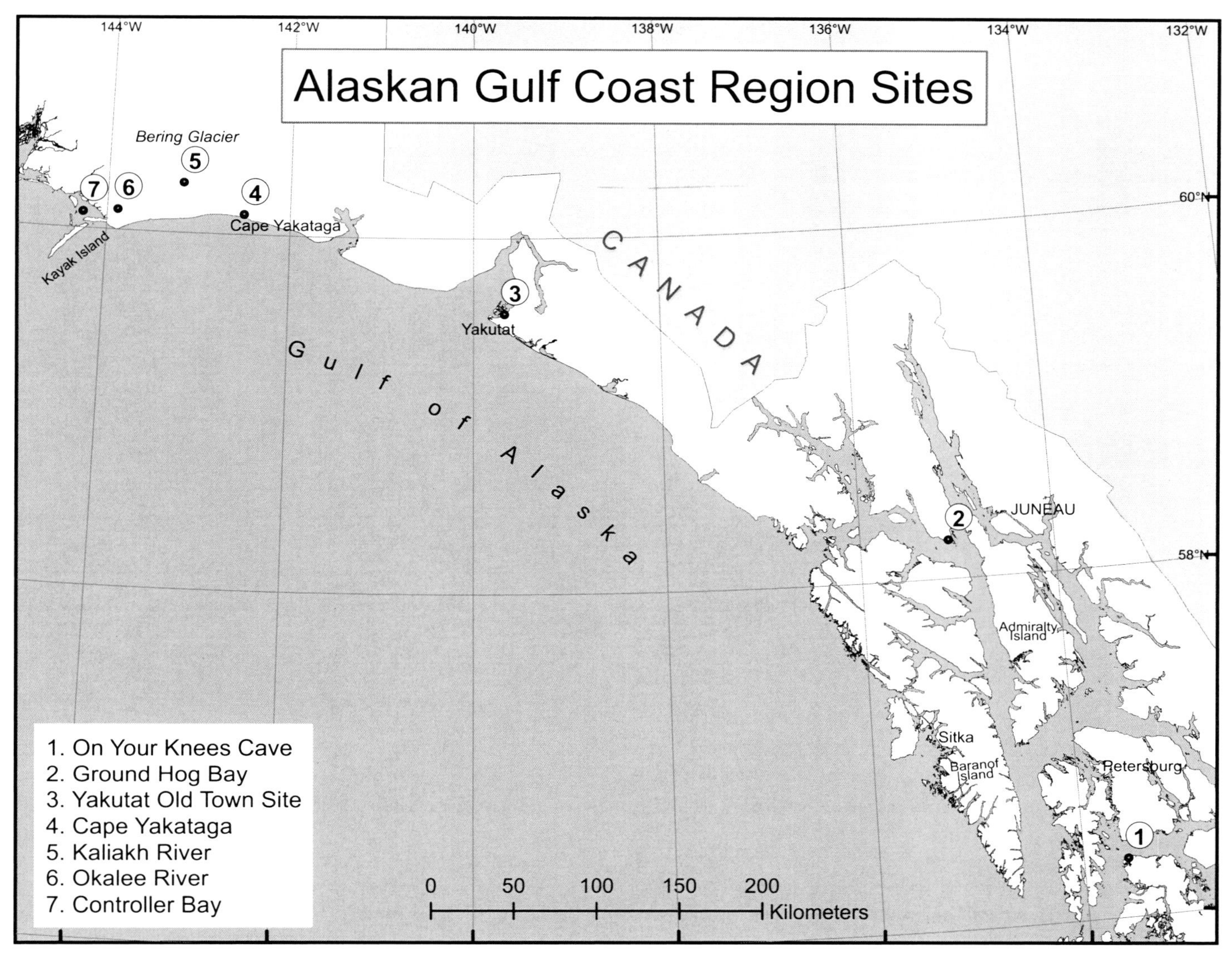

Figure 1. Map of the Gulf Coast of Alaska with known archaeological and historic site locations.

Bay east to just beyond Yakutat and were composed of four regional groups (Birket-Smith and De Laguna, 1938; De Laguna, 1990). The first was the Eyak proper, who moved into the Cordova and Copper River region in the early 1800s, displacing the Chugach. The second was the Eyak of Controller Bay, also known as the Chilcats, who were heavily influenced by the Tlingit by the mid-1800s. Around Yakutat Bay the Eyak were more strongly affected by the Tlingit and were ultimately absorbed by them before the Eyak farther west. Finally, the Eyak of the Bering Glacier Region were also known as the Yakatags, who occupied the coast between Cape Suckling and Cape Yakataga.

Near the Bering Glacier, two of these groups, the Chilcat and the Yakatag Eyak had at least three large villages along the coast (De Laguna, 1972, 1990). Their major villages were at the mouth of the Okalee River (Site 6, Fig. 1), inland on the Kaliakh River (Site 5, Fig. 1) and near the point of Cape Yakataga (Site 4, Fig. 1). Because these people were either absorbed by the Tlingit or killed in subsequent outbreaks of disease, very little is known about their history or cultural landscape. One of these surviving Controller Bay (Site 7, Fig. 1) natives can be seen carving a traditional canoe at Katalla near the beginning of the twentieth century in a photo from about 1913 (Fig. 2). The oral history of the Yakutat Tlingit, who absorbed many of the Eyak from this region, may include some of the Eyak's history in the Bering Glacier area as well as their own.

The Yakutat Tlingit have an extensive oral history of the region, which includes former Eyak and Tlingit villages, local glacial history, and trade routes near the Bering Glacier. Several generations of the Yakutat Galyax Kwaan Clan owned the territory and subsistence resources from Controller Bay to just west

Figure 2. Katalla Native standing next to a canoe he is carving ca. 1913. Photo courtesy of the Alaska State Library.

of Icy Bay (Ramos, 2005; De Laguna, 1972). This clan is said to have been partly descended from Ahtna Athapaskan Natives, who crossed the Bagley Ice Field from Chitina to Icy Bay after a dispute over the property of the deceased Raven Chief. The Galyax Kwaan Clan's first village on the west side of Icy Bay was overrun by the Malaspina Glacier between 1700 and 1791 and destroyed (De Laguna, 1958; Ramos, 2005). Thereafter their main village, Galyax (Site 5, Fig. 1), was relocated a short distance upstream of the Kaliakh River's mouth (De Laguna, 1990; Ramos, 2005). Here they harvested salmon, hooligan, and sea otters. Farther upstream near the headwaters of the Kaliakh, another small camp or village was established, where mink, land otter, fox, and brown and black bears were harvested. They also had a number of smaller or seasonally used villages or camp sites, which were used for a variety of other subsistence activities. A camp near Cape Yakataga, called Yaakw Deiyi or "Ravens Canoe Road," was occupied by the Eyak Yakatags and the Tlingit as a sea otter hunting camp (Site 4, Fig. 1).

An important part of the Tlingit economy was trade with both the Eyak as well as the Ahtna of the interior. Although the Ahtna traded copper and furs down the Copper River with the Eyak of Cordova and Controller Bay, the main trade route for the Tlingit and for the Yakatag Eyak was a cross-glacier route over the Bagley Ice Field. This was also the approximate route used by the ancestors of the Galyax Kwaan Clan to reach Icy Bay. The route was traveled on a regular basis by the Ahtna to trade lumps of native copper. The Duktoth River, near Kaliakh Village, was followed as the final portion of the trading trail to the coast. By the time Europeans arrived on the Gulf of Alaska, many of the Eyak they met were more culturally similar to the Tlingit than the Eyak of the Copper River region.

Europeans discovered the Alaskan Gulf coast in July 1741, when Danish Commander Vitus Bering's ship, the *St. Peter,* spotted a land with high mountains and snowy peaks after being at sea for over a month and a half (Lethcoe and Lethcoe, 2001). His expedition had been undertaken at the request of Peter the Great of Russia, who was interested in any lands between Siberia and the North American continent. Bering sent a landing party ashore at Kayak Island south of Controller Bay on 19 July 1741, where the noted naturalist Georg Wilhelm Steller, among others, observed that this newly discovered land was neither the Asian continent nor the mythical Juan de Gama land they had anticipated. Catastrophically, the *St. Peter* was wrecked on their return home. Bering and 19 of his men died before the remaining crew members constructed a small boat from the ship's wreckage and sailed back to Russia. His maps and other information accompanied the survivors back to Russia and had an influence on other early European explorers.

Bering's maps were instrumental to both the British and Spanish navies exploring Alaska's coast, discovering Prince

William Sound, and searching for the Northwest Passage as well as a promising trade route between Europe and Asia (Lethcoe and Lethcoe, 2001). Captain James Cook left England in July 1776 with two ships, the *Discovery* and the *Resolution,* taking almost two years to sail his ships into Prince William Sound, where he anchored behind his newly named *Cape Hitchenbrook*. Cook named this island and cape after the father of his patron, Viscount Hitchenbrook. The sheltered water behind this island was named *Sandwich Sound* after the Earl of Sandwich, his patron. However, while the name of Hitchenbrook Island has survived, Sandwich Sound was renamed by Cook's map editors after Prince William the Fourth. A year later, in 1779, a Spanish expedition, led by Lieutenant Inacio Aretega, reached Prince William Sound, where they visited with Natives who displayed artifacts left behind by Cook's expedition.

The legacy of Bering's expedition also fueled the first Alaskan stampede before Britain's two vessels even returned home from their explorations in 1780 (Lethcoe and Lethcoe, 2001). Survivors of the Bering expedition brought home sea otter pelts, which at that time had tremendous value in China. The result was that Russians returned to the Aleutian Islands en masse and began harvesting sea otters for their pelts to the point at which serious population declines began within a few years. Russian fur hunters expanded farther north and east until they reached Prince William Sound in 1783, when in August of that year Potap Zaikov of the Panov Company landed on Kayak Island. A small Russian trading post was established in 1785 at Nuchek, where Russians, Aleuts, and local Chugach Natives began harvesting pelts in earnest. They were joined by English fur traders in 1786.

The Russian fur companies dominated the fur trade in Alaska. They established a number of large, permanent trading posts along the coast, including a post opened on Hitchenbrook Island at Constantine Harbor in 1793 (Lethcoe and Lethcoe, 2001). The Russians next established a fort and agricultural colony at Yakutat in 1796 (De Laguna, 1990). Until this time the Russians had limited contact with the Eyak, nor had they explored the Gulf of Alaska coast. Later the same year Tarkhanov, a representative of the Shelikov Company of hunters and traders, left Yakutat Bay on foot seeking a route to Hudson Bay up the Copper River (Lethcoe and Lethcoe, 2001). He was the first European to cross the Bering Glacier's moraines and the oil seeps at Katalla, before ascending the Copper River.

The Russians' trading practices from 1783 through the 1860s had a devastating effect on the area's Natives (Lethcoe and Lethcoe, 2001). Trading strategies were aimed at making them more dependent on the Russian trading posts, prohibiting trade with independent American or English traders. The Russian traders were in the habit of taking hostages at each trading post and killing Natives to ensure compliance with their wishes. They encouraged hunting seals at the expense of subsistence activities, creating even greater dependence on the trading posts for purchased foods and goods. The result was that Prince William Sound's Native populations plummeted. This trend spurred a violent reaction in the area from the local tribes.

The Eyak from Cape Suckling and Tlingitized Yakutat Eyak attacked Alexsandr Baranov's parties in Prince William Sound about 1792 (De Laguna, 1990). The Russians responded by taking hostages from several villages, including those at the Kaliakh River and at Yakutat Bay in 1794 and 1796. The Eyak at Cape Suckling massacred a hunting party returning from Sitka in 1799 in reprisal for Russian brutality and poaching by the Russian's Aleut seal hunters. The natives of Yakutat then helped plan the destruction of the fort at Sitka in 1802. The Eyak finally destroyed the Russian fort and colony at Yakutat in 1805, permanently ending the Russian presence in their territory.

A series of subsequent events nearly wiped out this area's coastal natives. Shortly after the destruction of the Russian Fort at Yakutat, the Tlingit from Dry Bay and Southeast Alaska conquered the Yakutat area and absorbed the remaining Yakutat Eyak (De Laguna, 1990). In the years 1837 and 1838 smallpox ravaged the coast, killing about half of the remaining Tlingit and Eyak. After that, except for occasional interactions with the Russians, the natives of the Gulf Coast remained relatively isolated until the 1880s.

The United States purchased Alaska from Russia in 1867. Unfortunately, Congress refused to spend taxpayer dollars on the development of "Seward's Folly," and few Americans, official or otherwise, visited or explored the area (Lethcoe and Lethcoe, 2001), leaving the region to Native hunters and a small handful of fur traders. The Department of the Army eventually sent men to explore the coast, find a Copper River route into the interior, and to assess Alaska's coast resources. The first was Lieutenant W.R. Abercrombie's expedition of 1884, which landed at Nuchek on Hitchenbrook Island. Abercrombie heard about a route over the Valdez Glacier and proceeded to Valdez harbor to search for such a route. A year later, Lieutenant Henry Allen, who traveled with Abercrombie the previous year, returned to Nuchek with the goal of leading an expedition up the Copper River itself (Allen, 1887). He crossed to the mouth of the Copper River and enlisted aid from local Eyak, who accompanied him partway up the river. Another government-sponsored expedition in 1887 by Samuel Applegate of the U.S. Bureau of Fisheries surveyed the economic potential of the sound's fisheries (Lethcoe and Lethcoe, 2001). This short period of American exploration was followed by a small fleet of fishermen after Applegate's fisheries report reached markets in San Francisco and Seattle.

Between 1887 and 1889, two fish canneries were established in Prince William Sound, leading to additional shifts in local populations and hardships for the Eyak (Lethcoe and Lethcoe, 2001). These canneries were operated in Controller Bay near Eyak Village and had three major effects on Native populations. First, the opening of two canneries near Eyak village drew native populations from here as well as from nearby Nuchek, where declines in sea otter hunting had removed an important part of their economy. This shifted the region's Natives toward a local wage-based economy centered on the canneries and away from their traditional subsistence economies, increasing their dependence on store-bought trade goods. Second, uncontrolled exploitation of the Copper River Delta's salmon threatened the main subsistence food source of the Eyak. Finally, disease spread among the Natives, killing a large part of the remaining population. The survivors abandoned the canneries and moved to the village on Eyak Lake by 1892. The changes in population from the construction of the canneries were a mere trickle compared to the flood of humanity that poured over the coast when oil, coal, and gold were discovered in Alaska and the Canadian Klondike.

During the late 1890s, local promoters in Seattle had created a rush for an "All American" route to the Klondike gold fields from Prince William Sound's harbors, luring thousands of prospectors northward for a chance at quick riches. However, these gold prospectors were only a part of the wave of men searching for resources previously only known to Alaska's Natives. Oil was discovered at Katalla in 1896, and by 1904 the Controller Bay area was covered with >1100 coal and oil placer claims (Rakestraw, 1994). Katalla skyrocketed in population, with between 5000 and 10,000 people living in this small coastal boomtown by 1907 (Shaw, 2001). One of Katalla's first coastal oil camps and a test well can be seen in Figure 3. During this time many local Natives sought wage labor alongside Americans from the south in the oil fields and camps. Katalla's proximity to valuable coal fields along the Bering River and its competition with nearby Cordova to build a railway line to the Kennicott copper deposits drew ambitious people and financial backing from across the country. However, a large storm in November of that same year washed out the newly constructed breakwater and dock at Katalla, leaving Cordova with its natural harbor as the final candidate for the Kennicott and Northwestern rail line. The final blow for this area's early development was the federal government closing Alaska to oil and gas entry in 1910, except for a single oil claim at Katalla. The Bering River coalfields were reopened in 1914, and a small railroad, the Alaska Anthracite Railroad, was constructed to the coalfields in 1918 (Antonson, 1989). The coalfields and railroad operated only for three years before a lack of profits drove the company out of business. The Katalla oil fields remained closed until the Oil and Gas Lease Act of 1920 reopened federal lands to oil and gas entry.

Development at nearby Cape Yakataga didn't begin until just before the outbreak of World War II. During the late 1930s and early 1940s the United States was becoming increasingly concerned about an invasion of Alaska by either Japan or the Union of Soviet Socialist Republics (U.S.S.R.), which resulted in a program to construct defenses and support airfields along the Alaskan coast (Cloe, 1984). A large airfield was constructed at Yakutat in October of 1941, and a smaller, unimproved air strip was constructed at Cape Yakataga in September of the same year (Jangala, 2001). The Cape Yakataga airfield was augmented with a weather and low-frequency radio navigation facility in 1942, whose function throughout the war was to provide weather and navigation information to military and civilian aviators along the rough and unpredictable coast. After the war the Civilian Aeronautics Agency expanded the Cape Yakataga station for civilian air traffic as a manned Flight Service Station in 1945.

However, tensions between the United States and the U.S.S.R. continued to escalate after the war. The development of atomic weapons, plus the means of delivering them, required increased vigilance from Alaskan military forces. The U.S. Air Force needed a way to link its large radars, also known as the Distant Early Warning (DEW) line system or the Ballistic Missile Early Warning System (BMEWS), farther north with the North American Air Defense Command (NORAD) in Colorado (Lewis-Reynolds, 1988). The result was a system of large billboard antennae that bounced ultrahigh frequency radio signals off the high atmosphere, which was called the White Alice Communication System (WACS). One of these facilities was completed at Cape Yakataga in 1960 and communicated with similar antennae at Ocean Cape and Boswell Bay in Alaska. A move toward satellite communications resulted in this facility being decommissioned in 1976, less than a decade after the Federal Aviation Administration (FAA) closed the nearby Flight Service Station in 1967. Since this time the old Air Force and FAA facilities have been bought and sold by RCA Alascom, and now only a small handful of beach miners use the area.

Currently the area is used by a variety of non-Natives and Natives. There are two small beach mining operations near

Figure 3. Oil prospecting camp at Katalla. Photo courtesy of the Alaska and Polar Regions Collections, Elmer E. Rasmuson Library.

Cape Yakataga, where gold is often found near the mouth of the Yakataga River. Commercial fishermen from Cordova and Yakutat fish the mouths of the Kiklukh, Seal, and Kaliakh Rivers (Mark King, 2007, personal commun.). Sport hunters from area lodges hunt the Robinson Mountains for mountain goats, moose, and brown bear. The Yakutat-Tlingit people still harvest bird eggs and noncommercial salmon to the east near Icy Bay (Ramos, 2005). People from Yakutat also subsistence-fish the Kaliakh River for silver salmon after the commercial fishing season closes each fall (Mark King, 2007, personal commun.). Eyak Village members from Cordova use the area occasionally for both subsistence and commercial use, fishing for salmon or hunting for moose and seals. Frequently subsistence and commercial fishing take place concurrently as fisherman save a few fish from their commercial catch for their personal use.

SUMMARY AND DISCUSSION

A late Pleistocene migration route for early Native Americans through an ice-free North American interior corridor prior to 11,000 a ago appears unlikely, given recent information about the timing and extent of interior glaciers (Dixon, 2001; Jackson and Wilson, 2004) as well as the number of reliably dated older sites spread across the Americas (Haynes, 1992; Dillehay, 1997). Therefore, a coastal migration route along the Gulf of Alaska, which existed sometime between ~16,800 and 11,000 a ago (Dixon, 1999, 2001), is the most likely passage for early peoples entering the continent. Knowledge about the prehistory of the region, however, has numerous gaps caused by a lack of systematic archaeological work. Finding prehistoric sites older than a few hundred years is difficult and requires the incorporation of paleoenvironmental and geomorphological information that is specific to the area being investigated (Goldberg and Macphail, 2006). These more ancient archaeological sites are more likely to be found on older landforms that attracted humans and are stable enough to preserve older archaeological sites. In the Bering Glacier Region, older landforms such as elevated wave-washed terraces or beach ridges that are intersected by abandoned river channels may yield prehistoric sites of maritime-adapted peoples older than those found near modern channels of the Kaliakh or Duktoth Rivers.

The Gulf of Alaska's coast also remains relatively untouched by archaeological research but has a number of more recent ethnographically reported village sites and travel routes. Only one of the reported Eyak or Tlingit villages in the area has received any archaeological investigation so far. Additionally, the high alpine areas surrounding the Bering Glacier and Bagley Ice Field offer some of the highest probability areas for finding cultural resources. Archaeological work on semipermanent alpine ice patches in the Canadian Yukon (Farnell et al., 2004), and from the closer Wrangell and Amphitheater Mountains (Dixon et al., 2005; Vanderhoek et al., 2007), demonstrates the ability of these areas to preserve organic artifacts composed of wood, bone, and antler for several thousand years. Any long-term ice patches near the mountain passes used as part of the native-copper trade route over the Bagley Ice Field have the potential to preserve related organic artifacts, such as the wooden staves reportedly seen by miners traveling through the area in 1905 or 1906 (Moffit, 1918).

The period after the arrival of Europeans in the mid to late 1700s is much better documented. During this period the region's Native American populations changed dramatically from a subsistence lifestyle to one that increasingly relied upon wages and commercial goods. At the same time the area changed demographically to include larger numbers of European and Euro-American fur traders, fishermen, and miners. Unfortunately, among modern populations of rural coastal Alaska, significant information gaps exist about their use of the Bering Glacier Region. It is clear, however, that some amount of subsistence fishing and hunting continues in the area.

In summary, the Bering Glacier vicinity and the Gulf of Alaska have a long but poorly documented history of human use. There is very little direct evidence for the prehistoric occupation of the area prior to the arrival of Europeans. However, there is good evidence for a late Pleistocene human migration route as well as ancient maritime Native American cultures along the Alaskan coast. Therefore any potentially ancient landforms around the Bering Glacier could retain archaeological remains that could fill in the current sketchy picture of the prehistoric use of the Gulf of Alaska. Additionally, because even information about the modern coastal populations' subsistence use of the area's resources is anecdotal and poorly known, anthropological research on modern subsistence practices in this region is necessary for future planning and wise resource use.

REFERENCES CITED

Ackerman, R.E., 1996, Early maritime cultural complexes of the northern Northwest Coast, *in* Carlson, R., and Bona, L.D., eds., Early Human Occupation in British Columbia: Vancouver, University of British Columbia Press, p. 123–132.

Ackerman, R.E., Hamilton, T.D., and Struckenrath, R., 1979, Early cultural complexes of the northern Northwest Coast: Canadian Journal of Archaeology, v. 3, p. 195–209.

Allen, H., 1887, Report of an Expedition to the Copper, Tanana, and Koyukuk Rivers, in the Territory of Alaska, in the Year 1885: Washington, D.C., U.S. Government Printing Office, 172 p.

Antevs, E., 1935, The spread of aboriginal man to North America: Geographical Review, v. 25, p. 302–309, doi: 10.2307/209605.

Antonson, J., 1989, Bering River Train: Anchorage, U.S. Forest Service, 21 p.

Birket-Smith, K., and De Laguna, F., 1938, The Eyak Indians of the Copper River Delta, Alaska: Copenhagen, Levin and Munksgaard, 567 p.

Clark, D.W., 1984, Pacific Eskimo: Historical ethnography, *in* Damas, D., ed., Handbook of North American Indians: Arctic: Washington, D.C., Smithsonian Institution, v. 5, p. 185–197.

Cloe, J.H., 1984, Top Cover for America: The Air Force in Alaska 1920–1983: Missoula, Montana, Anchorage Chapter, Air Force Association and Pictorial Histories Publishing Co., 258 p.

Cruikshank, J., 2001, Glaciers and climate change: Perspectives from oral tradition: Arctic, v. 54, p. 377–393.

Davis, S.D., 1990, Prehistory of southeastern Alaska, *in* Suttle, W., ed., Handbook of North American Indians: Northwest Coast: Washington, D.C., Smithsonian Institution, v. 7, p. 197–202.

De Laguna, F., 1958, Geological confirmation of native traditions, Yakutat, Alaska: American Antiquity, v. 23, p. 434, doi: 10.2307/276497.

De Laguna, F., 1972, Under Mount Saint Elias, Smithsonian Contributions to Anthropology: Washington, D.C., Smithsonian Institution, v. 7, 1394 p.

De Laguna, F., 1990, Eyak, *in* Suttles, W., ed., Handbook of North American Indians: Northwest Coast: Washington, D.C., Smithsonian Institution, v. 7, p. 189–196.

De Laguna, F., Riddell, F.A., McGeein, D.F., Lane, K.S., Freed, J.A., and Osborne, C., 1964, Archaeology of the Yakutat Bay Area, Alaska: Washington, D.C., Bureau of American Ethnology Bulletin, v. 192, 244 p.

Dillehay, T.D., 1997, Monte Verde, a Late Pleistocene Settlement in Chile: The Archaeological Context and Interpretation: Washington, D.C., Smithsonian Institution Press, v. 2, 306 p.

Dixon, E.J., 1999, Bones, Boats, and Bison: Archeology and the First Colonization of Western North America: Albuquerque, University of New Mexico Press, 322 p.

Dixon, E.J., 2001, Human colonization of the Americas: Timing, technology and process: Quaternary Science Reviews, v. 20, p. 277–299, doi: 10.1016/S0277-3791(00)00116-5.

Dixon, E.J., Manley, M.F., and Lee, C.M., 2005, The emerging archaeology of glaciers and ice patches: Examples from Alaska's Wrangell–St. Elias National Park and Preserve: American Antiquity, v. 70, p. 129–143.

Farnell, R., Hare, P.G., Blake, E., Bowyer, V., Schweger, C., Greer, S., and Gotthardt, R., 2004, Multidisciplinary investigations of alpine ice patches in Southwest Yukon, Canada: Paleoenvironmental and paleobiological investigations: Arctic, v. 57, p. 247–259.

Goldberg, P., and Macphail, R., 2006, Practical and Theoretical Geoarchaeology: Cambridge, Massachusetts, Blackwell Science, 454 p.

Haynes, C.V., 1992, Contributions of radiocarbon dating to the geochronology of the peopling of the New World, *in* Taylor, R.E., Long, A., and Kra, R.S., eds., Radiocarbon after Four Decades: New York, Springer-Verlag, p. 355–374.

Jangala, J.W., 2001, Yakataga River Bridge (XBG-119): National Register evaluation: Glennallen, Alaska, U.S. Bureau of Land Management.

Kemp, B.M., Malhi, R.S., McDonough, J., Bolnick, D.A., Eshleman, J.A., Rickards, O., Martinez-Labarga, C., Johnson, J.R., Lorenz, J.G., Dixon, E.J., Fifield, T.E., Heaton, T.H., Worl, R., and Smith, D.G., 2007, Genetic analysis of early Holocene skeletal remains from Alaska and its implications for the settlement of the Americas: American Journal of Physical Anthropology, v. 132, 17 p.

Lethcoe, J., and Lethcoe, N., 2001, A History of Prince William Sound: Valdez, Alaska, Prince William Sound Books, 275 p.

Lewis-Reynolds, G., 1988, Historical overview and inventory: White Alice Communication System: Anchorage, U.S. Air Force, Alaska Command, p. 19–65.

Moffit, F.H., 1918, The Upper Chitina Valley, Alaska: U.S. Geological Survey Bulletin 675, 675 p.

Rakestraw, L., 1994, A History of the United States Forest Service in Alaska: Anchorage, U.S. Forest Service, 221 p.

Ramos, J., 2005, Historical and subsistence use of the Yakutat Galyax (Kaliakh) Kwaan Clan, Letter to the Administrative Record of the East Alaska Resource Management Plan, H.C. 1926–7: Glennallen, Alaska, U.S. Bureau of Land Management.

Shaw, R., 2001, Katalla Oil Field Claim 1: A historic field survey 100 years after drilling the first well: Anchorage, Cassandra Energy Corporation, p. 1–28.

Vanderhoek, R., Tedor, R., and McMahan, D., 2007, Cultural materials recovered from ice patches in the Denali Highway region, Central Alaska, 2003–2005: Alaska Journal of Anthropology, v. 5, p. 185–200.

Manuscript Accepted by the Society 02 June 2009

Printed in the USA

New Investigator Activities at Bering Glacier

An important aspect of the field investigations at the Bering Glacier has involved the participation of high school, undergraduate, and graduate students. The level of student participation has varied from simply assisting a more senior researcher collect field data to generation of original ideas that have resulted in a PhD dissertation or Master thesis. Other activities reported in this section of the monograph include faculty-supervised, student-run scientific investigations that could be accomplished within a short time duration set of field observations. Also reported in this section is the participation of minority students funded by the U.S. Department of Interior. Each of these reports has been reviewed by the supervising faculty or institution sponsoring the student activities.

New Investigator 1: **Chapter 17: Acoustic and seismic observations of calving events at Bering Glacier, Alaska**
Joshua P. Richardson, Katelyn A. FitzGerald, Gregory P. Waite, and Wayne D. Pennington

New Investigator 2: **Chapter 18: Ice-generated seismic events observed at the Bering Glacier**
Katelyn A. FitzGerald, Joshua P. Richardson, and Wayne D. Pennington

New Investigator 3: **Chapter 19: Mapping the fresh-water–salt-water interface in the terminal moraine of the Bering Glacier**
Austin B. Andrus, Kevin A. Endsley, Silvia Espino, and John S. Gierke

New Investigator 4: **Chapter 20: Satellite-derived turbidity monitoring in the ice marginal lakes at Bering Glacier**
Liza K. Jenkins

New Investigator 5: **Chapter 21: Investigating biological productivity in Vitus Lake, Bering Glacier, Alaska**
Nancy A. Auer

New Investigator 6: **Chapter 22: Bryophytes and bryophyte ecology of the Bering Glacier Region**
Nancy G. Slack and Diana G. Horton

New Investigator 7: **Chapter 23: Birds of the Bering Glacier Region: A preliminary survey of resident and migratory birds**
Carol Griswold

New Investigator 8: **Chapter 24: Bering Glacier, climate change, and the Southern University research experience**
Michael Stubblefield, Revathi Hines, and Lionel D. Lyles

The Geological Society of America
Special Paper 462
2010

Acoustic and seismic observations of calving events at Bering Glacier, Alaska

Joshua P. Richardson
Katelyn A. FitzGerald
Gregory P. Waite
Wayne D. Pennington
Department of Geological and Mining Engineering and Sciences, Michigan Technological University, 630 Dow Environmental Sciences and Engineering Building, 1400 Townsend Drive, Houghton, Michigan 49931, USA

ABSTRACT

The Bering Glacier, located in Southeastern Alaska, extends from the Bagley Ice Field to Vitus Lake, a tidally influenced fresh-water lake draining into the Gulf of Alaska. Calving events from the grounded and floating portions of the terminus are shown to produce both acoustic and seismic signals measurable with infrasound detectors and geophones, respectively. Based on the complex, emergent seismic signals recorded from calving events during a short-term experiment conducted at the Bering Glacier in the summer of 2007, we sought another technique for accurately locating these events. In August 2008 we deployed three small-aperture arrays of infrasound detectors to test their utility at determining the locations of subaerial calving events. Despite the complex nature of both the seismic and acoustic signals generated by calving, through the determination of azimuth from three small arrays of infrasound detectors, we were able to accurately locate both terminus calving and iceberg breakup events without relying on first motion picks for hypocenter locations.

INTRODUCTION

The Bering Glacier, the largest glacier in North America by surface area at 5200 km^2, is subject to mass loss from melting and calving in response to atmospheric and stress conditions. Calving is known to occur on glaciers and ice shelves throughout the world, and has been studied recently using passive seismology both on a large scale, by looking at global earthquake catalogues for undetected Greenland glacier calving events (Wilson, 2008), and on a smaller scale, at the Columbia Glacier in Southeastern Alaska (O'Neel et al., 2007). The focus of this study is to investigate the utility of infrasound detectors to record and locate calving events and make observations in conjunction with the seismic record. Using techniques frequently employed in volcanic seismology (e.g., Johnson et. al, 2004), the acoustic records of infrasound detectors in small triangular arrays were used to produce back-azimuths to accurately locate calving events along the terminus.

METHODS

We used both seismic and acoustic detectors, as well as direct visual and audio observations made near the terminus of the glacier, to study the range of signals generated by iceberg and glacier calving events. Several seismic stations operated (without infrasound) for a two-month period, and the infrasound detectors and seismic stations operated simultaneously for a five-day period. We made visual observations of calving during a six-hour time period on August 2, 2008, during the operation of the seismic stations, confirming the nature of some of the events observed in the seismic records.

Richardson, J.P., FitzGerald, K.A., Waite, G.P., and Pennington, W.D., 2010, Acoustic and seismic observations of calving events at Bering Glacier, Alaska, *in* Shuchman, R.A., and Josberger, E.G., eds., Bering Glacier: Interdisciplinary Studies of Earth's Largest Temperate Surging Glacier: Geological Society of America Special Paper 462, p. 327–335, doi: 10.1130/2010.2462(17). For permission to copy, contact editing@geosociety.org.

Seismic Array

Seven Mark Products L22 3-component geophones (2 Hz) equipped with RefTek 130 data acquisition systems and global positioning system (GPS) time clocks were deployed in early June 2008. The seismograph stations were powered with solar panels and 12 V sealed lead-acid batteries. These stations were deployed on small islands within Vitus Lake and on its margins, with site selection based on proximity to the ice edge and triangulation potential (Fig. 1). No stations were deployed on the ice directly above the terminus owing to the risk of unpredictable large calving events displacing stations into the lake. These stations recorded at a sampling rate of 125 Hz. In accordance with daylight and the weather, the sites were deployed over several days. Additional seismic stations were deployed up-glacier to study basal and local tectonic seismic energy as part of a larger experiment (FitzGerald et al., this volume) (Fig. 1), but these stations only rarely recorded signals produced along the calving front. Excluding time lost to equipment failure and animal disturbance, these sites were operational through early August, at which time infrasound arrays were added at three of the sites.

Infrasound Detectors

In early August, the existing seismic stations were serviced, and stations 03, 06, and 07 were outfitted with infrasound arrays consisting of three Honeywell differential pressure sensor microphones (0.05–20 Hz) designed to study acoustic signals produced by volcanic eruptions. Each array was designed to be an azimuthally independent pointer, with channel C coincident with the existing geophone, channel N, 30.5 m to magnetic north, and channel E, 30.5 m to magnetic N 60° E (Fig. 2). The infrasound detector signals were also recorded with RT130 digitizers at a sampling rate of 125 Hz.

Direct Visual and Audio Observations

To verify the assumptions made about the signals in the acoustic and seismic records, direct visual and audio range observations were made for more than six hours on the island at which station 03 was located. Calving parameters included event time, approximate location, size, and time duration noted by two observers on the island.

OBSERVATIONS

Seismic Record and Direct Observations

The characteristic seismic calving signal as recorded during the deployment exhibited an emergent onset, often several seconds long, containing frequencies of 1–30 Hz, decaying into a narrow-band <5 Hz signal. Coincident with part of this narrow-band signal, or shortly after its decay, depending on the proximity of the event to the station, were typically several higher frequency

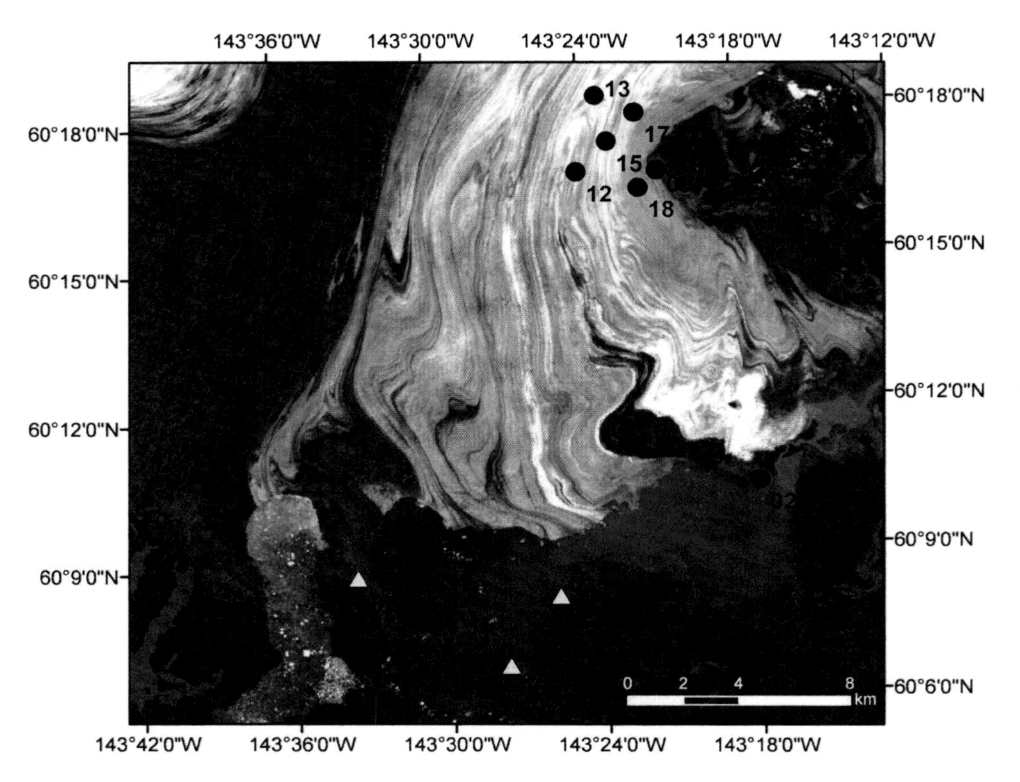

Figure 1. Map showing the locations of seismic stations on and around the Bering Glacier. Stations 03, 06, and 07, indicated by yellow triangles, are stations that were also implemented with infrasound arrays. Michigan Tech Research Institute (MTRI) 2007 Landsat image.

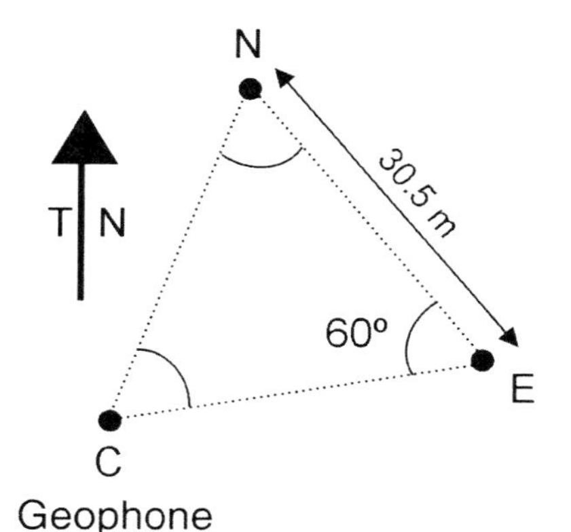

Figure 2. Diagram of the infrasound detector array configuration, with the central detector (C) at the position of the existing geophone at that station.

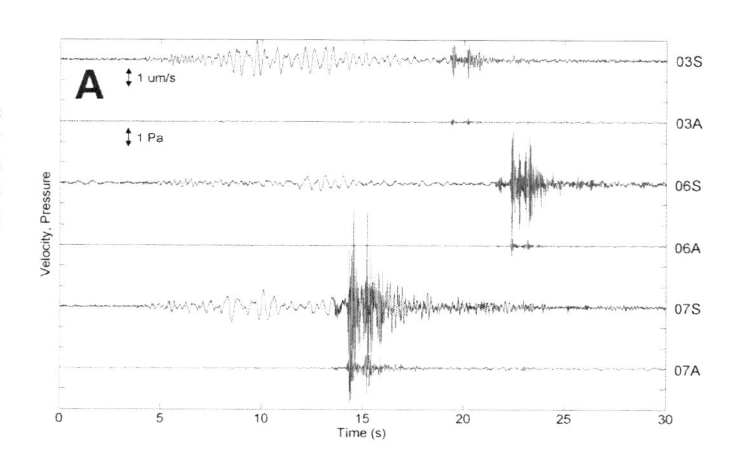

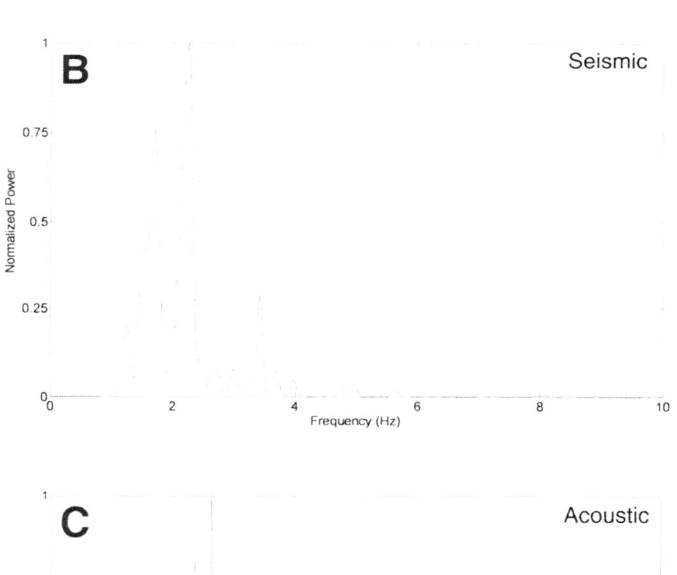

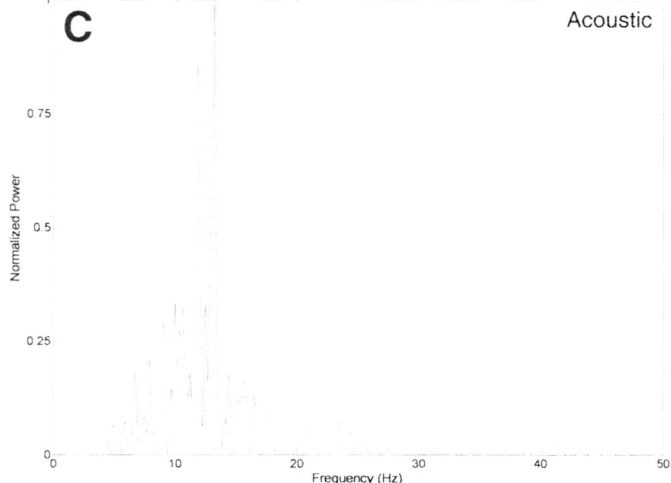

Figure 3. Seismic and acoustic (S and A) traces recorded at each array station from calving event (70) observed on 7 August 2008, 08:06 Coordinated Universal Time (UTC) (A). All waveforms were band-pass filtered from 1 Hz to 60 Hz. The normalized spectra of seismic (B) and acoustic (C) signals windowed respectively from 5–13 s and 14–18 s, as recorded at station 07.

arrivals (Fig. 3). Some seismic signals lacked the high-frequency arrivals, and others showed only those arrivals.

Prior to the installation of the infrasound detectors, visual observations were made over a six-hour period in an effort to identify the sources of the different arrivals observed in the seismic traces. The observed events ranged from single blocks breaking free from the glacier face, producing a short-lived acoustic signal, to a five-minute-long continuous calving sequence producing many acoustic ruptures and a large amplitude, long lived seismic signal. Most calving events were too small to be accurately located visually, although the sequences of several long-duration events produced large, freshly exposed blue-ice faces hundreds of meters long. A total of 70 discernible calving events were recorded during the single observation period on the afternoon of 2 August, consisting of events either seen, heard, or both. Through comparison with the seismic data, it was found that 27 of these events did not produce strong enough signals to exceed the average noise level and were not visible in the records at the coincident seismic station. Additionally, 12 observations were found in the seismic data, but either they were missed by the observers, or they could be attributed to undetected submarine calving or inaudible ruptures.

The direct observations made from the island were not a complete match to the seismic data, but they did confirm the higher frequency arrivals from the compression waves traveling through the air on the seismograms to within several seconds of the acoustic observations. In the seismograms, these acoustic arrivals were often preceded by emergent seismic arrivals from paths through the sedimentary glacial till and through the water, but the path through the air was the only audibly detectable signal.

The correlation of audible events with the high-frequency arrivals on the seismic trace provides convincing evidence that these arrivals are ground-coupled airwaves. Recent studies conducted on nearby Columbia Glacier, Alaska, suggested a crack tip enlargement model for the 1–3 Hz monochromatic waveforms frequently observed prior to or continuing on through the airwave arrival (O'Neel et al., 2007). If this model applies to the events we recorded on the Bering Glacier, the higher frequency airwave suggests that the initial rupture may also contain higher frequencies that were attenuated in the seismic record or were overwhelmed by the dominant 1–3 Hz signal (Richardson et al., 2010). The seismic traces that lack an observable airwave may be attributed to submarine calving; the events with only a ground-coupled airwave may correspond to smaller, entirely subaerial events with poor initial ground coupling. Through comparison of the visual observations with the seismic record, each high-frequency arrival seems to correspond to a unique rupture of the ice front in the calving sequence.

Infrasound Detector Record

The acoustic signal recorded by the infrasound detectors was dominated by atmospheric noise below 2 Hz but was found to be extremely effective at distinguishing each of the higher frequency acoustic rupture events that occur during significant subaerial calving sequences. Although the seismic arrivals were always emergent, the infrasound detectors recorded high-amplitude, impulsive arrivals nearly coincident with the ground-coupled airwave arrivals in the seismic records (Fig. 3A). The presence of a broader range of frequencies (strong signals >4 Hz) in the acoustic arrivals also suggests that these frequencies are indeed present in the initial rupture but are attenuated or overwhelmed by ice filtering or amplifying effects in the seismic record (O'Neel and Pfeffer, 2007). The waveforms in Figure 4A also suggest that each subsequent rupture as seen in the airwave arrivals as part of the calving event occurs at approximately the same location, as the plotted semblance-azimuth of 30 sample windows during the event produced a constant propagation azimuth direction (Fig. 4B). This supports the idea that one calving rupture triggers subsequent ruptures by destabilizing a nearly detached block of

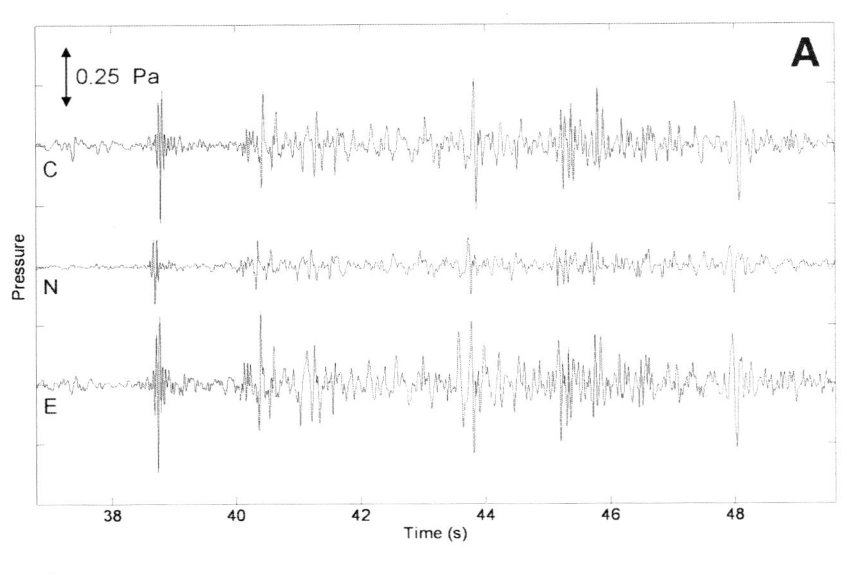

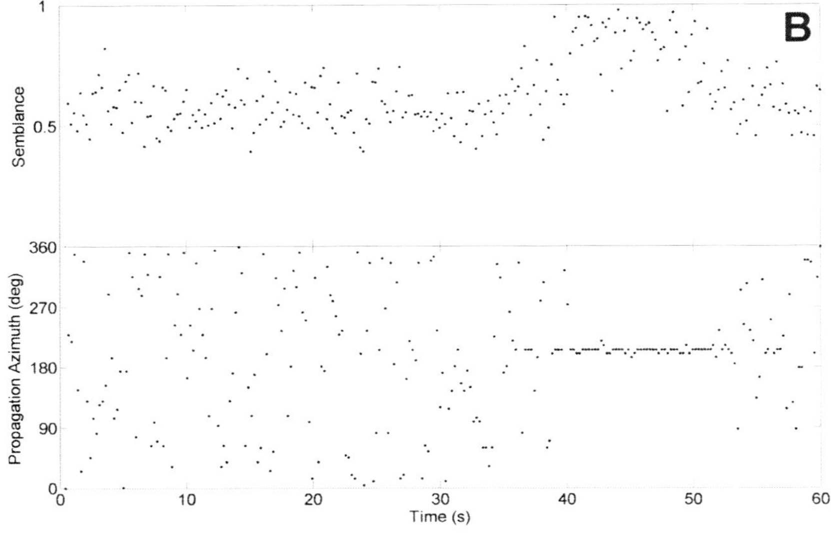

Figure 4. Acoustic arrivals at all three infrasound elements of the station 07 array from event (118) that occurred at 00:51 UTC on 9 August 2008 (A), and a plot of semblance quality and azimuth leading up to the same event (B).

ice (O'Neel and Pfeffer, 2007), although not all events produced a series of arrivals attributed to multiple, triggered ruptures. Individual event sequences ranged from one discrete acoustic arrival lasting less than a second to five discrete acoustic arrivals.

Data Processing

In order to test the utility of infrasound detectors to readily locate the hypocenters of calving events, a total of 126 events were selected over the five-day infrasound detector recording period, based qualitatively on the strength of the airwave arrivals and the presence of arrivals at more than one array. Using time windows (0.5–5 s) surrounding each acoustic arrival, or multiple arrivals if the event consisted of several rupturing events at the same location (Fig. 4A), a wide range of time shifts for the record of each detector in the array was searched to obtain the best-fit time delay across the array. Each time delay corresponded to the best azimuth and slowness of the airwave produced by the calving event. The slowness bounds were computed from zero to slightly faster than the speed of sound in air, and the azimuth bounds were computed at all azimuths at half-degree increments. We used the semblance method (Neidell and Taner, 1971) to measure the similarity of the time-shifted waveforms at each azimuth and slowness combination. Owing to the character of most calving acoustic signals, the plotted semblance of most grid searches produced convincing sharp peaks at the best slowness-azimuth combination consistent with the slowness of 3 s/km (Fig. 5A), and poor correlation or electronic noise spikes produced characteristically incorrect velocities that were apparent and not included in the solution. Additionally, calving signals impulsive at proximal stations but noisy at more distal stations were so similar across each tight array that even weak arrivals were well correlated, leading to a well-resolved solution for even the weakest signals. To account for the poor signal-to-noise ratios for some events, the semblances were normalized. Figure 5B displays the subset of the slowness-azimuth grid search at the slowness corresponding to a horizontally propagating acoustic wave.

In order to combine the results from each array, a grid consisting of 10 m^2 blocks was created over the region of potential event locations, and the summation was taken of each normalized semblance for each array at the corresponding azimuth to the center of each block. To retain relative confidence levels between events, the grid of summed values was also normalized. A 95% contour polygon was used as the minimum quality cutoff to find the best solution, representing 95% of the maximum summed semblance.

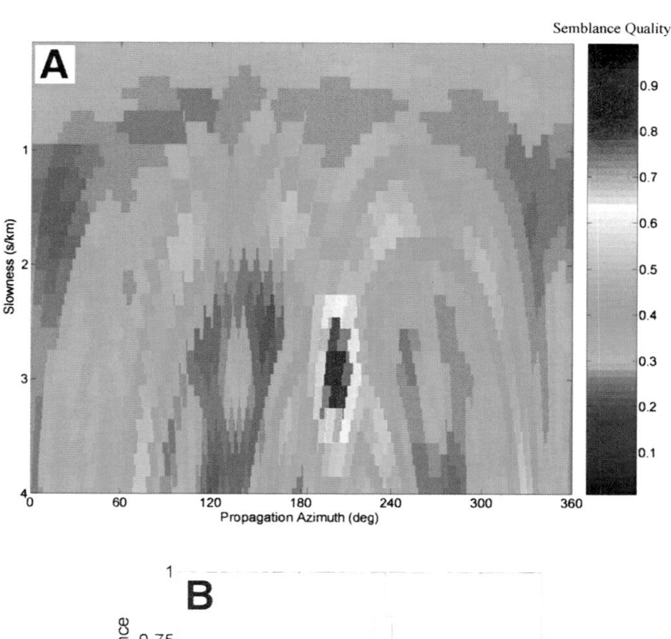

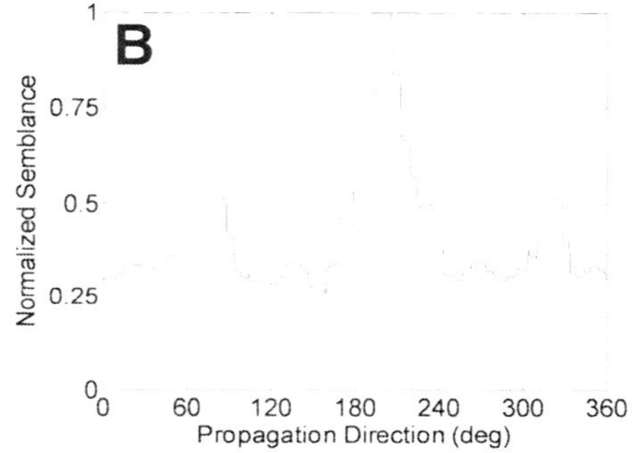

Figure 5. Representative example (event 118) of unnormalized semblance at each point in the slowness-azimuth grid (A), and raw semblance cross section at the correct slowness value of 3 s/km corresponding to a velocity of 333 m/s (B).

RESULTS

Most of the calving events we observed with the infrasound detector arrays produced arrivals on only one array. Of the 126 hand-picked events, 43 were located with back-azimuths from all three arrays. Three of these events were located with large errors outside of the reasonable region of confidence and were removed from the final solution set. An example location is shown in Figure 6 for probable calving event 118 that occurred at 00:51 UTC on 9 August 2008. The locations of each resulting hypocenter location revealed the presence of both terminus calving events and iceberg breakup events (Fig. 7). Note that the Landsat image in Figure 7 was taken in 2007, and the glacier front had retreated farther north during our observation period. Most event locations were away from the active calving front of the glacier. We observed numerous icebergs in the areas in which these events were concentrated, however, suggesting that the breakup of icebergs was the most probable source. The events that were located in the wide-aperture region of the infrasound array configuration were located with a high degree of confidence. Triangulation potential outside of the array was limited and produced elongated confidence regions (Fig. 7). Of the 40 events containing well-correlated acoustic arrivals across all three arrays, 30 produced significant arrivals in the seimic records other than the ground-coupled airwave. The remaining 10 events could correspond to mechanisms that do not couple well with the ground, perhaps including subaerial calving

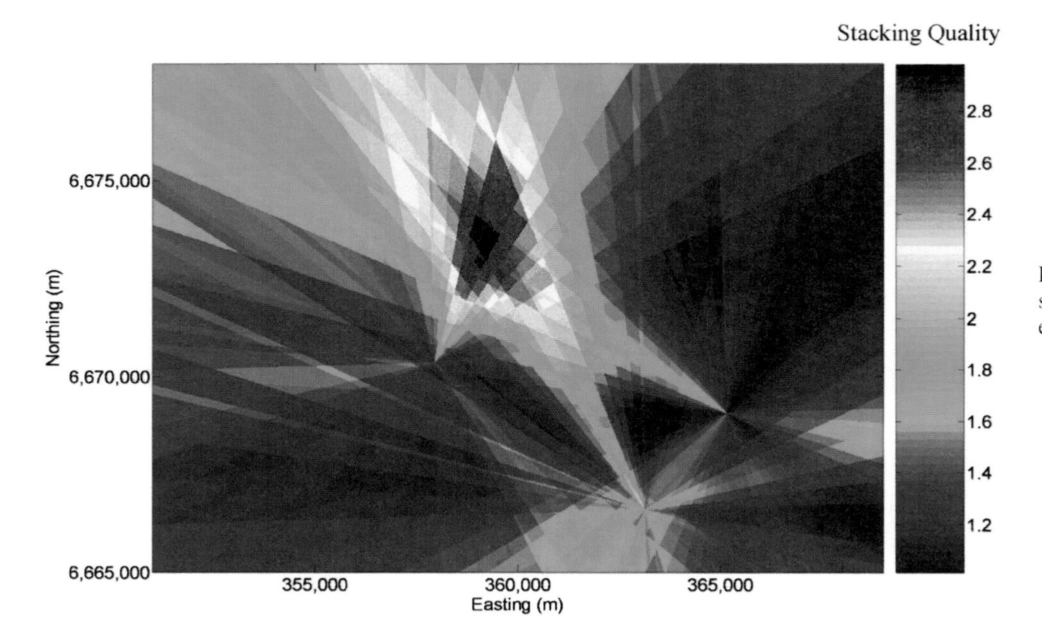

Figure 6. Resulting grid produced by summing all semblance qualities from event (118).

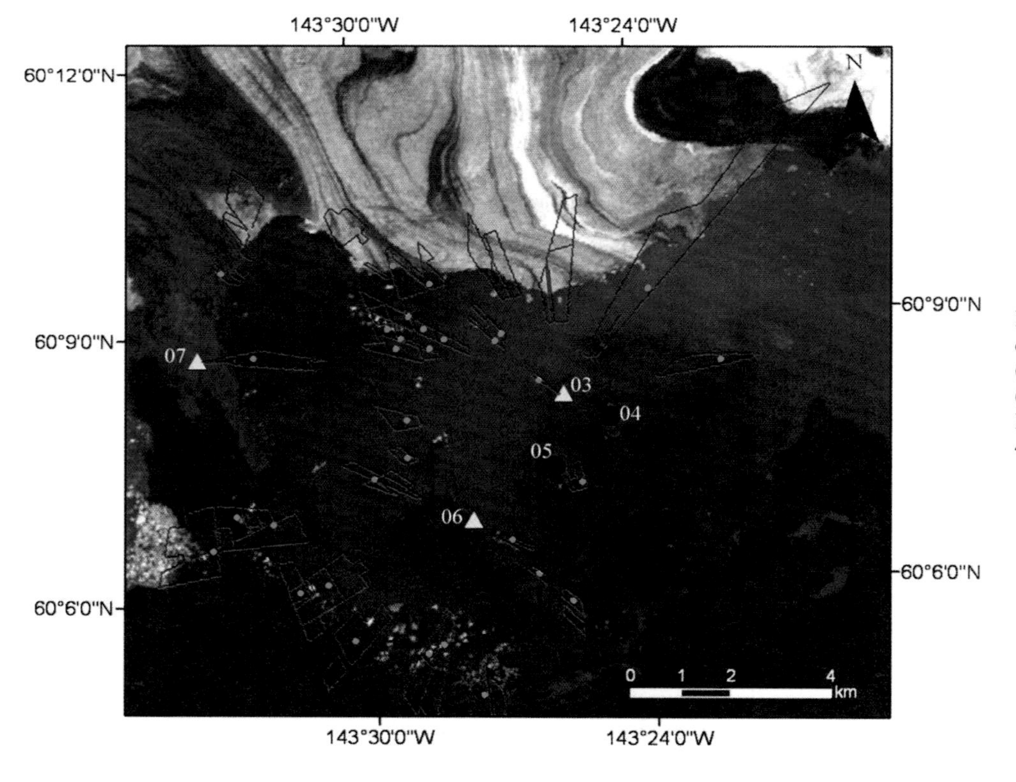

Figure 7. Seismic stations (black circles), seismic and acoustic stations (yellow triangles), hypocenter solutions (green circles), and 95% contours (red lines) for each hypocenter solution for 40 events. MTRI 2007 Landsat image.

of floating portions of the terminus or floating icebergs, or internal fracturing episodes not leading to calving.

DISCUSSION

Testing the Method

We employed two independent techniques in order to confirm the validity of the event location procedure. To verify the accuracy of the back-azimuths generated from the infrasound grid search, we conducted a particle motion analysis at the geophones in each of the three infrasound arrays. We analyzed horizontal particle motions recorded at the time of each acoustic arrival for the twelve best-recorded events on 7 August and compared the resultant direction with the direction obtained from the acoustic semblance method. The difference in azimuths for the two techniques ranged from −10° to 10° and −30° to 30° for stations 03 and 07, respectively, with means nearly zero for both stations. Station 06 had a faulty horizontal component, making this analysis impossible. The analysis demonstrates some degree of consistency between the direction of the ground-coupling into the geophone and the direction of propagation of the pressure wave through the air as observed by the pressure sensors.

Once the entire list of hypocenters was calcultated, theoretical travel times for the corresponding airwaves to each of the infrasound arrays were computed, and travel time differences computed between each array. Using only the events that correlate well at all three arrays, these theoretical travel time differences were compared with the actual travel time differences, assuming 333 m/s average speed of sound at 3.3 °C (38 °F). Although the acoustic arrival picks generally contained large uncertainties themselves, the travel time differences were in agreement with the acoustic arrival time differences where picks were possible. Additional errors likely occurred owing to dissimilarities in arrivals at different azimuths relative to the source and actual error in hypocenter locations in the narrow aperture regions outside of the array.

Assumptions

The underlying assumption made in the computation of back-azimuths from the infrasound arrays is that the pressure wave is distant enough from the source to be considered a plane wave, which holds for nearly all computed hypocenters as found in this experiment owing to the small spacing between components of each infrasound array. Significant deviations from this assumption would have become evident in errant inverse velocity solutions.

An additional necessary assumption was that the long wavelengths (~20–100 m) are not significantly affected by wind or topographical effects to within the resolution claimed. Also, although echoes were audibly observed as reflecting off the terminus wall, it was assumed that this energy was significantly reduced from the direct-path energy and did not contribute to altering the best back-azimuth solution (Johnson et al., 2004).

Possible Source Mechanisms

In addition to the expected calving events along the current terminal face of the Bering Glacier, we discovered many sources from floating and grounded icebergs within Vitus Lake (Fig. 7). Most of these sources produced both lower frequency (1–5 Hz) seismic energy recorded in the geophones (Fig. 8A, 8C) and higher frequency acoustic energy (>4 Hz) in both the geophones and infrasound detectors (Fig. 8A–8D), but 13 events with acoustic signals had very weak to no discernible seismic arrivals across multiple stations within the array. We found that the events on or adjacent to the present terminus consistently had stronger seismic signatures than events elsewhere within Vitus Lake, even though this conclusion is somewhat subjective. The infrasound signals, whether with or without significant associated seismic energy, were attributed to both calving and iceberg events (depending on their location), implying that a similar source is present in the case of both types of events. Source mechanisms that have been suggested for the seismic-waveform details of calving and iceberg events include cliff-face vibrations, single-force landslides, enlargement of a fluid-filled crack-tip (O'Neel and Pfeffer, 2007), and iceberg grounding (Müller et al., 2005).

Calving events along the terminus generally provided coherent recorded seismic energy, indicating that events along that face are either well-coupled to the ground, or perhaps the total energy intensity is greatest at the calving face of the large glacier, allowing significant seismic energy in spite of poor coupling. In addition, a weak correlation exists between iceberg events without significant seismic energy and location in deeper water, whereas most icebergs that appear to be grounded or are in shallow water apparently are well-coupled seismically. Another mechanism that may be responsible for seismic events, in addition to iceberg calving, could be the sudden grounding of moving icebergs, as seen on a larger scale in Antarctica by Müller et al. (2005). Additionally, the isolated acoustic arrivals that were observed from events in deeper water may be due to subaerial crack-tip propagation, or as a result of melting and buoyancy stresses causing conduit resonance induced by rolling over or simply repetitive pressure pulses from equilibrium buoyancy restoration. The presence or lack of seismic energy also simply may be due to a wide range of magnitudes where only larger events overcome the seismic noise level.

CONCLUSIONS

We accurately located 40 events that occurred within five days in relation to glacier and iceberg calving or breakup. These events were located using triangular infrasound detector arrays along the terminus of the Bering Glacier and within Vitus Lake from grounded and floating icebergs. The acoustic records

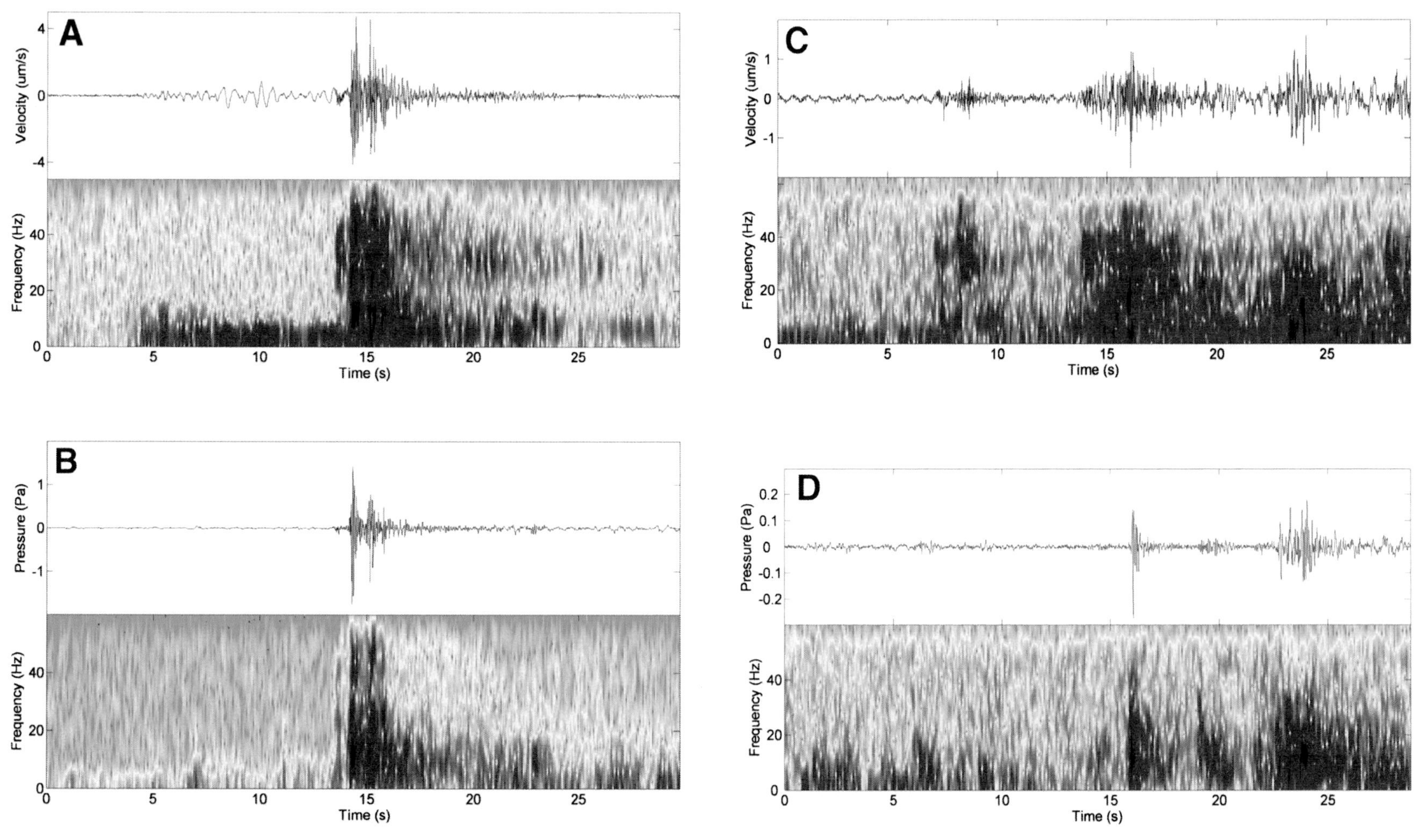

Figure 8. Seismic and acoustic records and spectrograms from calving event (70) (A and B) and from iceberg event (108) (C and D); filtered band-pass from 1 to 60 Hz.

demonstrate the wide range of acoustic and seismic sources that are present away from the terminus, which are affiliated with iceberg events. The acoustic arrivals are distinct from the low-frequency signals recorded seismically and provide improved hypocentral locations over those obtained from seismic data alone. We have demonstrated the utility of infrasound detector arrays for locating calving events without the need for direct visual observations.

ACKNOWLEDGMENTS

The authors would like to thank the following people for their assistance in making this project a success: U.S. Bureau of Land Management: Scott Guyer and Nathan Rathbun; Michigan Tech Research Institute: Bob Shuchman and Liza Jenkins; and Michigan Tech: John Gierke and Kevin Endsley. The seismic instruments were provided by the Incorporated Research Institutions for Seismology (IRIS) through the PASSCAL Instrument Center at New Mexico Tech. Data collected will be available through the IRIS Data Management Center. The facilities of the IRIS Consortium are supported by the National Science Foundation (NSF) under Cooperative Agreement EAR-0552316, the NSF Office of Polar Programs, and the DOE National Nuclear Security Administration. We would also like to thank the Michigan Tech Remote Sensing Institute, the Office of the Vice President for Research, and the Michigan Tech Fund for their support.

REFERENCES CITED

FitzGerald, K.A., Richardson, J.P., and Pennington, W.D., 2010, this volume, Ice-generated seismic events observed at the Bering Glacier, *in* Shuchman, R.A., and Josberger, E.G., eds., Bering Glacier: Interdisciplinary Studies of Earth's Largest Temperate Surging Glacier: Geological Society of America Special Paper 462, doi: 10.1130/2009.2462(18).

Johnson, J.B., Aster, R.C., and Kyle, P.R., 2004, Volcanic eruptions observed with infrasound: Geophysical Research Letters, v. 31, L14604, doi: 10.1029/2004GL020020.

Michigan Tech Research Institute, 2007, Landsat ETM+ scene L71064018_01820070810, SLC-Off, USGS, Sioux Falls, South Dakota, 8/10/2007.

Müller, C., Schlindwein, V., Eckstaller, A., and Miller, H., 2005, Singing icebergs: Science, v. 310, p. 1299, doi: 10.1126/science.1117145.

Neidell, N.S., and Taner, M.T., 1971, Semblance and other coherency measures for multichannel data: Geophysics, v. 36, p. 482–497, doi: 10.1190/1.1440186.

O'Neel, S., and Pfeffer, W.T., 2007, Source mechanics for monochromatic icequakes produced during calving at Columbia Glacier, AK: Geophysical Research Letters, v. 34, doi: 10.1029/2007GL031370.

O'Neel, S., Marshall, H.P., McNamara, D.E., and Pfeffer, W.T., 2007, Seismic detection and analysis of icequakes at Columbia Glacier, Alaska: Journal of Geophysical Research, v. 112, doi: 10.1029/2006JF000595.

Wilson, M., 2008, Calving icebergs may cause glacial earthquakes in Greenland: Physics Today, v. 9, p. 17–20.

Richardson, J.P., Waite, G.P., FitzGerald, K.A., and Pennington, W.D., 2010, Characteristics of seismic and acoustic signals produced by calving, Bering Glacier, Alaska: Geophysical Research Letters, v. 37, L03503, doi: 10.1029/2009GL041113.

MANUSCRIPT ACCEPTED BY THE SOCIETY 02 JUNE 2009

The Geological Society of America
Special Paper 462
2010

Ice-generated seismic events observed at the Bering Glacier

Katelyn A. FitzGerald
Joshua P. Richardson
Wayne D. Pennington
Department of Geological and Mining Engineering and Sciences, Michigan Technological University, 630 Dow Environmental Sciences and Engineering Building, 1400 Townsend Drive, Houghton, Michigan 49931, USA

ABSTRACT

The Bering Glacier progresses downhill through the mechanisms of plastic crystal deformation and basal sliding. Two summer field campaigns involving seismic monitoring in 2007 and 2008 were conducted in order to investigate basal processes near the terminus of the glacier. Many events were observed at stations deployed on the ice in 2007 near the Grindle Hills, but owing to the large distance between stations and the short recording period, few events were large enough to be recorded on sufficient stations to be accurately located. During August 2008, five stations were deployed in the same general area on the ice with closer spacing. Using this improved array, along with stations to the south of the glacier, four events were located. Of these, three appear to have occurred at or near the base of the glacier, at a point near the terminus, where the ice is severely folded, above the possible location of the Hope Creek fault. The fourth event was farther upstream beneath or within the glacier, but its location is poorly constrained.

BACKGROUND

The Bering Glacier progresses downhill through the mechanisms of internal plastic crystal deformation and basal sliding from its source in the Bagley Ice Field (or Bagley Ice Valley) to its terminus in Vitus Lake, which drains into the Gulf of Alaska. The Bering Glacier is also known to have periodic surges, likely enabled by these subglacial processes, which occur approximately every 20 a (Molnia and Post, 1995).

The purpose of this study was to investigate these processes near the terminus of the glacier. Because of the short period of recording and an insufficient array in the season of 2007, events were identified but were unlocatable. In the summer of 2008, a more robust array was deployed for a longer time, and a related seismic experiment studying calving was conducted simultaneously (Richardson et al., this volume).

Although the glacier was not surging during either experiment, this study may provide information on the nonsurging, subglacial base level activity. It is generally expected that another surge will take place within a few years (Molnia and Post, 1995), and baseline studies such as this may assist in understanding the nature of surges. Our study, as reported in this short note, was intended to provide background information for development of improved monitoring techniques, under future programs.

METHODS

In the summer of 2008, 13 Mark Products L-22 3-component short period geophones and RefTek 130 data acquisition systems, equipped with global positioning system (GPS) clocks, were deployed on and around the Bering Glacier. The array consisted of five stations on the ice, which were centered about the final turn of the glacier to the northwest of the Grindle Hills, one station on the bedrock at the western edge of the Grindle Hills, and seven stations south of the terminus on glacial till (Fig. 1).

The five-station array on the ice, stations 12, 13, 15, 17, and 18, was designed under the assumption that there would be more events near this turn in the glacier owing to the high stress

FitzGerald, K.A., Richardson, J.P., and Pennington, W.D., 2010, Ice-generated seismic events observed at the Bering Glacier, *in* Shuchman, R.A., and Josberger, E.G., eds., Bering Glacier: Interdisciplinary Studies of Earth's Largest Temperate Surging Glacier: Geological Society of America Special Paper 462, p. 337–340, doi: 10.1130/2010.2462(18). For permission to copy, contact editing@geosociety.org.

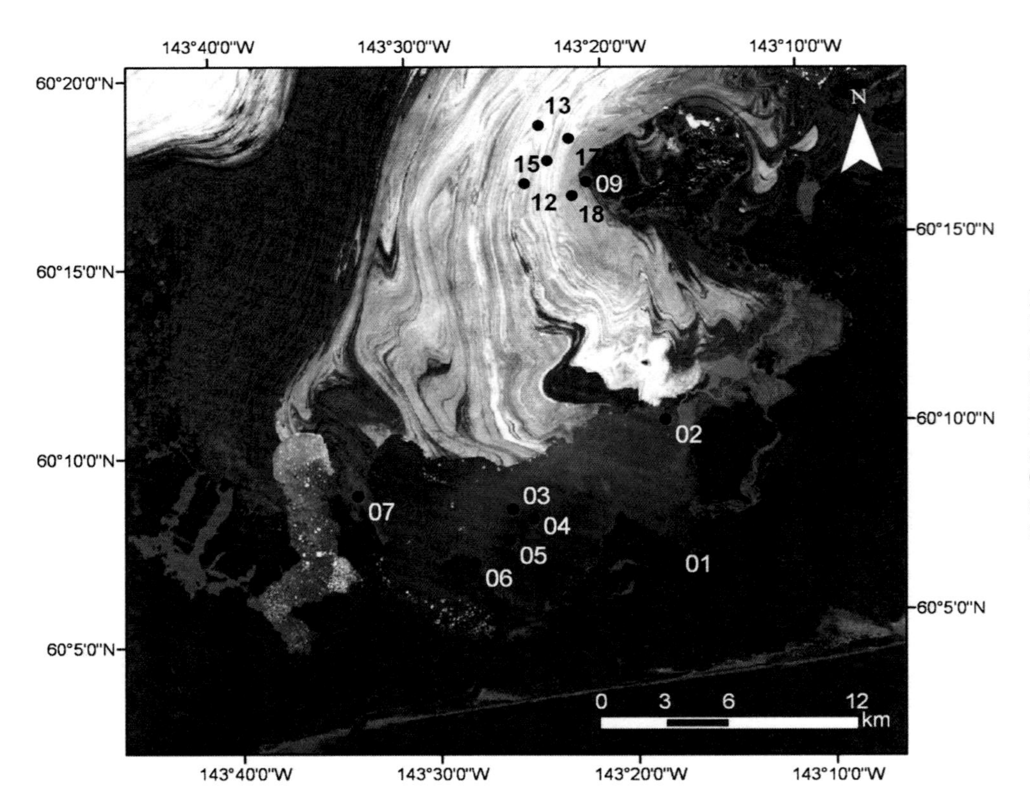

Figure 1. Map of the terminus of the Bering Glacier, showing seismograph station locations (Michigan Tech Research Institute [MTRI], 2007 Landsat image). The stations situated on ice are labeled in black, and the stations on land are labeled in white. Stations 01 through 07 are positioned on glacial till to the south of the glacier, and Station 09, in the Grindle hills, is on bedrock.

and deformation rate expected there. The stations were placed nearly a kilometer apart in order to have a wide enough array to locate various events, but still small enough to record the signals at multiple stations. These stations were deployed for only nine days, 1–9 August, because of melting and crevassing concerns on the glacier.

The seven southernmost stations also served to monitor calving events from the terminus of the glacier. These stations operated from 2 June, and nearly all were in continuous operation while the ice stations were operating, excluding instrument and maintenance down time. All stations were powered by 12 V sealed lead-acid batteries, and the eight stations on land were supplemented with solar panels for their longer operational period.

DATA ANALYSIS

During the nine days of nearly continuous operation of stations on the ice, thousands of events were observed. Most of these were difficult to locate, because the first arrivals we recorded were emergent, low-amplitude signals. The frequency content of events that were unable to be located seems to be similar to that reported by Leblanc et al. (2008) farther upstream on the glacier. Events were chosen for analysis on the basis of arrivals observed across three or more stations in the on-ice array. Forty events with good array continuity were chosen for study and reporting here. Ten of the events were confirmed to be global or regional tectonic events, through analysis of catalogue data (including that maintained by the Alaska Earthquake Information Center), and one other event was determined to be a large calving event (recognized by the presence of high-amplitude airwave arrivals seen at several stations south of the terminus). The remaining 29 events were then assumed to be internal ice events or glacial-rock interface events. Seismograms from 25 of those 29 icequakes exhibited first arrivals that were extremely emergent, had low signal-to-noise ratios, and/or had poor azimuthal station coverage, and could not be located with confidence. The four remaining events exhibited impulsive P-wave onsets and high signal-to-noise ratios at stations located at a variety of azimuths (see Fig. 2).

VELOCITY MODEL

A simple, laterally homogeneous layered model was used to determine the locations of the four apparent icequakes. The model extends from 0.4 km above sea level to 10 km depth (Fig. 3) and simulates the glacial ice and rock underlying station 13. The top layer is 600 m of glacial ice, and beneath this is bedrock with an underlying sequence of increasingly dense and higher velocity media. The top 300 m of the glacial ice is assumed to be highly fractured with a P-wave velocity of 3.80 km/s, whereas the lower, less-fractured ice has a velocity of 3.87 km/s (Deichmann et al., 2000). The underlying bedrock was assumed to be graywacke, which crops out in the Grindle Hills, and was assigned a velocity of 5.00 km/s, with increasing velocity to 6.00 km/s at 7 km depth (Plafker, 1967).

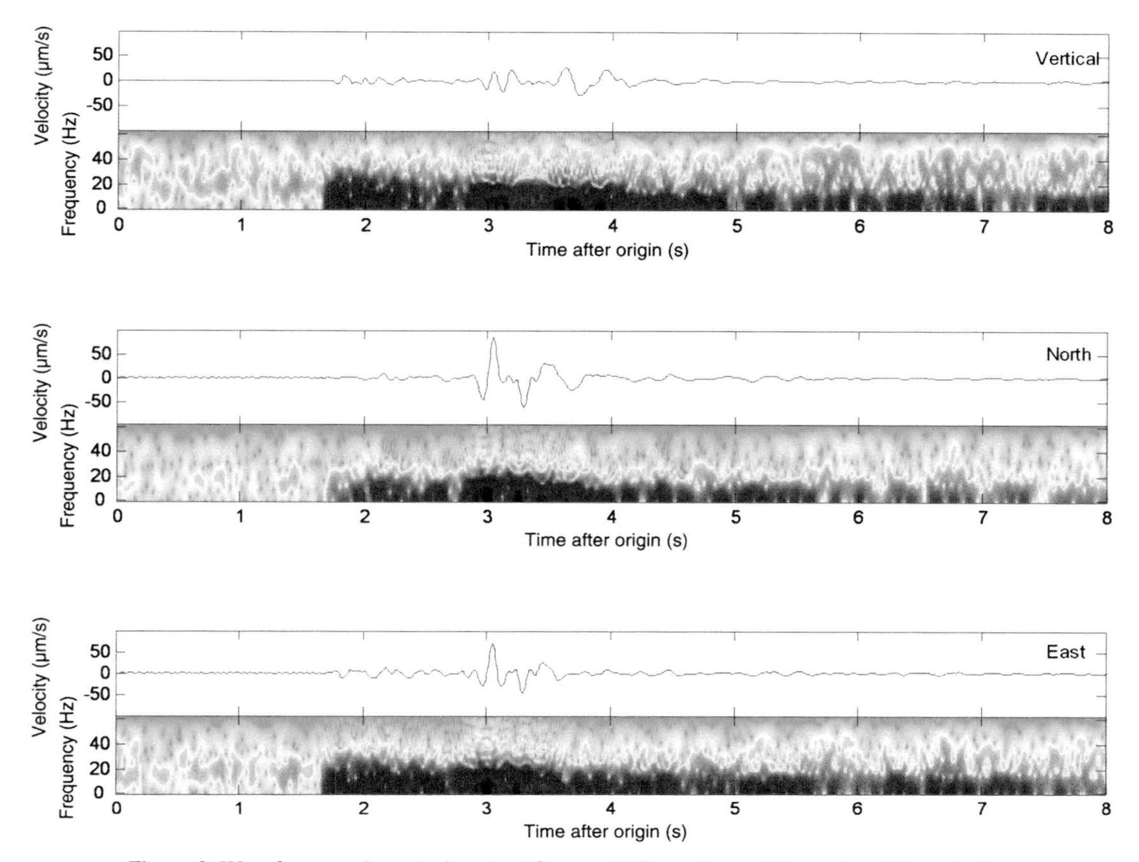

Figure 2. Waveforms and spectral content for event E3, as observed by station 18, on the ice.

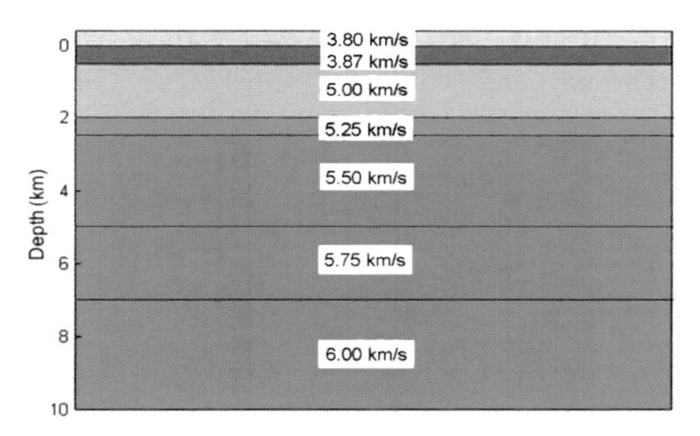

Figure 3. Cross section of the layered velocity model used in determining event locations.

EVENT LOCATIONS

Geiger's method was used to invert the P-wave relative arrival times for hypocenter locations using the velocity model and equally weighted picks. Each of the four hypocenter locations lies in or near the base of the glacier within reasonable error bounds. Events E2, E3, and E4 were located precisely with very low residual times, but event E1 had a weaker location and depth constraint owing to the poorer azimuthal coverage. Computed depths below sea level for the well-constrained events are E2 (0.00 km), E3 (0.70 km), and E4 (0.63 km). Events E2 and E3 were found in the Alaska Earthquake Information Center database, but their locations and depths were poorly constrained in the catalogue data.

The three well-located events (E2, E3, and E4) fall along a line separating the portion of the glacier where the striations are smooth and straight upstream, but become distorted and highly folded farther downstream (Fig. 4). This location is also coincident with the possible (unmapped) extension of the Hope Creek fault mapped by Plafker (1967) to the east of the glacier.

CONCLUSIONS

A large number of events were found, but to date only four have been located with any confidence. The located events all appeared to be shallow, glacially related events. Depths for events E2, E3, and E4 were consistent with the depth to the base of the glacier, but the depth for E1 was too weakly constrained to determine whether its position was within or beneath the glacier. The three well-constrained events all lie along a line that appears to separate regimes of the glacier that, in map view, contain smooth, laminar flow upstream of the line, and distorted, highly strained flow downstream of the line. This location also corresponds to a

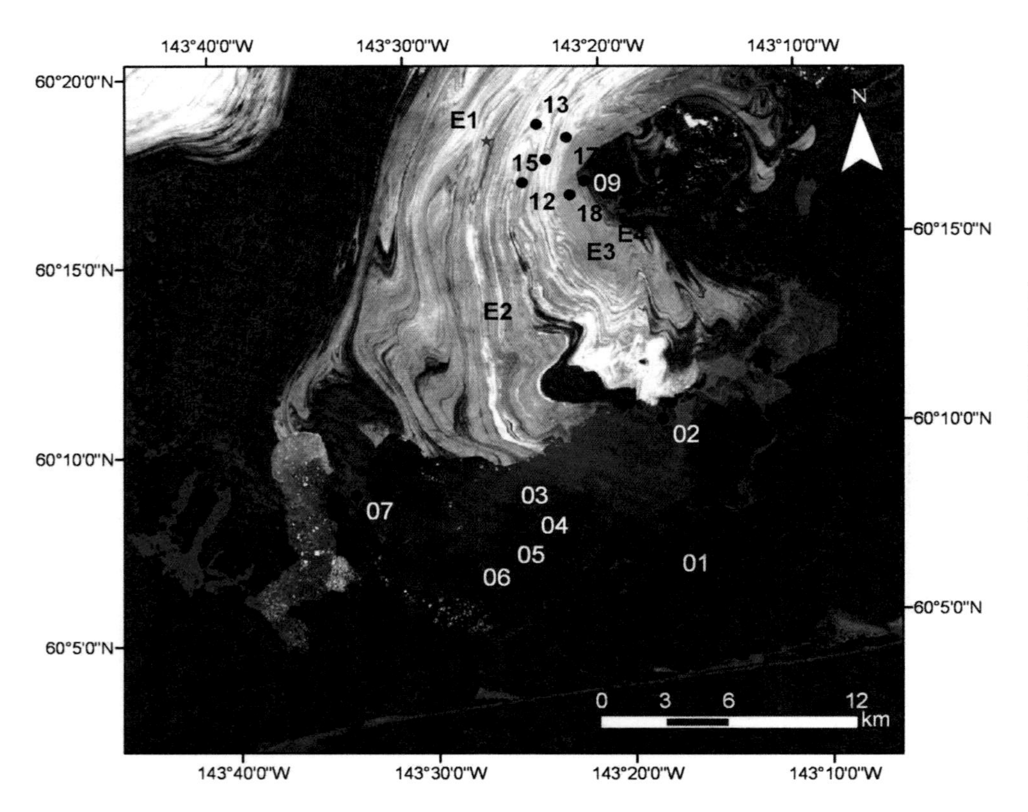

Figure 4. Map of the terminus of the Bering Glacier showing seismograph station locations as black circles and approximate event locations as stars, with the green events being better constrained (MTRI 2007 Landsat image). Note that the three well-located events fall along a line that separates smoothly flowing ice to the north (upstream) from highly contorted and folded ice to the south.

possible extension of the Hope Creek fault beneath the glacier. One possible explanation for these events is that the ice encounters a topographic obstacle at depth, whether it is caused by a fault, by an old moraine, or both, and that this obstacle causes stresses sufficient to result in seismic slip at or near the base of the ice. There are likely other possible explanations as well.

In general the emergent nature of the recorded arrivals and the low signal-to-noise ratios make these events very difficult to locate. One of the goals of this study was to optimize a method for obtaining event information from events on, in, or near the glacier. Another goal was to gain sufficient experience in designing new monitoring techniques that are more robust under the difficult field conditions posed by rapidly melting and ablating glaciers in the summer. We are currently investigating such field techniques as a result of the field work reported in this short note.

ACKNOWLEDGMENTS

The authors would like to thank the following people for their assistance in making this project a success: Scott Guyer and Nathan Rathbun, U.S. Bureau of Land Management; Robert Shuchman and Liza Jenkins, Michigan Tech Research Institute; Greg Waite and John Gierke, Michigan Tech; and the helpful staff at PASSCAL Instrument Center who assisted with equipment and computing support. We would also like to thank the Michigan Tech Remote Sensing Institute, the Office of the Vice President for Research, and the Michigan Tech Fund for their support. The seismic instruments were provided by the Incorporated Research Institutions for Seismology (IRIS) through the PASSCAL Instrument Center at New Mexico Tech. Data collected will be available through the IRIS Data Management Center. The facilities of the IRIS Consortium are supported by the National Science Foundation (NSF) under Cooperative Agreement EAR-0552316, the NSF Office of Polar Programs, and the DOE National Nuclear Security Administration.

REFERENCES CITED

Deichmann, N., Ansorge, J., Scherbaum, F., Aschwanden, A., Bernardi, F., and Gudmundsson, G.H., 2000, Evidence for deep icequakes in an Alpine glacier: Annals of Glaciology, v. 31, p. 85–90, doi: 10.3189/172756400781820462.

Leblanc, L.E., Larsen, C., West, M., O'Neel, S., and Truffer, M., 2008, Time-series analysis of icequakes and ice motion, Bering Glacier, AK: American Geophysical Union, Fall Meeting 2008, abstract #C11D-0535.

Molnia, B.F., and Post, A., 1995, Holocene history of Bering Glacier, Alaska: A prelude to the 1993–1994 Surge: Physical Geography, v. 16, p. 87–117.

NASA Landsat Program, 2007, Landsat ETM+ scene L71064018_01820070810, SLC-Off gap-fill: Ann Arbor, Michigan Tech Research Institute, 8/10/2007.

Plafker, G., 1967, Geologic map of the Gulf of Alaska tertiary province, Alaska: U.S. Geological Survey Miscellaneous Investigations Map I-484, scale 1:500,000, 1 sheet.

Richardson, J.P., FitzGerald, K.A., Waite, G.P., and Pennington, W.D., 2010, this volume, Acoustic and seismic observations of calving events at Bering Glacier, Alaska, *in* Shuchman, R.A., and Josberger, E.G., eds., Bering Glacier: Interdisciplinary Studies of Earth's Largest Temperate Surging Glacier: Geological Society of America Special Paper 462, doi: 10.1130/2009.2462(17).

Manuscript Accepted by the Society 02 June 2009

The Geological Society of America
Special Paper 462
2010

Mapping the fresh-water–salt-water interface in the terminal moraine of the Bering Glacier

Austin B. Andrus
Kevin A. Endsley
Silvia Espino
John S. Gierke
Geological and Mining Engineering and Sciences, Michigan Technological University, 1400 Townsend Drive, Houghton, Michigan 49931, USA

ABSTRACT

The principal objective of this investigation was to delineate the geometry of the subterranean discharge of fresh melt water from Vitus Lake (VL) at the terminus of the Bering Glacier into the Gulf of Alaska (GoA). It was hypothesized that during the seasons where the glacier was exhibiting melting, increases in VL elevation would push freshwater through the moraine toward the GoA. Because of the proximity to the GoA, intruded seawater would lie below the freshwater. The fresh and salt waters were delineated by measuring variations in apparent resistivity—as a proxy for salinity—with depth using two-dimensional electrical resistivity tomography at a variety of locations between the GoA and VL. The surveys were conducted using Schlumberger and pole-dipole arrays, along with temporal surveys to examine tidal effects. The results of field surveys conducted in August 2007 and 2008 showed that an interface between the fresh and salt water exists within the moraine at depth and that the interface is deeper progressing from the GoA toward VL. Although the fresh/salt water interface was readily apparent proximal to the GoA, a distinct interface was not apparent closer to VL. The inability to measure a distinct interface could be due to depth limitations (~68 m) for the maximum array length (330 m) and/or mixing of fresh and seawater due to the seasonal effects on melting, which would likely result in seasonal changes in the interface.

INTRODUCTION

Water budget analyses (Josberger et al., 2006) of the annual melting of the Bering Glacier in Alaska and of the surface hydrology of Vitus Lake and the Seal River, which discharges into the Gulf of Alaska (Fig. 1), suggest that significant amounts of fresh water are discharging to the Gulf as subterranean groundwater flow. This hypothesis is derived from reasonable estimates of evaporation from Vitus Lake and measured discharge of the Seal River.

These estimates indicate that the Seal River alone cannot discharge the amount of fresh water produced by glacial melting. The discovery of salt water at ~40 m depth in Vitus Lake (Josberger et al., 2006) further suggests that groundwater is moving through the area's unconsolidated end-moraine deposits and mixing with intruding salt water (Fig. 2).

OBJECTIVE

This investigation attempted to map whether an interface exists between the fresh-water table and an inferred salt-water wedge that potentially exists at depth at the terminal moraine of the Bering Glacier between Vitus Lake and the Gulf of

Andrus, A.B., Endsley, K.A., Espino, S., and Gierke, J.S., 2010, Mapping the fresh-water–salt-water interface in the terminal moraine of the Bering Glacier, *in* Shuchman, R.A., and Josberger, E.G., eds., Bering Glacier: Interdisciplinary Studies of Earth's Largest Temperate Surging Glacier: Geological Society of America Special Paper 462, p. 341–350, doi: 10.1130/2010.2462(19). For permission to copy, contact editing@geosociety.org.

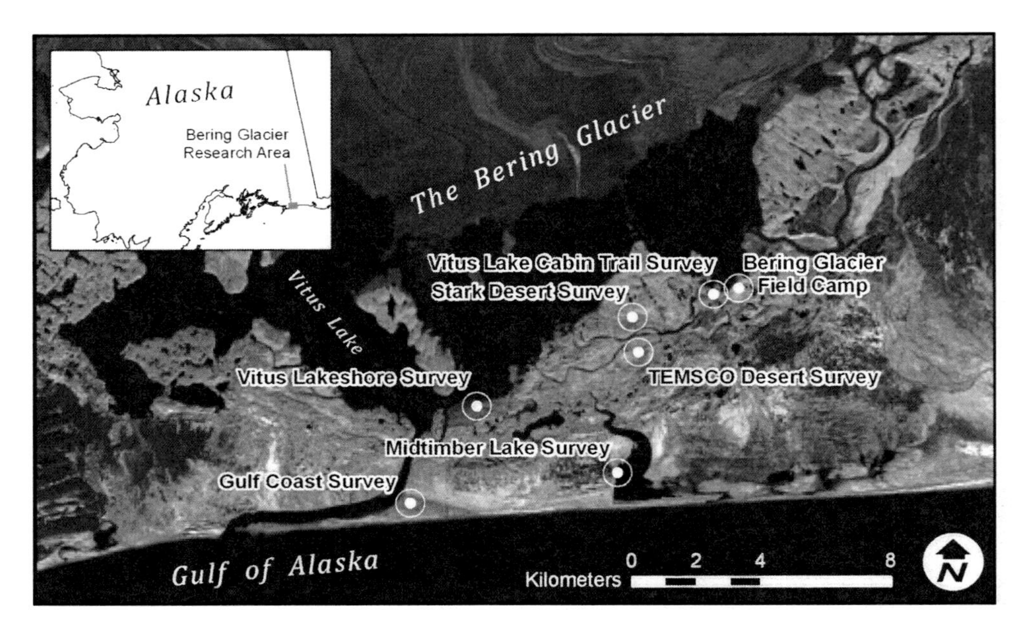

Figure 1. Site location of survey area. Specific survey locations are identified with circles.

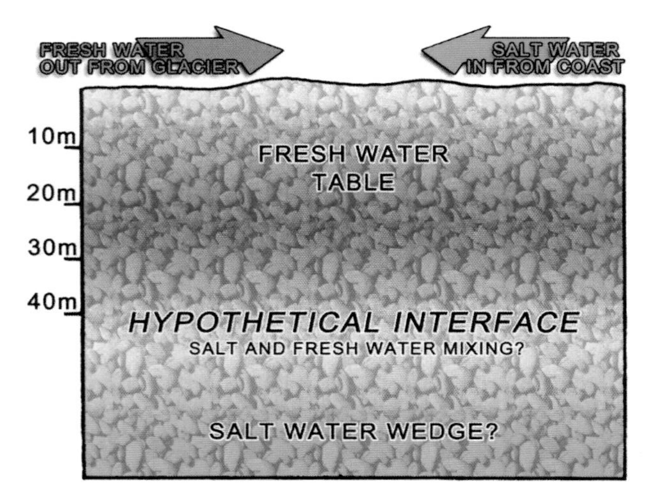

Figure 2. Model of hypothetical groundwater conditions (a hypothesis only).

Alaska. We also attempted to determine how fresh water from Vitus Lake might flow through the terminal moraine and subsequently to identify the influence of the tides on groundwater movement in this area.

METHODS AND MATERIALS

In a hydrogeological setting, as depicted in Figure 2, the operative property of interest for delineating variations in salinity with depth is electrical resistivity. Since salt water is denser than fresh water, salt-water intrusion from the sea will occur below fresh-water discharge from higher topographic elevations along a coastal region. Increased salinity decreases the electrical resistivity of water, and thus resistivity decreases with depth where both fresh water and seawater occur. Electrical resistivity sounding was chosen as a method of measuring electrical resistivity at different depths and attempt to identify the fresh-water–salt-water interface.

Investigations were conducted during early August field seasons in 2007 and 2008 at the Bering Glacier field camp (operated by the U.S. Bureau of Land Management). Survey locations were chosen on the basis of their relative locations between Vitus Lake and the Gulf of Alaska, and a lack of vegetation to enable the deployment of the survey arrays without clearing brush and trees. Using the Advanced Geosciences, Inc., SuperSting R1/IP multiple electrode resistivity system, 56 electrodes were placed on stakes in a straight array. All of the stakes were separated by an equal interval and inserted ~10–15cm deep into the ground. The SuperSting R1/IP selects two current and two potential electrodes from among the available 56 for each measurement. Any electrode selected by the instrument can be used either as a current or as a potential electrode. Depending on the survey, spacing between the electrodes varied between 4 and 6 m. Using an electrode separation of 4 m, an array of 56 electrodes spanned a length of 220 m, whereas an electrode separation of 6 m expanded it to 330 m long. A smaller electrode separation of 4 m (and consequently less depth penetration) was chosen in areas where the salt-water–fresh-water interface was expected to be close to the surface (for better resolution) or, in some cases, where the array could not be expanded with all 56 electrodes at 6-m spacing. An electrode spacing of 6 m was the maximum spacing used with this equipment.

Two array geometries were dominantly used: the Schlumberger array and the pole-dipole array. Both are excellent for electrical resistivity sounding. For the pole-dipole array, a current electrode was placed off-end but in line with the survey at a distance sufficiently large to represent infinity (greater than or equal to about twice the total length of the survey) and connected to the SuperSting R1/IP with wire. Because current depth penetration is directly related to current electrode spacing (Burger et al., 2006), the pole-dipole array allows visualization of resistivity contrasts at greater depths than the Schlumberger array. The depth penetration offered by a Schlumberger array is limited by the length of the array (when the first and last electrode are selected as the current electrodes), but a survey using this array rather than the pole-dipole array takes significantly less time to conduct. Surveys using the Schlumberger array required at most 90 min to complete, whereas surveys using the pole-dipole survey required ~150 min to complete. Two dipole-dipole surveys were conducted (one in 2007, the other in 2008) but were not considered for this chapter because the signal-to-noise ratio was too low.

The SuperSting R1/IP is fully automated once the survey is set up. A "command file" that instructs the device on how to run a survey is uploaded to the SuperSting R1/IP instrument from a field laptop computer. For both device operation and providing current for subsurface injection, the SuperSting was powered by two 12 V gel-cell batteries. Before beginning the survey, two contact resistance tests were performed: once before and once after pouring salt water on the ground contact at the base of each stake. After a reliable contact between each stake and the ground was established, the survey was implemented.

While the survey was running in automatic mode, terrain elevations at the stake locations were measured with a TruPulse 200 laser rangefinder/inclinometer (Laser Technology, Inc., Centennial, Colorado). Initially, these terrain data were used to improve the inversion of resistivity data. However, it was evident that the effect of terrain on resistivity measurements was negligible, because all surveys were conducted in relatively flat areas.

All survey data were imported into AGI EarthImager 2D, proprietary inversion software developed for use specifically with the SuperSting R1/IP or SuperSting R8/IP. The Schlumberger surveys conducted in August 2007 were inverted using the Smooth Inversion Method with four or five mesh divisions. The Smooth Inversion Method was chosen for Schlumberger arrays because better data values can be obtained from smoothing the overlapping curve segments that are generated (Burger et al., 2006). For pole-dipole surveys conducted in August 2008 the Robust Inversion Method was used with four or five mesh divisions.

DATA SUMMARY

Table 1 lists the Schlumberger surveys conducted during both field seasons and details their location, the times they were conducted, the root mean square (rms) error, and their maximum depth penetration. Table 2 lists the pole-dipole surveys completed during the 2008 field season with details on locations, date and time of survey, rms error, and maximum depth penetration.

TABLE 1. SCHLUMBERGER SURVEYS AND INVERSION DETAILS

Survey name	Date–time (ADT)	Position of no. 1 electrode	Position of no. 56 electrode	rms error	Maximum depth (m)
Vitus Lake Cabin Trail	July 31, 2007 13:00–14:16	60.11803°N, 143.29190°W	60.11732°N, 143.29556°W	1.40%	49.4
Midtimber Lake	August 2, 2007 14:30–15:50	60.065278°N, 143.34611°W	60.068056°N, 143.34583°W	11.09%	70.0
Midtimber Lake	August 2, 2008	60.06583°N, 143.34583°W	60.06889°N, 143.34639°W	8.29%	64.0
Stark Desert E-W	August 3, 2007			6.03%	68.0
Stark Desert N-S	August 3, 2007			5.33%	68.0
Gulf Coast 1	August 4, 2007 12:13–13:40	60.054444°N, 143.460278°W	60.055556°N, 143.463611°W	6.81%	45.4
Gulf Coast 2	August 4, 2007 13:41–15:07	60.054444°N, 143.460278°W	60.055556°N, 143.463611°W	7.39%	45.4
Gulf Coast 3	August 4, 2007 15:09–16:40	60.054444°N, 143.460278°W	60.055556°N, 143.463611°W	5.81%	45.4
Gulf Coast	August 9, 2008	60.05305°N, 143.42979°W	60.05313°N, 143.43316°W	2.22%	45.4
TEMSCO Desert E-W	August 5, 2007 10:44–12:14			3.86%	68.0
TEMSCO Desert N-S	August 5, 2007 13:26–14:57			4.59%	68.0
Vitus Lake shore	August 7, 2007 11:49–20:20	60.083056°N, 143.425278°W	60.082222°N, 143.431111°W	5.09%	68.0

Note: ADT—Alaska Daylight Time; rms—root mean square.

TABLE 2. POLE-DIPOLE SURVEYS AND INVERSION DETAILS

Survey name	Date–time (ADT)	Position of no. 1 electrode	Position of no. 56 electrode	Position of infinity electrode	rms error	Maximum depth (m)
Midtimber Lake	August 2, 2008	60.06583°N, 143.34583°W	60.06889°N, 143.34639°W	60.07000°N, 143.34583°W	22.33%	79.0
Bering Glacier field camp	August 5, 2008	60.11870°N, 143.28416°W	60.11895°N, 143.28004°W	60.12008°N, 143.27446°W	21.76%	102.0
Stark Desert	August 6, 2008	60.11143°N, 143.34045°W	60.10996°N, 143.34365°W	60.10703°N, 143.35123°W	3.21%	91.0
TEMSCO Desert	August 8, 2008	60.10139°N, 143.33106°W	60.09969°N, 143.34040°W	60.09543°N, 143.34040°W	19.99%	102.0

Note: ADT—Alaska Daylight Time; rms—root mean square.

The figures in this section were generated in AGI EarthImager 2D and feature three cross sections of the subsurface in each image. The top cross section, which displays scattered black dots representing every measurement, shows the apparent resistivity values at different depths. The middle cross section displays a computer-generated model of apparent resistivity values corresponding to different subsurface conditions which is checked against the real measurements (again, displayed in the top cross section). The bottom cross section is considered to be a "real" cross section of the subsurface showing real resistivity values that are inverted (calculated) from the model. Subsequent comments on resistivity contrasts in this section will be concerned with this bottom cross section, the inverted resistivity cross section.

The first survey conducted during the 2007 field season was along the Vitus Lake Cabin hiking trail, an area chosen for its close proximity to the field camp. In this paper, given trends are measured from the no. 1 electrode to the no. 56 electrode. The array was set up in a trend line of N 50° E. The field data and inversion results are displayed in Figure 3. From the inversion of this survey's measurements, there is a clear drop in resistivity from 1000 to 400 ohm-m at ~10–13 m depth.

Ground reconnaissance of the terminal moraine determined the location of the next survey, conducted at Midtimber Lake. The array line followed a trend of N 20° W. Figure 4 shows the inversion results of this survey from EarthImager 2D. At ~17–20 m depth there is a distinct drop in resistivity from 200 to 50 ohm-m. Then, at ~35-m depth, there is another sudden drop in resistivity from 50 to 10 ohm-m. The contrast between layers is easier to visualize in plots with more color bins than are displayed in Figure 4; however, a logarithmic scale is most appropriate for resistivity data, and the custom colors ensure scale consistency throughout all figures.

A series of surveys using the Schulmberger array were conducted near the Seal River and the Gulf of Alaska coast with the intention of observing tidal effects on groundwater movement. The site is not unlike the desert survey sites in terms of its sparse ground cover and flat terrain; at this site in particular there is no ground cover whatsoever. These surveys were conducted in the same location, without moving any of the electrodes in the array.

In this manner, three "time-lapse" resistivity soundings of groundwater near the coast were obtained with the array trending N 10° E. These three surveys are displayed in order of their execution in Figures 5–7. All three surveys indicate a drop in resistivity on the order of 1000–1500 ohm-m at 4–6m depth. They also indicate a second, deeper interface where the resistivity again drops by ~250–300 ohm-m; this interface becomes closer to the surface in the third and final survey conducted that day.

In the interior of the terminal moraine, away from the coast or lakeshore, two sparsely vegetated areas are commonly referred to as "deserts." These large, flat areas are covered only by scattered, small, low-lying bushes; no trees or grasses grow in these areas. Consequently, they are ideal for setting up and conducting resistivity surveys. In both areas, orthogonal Schlumberger arrays were set up—one survey was conducted with the Schlumberger array in an east-west orientation, and one was conducted in a north-south orientation. These orthogonal arrays shared the same midpoint: electrode no. 28 was effectively in the same position for both surveys (and by design, the SuperSting R1/IP instrument and batteries were as well). The orthogonal array pairs for the Stark Desert are not displayed.

The TEMSCO Desert orthogonal surveys (see Figs. 8 and 9), conducted in August 2007 using the Schlumberger array, indicate only a shallow interface at a depth of 4 m, where the drop in resistivity is on the order of 200 ohm-m. The Stark Desert orthogonal surveys conducted the same year using the Schlumberger array show the same interface at ~4 m, but the drop in resistivity across it is close to 800 ohm-m. Both the north-south and east-west surveys in each desert display elliptical pockets of high or low resistivity near the surface.

The final survey conducted in the August 2007 field season, at the shore of Vitus Lake, shows very low resistivity throughout the subsurface, with a maximum inverted resistivity value of 300 ohm-m at the surface.

INTERPRETATIONS

Beginning with the Vitus Lake Cabin Trail survey in August 2007, the fresh-water table appears to be at a depth of 13–15 m. At this depth a distinct drop in resistivity from 1000 ohm-m to

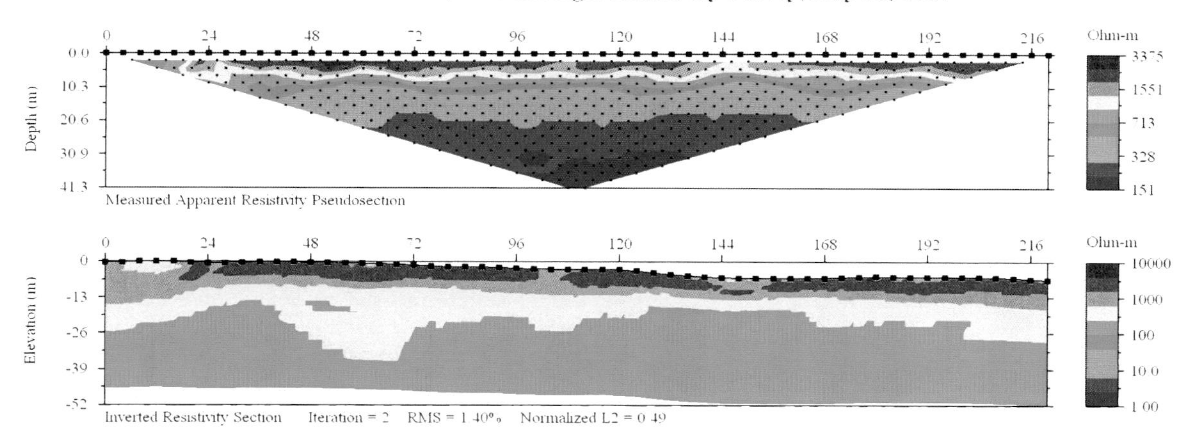

Figure 3. EarthImager 2D result from Vitus Lake Cabin Trail survey, a Schlumberger array that utilized 4-m electrode spacing. There is a distinct drop in resistivity from 1000 to 400 ohm-m between 10 and 13 m depth.

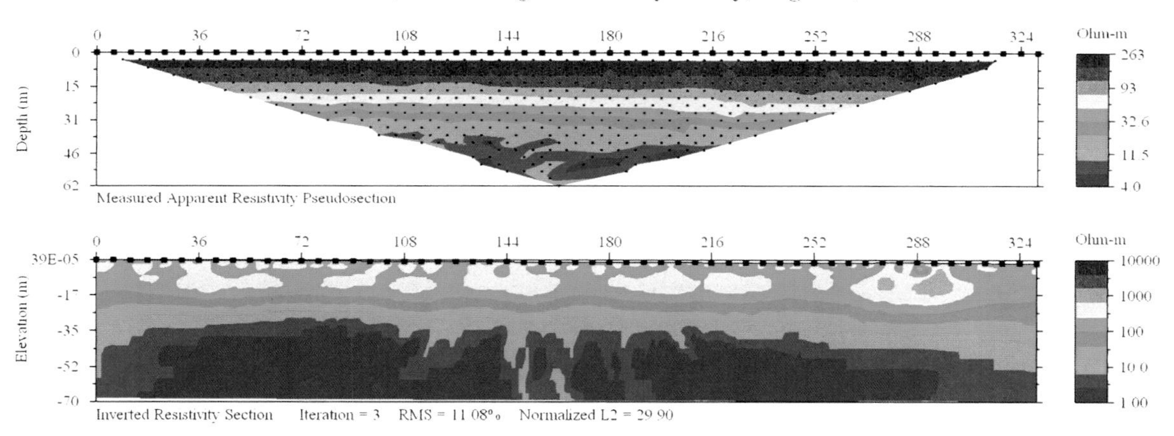

Figure 4. Earth Imager 2D results from Midtimber Lake survey, a Schlumberger array that utilized 6-m electrode spacing. At ~17–20-m depth there is a noted drop in resistivity from 200 to 50 ohm-m. At ~35-m depth, there is another sudden drop in resistivity from 50 to 10 ohm-m.

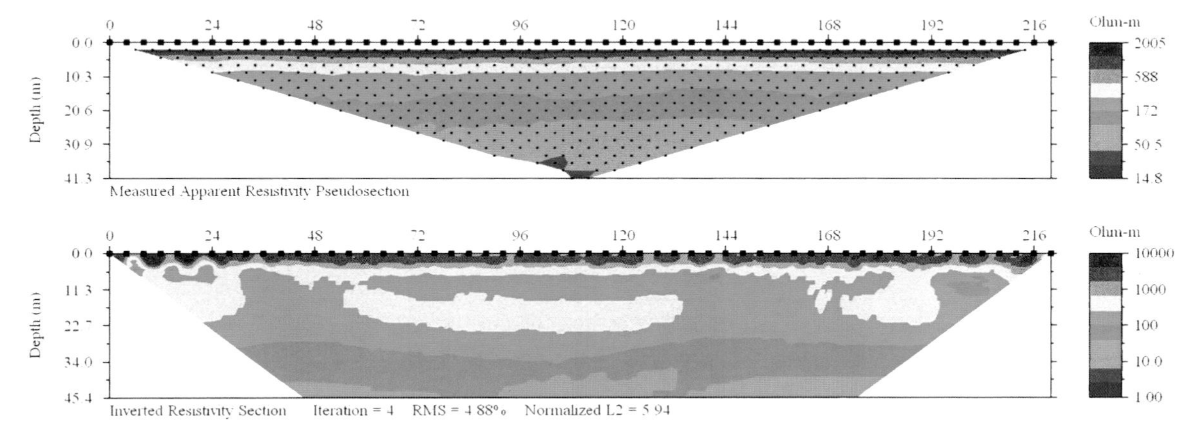

Figure 5. EarthImager 2D results of Gulf Coast Survey no. 1, a Schlumberger array utilizing 4-m electrode spacing. At ~4–6-m depth there is a noted drop in resistivity of ~1000 ohm-m. Also, at ~23–30 m depth, there is an undulating interface across which the resistivity drops ~250 ohm-m.

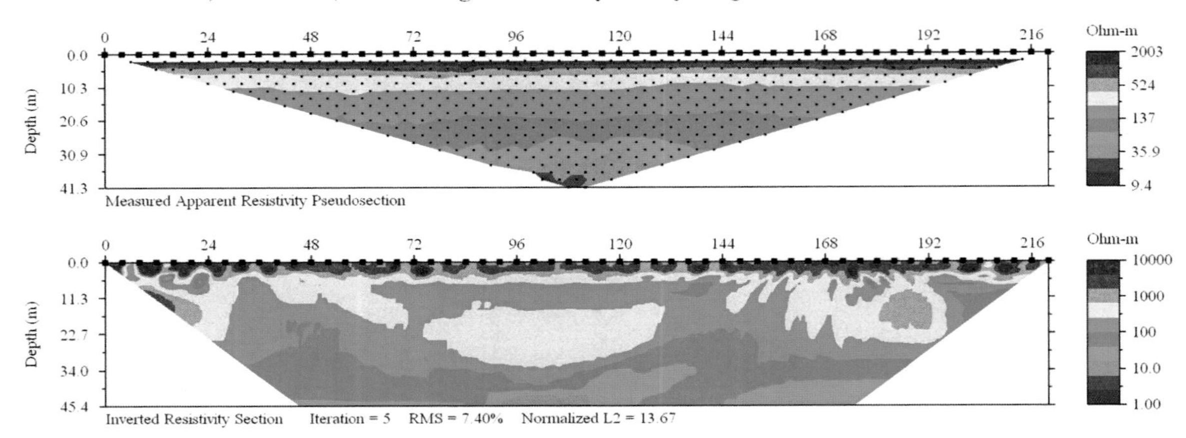

Figure 6. EarthImager 2D results of Gulf Coast Survey no. 2, a Schlumberger array utilizing 4-m electrode spacing. At ~4–6 m depth there is a noted drop in resistivity of ~1000 ohm-m. Also, at ~23–34 m depth there is an undulating interface across which there is a drop in resistivity of ~250 ohm-m.

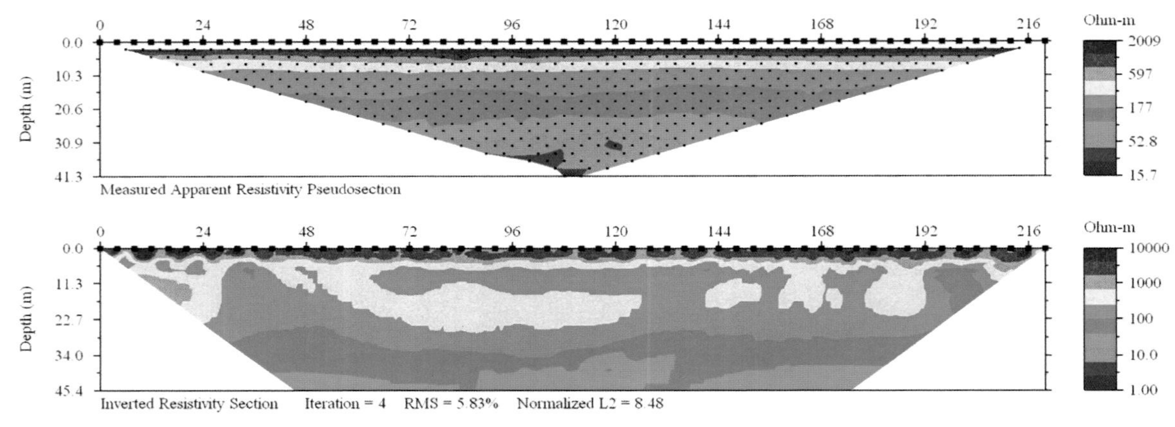

Figure 7. EarthImager 2D results of Gulf Coast Survey no. 3, a Schlumberger array utilizing 4-m electrode spacing. At 4–6 m there is a noted drop in resistivity from 1500 ohm-m. Also, at ~15–23 m there is an undulating interface across which there is a drop in resistivity of ~300 ohm-m.

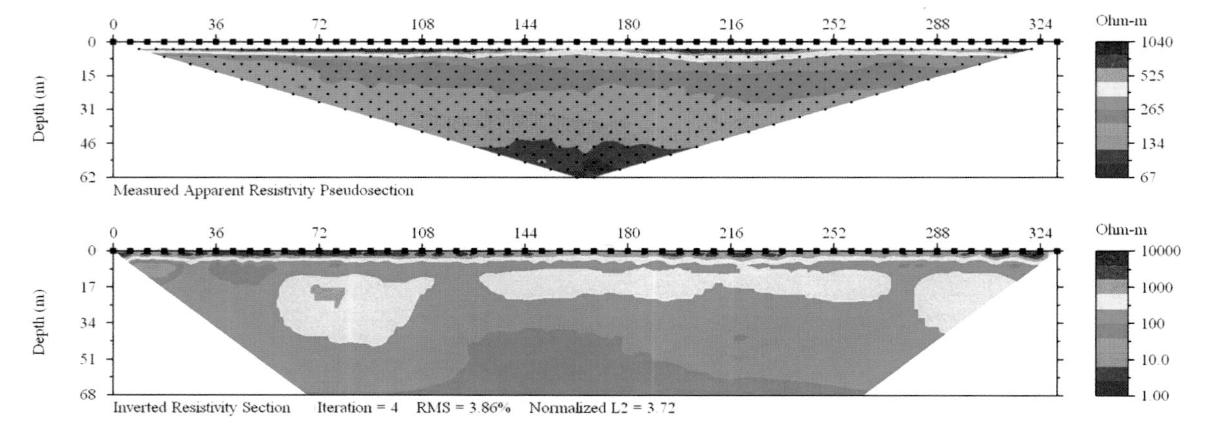

Figure 8. EarthImager 2D result from the TEMSCO Desert east-west survey, an orthogonal Schlumberger array using 6-m electrode spacing. At ~4 m there is an uneven drop in resistivity of ~200 ohm-m.

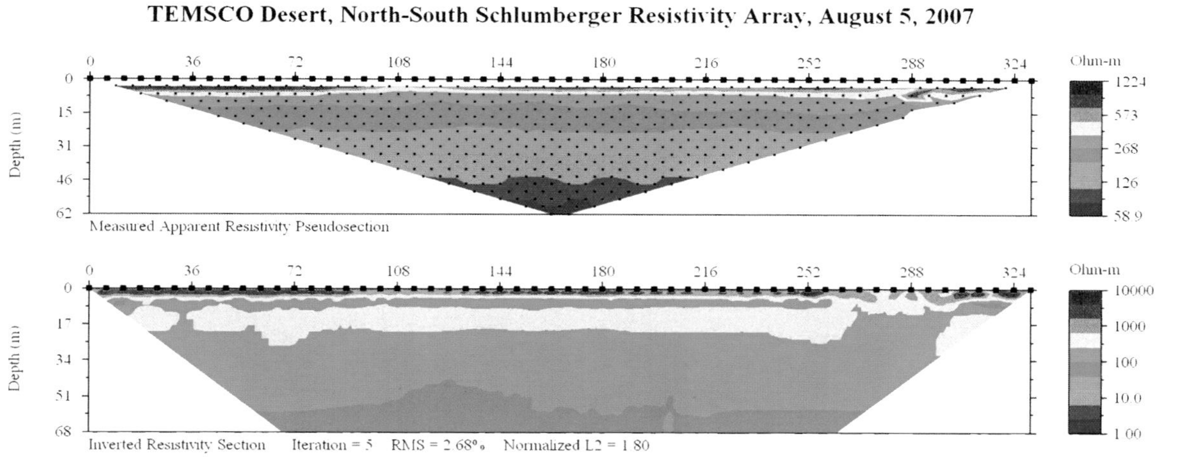

Figure 9. EarthImager 2D result from the TEMSCO Desert north-south survey, an orthogonal Schlumberger array using 6-m electrode spacing. At ~8 m there is a drop in resistivity of ~300 ohm-m.

400 ohm-m likely indicates a transition from unsaturated sediments near the surface to saturated sediments below the water table. At Midtimber Lake the fresh-water table appears to be very close to the surface, although at the site the ground was quite dry. In the inversion image from this survey (Fig. 4) there appears to be an interface at ~35–40 m deep, below which resistivity values drop to ~6–40 ohm-m. Resistivity values from 1 to 10 ohm-m are sufficiently low for this zone to be considered likely to contain salt water.

The series of Schlumberger surveys conducted at the Gulf of Alaska coast indicates that the fresh-water table in that area in August 2007 was at a depth of ~5–6 m. Salt water is apparent at a depth of ~25–35 m, where the resistivity drops to below 85 ohm-m and continues to decrease to a minimum resistivity of ~5 ohm-m where the survey bottoms out. It was superficially observed that the tide was coming in the afternoon these surveys were conducted. From the rise and fall of the undulating interface between 25 and 35 m deep in the inversion results, it appears that the fresh-water–salt-water interface is descending between 12:13 and 15:07 ADT (Alaska Daylight Time), but then rising between 15:09 and 16:40 ADT.

The Schlumberger surveys conducted in the TEMSCO Desert seem to resolve no fresh-water–salt-water interface, although the resistivity values from the inversion of the survey oriented north-south drop as low as 67 ohm-m. The depth of the fresh-water table here is variable, residing at a depth as shallow as 3 m or as deep as 10 m. In the Stark Desert as well there appears to be no fresh-water–salt-water interface from the Schlumberger surveys conducted. There is very low resistivity overall in the subsurface according to the inversion of the east-west survey. Zones of low resistivity on the order of 40–100 ohm-m extend from 68 m depth toward 10 m below the surface in discrete vertical paths. In between these zones and near the surface, elliptical patches of higher resistivity, from 600 to 2000 ohm-m, have been interpreted to be either boulders or old river channels that have been filled in with more resistive material. The north-south survey from the Stark Desert shows a more homogeneous subsurface with lower resistivity overall, but also shows near-surface anomalies with resistivity higher or lower than the background.

From the Schlumberger resistivity surveys conducted in August 2007 it appears that salt water can be imaged only at sites near the coast. Only at Midtimber Lake and the Gulf of Alaska coast were inverted resistivity values obtained that correspond to the expected resistivity of salt-water–saturated sediments.

In August 2007 three surveys were conducted over time at the Gulf Coast with the array in the same position. The survey was repeated once in 2008 to compare with the previous year's results.

The first survey completed during the August 2008 field season was along the western side of Midtimber Lake (see Figure 10). This survey was the first of a series of pole-dipole arrays. Owing to space constraints, the electrode at infinity could not be placed far enough from the end of the array, and the survey did not provide significantly deeper resolution. However, at this location, sufficient changes in resistivity with depth appear to suggest the presence of salt water. The inversion of the 2008 Midtimber Lake pole-dipole survey is presented in Figure 11. A Schlumberger survey of this area yielded results similar to the previous year's survey.

The August 2008 pole-dipole survey of the Stark Desert site yielded a similar subsurface pseudosection as the 2007 results, with an increased depth penetration of ~23 m. However, the deeper resistivity readings do not show a second significant drop in resistivity. Figure 12 shows the inverted resistivity data for this survey.

A single survey along the Gulf Coast was completed for comparison with the 2007 results. The survey site was farther east and ran east-west along an abandoned aircraft runway. The

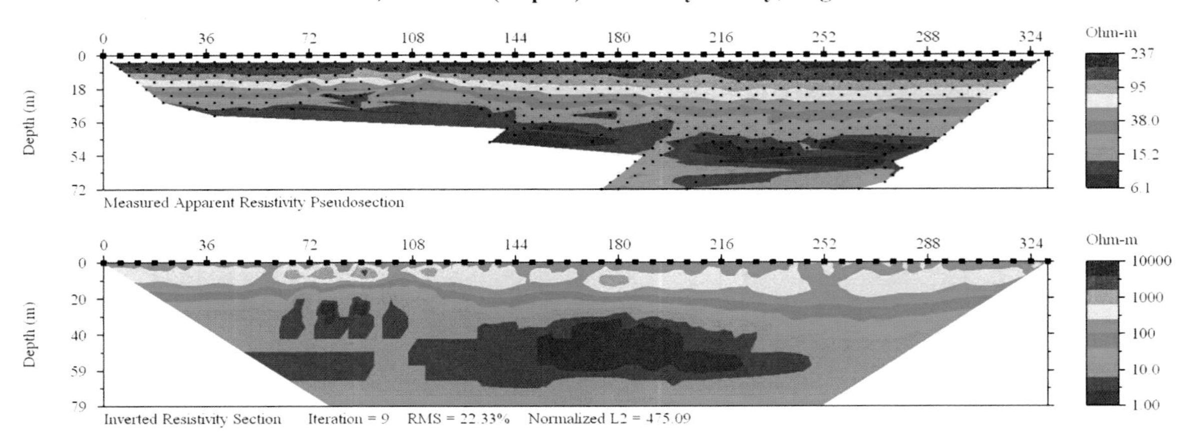

Figure 10. EarthImager 2D result from the 2008 north-south Midtimber Lake survey, a pole-dipole array using 6-m electrode spacing. At ~12 m there is a drop in resistivity of ~500 ohm-m. Between 40 and 50 m there is another drop in resistivity from ~100 ohm-m to 1–10 ohm-m.

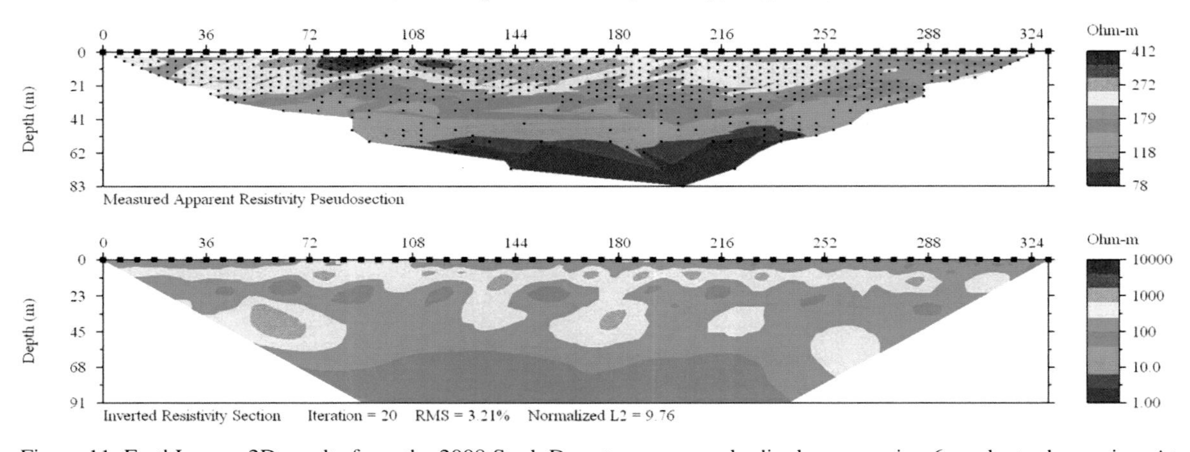

Figure 11. EarthImager 2D results from the 2008 Stark Desert survey, a pole-dipole array using 6-m electrode spacing. At ~20 m there is a resistivity drop of ~300 ohm-m. There is a second, slight drop in resistivity (~100 ohm-m) at ~70 m depth.

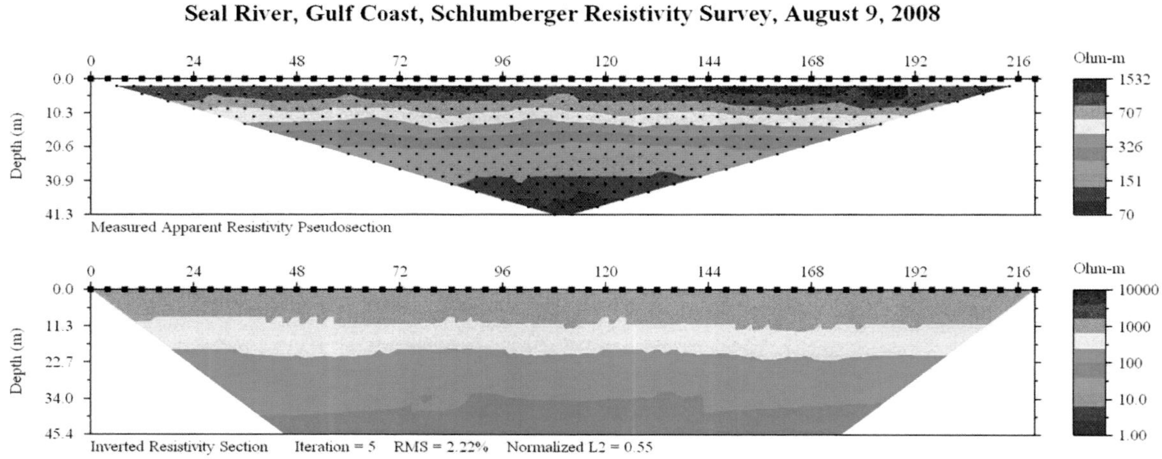

Seal River, Gulf Coast, Schlumberger Resistivity Survey, August 9, 2008

0 24 48 72 96 120 144 168 192 216

Ohm-m 1532 707 326 151 70

Depth (m) 0.0 10.3 20.6 30.9 41.3

Measured Apparent Resistivity Pseudosection

Ohm-m 10000 1000 100 10.0 1.00

Depth (m) 0.0 11.3 22.7 34.0 45.4

Inverted Resistivity Section Iteration = 5 RMS = 2.22% Normalized L2 = 0.55

Figure 12. EarthImager 2D results for the 2008 Gulf Coast survey, a Schlumberger array using 4-m electrode spacing. Horizontal layers suggest a saturated zone with salinity increasing with depth.

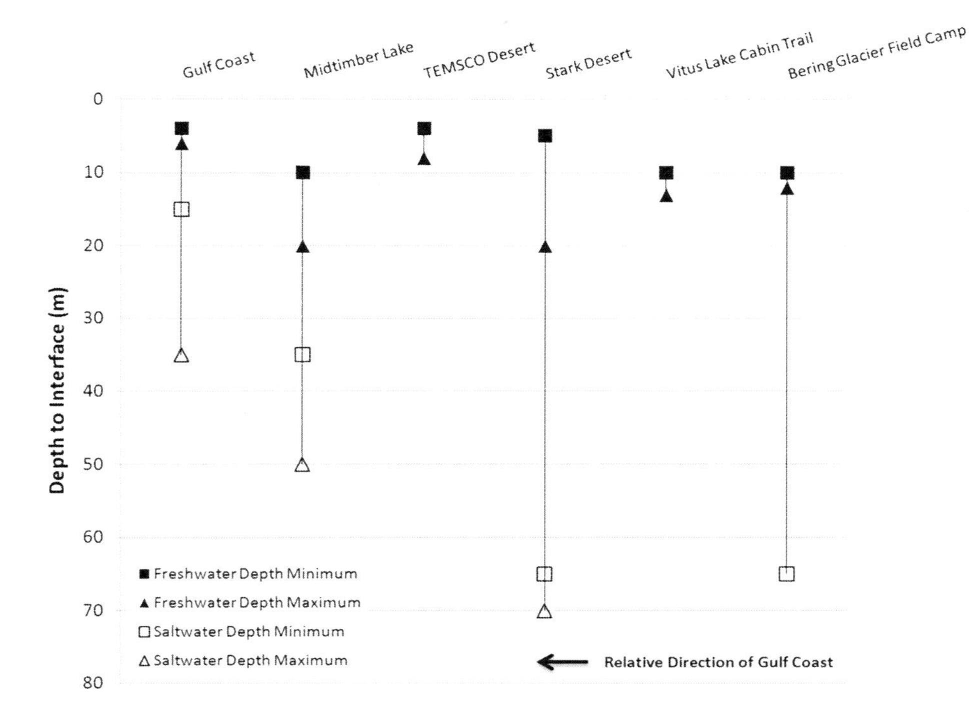

Figure 13. Plot of the minimum and maximum depths to various interfaces. The landward direction is generally to the right, and the coast generally to the left.

survey was conducted with a Schlumberger array and 4-m electrode spacing, and displayed an even layering that shows a relatively gradual change in resistivity from ~1000 ohm-m to 100 ohm-m. Figure 13 presents these 2008 results for the Gulf Coast.

RESULTS AND DISCUSSION

The subsurface setting of the terminal moraine of the Bering Glacier suggests that reasonable amounts of water may flow within it. Ground resistivity surveys suggest that a layer of salt water lies beneath the fresh-water table, which is indicated by a sharp decrease in resistivity values below an inferred fresh-water table. Lateral alignment of the data shows that the salt-water influence appears to diminish farther landward. Depth to the salt-water table at survey sites nearest to Vitus Lake appears too great to suggest correspondence with the salt water at 40 m depth in the lake. The following is a summary of our interpretations (which are quantified in Table 3 and Fig. 13).

Gulf Coast

The Schlumberger surveys in this area indicate that the seasonal fresh-water table exists at a depth of 4–6 m. The fresh-water–salt-water interface seems to be anywhere between 15 and 35 m, depending on the tides. The apparent change in depth to this interface over time is likely a result of the tides.

Midtimber Lake

The Schlumberger surveys at this site suggest that the fresh-water table was at a depth of 17–20 m in August 2007; the fresh-water–salt-water interface was at a depth of ~35 m. The

TABLE 3. INFERRED DEPTHS TO FRESH-WATER TABLE AND FRESH-WATER–SALT-WATER INTERFACE AT EACH SITE, ORDERED FROM SOUTH TO NORTH

Site name	Depth to fresh-water table in August (m)		Depth to fresh–salt-water interface in August (m)	
	Minimum	Maximum	Minimum	Maximum
Gulf Coast	4	6	15	35
Midtimber Lake	10	20	35	40–50
TEMSCO Desert	4	8		
Stark Desert	5–10	20	65?	70
Vitus Lake Cabin Trail	10	13		
Bering Glacier field camp	10	12	65–85?	

bipole-dipole survey suggests that the fresh-water table is present ~15 m below the surface. The fresh-water–salt-water interface is present at ~30–35 m depth.

TEMSCO Desert

The fresh-water table here appears to correspond with the level of that of most other sites, with a likely depth of 8–10 m. The survey results were inconclusive for a salt-water interface, but the data suggest that it may not have been reached even with the additional depth penetration of the pole-dipole array.

Stark Desert

The fresh-water table lies at ~6 m depth. Salt water appears absent, but a contrast in resistivity is present at ~65 m below the surface, which may suggest fresh water and salt water mixing. However, even at the 90-m depth seen in the bipole-dipole survey, resistivity values are not low enough to suggest salt water.

Vitus Lake Cabin Trail

The fresh-water table was determined by the Schlumberger survey to be at ~10–13 m depth. No possible salt-water interface was found from surveys completed at this location.

Bering Glacier Field Camp

Data from this survey suggest a depth of ~10 m for the fresh-water table. Pole-dipole survey results were inconclusive for the salt-water interface but may suggest a considerable resistivity change between 65 and 85 m depth.

CONCLUSIONS

Schlumberger surveys from 2007 compare very well with those from 2008. The increased depth penetration (as deep as 90–100 m) offered by the bipole-dipole surveys is a significant improvement over the penetration of Schlumberger surveys (60–65 m depth with 6 m spacing). At depths of 90–100 m, however, there is still no evidence of salt water in the Stark or the TEMSCO Desert. This does not support the hypothesis of salt water intruding through the subsurface.

The slight increase in depth penetration of the 2008 Midtimber Lake surveys over the 2007 surveys allows the resolution of deeper salt water. Overall, the bipole-dipole surveys of 2008 indicate the same interfaces as the Schlumberger surveys of 2007 did for the Midtimber Lake site, reinforcing the interpretation of the salt-water–fresh-water interface being confined to a depth of 30–35 m at this site. Additionally, both the 2007 and 2008 surveys at this site indicate that this interface dips toward the north (away from the Gulf of Alaska).

In the Stark Desert the 2008 bipole-dipole survey and the 2007 east-west Schlumberger survey show similar subsurface conditions: large, elliptical pockets of higher resistivity. This would seem to indicate that bipole-dipole surveys are as reliable as Schlumberger surveys in this region.

Conclusively, however, the data suggest that the salt-water wedge diminishes toward Vitus Lake, which would suggest a disconnect in the hydrologies of the Gulf of Alaska and Vitus Lake. This is shown by the slight dip of the salt-water–fresh-water interface toward the north at Midtimber Lake. Regardless, salt water has been recorded at a depth of ~40 m in Vitus Lake. The results of the 2007 and 2008 surveys lead to the assumption that the lake and the ocean are not hydraulically connected in the subsurface. However, the presence of salt water in Vitus Lake remains to be explained.

ACKNOWLEDGMENTS

The authors would like to thank the following people for their valuable assistance: from Michigan Tech: Wayne Pennington, Jimmy Diehl, and Nancy Auer; from the U.S. Bureau of Land Management: Scott Guyer, Chris Noyles, and Nathan Rathbun; from the Michigan Tech Research Institute: Robert Shuchman, Liza Jenkins, Erik Josberger, and Luke Spaete; Alaa Shams and students and faculty of the Southern University research team; and from the State University of New York at Oneonta: Jay Fleisher.

REFERENCES CITED

Burger, H.R., Sheehan, A.F., and Jones, C.H., 2006, Introduction to Applied Geophysics: Exploring the Shallow Subsurface: New York, W.W. Norton, 554 p.

Josberger, E.G., Shuchman, R.A., Meadows, G.A., Savage, S., and Payne, J., 2006, Hydrography and circulation of ice-marginal lakes at Bering Glacier, Alaska, U.S.A.: Arctic, Antarctic, and Alpine Research, v. 38, p. 547–560, doi: 10.1657/1523-0430(2006)38[547:HACOIL]2.0.CO;2.

Manuscript Accepted by the Society 02 June 2009

The Geological Society of America
Special Paper 462
2010

Satellite-derived turbidity monitoring in the ice marginal lakes at Bering Glacier

Liza K. Jenkins
Michigan Tech Research Institute, 3600 Green Court, Suite 100, Ann Arbor, Michigan 48105, USA

ABSTRACT

From an evolutionary perspective, glacial lakes at the Bering Glacier System are highly immature and are classified as extremely oligotrophic, resulting from their relatively recent formation and the surrounding harsh, northern climate. Unlike temperate or tropical lakes, northern glacial lakes do not contain significant amounts of biological material. Instead, these lakes are dominated by rock flour, suspended sediment originating from glacial rock weathering. This lack of biological influence makes satellite turbidity mapping and prediction more straightforward and potentially more accurate than similar efforts in temperate or tropical environments, where biology typically drives these systems and strongly affects the remotely sensed, electro-optical signal.

In-situ turbidity data, collected using an autonomous robot buoy, were used to develop a model-based turbidity algorithm. Multiple linear regression analyses were conducted using different Landsat 7 ETM+ bands to determine the best predictor(s) of turbidity in glacial lakes. The final algorithm utilized Landsat 7 ETM+ band 3 (red portion of the electromagnetic spectrum) and band 4 (near-infrared portion of the electromagnetic spectrum) data to predict turbidity concentrations.

Turbidity maps created using the algorithm can be used to help determine inter- and intra-annual sediment dynamics of Vitus Lake. This information could be used to help researchers predict significant glacial events such as outburst floods or surge events. The turbidity maps could also provide insight into the hydrologic routing of the Bering Glacier System by showing where the Glacier is discharging sediment-laden fresh water into Vitus Lake through subsurface conduits. The turbidity algorithm also has broader applicability to other glacial lakes in south-central Alaska and potentially to glacial lakes worldwide.

INTRODUCTION

Recession of the Bering Glacier has resulted in the formation of several ice marginal lakes. The largest of these lakes is Vitus Lake, which lies at the terminus of the Bering Glacier. As the Glacier has retreated, Vitus Lake has expanded rapidly in both area and volume. From 1995 to 2004, this lake has expanded 95.3% in area and 163.1% in volume (Josberger et al., 2006 and this volume). As of 2007, the lake has an area of 138 km^2 (Shuchman et al., this volume). This is a 136.3% change in area since 1995. From an evolutionary perspective, Vitus Lake is considered highly immature, as it was almost nonexistent 10 a ago.

Vitus Lake is very cold and deep. Surface temperatures range from near 0 to +2 °C, and depths exceed 240 m at some localities (Josberger et al., 2006 and this volume). Owing to the relative immaturity of the lake, cold water temperatures, and harsh climate

Jenkins, L.K., 2010, Satellite-derived turbidity monitoring in the ice marginal lakes at Bering Glacier, *in* Shuchman, R.A., and Josberger, E.G., eds., Bering Glacier: Interdisciplinary Studies of Earth's Largest Temperate Surging Glacier: Geological Society of America Special Paper 462, p. 351–360, doi: 10.1130/2010.2462(20).

of the region, almost no phytoplankton, zooplankton, or other lower food web organisms exist in Vitus Lake. This was confirmed through a series of plankton and larval fish tows conducted during the 2007 summer field season (Auer, this volume). Higher food web organisms such as harbor seals, waterfowl, and fish have been observed in Vitus Lake. Interestingly, scat studies of the harbor seals *(Phoca vitulina richardsii)* have revealed only marine prey in their diets, thus indicating only Gulf of Alaska foraging despite their observed residence in Vitus Lake (Savarese, 2004). It is also not known whether the fish reside more permanently in Vitus Lake or if they only seasonally enter the system from the Gulf of Alaska.

The hydrologic routing of the Vitus Lake system is poorly understood because of subglacial and subsurface water movement. Evidence exists, in the form of rock flour—rock crushed to a powder by a moving glacier—in Vitus Lake and moulins on the surface of the glacier of subglacial and subsurface water movement, but specific areas and amounts of discharge into the lake are unknown. Known and quantifiable water contributions into Vitus Lake result from glacier melt, overland flow off the surface of the glacier, iceberg calving from the glacier terminus, runoff from the surrounding environment, and discharge from the Tsiviat Lake basin either through the Abandoned River (periodically and currently dry) or other channel. Known water discharge from Vitus Lake occurs primarily through the Seal River, which connects Vitus Lake with the Gulf of Alaska.

Vitus Lake receives large contributions of sediment from glacial processes, and this sediment is clearly visible from space and easily measured in-situ. The suspended sediment is principally rock flour, and this is virtually the only substance contributing to the high values of turbidity found in this lake. Since there are little other materials, biological or physical, confounding the remotely sensed optical signal, delineation of turbidity concentrations in glacial environments is straightforward.

There is little agreement among researchers on the strength, form, or even optimum wavelengths to be used in determining the relationship between in-situ turbidity measurements and satellite data (Novo et al., 1989). The lack of consensus in previous studies is due in part to differences in sediment type, notably particle size and mineral color (Novo et al., 1989). Water reflectance is affected by the absorption and scattering properties of sediment. Fine-grained material contains more particles and thus scatters more light than would an equal weight of coarse-grained material (Novo et al., 1989). The rock flour in glacial environments is very fine-grained and thus produces high reflectance. As all rock flour is extremely fine-grained, and fine-grained material results in a spectrally more uniform strength of correlation between suspended sediment concentration and reflectance (Novo et al., 1989), glacial environments are good candidates for development of a satellite-based algorithm to measure turbidity.

METHODS

In-situ turbidity data were collected during the first week of August in both 2006 and 2007. Secchi disk data were collected concurrently with the turbidity sampling in 2007. Turbidity data were collected using ALWAS (Automated Lagrangian Water Quality Assessment System), an autonomous drifter buoy (Fig. 1), and Secchi disk data were collected using a Wildco 20 cm diameter professional Secchi disk.

ALWAS is a free-drifting, sail-powered or electrically propelled water-quality measuring and watershed evaluation buoy. It is capable of measuring a data point with multiple parameters as rapidly as every 40 s. Data are transmitted for real-time viewing and are stored for future retrieval and analysis. The stored data are easily downloaded into geographic database (ESRI shapefile) and spreadsheet formats. ALWAS uses advanced sensors to measure water quality parameters and global positioning system (GPS) data. ALWAS is the result of a joint partnership between the Michigan Tech Research Institute (MTRI) and the University of Michigan Marine Hydrodynamics Laboratory.

At the time of data collection the ALWAS buoy was configured to measure and store onboard depth, temperature, conductivity, salinity, total dissolved solids, dissolved oxygen,

Figure 1. ALWAS is an autonomous drifter buoy that measures various water quality parameters, including turbidity, as a function of location. ALWAS was deployed in Vitus Lake during the 2006 and 2007 summer field seasons.

pH, oxidation-reduction potential, turbidity, chlorophyll, blue-green algae, nitrates, ammonium, and chlorides. In addition, the ALWAS buoy also provided the full GPS position and navigation suite, including not only position but also course and speed over ground.

The sensor head in ALWAS is a YSI 6600 sonde. The turbidity sensor is an optical YSI 6136 Turbidity Probe, and the range of measurement of the sensor is from 0 to 1000 NTU (Nephelometric Turbidity Unit), with a resolution of 0.1 NTU (YSI, 2007). The accuracy of the sensor is ±2% of the reading or 0.3 NTU, whichever is greater (YSI, 2007). The performance of the turbidity sensor has been verified through the U.S. Environmental Protection Agency's Environmental Technology Verification Program, and extensive empirical field and laboratory tests document close agreement between in-situ measurements made with the sensor and data from a Hach 2100AN, a laboratory instrument recognized as the standard for turbidity measurement (YSI, 2007).

ALWAS and Secchi disk data were collected at a set of historical sampling stations across Vitus Lake (Fig. 2), where conductivity, temperature, and depth (CTD) measurements were recorded for the past 7 a. These stations were evenly distributed across Vitus Lake and represented the entire range of sediment distribution found in the lake. In 2007, data were collected at sites 1–9 and sites 14–19. Too many icebergs were present in the western part of Vitus Lake (Tashalich Arm) to sample at sites 10–13. In 2006, sites 1–12 were sampled.

ALWAS data were collected by deploying the buoy upon arrival on station, allowing the buoy to float and collect data while other sampling was occurring, and retrieving the buoy upon station departure. Secchi disk data were collected by lowering the disk at the station until it was no longer visible, raising the disk, and taking a reading at the point at which the black and white quadrants became visible. This procedure was repeated until three measurements were recorded for each station, and the mean was calculated for the overall station measurement.

Two Landsat 7 ETM+ images from path 64, row 18, were used in this analysis. The 2006 image was acquired on 7 August 2006, and the 2007 image was acquired on 10 August 2007. The images used in this study were acquired from the U.S. Geological Survey (USGS) and were Level 1T (standard terrain correction) products. No additional processing or corrections were applied to the images. Bands 1–5 and band 7 were used in the analysis. Band 6, the thermal band, and band 8, the panchromatic band, were omitted.

Both of the images used in this study were affected by the scan line corrector (SLC) error, a technical problem the Landsat 7 satellite is currently undergoing. The SLC compensates for the forward motion of the satellite, and without the SLC a series of zigzag patterns (i.e., no data) are visible across the satellite

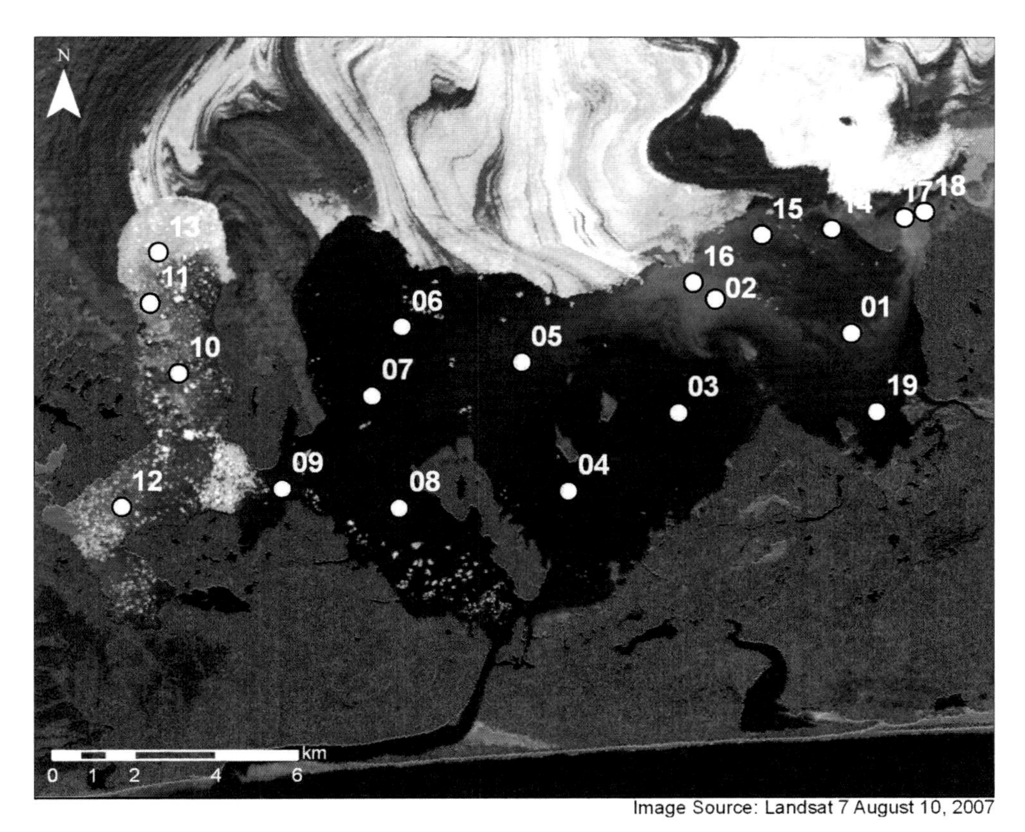

Figure 2. Turbidity and Secchi disk data were collected at a set of historical sampling sites across Vitus Lake. These sites are evenly distributed and represent the entire range of turbidity values found in Vitus Lake.

ground track. Despite this technical issue, the sensor still acquires 75% of the data for any given scene, and the center of the image scene is not affected, according to the U.S. National Aeronautics and Space Administration (NASA). In our images, most of Vitus Lake fell within the unaffected portion of the image scene. Only a small part of the Tashalich Arm section of the lake (the far western part) was affected by the SLC error.

It is important to note that the in situ-data and the Landsat satellite data were collected almost concurrently. The 2006 turbidity data were collected 8 and 9 August, and the Landsat overpass occurred 7 August. In 2007 the turbidity and Secchi disk data were collected 9 August, and the Landsat overpass was 10 August.

An average turbidity measurement and an average radiance (as defined by the digital number [DN] in this study) for each Landsat 7 band were determined for each station location. Recall, radiance is the amount of electromagnetic radiation leaving or arriving at a point on a surface and is proportional to the DN. At every station, the number of individual turbidity measurements ranged from 6 to 42, and these measurements spanned 1–11 pixels. Radiance was determined for each individual turbidity measurement, and these turbidity-radiance measurements were averaged to obtain a single average turbidity and a single average radiance for each band at each station. It is important to note that the spatial variation of turbidity and the average radiance at each site were small (Table 1).

Multiple linear regression methods were used to create a model-based algorithm to predict turbidity. The 2007 data were used to create the algorithm, and the algorithm was validated using 2006 data. Before the algorithm was applied, the Vitus Lake boundary had to be digitized. The icebergs floating in Vitus Lake and the islands in Vitus Lake also had to be identified and excluded from the area of analysis. Several techniques were used for iceberg and island removal, including a slice tool and object-based image classification, but manual identification produced the best results. The automated techniques consistently confused the highly reflective icebergs with highly reflective, sediment-laden water in the eastern part of Vitus Lake, and therefore the automated techniques were not used in this analysis.

RESULTS

Average turbidity at each site ranged from 3.3 to 663.9 NTU in 2007 and from 2.2 to 997.0 NTU in 2006 (Table 1). Average Secchi disk depths ranged from 0 to 13.8 ft in 2007. Average radiance ranged from 51 to 75 DN in 2006 and 47–75 DN in 2007 for band 1, 33–57 DN in 2006 and 34–58 DN in 2007 for band 2, 20–56 DN in 2006 and 18–52 DN in 2007 for band 3, 10–33 DN for 2006 and 10–30 DN in 2007 for band 4, 10–12 DN in 2006 and 9–11 DN in 2007 for band 5, and 9–11 DN in 2006 and 8–11 DN in 2007 for band 7.

ALWAS turbidity values were log (base 10) transformed, and their Pearson correlation with radiance evaluated (Table 2). In 2007 the highest correlation was between log average turbidity and band 3 (0.949), and in 2006 the highest correlation was between log average turbidity and band 4 (0.950). Band 2 also showed relatively high correlations (0.808 for 2007 and 0.882 for 2006) with log average turbidity, but bands 1, 5, and 7 showed very little if any correlation.

TABLE 1. SUMMARY OF THE VALUES OBTAINED FOR AVERAGE RADIANCE, DN, FOR LANDSAT BANDS 1–5 AND BAND 7

	2006							2007							
Site	B1	B2	B3	B4	B5	B7	Avg. turb.	B1	B2	B3	B4	B5	B7	Avg. turb.	Avg. Secchi
01	68	55	50	20	10	10	42.7	69	48	28	10	9	9	5.4	6.1
02	69	55	52	22	11	10	149.5	68	51	46	15	9	8	101.1	1.9
03	75	50	29	11	10	9	12.0	58	34	18	10	10	9	3.3	13.8
04	66	39	20	10	10	10	6.8	57	34	18	10	10	11	7.0	7.2
05	69	46	26	12	10	10	13.7	72	45	23	10	10	10	9.6	5.1
06	67	40	21	10	10	9	6.7	59	36	19	10	11	8	4.5	9.1
07	66	40	21	10	10	10	5.7	64	40	21	10	10	9	9.2	7.5
08	63	39	21	10	10	10	4.2	62	38	21	10	11	10	5.6	9.2
09	55	42	32	12	10	9	18.1	47	35	22	11	10	9	7.5	3.1
10	51	33	21	10	10	10	3.0	NA	NA	NA	NA	NA	NA	NA	NA
11	52	34	20	10	10	10	2.2	NA	NA	NA	NA	NA	NA	NA	NA
12	58	45	43	17	10	9	51.9	NA	NA	NA	NA	NA	NA	NA	NA
14	NA	NA	NA	NA	NA	NA	NA	68	48	30	10	10	9	15.9	1.7
15	NA	NA	NA	NA	NA	NA	NA	69	52	37	12	9	10	25.0	2.4
16	NA	NA	NA	NA	NA	NA	NA	68	51	46	15	9	8	87.0	2.4
17	NA	NA	NA	NA	NA	NA	NA	72	57	52	30	10	10	663.9	0
18	NA	NA	NA	NA	NA	NA	NA	75	58	51	18	10	11	130.5	0.9
19	NA	NA	NA	NA	NA	NA	NA	68	46	25	10	9	9	4.7	7.1
Plug	71	57	56	33	12	11	997.0	NA	NA	NA	NA	NA	NA	NA	NA

Note: Average turbidity (NTU), and average Secchi disk depth (ft) are presented for each site.

TABLE 2. PEARSON CORRELATION COEFFICIENTS FOR LOG AVERAGE TURBIDITY AND LANDSAT BAND AVERAGE RADIANCE SHOW CONSISTENTLY HIGH CORRELATIONS FOR BANDS 3 AND 4 FOR THE TWO YEARS

	2007 log average turbidity	2006 log average turbidity
Band 1	0.566	0.479
Band 2	0.808	0.882
Band 3	0.949	0.938
Band 4	0.904	0.950
Band 5	–0.185	0.909
Band 7	0.039	0.482

A relatively high Pearson correlation (–0.871) was observed between the log Secchi data and the log turbidity data. Despite the high correlation between these two variables, only the turbidity data were used in this analysis. Secchi disk measurements were deemed to be inherently more subjective and thus more prone to inconsistencies and reduced accuracy. The high correlation observed between these variables did indicate the potential use of Secchi disk data in the absence of future or historical turbidity data.

Regression Modeling

Average radiance values from bands 3 and 4 were both used as independent variables in a multiple linear regression (MLR) analysis. The resultant model was $Y = -0.441 + 0.0340*B3 + 0.0507*B4$. The MLR model produced an excellent fit with an r^2 value of 0.956 and an adjusted r^2 value of 0.949. Other band combinations were used as inputs into MLR models, and while I was able to achieve higher r^2 values, I was unable to achieve a higher adjusted r^2 value than with the combined band 3 and band 4 input. Additionally, an MLR model using all of the Landsat bands resulted in *p*-values statically insignificant for bands 1, 2, 5, and 7. Thus these bands did not make a significant contribution to estimating log average turbidity and were not included in the final model.

The 2006 turbidity and Landsat data were not used to create the models. These data were intentionally set aside for validation after the models were created. The MLR model and the 2006 Landsat bands 3 and 4 data were used to determine how well the model predicted the known 2006 turbidity values. Residual and standardized residual plots for the model validation analysis showed that all the residuals fall within +0.2 and –0.3 log NTU from the actual values. Standardized residuals were within ±1.5 standard deviations from the mean. These small deviations between actual and predicted values indicated that the model fit the 2006 data very well.

Turbidity Maps

The model-based algorithm was used to predict log turbidity values at every pixel in Vitus Lake from the 2006 and 2007 Landsat images. The resultant turbidity maps (Figs. 3, 4) showed definite turbidity gradients and turbidity hotspots across Vitus Lake. In general, the maps indicated more sediment loading in 2006.

The 2006 turbidity map showed approximately four locations of high turbidity concentrations in Vitus Lake. These areas were in the far northeastern part of the lake, the far southeastern part of the lake, along the glacier terminus in the northern part of the lake, and in the far southwestern part of the lake. The 2007 map showed approximately two areas of high turbidity, and although these areas also displayed high values in 2006, the spatial extents of these turbidity plumes decreased. The high turbidity concentrations from 2007 were located in the far northeastern part of the lake and along the glacier terminus in the northern part of the lake. The 2007 turbidity map did not contain the far western part of Vitus Lake (the Tashalich Arm section) because in 2007 too many icebergs were present to run the algorithm in this area of the lake.

DISCUSSION AND CONCLUSIONS

In the future, the developed algorithm can be used both forward and backward in time, taking advantage of the long-term historical Landsat record, without the need for further in-situ data inputs. All that is required for the algorithm is a Landsat image scene for the area of interest. The algorithm was developed for use in Vitus Lake, but it also has broader applicability to other lakes within the Bering Glacier Region, other glacier-influenced lakes in coastal and south-central Alaska, and even to similar glacier systems worldwide.

All of the Landsat bands were tested for their suitability in a final turbidity algorithm, but bands 3 and 4 as independent variables in a multiple linear regression equation produced the best fit. Recall that band 3 senses energy in the red portion of the electromagnetic spectrum and band 4 senses energy in the near infrared portion. Previous studies have identified these portions of the electromagnetic spectrum as useful for detection of sediment in aquatic environments, but previous studies also identified other bands and band combinations as useful without much consensus among the different studies. Further confounding consensus is that previous research activities were conducted in various geographical areas with different factors that affected the aquatic system and the remotely sensed electro-optical signal. Very little evidence exists of aquatic remote sensing applications in northern environments where conditions are typically more spatially homogeneous, less affected by biological influences, and thus more likely to produce more consistent and accurate results.

The Pearson correlation coefficients and the adjusted r^2 values from this analysis are much higher than those reported in other studies. These high values were obtained despite any attempt at a window of analysis or additional atmospheric correction of the Landsat data. In my search of the literature I found that most if not all of the other studies involving aquatic remote sensing applications utilize some sort of atmospheric correction, window of analysis, or both, to achieve a higher correlation between remote sensing data and ground reference data. Baban (1993)

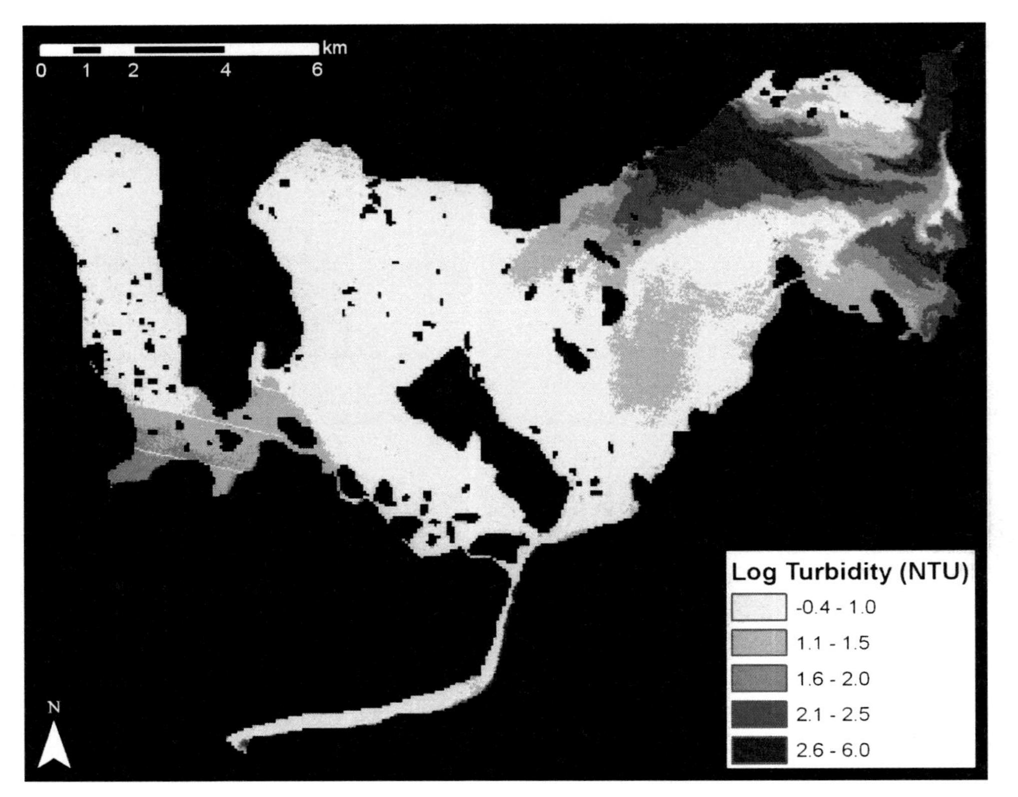

Figure 3. The 2006 Vitus Lake turbidity map shows approximately four turbidity hotspots in Vitus Lake. These hotspots occur in the far northeastern part of the lake, the far southeastern part of the lake, along the glacier terminus in the northern part of the lake, and in the far southwestern part of the lake.

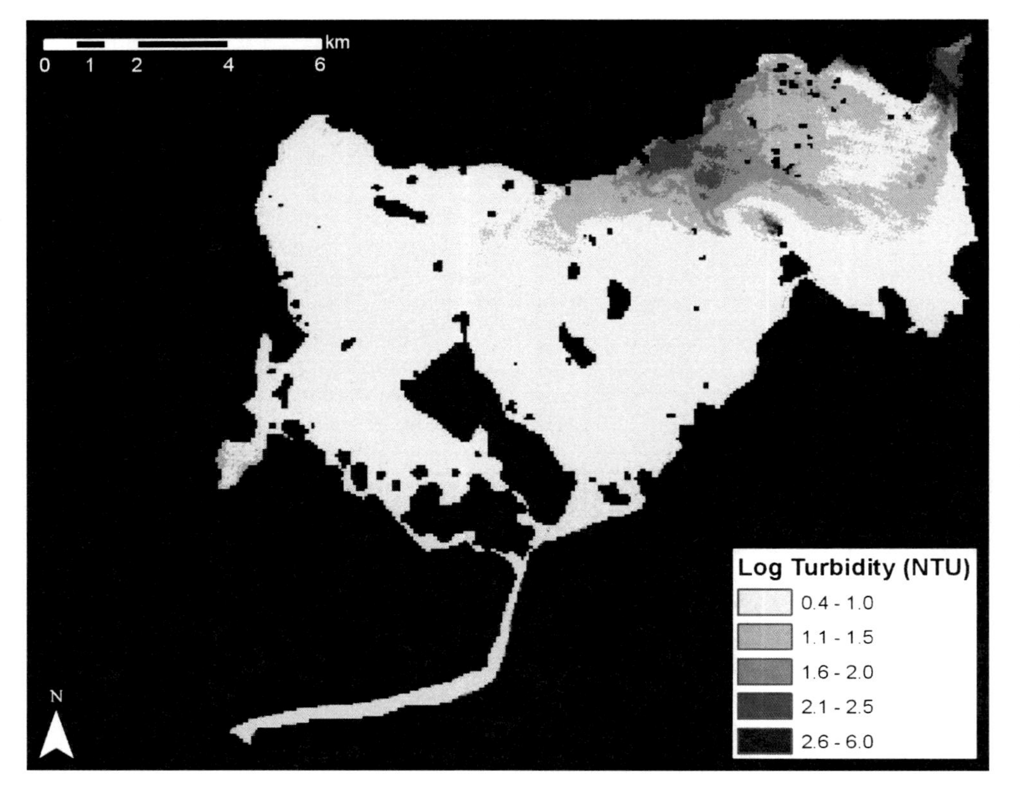

Figure 4. The 2007 Vitus Lake turbidity map shows approximately two turbidity hotspots in Vitus Lake, which occur in the far northeastern part of the lake and along the glacier terminus in the northern part of the lake.

tested windows of 1 by 1, 2 by 2, and 3 by 3 in his research, and found that the 3 by 3 window was the most representative in correlation with his ground reference data. Wang et al. (2006) conducted elaborate atmospheric corrections consisting of both radiometric correction and the application of an improved image-based COST (cosine of the solar zenith angle) model before correlating reflectance data and water quality parameters.

Preprocessing of my image scene or utilizing a window of analysis was not necessary for two main reasons. First, as discussed above, the study site was in a northern environment. Second, the spatially extensive ALWAS data provided high-accuracy data for creation of the algorithm.

The ALWAS data had many different benefits in this analysis. First, the ALWAS data collected GPS information for each data point. This meant that I was able to accurately assign turbidity data points to specific Landsat pixels and thus eliminate any errors associated with wrong or slightly incorrect GPS information. Second, the ALWAS buoy, as it drifted, provided many data points for each site. With many data points, I was able to take the average and get a more accurate measurement of ground truth. Third, the ALWAS buoy typically covered more than one Landsat pixel. By having several data points in one pixel and/or data scattered among several pixels, the potential of problems associated with mixed pixels or edge contamination was reduced. In the literature, most studies relied upon single point measurements of ground reference data. This type of research design does not take into account the possibility of the point measurement occurring on the pixel edge, nor does it provide enough data to obtain an accurate representation of ground truth.

The developed turbidity algorithm produced an adjusted r^2 of 0.949, which indicated that the algorithm fit the data very well. Site 17 was the only statistically identified outlier, and this outlier was present in all three of the regression analyses. This site had the highest observed turbidity in 2007 with an average turbidity of 663.9 NTU, and turbidity measurements at this site were much higher than those of any of the other sites. The next highest average turbidity was 130.5 NTU, which was observed at site 18. Statistically, it might have made sense to remove this outlier, but from a prediction sense, it was important to include this point in the analysis because it represented the upper bound of turbidity values found in Vitus Lake. For this reason, this point was not removed, and a log transform was applied to the turbidity data.

The identification of site 17 as an outlier leads to the conclusion that the algorithm has better predictive power at lower turbidity values. The algorithm could most likely be improved through the inclusion of more points with higher turbidities, but the first step would be to test the prediction power of the current algorithm using more known, high-value turbidity points. The algorithm might also be improved by looking into nonlinear fitting techniques or spline fitting.

The turbidity maps of Vitus Lake provide unique insight into sediment dynamics at the Bering Glacier. Mapping of the sediment distribution in this lake is important because it provides information about the hydrologic routing of the Bering Glacier System. Rock flour is transported to the ice marginal lakes through the glacier's internal plumbing system, and at this point in time the plumbing is largely unknown. Moulins, which are narrow, tubular chutes or crevasses through which water enters a glacier from the surface, carry fresh water deep into the bowels of the Bering System, but no further evidence of the course or final point of discharge exists. Currently, the best indication of where subglacial discharge occurs is the sediment concentrations in Vitus Lake and the other surrounding ice marginal lakes.

The 2006 and 2007 turbidity maps show several areas of high suspended sediment concentration. These hotspots are most likely to be where the glacier is discharging sediment-laden fresh water through subsurface conduits. Interestingly, the 2006 map shows more sediment and more sediment hotspots than the 2007 map. This indicates that the hydrologic routing of the glacier changed between the acquisitions of these two images.

A significant glacial event occurred in late August 2006 that explains many of the interannual changes observed in the turbidity maps. While out in the field in early August 2006, we noticed an upwelling of high turbidity, super-cooled water at the terminus of the glacier in the far northeastern portion of Vitus Lake. Jay Fleisher and his team reported that a gradual, yet progressive leakage occurred at this "plug" location beginning in late June 2006, and that the failure of this plug eventually occurred before the end of the summer melt season (Fleisher et al., 2007 and this volume). The failure of the plug resulted from pressure behind the ice dam separating the higher elevation Tsiviat Lake basin from the lower elevation Vitus Lake. Before the failure event, the Tsiviat Lake basin drained through the Abandoned River, but during our August 2007 field work we noticed that the Abandoned channel had dried up and is currently abandoned. This was also confirmed with the 2007 Landsat image (Fig. 5).

The turbidity maps showed evidence of the plug failure event. First, the Abandoned River discharge location in the far southeast part of the lake was one of the turbidity hotspots in the 2006 map. The 2007 map did not show any turbidity at the Abandoned River discharge location, thus providing evidence for a change in the hydrologic routing of the system after the failure event. Second, the hotspot areas in the far northeastern and northern parts of the lake changed in both sediment concentration and spatial extent between the two years. This was also indicative of a significant change in the hydrologic routing of the system. It appeared as if sediment were being vigorously discharged about the time of the plug failure, but one year after the failure the discharge had lessened. This could have been due to a reduction in pressure associated with the outburst. Recall from the Shuchman et al. Remote Sensing and Ablation chapters (this volume) that 2006 was a record warm summer with resulting increase in glacial melt. More melt results in increased water discharge from the glacier and thus subsequently more sediment discharge. This is also a contributing factor in the difference between the 2006 and 2007 turbidity maps.

It is important to monitor changes in the sediment distribution in Vitus Lake, because changes can be both a precursor and a

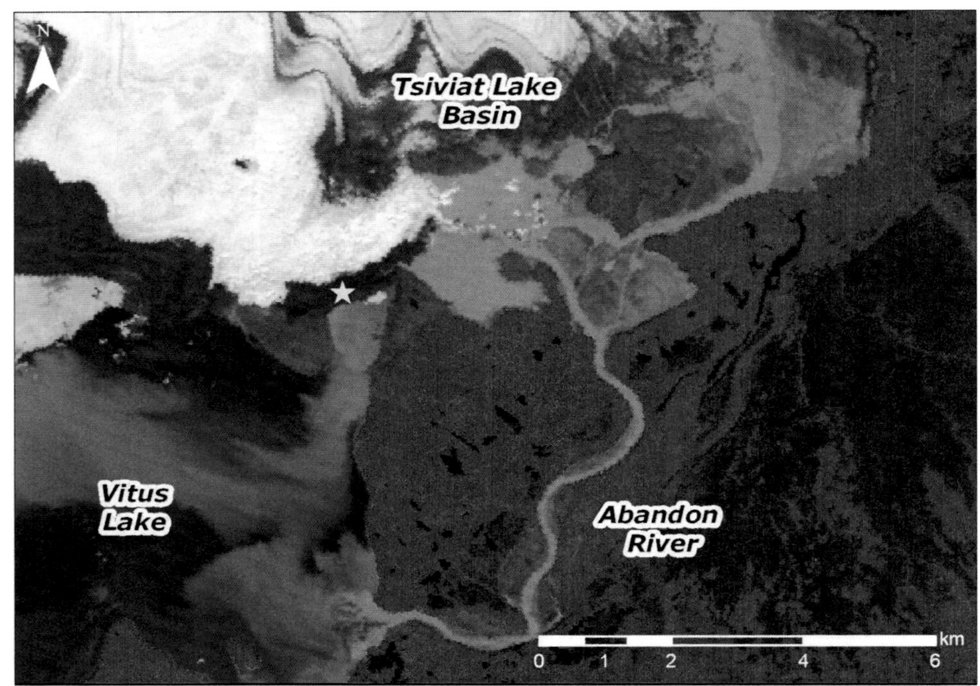

Image Source: Landsat 7 August 7, 2006

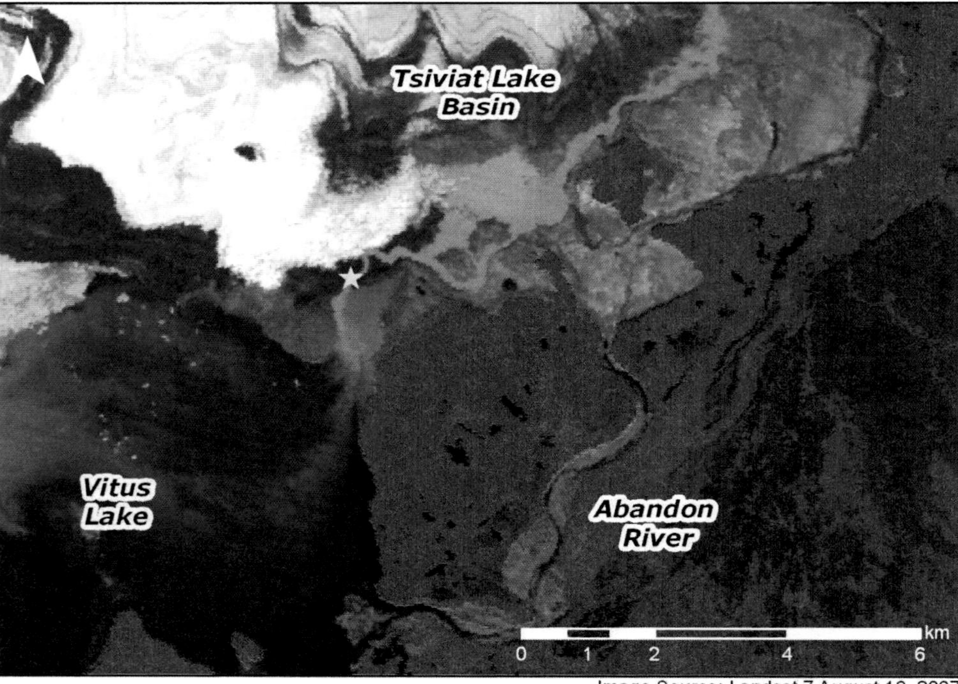

Image Source: Landsat 7 August 10, 2007

Figure 5. The "plug" failure that occurred before the end of the 2006 melt season resulted in several changes in the landscape and the hydrologic routing of the glacier system. The plug is marked with a yellow star in both images. The 2006 image (top) was acquired before the event, and this image shows the conditions prior to the failure event. The event changed the outflow location of the Tsiviat Lake basin, which drained in 2006 through the Abandoned River and is currently draining through the plug location (2007 bottom image). Note that the Abandoned River is no longer discharging sediment into Vitus Lake (bottom image).

successor of significant glacial events. As described above, vigorous sediment discharges are associated with plug failure events. Similar conditions might occur before a glacier surge event or an outburst flood event. More observations like those reported here will be needed to help identify and differentiate early warning signals of significant glacial processes.

The user of these turbidity maps and the turbidity algorithm must be cautioned about a few potential caveats. The first is that some pixels at the edge of the lake boundary may show elevated turbidity values. This could be a result of bottom reflectance occurring in areas of shallow, clear water or the result of shoreline contamination where the pixel may not actually contain lake area (owing to poor digitization of the lake boundary), or the pixel may contain a mixture of land and lake area (owing to the spatial resolution of the satellite sensor). With any of the above scenarios, the pixel is not reflecting an actual turbidity measurement and should be ignored. Another potential problem, similar to the first, is the possibility of bottom reflectance in shallow areas of the lake. Vitus Lake is extremely deep in most areas, but this could be a concern in the far southwestern part of the lake. The 2006 turbidity map (Fig. 3) shows a turbidity hot spot in this section of the lake, and it is difficult to determine if this is the result of subglacial discharge or shallow water. From the Vitus Lake bathymetric data (Josberger et al., 2006 and this volume) we know that this part of the lake is shallower, but it is difficult to know whether bottom reflectance is occurring. My experience indicates that bottom reflectance is not occurring, and that the turbidity hotspot here is due to glacial discharge.

FUTURE RESEARCH

The next step for further research in this field would be to apply the turbidity algorithm to other lakes in the Bering Glacier Region. From satellite images I have observed definite differences in turbidity among the lakes, and, interestingly, I have observed stark turbidity differences among lakes in close proximity to one another despite any obvious surface connections (Fig. 6). Information obtained from running the turbidity algorithm on these lakes may provide insight into the hydrologic routing of the glacier. At this point it appears as if the glacier is discharging sediment-laden water into some of the lakes through underground conduits. If this is the case, some of these conduits would have to cover relatively large distances and have highly localized discharge areas.

Future research activities could be focused on further inter- and intra-annual comparisons of turbidity maps. By running the algorithm on several Landsat images throughout a single year, we would be able to gain insight into the seasonal dynamics of sediment distribution. Through yearly comparisons of turbidity maps created using historical Landsat data, we would be able to quantify and better understand how the sediment distribution changes over a larger time period. We could also look into how sediment loading is affected by glacier surge and retreat.

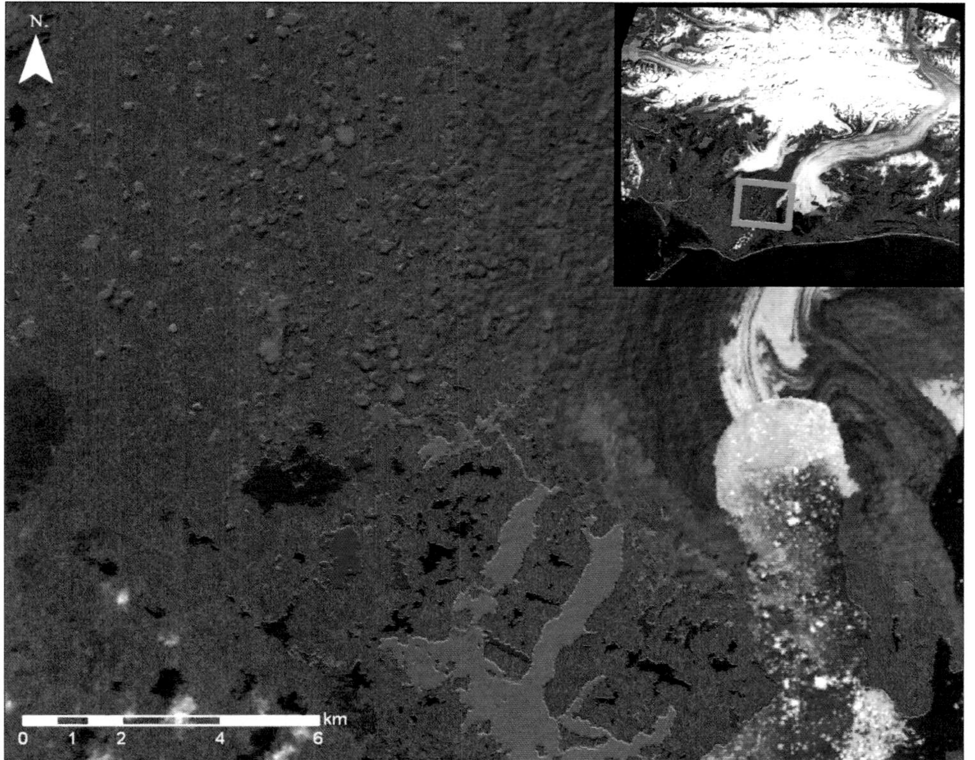

Figure 6. Large turbidity differences exist among surrounding lakes in the Bering Glacier Region. These lakes have very different turbidity concentrations despite their close geographical proximity. These differences in turbidity could be indicative of subsurface glacial discharge through underground conduits.

Both Landsat 5 and 7 are undergoing technical difficulties, and both satellites have performed beyond their expected life span. The Landsat Data Continuity Mission is currently seeking replacement options for the old and ailing Landsat satellites in orbit, but the next-generation Landsat satellite is not expected to be launched until 2011 (USGS). Despite the uncertainties surrounding the Landsat program, other platforms exist that could provide an input data replacement for the algorithm. For example, ASTER (Advanced Spaceborne Thermal Emission and Reflection Radiometer) is a viable alternative. ASTER is an instrument on the Terra EOS (Earth Observing System) satellite launched December 1999 (Jet Propulsion Laboratory: JPL). ASTER captures data in 14 bands ranging from the visible to the thermal infrared wavelengths. ASTER bands 2 and 3 (0.63–0.69 μm and 0.76–0.86 μm, respectively) are almost exactly the same as Landsat EMT+ bands 3 and 4. The spatial resolution of ASTER bands 2 and 3 (15 m) are also similar to Landsat EMT+ bands 3 and 4 (30 m) (JPL).

The algorithm developed as part of this research is applicable to other glaciers in coastal and south-central Alaska. Along the coast there are many other glacial systems similar to the Bering Glacier, and the algorithm could be useful in determining glacier dynamics and sediment loading from a regional perspective. This algorithm is also applicable to all lake systems affected by sediment-laden glacial discharge.

The algorithm may have to be slightly modified (i.e., the coefficients adjusted) in order to be more broadly applicable to regional and global systems, and this would be accomplished through the inclusion of more data points. Ideally, ALWAS turbidity data points would be best, but in the situations where this type of data is not available, turbidity data collected using traditional sampling techniques or Secchi disk data could be substituted. Recall from the Results section of this paper that a strong correlation was observed between the turbidity and the Secchi disk data. Secchi disk data would not be ideal but would offer a low-cost alternative. Additionally, as the Secchi disk has been available since 1865 (Cole, 1994), historical Secchi data may be obtainable for some areas. Historical data would provide important data points from the perspective of a time series analysis.

In the face of a changing climate and political uncertainty it is more important than ever to monitor glaciers and their associated environments. Glacial environments will be the first affected by a warmer climate, and these environments have the potential to act as early warning systems, but we need more data to support scientific claims. The developed algorithm has the potential to provide data, and to provide these data remotely and at reduced cost over traditional, field-based data collection efforts.

ACKNOWLEDGMENTS

This work was originally presented as my master's thesis from the University of Michigan. I would like to thank my thesis committee, Mike Wiley, Guy Meadows, and Robert Shuchman, for their input and direction. The data collection for this study would not have been possible without the U.S. Bureau of Land Management (BLM), and specifically John Payne, Scott Guyer, Chris Noyles, and Nathan Rathbun from the BLM. Their smooth operation of the Bering Glacier camp is greatly appreciated, and their behind-the-scenes efforts make the research carried out at camp a success. I would also like to thank those that have helped in the field with data collection, including Bob Shuchman, Luke Spaete, and Erik Josberger from the Michigan Tech Research Institute (MTRI); Guy Meadows and Hans Van Sumeran from the University of Michigan; and Scott Guyer from the BLM.

REFERENCES CITED

Baban, S.M.J., 1993, Detecting water quality parameters in the Norfolk Broads, U.K., using Landsat imagery: International Journal of Remote Sensing, v. 14, p. 1247–1267, doi: 10.1080/01431169308953955.

Cole, G.A., 1994, Textbook of Limnology, 4th edition: Prospect Heights, Illinois, Waveland Press, 412 p.

Fleisher, P.J., Bailey, P.K., and Natel, E.M., 2006, Ice dam breakout at Bering Glacier, Alaska: Philadelphia, Pennsylvania, Geological Society of America 2006 Annual Meeting, Abstracts with Programs, v. 38, no. 7, p. 236.

Josberger, E.G., Shuchman, R.A., Meadows, G.A., Savage, S., and Payne, J., 2006, Hydrography and circulation of ice-marginal lakes at Bering Glacier, Alaska, U.S.A.: Arctic, Antarctic, and Alpine Research, v. 38, p. 547–560, doi: 10.1657/1523-0430(2006)38[547:HACOIL]2.0.CO;2.

Jet Propulsion Laboratory, NASA, 2007, ASTER: Advanced Spaceborne Thermal Emission and Reflection Radiometer: http://asterweb.jpl.nasa.gov/index.asp.

National Aeronautics and Space Administration (NASA), The Landsat Program: http://landsat.gsfc.nasa.gov/.

Novo, E.M.M., Hansom, J.D., and Curran, P.J., 1989, The effect of sediment type on the relationship between reflectance and suspended sediment concentration: International Journal of Remote Sensing, v. 10, p. 1283–1289, doi: 10.1080/01431168908903967.

Savarese, D.M., 2004, Seasonal trends in harbor seal (*Phoca vitulina richardsii*) abundance at the Bering Glacier in south central Alaska [M.S. thesis]: University of Alaska, Anchorage, 70 p.

Shuchman, R.A., Josberger, E.G., Jenkins, L.K., Payne, J.F., Hatt, C.R., and Spaete, L., 2010a, this volume, Remote sensing of the Bering Glacier region, *in* Shuchman, R.A., and Josberger, E.G., eds., Bering Glacier: Interdisciplinary Studies of Earth's Largest Temperate Surging Glacier: Geological Society of America Special Paper 462, doi: 10.1130/2010.2462(03).

Shuchman, R., Josberger, E., Hatt, C., Roussi, C., Fleisher, J., and Guyer, S., 2010b, this volume, Bering Glacier Ablation Measurements, *in* Shuchman, R.A., and Josberger, E.G., eds., Bering Glacier: Interdisciplinary Studies of Earth's Largest Temperate Surging Glacier: Geological Society of America Special Paper 462, doi: 10.1130/2010.2462(05).

U.S. Geological Survey (USGS), 2007, Landsat Data Continuity Mission: http://ldcm.usgs.gov/LDCMHome.php.

Wang, F., Han, L., Kung, H.-T., and Van Arsdale, R.B., 2006, Applications of Landsat-5 TM imagery in assessing and mapping water quality in Reelfoot Lake, Tennessee: International Journal of Remote Sensing, v. 27, p. 5269–5283, doi: 10.1080/01431160500191704.

YSI, 2007, YSI 6136 Turbidity Sensor Spec Sheet: http://www.ysi.com/DocumentServer/DocumentServer?docID=EMS_E56_6136TURBIDITY.

Manuscript Accepted by the Society 02 June 2009

The Geological Society of America
Special Paper 462
2010

Investigating biological productivity in Vitus Lake, Bering Glacier, Alaska

Nancy A. Auer*
Department of Biological Sciences, Michigan Technological University, Houghton, Michigan 49931, USA

ABSTRACT

The fresh water from melting glaciers gradually collects in lake systems; however, global warming accelerates this input. How does vigorous melt influx impact biological systems within the fresh-water lake environment? A freshman biology student assisted in this research experience. Using benthic grab samplers and seines, macroinvertebrates and fish were collected from nearshore areas of Vitus Lake, Bering Glacier, Alaska. Water <1 m near the southern shoreline proved to be important habitat for several fish species and benthic invertebrates consumed by those fish. This nearshore zone of productivity may be important nursery areas for several species of fish, and these fish in turn may be important to nesting and migrating waterfowl.

INTRODUCTION

Vitus Lake, a fresh-water lake now expanding at the base of Bering Glacier, was 114 km^2 in 2003 (Josberger et al., 2006). Investigations of the biological components of Vitus Lake are limited (von Hippel and Weigner, 2004), and there has been no study of the food web in this lake. In 2007, open lake sampling for zooplankton indicated low to no productivity owing to suspended rock flour and cold temperatures. In 2008, investigations of biologic components focused on the very nearshore region of Vitus Lake, which has some salt-water intrusion at depth (Josberger et al., 2006). The Seal River drains Vitus Lake, and some tidal fluctuation is ongoing in that region. Salinity in Vitus Lake averaged 2.5 practical salinity units (psu) for the upper 10 m in 2001–2003 and rose to 17.5 psu at 60 m or more in depth (Josberger et al., 2006). The objective of this study was to determine if locations farthest from the glacier were more biologically productive than newer sites near the glacier. In 2008, four fresh-water stations were selected on the basis of a gradient of productivity observed along the southern shoreline, a region least impacted by glacial surge and retreat (Fig. 1). The research experience included a freshman biology major from Michigan Technological University, who assisted in the field and in processing samples. The student was exposed to several unique field sampling experiences, sampling in extreme environments, how organization and planning are important to obtain adequate data, and the importance of team effort and concern. By exposing a young Midwestern student to such diverse experiences, I hoped to instill excitement and concern for the environment and future scientific pursuits.

METHODS

In early August 2007, preliminary investigation revealed a small zone of productivity primarily along the southern edge of Vitus Lake. This narrow zone produced green algae, periphyton, invertebrate amphipods, and some small fish. In early August 2008, a greater effort was made to sample macroinvertebrates and fish. At each station, invertebrates were sampled using a Wildco benthic "Mighty Grab" sampler (277.5 cm^2) with 210 µm mesh screen adaptors, and washed through a 210 µm sieve. Fish were captured using a 1.22 × 4.57 m, 0.32 cm mesh seine, and four 20-hook (no. 6) set-lines baited with pieces of frozen fish. All captured fish were weighed and measured on site, and all but voucher specimens were released alive. Samples were preserved

*naauer@mtu.edu

Auer, N.A., 2010, Investigating biological productivity in Vitus Lake, Bering Glacier, Alaska, *in* Shuchman, R.A., and Josberger, E.G., eds., Bering Glacier: Interdisciplinary Studies of Earth's Largest Temperate Surging Glacier: Geological Society of America Special Paper 462, p. 361–364, doi: 10.1130/2010.2462(21).

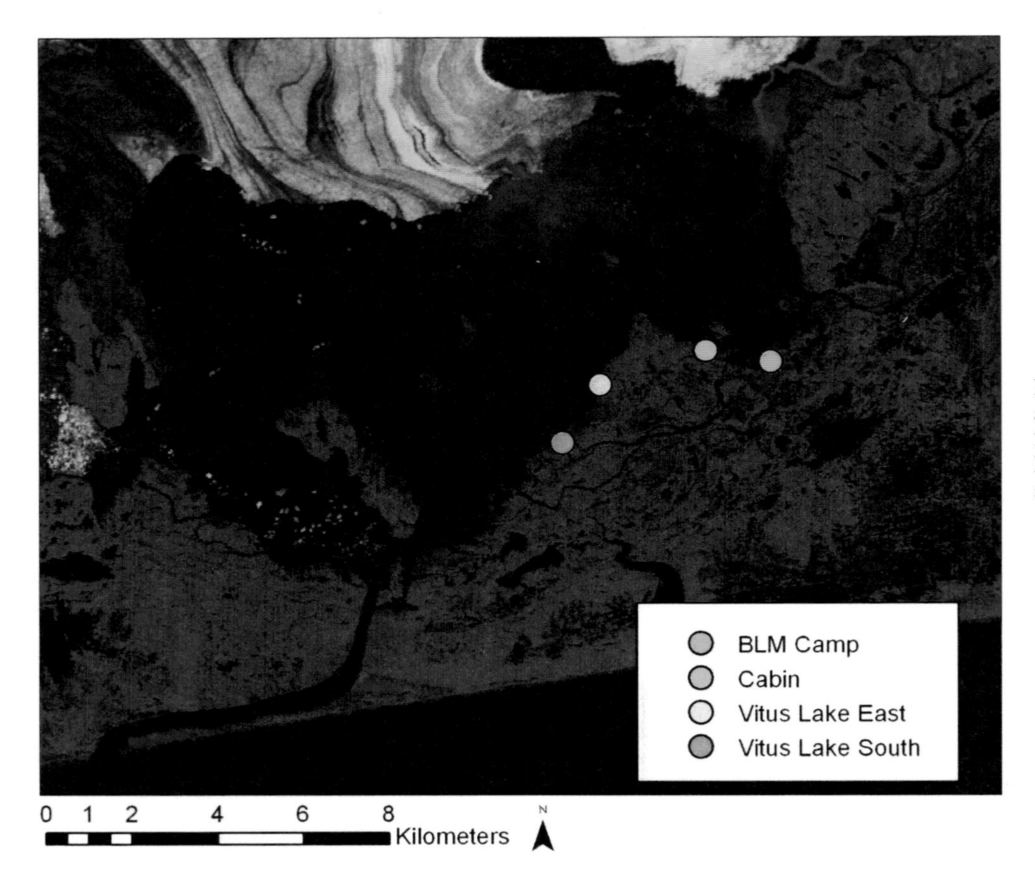

Figure 1. Map of Vitus Lake, Alaska, and the four sampling locations for 2008. (Supplied by R. Shuchman and L. Jenkins, Michigan Technological Research Institute, Ann Arbor.)

either frozen or in alcohol and were later examined in a laboratory. Gut samples from four fish were analyzed for content.

RESULTS AND DISCUSSION

Biological activity was confined to a very narrow region in comparison with the vast surface area of Vitus Lake (Table 1). A complete survey of the Vitus Lake shoreline has not yet been accomplished; however, an area within 15 m of the shoreline and <1 m in depth along some areas of the southern edge of the lake provides habitat for a variety of benthic organisms and algae as well as for some fish species (Table 2). Sampling using a seine collected adults and juveniles of threespine stickleback, *Gasterosteus aculeatus*; and juveniles of sockeye salmon, *Onchorhynchus nerka*; starry flounder, *Platichthys stellatus*; and speckled sanddab, *Citharichthys stigmaeus*. On the set-lines, only Pacific staghorn sculpin, *Leptocottus armatus*, both adult and juveniles, were collected (Table 3). Many of these species are known for their ability to live in fresh and brackish waters (Mecklenburg et al., 2002). One of the fish species, speckled sanddab, encountered in this study had not been reported for Vitus Lake previously. Major invertebrates included snails, amphipods, and chironomids, not yet identified beyond family level.

Gut analysis of four threespine sticklebacks, the most abundant fish seen in the lake, revealed stomachs filled with amphipods and chironomids, benthic invertebrates known to utilize nearshore lake habitats and common to the diet of fishes (Nalepa, 1987; Pothoven and Nalepa, 2006). These fish are also important

TABLE 1. MAXIMUM DEPTH AND DISTANCE FROM SHORE OF BIOLOGICAL PRODUCTIVITY, VITUS LAKE, 1–8 AUGUST 2008

Site, 2008	Maximum depth (m)	Distance from shore (m)	Total number of invertebrates/m^2	Water temp. (°C)	Seined fish/m^2
Base Camp	<0.3	< 1	10.7	n.a.	0
Cabin	0.64	12.35	167.8	6	0
Vitus Lake East	0.33	6.43	3953.6	8	0
Vitus Lake South	0.28	15.71	2223.2	9	0.48

TABLE 2. SPECIES OF FISH COLLECTED ON SET-LINES AND WITH SEINE, VITUS LAKE, 1–8 AUGUST 2008

Fish gear	Common and species name	Number and life stage	Range total length (mm)	Range weight (g)
Baited set-line	Pacific staghorn sculpin *Leptocottus armatus*	10 adult	242–135	178–28
Seine	Sockeye salmon *Onchorhynchus nerka*	1 smolt	70	2
Seine	Starry flounder *Platichthys stellatus*	1 juvenile	70–53	4–1
Seine	Threespine stickleback *Gasterosteus aculeatus*	18 adult	81–50	4–1
Seine	Threespine stickleback *Gasterosteus aculeatus*	198 juveniles	25–16	4–<1
Seine	Speckled sanddab *Citharichthys stigmaeus*	2 juveniles	35–32	4–1

TABLE 3. FISH COLLECTED DURING SEINING AT VITUS LAKE SOUTH, 2008

Date	33 m	Stickleback adults	Stickleback <1 in.	Flounder	Salmon	No. of fish/m^2
7 Aug.	1	3	130	0	1	0.89
7 Aug.	2	0	30	1	0	0.20
7 Aug.	3	15	38	0	0	0.35
Total		18	198	1	1	Avg. 0.48

TABLE 4. NUMBER OF MAJOR INVERTEBRATES IN BENTHOS SAMPLES TAKEN FROM VITUS LAKE 1–8 AUGUST 2008

Macroinvertebrates number/m^2	Snails	Amphipods	Chironomids	Other	Total number/m^2
Base Camp	0	12	0	0	10.7
Cabin	0	47.6	119	0	167.8
Vitus Lake East	690	1035.7	199	238	3953.6
Vitus Lake South	44	1107	652	8.9	2223.2

in the diet of some waterfowl, such as loons, especially when they are young (Jackson, 2003). Owing to rock flour suspension and cold water temperatures, the most biologically productive areas of Vitus Lake are within a narrow band along the southern shoreline, farthest from the glacial melt and often near stream outflows. Fish use these areas to feed on benthic invertebrates, and such regions can be important nursery areas for some species as well. The station nearest the base camp site seems to be the farthest and least disturbed station from glacial meltwater currents that actually move icebergs, fresh meltwater, and rock flour into this small bay, which may reduce productivity (Table 4).

Future research will incorporate mapping water movement within the lake to track rock flour deposition. Also, investigation of the fish-benthos-waterbird food web in the nearshore region of Vitus Lake will continue incorporating phytoplankton and periphyton. More students will be invited into future field work to become exposed to and learn more about our changing world.

ACKNOWLEDGMENTS

I am very grateful for the help and encouragement of Robert Shuchman, Liza Jenkins, Scott Guyer, Nathan Rathbun, and all the people who made research at the U.S. Bureau of Land Management Bering site safe and productive. I also wish to thank my student assistant in 2008, Sharon Rayford.

REFERENCES CITED

Jackson, D.B., 2003, Between lake differences in the diet and provisioning behavior of black-throated divers *Gavia arctica* breeding in Scotland: The Ibis, v. 145, p. 30–44, doi: 10.1046/j.1474-919X.2003.00119.x.

Josberger, E.G., Shuchman, R.A., Meadows, G.A., Savage, S., and Payne, J., 2006, Hydrography and circulation of ice marginal lakes at Bering Glacier, Alaska, U.S.A.: Arctic, Antarctic, and Alpine Research, v. 38, p. 547–560, doi: 10.1657/1523-0430(2006)38[547:HACOIL]2.0.CO;2.

Mecklenburg, C.W., Mecklenburg, T.A., and Thorsteinson, L.K., 2002, Fishes of Alaska: Bethesda, Maryland, American Fisheries Society, 1037 p.

Nalepa, T.F., 1987, Long-term changes in the macrobenthos of Southern Lake Michigan: Canadian Journal of Fisheries and Aquatic Sciences, v. 44, p. 515–524, doi: 10.1139/f87-064.

Pothoven, S.A., and Nalepa, T.F., 2006, Feeding ecology of lake whitefish in Lake Huron: Journal of Great Lakes Research, v. 32, p. 489–501, doi: 10.3394/0380-1330(2006)32[489:FEOLWI]2.0.CO;2.

Von Hippel, F.A., and Weigner, H., 2004, Sympatric anadromous-resident pairs of threespine stickleback species in young lakes and streams at Bering Glacier, Alaska: Behaviour, v. 141, p. 1441–1464, doi: 10.1163/1568539042948259.

Manuscript Accepted by the Society 02 June 2009

The Geological Society of America
Special Paper 462
2010

Bryophytes and bryophyte ecology of the Bering Glacier Region

Nancy G. Slack
Biology Department, The Sage Colleges, Troy, New York 12180, USA

Diana G. Horton
Department of Biological Sciences, University of Iowa, Iowa City, Iowa 52240, USA

INTRODUCTION

As a bryophyte ecologist, I was invited to survey bryophytes for the Bering Glacier study funded by the U.S. Bureau of Land Management (BLM). We worked out of the BLM Bering Glacier field camp during August of 2001. Bryologist Diana Horton, of the University of Iowa, came with me. We were attached to the botanical team, which included Marilyn Barker (see Chapter 7 of this volume) and others who had already spent time in previous summers collecting and identifying vascular plants at this and surrounding sites. The 2001 botanical group was able to survey a number of varied and interesting sites for both vascular plants and bryophytes because of the helicopter access.

PREVIOUS BRYOPHYTE COLLECTIONS

Little survey work has been done and published on mosses or liverworts in this part of Alaska. The *Sphagnum* flora of Alaska is probably better known than any other group of bryophytes because of the collections of Richard Andrus and associates. A larger Alaskan *Sphagnum* survey was due to be made in 2008 in Southeast Alaska. There has been little previous collecting of bryophytes in the Bering Glacier Region. A few were collected along with vascular plants by the botany team in prior seasons. A spreadsheet from the University of Alaska herbarium at Fairbanks lists barely 20 species from the Bering Glacier Region, most with little or no data on the collectors or the dates of the collections. Prior to 2001 none seem to have been collected by bryologists other than O.M. Afonina, who collected bryophytes for the herbarium in many parts of Alaska, particularly in the 1990s. The bryophytes collected at the Glacier Bay sites together with site descriptions are given Appendix 1. The Alaskan bryophytes in the University of Alaska herbarium collected by Nancy G. Slack are listed in Appendix 2.

Bryophyte floras for other parts of Alaska have been previously reported by W.B. Schofield and Stephen and Sandra Talbot (Schofield et al., 2004) and earlier for Kodiak Island, the Alexander Archipelago, and Chisik Island (see references in the above paper). Most recently, in 2005, Eve Langer, field associate, California Academy of Sciences, made extensive collections of bryophytes on Kodiak Island as part of a botanical expedition to inventory the late glacial refugium region. (Laeger, 2006) Nancy Slack's Bering Glacier collections are at the University of Alaska at Fairbanks, with a duplicate set at the Russell Sage College Herbarium, Troy, New York. These specimens, collected 3 to 15 August 2001, also include bryophytes collected in the Anchorage area At Chugach State Park and at Jewel Lake fen and at two sites near Cordova, Alaska. The latter collections are useful for comparison with the bryoflora found at the Bering Glacier sites.

BRYOPHYTE ECOLOGY

Among the unusual sites found in our Bering Glacier work are intermediate and rich fens not previously known. These had interesting floras, described below. We also found early successional sites, some with very high percentages (within the quadrats surveyed) of the pioneer liverwort *Blasia pusilla.* In addition, a preliminary study was carried out on Weeping Peat Island, where the retreat of the ice has been well documented. Here bryophytes are the major pioneer species in the first stages of succession after ice retreat. The alpine areas have a diverse bryophyte flora; the mosses and liverworts present, as well as the associated vascular plants, are similar to alpine flora in the Anchorage area. Coastal forests have high biomass, particularly of epiphytic mosses. They are similar to forests visited near Cordova, also in 2001. Their bryophyte flora deserves further study.

Berg Lake Site

A recently flooded area was studied here. Plots with dimensions of 2 × 5 dm were measured in both mud and gravel areas.

Slack, N.G., and Horton, D.G., 2010, Bryophytes and bryophyte ecology of the Bering Glacier Region, *in* Shuchman, R.A., and Josberger, E.G., eds., Bering Glacier: Interdisciplinary Studies of Earth's Largest Temperate Surging Glacier: Geological Society of America Special Paper 462, p. 365–371, doi: 10.1130/2010.2462(22).

In quadrats on mud, almost no vascular plants were present, only seedlings of *Mimulus guttatus* and *Chamerion latifolium* and small *Equisetum arvense.* Although several moss species were present, notably *Barbula unguiculata* and *Pohlia drummondii*, and more common mosses, *Ceratodon purpureus* and *Plagiomnium cuspidatum*, the plots were dominated by the pioneer liverwort *Blasia pusilla*. In some plots it was the only species, covering up to 75% of the plot. The liverwort is unusual among bryophytes in having pockets of cyanobacteria, which fix nitrogen. It has several stages of growth and reproduction and a relatively short life span. Its old leaves, however, provide shelter for spores of other bryophyte species to develop. Plots in gravel substrates contained six or more bryophyte species and invading *Alnus viridis*, *Salix sitchensis*, and *Picea sitchensis* seedlings.

Weeping Peat Site

In seven quadrats in an area that had been ice free only since 1999, six species of bryophytes were present; the only vascular plant pioneer was an *Alnus* seedling. Mosses of note were *Aongstroemia longipes*, *Dicranella varia*, and *Pohlia filum*. Old dead leaves of *Blasia pusilla* (see above) were present in some of the quadrats as well, apparently facilitating germination of moss spores. Quadrats in areas that had been ice free for a longer period (probably 1995–1998) were still dominated by *Dicranella* and *Aongstroemia*, but more widespread mosses such as *Ceratodon purpueus*, *Bryum pseudotriquetrum*, and *Racomitrium canescens* had invaded, as well as a variety of vascular plants, including *Alnus viridis*, *Equisetum variegatum*, and *Salix barclayi*.

Rich Fen, Robinson Mountains

Parts of the Robinson Mountains are believed to be an unglaciated refugium. We found a rich fen in a graben here, with a very interesting bryophyte flora. Some of the notable moss species were *Calliergon stramineum*, *Hygrohypnum luridum*, *Paludella squarrosa*, *Sarmenthypnum sarmentosum*, and a member of the Splachnaceae, *Tayloria lingulata*. A number of sphagnum species characteristic of minerotrophic waters were also found in this fen: *S. squarrosum*, *s. teres*, *and S. warnstorfii*.

Intermediate Fen, Robinson Mountains

At another site in the Robinson Mountains we found an intermediate fen in an alpine cirque at nearly 2000 ft elevation. Notable in this fen were very large patches of *Paludella squarrosa* and *Pseudobryum cinclidioides*. Two liverworts were also found here, *Anthelia juratzkana* and *Lophozia alpestris*. Here the *Sphagnum* species were *S. subsecundum*, *S. tenellum*, and *S. teres*. The *Sphagnum* species found here and in the rich fen above are very different from those I have found in poorer sites. (See complete list of *Sphagnum* species below.)

The rich and intermediate fens discovered and surveyed are very interesting and deserve further study; here are surely additional ones in the Bering Glacier area. They contain many interesting bryophyte and vascular plant species, and in my experience they are very different in both respects from the fens studied in the Glacier Bay area. These fens would be worth a special study in comparison with fens found elsewhere (Slack et al., 1980; Slack, 1994).

Alpine Tundra, Khitrov Hills, and Grindle Hills

These alpine sites in the Bering Glacier Region (to 3000 ft) are similar to such sites elsewhere in Alaska, e.g., in the Anchorage region, but also show similarities with the vascular and bryophyte species even in the eastern United States on Mt. Washington (Slack and Bell, 2006) and on the Gaspé Peninsula, Quebec, with species of *Empetrum*, *Phyllodoce*, *Cassiope*, and *Vaccinium uliginosum*. Bryophyte species are similar as well, including those of *Racomitrum*, *Pogonatum*, *Dicranum*, and *Conostomum tetragonum*. Species of special interest are *Kiaeria falcata*, *Bryum articum*, and the rare moss *Tetradontium brownianum*, found by Diana Horton in a small cave at 2400 ft in the Khitrov Hills. The many Bering Glacier alpine areas would repay quantitative study, including bryophytes and vascular plants as well as lichens.

BLM Bering Glacier Field Camp

This was the only site for which we did not need helicopter service. It is also the only site where there has been major human as well as natural disturbance. The diversity of both bryophytes and vascular plants on the sandy shore and the recessional moraines was low. Amidst the patches of *Alnus viridis*, three *Salix* species, and considerable *Lupinus nootkatensis*, the bryophyte cover was unusually high, consisting mainly of *Racomitrium canescens*, a species that is a pioneer after glacial recession elsewhere in Alaska, e.g., Glacier Bay. *Chamerion angustifolium* is another such species also common at the BLM camp. The most interesting and surprisingly common species was *Tetraplodon mnioides*, a member of the Splachnaceae, which has a wide distribution in the United States and Canada, but is often rare. A greater variety of bryophytes was found in niches along the sandy shore.

The following species of *Sphagnum* were found during this study: *S. angustifolium*, *S. fimbriatum*, *S. girgensohnii*, *S. magellanicum*, *S. papillosum*, *S. squarrosum*, *S. subsecundum*, *S. tenellum*, *S. teres*, and *S. warnstorfii.*

These species represent a range of *Sphagnum* species from acid conditions to high pH-rich fen conditions. The Bering Glacier species are almost completely different from the 15 *Sphagnum* species found in two fens near Anchorage, neither of which was a rich fen. More exploration of Bering Glacier would probably yield additional bog and fen types and *Sphagnum* species.

The Bering Glacier sites provide an excellent opportunity for studying primary succession after glacial retreat. The bryophytes are the most important plants in the early successional stages. The liverwort *Blasia pusilla* with its nitrogen-fixing cyanobacteria is important in early succession. Its role was documented and is of interest to population and community ecology.

The forest sites were visited only briefly. They are similar in bryophyte flora to forest sites studied elsewhere in southern Alaska and British Columbia and deserve further investigation. Species such as *Plagiothecium undulatum*, *Hypnum plicatum*, and *Oligotrichum aligerum* that Slack found in Orca Sound and Cordova forests should also be in Bering Glacier forest habitats.

The same is true of the higher elevation sites. They contain similar species to those I found on Little O'Malley Peak, Chugach State Park, but additional species found there, especially liverworts, may be present at additional higher elevation Bering Glacier sites. Further exploration is needed.

The Bering glacier sites also contained a great many lichen species, which to my knowledge have not been surveyed. This can only be a preliminary report on the bryophytes of the Bering Glacier Region, because both the time spent there and the number of sites visited was limited. However, these data should be valuable to future bryologists visiting the Bering Glacier. Therefore the species are listed according to sites visited and with basic vascular plant data for the particular habitats studied. See Appendix 1 for a list of sites visited, with their associated bryophyte flora. Appendix 2 is an alphabetical checklist of bryophytes found in the Bering Glacier Region.

APPENDIX 1

The Bering Glacier sites are described, and their bryophytes are listed below, in the order in which collections were made.

Ancient Forest Site
6 August 2001
60°08′22.51″N, 143°37′10.65″W, 1 ft elevation
Bering Glacier Main Lobe, SW margin, N of Tashalich River
Ancient forest site with vernal pools. Forest trees dated at A.D. 800; ancient vertical and tipped tree stumps. Scattered *Salix sitchensis*, *S. barclayi*, *S. alaxensis*, *Chamerion latifolium*, *Equisetum variegatum*.
N. Slack accession numbers 2001-85 to 2011-114.
MOSSES:
Bryum argenteum Hedw.
Bryum pseudotriquetrum (Hedw.) P. Gaertn, B. Mey. & Scherb.
Ceratodon purpureus (Hedw.) P. Beauv.
Limprichtia revolvens (Sw.) Loeske
Philonotis fontana (Hedw.) P. Beauv.
Racomitrium canescens (Hedw.) Bruch & Schimp.
LIVERWORTS:
Gymnocolea inflata (Huds.) Dum.
Marchantia polymorpha L.

Grindle Hills
7 August 2001
60°16′18″N, 143°37′16.17″W
Bering Glacier Main Lobe, Grindle Hills, W part
Stop 1. 1359 ft elevation
Low tundra with stream, tarn, and waterfall over siliceous outcrops; patches of *Alnus* and *Salix*. *Solenostoma* abundant over waterfall rocks.
N. Slack accession numbers 2001-115 to 2001-151
MOSSES:
Cirriphyllum cirrosum (Schwaegr. *in* Schultes) Grout
Conostomum tetragonum (Hedw.) Lindb.
Dicranum acutifolium (Lindb. & Arnell) C. Jens. X Weinm.
Dicranum spadiceum Zett.
Dicranum tauricum Sapeh.
Kiaeria falcata (Hedw.) Hag.
Pseudoleskea baileyi Best & Grout *in* Grout
Pleurozium schreberi (Brid.) Mitt.
Pogonatum urnigerum (Hedw.) P. Beauv.
Pohlia nutans (Hedw.) Lindb.
Pseudoleskea baileyi Best & Grout *in* Grout
Racomitrium canescens (Hedw.) Brid.
Racomitrium heterostichum (Hedw.) Brid.
Rhytidiadelphus loreus (Hedw.) Warnst.
Rhytidiadelphus squarrosus (Hedw.) Warnst.
Rhytidiopsis robusta (Hook.) Broth.
Sanionia unciata (Hedw.) Loesk.
LIVERWORTS:
Barbilophozia floerkii (Web. et Mohr) Loeske
Barbilophozia lycopodioides (Wallr.) Loeske
Ptilidium ciliare (L.) Hampe
Stop 2. On ridge, 60°16′25″N, 143°15′26″W, 2284 ft elevation
N. Slack accession numbers 2001-152 to 2001-1
MOSSES:
Dicranum spadiceum Zett.
Hypnum subimponens Lesq.
Oncophorus wahlenbergii Brid.
Pogonatum urnigerum (Hedw.) P. Beauv.
Polytrichum juniperinum Hedw.
Polytrichum piliferum Hedw.
Pseudoleskea baileyi Best & Grout *in* Grout
Psilopilum cavifolium (Wils.) Hag.
Racomitrium canescens (Hedw.) Brid.
Racomitrium lanuginosum (Hedw.) Brid.
Rhytidiadelphus squarrosus (Hedw.) Warnst.
Sanionia uncinata (Hedw.) Loesk.
LIVERWORTS:
Ptilidium ciliare (L.) Hampe
Stop 3. 60°16′38.32″N, 143°15′26″W, 1358 ft elevation
Lakeshore; also stream and waterfall; lake site (BG-01-06)
N. Slack accession numbers 2001-174 to 2001-191
MOSSES:
Abietinella abietina (Hedw.) Fleisch.
Brachythecium turgidum (Hartm.) Kindb.
Brachythecium rivulare Schimp.
Bryum pseudotriquetrum (Hedw.) P. Gaertn., B. Mey. & Scherb.
Dicranoweisia crispula (Hedw.) Lindb. *ex* Milde
Hygrohypnum ocraceum (Turn. *ex* Wils.) Loeske
Pogonatum urnigerum (Hedw.) P. Beauv.
Pohlia wahlenbergii (Web. & Mohr) Andrews
Pseudoleskea baileyi Best & Grout *in* Grout
Racomitrium fasciculare (Hedw.) Brid.
Racomitrium heterostichum (Hedw.) Brid.
Rhytidiadelphus loreus (Hedw.) Brid.
Phytidiadelphus squarrosus (Hedw.) Warnst.
Rhytidiopsis robusta (Hook.) Broth.
Sanionia unciata (Hedw.) Loesk.
LIVERWORTS:
Anthelia julacea (L.) Dum.
Jungermannia exsertifolia Steph. ssp. *cordifolia* (Dum.) Vana
Scapania undulata (L.) Dum.

Berg Lake Site
8 August 2001
60°25′30″N, 143°50′40″W, 459 ft elevation

Bering Glacier Stellar Lobe, Berg lake, NW corner
Stream from head of alluvial fan, unsorted material, gravel to boulder size; some sandy mud areas; sparse vegetation with *Epilobium latifolium*, *Salix alaxensis*, *Mimulus guttatus*; 5 × 5 m plots, 50 m apart, subject to flooding, with very high percentages (to >75%) in plots of the pioneer liverwort *Blasia pusilla*.
Stops 1–4, head of stream
N. Slack accession numbers 2001-192 to 2001-216
MOSSES:
Barbula unguiculata Hedw.
Ceratodon purpueus (Hedw.) Brid.
Plagiomnium cuspidatum (Hedw.) T. Kop.
Pohlia drummondii (C. Mull.) Andrews
LIVERWORTS:
Blasia pusilla L.
Cephalozia sp.

Khitrov Hills Sites
Between Bering Glacier Main Lobe and Stellar Lobe
60°26′23″N, 143°24′59″W
Stop 1. 2399 ft elevation
Tundra with sandstone outcrops and latye snowmelt patches. *Cassiope stelleriana*, *Phyllodoce*,
Empetrum, *Luetkia*, *Vaccinium uliginosum*. Seeps and waterfalls with *Pellia*, *Marsupella*,
Philonotis, and *Pohlia wahlenbergii*. Outcrops with *Andreaea*.
Stop 1. 2399 ft
N. Slack accession numbers 2001-217 to 2001-270
MOSSES:
Blindia acuta (Hedw.) Bruch & Schimp *in* B.S.G.
Calliergon stramineum (Brid.) Kindb.
Claopodium bolanderi Best
Dicranum fuscescens Turn.
Dryoptodon patens (Hedw.) Brid.
Philonotis fontana (Hedw.) Brid.
Plagiothecium denticulatum (Hedw.) Schimp. *in* B.S.G.
Platydictya jungermannioides (Brid.) Crum
Pohlia cruda (Hedw.) Lindb.
Pseudoleskea baileyi Best & Grout *in* Grout
Racomitrium canescens (Hedw.) Brid.
Racomitrium heterostichum (Hedw.) Brid.
Racomitrium sudeticum (Funck) Bruch. & Schimp. *in* B.S.G.
Rhizomnium punctatum (Hedw.) T. Kop
Rhytidiopsis robusta (Hook) Broth.
Sphagnum teres (Schimp.) Angstr. *in* Hart
Tetradontium brownianum (Dicks.) Schwaegr.*
Tortella tortuosa (Hedw.) Limpr.
*This is a rare moss found in a small cave by Diana Horton. Slack collected it also; deposited in the herbaria.
LIVERWORTS:
Barbilophozia floerkei (Web. et Mohr) Loeske
Blepharostoma trichophyllum (Lightf.) Corda
Marchantia polymorpha L.
Marsupella sp.
Pellia sp.
Stop 2. Top of peak, 3003 ft elevation
MOSSES:
Brachythecium albicans (Hedw.) Schimp. *in* B.S.G.
Bryum arcticum (Bruch & Schimp *in* B.S.G.
Dicranum fuscescens Turn. (Coll. Anne Pasch)
Dicranum tauricum Sapeh. (Coll. Anne Pasch)
Pleurozium schreberi (Brid.) Mitt.
Pogonatum urnigerum (Hedw.) P. Beauv.
Racomitrium canescens (Hedw.) Brid.
Racomitrium heterostichum (Hedw.) Brid.
LIVERWORTS:
Barbilophozia floerkii (Web. & Mohr) Loeske
Fossombronia sp.

BLM Bering Glacier Field Camp
6 and 10 August 2001
60°07′15″N, 143°17′02″W, 39 ft elevation
Bering Glacier Main Lobe, S of SE margin, W of Tsiu River
Sandy shore under alders; recessional moraine with patches of *Salix sitchensis*, *S. barclayi*, *Alnus viridis*, *Lupinus nootkatensis*, *Epilobium latifolium*; dominant moss *Racomitrium canescens*, but *Tetraplodon mnioides*, an often rare member of the Splachnaceae common.
N. Slack accession numbers 2001-185 to 2001-189, and 2001-271 to 2001-277
MOSSES:
Aongstroemia longipes (Somm. Bruch & Schimp. *in* B.S.G.
Brachythecium frigidum (C.Mull.) Besch.
Bryum argenteum Hedw.
Pogonatum urnigerum (Hedw.) P. Beauv.
Racomitrium canescens (Hedw.) Brid.
Tetraplodon mnioides (Hedw.) Bruch and Schimp.

Weeping Peat Island Site
10 August 2001
60°12′02″N, 143°12′23″W, 213 ft elevation
Bering Glacier Main Lobe, island called *Weeping Peat* near SE margin. Recessional moraines from 1999 and 1996. Early successional barren with sparse *Barbula*, *Pohlia*, *Bryum*, *Racomitrium canescens*. Later successional with dense *Alnus viridis*, *Salix Barclay*, *S. sitchensis*, *Pyrola secunda*. Ephemeral pools with *Spiranthes romanzoffiana*, *Platanthera dilatata*, *Juncus* spp. Grasses and bryophytes: *Drepanocladus*, *Barbula*, *Bryum pseudotriquetrum*, *Brachythecium*, *Philonotis*.
N. Slack accession numbers 2001-278 to 2001-285
MOSSES:
Aongstroemia longipes (Somm.) Bruch & Schimp *in* B.S.G.
Brachythecium frigidum (C. Mull.) Besch.
Bryum pseudotriquetrum (Hedw.) Gaertn. et al.
Campylium stellatum (Hedw.) C. Jens
Dichodontium pellucidum (Hedw.) Schimp.
Racomitrium canescens (Hedw.) Brid.
Sanionia uncinata (Hedw.) Loeske
Weeping Peat successional transects
N. Slack accession numbers 2001-286 to 2001-301
MOSSES:
Aongstroemia longipes (Somm.) Bruch & Schimp. *in* B.S.G.
Bryum pseudotriquetrum (Hedw.) P. Gaern., B. Mey. & Scherb.
Dicranella varia (Hedw.) Brid.
Pogonatum urnigerum (Hedw.) P. Beauv.
Pohlia filum (Schimp.) Mart.
HEPATICS:
Blasia pusilla L.
Cephalozia sp.

Hanna Lake Site
12 August 2001
60°13′51″N, 143°08′22″W, 92 ft elevation
Bering Glacier Main Lobe, eastern Grindle Hills, lake S of Hanna Lake, NW shore.
Dry, rocky streambed with *Chamerion latifolium*, small patches of *Racomitrium canescens*,
associated forest of *Picea sitchensis*, *Populus trichocarpa*, *Alnus viridis*, *Oplopanax*, *Pyrola secunda*, *Tiarella trifoliata*, *Lycopodiums*; dominant bryophytes: *Rhytidiadelphus loreus*, *Dicranum*; epiphytes: *Antitrichia*, *Ulota*.

Stop 1. Along a dry, rocky streambed and in rain forest
N. Slack accession numbers 2001-302 to 2001-314
MOSSES:
Antitrichia curtipendula (Hedw.) Brid.
Aulacomnium androgynum (Hedw.) Schwaegr.
Ceratodon purpureus (Hedw.) Bid.
Dicranum fuscescens Turn.
Polytrichastrum alpinum (Hedw.) G.I. Sm.
Rhizomnium glabrescens (Kindb.) Kop.
Rhytidiadelphus triquetrus (Hedw.) Warnst.
Ulota crispa (Hedw.) Brid.
Ulota obtusiuscula C. Mull & Kindb. *in* Mac & Kindb.
LIVERWORTS:
Pellia sp.
Ptilidium pulcherrimum (G. Web.) Hampe
Stop 2. Sloping intermediate fen, interspersed Sitka spruce and fen patches on steep slope, southeast of Hanna Lake, Donald Ridge
N. Slack accession numbers 2001-315 to 2001-325
MOSSES:
Aulacomnium palustre (Hedw.) Schwaegr.
Calliergon stramineum (Brid.) Kindb.
Dicranoweisia crispula (Hedw.) Lindb. *ex* Milde
Pleurozium schreberi (Brid.) Mitt.
Pogonatum urnigerum (Hew.) P. Beauv.
Rhytidiadelphus squarrosus (Hedw.) Warnst.
Schistidium agassizii Sull. & Lesq. *in* Sull.
Schistidium apocarpum (Hedw.) Bruch & Schimp.
Sphagnum angustifolium (C. Jens. *ex* Russ) *in* Tolf
Sphagnum fimbriatum Wils. *in* Wils. & Hook.
Sphagnum girgensohnii Russ.
Sphagnum magellanicum Brid.
Sphagnum warnstorfii Russ.
Sphagnum papillosum Lindb.

McIntosh Mountain "Alpine" Site—Unglaciated Refugium

13 August 2001
60°18′17″N, 143°01′10″W, 1978–2078 ft elevation
Stop 1. Rich fen in a graben
Bering Glacier Main Lobe, McIntosh Hills, McDougall Ridge
N. Slack accession numbers 2001-326 to 2001-340
MOSSES:
Aulacomnium palustre (Hedw.) Schwaegr.
Calliergon stramineum (Brid.) Kindb.
Drepanocladus aduncus (Hedw.) Warnst.
Hygrohypnum luridum (Hedw.) Warnst.
Oncophorus virens (Hedw.) Brid.
Paludella squarrosa (Hew.) Brid.
Philonotis fontana (Hedw.) Brid.
Pseudoleskea baileyi Best & Grout *in* Grout
Rhytidiadelphus squarrosus (Hedw.) Warnst.
Sarmenthypnum sarmentosum (Wahlenb.) Toum. & T. Kop.
Sphagnum squarrosum Crome
Sphagnum teres (Schimp.) Angstr. *in* Hartm.
Sphagnum warnstorfii Russ.
Sanionia uncinata (Hedw.) Loeske
Tayloria lingulata (Dicks.) Lindb.
Stop 1. Not in fen, but in stream and alpine
N. Slack accession numbers 2001-341 to 2001-358
MOSSES:
Bryum pseudotriquetrum (Hedw.) P. Gaertn., B. Mey. & Scherb.
Bryum callophyllum R. Br.
Campylium stellatum (Hedw.) C. Jens.
Conostomum tetragonum (Hedw.) Lindb.
Dicranum spadiceum Zett.
Hygrohypnum ocraceum (Turn. *ex* Wils.) Loeske
Kiaeria blytii (Schimp.) Broth.
Pleurozium schreberi (Brid.) Mitt.
Pseudoleskea atrichia (Kindb. in Mac. & Kindb.) Kindb.
Racomitrium affine (Schleich ex Web & Mohr) Lindb.
Racomitrium heterostichum (Hedw.) Brid.
Rhizomnium nudum (Britt. & Williams) T. Kop.
LIVERWORTS:
Jungermannia pumila With.
Marsupella spacelata (Goesele) Dum.
Ptilidium ciliare (L.) Hampe
Scapania undulata (L.) Dum.
Stop 2. Intermediate fen in alpine cirque, 1978 ft
Salix polaris, *Fauria crus-gallii*, *Geum calthifolium*, *Chamerion angustifolium*, *E. scheuchzeri* var. *tenuifolium*, *Carex enanderi*, and *Muhlenbergia glomerata*; also large populations of the mosses *Paludella squarrosa* and *Pseudobryum cinclidioides*.
N. Slack accession numbers 2001-359 to 2001-370
MOSSES:
Calliergon stramineum (Brid.) Kindb.
Campylium stellatum (Hedw.) C. Jens.
Drepanocladus aduncus (Hedw.) Warnst.
Limprichtia revolvens (Sw.) Loeske
Paludella squarrosa (Hedw.) Brid.
Pseudobryum cinclidiodes (Hub.) T. Kop.
Sphagnum subsecundum Mess *in* Sturm
Sphagnum tenellum (Brid.) Bory
Sphagnum teres (Schimp.) Angstr. *in* Hartm.
LIVERWORTS:
Anthelia juratzkana (Limpr.) Trev.
Lophozia alpestris (Schleich. *ex* Web.) Evans
Ridge top, 2078 ft. Collected by M. Barker; identified by N. Slack
N. Slack accession numbers 2001-371 to 2001-374
MOSSES:
Dicranoweisia crispula (Hedw.) Lindb. *ex* Milde
Paraleucobryum (Thed.) Loeske
Pleurozium schreberi (Brid.) Mitt.
Polytrichum piliferum Hedw.
Racomitrium heterostichum (Hedw.) Brid.

APPENDIX 2. BRYOPHYTES OF ALASKA
(collected by Nancy G. Slack, now in the University of Alaska herbarium)

Mosses
Abietinella abietina
Andreaea alpestris
Andreaea rupestris
Antritrichia curtipendula
Aongstroemia longipes
Aulacomnium turgidum
Aulacomnium androgynum
Aulacomnium palustre
Aulacomnium palustre
Barbula unguiculata
Bartramia ithyphylla
Blindia acuta
Brachythecium albicans
Brachythecium frigidum
Brachythecium rivulare
Brachythecium turgidum
Bryum arcticum
Bryum argenteum

Bryum callophyllum
Bryum pseudotriquetrum
Calliergon cordifolium
Calliergon stramineum
Campylium stellatum
Ceratodon purpureus
Cirrophyllum cirrosum
Cladopodium bolanderi
Conostomum tetragonum
Dichodontium pellucidum
Dicranella varia
Dicranoweisia crispula
Dicranum acutifolium
Dicranum fuscescens
Dicranum spadiceum
Dicranum tauricum
Drepanocladus aduncus
Dryptodon patens
Grimmia torquata
Hygrohypnum luridum
Hygrohypnum ocraceum
Hylocomium splendens
Hypnum plicatulum
Hypnum subimponens
Kiaeria blytii
Kiaeria falcata
Limprichtia revolvens
Oligotrichum aligerum
Oncophorus virens
Oncophorus wahlenbergii
Paludella squarrosa
Paraleucobryum enerve
Paraleucobryum enerve
Philonotis fontana
Plagiomnium cuspidatum
Plagiothecium denticulatum
Plagiothecium undulatum
Platydictya jungermannioides
Pleurozium schreberi
Pogonatum urnigerum
Pohlia cruda
Pohlia drummondii
Pohlia filum
Pohlia natans
Pohlia wahlenbergii
Polytrichastrum alpinum
Polytrichum juniperinum
Polytrichum piliferum
Pseudobryum cinclidioides
Pseudoleskea atricha
Pseudoleskea baileyi
Psilopilum cavifolium
Racomitrium affine
Racomitrium canescens
Racomitrium fasciculare
Racomitrium heterostichum
Racomitrium lanuginosum
Racomitrium sudeticum
Rhizomnium glabrescens
Rhizomnium nudum
Rhizomnium punctatum
Rhytidiadelphus loreus
Rhytidiadelphus squarrosus
Rhytidiadelphus triquetrus
Rhytidioipsis robusta
Sanionia uncinata
Sarmenthypnum sarmentosum
Schistidium agassizii
Schistidium apocarpum
Sphagnum alaskensis
Sphagnum andersonianum
Sphagnum angustifolium
Sphagnum aongstroemii
Sphagnum balticum
Sphagnum brevifolium
Sphagnum compactum
Sphagnum fimbriatum
Sphagnum fuscum
Sphagnum girgensohnii
Sphagnum inundatum
Sphagnum lenense
Sphagnum lindbergii
Sphagnum magellanicum
Sphagnum majus
Sphagnum obtusum
Sphagnum papillosum
Sphagnum russowi
Sphagnum squarrosum
Sphagnum subobesum
Sphagnum subsecundum
Sphagnum tenellum
Sphagnum teres
Sphagnum warnstorfii
Tayloria lingulata
Tetradontium brownianum
Tetraplodon mnioides
Tomentypnum nitens
Tortella tortuosa
Tretraplodon mnioides
Ulota crispa v. alaskana=
Ulota obtusiuscula
Warnstorfia exannulata

Hepatics

Anastrophyllum michauxii
Anthelia julacea
Anthelia juratzkana
Barbilophozia floerkei
Barbilophozia lycopodioides
Blasia pusilla
Blepharostoma trichophyllum
Diplophyllum albicans
Fossombronia sp.
Gymnocolea inflata
Gymnomitrion concinnatum
Gymnomitrion coralloides
Jungermannia exsertifolia
Jungermannia pumila
Lophozia alpestris
Marchantia polymorpha
Marsupella sp.
Marsupella spacelata
Mylia taylori
Odontoschisma macounii
Pellia neesiana
Pleurocladia albescens
Preissia quadrata
Ptilidium ciliare

Ptilidium pulcherrimum
Radula obtusiloba
Scapania undulata
Tetralophozia setiformis

ACKNOWLEDGMENTS

I would like to thank the following: the Bureau of Land Management, Marilyn Barker, and the botanical team for including bryologists in their 2001 explorations; this group, as well as Anne Pasch, for their companionship and help, especially in choosing field sites; the helicopter pilots for transporting us to these sites; Jim Sisk for help in the field at several sites. Special thanks go to Wilfred B. Schofield for helpful consultation and the determination of several difficult specimens and to Janice Glime for a helpful review of the manuscript.

REFERENCES CITED

Laeger, E., 2006, Bryophytes of the Kodiak Late Glacial Refugium, Kodiak National Wildlife Refuge, Kodiak Island, Alaska: Fairbanks, Report and specimens at the California Academy of Sciences at the University of Alaska Museum of the North Herbarium.

Schofield, W.B., Talbot, S.S., and Talbot, S.L., 2004, Bryophytes from Simeonof Island in the Shumagin Islands, Southwestern Alaska: Journal of the Hattori Botanical Laboratory, v. 95, p. 155–198.

Slack, N.G., 1994, Can one tell the mire type from the bryophytes alone?: Journal of the Hattori Botanical Laboratory, v. 75, p. 149–159.

Slack, N.G., and Bell, A.W., 2006, AMC Field Guide to the New England Alpine Summits (2nd edition): Boston, Appalachian Mountain Club Books, 104 p.

Slack, N.G., Vitt, D.H., and Horton, D.G., 1980, Vegetation gradients of minerotrophically rich fens in western Alberta: Canadian Journal of Botany, v. 58, p. 330–350.

MANUSCRIPT ACCEPTED BY THE SOCIETY 02 JUNE 2009

The Geological Society of America
Special Paper 462
2010

Birds of the Bering Glacier Region: A preliminary survey of resident and migratory birds

Carol Griswold
University of Alaska Anchorage, P.O. Box 1342, Seward, Alaska 99664, USA

ABSTRACT

A preliminary survey of birds in the Bering Glacier Region was conducted over a 12-day period from 30 July to 10 August 2007. The habitats ranged from alpine to subalpine mountains, coastal temperate rain forest, glacial forelands, wetlands including fresh- and salt-water marshes, coastal beaches, and the nearshore waters of the Gulf of Alaska. The objective was to add to the existing database of bird species in specific locations and unique habitats, recognizing that birds are an important component of this dynamic ecosystem.

Prior to this study, Kevin Lynch compiled bird observations while primarily focusing on bear safety in two two-week periods in 2002 and 2003. His surveys recorded 40 species of 17 families. Eight species were recorded that were not observed or verified in the 2007 survey. The 2007 survey recorded 51 species, representing 22 families, and 4 others noted as unidentified species. The study areas overlapped only at the Bering Glacier camp, Midtimber Lake Trail, Pacific coast, Seal River, and Vitus Lake.

INTRODUCTION

The Bering Glacier Region is strategically located in a transition zone where breeding birds from southeast Alaska mix with those from south-central Alaska. Migratory birds are funneled around the high mountains to the north and through the passes. Many stop over to feed in the rich estuarine lagoons, rivers, and ocean on their long journey to and from the breeding grounds.

As the ice retreats, new habitat is exposed for exploitation by birds, finding new niches as the succession of plants proceeds from pioneer to climax species. The bird survey thus is intimately connected to the flora of the region and so this was done for the most part in conjunction with the botanical survey.

The objectives of this inventory were to (1) compile data on the bird species of the region, (2) identify areas of interest for future bird surveys, and (3) assist land managers in policy decisions. (See Appendix 1 for the 2007 bird survey, and Appendix 2 for bird surveys in 2002 and 2003.)

METHODS

As a member of the botany team, I noted birds and bird sign in the Bering Glacier Region from 30 July to 10 August 2007. We flew by helicopter to remote locations and hiked to sites near the Bering Glacier camp. Once on site, I walked through the area with the team of botanists, writing field notes and taking digital photographs when possible to document any birds, bird activity, bird sign, and significant bird habitat. I also birded a short time from a Zodiac inflatable boat on Vitus Lake.

Since birding was a secondary activity to the botanical survey, not all sites were studied equally, and bird sites of interest were limited to the time and locations allotted for the plants. The only place a scope and tripod was set up was at camp, focused on Vitus Lake.

Weather is briefly noted in the Site Reports, particularly because precipitation on eyeglasses and binoculars impairs observations. Birds tend to be less active as well.

Griswold, C., 2010, Birds of the Bering Glacier Region: A preliminary survey of resident and migratory birds, *in* Shuchman, R.A., and Josberger, E.G., eds., Bering Glacier: Interdisciplinary Studies of Earth's Largest Temperate Surging Glacier: Geological Society of America Special Paper 462, p. 373–380, doi: 10.1130/2010.2462(23). For permission to copy, contact editing@geosociety.org.

SITE REPORTS

Bering Glacier Camp

Observations were taken over the entire period of the survey, from 30 July to 10 August, with a longer period of observation on 8 August.

Birding in camp was excellent: one evening an adult bald eagle flew over camp hauling a fish for dinner in its talons; two common ravens passed overhead, croaking loudly and demonstrating their aerial skills with a quick tuck, roll, and flip; one afternoon, a stealthy gray male merlin flew low, right through camp and quietly vanished; a noisy male belted kingfisher rattled his way from the lake, landed awkwardly on top of the yurt's peak, then clattered off. Small flocks of redpolls dipped overhead, calling their distinctive "chit-chit-chit-chit-chit" but never landed to confirm whether they were common or hoary. Glaucous-winged gulls, including one with a dangling left leg, and the smaller mew gulls often flew over camp, between the lake and ocean.

A trio of warbler species flitted quickly from branch to twig in the alders and willows bordering the camp, gleaning tiny insects from the leaves: the bright yellow male Wilson's warblers with a contrasting black cap, equally bright yellow warblers, and drab olive-yellow orange-crowned warblers. The larger, dark brown fox sparrows with a yellow lower bill and dark upper bill foraged for insects and seeds in the branches and on the ground. A Lincoln's sparrow sang its bubbling melody nearby. Golden-crowned sparrows ("Oh, dear me!") probably live here; a desiccated juvenile found behind the tents documented their presence.

Researchers from previous years reported nesting willow ptarmigan and other sightings near the scrubby willows near camp. As reported, I finally heard one bursting away in a flurry of feathers and startled clucks. Thanks to an open outhouse door, I received the treat of watching a majestic great blue heron wing from the lake to an unknown roosting site late in the evening.

In June, a reported semipalmated plover chose to nest near the airfield in the mowed, open ground instead of the usual gravel bar along a river or pond. An outline of bright surveyor's tape guarded the camouflaged nest long after she was gone.

If trolling for hummingbirds, it pays to wear red while in camp or in the field. My red jacket attracted a hummer that buzzed me loudly, regrettably from the back, so I didn't get even a glimpse before it shot away. The rufous hummingbird generally leaves Alaska in mid to late July, so any late hummers could be an unusual species, such as Anna's. It is difficult to get a good look at these tiny, erratic, and speedy wonders. One also buzzed the red detergent bottle at the wash-up barrel; it looked larger than a rufous, but no ID was confirmed.

Nearby Vitus Lake provides habitat for fresh-water divers, dabblers, and probers. Common loons yodeled their haunting wilderness cry at all hours of the day and night. Unusually large flotillas were often seen mingling near camp. Once a group of five swam slowly over to another group of six and merged to swim and feed together. Another day, a flock of nine swam together. Normally territorial, these large groups may be gathering prior to migrating to the coast for winter (Fig. 1).

Figure 1. Unusually large raft of nine common loons gather in front of camp on Vitus Lake.

The smaller red-throated loons often flew over camp, calling loudly, a bizarre, raucous, goose-like "kak-kak-kak" cry or wails and shrieks that continually baffled the listener. They, too, gathered in rafts of five to seven near the camp.

A small flock of 11 Canada geese and five white-winged scoters visited the lake, almost out of binocular range, but within scope range. The short boat ride across the lake to the impressive glacier face added five red-necked grebes to the species list.

Mixed flocks of shorebirds probed the shoreline, the tiny western sandpipers with their short black legs (16), and the larger spotted sandpipers (five), without spots in juvenile and nonbreeding adult fall plumage. A lone mallard hen dabbled on an abundant green algal bloom, undisturbed by the thundering helicopter overhead while the skittish shorebirds raced away and then back. The characteristic "tew-tew-tew" cry announced the greater yellowlegs, come to feed.

OUT IN THE FIELD

Midtimber Lake Trail, 31 July 2007

Fog, mist, and light rain.

The 11-mile round-trip hike west from camp along a generally well defined trail led through a variety of habitats from alder and willow shrubs, puddles, shrinking ponds, and small streams, across the sandy "Stark Desert," through a meadow of spectacular magenta fireweed and blue lupine, past salmonberry bushes loaded with juicy berries, near scattered large spruce trees, and across one large stream.

Two song sparrows, one Lincoln's sparrow, a few orange-crowned and Wilson's warblers fed in the willows, and one black-billed magpie chatted conversationally from the alders. A parent black-capped chickadee carried a beak-full of food to an unseen fledging. At Drosera Creek, five glaucous-winged gulls flew overhead following the winding course, and two greater yellowlegs and three least sandpipers probed for invertebrates along the shore.

The insistent rattling and diving of a female belted kingfisher drew attention to a lone spruce tree off in the distance. Closer inspection with binoculars showed the object of her anger, a great horned owl, perched at the top, its "horns" silhouetted against the sky. She continued to dive and harass this bird predator until it finally flew off, and she also left, satisfied.

Swan Lake, 1 August 2007

Helicopter trip east of camp. Overcast.

Appropriately enough, two trumpeter swans graced this small lake ringed with pond lilies. Two common loons floated tranquilly nearby. No young of either were seen. An adult bald eagle supervised the botany team at work for several hours from a perch high in a spruce tree, occasionally crying out loudly.

A big surprise was a perfectly preserved young bank swallow, its tiny beak still yellow at the base. It was discovered at the base of a log at the water's edge. We speculate that it accidentally hit the lake and swam to shore, where it died of exhaustion. Juvenile mortality is high, as the risks are great (Fig. 2).

Three widely separated Lincoln's sparrows sang from the meadow grasses and sedges. Two song sparrows, so very dark with long tails, hopped around the alders and willows together.

Too high to properly identify, a mixed flock of ventriloquist chickadees and kinglets flitted among the dense branches of the nearby spruce-hemlock forest. A Steller's jay flashed below. On 20 July 1741, this blue and black crested jay confirmed the discovery of the New World to naturalist Georg Wilhelm Steller on Vitus Bering's expedition. They landed on Kayak Island, ~100 miles to the east in Prince William Sound. The crew got fresh water while he discovered plants and animals not found in Asia or Europe and documented a whole new continent. Now, 266 years later, scientific collections are still being made, and new discoveries documented, at the Bering Glacier, which honors that remarkable explorer.

Well-trod bear trails transected the sedge meadow. Water striders skated over the calm water, damselflies, dragonflies, and mayflies hovered. A long, thin, dark horsehair worm collected inadvertently with the pond plants, tied itself in intricate loops and knots. The usual host of hungry mosquitoes and small biting flies enjoyed our visit very much.

Weeping Peat Island, 2 August 2007

Gray sky clearing to blue with scattered clouds.

Almost immediately after landing and walking to inspect the 1200-year-old trees of the Ancient Forest, I watched a peregrine falcon fly swiftly upstream to land in a snag protruding from a steep bank, where it seemed content to stay. The steep, sandy banks of Weeping Peat Island are riddled with holes, most likely from bank swallow burrows, and possibly belted kingfishers that also burrow 2.5–3 feet into such banks (Fig. 3).

Orange-crowned warblers again worked the fringe of dense alders and willow thickets. An American robin flew past the Abandoned River to the bushes fringing the island's top. Another thrush, the hermit thrush and one scruffy juvenile, watched us quietly from the alders.

Few birds were seen, but many shorebird tracks wove around the two small ponds nestled in the alders and willows at the top of the island. More peep tracks stitched outlines of shallow puddles left in the sand and silt below (Fig. 4). Brown bear, moose, wolf,

Figure 3. Likely bank swallow nesting holes, larger ones possibly belted kingfisher, in the bank of Weeping Peat Island.

Figure 2. Bank swallow that drowned in Swan Lake. Note the yellow bill of this juvenile.

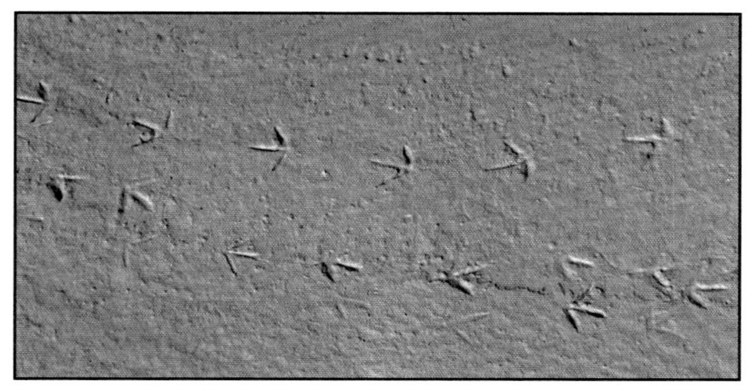

Figure 4. Shorebird tracks embossed in the silt west of Weeping Peat Island.

and coyote tracks also etched the sand and silt, often accompanied by scat overgrown with specialized green moss in an otherwise barren landscape.

An empty mew gull egg left in the sandy Abandoned River channel offered mute testimony to the cacophony of the June nesting season. Additionally, the egg showed evidence of predation; normally hatched eggs fall open in two parts, whereas this one was attacked and opened from the side (Fig. 5).

Figure 5. Mew gull egg, showing predator attack from sides, with a dime for scale.

Tashalich River, 3 August 2007

Fog and low clouds, improving to high overcast.

Over 30 phalaropes danced over the waters of the Seal River, small, light flyers in a tight flock, likely red-necked. Brown bear trails crisscrossed the vibrant carpet of magenta fireweed rolling for miles over the coastal dunes; we spotted several brown bears, including one sow with two cubs, and one moose from the helicopter.

Upon landing in a wet coastal meadow with copses of sweet gale, willows, and alders, a very upset greater yellowlegs immediately greeted us with its ringing "tew-tew-tew!!!" and flew nearby, landing in the tops of flimsy shrubs. Although nesting season was over, its young may have been nearby, and it was guarding the territory. It was easy to know when we were near our stashed gear when we returned, as it again raised the alarm.

A large coastal lagoon provided habitat for two trumpeter swans, a flock of mallards, and a lone male canvasback. Goose poop lined the muddy shore, but no geese were seen here. A few dead sticklebacks and empty fresh-water snail shells provided some evidence of what the birds were eating, in addition to the abundant pond vegetation. A sudden call of a red-wing blackbird sitting on a dead willow branch at the edge of the lagoon added another species. Six blackbirds called to each other before winging away. A large moose grazed peacefully on the far side of the lagoon.

A fox sparrow with its bicolored bill, seven dark song sparrows, and six orange-crowned warblers gleaned insects from the numerous willows. Lincoln's sparrows sang their sweet song from the dune grass, while savannah sparrows, with their bubblegum-pink legs, watched us nervously, flitting off to hide in the grasses as we approached.

Two juvenile semipalmated plovers with dark bills and eye patches did the typical plover stop and start along the sandy patches paralleling the dunes. Numerous brown bear tracks and scat decorated the terrain. One peculiar brown bear scat was composed completely of Styrofoam, perhaps the fate of a cooler, no doubt accompanied by some indigestion and a bad attitude. Coyote tracks also dotted the sands. Two piles of bird feathers proclaimed hunting success, from either a winged or furred predator.

The rhythmic surf boomed as we crested the sand dunes. An adult bald eagle cruised low over the coast. Dark parasitic jaegers patrolled the crying flocks of glaucous-winged gulls and black-legged kittiwakes, winging over the ocean swells, hoping to steal their food. Two juvenile common loons dove for fish amongst the shrinking bergy bits as gray ghosts of fog drifted in and out.

Five marbled murrelets dove for fish, but also took time out for a little courtship behavior, the pairs swimming closely side by side, heads held high, mirroring each other's actions. I wonder where this little alcid nests, as it prefers a moss-covered branch of old-growth spruce or hemlock, a scarce commodity in this new land. It will commute up to 45 miles one way to its solitary nest rather than join its cousins, the murres and puffins, in a noisy, congested seabird colony.

A Steller's sea lion also cruised the rich shoreline in search of fish. The pink salmon are spawning at the mouths of coastal streams, while the silvers wait just offshore, getting ready to complete their life's work, returning to spawn in the same stream where they were hatched up to five years ago. It's a feast for both furred and feathered.

Robinson Mountains, 4 August 2007

Cool and overcast with sprinkles.

Up to 2100 feet to inventory the alpine zone. An abundance of beautiful flowers, but very few birds today.

Al and Tony flushed a ptarmigan with two chicks in a mossy wet meadow, but there was no verified identification. Redpolls chirred overhead. Goat tracks marked the ridge, and marmots whistled alarms. One large brown bear glided effortlessly up a steep gulley full of snow toward our pick-up spot, fortunately just after we were whisked away.

Oaklee River, 5 August 2007

Overcast with sprinkles and fog.

Back to the coast to a large estuary west of camp. From the helicopter we spotted 12 brown bears plus one black bear and a sow with three cubs. We also found three pairs of trumpeter swans, one with a cygnet. I haven't seen many of the gray cygnets yet. The pilot reported 29 swans together on the Seal River, perhaps gathering during this stretch of bad weather.

As usual in the grasses and sedges of the marsh, Lincoln's and savannah sparrows chipped and played hide and seek. Small flocks of peeps, five to 12 birds each, and a larger flock of about 30, flew about, eluding positive ID through the binoculars' spots and dots on this drippy day. A determined merlin stroked after one of these nervous flocks with a rapid wing beat, but failed to nail dinner. A small group of mallards, northern pintails, and gadwalls dabbled in the river. An adult bald eagle flew past, and a glaucous-winged gull surveyed the river upstream while two common loons flew downstream to feed in the ocean. When the tide is out, the table is set for shorebirds along the extensive mudflats.

Abandoned River, 6 August 2007

Dense fog clearing to scattered clouds and blue sky.

Only one bird observed, a song sparrow in an alder near the Abandoned River. This species is generally nonmigratory, so one can expect to find it here year-round.

Khitrov Mountains, 7 August 2007

Glorious sunny day!

Back up to alpine to 3400 feet via the magic of the helicopter. Stunning scenery with bird's-eye views of both the Steller and Bering Lobes and the snowy coastal mountains just to the north.

The first bird, a familiar song sparrow hopping along the rocks near the landing site. Then three American pipits calling "pi-pit!" flying past, flashing their distinctive white outer tail feathers, a more likely alpine species that nests here. A marmot whistled its alarm as the ominous dark shadow of a golden eagle dipped up and down following the contours of the mountainside, while the great raptor glided purposefully on outstretched wings on its mission. Five juvenile gray-crowned rosy-finches in a loose flock preened in the warm sunshine on the steep, rocky slope.

The hardy resident grouse of high, barren tundra, a hen rock ptarmigan, stayed close to her single chick near a flower-covered rocky alpine ridge. Down to one from a normal clutch size of seven to nine eggs, mortality was high. Later in the afternoon, a male rock ptarmigan popped his curious head over a rocky outcrop to investigate me, his red eyebrow and light brown body framed against a brilliant blue sky (Figs. 6–9).

Five American pipits wagged their tails goodbye as the last birds seen at the magnificent and majestic Khitrovs.

Bering Glacier Camp, 8 August 2007

Partly sunny, mostly overcast.

See initial field report entry.

Figure 6. Female rock ptarmigan in her magnificent but harsh world.

Figure 7. From an average clutch size of seven to nine eggs, only one rock ptarmigan chick survived.

Cape Suckling, 9 August 2007

Foggy conditions return, socking in much of Bering Glacier, partly sunny at coast, with occasional sprinkles.

We explored for a very short time at the fabled Cape, infamous on marine forecasts: "Cape Suckling to Gore Point: gale

Figure 8. Male rock ptarmigan patrols the alpine zone of the Khitrov Mountains.

Figure 9. Eagle feather, wild and free on the Khitrovs.

force winds and hazardous seas." Today, the fierce winds only whispered, and the mighty Gulf of Alaska rolled against the sandy beach in a steady, unruffled rhythm.

Surprisingly tall spruce trees sprouted out of the rocky Cape headlands, providing a fringe of forest habitat adjacent to the beach ryegrass. Chestnut-backed chickadees sang their hoarse, cheerful song in the branches. A black-billed magpie chattered nearby. Bald eagles soared overhead. Shortly after noon we moved to another site to the east.

Tsiu River, 9 August 2007

The broad, salmon-spawning Tsiu River was loaded with birds. Over one hundred ducks, just out of binocular range, rafted up among the sedges across the river. Eleven Canada geese, probably dusky, flew in stately formation, joining another 20 already floating downstream. Across the river, two trumpeter swans tipped up, feeding on aquatic vegetation. Fifteen semipalmated plovers, including adults and juveniles, did their stop-start action hunting for tiny insects and invertebrates along the sandy shore. A solitary western sandpiper probed the water's edge. Five greater yellowlegs stalked the shallows. A bank swallow flew overhead, hopefully gorging on the abundant insect life that we would gladly share.

Paul spied a new bird, a northern harrier, the raptor with the white rump and tilted wings, weaving its way over the beach ryegrass carpeting the sand dunes, hunting for rodents such as voles. A Lincoln's sparrow again graced the grasses with its lilting melody. An adult bald eagle, its back to the blazing fireweed, watched for lunch from the edge of an eroding dune, a secret hiding place until it flew off to the river to terrorize the ducks.

Yet another raptor, a peregrine falcon, flap-glided above the sand dunes and quickly disappeared.

An American robin clucked, a Steller's jay rat-tatted, and a red squirrel scolded from the edge of a dense spruce forest crowding the alder and willow fringe adjacent to a large wet meadow. An incongruous airboat lay nosed into the alders awaiting the upcoming salmon fishing and bear hunting season.

No birds graced the gloomy, ghostly spruce forest where moss grew 50 feet or more up the skinny trunks, thanks to high humidity. Almost nothing grew below the dense canopy, save for the occasional opening where the needle-strewn forest floor erupted into green devil's club and salmonberry bushes. Shortly after we found a weather-beaten hunting shack sagging into the forest we heard the odd sounds of "civilization": a radio and the regular pinging of target shooting. A well-established ATV trail quickly led us back to the beach.

A sharp-shinned hawk erupted from the forest edge ahead to land on a driftwood post. A large female, she lingered long enough for us to admire her long, banded tail and catch a glimpse of her horizontally striped rusty breast before rushing away.

Back to the coastal dunes, three young, frowzy-looking juvenile Steller's jays watched us with interest from their alder sanctuary. One repeatedly imitated the cry of a hawk, a very good mimic indeed!

CONCLUSIONS

Despite the generally uncooperative weather and secondary focus on birds, I identified 51 species representing 22 families, and 4 others as unidentified species. Owing to the advanced season, many of the songbirds had already migrated. There was much more bird activity primarily in the rivers, lagoons, and in the ocean.

A survey focused on birds using a scope in addition to binoculars from mid-May to September would definitely reveal many more birds and species, particularly in mid-June during the courting and nesting season. Boat access to the lake and to known birding hot spots such as Whale Back Island, a reported location of a Caspian tern nesting colony, would be productive. I would also recommend more time along the larger rivers to survey ducks, and along the beach to survey oceanic species.

The Bering Glacier Region provides outstanding and varied habitats for a diverse number of species. Some seem protected by their remote location and challenging weather. The Pacific coast area, especially near the Tsiu River, shows the most human impact. There are ATV tracks running through the coastal vegetation, an airboat and other gear stored near the river, two lodges specializing in salmon fishing and bear hunting served by rudimentary airstrips, and hunting shacks dotting the dunes.

A thorough baseline survey of bird species and behavior is essential to document crucial information as the dynamic landscape responds to climate change, and as the land is modified by human impact. Land managers need this information to protect and preserve this deceptively fragile and magnificent ecosystem.

ACKNOWLEDGMENTS

Thanks to Marilyn Barker for including me on this phenomenal trip, and to the other members of the botany team, Alan Batten, Tony Reznicek, and Paul Sharaba, who faithfully carried the heavy bear gun to protect us. It was an honor to be a member of this dedicated and knowledgeable team. I appreciate the confirmations from birder Joe Staab of Seward, and Steve Heinl of Juneau, who helped me with the tricky juvenile least sandpiper identification.

Manuscript Accepted by the Society 02 June 2009

APPENDIX 1. BIRDS OF THE BERING GLACIER, 31 JULY–10 AUGUST 2007

Family	Common name	Locations observed
Swans, geese	Canada goose	TSIU, VL
	Trumpeter swan	H, SL, TR, TSIU
Ducks	Gadwall	OR, VL
	Mallard	OR, TR, TSIU, VL
	Northern pintail	OR
	Canvasback	TR
	White-winged scoter	VL
Ptarmigan	Willow ptarmigan	BGC
	Rock ptarmigan	KM
Loons	Red-throated loon	VL daily
	Common loon	OR, PC, SL,VL daily,
Grebes	Red-necked grebe	VL
Herons	Great-blue heron	VL
Hawks, falcons	Bald eagle	BGC, CS, OR, PC, SL, TSIU
	Northern harrier	TSIU
	Sharp-shinned hawk	TSIU
	Golden eagle	KM
	Merlin	BGC, OR
	Peregrine falcon	TSIU, WPI
Plovers	Semipalmated plover	TR, TSIU
Sandpipers	Greater yellowlegs	MTLT, TR, TSIU, VL
	Spotted sandpiper	VL
	Western sandpiper	TSIU, VL
	Least sandpiper	MTLT
Jaegers, gulls	Parasitic jaeger	PC
	Mew gull	BGC
	Glaucous-winged gull	MTLT, PC, VL
	Black-legged kittiwake	PC
Puffins, murres	Marbled murrelet	PC
Owls	Great-horned owl	MTLT
Kingfishers	Belted kingfisher	BGC, MTLT
Jays, crows	Steller's jay	SL, TSIU
	Black-billed magpie	BGC, CS
	Northwestern crow	PC
	Common raven	BGC, TR
Swallows	Bank swallow	SL, TSIU
Chickadees	Black-capped chickadee	MTLT
	Chestnut-backed chickadee	CS
Thrushes and allies	Hermit thrush	WPI
	American robin	TSIU, WPI
Pipits	American pipit	KM
Wood-warblers	Orange-crowned warbler	BGC, MTLT, PC, TR, WPI
	Yellow warbler	BGC
	Wilson's warbler	BGC, MTLT
Sparrows	Savannah sparrow	OR, TR
	Fox sparrow	BGC, TR
	Song sparrow	AR, KM, MTLT, SL, TR
	Lincoln's sparrow	BGC, MTLT, OR, SL, TR, TSIU
	Golden-crowned sparrow	BGC
Blackbirds	Red-winged blackbird	TR
Finches	Gray-crowned rosy-finch	KM
Unidentified species		
Sandpipers	Phalarope sp., likely red-necked	SR
Hummingbirds	Hummingbird sp.	BGC 8/06, 8/08
Thrushes and allies	Kinglet sp., likely golden-crowned	SL
Finches	Redpoll sp., likely common	BGC, MTLT, RM

(*continued*)

APPENDIX 1. BIRDS OF THE BERING GLACIER, 31 JULY–10 AUGUST 2007 (*continued*)

Date	Location	Key	GPS	Elevation
8/06	Abandoned River	AR	No data	
7/31–8/10	Bering Glacier Camp	BGC	No data	
8/09	Cape Suckling	CS	60 00.125′ N 143 58.070′ W	1 m
8/01–8/09	Helicopter	H	Various	
8/07	Khitrov Hills	KH	60 26.755′ N 143 15.515′ W	1018 m
7/31	Midtimber Lake Trail	MTLT	60 06.281′ N 143 20.965′ W	4 m
8/05	Oaklee River	OR	60 02.238′ N 144 03.621′ W	4 m ± 6 m
8/03	Pacific Coast	PC	No data	
8/04	Robinson Mountains	RM	60 18.522′ N 143 01.701′ W	653 m
8/01	Seal River	SR	No data	
8/01	Swan Lake	SL	60 08.148′ N 142 56.683′ W	13 m ± 6 m
8/03	Tashalich River	TR	60 02.515′ N 143 37.076′ W	6 m ± 12 m
8/09	Tsiu River	TSIU	60 05.349′ N 142 59.956′ W	10 m
8/02	Weeping Peat Island	WPI	60 11.446′ N 143 11,992′ W	13 m
7/31–8/10	Vitus Lake	VL	No data	

Note: Common names for birds, unlike those for plants, are generally very specific and stable, so the scientific names are not listed.

APPENDIX 2. BIRDS OF THE BERING GLACIER, AUGUST 2002 AND 2003

Family	Common name	Locations observed
Swans, geese	Canada goose (dusky)	BL
	Trumpeter swan	BGC, TR, HL
Ducks	Gadwall	BGC
	Mallard	MTLT
Ptarmigan	Willow ptarmigan	BGC
	Rock ptarmigan	MP
Loons	Red-throated loon	BL, BLC, HL
	Common loon	BLC, PC
Hawks, falcons	Bald eagle	BGC
	Northern harrier	SR
	Northern goshawk	SR
	Merlin	BGC, DR, HP, KH
Plovers	Semipalmated plover	BGC
	Killdeer	BGC
Sandpipers	Greater yellowlegs	MTLT, TR
	Spotted sandpiper	BGC, TR
	Red-necked phalarope	VL
Jaegers, gulls	Parasitic jaeger	SR
	Mew gull	SR
	Herring gull	MTLT, SR,TR
	Caspian tern	SR, VL
Hummingbirds	Rufous hummingbird	BGC
Kingfishers	Belted kingfisher	HL, TR
Woodpeckers	Hairy woodpecker	DF
Jays, crows	Steller's jay	DF, HP, SH
	Black-billed magpie	AF, BGC, SH
	Common raven	AF, BGC, SH
Chickadees	Black-capped chickadee	MTLT
Kinglets	Ruby-crowned kinglet	MTLT
Thrushes and allies	Hermit thrush	DF
Pipits	American pipit	AF, DR, GH, MP, SH
Wood-warblers	Orange-crowned warbler	BGC
	Yellow warbler	BGC
	Wilson's warbler	BGC
Sparrows	Savannah sparrow	BGC, SR
	Fox sparrow (sooty)	BGC, DF, KH, MTLT, SH, TR
	Song sparrow	BGC, SR
	Lincoln's sparrow	MTLT
	Golden-crowned sparrow	KH
Finches	Pine siskin	DF

Note: Common names for birds, unlike those for plants, are generally very specific and stable, so the scientific names are not listed. Source: Kevin Lynch.

Locations: AF—Ancient Forest; BL—Berg Lake; BGC—Bering Glacier camp; DR—Donald Ridge; DF—Donald Fen; GH—Grindle Hills; HL—Hanna Lake; HP—Hanna Peak; KH—Khitrov Hills; MP—McIntosh Peak; MTLT—Midtimber Lake Trail; PC—Pacific Coast; SR—Seal River; SH—Sucking Hills; TR—Tashalich River; VL—Vitus Lake.

Printed in the USA

The Geological Society of America
Special Paper 462
2010

Bering Glacier, climate change, and the Southern University research experience

Michael Stubblefield
Revathi Hines
Lionel D. Lyles
Nelson Mandela School of Public Policy, Department of Public Policy, P.O. Box 9656, Southern University, Baton Rouge campus, Baton Rouge, Louisiana 70803, USA

INTRODUCTION

The United Nation's Intergovernmental Panel on Climate Change (UNIPCC) has ranked the years 1995–2006 as the warmest years since 1850. This increase in temperature is global in perspective and is greater at higher northern latitudes. The land regions seem to have warmed at a faster rate than the oceans, and the average temperature of the Arctic has been double that of the global average of climatic warming. The continued warming of glaciers and ice caps are contributing to an increased melting rate. Glaciers and ice caps are shedding many cubic kilometers of fresh water into the seas (Josberger et al. and Shuchman et al., this volume).

One such affected glacier is the Bering Glacier, the largest glacier in continental North America. It has an area of ~5,175 km^2 and a length of 190 km. It is located in coastal south-central Alaska at 60°–61° north latitude, and 141°–145° west longitude; it is bounded on the north by the Saint Elias Mountains and on the south by the Gulf of Alaska. Added to this formidable ice mass is the fact that in some places the Bering Glacier's thickness exceeds 800 m. It constitutes 6% of the total glacier-covered area of Alaska (Molnia and Post, 1995).

The Bering Glacier is not stationary. It is also the largest surging glacier in North America. The terminus of this 5000 km^2 glacier has receded 10–12 km from its terminus position 100 yr ago. Since 1995, a warmer polar climate has created conditions for increased melting as evidenced by the appearance of surface water and moulins.

During the twentieth century the Bering Glacier underwent six major surges, the most recent of which occurred during the 1993–1995 event (Molnia et al., this volume). According to various authors of this volume, throughout this period there have been several hundred meters of glacial thinning, and ~12 km of terminus retreat. The Bering Glacier's last surge occurred in 1995, and ever since this time it has been undergoing rapid *retreat*, namely, melting. Interestingly, it might be expected that the largest gains in temperature would have occurred in nonpolar regions. On the contrary, the largest increases have occurred in the northern latitudes where the Bering Glacier is located. As a result of surge-retreat action over time, the Bering Glacier carved out an unusually large lake known today as Vitus Lake, which bears Vitus Jonassen Bering's name, including that of the glacier as well. Bering, a Dane, was one of the earliest Arctic explorers to visit the southeastern Alaskan coast.

During the summer of 2007, a team of faculty and students from Southern University and A&M College, Baton Rouge, Louisiana, engaged in a research expedition to the Bering Glacier. Southern University researchers were interested in exploring the question, "How has climate change affected glacial melting at the Bering Glacier and what has it meant to the surrounding ecosystem, habitat, and sea level rise?" Additionally, the research team wanted to draw some conclusions on what a future-based sea-level rise could mean to the state of Louisiana.

BERING GLACIER AND SOUTHERN UNIVERSITY (SUBR) RESEARCH

The SUBR 2007 Research Team

Before 2007, of the 100 Historically Black Colleges and Universities (HBCU) in the United States, none had ever sent a research team to the Bering Glacier to conduct climate change research. However, in March 2007, Dr. Michael Stubblefield, who is currently the Vice Chancellor for the Office of Research and Strategic Initiatives (ORSI), was contacted by Dr. John

Stubblefield, M., Hines, R., and Lyles, L.D., 2010, Bering Glacier, climate change, and the Southern University research experience, *in* Shuchman, R.A., and Josberger, E.G., eds., Bering Glacier: Interdisciplinary Studies of Earth's Largest Temperate Surging Glacier: Geological Society of America Special Paper 462, p. 381–384, doi: 10.1130/2010.2462(24). For permission to copy, contact editing@geosociety.org.

Payne and Scott Guyer, both U.S. Bureau of Land Management (BLM) administrators in the Anchorage, Alaska, office, to discuss the possibility of forming a research team to travel to BLM's Bering Glacier camp during the summer of 2007. The purpose of the invitation was twofold: (1) to receive an orientation related to how the Bering Glacier camp operates, and (2) to meet with other research teams to discuss their research activities and also to participate in their planned summer 2007 research activities. A Memorandum of Understanding was created, establishing a partnership between Southern University and BLM.

After this initial contact was made, Dr. Stubblefield contacted Dr. Lionel D. Lyles, Associate Professor, Geography/GIS, to determine if there was any interest in conducting research at the Bering Glacier in Alaska. During the brief meeting, and after listening to Dr. Stubblefield outline the pros and cons of the idea, Dr. Lyles informed him that the latter was an opportunity we had to accept, not only because of what we could learn about the Bering Glacier, but, equally important, what an opportunity this would be for SUBR's graduate and undergraduate students to view and walk on the Bering Glacier itself! Of all the African-Americans in the United States, less than half of one percent will ever have such an experience!

Dr. Stubblefield, feeling the same as Dr. Lyles did about the opportunity, instructed him to immediately put together what has come to be known as the Southern University Bering Glacier Research Team (SUBGRT). Imagine SUBR's challenge: How does the university convince African-Americans from the Deep South, and from lowland, flat, humid Louisiana, where the temperature seldom falls below 32 °F, to get excited about traveling to icy, cold Bering Glacier, Alaska? To the pleasant surprise of both Dr. Stubblefield and Dr. Lyles, the task was easy.

Dr. Lyles' first act was to get another professor involved who could inspire and motivate female graduates and undergraduates to join the SUBGRT research enterprise. He approached Dr. Revathi Hines to join our team, and she immediately accepted. Next, he got the University's GIS Computer Lab Technician, Alaa Shams, to commit. Drs. Stubblefield, Hines, and Lyles, along with Alaa Shams, formed the leadership division of the SUBGRT.

With a strong, determined leadership team, the next task was to recruit undergraduate and graduate students. Through the Center for Coastal Zone Assessment and Remote Sensing (CCZARS), the task of recruiting our first group of undergraduate students was relatively easy. The summer 2007 SUBGRT consisted of Drs. Stubblefield, Hines, and Lyles; Alaa Shams, GIS Lab Technician; Jacquole Landry and Mykel Delandro, both Civil Engineering undergraduate students; and Pamela Brue-Johnson, SUBGRT Program Coordinator. The 2007 SUBGRT was such a success that the following year, summer 2008, it increased from 7 to 11, including an addition of 5 new students.

SUBR-Bering Glacier Research Initiatives

Having made all the logistical arrangements—clothing, research equipment, etc.—the SUBGRT embarked on its journey to the Bering Glacier camp in summer 2007 and 2008. In both instances the team left the Baton Rouge Airport on 31 July and arrived in Cordova, Alaska, where they stayed one night at the Orca Lodge. Weather permitting ("hurry up and wait"), they arrived at the Bering Glacier camp 2 August via a 45 min bush plane flight. For the undergraduate and graduate students, this plane ride was unique because it was possible for them to observe landforms and land cover at low elevation, and grizzly bears and moose in their natural habitat. Upon arriving at the Bering camp, everyone on the team was assigned to living quarters according to gender. The mess hall was a popular stop, where all of the researchers from the participating universities (Michigan Tech, University of Alaska, Anchorage, etc.) got a chance to sit down, interact, and finalize research plans for the day, such as the following:

1. Bering Glacier Terminus Research Project: During the summer 2007, BLM officials and Michigan Tech Research Institute (MTRI) scientists thought mapping the Bering Glacier Terminus would be an added value to the body of interdisciplinary research going on at the Bering camp. As such, Alaa Shams and Dr. Lyles took this research project on, using the Redhen Technology, which makes it possible to continuously collect lat/long data while flying the Bering Glacier terminus in a helicopter. After the helicopter flights, they downloaded the lat/long data into our computer, and created a 2007 map of the Bering Glacier terminus. When the team overlaid our 2007 map on a 2006 Landsat 7 image, they discovered that the Bering Glacier terminus had retreated ~0.4–0.5 km between 2006 and 2007. Repeating this process during summer 2008, they also observed a similar retreat (see Figs. 1, 2).

2. Water Quality, Seismic, and Vegetation Classification: Dr. John Payne's vision of the Bering Glacier Research Initiative, being primarily interdisciplinary, was observed as successful because the SUBGRT members joined other research teams to assist and learn more about the Bering Glacier ecosystem. Specifically, some of the team members assisted simultaneously with taking water measurements to monitor for salinity, temperature as a function of depth, and to perform water quality observations (see Jenkins, this volume). Others assisted with placement of seismic units, which record Bering Glacier's calving and cracking events (see FitzGerald et al. and Richardson et al., this volume). Still others got to learn how to carry out resistivity research, and take measurements to determine the presence of salinity in the subsurface water table (see Andrus et al., this volume).

IMPLICATIONS FOR LOUISIANA'S GULF COAST

The glacial melting and sea level rise in the Pacific can be generalized to what climate change and sea level rise can mean to the coastal state of Louisiana. Sea level rise can significantly impact Louisiana in two ways. Firstly, Louisiana is a coastal state and borders the Gulf of Mexico, and secondly, the Mississippi River (which originates from Lake Itaca in Minnesota) flows via Louisiana and empties into the Gulf of Mexico. Therefore, Louisiana is susceptible both to coastal flooding (Gulf of Mexico)

Figure 1. Retreat of the Bering Glacier terminus, 2006–2007, where the red line is the 2007 terminus location superimposed on the Landsat 2006 image. Source: Southern University Bering Glacier Research Team, Dr. Lionel D. Lyles, Co-PI, 2007.

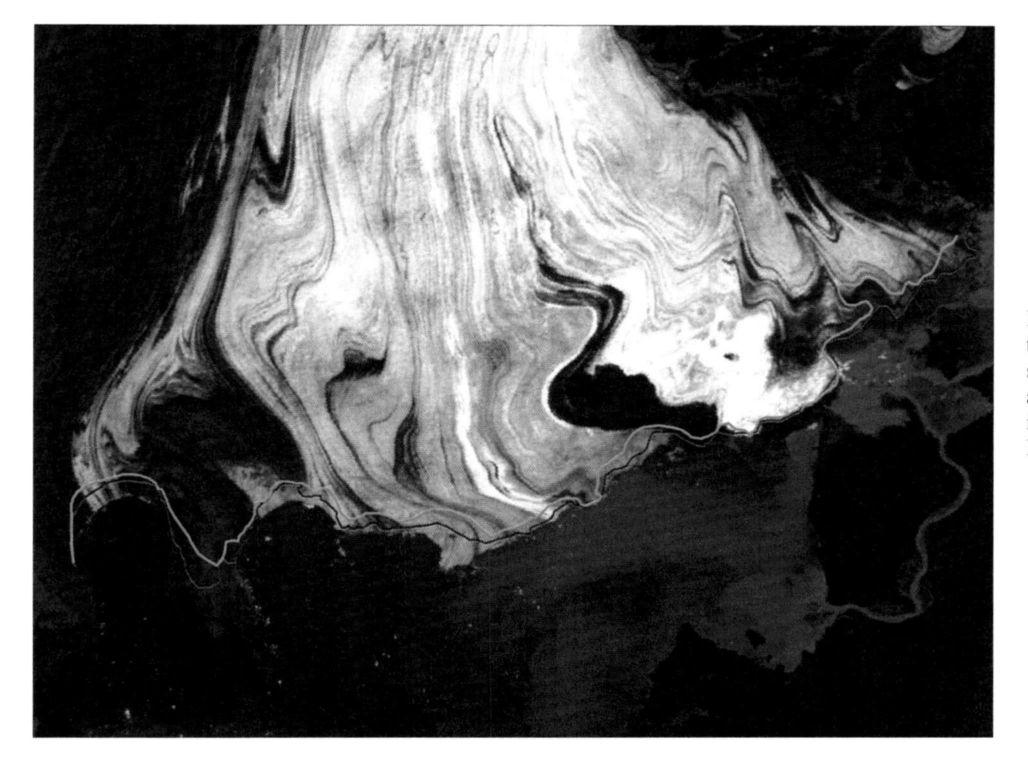

Figure 2. Retreat of the Bering Glacier terminus, 2007 (red) to 2008 (green) superimposed on the Landsat 2006 image. Source: Southern University Bering Glacier Research Team, Dr. Lionel D. Lyles, Co-PI, August 2008.

and mainland flooding (Mississippi River and its tributaries). Sea level rise therefore will affect Louisiana and cause flooding, salt-water intrusion into aquifers, land erosion, and destruction of fresh-water habitats. Even minimal sea level rise in the twentieth century led to soil erosion and loss of 100 km^2 of wetlands per year in the Mississippi River Delta region.

Melting of the other ice sheets, such as the Greenland Ice Sheet and the West Antarctic Ice Sheet, will lead to a significant rise in sea level. If current levels of melting of these ice sheets continue, the U.S. Geological Survey predicts a sea level rise of ten or more meters by the end of the century. This rise will affect ~25% of the U.S. population, with the major impact being borne by Eastern and Gulf Coast states, including Louisiana.

Another equally worrisome concern is an increasing imbalance in the ocean "conveyor belt." The latter maintains a delicate balance in the flow of very cold water from the North Atlantic to the southern Atlantic Basin, forcing warmer Caribbean water through the Gulf Stream back to the North Atlantic. This ocean conveyor belt regulates air temperatures, which maintains a balance in the European climate. Cold heavy water from around Greenland sinks and cycles south, pushing water from the tropics north to warm Europe. With each drop of lightweight ice melt pouring off Greenland, that belt frays a little more. The last time it was disrupted, much of Europe was buried under glaciers.

CONCLUSIONS

The Bering Glacier is best understood from data recorded over five-year time periods and longer. The SUBGRT has documented the retreat of the Bering Glacier terminus, showing that the latter is still undergoing significant calving. As mentioned earlier, the retreat of the Bering Glacier is releasing, via Vitus Lake and the Seal River, an estimated minimum of 6–8 km^3 of fresh water into the Gulf of Alaska each summer. This amount has a direct influence on sea level rise along the Gulf Coast of Louisiana, which is already at or below sea level. Given this fact, the SUBGRT is engaged in scientific investigations at the Bering Glacier to determine how significant, or insignificant, is the threat of climate change to the Gulf of Alaska and the Gulf of Mexico (Louisiana in particular), although, geographically speaking, these two physical features, including human and animal life, are several thousand miles apart. It is a well-known fact that a mere 3 m rise in the Gulf of Mexico will bring catastrophic flooding to New Orleans and other low-lying coastal cities, both in the United States and worldwide.

Lastly, the Bering Glacier has become, for our SUBGRT, an interdisciplinary laboratory in which professional scientists carry out their research on varying aspects of the glacier with the direct, hands-on assistance of undergraduate and graduate students.

REFERENCES CITED

Andrus, A.B., Endsley, K.A., Espino, S., and Gierke, J.S., 2010, this volume, Mapping the fresh-water–salt-water interface in the terminal moraine of the Bering Glacier, *in* Shuchman, R.A., and Josberger, E.G., eds., Bering Glacier: Interdisciplinary Studies of Earth's Largest Temperate Surging Glacier: Geological Society of America Special Paper 462, doi: 10.1130/2010.2462(19).

FitzGerald, K.A., Richardson, J.P., and Pennington, W.D., 2010, this volume, Ice-generated seismic events observed at the Bering Glacier, *in* Shuchman, R.A., and Josberger, E.G., eds., Bering Glacier: Interdisciplinary Studies of Earth's Largest Temperate Surging Glacier: Geological Society of America Special Paper 462, doi: 10.1130/2010.2462(18).

Josberger, E.G., Shuchman, R.A., Meadows, G.A., Savage, S., and Payne, J., 2010, this volume, Hydrography and circulation of ice-marginal lakes at Bering Glacier, Alaska, USA, *in* Shuchman, R.A., and Josberger, E.G., eds., Bering Glacier: Interdisciplinary Studies of Earth's Largest Temperate Surging Glacier: Geological Society of America Special Paper 462, doi: 10.1130/2010.2462(04).

Molnia, B.F., and Post, A., 1995, Holocene history of Bering Glacier, Alaska: A prelude to the 1993–1994 surge: Physical Geography, v. 16, p. 87–117.

Molnia, B.F., and Post. A., 2010, this volume, Surges of the Bering Glacier, *in* Shuchman, R.A., and Josberger, E.G., eds., Bering Glacier: Interdisciplinary Studies of Earth's Largest Temperate Surging Glacier: Geological Society of America Special Paper 462, doi: 10.1130/2010.2462(15).

Richardson, J.P., FitzGerald, K.A., Waite, G.P., and Pennington, W.D., 2010, this volume, Acoustic and seismic observations of calving events at Bering Glacier, Alaska, *in* Shuchman, R.A., and Josberger, E.G., eds., Bering Glacier: Interdisciplinary Studies of Earth's Largest Temperate Surging Glacier: Geological Society of America Special Paper 462, doi: 10.1130/2010.2462(17).

Shuchman, R.A., Josberger, E.G., Jenkins, L.K., Payne, J.F., Hatt, C.R., and Spaete, L., 2010, this volume, Remote sensing of the Bering Glacier region, *in* Shuchman, R.A., and Josberger, E.G., eds., Bering Glacier: Interdisciplinary Studies of Earth's Largest Temperate Surging Glacier: Geological Society of America Special Paper 462, doi: 10.1130/2010.2462(03).

Manuscript Accepted by the Society 02 June 2009